2 projects: 1- defender: currently owned
2- challenger: alternat

① find economic service life of challenger:
AEC is lowest !!

② Defender (use Market Value → P)
find AEC of defender

∴ keep defender until AEC_D is greater than
AEC_C then switch.

Depreciation: $D = \dfrac{(P-S)}{N}$

$\qquad B_n = P - n\dfrac{(P-S)}{N}$ $\qquad$ N = useful life
$\qquad\qquad\qquad\qquad\qquad\qquad\qquad$ n = book value

SECOND CANADIAN EDITION

Contemporary Engineering Economics

A Canadian Perspective

Chan S. Park • Ronald Pelot • Kenneth C. Porteous • Ming J. Zuo

Addison
Wesley
Longman

Toronto

Canadian Cataloguing in Publication Data

Main entry under title:
  Contemporary engineering economics: a Canadian perspective

2nd Canadian ed.
Includes index.
ISBN 0-201-61390-5

1. Engineering economy   2. Engineering economy—Canada   I. Park, Chan S.

TA177.4C66 2001   658.15'024'62   C00-931929-8

ISBN 0-201-61390-5

Vice President, Editorial Director: Michael Young
Acquisitions Editor: Dave Ward
Marketing Manager: Cathleen Sullivan
Developmental Editor: Laurie Goebel
Production Editor: Jennifer Therriault
Copy Editor: Tally Morgan
Production Coordinator: Deborah Starks
Page Layout: Heidy Lawrance Associates; Bill Renaud
Art Director: Mary Opper
Cover Design: Alex Li
Cover Image: Eye Wire/E010286

  2 3 4 5   05 04 03 02

Printed and bound in USA

In memory of my mother, Hong Yong Hee, whose wisdom as a parent has inspired those she influenced.

Chan S. Park

To my wife, Milica, in appreciation of all your support and encouragement.

Ronald Pelot

To my wife, Helen, and daughters Elizabeth and Suzanne.

Kenneth Porteous

To my wife, Ninghe Hu, for her love, support, and understanding.

Ming Zuo

Table of Contents

PREFACE

What's "Contemporary" about Engineering Economics?

Decisions made during the engineering design phase of a project can have a major impact on the total cost of the project. In the case of a new product development, its design may determine the majority (some say 85%) of the costs of manufacturing the product. And, as design and manufacturing processes become more complex, the engineer, increasingly, will be called upon to make decisions that involve money. In the twenty-first century, the competent and successful engineer will need an improved understanding of the principles of science, engineering, and economics, coupled with relevant design experience. This is because, in the new world economy, successful businesses will rely on engineers with this type of expertise more and more.

In the product/service life cycle, economic and design issues are inextricably linked. One of the strongest motivations in writing this text was, therefore, to bring the realities of economics and engineering design into the classroom and to help students integrate these issues when contemplating an engineering problem or making an engineering decision.

Another compelling motivation was, once students had mastered fundamental concepts, to introduce the computer as a productivity-enhancing tool for modeling and analyzing engineering decision problems. Spreadsheets are currently the undisputed standard for automating complex engineering economic problems in the industry, and they are used extensively in the classroom. At the ends of many chapters, this text introduces the topic of spreadsheets (Microsoft Excel) in sections dedicated to spreadsheet use.

In addition, the World Wide Web site for this edition of *Contemporary Engineering Economics: A Canadian Perspective* includes a revised version of the **EzCash** software which students can download at no cost. **EzCash** was developed by Chan Park, with a grant from the National Science Foundation, to open visually the economic computing environment to the student's understanding. **EzCash** is an integrated package that includes the most frequently used methods of economic analysis. **EzCash** organizes information via graphically-based structures that can be explored independently by students. **EzCash** runs with Microsoft Windows 95, Microsoft Windows 98 and higher versions.

The underlying motivation for writing this book was not only to simply address contemporary needs, but also to address the timeless goal of all educators: To help students learn. Thus, thoroughness clarity, and accuracy of presentation of essential engineering economics were our aims at every step in the development of the text.

Although the major focus of this book is the economic evaluation of engineering projects, the methodology is directly applicable to any financial situation. Hence, a secondary theme in the text is personal financial management and the analysis of personal investments.

Changes in This Second Edition

We are living in the age of information, within the complex and changing world of a global economy. The practice of engineering economics is therefore dynamic, and new developments as they occur, should be incorporated into a text such as this. Along with the publisher, we, as teachers, are constantly seeking ways to improve *Contemporary Engineering Economics: A Canadian Perspective* in terms of its clarity and provision of understanding. As a result, several important changes were made to this second edition.

1. The text was completely updated to reflect the latest tax laws, interest rates, and other financial developments.

2. All spreadsheet examples are now in Microsoft Excel, the market leader for Windows-based environments.

3. A revised version of **EzCash** is available from the book's web site. This allowed us to remove **EzCash** discussions from the main text. The new version incorporates a scrolling feature and resolves the application operating and printing problems encountered with the original version.

4. The number of end-of-chapter problems was increased and categorized by degree of difficulty. Level 1 problems are the easiest and generally involve the straightforward application of formulas in the text. Level 3 problems are the most difficult by virtue of the complexity of the concepts, data or both. Level 2 problems fall between these extremes. The solution to starred problems are provided at the end of the book.

5. Approximately one third of the examples are either new, or revised, and reflect the contemporary nature of economic decision problems.

6. Examples and problems which have a personal finance context (loans, mortgages, bonds, stocks, RRSP's etc) are identifiable by the $ sign preceding the number.

7. A French/English glossary has been added and a French version of the Table of Contents is available on the book's web site.

8. In Chapter 1, most of the real-world engineering economic decision problems have been replaced with more timely examples from *The New York Times*. At the suggestion of reviewers, a section on short-term operational economic decisions (commonly known as present economic studies) and estimation of project costs and benefits were added to this chapter.

9. The material on the measurement of investment attractiveness in Chapters 4, 5, and 6 has been completely reorganized. Present, future and annual equivalent criteria and rate of return are introduced in Chapter 4 and applied to independent investments. The application of these measures to situations involving

mutually exclusive investments is the topic of Chapter 5. In Chapter 6, the use of these measures is extended to engineering design and replacement analysis.

10. The introduction of the before tax treatment of replacement analysis in Chapter 6 allows this important topic to be covered earlier in the course and without the additional complications of income taxes. The effect of income taxes in this analysis is included in Chapter 11.

11. In Chapters 4, 8, 9, and 13 some of the advanced topics (or optional materials), such as multiple rate of return problems, the more rigorous calculation of disposal tax effects, capital tax factors and risk simulation, have been removed from the main text and placed in chapter appendices. This separation allows instructors to budget their lecture hours more effectively, according to both their audience and the curriculum.

12. Personal finance topics such as bond and stock investments and RRSP's with the associated income tax implications are included in Chapter 11.

13. At the suggestion of reviewers, project financing is no longer covered as a separate chapter. Chapter 11 and Chapter 13 of the first edition were combined to give a unified treatment to project financing and capital budgeting topics.

Overview of the Text

Although containing little advanced math and few truly difficult concepts, the introductory engineering economics course is often curiously challenging for engineering students at all levels. There are several likely explanations for this difficulty.

1. The course may be the first time a student is asked to make an analytical consideration of money (a resource with which he, or she, may have had little direct contact, beyond paying for tuition, housing, food, and textbooks).

2. An emphasis on theory, while critically important to forming the foundation of a student's understanding, may obscure for the student the fact that the aim of the introductory engineering economics course, among other things, is to develop a very practical set of analytical tools for measuring a project's worth. This is unfortunate since at one time or another virtually every engineer, not to mention every individual, is responsible for allocating limited financial resources wisely.

3. The mixture of industrial, civil, mechanical, electrical, manufacturing, and other engineering undergraduates who take the course often fail to "see themselves" as needing to use, or develop, the skills the course and text are intended to foster. This is perhaps less true for industrial engineering students but students in other disciplines are often motivationally shortchanged by a text's lack of applications that have direct appeal.

Goal of the Text

This text aims not only to build a sound and comprehensive coverage of the concepts of engineering economics, but also to address the basic difficulties experienced by the types of students described above–all of which have their basis in a lack of appreciation of the practical concerns of engineering economics. More specifically, this text has the following major goals:

1. To build a thorough understanding of the theoretical and conceptual basis on which the practice of financial project analysis is built.

2. To satisfy the very practical need that engineers will be called upon to make informed financial decisions when acting as team members or as project managers of engineering projects.

3. To incorporate all critical decision-making tools–including the most contemporary, computer-oriented ones–that engineers bring to the task of making informed financial decisions.

4. To appeal to the full range of engineering disciplines including industrial, civil, mechanical, electrical, computer, aerospace, chemical, and manufacturing engineering, as well as engineering technology.

5. To link the financial analysis concepts to common personal finance situations.

Prerequisites

The text is intended for undergraduate engineering students at the sophomore level or above. The only mathematical background required is elementary calculus. For Chapter 13, a first course in probability or statistics would be helpful, but not necessary, because the treatment of basic topics in this chapter is essentially self-contained.

Content and Approach

Educators generally agree upon the proper content and organization of an engineering economics text. A glance at the table of contents demonstrates that this text addresses the standard embraced by most instructors and that is reflected in competing texts. However, one of the driving motivations in this text was to supersede the standard in terms of the depth of coverage and the care with which difficult concepts were presented. Accordingly, the content and approach of the second edition of *Contemporary Engineering Economics: A Canadian Perspective* was designed to provide the following:

Thorough Development of the Concept of the Time Value of Money

The notion of the time value of money and the interest formulas that model it form the foundation upon which all other topics in engineering economics are built. Because of their great importance, and because many students are being exposed to an analytical approach to money for the first time, interest topics are carefully and thoroughly developed in Chapters 2 and 3.

1. Chapter 2 carefully examines the conceptual understanding of interest–the time

value of money—more "what if" and more graphical explorations are provided than in any other current text.

2. Chapter 3 extends an understanding of the time value of money via its real world complexities—effective interest, noncomparable payment, and compounding periods, etc.

Thorough Coverage of Major Analysis Methods

The equivalence methods–present worth, annual worth, and future worth–and rate of return analysis are the bedrock of project evaluation and comparison. This text carefully develops these topics in Chapters 4, 5, and 6, where they are paced for maximum student comprehension of the subtleties, strengths, and weaknesses of each method.

1. Independent investments and mutually exclusive investments are dealt with in Chapters 4 and 5 respectively.

2. The difficulties and exceptions associated with rate of return analysis are thoroughly covered in Chapter 4. Coverage of internal rate of return for nonsimple projects, which is located in Appendix 4A, is optional for those who wish to avoid teaching this topic in an introductory course. The appendix can be omitted without disrupting the flow of topics.

3. A separate chapter (Chapter 6) extends the coverage to related topics including unit cost, design economics and replacement analysis.

Increased Emphasis on Developing After-Tax Cash Flows

For most practicing engineers, estimating and developing project cash flows are the first critical steps they must take in conducting an engineering economic analysis. A particularly important goal of this text is to build confidence in developing after-tax cash flows. Further analysis, comparison of projects, and decision-making all depend on intelligently developed project cash flows. This text provides more emphasis on this topic than does any competitive text.

1. Chapter 9 provides a synthesis of previously developed topics (analysis methods, depreciation, and income taxes) and is dedicated to building skill and confidence in developing after-tax cash flows for a series of fairly complex projects.

2. To account for the ever-changing nature of tax systems, the *Contemporary Engineering Economics: A Canadian Perspective* web site was created. It will be maintained so that changes in tax regulations and rates can be posted on this increasingly popular Internet tool.

Complete Coverage of Special Topics

A number of special topics are important to a comprehensive understanding of introductory engineering economics. Chapters 10 through 14 cover topics such as (1) capital budgeting, (2) income tax effects on personal investments and replacement analysis, (3) inflation, (4) project risk and uncertainty, and (5) public sector analysis.

Recognizing that availability of time and priorities vary from course to course, and from instructor to instructor, each one of these chapters is sufficiently self-contained so that it may be skipped or covered out of sequence, as needed.

Addressing Educational Challenges

The features of *Contemporary Engineering Economics: A Canadian Perspective* were selected and shaped to address key educational challenges. It is our observation and that of the publisher–based on many conversations with engineering educators–that, across the engineering curriculum,,certain challenges consistently frustrate both instructors and students alike. Low student motivation and enthusiasm, student difficulty in developing problem-solving skills and intuition, challenges to integrate technology without short-changing fundamental concepts and traditional methods, and student difficulty in prioritizing and remembering enormous amounts of information are among the key educational challenges that drove the features of this second edition of *Contemporary Engineering Economics: A Canadian Perspective.*

Building Problem-Solving Skills and Confidence

The examples in the text are formatted to maximize their usefulness as guides to problem solving. Further, they are intended to stimulate student curiosity to look beyond the mechanics of problem solving to "what if" issues, alternative solutions, and interpretation of solutions. Each example in the text is formatted as follows:

- **Example titles** promote ease of student reference and review
- **Discussion sections** at the beginning of complex examples help students begin organizing a problem-solving approach.
- **Given** and **Find** heads in **Solution** sections of Examples help students identify critical data. This convention is primarily employed in Chapters 2 through 9, and then generally omitted in Chapters 10 through 14, after student confidence in setting up solution procedures has been established.
- **Comments sections** at the end of examples add additional insights–these could take the form of an alternative solution method, a short-cut, or an interpretation of the numerical solution, thereby extending the educational value of the example.
- **Problems** are categorized by level of difficulty thereby allowing students to build confidence as they master the concepts and solve increasingly difficult problems.

Capturing the Student's Imagination

Students want to know how the conceptual and theoretical knowledge they are acquiring will be put to use. To stimulate student enthusiasm and imagination,

Contemporary Engineering Economics: A Canadian Perspective incorporates real-world applications and contexts in a number of ways.

1. *The real-world, conceptual overview of engineering economics established in Chapter 1:* This provides an engaging introduction to engineering economics via examples of its practical use.
2. *The chapter-opening scenarios:* These establish an interest in the need-to-know chapter concepts within the context of a practical application.
3. *The many homework problems involving real engineering projects:* These stimulate student interest and motivation with actual engineering investment projects, many taken from today's headlines.
4. *The full range of engineering disciplines represented in problems, examples, chapter openers, and case studies:* The diversity of these elements illustrates that many disciplines require engineering economics. Industrial, chemical, civil, electrical, mechanical, manufacturing, and other areas are all represented.
5. *Personal finance applications:* Topics relating to personal financial management demonstrate the usefulness of the concepts to the individual.

Harnessing the Power of the Computer

The integration of the computer into the course and text is another important features of *Contemporary Engineering Economics: A Canadian Perspective*. As a consequence, students will have greater access to, and familiarity with, various spreadsheet tools, and instructors will have a greater inclination either to treat these topics explicitly in the course or to encourage students to experiment independently.

A concern may be that the computer will undermine true understanding of course concepts. However, this text does not promote the trivial or mindless use of computers as a replacement for genuine understanding of, and skill in, applying traditional solution methods. Rather, it focuses on the computer's productivity-enhancing benefits for complex project cash flow development and analysis. Specifically, *Contemporary Engineering Economics: A Canadian Perspective* includes a robust introduction to computer automation in the form of the Computer Notes sections, which appear at the end of most chapters.

Spreadsheets are introduced via Microsoft Excel examples. In terms of spreadsheet coverage, the emphasis is on demonstrating that the more complex chapter concepts can be much more efficiently resolved by a computer than by traditional longhand methods. In Appendix B, conversion tables are included so that the built-in financial functions of both Lotus and Quattro Pro can be compared to Excel for case in "translating" examples to other software programs.

An Internet Tool, the *Contemporary Engineering Economics: A Canadian Perspective* web site (located at www.pearsoned.ca/park) has been created and will be maintained. This text takes advantage of the Internet, which will become increasingly important as a resource to access a variety of information published in cyberspace.

The web site provides instructors and students with additional and more up to date information which can be used to make the course and student learning more effective. Material which is available includes current tax rates, links to other web sites such as the Canada Customs and Revenue Agency, Statistics Canada, the Globe and Mail which provide information on tax regulations, the economy, general business conditions, market quotations, and a French Table of Contents. In addition, a revised Canadian version of the **EzCash** software can be downloaded from the web site free of charge.

A great deal of care and attention has gone into the preparation of this text to make it virtually error free. When instructors do encounter errors (misspelled words, arithmetic mistakes, etc) or have suggestions, they are encouraged to contact the authors using the direct email link.

For those instructors who are considering adopting *Contemporary Engineering Economics: A Canadian Perspective* as their course text, the web site includes the Table of Contents and the Preface describing the text, biographical information on the authors, and a Pearson Education, Canada link to order a copy.

Flexibility of Coverage

For a typical three-credit-hour, one-semester course, the majority of topics in this text can be covered by taking advantage of the depth and breadth in which they are presented. For other arrangements—quarter terms or fewer credit-hours—chapters 1 through 9 present the essential topics, and subsequent chapters present optional coverage. By varying the depth of coverage, and supplementing the reading with case studies, enough materials are provided for a continuing, two-term engineering economics course.

Because the topics of the time value of money and interest relationships are so basic to the overall subject of engineering economics, these are treated in depth in Chapters 2 and 3. For those wishing a briefer coverage of these topics we suggest covering Chapter 2 in its entirety, and Sections 3.1 and 3.2 of Chapter 3. The remaining topics in Chapter 3 may be omitted entirely or assigned as additional readings.

Supplement

An Instructor's Manual is available to adopters of this text. In addition to complete solutions to all problems, it contains:

- A problem key for each chapter where problems are categorized by topic and degree of difficulty.
- Transparency masters for key cash flow diagrams from the text.

ACKNOWLEDGMENTS
BY CHAN PARK

This book reflects the efforts of a great many individuals who reviewed and contributed to the first edition. First, I would like to thank each of them once again:

Kamran Abedini, California Polytechnic – Pomana; James Alloway, Syracuse University; Mehar Arora, U. Wisconsin – Stout; Joel Arthur, California State University – Chico; Robert Baker, University of Arizona; Robert Barrett, Cooper Union and Pratt Institute; Tom Barra, Iowa State University; Charles Bartholomew, Widener University; Richard Bernhard, North Carolina State University; Bopaya Bidanda, University of Pittsburgh; James Buck, University of Iowa; Philip Cady, The Pennsylvania State University; Tom Carmichal, Southern College of Technology; Jeya Chandra, The Pennsylvania State University; Max C. Deibert, Montana State University; Stuart E. Dreyfus, University of California – Berkeley; W.J. Foley, RPI; Jane Fraser, Ohio State; Bruce Hartsough, University of California – Davis; Carl Hass, University of Texas – Austin; John Held, Kansas State University; T. Allen Henry, University of Alabama; R.C. Hodgson, University of Notre Dame; Philip Johnson, University of Minnesota; Harold Josephs, Lawrence Tech; Henry Kallsen, University of Alabama; W.J. Kennedy, Clemson University; Oh Keytack, University of Toledo; Wayne Knabach, South Dakota State University; Stephen Kreta, California Maritime Academy; John Krogman, University of Wisconsin – Platteville; Dennis Kroll, Bradley University; Michael Kyte, University of Idaho; William Lesso, University of Texas – Austin; Martin Lipinski, Memphis State University; Robert Lundquist, Ohio State University; Richard Lyles, Michigan State University; Abu S. Masud, The Wichita State University; James Milligan, University of Idaho; Richard Minesinger, University of Massachusetts – Lowell; James S. Noble, University of Missouri, Columbia; Wayne Parker, Mississippi State University; Elizabeth Pare-Cornell, Stanford University; Cecil Peterson, GMI; george Prueitt, U.S. Naval Postgraduate School; J.K. Rao, California State University – Long Beach; Susan Richards, GMI; Mark Roberts, Michigan Tech; John Roth, Vanderbilt University; Bill Shaner, Colorado State University; Fred Sheets, California Polytechnic – Pomona; Dean Shup, University of Cincinnati; Milton Smith, Texas Tech; Charles Stavridge, FAMU/PSU; Junius Storry, South Dakota State University; Frank E. Stratton, San Diego State University; George Stukhart, Texas A & M University; Donna Summers, University of Dayton; Joe Tanchoco, Purdue University; Deborah Thurston, University of Illinois – UC; L. Jackson Turaville, Tennessee Technological University; Thomas Ward, University of Louisville, Theo De Winter, Boston University.

In addition, the following individuals reviewed the first edition or the revised manuscript and provided detailed comments and suggestions for improving the second edition:

Kamran Abedini, California State Polytechnic University, Pomona
Bopaya Bidanda, University of Pittsburgh
Stuart Dreyfus, University of California, Berkeley
William J. Foley, Renessclaer Polytechnic Institute
Anil K. Goyal, Renessclaer Polytechnic Institute
Bruce Hartsough, University of California, Davis
Scott Iverson, University of Washington
Peter Jackson, Cornell University
William J. Kennedy, Clemson University
Sue McNeil, Carnegie – Mellon University
Gary Moynihan, The University of Alabama
Bruce A. Reichert, Kansas State University
Susan E. Richards, GMI Engineering & Management Institute
Mark C. Roberts, Michigan Technological University
Paul L. Schillings, Montana State University
David C. Slaughter, University of California, Davis
Donna C.S. Summers, University of Dayton

Personally, I would like to thank Barbara Brown, who served as copy editor and editorial coordinator during the preparation of this second edition of *Contemporary Engineering Economics*. Her writing skills and precise knowledge of the English language were of immense value for helping ensure that the written component of the text is as clear, accurate, and relevant as possible. The Addison Wesley book team, especially, Melanie van Rensburg, Rob Merino, Nancy Smith, and Dan Joranstaad helped greatly with all phases of the revision. Thanks are also due to James Treharne, who worked closely with me at every stage of reviewing the galley proofs, and Venkat Narayanan who helped me in preparing earlier drafts of the manuscript.

Finally, I would like to thank Ed Unger, Head of Industrial & Systems Engineering at Auburn University, who provided me with the resources and the constant encouragement that allowed me to revise this second edition of the book.

Acknowledgments by the Canadian Authors

A number of individuals provided assistance in the preparation of the first edition. Their contributions have carried over directly or indirectly into the second edition. We again acknowledge the following people:

John R. McDougall, Alberta Research Council; Douglas A. Hackbarth, Hackbarth Environmental Consultants; Ron G. Bryant, RBC Dominion Securities; Barry J. Walker, Peterson Walker Chartered Accountants; Carl F.

Hunter, Dalcor Consultants Ltd.; Tony D'Andrea, Toronto Dominion Bank; Glen A. Murney, University of Alberta; Peter Wilson, Technical University of Nova Scotia; Scott Dunbar, University of British Columbia; Eldon Gunn, Dalhousie University; Ken Rose, Queen's University; Milica Saagh of St. Mary's University; Kenneth Sadler, an author of the first edition; Gervais Soucy, University of Sherbrooke; Sastry VanKamamidi, Saint Mary's University, and Jerry Ward, University of New Brunswick suggested changes or improvements which were incorporated in the second edition and/or reviewed the final manuscript. We appreciate their efforts in helping us improve the second edition.

Pearson Education Canada would like to acknowledge the contribution of Professor Kenneth F. Sadler from Dalhousie University (Previously Technical University of Nova Scotia). Professor Sadler was a co-author of the first Canadian edition and continued to actively participate in this edition of the text as a reviewer. We would like to sincerely thank Professor Kenneth F. Sadler for his contribution as an author and his continued support as a reviewer of the text.

Each of us sought help from various people in the preparation of this second edition. The kind of assistance included clerical help with the manuscript, specialized advice on financial or income tax matters and programming changes to **EzCash**. The contributions of the following individuals were greatly appreciated: Val Arychuk, Nelson Epp, Jill Stanton, Cara Wood, Lina Xu from the University of Alberta; Sandy Wishloff of KPMG.

The possibility of a second edition was first presented to us by Brian Henderson, who was then Publisher, College Division, Addison Wesley Longman, Canada. When we started the project, we were again working with Linda Scott who had done such an excellent job as Managing Editor of the first edition. Unfortunately, Linda's health did not permit her to continue. Carol Ring did the copy editing on the first draft of all the chapters. Dave Ward, Senior Acquisitions Editor of Pearson Education Canada, assumed responsibility for the overall project and saw it through to completion. Laurie Goebel, Development Editor worked with us to get the manuscript to the production stage. Jennifer Therriault; Production Editor, and the production staff at Pearson Education Canada then published the book on a tight schedule. We thank all of these people for their belief in and commitment to this project.

Finally, the preparation of this manuscript would not have been possible without the support and understanding of our families.

Chan S. Park
Auburn University
Auburn Alabama

Kenneth C. Porteous
University of Alberta
Edmonton, Alberta

Ronald P. Pelot
Dalhousie University
Halifax, Nova Scotia

Ming J. Zuo
University of Alberta
Edmonton, Alberta

CHAPTER 1

Engineering Economic Decisions

Most consumers abhor lukewarm beverages, especially during the hot days of summer, but throughout history, necessity has been the mother of invention. Several years ago, Sonya Talton, an electrical engineering student, had a revolutionary idea—a self-chilling pop can!

Picture this. It's one of those sweltering, hazy August afternoons. Your friends have finally gotten their acts together for a picnic at the lake. Together you go over the list of stuff you need—blankets, radio, sunscreen, sandwiches, chips, and pop. You wipe the sweat from your neck, reach for a pop, and realize that it's about the same temperature as the 30°C afternoon. Great start. Everyone's just dying to make another trip back to the store for ice. Why can't they come up with a pop container that can chill itself, anyway?

Sonya decided to take on the topic of a pop container that can chill itself as a term project in her engineering graphics and design course. The professor stressed innovative thinking and urged students to consider unusual or novel concepts. The first thing Sonya needed to do was to establish some goals for the project:

- Get the pop as cold as possible in the shortest possible amount of time.
- Keep the container design simple.
- Keep the size and weight of the newly designed container similar to that of the traditional pop can. (This would allow beverage companies to use existing vending machines and storage equipment.)
- Keep the production cost low.
- Make the product environmentally safe.

With these goals in mind, Sonya had to think of a practical yet innovative way of chilling the can. Ice was the obvious choice—practical, but not innovative. Sonya had a great idea—What about a chemical ice pack? Sonya's next question was, What's inside? The answer was ammonium nitrate (NH_4NO_3) and a water pouch. When the needed pressure is applied to the chemical ice pack, the water pouch breaks and mixes with the NH_4NO_3, creating an endothermic reaction (the absorption of heat). The NH_4NO_3 draws the heat out of the pop, causing it to chill. How much water is in the water pouch? Sonya measured it: 135 ml. She wondered what would happen if she reduced the amount of water. After several trials involving different amounts of water, Sonya found that she could chill the pop can from 27°C to 9°C in a 3-minute period. At this point, she needed to determine how cold a refrigerated pop gets. She put a can in the fridge for 2 days and found out it chilled to 5°C. Sonya's idea was definitely feasible. But was it economically marketable?

In her engineering graphics and design course, the topic of how economic feasibility plays a major role in the engineering design process was discussed. The professor emphasized the importance of marketing surveys and cost/benefit analyses as ways to gauge a product's potential. Sonya surveyed approximately 80 people. She asked them only two questions: Their age and how much would they be willing to pay for a self-chilling can of pop. The under-21 group was willing to pay the most, 84 cents. The 40-plus bunch only wanted to pay 68 cents. Overall, the surveyed group would be willing to shell out 75 cents for a self-chilling pop. (It was hardly a scientific market survey, but it did give Sonya a feel for what would be a reasonable price for her product.)

The next hurdle was to determine the existing production cost of one can of pop. How much more would it cost to produce the self-chiller? Would it be profitable? She went to the library, and there she found the bulk cost of chemicals and materials she would need. Then she calculated how much she would require for one unit. She couldn't believe it! It only costs 12 cents to manufacture one can of pop, including transportation. Her can would cost 2 or 3 cents more. That wasn't bad, considering the average consumer was willing to pay up to 25 cents more for the self-chilling can than for the traditional one.

The only two constraints left to consider were possible chemical contamination and recyclability. Theoretically it should be possible to build a machine that would drain the solution from the can and recrystallize it. The ammonium nitrate could then be reused in future pop cans, including the plastic outer can. Chemical contamination, the only remaining restriction, was a big concern. Unfortunately, there was absolutely no way to ensure that the chemical and the pop would never come in contact with one another inside the cans. To ease consumer fears, Sonya decided a color or odor indicator could be added to alert the consumer to contamination if it occurred.

Sonya's conclusion?—The self-chilling beverage can would be an incredible technological advancement. The product would be convenient for the beach, picnics, sporting events, and barbecues. Its design would incorporate consumer convenience while addressing environmental concerns. It would be innovative, yet inexpensive, and it would have an economic, as well as a social, impact on society ...[1]

1.1 Economic Decisions

The economic decisions that engineers make in business differ very little from those made by Sonya, except for the scale of the concern. Suppose, for example, that a firm is using a lathe that was purchased 12 years ago to produce pump shafts. As the production engineer in charge of this product, you expect demand to continue into the foreseeable future. However, the lathe has begun to show its age: It has broken frequently during the last 2 years and has finally stopped operating altogether. Now you have to decide whether to replace or repair it. If you expect a more efficient lathe to be available in the next 1 or 2 years, you might repair it instead of replacing it. The major issue is whether you should make the considerable investment in a new lathe now or later. As an added complication, if demand for your product begins to decline, you may have to conduct an economic analysis to determine whether declining profits from the project offset the cost of a new lathe.

Let us consider a real world engineering decision problem of much larger scale. Public concern about poor air quality is increasing, particularly that caused by gasoline-powered automobiles. With requirements looming in California, New York, and Massachusetts for automakers to produce electric vehicles, General Motors Corporation has decided to build an advanced electric car to be known as Impact.[2] The biggest question remaining about the feasibility of the vehicle concerns its battery. With its current experimental battery design, Impact's monthly operating cost would be roughly twice that of a conventional automobile. The primary advantage

---

[1] Background material from 1991 Annual Report, GWC Whiting School of Engineering, Johns Hopkins University (with permission).
[2] "G.M. to Begin Production of a Battery-Powered Car," *The New York Times*, April 19, 1990; "G.M. Displays the Impact Car, an Advanced Electric Car," *The New York Times*, January 3, 1990; and "Make-or-Break Year for Electric Cars," *The New York Times*, January 3, 1994. (Reprinted with permission.)

of the design, however, is that Impact does not emit any pollutants, a feature that could be very appealing at a time when government air-quality standards are becoming more rigorous, and consumer interest in the environment is strong.

Engineers at General Motors have stated that California would be the primary market for Impact, but they added that a nationwide annual demand of 100,000 cars would be necessary to justify production. Despite General Motors management's decision to build the battery-powered electric car, the engineers involved in making the engineering economic decision were still uncertain whether the demand for such a car would be sufficient to justify its production.

Obviously, this level of engineering decision is more complex and more significant to the company than a decision about when to purchase a new lathe. Projects of this nature involve large sums of money over long periods of time, and it is difficult to predict market demand accurately. An erroneous forecast of product demand can have serious consequences: With any overexpansion, unnecessary expenses will have to be paid for unused raw materials and finished products. In the case of Impact, if the improved battery design never materializes, demand may remain insufficient to justify the project.

In this book, we will consider many investment situations, personal investments as well as business investments. The focus, however, will be on evaluating engineering projects on the basis of economic desirability and on investment situations that face a typical firm.

1.2 Predicting the Future

Economic decisions differ in a fundamental way from the types of decisions typically encountered in engineering design. In a design situation, the engineer utilizes known physical properties, the principles of chemistry and physics, engineering design correlations, and engineering judgment to arrive at a workable and optimal design. If the judgment is sound, the calculations are done correctly, and we ignore technological advances, the design is time invariant. In other words, if the engineering design to meet a particular need is done today, next year, or in five years' time, the final design would not change significantly.

In considering economic decisions, the measurement of investment attractiveness, which is the subject of this book, is relatively straightforward. However, information required in such evaluations always involves the prediction or forecasting of product sales, product selling price, and various costs over some future timeframe— 5 years, 10 years, 25 years, etc.

All such forecasts have two things in common. First, they are never completely accurate when compared with the actual values realized at future times. Secondly, a prediction or forecast made today is likely to be different than one made at some point in the future. It is this ever-changing view of the future which can make it necessary to revisit and even change previous economic decisions. Thus, unlike engineering design, the conclusions reached through economic evaluation are not necessarily time invariant. Economic decisions have to be based on the best information available at the time of the decision and a thorough understanding of the uncertainties in the forecasted data.

1.3 Role of Engineers in Business

Apple Computer, Microsoft Corporation, and Sun Microsystems produce computer products and have a market value of several billion dollars. These companies were all started in the late 1970s or early 1980s by young college students with technical backgrounds. When they went into the computer business, these students initially organized their companies as proprietorships. As the businesses grew, they became partnerships and were eventually converted to corporations. This chapter will introduce the three primary forms of business organization and briefly discuss the role of engineers in business.

1.3.1　Types of Business Organization

As an engineer, it is important to understand the nature of the business organization with which you are associated. This section will present some basic information about the type of organization you should choose should you decide to go into business for yourself.

The three legal forms of business, each having certain advantages and disadvantages, are proprietorships, partnerships, and corporations.

Proprietorships

A **proprietorship** is a business owned by one individual. This person is responsible for a firm's policies, owns all its assets, and is personally liable for its debts. A proprietorship has two major advantages. First, it can be formed easily and inexpensively. No legal and organizational requirements are associated with setting up a proprietorship, and organizational costs are therefore virtually nil. Second, the earnings of a proprietorship are taxed at the owner's personal tax rate, which may be lower than the rate at which corporate income is taxed. Apart from personal liability considerations, the major disadvantage of a proprietorship is that it cannot issue stocks and bonds, making it difficult to raise capital for any business expansion.

Partnerships

A **partnership** is similar to a proprietorship except that it has more than one owner. Most partnerships are established by a written contract between the partners, which normally specifies salaries, contributions to capital, and the distribution of profits and losses. A partnership has many advantages, among which are its low cost and ease of formation. Because more than one person makes contributions, a partnership typically has a larger amount of capital available for business use. Since the personal assets of all the partners stand behind the business, a partnership can borrow money more easily from a bank. Each partner pays only personal income tax on his or her share of a partnership's taxable income.

On the negative side, under partnership law, each partner is liable for a business's debts. This means that the partners must risk all their personal assets, even those not invested in the business. And while each partner is responsible for his or her portion of the debts in the event of bankruptcy, if any partner cannot meet his or her pro rata claim, the remaining partners must take over the unresolved claims. Finally, a partnership has a limited life insofar as it must be dissolved and reorganized if one of the partners quits.

Corporations

A **corporation** is a legal entity created under provincial or federal law. It is separate from its owners and managers. This separation gives the corporation four major advantages: (1) It can raise capital from a large number of investors by issuing stocks and bonds; (2) it permits easy transfer of ownership interest by trading shares of stock; (3) it allows limited liability—personal liability is limited to the amount of the individual's investment in the business; and (4) it is taxed differently than proprietorships and partnerships, and under certain conditions, the tax laws favor corporations. On the negative side, it is expensive to establish a corporation. Furthermore, a corporation is subject to numerous governmental requirements and regulations.

As a firm grows, it may need to change its legal form because the form of a business affects the extent to which it has control of its own operations and its ability to acquire funds. The legal form of an organization also affects the risk borne by its owners in case of bankruptcy and the manner in which a firm is taxed. Apple Computer, for example, started out as a two-man garage operation. As the business grew, the owners felt constricted by this form of organization: It was difficult to raise capital for business expansion; they felt that the risk of bankruptcy was too high to bear; and as their business income grew, their tax burden grew as well. Eventually, they found it necessary to convert the partnership into a corporation.

In Canada, the overwhelming majority of business firms are proprietorships, followed by corporations and partnerships. However, in terms of total business volume (dollars of sales), the quantity of business done by proprietorships and partnerships is several times less than that of corporations. Since most business is conducted by companies of the corporation form, this text will generally address economic decisions encountered in corporations.

1.3.2   Engineering Economic Decisions

What role do engineers play within a firm? What specific tasks are assigned to the engineering staff, and what tools and techniques are available to it for improving a firm's profits? Engineers are called upon to participate in a variety of decision-making processes, ranging from manufacturing, through marketing, to financing decisions. We will restrict our focus, however, to various economic decisions related to engineering projects. We refer to these decisions as **engineering economic decisions**.

In manufacturing, engineering is involved in every detail of a product's production, from the conceptual design to the shipping. In fact, engineering decisions account for the majority (some say 85%) of product costs. Engineers must consider the effective use of capital assets such as buildings and machinery. One of the engineer's primary tasks is to plan for the acquisition of equipment (**capital expenditure**) that will enable the firm to design and produce products economically.

With the purchase of any fixed asset, equipment for example, we need to estimate the profits (more precisely, cash flows) that the asset will generate during its service period. In other words, we have to make capital expenditure decisions based on predictions about the future. Suppose, for example, you are considering the purchase of a deburring machine to meet the anticipated demand for hubs and sleeves used in the production of gear couplings. You expect the machine to last 10 years. This purchase decision thus involves an implicit 10-year sales forecast for the gear couplings, which means that a long waiting period will be required before you will know whether the purchase was justified.

An inaccurate estimate of asset needs can have serious consequences. If you invest too much in assets, you incur unnecessarily heavy expenses. Spending too little on fixed assets is also harmful, for then the firm's equipment may be too obsolete to produce products competitively and, without an adequate capacity, you may lose a portion of your market share to rival firms. Regaining lost customers involves heavy marketing expenses and may even require price reductions and/or product improvements, all of which are costly.

1.3.3   Personal Economic Decisions

In the same way that an engineer can play a role in the effective utilization of corporate financial assets, each of us is responsible for managing our personal financial affairs. After paying for nondiscretionary or essential needs such as housing, food, clothing, and transportation, any remaining money is available for discretionary expenditures on items such as entertainment, travel, investments, etc. For money which we choose to invest, we want to maximize the economic benefit at some acceptable risk. The investment choices are unlimited, including savings accounts, guaranteed investment certificates, stocks, bonds, mutual funds, registered retirement savings plans, rental properties, land, business ownership, etc.

How do you choose? Analysis of personal investment opportunities utilizes the same techniques used for engineering economic decisions. Again, the challenge is predicting the performance of an investment into the future. Choosing wisely can be very rewarding while choosing poorly can be disastrous. Some investors in the gold mining stock BreX who sold prior to the fraud investigation became millionaires. Others, who did not sell, lost everything.

A wise investment strategy is one which manages risk by diversification of investments. You have a number of different investments which range from very low to very high risk and are in a number of business sectors. Since you do not have all your money in one place, the risk of losing everything is significantly reduced.

1.4 Large-Scale Engineering Projects

In the development of any product, a company's engineers are called upon to translate an idea into reality. A firm's growth and development largely depends upon a constant flow of ideas for new products, and for the firm to remain competitive, it has to make existing products better or produce them at a lower cost. Traditionally, a marketing department would propose a product and pass the recommendation to the engineering department. The engineering department would work up a design and pass it on to manufacturing, which would make the product. With this type of product development cycle, a new product normally takes several months (or even years) to reach the market. A typical tool maker, for example, would take 3 years to develop and market a new machine tool. However, the Ingersoll-Rand Company, a leading tool maker, was able to cut down the normal development time of its products by one-third. How did the company do this? A group of engineers examined the current product development cycle to find out why things dragged on. They learned how to compress the crippling amount of time it took to bring products to life. In the next section, we will present an example of how a design engineer's idea eventually turned into a popular consumer product.

1.4.1   How a Typical Project Idea Evolves

In many ways, the Gillette Sensor razor (Figure 1.1), introduced in January 1990, is not a "typical" project. It represents the single most expensive project the Gillette Company has ever developed, and its success has been nothing short of phenomenal—it is currently the bestselling razor in the 39 countries in which it has been launched. Less than 2 years after its introduction, the one billionth Sensor cartridge was manufactured by Gillette.

Product Development Summary (Sensor)	
Patents	18 granted, 4 pending
Parts	23
Assembly process	34 steps
Development time	13 years
Capital investment	$125 million*
Research and development	$75 million*
1990 advertising	$110 million*
Retail price	$3.75 for razor and three blades

*US dollars

A look at the evolution of the investment project that produced Sensor reveals a number of stages and events that are common to most engineering investment projects (see above). The following excerpted passages are from *Business Week*, January 29, 1990. Additional information has been provided by Gillette.

There are 40 engineers, metallurgists, and physicists at Gillette Company's Reading (Britain) research facility who spend their days thinking about shaving and little else. In 1977, one of them had a bright idea. John Francis had already figured out how to create a thinner razor blade that would make Gillette's cartridges easier to clean. Then, the design engineer remembered a notion he had toyed with for years: He could set the thinner blades on springs so that they would follow the contours of a man's face. He built a simple prototype, gave it a test, and thought, "This is pretty good." He passed the idea to his boss, then went on to the next project.[3]

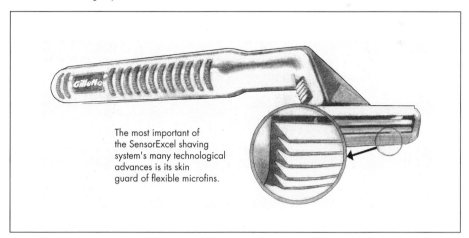

The most important of the SensorExcel shaving system's many technological advances is its skin guard of flexible microfins.

Figure 1.1   Gillette Corporation's new razor — Sensor (Courtesy of The Gillette Company.)

---

[3] *Business Week*, January 29, 1990. (Reprinted with permission of *Business Week Magazine* and The Gillette Company.)

The initial idea for an investment project may be born of inspiration or necessity and may entail an entirely new venture for a firm or the expansion or improvement of current ventures. In the case of Sensor, John Francis's inspiration was for an improvement so profound that it caused consumers and competitors to view Sensor as the next generation in razors.

The case of Sensor also demonstrates that a long period of time may elapse between the moment an idea is articulated and the moment it is fully implemented as a completed, for-sale product. This time interval may vary widely from industry to industry. Computer manufacturers, for example, place a premium on swift implementation of investment projects due to the rapidly evolving technologies with which they work and compete.

Selection of materials and processes are some of the most fundamental decisions affecting project costs. Gillette faced some particularly sticky issues with Sensor:

> Gillette used styrene plastic to mold blade cartridges for all its razors, because it was inexpensive and easy to work with. But a styrene spring, tests showed, lost some of its bounce over time. The engineers turned to a resin called Noryl, a stronger material that kept its bounce.[4]

In addition to innovating the material with which Sensor's blade cartridge would be constructed, Gillette made a remarkable production decision:

> Sensor's blades were to "float" on the springs independently of each other. That meant the blades had to be rigid enough to hold their shape—though each was no thicker than a sheet of paper. Engineers decided to attach each blade to a thicker steel support bar.
>
> The question was, how? For mass manufacturing, glue was too messy and too expensive. The answer was lasers. Engineers built a prototype laser that spot-welded each blade to a support without creating heat that would damage the blade edge, relying on a process more commonly used to make things such as heart pacemakers.[5]

As with any investment project, the materials and manufacturing processes that Sensor required had cost implications. Estimating costs is one of the more critical tasks involved in any investment project:

> Sensor was the single most expensive project Gillette had ever taken on: By the time the razor hit the stores, the company had spent an estimated $200 million in research, engineering, and tooling.[6]

---

[4] Ibid.
[5] Ibid.
[6] Ibid.

Engineers conducting economic analyses typically consider equipment, material, tooling, and labor costs. However, they may not be used to factor in expenses such as advertising, promotion, and public relations. In the case of Sensor, such costs added another $110 million to the project's overall cost. In making an accept or reject decision about a project, after estimating costs, the next concern is revenues:

> Gillette will need a huge win to justify its investment in Sensor. The new razor must add about four percentage points to Gillette's market share in the United States and Europe, just to recoup its ad budget.
>
> Those pressures also help explain why it took so long to get Sensor out the door. Gillette executives were reluctant to make the huge investment in manufacturing and marketing at a time when the company could still count on prodigious profits from its existing razors.[7]

Clearly, with a sizable investment, a firm needs a sizable return to justify undertaking an investment project. In the case of Sensor, the advertising budget alone was a significant expense; recovering that cost required that the company increase its market share by 4%. Note also that if existing ventures are already generating adequate income, a firm may be reluctant to undertake a risky new venture, which could fail or which might jeopardize existing projects by taking resources and sales away from them. (In fact, in Sensor's case, the new product took sales away from other Gillette products, namely Trac II and Atra, but to a lesser extent than the company expected.)

Even with the most sophisticated market research to support it, no product can be assured of success. Estimating project costs and revenues as exactly as possible is critical to a firm's decision to pursue a project. But finally the success or failure of a project cannot be judged until it is implemented and evaluated in comparison with predictions of its success. In the case of Sensor, success has been well documented, and the risk undertaken by Gillette has been justified.

Virtually every employee of a company is a potential source of ideas for a new product or the improvement of existing products, and many companies encourage their employees to present new ideas for evaluation. Many good ideas for product improvement come from the engineers actually involved in production or marketing. The process, of course, may take a long time. As our discussion illustrated, it took 13 years to realize John Francis's idea. Fortunately, not all engineering ideas take such a long time to develop into products.

---

[7] Ibid.

1.4.2 Impact of Engineering Projects on Financial Statements

Engineers must understand the business environment in which a company's major business decisions are made. It is important for an engineering project to generate profits, but it also must strengthen the firm's overall financial position. How do we measure The Gillette Company's success in the Sensor project? Will enough Sensor razors be sold, for example, to keep the blade business as Gillette's biggest source of profits? While the Sensor project will provide comfortable, reliable, low-cost shaving for the customers, the bottom line is its financial performance over the long run.

Regardless of a business's form, each company has to produce basic financial statements at the end of each operating cycle (typically a year). These financial statements provide the basis for future investment analysis. In practice, we seldom make investment decisions based solely on an estimate of a project's profitability because we must also consider its overall impact on the financial strength and position of the company.

Suppose that you were the president of The Gillette Company. Let us further suppose that you even hold some shares in the company, which also makes you one of the company's many owners. What objectives would you set for the company? While all firms are in business in hopes of making a **profit**, what determines the market value of a company is not profits per se, but **cash flow**. It is, after all, available cash that determines the future investments and growth of the firm. Therefore, one of your objectives should be to increase a company's value to its owners (including yourself) as much as possible. The **market price** of your company's stock to some extent represents the value of your company. Many factors affect your company's market value: Present and expected future earnings, the timing and duration of these earnings, as well as risks associated with them. Certainly, any successful investment decision will increase a company's market value. Stock price can be a good indicator of your company's financial health and may also reflect the market's attitude about how well your company is managed for the benefit of its owners.

In the case of The Gillette Company, the firm's financial position prior to, and after, introducing the Sensor to the market was as follows.

The Gillette Company at a Glance*				
Year ended (Dec. 31)	1997	1995	1990	1989
Revenues	$10,062	$6795	$4345	$3819
Net income	$1427	$824	$368	$285
Earnings per share	$1.27	$0.97	$0.36	$0.29
Common shares outstanding†	1118	1100	906	904
Total assets	$10,804	$6340	$3671	$3114
Total liabilities	$6023	$3827	$2805	$2444
Net worth	$4841	$2513	$866	$670

*Millions of US dollars, except share amounts (Courtesy of The Gillette Company.)
†Millions of shares

Now let's examine how The Gillette Company's financial situation changed from 1989 to 1990. Gillette's 1990 sales were almost $4.345 billion. The $3.671 billion of assets (what it owns) and the $2.805 billion of liabilities (what it owes to creditors) were necessary to support these sales. The $866 million of net worth indicates that portion of the company's assets provided by the investors (owners or stockholders). We can see that the company had earnings of $368 million, but only $311 million available to common stockholders (after paying out $57 million in cash dividends to its preferred stockholders). Gillette had about 906 million shares of common stock, so the company earned $0.36 per share of stock outstanding. (We calculate this **earnings per share** by dividing the net income available to common stockholders by the number of common stock shares outstanding.) This earnings per share indicates an increase of 18% compared with that of 1989.

Investors liked the new product, which resulted in an increased demand for the company's stock. This, in turn, caused stock prices, and hence shareholder wealth, to increase. The Sensor shaving system franchise has continued to build substantially and has accounted for one-third of the total Gillette blade and razor sales dollars in year 1994. This progress has been accelerated by the success of the Sensor for Women (introduced in mid-1992) and the SensorExel (introduced in 1994) shaving systems. In fact, this new, heavily promoted, high-tech Sensor razor turned out to be a smashing success and contributed to sending The Gillette Company's stock to an all-time high in late 1995. This caused Gillette's market value to continue to increase well into 1997.

Year (Dec. 31)	Stock Prices	Market Value
1997	$38.88	$43.040 billion
1996	$26.06	$28.666 billion
1995	$26.06	$23.143 billion
1994	$18.72	16.692 billion
1993	$14.91	13.117 billion
1990	$7.84	3.043 billion
1989	$6.13	2.364 billion

Any successful investment decision on Sensor's scale will tend to increase a firm's stock prices in the marketplace and promote long-term success. Thus, in making a large-scale engineering project decision, we must consider its possible effect on the firm's market value.

1.5 Types of Strategic Engineering Economic Decisions

The story of how The Gillette Company successfully introduced a new product and regained the razor market share previously lost to competitors, is typical: Someone had a good idea, executed it well, and obtained good results. Project ideas such as the Sensor can originate from many different levels in an organization. Since some ideas will be good, while others will not, we need to establish procedures for screening projects. Many large companies have a specialized project analysis division that actively searches for new ideas, projects, and ventures. Once project ideas are identified, they are typically classified as (1) equipment and process selection, (2) equipment replacement, (3) new product and product expansion, (4) cost reduction, and (5) service or quality improvement. This classification scheme allows management to address key questions. Can the existing plant, for example, be used to achieve the new production levels? Does the firm have the knowledge and skill to undertake this new investment? Does the new proposal warrant the recruitment of new technical personnel? The answers to these questions help firms screen out proposals that are not feasible given a company's resources.

Gillette's Sensor project represents a fairly complex engineering decision that required the approval of top executives and the board of directors. Virtually all big businesses at some time face investment decisions of this magnitude. In general, the larger the investment, the more detailed is the analysis required to support the expenditure. For example, expenditures to increase output of existing products, or to manufacture a new product, would invariably require a very detailed economic justification. Final decisions on new products and marketing decisions are generally made at a high level within the company. On the other hand, a decision to repair damaged equipment can be made at a lower level within a company. In this section, we will provide many real examples to illustrate each class of engineering economic decisions. At this point, our intention is not to provide the solution to each example, but rather to describe the nature of decision problems that a typical engineer would face in the real world.

1.5.1   Equipment and Process Selection

What we have just described is a class of engineering decision problems that involve selection of the best course of action when several ways to meet a project's requirements exist. Which of several proposed items of equipment shall we purchase for a given purpose? The choice often turns on which item is expected to generate the largest savings (or return on the investment). Example 1.1 illustrates a process selection problem.

EXAMPLE 1.1   Material selection for an automotive exterior body

Engineers at General Motors want to investigate alternative materials and processes for the production of automotive exterior body panels. The engineers have identified two types of material: Sheet metal and glass fiber reinforced polymer, known as plastic sheet molding compound (SMC), as shown in Figure 1.2. Exterior body panels are traditionally made of sheet steel. With a low material cost, this sheet metal also lends itself to the stamping process, a very high-volume, proven manufacturing process. On the other hand, reinforced polymer easily meets the functional requirements of body panels (such as strength and resistance to corrosion). There is considerable debate among engineers as to the relative economic merits of steel as opposed to plastic panels. Much of the debate stems from the dramatically different cost structures of the two materials.

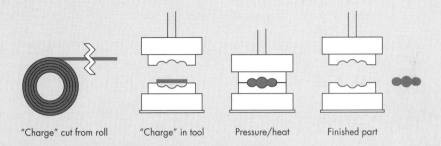

"Charge" cut from roll   "Charge" in tool   Pressure/heat   Finished part

Figure 1.2   Sheet molding compound process (Courtesy of Dow Plastics, a business group of the Dow Chemical Company.)

Description	Plastic SMC	Steel Sheet Stock
Material cost ($/kg)	$1.65	$0.77
Machinery investment	$2.1 million	$24.2 million
Tooling investment	$0.683 million	$4 million
Cycle time (minute/part)	2.0	0.1

Since plastic is petroleum based, it is inherently more expensive than steel, and because the plastic-forming process involves a chemical reaction, it has a slower cycle time. However, both machinery and tool costs for plastic are lower than for steel due to relatively low forming pressures, lack of tool abrasion, and single-stage pressing involved in handling. Thus, the plastic would require a lower initial investment, but would incur higher material costs. Neither material is obviously superior economically. What, then, is the required annual production volume that would make the plastic material more economical?

Comments: The choice of material will dictate the manufacturing process for the body panels. Many factors will affect the ultimate choice of the material, and engineers should consider all major cost elements, such as machinery and equipment, tooling, labor, and material. Other factors may include press and assembly, production and engineered scrap, the number of dies and tools, and the cycle times for various processes.

1.5.2 Equipment Replacement

This category of investment decisions involves considering the expenditure necessary to replace worn-out or obsolescent equipment. For example, a company may purchase ten large presses expecting them to produce stamped metal parts for 10 years. After 5 years, however, it may become necessary to produce the parts in plastic, which would require retiring the presses early and purchasing plastic molding machines. Similarly, a company may find that, for competitive reasons, larger and more accurate parts are required, which will make the purchased machines obsolete earlier than expected. Example 1.2 will provide a real-world example in this category.

EXAMPLE 1.2 **Replacing PCB transformers**

A large power generation and transmission company has a nuclear power plant. This facility was constructed with 44 electrical transformers that utilize a fluid containing polychlorinated biphenyl (PCB) as the liquid portion of the transformer insulation systems. The transformers use the fluid containing PCB as a coolant. As illustrated in Figure 1.3, the liquid, driven by convection currents, flows around the transformer coils, and the "hot" liquid travels outside of the transformer, where it is

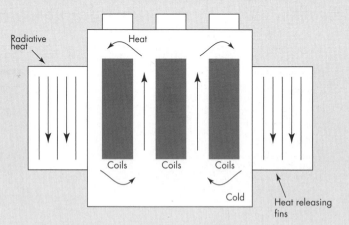

Figure 1.3 A liquid-cooled transformer (the arrows indicate the flow of the liquid and the direction of convection currents)

cooled; it then returns to the coils and the process is repeated. In the event of a fire or a transformer explosion, PCBs may form carcinogenic compounds which are extremely dangerous to human health. Government regulations now ban the production of PCBs. PCB transformers are being replaced or converted to alternate coolants. An internal risk assessment study by the company indicated a slight probability of an explosion with existing PCB transformers. The potential clean-up costs of a PCB transformer fire would range from $15 to $100 million. The study also identified the transformers offering the greatest risk to be those in the turbine building due to a large open area, which could be contaminated in a violent failure.[8]

The company was faced with making a decision as to whether it should continue to live with the well known PCB environmental risks or to reduce the risks by implementing one of the following options:

- Option 1: Replace all 44 transformers with nonhazardous liquid filled transformers (LFTs). The high dielectric strength of LFTs provides greater design flexibility in optimizing specific load requirements so that lower operating costs can be achieved.

- Option 2: Replace 24 transformers in the open area with LFTs and retain the remaining transformers. By only replacing transformers in open areas, 20 transformers in other areas still remain at risk.

- Option 3: Reclassify all the transformers by replacing the PCB fluid with a silicone fluid, which is nontoxic and environmentally safe. The challenge associated with reclassification is the possibility of derating the transformers.

The LFTs would cost $30,000 each, and the reclassified transformers $14,000 each. Annual maintenance costs including parts and labor are estimated at $500 per reclassified transformer and $300 per replaced transformer. A conservative estimate for clean-up in the event of an accident was made at $80 million over a period of 20 years. Knowing that an individual transformer failure rate is estimated at 0.6% over its life time, engineers predict that it is highly likely that one incident would be encountered over a 20-year operation.

Comments: This example involves weighing an expenditure to replace serviceable, but potentially hazardous, equipment in case of an accident. In this problem, we need to consider the economic consequences in the case of an accident, even though its probability is relatively small. With ever-increasing awareness of transformer safety, we need to consider environmental as well as safety elements in an engineering decision. No adverse environmental effects associated with silicon liquid use, production, or disposal have been identified. In fact, silicon liquid is used extensively in hair spray, skincare products, antacids, and even in the food processing industry. In addition, silicon liquid can be reconditioned and reused.

---

[8] Example provided by Rob Wright, Mike Whitt, and Philip Wilson.

1.5.3   New Product and Product Expansion

Investments in this category are those that increase the revenues of a company if output is increased. A description of two common types of expansion decision problems follows: The first category includes decisions about expenditures to increase the output of existing production or distribution facilities. In these situations, we are basically asking, "Shall we build or otherwise acquire a new facility?" The expected future cash inflows in this investment category are the profits from the goods and services produced in the new facility. The second type of decision problem includes considering expenditures necessary to produce a new product or to expand into a new geographic area. These projects normally require large sums of money over long periods. We will provide two examples to illustrate the types of capacity expansion problems.

EXAMPLE 1.3   **Marketing a rapid prototype system**

E.I. DuPont de Nemours, a leading chemical company, has set up a separate venture, Somos, to develop and market a rapid prototype system. This computer-driven technology allows design engineers to design and create plastic prototypes of complicated parts in just a few hours. An example of a rapid prototype system is shown in Figure 1.4.

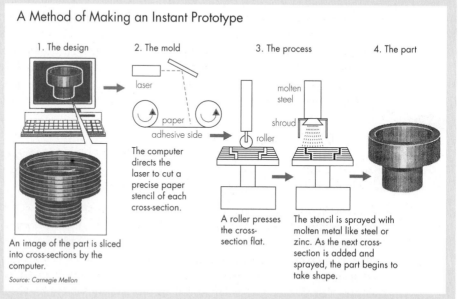

A Method of Making an Instant Prototype

1. The design

An image of the part is sliced into cross-sections by the computer.

Source: Carnegie Mellon

2. The mold

laser

paper
adhesive side

The computer directs the laser to cut a precise paper stencil of each cross-section.

3. The process

molten steel

shroud

roller

A roller presses the cross-section flat.

The stencil is sprayed with molten metal like steel or zinc. As the next cross-section is added and sprayed, the part begins to take shape.

4. The part

Figure 1.4   A typical rapid prototype system (*The New York Times*, April 7, 1993. Reprinted with permission.)

Rapid prototyping has allowed designers to complete projects that formerly took 6 months, or more, in less than 3 weeks. Assignments that took several weeks may now be done in a day or two. Rapid prototyping also saves tens of thousands of dollars per part in modeling costs, compared to traditional methods.

The technology might also give birth to true just-in-time manufacturing for some businesses. An auto parts replacement shop, for example, might simply stock metal and plastic powders along with a library of computer programs that would allow it to build any part a customer needs on the spot. The venture will require an investment of $40 million on the part of DuPont, and the prototype system once developed would be sold for $385,000.

Comments: In this example, we are asking the basic questions involved in introducing a new product: Is it worth spending $40 million to market the rapid prototyping? How many sales are required to recover the investment? The future expected cash inflows in this example are the profits from the goods and services produced with the new product. Are these profits large enough to warrant the investment in equipment and the costs required to make and introduce the product?

EXAMPLE 1.4

Videos from heaven

In 1995, Hughes Communications started to beam 150 television channels directly from space to homes in the continental United States. Hughes offers personalized TV by giving customers the chance to pick what they want and pay for only what they view. As shown in Figure 1.5, customers can now receive Hughes's signals over 45 centimetre satellite dishes—about the size of a large pizza—rather than the 3 metre backyard dishes common in rural areas.[9] Hughes has already set fees for its initial service, though executives expect to match monthly rates currently charged for cable service in the years to come. By using the computer precision of digital signals, the service provides sound and pictures with the fidelity and crispness of laser-disc movies. Still, the risks are huge—and not just because of the project's $1 billion price tag. For all its space age luster, the satellite system could be considered to be old fashioned in just 2 or 3 years, eclipsed by new-age cable television.

To attract customers, Hughes budgeted $40 million for advertising in 1996. It has also sold franchise rights totaling about $125 million to members of rural telephone and electric utilities. These utilities will market the satellite dishes and services within their regions. As planned, the system was in full swing by the Fall of 1995. Hughes expected to break even by the end of 1996, with 3 million customers who are expected to spend an average of about $30 a month. By the year 2000, Hughes expects to capture 10 million homes, or 10 percent of the television market. But some industry experts remain skeptical.[10] At $700 apiece, the satellite receivers and set-top

[9] "Betting Big on Small-Dish TV." *The New York Times*, December 15, 1993. (Reprinted with permission.)
[10] The subscriber base in mid-1999 was 7.4 million and the service was available in Canada.

converters are expensive, and customers will still have to pay monthly fees comparable to those of ordinary cable television. Beyond that, the system is unable to deliver local news and programming because it is beamed nationally. And unless customers buy deluxe receivers for $900, they will not be able to hook up a second television in a different room.

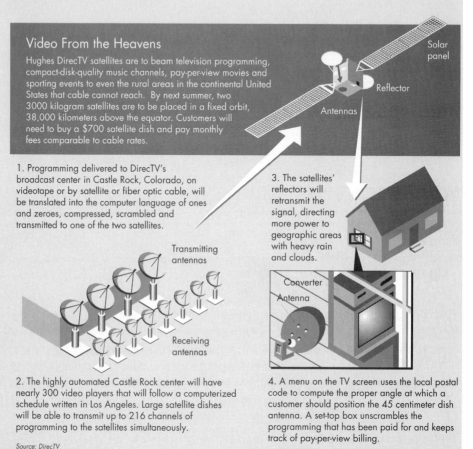

Video From the Heavens

Hughes DirecTV satellites are to beam television programming, compact-disk-quality music channels, pay-per-view movies and sporting events to even the rural areas in the continental United States that cable cannot reach. By next summer, two 3000 kilogram satellites are to be placed in a fixed orbit, 38,000 kilometers above the equator. Customers will need to buy a $700 satellite dish and pay monthly fees comparable to cable rates.

1. Programming delivered to DirecTV's broadcast center in Castle Rock, Colorado, on videotape or by satellite or fiber optic cable, will be translated into the computer language of ones and zeroes, compressed, scrambled and transmitted to one of the two satellites.

3. The satellites' reflectors will retransmit the signal, directing more power to geographic areas with heavy rain and clouds.

2. The highly automated Castle Rock center will have nearly 300 video players that will follow a computerized schedule written in Los Angeles. Large satellite dishes will be able to transmit up to 216 channels of programming to the satellites simultaneously.

4. A menu on the TV screen uses the local postal code to compute the proper angle at which a customer should position the 45 centimeter dish antenna. A set-top box unscrambles the programming that has been paid for and keeps track of pay-per-view billing.

Source: DirecTV

Figure 1.5   Betting on small-dish TV: Hughes DirecTV satellites (*The New York Times*, December 15, 1993. Reprinted with permission.)

Hughes's executives contend they can make money even if they do not grab much of the cable market. More than 9 million American homes do not have access to cable television, and about 5% do not receive major broadcast networks. About 2 million homes already own a large backyard satellite dish. It remains unclear how many of them would switch, or how many of the households that have refused to spend $2000 on the bigger dishes could be enticed to spend $700 on the Hughes equipment. Industry analysts say Hughes has a good chance of reaching its goal of 10 million customers, but they are more skeptical about a broader reach.

Comments:　　　This example illustrates an investment decision problem that requires expenditures necessary to expand into a geographic area not currently being served. These projects normally require large sums of money over long periods of time. This type of project also requires strategic decisions that could change the fundamental nature of a business. Invariably, a very detailed analysis is required, and final decisions on new market ventures are generally made by the board of directors as a part of a company's overall strategic plan.

1.5.4   Cost Reduction

A cost-reduction project is one that attempts to lower a firm's operating costs. Typically we need to consider whether a company should buy equipment to perform an operation now done manually or spend money now in order to save more money later. The expected future cash inflows on this investment are savings resulting from lower operating costs. Example 1.5 illustrates a cost-reduction project.

EXAMPLE

1.5

Lone Star Trucking Company

Lone Star Trucking, a Toronto company with 50 trucks, wishes to convert this fleet to use a fuel other than gasoline. If possible, it would like to have 30% or more of its fleet vehicles operating on an alternative fuel by September 1, 2000, 50% or more of the fleet operating on alternative fuel by September 1, 2002, and 90% or more of the fleet using alternative fuels by September 1, 2004.

Two different types of fuels are under consideration: Compressed Natural Gas (CNG) and Liquid Petroleum Gas (LPG). Both CNG and LPG engines are considered to be economically more efficient than conventional gasoline engines. However, the initial conversion cost for the CNG option will be $305,520. The use of CNG will save the company $16,520 in fuel costs per year for the first 2 years, $27,535 per year for the next 2 years, and $49,560 each year for the remaining 26 years of the project.

To use the LPG option will require an initial investment of $92,520. Every 3 years, each LPG-converted engine must be overhauled at a rate of $350. The savings in fuel costs for LPG are $11,440 a year for the first 2 years, $19,100 a year for the next 2 years, and $34,368 a year for the remaining 26 years.

The major drawback in the CNG project is the high cost of the required compressor and the relatively high cost of the conversion kits. Lone Star needs to decide whether the large initial cost of the CNG project more than offsets the increased fuel savings.

Comments: The choice of whether or not to change over to the alternative fuels and which fuel to switch to is affected by many factors. Among these factors are the number of trucks used by the company and the future outlook of the alternative fuel price. As the size of the fleet increases, so do the savings with the alternative fuel. Therefore, a small company with only a few trucks is less likely to change its fleet of trucks over to an alternative fuel. In other words, a small company does not travel enough cumulative miles to save the money required to recover the initial investment.

1.5.5   Service Improvement

All the examples in the previous sections were related to economic decisions in the manufacturing sector. The decision techniques we develop in this book are also applicable to various economic decisions involved in improving services. Example 1.6, which follows, will illustrate this.

EXAMPLE 1.6   Levi's personal pair of jeans

Levi Strauss & Company, the world's largest jeans maker, found that women complained more than men about difficulties in finding off-the-rack jeans that fit properly. This led to a consideration of whether to sell made-to-order Levi's for women. As shown in Figure 1.6, the idea is that sales clerks at Original Levi's Stores can use a personal computer and the customer's measurements to create what amounts to a digital blue jeans blueprint. When transmitted electronically to a Levi's factory in Tennessee, the computer file would instruct a robotic tailor to cut a bolt of denim precisely to the woman's measurements. The finished Levi's, about $10 more than a mass-produced pair, are shipped back to the store within 3 weeks—or directly to the customer by Federal Express for an additional $5 fee. The service being market-tested by Levi is the first of its type in the clothing industry. It has the potential to change the way people buy clothes, and it will allow stores to cut down on inventory. It is part of an emerging industrial trend toward so-called mass customization, in which computerized instructions enable factories to modify mass-market products one at a time to suit the needs of individual customers.

Levi Strauss, which has been offering this service for several months in Cincinnati, plans to offer the service at more than 30 Original Levi's Stores throughout the country. This national marketing thrust would require a considerable investment in robotic tailors and realignments of the current sewing operation and the assembly process.

Comments: Levi said each computer-designed pair of jeans was basically a customized product. Eventually, this could mean no inventory and no markdowns. A company following this route is not mass producing a product and hoping it sells.

From Data to Denim

A new computerized system being installed at some Original Levi's Stores allows women to order customized blue jeans. Levi Strauss declined to have its factory photographed, so here is an artist's conception of how the process works.

A sales clerk measures the customer using instructions from a computer as an aid.

The clerk enters the measurements, and adjusts the data based on the customer's reaction to samples.

The final measurements are relayed to a computerized fabric cutting machine at the factory.

Bar codes are attached to the clothing to track it as it is assembled, washed and prepared for shipment.

Figure 1.6 Making customized blue jeans for women (*The New York Times*, November 8, 1994. Reprinted with permission.)

This is no small consideration in the nation's apparel industry, in which an estimated $25 billion worth of manufactured clothing each year either goes unsold or sells only after severe markdowns. Levi's main problem is to determine how much demand the women's line would generate. How many more jeans would Levi need to sell to justify the cost of additional robotic tailors? This analysis should involve a comparison of the cost of operating the additional robotic tailors with the additional revenue generated by selling more jeans.

1.6 Cost/Benefit Estimation and Engineering Project Evaluation

Before the economics of an engineering project can be evaluated, it is necessary to have reasonable estimates of the various cost and revenue components which describe the project. Engineering projects may range from something as simple as the purchase of a new milling machine to the design and construction of a multibillion dollar process or resource recovery complex as in the case of Syncrude or Hibernia.

The engineering projects appearing in this book as examples and problems already include the necessary cost and revenue estimates. Developing adequate estimates for these quantities is extremely important and can be a time-consuming activity. Cost estimating techniques are not the focus of this book. However, it is worthwhile to mention some of the possible approaches in the context of simple projects which are straightforward and involve little or no engineering design, and complex projects which tend to be large and may involve many thousands of hours of engineering design. Obviously, there are projects which fall between these extremes and some combination of approaches may be appropriate in these cases.

Simple Projects

Projects in this category usually involve a single "off the shelf" component or a series of such components which are integrated in a simple manner. The acquisition of a new milling machine is an example.

The installed cost is the price of the equipment as determined from catalogues or supplier quotations, shipping and handling charges, and the cost of building modifications and changes in utility requirements. The latter may require some design effort to define the scope of the work which is the basis for contractor quotations.

Project benefits are in the form of new revenue and or cost reduction. Estimating new revenue requires agreement on the total units produced and the selling price per unit. These quantities are related through supply and demand considerations in the market place. In highly competitive product markets, sophisticated marketing studies are often required to establish price–volume relationships. These studies are undertaken as one of the first steps in an effort to define the appropriate scale for the project.

Estimates of cost reduction are in general relatively straightforward and include everything which is no longer required.

The on-going costs to operate and maintain the equipment can be estimated at various levels of detail and accuracy. Familiarity with the cost–volume relationships for similar facilities would allow the engineer to establish a "ball park" cost. Some of this information may be used in conjunction with more detailed estimates in other areas. For example, maintenance costs may be estimated as a percentage of the installed cost value. Such percentages are derived from historical data and are frequently available from equipment suppliers. Other costs such as manpower, energy, etc., may be estimated in detail to reflect specific local considerations. The most comprehensive and time-consuming type of estimate involves a detailed estimate around each type of cost associated with the project.

Complex Projects

The estimates developed for complex projects involve the same general considerations as discussed for simple projects. However, such projects usually include specialized equipment which is not "off the shelf" and must be fabricated from detailed engineering drawings. For certain projects such drawings are not even available until after a commitment has been made to proceed with the project. The typical phases of a project are:

- Development
- Conceptual design
- Preliminary design
- Detailed design

Depending upon the specific project, some phases may be combined. Project economics are performed during each of these phases to confirm the project attractiveness and the incentive to continue. The types of estimates and estimating techniques are a function of the stage of the project.

Benefits are usually well known at the outset in terms of price–volume relationships and/or cost reduction potential when the project involves the total or partial replacement of an existing facility. In the case of natural resource projects, oil and gas and mineral price forecasts are subject to considerable uncertainty.

At the development phase, work is undertaken to identify potential technologies and confirm the technical viability of these. Installed cost estimates and operating cost estimates are based on similar existing facilities or parts of these.

Conceptual design examines issues of project scale and technology alternatives. Again estimates tend to be based on large pieces of or processes within the overall project which correspond to similar facilities already in operation elsewhere.

Preliminary design takes the most attractive alternative from the conceptual design phase to a level of detail which provides specific sizing and layout for the actual equipment and associated infrastructure. Estimates at this stage tend to be based on individual pieces of equipment. The estimating basis is similar pieces of equipment in use elsewhere.

For very large projects, the cost of undertaking the detailed engineering is prohibitive unless the project is going forward. At this stage, detailed fabrication and construction drawings which would provide the basis for an actual vendor quotation become available.

The accuracy of the estimates available improves with each phase as the project becomes defined in greater detail. Normally, the installed cost estimate from each phase includes a contingency which is some fraction of the actual estimate calculated. The contingency at the development phase can be 50 to 500% and decreases to something in the order of 10% after preliminary design.

More information on cost estimating techniques can be found in reference books on project management and cost engineering. Industry-specific data books are also available for some sectors where costs are summarized on some normalized basis such as dollars per square metre of building space or dollars per tonne of material moved in operating mines. In these cases, the data are categorized by type of building and type of mine. Engineering design companies maintain extensive databases of such cost information.

1.7 Short-Term Operational Economic Decisions

Up to this point, the various engineering economic decision problems that we have presented have been long-term strategic investment problems. Typically they required investments in physical assets, and their benefits extended over several years. At the level of plant operations, however, engineers must make decisions involving materials, plant facilities, and the in-house capabilities of company personnel. Let us consider as an example the manufacture of food-processors. In terms of material selection, several of the parts could be made of plastic, while others must be made of metal. Once materials have been chosen, engineers must consider the production methods, the shipping weight, and the method of packaging necessary to protect the different types of material. In terms of actual production, parts may be made in-house or purchased from an outside vendor. The decision as to which parts to produce in-house depends on the availability of machinery and labor. All these operational decisions (commonly known as **present economic studies** in traditional engineering economic texts) require estimating the costs associated with various production or manufacturing activities. As these costs provide the basis for developing successful business strategies and planning future operations, it is important to understand how various costs respond to changes in levels of business activity.

1.7.1 Fundamental Cost–Volume Relationships

The operating costs of any company are likely to respond in some way to changes in its volume of activity. In studying cost behavior, we need to determine some measurable volume or activity that has a strong influence on the amount of cost incurred. The unit of measure used to define "volume" is called a **volume index**. A volume index may be based on production inputs, such as tons of coal processed, direct labor hours used, or machine-hours worked; or it may be based on production outputs, such as number of kilowatt-hours generated. Once we identify a volume index, we try to find out how costs change in response to changes in this volume index. To illustrate the relationships between costs and levels of activity, we will consider various types of cost behavior in a simple and familiar setting, the cost of operating an automobile. In this situation, we may use **kilometres driven** as the volume index of operating a vehicle.

Fixed Costs

The costs of providing a company's basic operating capacity are known as its **fixed cost** or **capacity cost**. For example, the annual insurance premium, property tax and license fee on a company automobile are fixed costs since they are independent of the number of kilometres driven.

Variable Costs

In contrast to fixed operating costs, **variable operating costs** have a close relationship to the level of volume. If, for example, volume increases 10%, a total variable cost will also increase by approximately 10%. In a typical manufacturing environment, direct labor and material costs are major variable costs. Gasoline is a good example of a variable automobile cost, since fuel consumption is directly related to kilometres driven. Similarly, the tire replacement cost will also increase as a vehicle is driven more.

Semi-Variable Costs

Some costs do not fall precisely into either the fixed or the variable category, but contain elements of both. These are generally referred to as **semi-variable costs**. A typical example of a semi-variable cost is the cost of electric power: Some components of a company's power consumption, such as lighting, are independent of operating volume, while other components may likely vary directly with volume (e.g., number of machine-hours operated). In our automobile example, **depreciation** (loss of value) is a semi-variable cost. Some depreciation occurs simply from passage of time, regardless of how many kilometres a car is driven, and this represents the fixed portion of depreciation. On the other hand, the more kilometres an automobile is driven a year, the faster it loses its market value, and this represents the variable portion of depreciation.

Average Unit Cost

The foregoing description of fixed, variable, and semi-variable costs was expressed in terms of total volume for a period. We often use the term **average cost** to express activity cost on a per unit basis. In terms of unit costs, the description of cost is quite different. The variable cost per unit of volume is a constant. Fixed cost per unit varies with changes in volume: As volume increases, the fixed cost per unit decreases. The semi-variable cost per unit also changes as volume changes, but the amount of change is smaller than that for fixed costs. To demonstrate the behavior of average cost, consider the somewhat simplified data to describe the cost of owning and operating a typical passenger car in the following:

Type of Cost	Amount
Fixed cost	
Insurance	$720 per year
Property taxes	$120 per year
License fee	$55 per year
Variable cost	
Gasoline, tire, oil, and lubrication	10 cents per km
Servicing	2 cents per km
Semi-variable cost	
Depreciation (fixed)	$1200 per year
Depreciation (variable)	4 cents per km

In the table above, we itemize the cost entries by fixed, variable, and semi-variable classes. We also summarize the costs of owning and operating the automobile at various annual operating volumes from 5000 to 20,000 kilometres in the table that follows.

	Cost Per Kilometre of Owning and Operating an Automobile			
Kilometres driven per year	5000	10,000	15,000	20,000
Costs				
Fixed cost	$895	$895	$895	$895
Variable cost	600	1200	1800	2400
Semi-variable cost				
Fixed portion	1200	1200	1200	1200
Variable portion	200	400	600	800
Total costs	$2895	$3695	$4495	$5295
Cost per km	$0.5790	$0.3695	$0.2997	$0.2648

Once the total cost figures are available at specific volumes, we can calculate the effect of volume on unit (per-kilometre) costs by converting the total cost figures to average unit costs.

To determine the estimated annual costs for any assumed number of kilometres, we construct a **cost–volume diagram** as shown in Figure 1.7(a). Further we can show the relation between volume (kilometres driven per year) and the three types of cost class separately as in Fig. 1.7. We can use these cost–volume graphs to estimate both the separate and combined costs of operating at other possible volumes. For example, an owner who expects to drive 12,000 kilometres in a given year may estimate the total cost at $4015, or 33.45 cents per kilometre. In Fig. 1.7(a), all costs in excess of those necessary to operate at the zero level are known as variable costs. Since the fixed cost is $2095 a year, the remaining $1920 are variable costs. By combining all the fixed and variable elements of cost, we can state simply that the cost of owning and operating an automobile is $2095 per year, plus 16 cents per kilometre driven during the year.

Figure 1.8 illustrates graphically the average unit cost of operating an automobile. The average fixed unit cost, represented by the height of the solid line in Figure 1.8, will decline steadily as the volume increases. The average unit cost is high when volume is low because the total fixed cost is spread over a relatively few units of volume. In other words, the total fixed costs remain the same regardless of the number of kilometres driven, but they decrease on a per-kilometre basis as the kilometres driven increase.

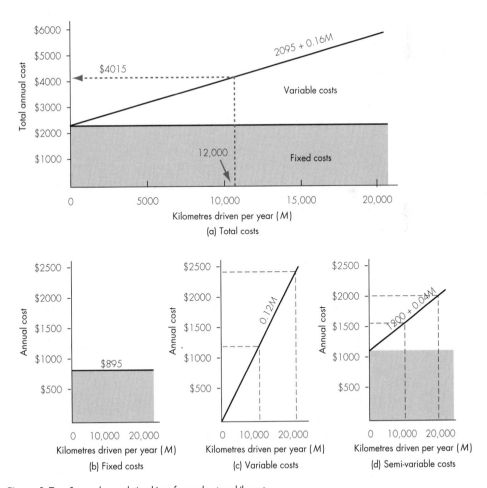

Figure 1.7   Cost–volume relationships of annual automobile costs

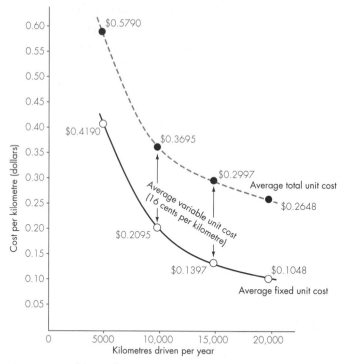

$0.5790

$0.4190

$0.3695

$0.2997

Average total unit cost

$0.2648

Average variable unit cost
(16 cents per kilometre)

$0.2095

$0.1397

$0.1048

Average fixed unit cost

Kilometres driven per year

0   5000   10,000   15,000   20,000

Cost per kilometre (dollars)

Figure 1.8   Average cost per kilometre for owning and operating a car

Marginal Costs

Another cost term useful in cost–volume analysis is marginal cost. We define **marginal cost** as the added cost that results from increasing the rate of output by a single unit. To illustrate, consider a company that has an available electric load of 37 horsepower and that purchases its electricity at the following rates:

kWh/Month	@$/kWh	Average Cost ($/kWh)
First 1500	$0.050	$0.050
Next 1250	0.035	$\dfrac{\$75.00 + 0.0350\,(X - 1500)}{X}$
Next 3000	0.020	$\dfrac{\$118.75 + 0.020\,(X - 2750)}{X}$
All over 5750	0.010	$\dfrac{\$178.75 + 0.010\,(X - 5750)}{X}$

In this rate schedule, the unit variable cost in each rate class represents the marginal cost per kilowatt-hours (kWh). On the other hand, we may determine the average costs in the third column by finding the cumulative total cost and dividing it by the total number of kWh (X). For example, if the current monthly consumption of electric power averages 3200 kWh, then the marginal cost of adding one more kWh is $0.020. However, for a given operating volume (3200 kWh), we determine the average cost per kWh as follows:

kWh	Rate ($/kWh)	Cost
First 1500	0.050	$75.00
Next 1250	0.035	43.75
Remaining 450	0.020	9.00
Total		$127.75

The average variable cost per kWh is

$$\frac{\$127.75}{3200 \text{ kWh}} = \$0.0399 \text{ kWh}.$$

Changes in the average variable cost per unit are the result of changes in the marginal cost. As shown in Figure 1.9, the average variable cost continues to fall because the marginal cost is lower than the average variable cost over the entire volume.

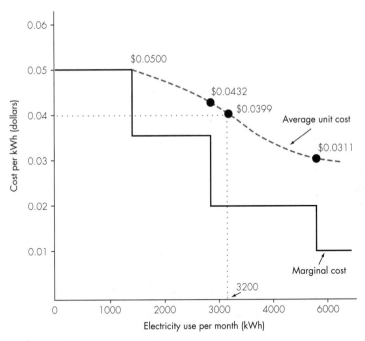

Figure 1.9 Marginal versus average cost per kWh

1.7.2 Short-Run Decisions

Cost–volume relationships find many engineering applications. They are useful in making a variety of short-term operational decisions. Many short-run problems have the following characteristics:

- The base case is the status quo (current operation or existing method), and we propose an alternative to the base case. If we find the alternative to have lower costs than the base case, we accept the alternative, assuming that nonquantitative factors do not offset the cost advantage. If several alternatives are possible, we select the one with the lowest cost. Problems of this type are often called trade-off problems because one type of cost is traded off for another.
- New investments in physical assets are not required.
- The planning horizon is relatively short (a week, a month, but less than a year).
- Relatively few cost items are subject to change by management decision. The best information about the future costs is often derived from an analysis of historical costs.

Some common examples include method changes, operations planning, make or buy decisions, and order quantities.

Method Change

The proposed alternative is to consider some new method of performing an activity. If the costs of the proposed method are significantly lower than those of the current method, we adopt the new method. For example, the engineering department at an auto-parts manufacturer comes up with the suggestion that replacing the current dies (base) with higher quality dies (alternative) will result in substantial savings in manufacturing one of the company's products. The higher cost of materials would be more than offset by the savings in machining time and electricity. Estimated monthly costs under the two alternatives are as follows:

	Current Dies	Better Dies
Variable costs		
Materials	$150,000	$170,000
Machining labor	85,000	64,000
Electricity	73,000	64,000
Fixed costs		
Supervision	25,000	25,000
Taxes	16,000	16,000
Depreciation	43,000	43,000
Total	$392,000	$382,000

The better dies cost less than the current dies. The cost savings are $392,000 − $382,000 = $10,000 per month. We should replace the current dies with the better dies.

Operations Planning

In a typical manufacturing environment, when demand is high, managers are interested in whether to use a one-shift plus overtime operation or to add a second shift. When demand is low, it is equally possible to explore whether to operate temporarily at very low volume or to shut down until operations at normal volume become economical. In a chemical plant, several routes exist for scheduling products through the plant. The problem is which route provides the lowest cost.

To illustrate how engineers may use cost–volume analysis in a typical operational analysis, let's consider Sandstone Corporation, which has one of its manufacturing plants operating on a single-shift 5-day week. The plant is operating at its full capacity (24,000 units of output per week) without the use of overtime or extra-shift operation. Fixed costs for single-shift operation amount to $90,000 per week. The average variable cost is a constant $30 per unit, at all output rates, up to 24,000 units per week. The company has received an order to produce an extra 4000 units per week beyond the current single-shift maximum capacity. Two options are being considered to fill the new order.

- Option 1: Increase the plant's output to 36,000 units a week by adding overtime or by adding Saturday operations, or both. No increase in fixed costs is entailed, but the variable cost is $36 per unit for any output in excess of 24,000 units per week, up to a 36,000-unit capacity.

- Option 2: Operate a second shift. The maximum capacity of the second shift is 21,000 units per week. The variable cost on the second shift is $31.50 per unit, and operation of a second shift entails additional fixed costs of $13,500 per week.

In this example, the operating costs related to the first shift operation will remain unchanged if one alternative is chosen instead of another. Therefore, these costs are irrelevant to the current decision and can safely be left out of the analysis. Therefore we need to examine only the increased total cost due to the additional operating volume under each option (known as **incremental analysis**). Let Q denote the additional operating volume:

- Option 1: Overtime and Saturday operation: $\$36 \times Q$

- Option 2: Second-shift operation: $\$13,500 + \$31.50 \times Q$

We can find the breakeven volume (Q_b) by equating the incremental cost functions and solving for Q:

$$\begin{aligned} 36Q &= 13,500 + 31.5Q \\ 4.5Q &= 13,500 \\ Q_b &= 3000 \text{ units.} \end{aligned}$$

If the additional volume exceeds 3000 units, the second-shift operation becomes more efficient than overtime or Saturday operation. A **break-even** (or cost–volume) **graph** based on the above data is shown in Figure 1.10. The horizontal scale represents additional volume per week. The upper limit of the relevant volume range for Option 1 is 12,000 units, whereas the upper limit for Option 2 is 21,000 units. The vertical scale is in dollars of costs. The operating savings expected at any volume may be read from the cost–volume graph. For example, the break-even point (zero savings) is 3000 units per week. Option 2 is a better choice, since the additional weekly volume from the new order exceeds 3000 units.

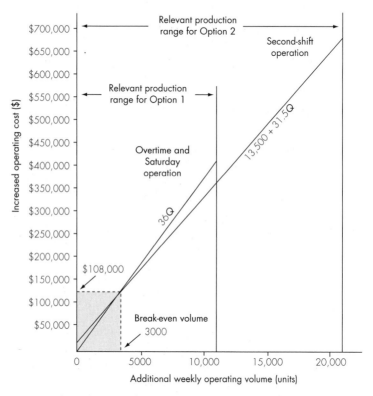

Figure 1.10 Cost–volume relationships of operating an overtime and a Saturday operation versus second-shift operation beyond 24,000 units

Make or Buy

In business, the **make-or-buy decision** arises on a fairly frequent basis. Many firms perform certain activities using their own resources, and they pay outside firms to perform certain other activities. It is a good policy constantly to seek to improve the balance between these two types of activities by asking whether we should outsource some function that we are now performing ourselves to other firms or vice versa.

To illustrate this type of trade-off decision, consider Danford Company, a farm-equipment manufacturer that currently produces 20,000 units of gas filters annually for use in its lawnmower production. The expected annual production cost of the gas filters is summarized below:

Variable cost	
Direct materials	$100,000
Direct labor	190,000
Power and water	35,000
Fixed costs	
Heating and light	20,000
Depreciation	100,000
Total cost	$445,000

Tompkins Company has offered to sell Danford 20,000 units of gas filters for $17.00 per unit. If Danford accepts the offer, some of the manufacturing facilities currently used to manufacture the gas filters could be rented to a third party at an annual rent of $35,000. Should Danford accept Tompkins' offer, and why?

This problem is unusual in the sense that the buy option would generate a rental fee of $35,000. In other words, Danford could rent out the current manufacturing facilities if it were to purchase the gas filters from Tompkins. To compare the two options, we need to examine the cost for each option.

	Make Option	Buy Option
Variable cost		
Direct materials	$100,000	
Direct labor	190,000	
Power and water	35,000	
Gas filters		$340,000
Fixed costs		
Heating light	20,000	20,000
Depreciation	100,000	100,000
Rental income		−$35,000
Total cost	$445,000	$425,000
Unit cost	$22.25	$21.25

The buy option has a lower unit cost and saves $1 for each gas-filter use. If the rental income were not considered, however, the decision would favor the make option.

Operational decision problems described in this section tend to have a relatively short-term horizon. That is, decisions do not commit a firm to a certain course of action over a considerable period into the future. If operational decision problems significantly affect the amount of funds that must be invested in a firm, fixed costs will have to increase. In our make–buy example, if the "make" alternative required the acquisition of machinery and equipment, then it is a long-term investment decision, not a short-run decision problem. (We will revisit the make–buy decision in Chapter 6 when we consider the effect of purchasing the equipment for the make option.) In the remaining chapters, we will study situations where we need to consider the effect of long-term investments.

1.8 Summary

- The factors of **time** and **uncertainty** are the defining aspects of investments.

- The three types of business organizations are (1) proprietorships, (2) partnerships, and (3) corporations. Although corporations make up only a small portion of businesses in Canada, they conduct a very large proportion of all business volume in sales dollars. Thus, our focus in future chapters will be primarily on the role of engineering economic decisions within corporations.

- The term **engineering economic decision** refers to all investment decisions relating to engineering projects. The facet of an economic decision of most interest from an engineer's point of view is the evaluation of costs and benefits associated with making a capital investment.

- The five main types of engineering economic decisions are:

  1. Equipment and process selection

  2. Equipment replacement

  3. New product and product expansion

  4. Cost reduction

  5. Service or quality improvement

- Operating costs fall into one of three categories: fixed, variable or semivariable. **Fixed costs** represent the costs of providing the basic operating capacity and are independent of production volume. **Variable costs** are directly proportional to production volume. **Semi-variable costs** contain fixed and variable cost components.

Problems

1 Compare the price of computer RAM 5 years ago with the price today. What would you expect the price to be in 5 years' time?

2 What factors have contributed to the decline in RAM pricing?

3 Many oil price forecasts in the early 1980s indicated that the price of oil in the year 2000 would be in excess of $100 per barrel. What is the price today? Why are these prices so difficult to predict?

4 Assume you purchased 100 shares of each of the following Canadian stocks on January 1, 1997: Nortel, Imperial Oil, and Bombardier. How much are they worth today? What things would you consider in deciding to keep or sell these stocks?

5 Give a specific example, from businesses in your geographical area, for each of the types of engineering economic decisions discussed in Section 1.5.

6 Determine the average cost per kilometre as a function of the kilometres driven per year for your own or your parents' automobile.

7 In planning the publication of a new engineering economics textbook, the publisher has identified the following fixed and variable costs.

Fixed Costs

Overhead	$10,000
Editing and typesetting	$100,000
Author's fee	$10,000

Variable Costs

Printing/binding	$25/copy on first 5000 copies
	$20/copy on copies above 5000
Author's royalty	$2/copy
Warehousing/distribution	$1/copy

If the publisher prints 4000 copies, determine the average and marginal cost per copy. What are these costs if the publisher prints 7500 copies?

8 Work in small groups and brainstorm ideas about how a common appliance, device or tool could be redesigned to improve it in some way. Identify the steps involved and the economic factors which you would need to consider prior to making a decision to manufacture the redesigned product. A detailed design and actual cost estimates are not required. Some items which could be considered for this redesign exercise are: a shopping cart, telephone, can opener, screwdriver …

CHAPTER 2

Equivalence and Interest Formulas

You may have already won $2 million!!! Just peel the game piece off the Instant Winner Sweepstakes ticket, and mail it to us along with your orders for subscriptions to your two favorite magazines. As a Grand Prize Winner you may choose between a $1 million cash prize paid immediately or $100,000 per year for 20 years—that's $2 million!!!

If you were the winner of the Super Prize described above, you might well wonder why the value of one prize, $1 million paid immediately, is so much lower than the $2 million paid in 20 installments. Isn't receiving the $2 million overall a lot better than receiving $1 million now? The answer to your question involves the principles we will discuss in this chapter, namely, the operation of interest and the time value of money.

Now suppose that you acted on your first impulse and selected as your prize annual payments totalling $2 million. You may be surprised by how your decision stands up under economic analysis. First, most people familiar with investments would tell you that receiving $1 million today is likely to prove a far better deal than taking $100,000 a year for 20 years. In fact, based on the principles you will learn in this chapter, the real present value of your earnings—the value that you could receive today in the financial marketplace for the promise of $100,000 a year for the next 20 years—can be shown to be worth considerably less than $1 million. And that is before we even consider the effects of inflation!

We can also use the techniques presented in this chapter to show that, if you save your winnings for the first 7 years and then spend every cent of your winnings in the remaining 13, you are likely to come out wealthier than if you do the reverse and spend for 7 years and then save for 13! (Both examples assume that the economy will remain stable.) The reason for this surprising result is the **time value of money**; that is, the earlier a sum of money is received, the more it is worth, because over time money can earn more money, or interest.

In engineering economic analysis, the principles discussed in this chapter are regarded as the underpinning for nearly all project investment analysis. This is because we always need to account for the effect of interest operating on sums of cash over time. Interest formulas allow us to place different cash flows received at different times in the same time frame and to compare them. As will become apparent, almost our entire study of engineering economic analysis is built on the principles introduced in this chapter.

2.1 Interest: The Cost of Money

Most of us are familiar in a general way with the concept of interest. We know that money left in a savings account earns interest so that the balance over time is greater than the sum of the deposits. We know that borrowing to buy a car means repaying an amount over time that includes interest and that is therefore greater than the amount borrowed. What may be unfamiliar to us is the idea that, in the financial world, money itself is a commodity, and like other goods that are bought and sold, money costs money.

The cost of money is established and measured by an **interest rate**, a percentage that is periodically applied and added to an amount (or varying amounts) of money over a specified length of time. When money is borrowed, the interest paid is the charge to the borrower for the use of the lender's property; when money is loaned or invested, the interest earned is the lender's gain from providing a good to another. **Interest**, then, may be defined as the cost of having money available for use. In this section, we examine how interest operates in a free-market economy and establish a basis for understanding the more complex interest relationships that follow later on in the chapter.

2.1.1   The Time Value of Money

The way interest operates reflects the fact that money has a time value. This is why amounts of interest depend on lengths of time; interest rates, for example, are typically given in terms of a percentage per year. The principle of the time value of money can be formally defined as follows: The economic value of a sum depends on when it is received. Because money has **earning power** over time (it can be put to work, earning more money for its owner), a dollar received today has a greater value than a dollar received at some future time.

When we deal with large amounts of money, long periods of time, or high interest rates, the change in the value of a sum of money over time becomes extremely significant. For example, at a current annual interest rate of 10%, $1 million will earn $100,000 in interest in a year; thus, to wait a year to receive $1 million clearly involves a significant sacrifice. When deciding among alternative proposals, we must take into account the operation of interest and the time value of money to make valid comparisons of different amounts at various times.

It is important to differentiate between the time value of money as we use it in this chapter and the effects of inflation, which we will study in Chapter 12. The notion that a sum of money is worth more the earlier it is received can refer to its earning potential over time, to its decrease in value due to inflation over time, or to both. The earning power of money and its loss of value due to inflation represent different analytical techniques. In this chapter, we will consider these issues separately.

2.1.2   Elements of Transactions Involving Interest

Many types of transactions involve interest—e.g., borrowing or investing money, or purchasing machinery on credit—but certain elements are common to all of these types of transactions.

1. An initial amount of money that, in transactions involving debt or investments, is called the **principal**.

2. The **interest rate** that measures the cost or price of money and that is expressed as a percentage per period of time.

3. A period of time, called the **interest period**, that determines how frequently interest is calculated. (Note that even though the length of time of an interest period can vary, interest rates are frequently quoted in terms of an annual

percentage rate. We will discuss this potentially confusing aspect of interest in Chapter 3.)

4. A specified length of time that marks the duration of the transaction and thereby establishes a certain **number of interest periods**.

5. A **plan for receipts or disbursements** (**payments**) that yields a particular cash flow pattern over a specified length of time. (For example, we might have a series of equal monthly payments that repay a loan.)

6. A **future amount of money** that results from the cumulative effects of the interest rate over a number of interest periods.

For the purposes of calculation, these elements are represented by the following variables:

A_n = A discrete payment or receipt occurring at the end of some interest period.

i = The interest rate per interest period.

N = The total number of payment or interest periods.

P = A sum of money at a time chosen for purposes of analysis as time zero, sometimes referred to as the **present value** or **present worth**.

F = A future sum of money at the end of the analysis period, sometimes referred to as the **future value** or **future worth**. This sum may be specified as F_N.

A = An end-of-period payment or receipt in a uniform series that continues for N periods. This is a special situation where $A_1 = A_2 = ... = A_N$.

V_n = An equivalent sum of money at the end of a specified period n that considers the time value of money. Note that $V_0 = P$ and $V_N = F$.

Because frequent use of these symbols will be made in this text, it is important that you become familiar with them. Note, for example, the distinction between A, A_n, and A_N. A_n refers to a specific payment or receipt, at the end of period n, in any series of payments. A_N is the final payment in such a series because N refers to the total number of interest periods. A refers to any series of cash flows where all payments or receipts are equal.

Example of an Interest Transaction

As an example of how the elements we have just defined are used in a particular situation, let us suppose that an electronics manufacturing company buys a machine for $25,000 and borrows $20,000 from a bank at a 9% annual interest rate. A $200 loan origination fee reduces the cash value of the loan to $19,800. The bank offers two repayment plans, one with equal payments made at the end of every year for the next 5 years, and the other with a single payment made after the loan period of 5 years. These payment plans are summarized in Table 2.1.

In Plan 1 the principal amount, P, is $20,000 and the interest rate, i, is 9%. The interest period is 1 year, and the duration of the transaction is 5 years, which

Table 2.1
Repayment Plans for Example Given in Text (for $N = 5$ years and $i = 9\%$)

| | Receipts | Payments | |
		Plan 1	Plan 2
Year 0	$20,000.00	$ 200.00	$200.00
Year 1		5141.85	0
Year 2		5141.85	0
Year 3		5141.85	0
Year 4		5141.85	0
Year 5		5141.85	30,772.48

$P = \$20,000$, $A = \$5141.85$, $F = \$30,772.48$

Note: You actually receive $19,800 in cash after paying the origination fee of $200, but you pay back based on the full loan value of $20,000.

means there are five interest periods ($N = 5$). It bears repeating that whereas 1 year is a common interest period, interest is frequently calculated at other intervals: monthly, quarterly, or semi-annually, for instance. For this reason, we used the term **period** rather than **year** when we defined the preceding list of variables. The receipts and disbursements planned over the duration of this transaction yield a cash flow pattern of five equal payments, A, of $5141.85 each, paid at year-end during years 1 through 5. (You'll have to accept these amounts on faith for now—the following section presents the formula used to arrive at the amount of these equal payments, given the other elements of the problem.)

Plan 2 has most of the elements of Plan 1, except that instead of five equal repayments we have a grace period followed by a single future repayment, F, of $30,772.48.

Cash Flow Diagrams

Problems involving the time value of money can be conveniently represented in graphic form with a cash flow diagram (Figure 2.1). **Cash flow diagrams** represent time by a horizontal line marked off with the number of interest periods specified. The cash flows over time are represented by arrows at relevant periods: Upward arrows represent positive flows (receipts) and downward arrows represent negative flows (disbursements). Note, too, that the arrows actually represent **net cash flows**: Two or more receipts or disbursements made at the same time are summed and shown as a single arrow. For example, $20,000 received during the same period as a $200 payment would be recorded as an upward arrow of $19,800. The lengths of the arrows can also suggest the relative values of particular cash flows.

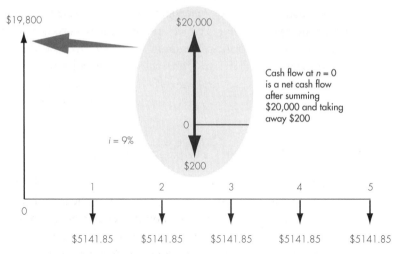

Figure 2.1 A cash flow diagram for plan 1 of the loan repayment example summarized in Table 2.1

Cash flow diagrams function in a manner similar to free body diagrams or circuit diagrams, which most engineers frequently use: Cash flow diagrams give a convenient summary of all the important elements of a problem as well as a reference point to determine whether a problem statement has been converted into its appropriate parameters. The text makes frequent use of this graphic tool, and you are strongly encouraged to develop the habit of using well-labeled cash flow diagrams as a means of identifying and summarizing pertinent information in a cash flow problem. Similarly, a table such as the one shown in Table 2.1 can help you organize information in another summary format.

End-of-Period Convention

In practice, cash flows can occur at the beginning or in the middle of an interest period, or at practically any point in time. One of the simplifying assumptions we make in engineering economic analysis is the **end-of-period convention**, which is the practice of placing all cash flow transactions at the end of an interest period. This assumption relieves us of the responsibility of dealing with the effects of interest within an interest period, which would greatly complicate our calculations.

It is important to be aware of the fact that, like many of the simplifying assumptions and estimates we make in modeling engineering economic problems, the end-of-period convention inevitably leads to some discrepancies between our model and real-world results.

Suppose, for example, that $100,000 is deposited during the first month of the year in an account with an interest period of 1 year and an interest rate of 10% per year. In such a case, the difference of 1 month would cause an interest income loss of $10,000. This is because, under the end-of-period convention, the $100,000 deposit made during the interest period is viewed as if the deposit were made at the end of the year as opposed to 11 months earlier. This example gives you a sense of why financial institutions choose interest periods that are less than 1 year, even though they usually quote their rate in terms of annual percentage.

Armed with an understanding of the basic elements involved in interest problems, we can now begin to look at the details of calculating interest.

2.1.3   Methods of Calculating Interest

Money can be loaned and repaid in many ways, and, equally, money can earn interest in many different ways. Usually, however, at the end of each interest period, the interest earned on the principal amount is calculated according to a specified interest rate. The two computational schemes for calculating this earned interest are said to yield either **simple interest** or **compound interest**. Engineering economic analysis uses the compound interest scheme almost exclusively.

Simple Interest

The first scheme considers interest earned on only the principal amount during each interest period. In other words, under simple interest, the interest earned during each interest period does not earn additional interest in the remaining periods, *even though you do not withdraw it.*

In general, for a deposit of P dollars at a simple interest rate of i for N periods, the total earned interest I would be

$$I = (iP)N. \tag{2.1}$$

The total amount available at the end of N periods, F, thus would be

$$F = P + I = P(1 + iN). \tag{2.2}$$

Simple interest is commonly used with add-on loans or bonds; these are reviewed in Chapter 3.

Compound Interest

Under a compound interest scheme, the interest earned in each period is calculated based on the total amount at the end of the previous period. This total amount includes the original principal plus the accumulated interest that has been left in the account. In this case, you are in effect increasing the deposit amount by the amount of interest earned. In general, if you deposited (invested) P dollars at interest rate i, you would have $P + iP = P(1 + i)$ dollars at the end of one period. If the entire amount (principal and interest) is reinvested at the same rate, i, for another period, you would have, at the end of the second period,

$$P(1 + i) + i[P(1 + i)] = P(1 + i)(1 + i)$$
$$= P(1 + i)^2.$$

Continuing, we see that the balance after period three is

$$P(1 + i)^2 + i[P(1 + i)^2] = P(1 + i)^3.$$

This interest-earning process repeats, and after N periods, the total accumulated value (balance) F will grow to

$$F = P(1 + i)^N. \tag{2.3}$$

EXAMPLE 2.1 $

Compound interest

Suppose you deposit $1000 in a bank savings account that pays interest at a rate of 8%, compounded annually. Assume that you don't withdraw the interest earned at the end of each period (year), but let it accumulate. How much would you have in the account at the end of year 3?

Solution

Given: P = $1000, N = 3 years, and i = 8% per year
Find: F

Applying Eq. (2.3) to our 3-year, 8% case, we obtain

$$F = \$1000(1 + 0.08)^3 = \$1259.71.$$

The total interest earned is $259.71, which is $19.71 more than was accumulated under the simple interest method (Figure 2.2). We can keep track of the interest accruing process more precisely as follows:

Period	Amount at Beginning of Interest Period	Interest Earned for Period	Amount at End of Interest Period
1	$1000.00	$1000(0.08)	$1080.00
2	1080.00	1080(0.08)	1166.40
3	1166.40	1166.40(0.08)	1259.71

Comments: At the end of the first year, you would have $1000 plus $80 in interest, or a total of $1080. In effect, at the beginning of the second year, you would be depositing $1080, rather than $1000. Thus, at the end of the second year, the interest earned would be 0.08($1080) = $86.40, and the balance would be $1080 + $86.40 = $1166.40. This is the amount you would be depositing at the beginning of the third year, and the interest earned for that period would be 0.08($1166.40) = $93.31. With a beginning principal amount of $1166.40 plus the $93.31 interest, the total balance would be $1259.71 at the end of year 3. If the total balance were then withdrawn, the net cash flow in this case would appear as

Year	Cash Flow
0	–$1000
1	0
2	0
3	1259.71

which is the same value worked out previously.

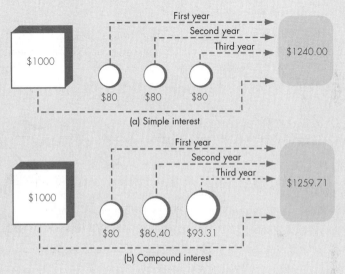

Figure 2.2 Two methods of calculating the balance when $1000 at 8% interest is deposited for 3 years (Example 2.1)

2.1.4 Simple Interest versus Compound Interest

From Eq. (2.3), the total interest earned over N periods is

$$I = F - P = P[(1 + i)^N - 1]. \qquad (2.4)$$

When compared with the simple interest scheme, the additional interest earned with compound interest is

$$\Delta I = P[(1 + i)^N - 1] - (iP)N \qquad (2.5)$$
$$= P[(1 + i)^N - (1 + iN)]. \qquad (2.6)$$

As either i or N becomes large, the difference in interest earnings also becomes large, so the effect of compounding is further pronounced. Note that, when $N = 1$, compound interest is the same as simple interest. Example 2.2 will illustrate the comparison.

EXAMPLE 2.2

Comparing simple to compound interest

In 1626, Peter Minuit of the Dutch West India Company paid $24 to purchase Manhattan Island from the Indians. In retrospect, if Mr. Minuit had invested the $24 in a savings account that earns 8% interest, how much would it be worth in 2000?

Solution

Given: $P = \$24$, $i = 8\%$ per year, $N = 374$ years
Find: F based on (a) 8% simple interest and (b) 8% compound interest

(a) With 8% simple interest:

$$F = \$24[1 + (0.08)(374)] = \$742.$$

(b) With 8% compound interest:

$$F = \$24(1 + 0.08)^{374} = \$75,979,388,482,896.$$

Comments: The significance of compound interest is obvious in this example. Many of us can hardly comprehend the magnitude of $76 trillion. Certainly, there is no way of knowing exactly how much Manhattan Island is worth today, but most real estate experts would agree that the value of the island is not anything near 76 trillion dollars.

2.2 Economic Equivalence

The observation that money has a time value leads us to an important question: If receiving $100 today is not the same as receiving $100 at any future point, how do we measure and compare various cash flows? How do we know, for example, whether we should prefer to have $20,000 today, and $50,000 ten years from now, or $8000 each year for the next 10 years (Figure 2.3)? In this section, we will describe the basic analytical techniques for making these comparisons. Then, in Section 2.3, we will use these techniques to develop a series of formulas that can greatly simplify our calculations.

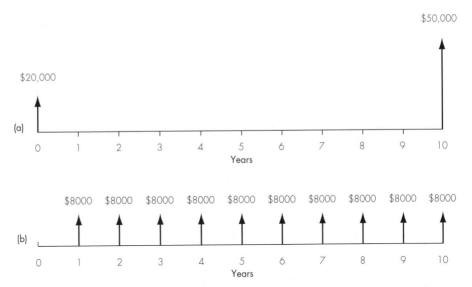

Figure 2.3 Which option would you prefer? (a) Two payments ($20,000 now and $50,000 at the end of 10 years) or (b) ten equal annual receipts in the amount of $8000

2.2.1 Definition and Simple Calculations

The central question in deciding among alternative cash flows involves comparing their economic worth. This would be a simple matter if, in the comparison, we did not need to consider the time value of money: We could simply add the individual cash flows, treating receipts as positive cash flows and payments (disbursements) as negative cash flows. The fact that money has a time value makes our calculations more complicated. We need to know more than just the size of a payment in order to determine its economic effect completely. In fact, as we will discuss in this section, we need to know several things:

- Its magnitude.

- Its direction—is it a receipt or a disbursement?

- Its timing—when does the transaction occur?

- The applicable interest rate during the time period under consideration.

It follows that to assess the economic impact of a series of cash flows, we must consider the impact of each individual cash flow.

Calculations for determining the economic effects of one or more cash flows are based on the concept of economic equivalence. **Economic equivalence** exists between cash flows that have the same economic effect and could therefore be traded for one another in the financial marketplace, which we assume to exist.

Economic equivalence refers to the fact that a cash flow—whether a single cash flow or a series of cash flows—can be converted to an *equivalent* cash flow at any point in time. For example, we could find the equivalent future value, F, of a present amount, P, at interest rate, i, at period N; or we could determine the equivalent present value, P, of N equal cash flows, A.

The strict conception of equivalence, which limits us to converting a cash flow into another equivalent cash flow, may be extended to include the comparison of alternatives. For example, we could compare the value of two proposals by finding the equivalent value of each at any common point in time. If financial proposals that appear to be quite different turn out to have the same monetary value, then we can be *economically indifferent* to choosing between them: In terms of economic effect, one would be an even exchange for the other so no reason exists to prefer one over the other in terms of their economic value.

A way to see the concepts of equivalence and economic indifference at work in the real world is to note the variety of payment plans offered by lending institutions for consumer loans. Table 2.2 extends the example we developed earlier to include three different repayment plans for a loan of $20,000 for 5 years at 9% interest. You will notice, perhaps to your surprise, that the three plans require significantly different repayment patterns and different total amounts of repayment. However, because money has time value, these plans are equivalent, and economically, the bank is indifferent to a consumer's choice of plan. We will now discuss how such equivalence relationships are established.

Table 2.2

Typical Repayment Plans for a Bank Loan of $20,000 (for $N = 5$ years and $i = 9\%$)

	Plan 1	Repayments Plan 2	Plan 3
Year 1	$5141.85	0	$ 1800.00
Year 2	5141.85	0	1800.00
Year 3	5141.85	0	1800.00
Year 4	5141.85	0	1800.00
Year 5	5141.85	$30,772.48	21,800.00
Total of payments (1)	$25,709.25	$30,772.48	$29,000.00
Total interest paid (2)	$5709.25	$10,772.48	$9000.00

Plan 1: Equal annual installments; Plan 2: End-of-loan-period repayment of principal and interest; Plan 3: Annual repayment of interest and end-of-loan repayment of principal
(1) Ignores time value of money effects
(2) Equals total of payments minus $20,000

Equivalence Calculations: A Simple Example

Equivalence calculations can be viewed as an application of the compound interest relationships we developed in Section 2.1. Suppose, for example, that we invest $1000 at 12% annual interest for 5 years. The formula developed for calculating compound interest, $F = P(1 + i)^N$ (Eq. 2.3), expresses the equivalence between some present amount, P, and a future amount, F, for a given interest rate, i, and a number of interest periods, N. Therefore, at the end of the investment period, our sums grow to

$$\$1000(1 + 0.12)^5 = \$1762.34.$$

Thus we can say that at 12% interest, $1000 received now is equivalent to $1762.34 received in 5 years, and that we could trade $1000 now for the promise of receiving $1762.34 in 5 years. Example 2.3 further demonstrates the application of this basic technique.

EXAMPLE 2.3 $ Equivalence

Suppose you are offered the alternative of receiving either $3000 at the end of 5 years or P dollars today. There is no question that the $3000 will be paid in full (no risk). Because you have no current need for the money, you would deposit the P dollars in an account that pays 8% interest. What value of P would make you indifferent to your choice between P dollars today and the promise of $3000 at the end of 5 years?

Discussion: Our job is to determine the present amount that is economically equivalent to $3000 in 5 years given the investment potential of 8% per year. Note that the problem statement assumes that you would exercise the option of using the earning power of your money by depositing it. The "indifference" ascribed to you refers to economic indifference, i.e., within a marketplace where 8% is the applicable interest rate, you could trade one cash flow for the other.

Solution

Given: $F = \$3000$, $N = 5$ years, $i = 8\%$ per year
Find: P

Equation: Eq. (2.3), $F = P(1+i)^N$

Rearranging to solve for P,

$$P = F/(1+i)^N.$$

Substituting,

$$P = \$3000/(1+0.08)^5 = \$2042.$$

We can summarize the problem graphically as in Figure 2.4.

Comments: In this example, it is clear that, if P is anything less than $2042, you would prefer the promise of $3000 in 5 years to P dollars today; if P were greater than $2042, you would prefer P. As you may have already guessed, at a lower interest rate, P must be higher to be equivalent to the future amount. For example, at $i = 4\%$, $P = \$2466$.

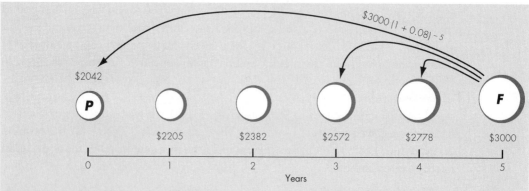

Figure 2.4 Various dollar amounts that will be economically equivalent to $3000 in 5 years, given an interest rate of 8% (Example 2.3)

2.2.2 Equivalence Calculations: General Principles

In spite of their numerical simplicity, the examples we have developed reflect several important general principles, which we will now explore.

Principle 1: Equivalence calculations made to compare alternatives require a common time basis

Just as we must convert fractions to common denominators to add them together, we must also convert cash flows to a common basis to compare their value. One aspect of this basis is the choice of a single point in time at which to make our calculations. In Example 2.3, if we had been given the magnitude of each cash flow, and had been asked to determine whether they were equivalent, we could have chosen any reference point and used the compound interest formula to find the value of each cash flow at that point. As you can readily see, the choice of $n = 0$ or $n = 5$ would make our problem simpler because we only need to make one set of calculations: At 8% interest, either convert $2042 at time 0 to its equivalent value at time 5, or convert $3000 at time 5 to its equivalent value at time 0. (To see how to choose a different reference point, take a look at Example 2.4.)

When selecting a point in time at which to compare the value of alternative cash flows, we commonly use either the present time, which yields what is called the **present worth** of the cash flows or some point in the future, which yields their **future worth**. The choice of the point in time often depends on the circumstances surrounding a particular decision, or it may be chosen for convenience. For instance, if the present worth is known for the first two of three alternatives, all three may be compared by simply calculating the present worth of the third.

EXAMPLE 2.4

Equivalent cash flows are equivalent at any common point in time

In Example 2.3, we determined that, given an interest rate of 8% per year, receiving $2042 today is equivalent to receiving $3000 in 5 years. Are these cash flows also equivalent at the end of year 3?

Discussion: This problem is summarized in Figure 2.5. The solution consists of solving two equivalence problems: (1) What is the future value of $2042 after 3 years at 8% interest (part (a) of the solution)? (2) Given the sum of $3000 after 5 years and an interest rate of 8%, what is the equivalent sum after 3 years (part (b) of the solution)?

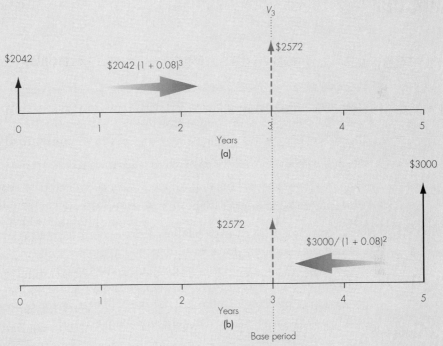

Figure 2.5 Selection of a base period for an equivalence calculation (Example 2.4)

Solution

Given:
(a) $P = \$2042$; $i = 8\%$ per year; $N = 3$ years
(b) $F = \$3000$; $i = 8\%$ per year; $N = 5 - 3 = 2$ years
Find: (1) V_3 for part (a); (2) V_3 for part (b); (3) Are these two values equivalent?

Equation:

$$\text{(a)} \quad F = P(1 + i)^N$$
$$\text{(b)} \quad P = F/(1 + i)^N$$

Notation: The usual terminology of *F* and *P* is confusing in this example, since the cash flow at *n* = 3 is considered a future sum in Part (a) of the solutions and a past cash flow in Part (b) of the solution. To simplify matters, we are free to arbitrarily designate a reference point, *n* = 3, and understand that it need not be now or the present. Therefore we assign a single variable, V_3, for the equivalent cash flow at *n* = 3.

1. The equivalent worth of $2042 after 3 years is

$$V_3 \quad = \quad 2042(1 + 0.08)^3$$
$$= \quad \$2572.$$

2. The equivalent worth of the sum $3000, 2 years earlier is:

$$V_3 \quad = \quad F/(1 + i)^N$$
$$= \quad \$3000/(1 + 0.08)^2$$
$$= \quad \$2572.$$

(Note that *N* = 2 because that is the number of periods over which discounting (Figure 2.10) is calculated in order to arrive back at year 3.)

3. While our solution doesn't strictly prove that the two cash flows are equivalent at any time, they will be equivalent at any time as long as we use an interest rate of 8%.

Principle 2: Equivalence depends on interest rate

The equivalence between two cash flow series is a function of the direction, magnitude and timing of individual cash flows and the interest rate or rates that operate on those cash flows. This is easy to grasp in relation to our simple example: $1000 received now is equivalent to $1762.34 received 5 years from now only at a 12% interest rate. Any change in the interest rate will destroy the equivalence between these two sums as we will demonstrate in Example 2.5.

EXAMPLE 2.5

Changing the interest rate destroys equivalence

In Example 2.3, we determined that, given an interest rate of 8% per year, receiving $2042 today is equivalent to receiving $3000 in 5 years. Are these cash flows equivalent at an interest rate of 10%?

Solution

Given: *P* = $2042; *i* = 10% per year, *N* = 5 years
Find: *F*: Is it equal to $3000?

We first determine the base period where an equivalence value is computed. Since we can select any period as the base period, let's select $N = 5$. Then, we need to calculate the equivalent value of $2042 today 5 years from now.

$$F = \$2042(1+0.10)^5 = \$3289.$$

Since this amount is greater than $3000, the change in interest rate destroys the equivalence between the two cash flows.

Principle 3: Equivalence calculations may require the conversion of multiple payment cash flows to a single cash flow

In all the examples presented thus far, we have limited ourselves to the simplest case of converting a single payment at one time to an equivalent single payment at another time. Part of the task of comparing alternative cash flow series involves moving each individual cash flow in the series to the same single point in time and summing these values to yield a single equivalent cash flow. We perform such a calculation in Example 2.6.

EXAMPLE 2.6 $

Equivalence calculations with multiple payments

Suppose that you borrow $1000 from a bank for 3 years at 10% annual interest. The bank offers two options: (1) Repaying the interest charges for each year at the end of that year and repaying the principal at the end of year 3 or (2) repaying the loan all at once (including both interest and principal) at the end of year 3. The repayment schedules for the two options are as follows:

Options	Year 1	Year 2	Year 3
Option 1: End-of-year repayment of interest and principal repayment at end of loan	$100	$100	$1100
Option 2: One end-of-loan repayment of both principal and interest	0	0	1331

Determine whether these options are equivalent, assuming that the appropriate interest rate for our comparison is 10%.

Discussion: Since we pay the principal after 3 years in either plan, the repayment of principal can be removed from our analysis. This is an important point: *We can ignore the common elements of alternatives being compared so that we can focus entirely on comparing the interest payments.* Notice that under Option 1, we will pay a total of $300 interest, whereas under Option 2, we will pay a total of $331. Before concluding that we prefer Option 1, remember that a comparison of the two cash flows is based on a *combination of payment amounts and timing of those payments*. To make our comparison we must compare the equivalent value of each option at a single point in time. Since Option 2 is already a single payment at $n = 3$ years, it is simplest to convert the Option 1 cash flow pattern to a single value at $n = 3$. To do this, we must convert the three disbursements of Option 1 to their respective equivalent values at $n = 3$. At that point, since they share a time in common, we can simply sum them in order to compare them to the $331 sum in Option 2.

Solution

Given: The cash flow diagrams in Figure 2.6; $i = 10\%$ per year
Find: A single future value, F, of the flows in Option 1

Equation: $F = P(1+i)^N$, applied to each disbursement in the cash flow diagram

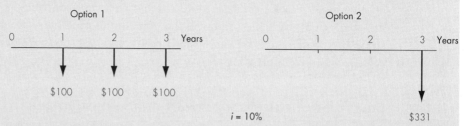

Figure 2.6 Equivalent cash flow diagram for Option 1 and Option 2 (excluding the common principal payment $1000 at the end of year 3) (Example 2.6)

N in Eq. (2.3) is the number of interest periods upon which interest is in effect; n is the period number (i.e., for year 1, $n = 1$). We determine its value by finding the interest period for each payment. Thus, for each payment in the series, N can be calculated by subtracting n from the total number of years of the loan (3). That is, $N = 3 - n$. Once the value of each payment has been found, we sum them:

$$F_3 \text{ for } \$100 \text{ at } n = 1 : \$100(1 + .10)^{3-1} = \$121$$
$$F_3 \text{ for } \$100 \text{ at } n = 2 : \$100(1 + .10)^{3-2} = \$110$$
$$F_3 \text{ for } \$100 \text{ at } n = 3 : \$100(1 + .10)^{3-3} = \$100$$
$$\text{Total} = \$331.$$

By converting the cash flow in Option 1 to a single future payment at year 3, we can compare it to Option 2. We see that the two interest payments are equivalent. Thus, the bank would be economically indifferent to a choice between the two plans. Note that the final interest payment in Option 1 does not accrue any compound interest.

Comments: If it is difficult to grasp the fact that Options 1 and 2 are equivalent, even though $31 more interest is paid under Option 2; focus again on the concept of the time value of money.

Note that Option 1 excludes the opportunity provided by Option 2 of earning interest on the deferred interest payments (Figure 2.7). Under Option 1, the interest must be paid at the end of each interest period. Because of the time value of money, the $100 paid at the end of period 1 has a future worth of $121 from the perspective of time at the end of period 3. A person under Option 2 can actually realize that future worth by depositing $100 in an account at 10% interest. Of course, the person may not be able to get 10% interest or could possibly earn a higher rate. This brings up the question, what is the appropriate interest rate to use in equivalence calculations? We will address this topic in Chapter 10. For example, if a person earns only 8%, the two options would not be equivalent.

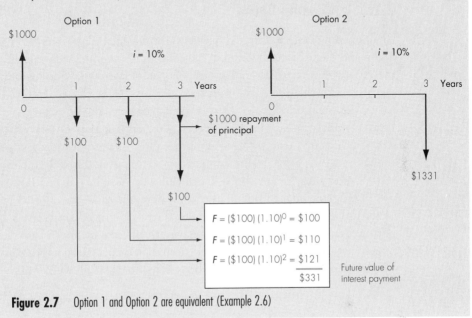

Figure 2.7 Option 1 and Option 2 are equivalent (Example 2.6)

Principle 4: Equivalence is maintained regardless of point of view

As long as we use the same interest rate in equivalence calculations, equivalence can be maintained regardless of point of view. In Example 2.6, the two options were equivalent at an interest rate of 10% from the banker's point of view. What about from a borrower's point of view? Suppose you borrow $1000 from a bank and deposit it in another bank that pays 10% interest annually. Then, you make future loan repayments out of this savings account. Under Option 1, your savings account at the end of year 1 will show a balance of $1100 after the interest earned during the first period has been credited. Now you withdraw $100 from this savings account (the exact amount required to pay the loan interest during the first year), and you make the first year interest payment to the bank. This leaves only $1000 in your savings account. At the end of year 2, your savings account will earn another interest payment in the amount of $1000(0.10) = $100, making an end-of-year balance of $1100. Now you withdraw another $100 to make the required loan interest payment. After this payment, your remaining balance will be $1000. This balance will grow again at 10%, so you will have $1100 at the end of year 3. After making the last loan payment ($1100), you will have no money left in either account. For Option 2, you can keep track of the yearly account balances in a similar fashion. You will find that you reach a zero balance after making the lump sum payment of $1331. If the borrower had used the same interest rate as the bank, the two options would be equivalent.

2.2.3   Looking Ahead

The preceding examples should have given you some insight into the basic considerations and calculations involved in the concept of economic equivalence. Obviously, the variety of financial arrangements possible for borrowing and investing money is extensive, as is the variety of time-related factors. In the case of alternative proposals for various engineering projects, the latter include items such as maintenance costs over time, increased productivity over time, etc. For personal investment alternatives this list would include payments to the investor over time (interest, dividends, share of profits), the future selling price, etc. It is important to recognize that even the most complex relationships incorporate the basic principles we have introduced in this section.

In the remainder of this chapter, we will represent all cash flow diagrams in the context of an initial deposit, with a subsequent pattern of withdrawals, or an initial borrowed amount with a subsequent pattern of repayments. A cash flow diagram representation of a more complicated equivalence calculation appears in Figure 2.8, which summarizes the payment options offered to the 1995 winner of the Publishers

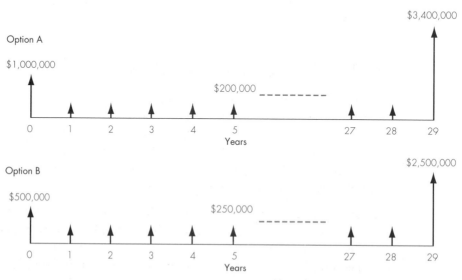

Figure 2.8   Two payment options provided by the 1995 Publishers' Clearing House, Ten Million Dollar SuperPrize: Option A, $1,000,000 cash now, $200,000 yearly plus a $3,400,000 final payment at the end of 29 years. Option B, $500,000 cash now, $250,000 a year thereafter, plus $2,500,000 at the end of 29 years

Clearing House Ten Million Dollar Super Prize. If we were limited to the methods developed in this section, a comparison between the two payment options would involve a large number of calculations. Fortunately, in the analysis of many transactions, certain cash flow patterns emerge that may be categorized. For many of these cash flow patterns we can derive formulas that can be used to simplify our work. In Section 2.3, we develop these formulas.

2.3 Development of Interest Formulas

Now that we have established some working assumptions and notations and have a preliminary understanding of the concept of equivalence, we will develop a series of interest formulas for use in more complex comparisons of cash flows.

As we begin to compare series of cash flows instead of single payments, the required analysis becomes more complicated. However, when patterns in cash flow transactions can be identified, we can take advantage of these patterns by developing concise expressions for computing either the present or future worth of the series. We will classify five major categories of cash flow transactions, develop interest formulas for them, and present several working examples of each type. Before we present the details, however, these five types of cash flows will be described briefly in the following section.

Before beginning this section, it is recommended that you review the notations in Section 2.1.2. These variables will be used consistently throughout the development of interest formulas.

2.3.1 The Five Types of Cash Flows

Whenever we identify patterns in cash flow transactions, we may use these patterns to develop concise expressions for computing either the present or future worth of the series. For this purpose, we will classify cash flow transactions into five categories: (1) Single cash flow, (2) uniform series, (3) linear gradient series, (4) geometric gradient series, and (5) irregular series. To simplify the description of various interest formulas, we will use the following notation:

1. **Single Cash Flow:** The simplest case involves the equivalence of a single present amount and its future worth. Thus, the single-cash flow formulas deal with only two amounts: A single present amount, P, and its future worth after N periods, F (Figure 2.9a). You have already seen the derivation of one formula for this situation in Section 2.1.3, which gave us Eq. (2.3):

$$F = P(1 + i)^N.$$

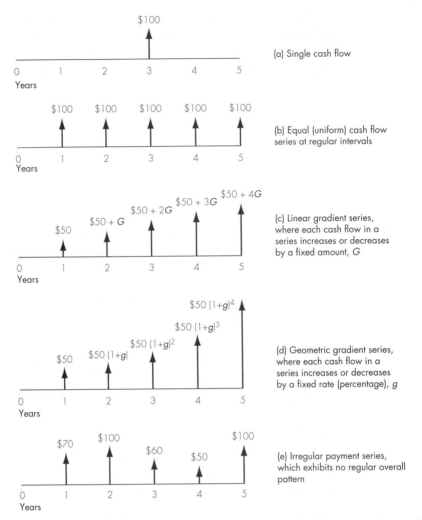

Figure 2.9 Five types of cash flows: (a) Single cash flow, (b) equal (uniform) cash flow series, (c) linear gradient series, (d) geometric gradient series, and (e) irregular payment series

2. **Equal (Uniform) Series:** Probably the most familiar category includes transactions arranged as a series of equal cash flows at regular intervals, known as an **equal cash flow series** (or **uniform series**) (Fig. 2.9b). For example, this describes the cash flows of the common installment loan contract, which arranges repayment of a loan in equal periodic installments. The equal cash flow formulas deal with the equivalence relations of *P*, *F*, and *A* (the constant amount of the cash flows in the series).

3. **Linear Gradient Series:** While many transactions involve series of cash flows, the amounts are not always uniform; they may, however, vary in some regular way. One common pattern of variation occurs when each cash flow in a series

increases (or decreases) by a fixed amount (Fig. 2.9c). A 5-year loan repayment plan might specify, for example, a series of annual payments that increase by $500 each year. We call this type of cash flow pattern a **linear gradient series** because its cash flow diagram produces an ascending (or descending) straight line, as you will see in Section 2.3.5. In addition to P, F, and A, the formulas used in such problems involve a *constant amount*, G, of the change in each cash flow from the preceding period.

4. **Geometric Gradient Series:** Another kind of gradient series is formed when the series cash flow is determined, not by some fixed amount like $500, but by some fixed *rate*, expressed as a percentage. For example in a 5-year financial plan for a project, the cost of a particular raw material might be budgeted to increase at a rate of 4% per year. The curving gradient in the diagram of such a series suggests its name which is a **geometric gradient series** (Fig. 2.9d). In the formulas dealing with such series, the rate of change is represented by a lower-case g.

5. **Irregular Series:** Finally, a series of cash flows may be irregular, in that it does not exhibit a regular overall pattern. Even in such a series, however, one or more of the patterns already identified may appear over segments of time in the total length of the series. The cash flows may be equal, for example, for five consecutive periods in a ten-period series. When such patterns appear, the formulas for dealing with them may be applied and their results included in calculating an equivalent value for the entire series.

Interest Tables

Interest formulas such as the one developed in Eq. (2.3), $F = P(1 + i)^N$, allow us to substitute known values from a particular situation into the equation and to solve for the unknown. Before the hand calculator was developed, solving these equations was very tedious. With a large value of N, for example, one might need to solve an equation such as $F = \$20,000(1 + 0.12)^{15}$. More complex formulas required even more involved calculations. To simplify the process, tables of compound interest factors were developed, and these allow us to find the appropriate factor for a given interest rate and the number of interest periods. Even with hand calculators, it is still often convenient to use these tables, and they are included in this text in Appendix C. Take some time now to become familiar with their arrangement and, if you can, locate the compound interest factor for the example just presented, in which we know P, and to find F we need to know the factor by which to multiply $20,000 when the interest rate, i, is 12%, and the number of periods is 15:

$$F = \$20,000 \underbrace{(1 + 0.12)^{15}}_{5.4736} = \$109,472.$$

Factor Notation

As we continue to develop interest formulas in the rest of this chapter, we will express the resulting compound interest factors in a conventional notation that can be substituted in a formula to indicate precisely which table factor to use in solving an equation. In the preceding example, for instance, the formula derived as Eq. (2.3) is $F = P(1 + i)^N$. In ordinary language, this tells us that to determine what future amount, F, is equivalent to a present amount, P, we need to multiply P by a factor expressed as 1 plus the interest rate, raised to the power given by the number of interest periods. To specify how the interest tables are to be used, we may also express that factor in a functional notation as follows: $(F/P, i, N)$, which is read as "Find F, given P, i, and N." This is known as the **single cash flow compound amount factor**. When we incorporate the table factor in the formula, it is expressed as follows:

$$F = P(1 + i)^N = P(F/P, i, N).$$

Thus, in the preceding example, where we had $F = \$20,000(1.12)^{15}$, we can write $F = \$20,000(F/P, 12\%, 15)$. The table factor tells us to use the 12% interest table and find the factor in the F/P column for $N = 15$. Because using the interest tables is often the easiest way to solve an equation, this factor notation is included for each of the formulas derived in the following sections.

2.3.2   Single Cash Flow Formulas

We begin our coverage of interest formulas by considering the simplest of cash flows: Single cash flows.

Compound Amount Factor

Given a present sum, P, invested for N interest periods at interest rate, i, what sum will have accumulated at the end of the N periods? You probably noticed right away that this description matches the case we first encountered in describing compound interest. To solve for F (the future sum) we use Eq. (2.3):

$$F = P(1 + i)^N = P(F/P, i, N).$$

Because of its origin in the compound interest calculation, the factor $(F/P, i, N)$ is known as the **single cash flow compound amount factor**. Like the concept of equivalence, this factor is one of the foundations of engineering economic analysis. Given this factor, all the other important interest formulas can be derived.

This process of finding F is often called the **compounding process**. The cash flow transaction is illustrated in Figure 2.10. (Note the time scale convention. The first period begins at $n = 0$ and ends at $n = 1$.) If a calculator is handy, it is easy enough to calculate $(1 + i)^N$ directly. However, the appropriate interest table can also be used, and a wide range of i and N values can be found there.

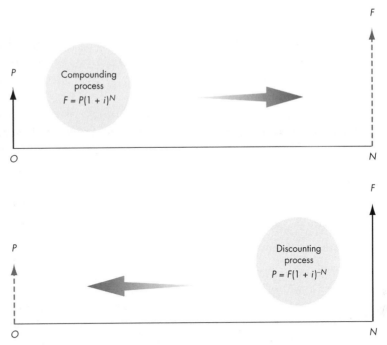

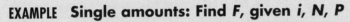

Figure 2.10   Equivalence relationship between *F* and *P*

EXAMPLE 2.7 $

Single amounts: Find *F*, given *i*, *N*, *P*

If you had $2000 now and invested it at 10%, how much would it be worth in 8 years (Figure 2.11)?

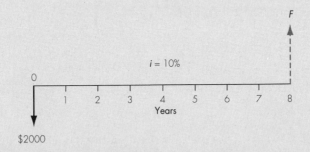

Figure 2.11   Cash flow diagram from the investor's point of view (Example 2.7)

Solution

Given: $P = \$2000$, $i = 10\%$ per year, and $N = 8$ years
Find: *F*

We can solve this problem in any of three ways:

1. Using a calculator: You can simply use a calculator to evaluate the $(1 + i)^N$ term. (Financial calculators are pre-programmed to solve most future-value problems.)

$$F = \$2000(1 + 0.10)^8$$
$$= \$4287.18.$$

2. Using compound interest tables: The interest tables can be used to locate the compound amount factor for $i = 10\%$ and $N = 8$. The number you get can be substituted into the equation. Compound interest tables are included as Appendix C of this book.

$$F = \$2000 \ (F/P,10\%,8) = \$2000(2.1436) = \$4287.20.$$

This is essentially identical to the value obtained by the direct evaluation of the single cash-flow compound amount factor. This slight deviation is due to rounding errors.

3. Using a computer: Many financial software programs for solving compound interest problems are available for use with personal computers. As summarized in Appendix B, many spreadsheet programs such as Lotus 1-2-3, Quattro Pro or Excel also provide financial functions to evaluate various interest formulas.

EXAMPLE
2.8
$

Single amounts: Find *N*, given *P*, *F*, *i*

You have just purchased 100 shares of Nortel stock at $60 per share. You will sell the stock when its market price has doubled. If you expect the stock price to increase 20% per year, how long do you expect to wait before selling the stock (Figure 2.12)?

Figure 2.12   Cash flow diagram (Example 2.8)

Solution

Given: $P = \$6000$, $F = \$12,000$, $i = 20\%$ per year
Find: N (years)

Using the single-cash flow compound amount factor, we write

$$F = P(1 + i)^N = P(F/P, i, N)$$
$$\$12{,}000 = \$6000(1 + 0.20)^N = \$6000(F/P, 20\%, N)$$
$$2 = (1.20)^N = (F/P, 20\%, N).$$

Again, we could use a calculator or a computer spreadsheet program to find N.

1. Using a calculator: Solving for N gives

$$\log 2 = N \log 1.20$$
$$N = \frac{\log 2}{\log 1.20}$$

$$= 3.80 \approx 4 \text{ years.}$$

2. Using a spreadsheet program: Within Excel, the financial function NPER($i,0,P,F$) computes the number of compounding periods it will take an investment (P) to grow to a future value (F), earning a fixed interest rate (i) per compounding period. In our example, the Excel command would look like this:

$$= \text{NPER}(20\%, 0, -6000, 12{,}000)$$
$$= 3.801784.$$

Comments: A very handy rule of thumb, called the Rule of 72, can determine approximately how long it will take for a sum of money to "double." The rule states that, to find the time it takes for the present sum of money to grow by a factor of 2, we divide 72 by the interest rate. For our example, the interest rate is 20%. Therefore, the Rule of 72 indicates 72/20 = 3.60 or roughly 4 years for a sum to double. This is, in fact, relatively close to our exact solution.

Present Worth Factor

Finding the present worth of a future sum is simply the reverse of compounding and is known as the **discounting process** as illustrated in Figure 2.10. In Eq. (2.3), we can see that if we were to find a present sum, P, given a future sum, F, we simply solve for P:

$$P = F\left[\frac{1}{(1 + i)^N}\right] = F(P/F, i, N). \tag{2.7}$$

The factor $1/(1 + i)^N$ is known as the **single cash flow present worth factor**, and is designated ($P/F, i, N$). Tables have been constructed for P/F factors and for various values of i and N. The interest rate i and the P/F factor are also referred to as the **discount rate** and **discounting factor**, respectively.

EXAMPLE
2.9
$

Single amounts: Find *P*, given *F*, *i*, *N*

Suppose that $1000 is to be received in 5 years. At an annual interest rate of 12%, what is the present worth of this amount (Figure 2.13)?

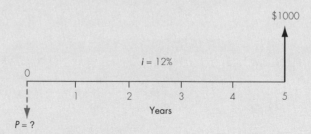

Figure 2.13 Cash flow diagram (Example 2.9)

Solution:

Given: *F* = $1000, *i* = 12% per year, and *N* = 5 years
Find: *P*

$$P = \$1000(1 + 0.12)^{-5} = \$1000(0.5674) = \$567.40.$$

Using a calculator may be the best way to make this simple calculation. To have $1000 in your savings account at the end of 5 years, you must deposit $567.40 now.

We can also use the interest tables to find that:

$$P = \$1000 \overbrace{(P/F,12\%,5)}^{(0.5674)} = \$567.40$$

Again, you could use a financial calculator or computer to find the present worth.

Compounding and Discounting Processes: Graphic Views

Figure 2.14 illustrates the characteristics of the *F/P* and *P/F* factors with variations in *i* and *n*. Fig. 2.14(a) shows how $1 (or any other sum) grows over time at various rates of interest. Note the rapid increases in the future value, *F*, with increases in either *n* or *i*. The higher the rate of interest, the faster the rate of growth. The interest rate is, in fact, a **growth rate**.

Figure 2.14(b) shows what $1 is worth today, if it is received *n* periods in the future. The curves in the figure show that the present worth of a sum to be received at some future time (*F*) decreases (1) as the payment period extended further into the future and (2) as the discount rate increases.

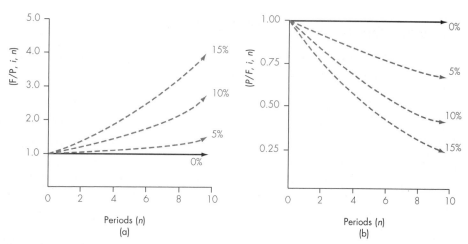

Figure 2.14 Characteristics of the F/P and P/F factors with variations in i and n

EXAMPLE
2.10
$

Effect of varying discounting rates

Consider a lump sum amount of $1 million which will be received 50 years from now. What is the equivalent present worth of this amount if your interest rate is (a) 5%, (b) 10% and (c) 25%, respectively?

Solution

Given: F = $1 million, N = 50 years, i = 5%, 10% or 25% per year
Find: P at each of the following interest rates

(a) at i = 5%:

$$P = \$1,000,000(P/F,5\%,50) = \$1,000,000(0.0872) = \$87,200;$$

(b) at i = 10%:

$$P = \$1,000,000(P/F,10\%,50) = \$1,000,000(0.008519) = \$8519;$$

(c) at i = 25%:

$$P = \$1,000,000(P/F,25\%,50) = \$1,000,000(0.00001427) = \$14.27.$$

As seen in the calculation, when relatively high discount rates are applied, funds due in the future are worth very little today. Even at relatively low discount rates, the present worth of funds due in the distant future is relatively quite small.

2.3.3　Uneven Cash Flow Series

A common cash flow transaction involves a series of disbursements or receipts. Familiar examples of series payments are payment of installments on car loans and home mortgage payments. Payments on car loans and home mortgages typically involve identical sums to be paid at regular intervals. However, when there is no clear pattern over the series, we call the transaction an **uneven cash-flow series**.

We can find the present worth of any uneven stream of payments by calculating the present value of each individual payment and summing the results. Once the present worth is found, we can make other equivalence calculations, e.g., future worth can be calculated by using the interest factors developed in the previous section.

EXAMPLE 2.11

Present value of an uneven series by decomposition into single cash flows

Wilson Technology, a growing machine shop, wishes to set aside money now to invest over the next 4 years in automating its customer service department. The company can earn 10% on a lump sum deposited now, and it wishes to withdraw the money in the following increments:

Year 1: $25,000 to purchase a computer and database software designed for customer service use;

Year 2: $3000 to purchase additional hardware to accommodate anticipated growth in use of the system;

Year 3: No expenses; and

Year 4: $5000 to purchase software upgrades.

How much money must be deposited now to cover the anticipated payments over the next 4 years?

Discussion:　This problem is equivalent to asking what value of P would make you indifferent in your choice between P dollars today and the future expense stream of ($25,000, $3000, $0, $5000). One way to deal with a series of uneven cash flows is to calculate the equivalent present value of each single cash flow and to sum the present values to find P. In other words, the cash flow is broken into three parts as shown in Figure 2.15.

Solution

Given: Uneven cash flow in Fig. 2.15, $i = 10\%$ per year
Find: P

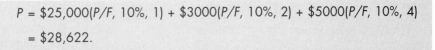

$$P = \$25,000(P/F, 10\%, 1) + \$3000(P/F, 10\%, 2) + \$5000(P/F, 10\%, 4)$$

$$= \$28,622.$$

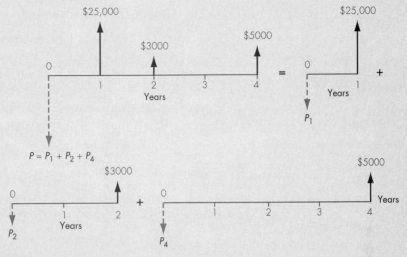

Figure 2.15 Decomposition of uneven cash flow series (Example 2.11)

Comments: To see if indeed $28,622 is sufficient amount, let's calculate the balance at the end of each year. If you deposit $28,622 now, it will grow to (1.10)($28,622) or $31,484 at the end of year 1. From this balance, you pay out $25,000. The remaining balance, $6484, will again grow to (1.10)($6484) or $7132 at the end of year 2. Now you make the second payment ($3000) out of this balance, which will leave you with only $4132 at the end of year 2. Since no payment occurs in year 3, the balance will grow to $(1.10)^2($4132) or $5000 at the end of year 4. The final withdrawal in the amount of $5000 will deplete the balance completely.

EXAMPLE
2.12
$

Calculating the actual worth of a contract

Troy Aikman, the Dallas Cowboys' quarterback, agreed to an 8-year, $50 million contract that will make him the highest-paid player in professional football history.[1] The contract includes a signing bonus of $11 million. The agreement calls for annual salaries of $2.5 million in 1993, $1.75 million in 1994, $4.15 million in 1995, $4.90 million in 1996, $5.25 million in 1997, $6.2 million in 1998, $6.75 million in 1999, and $7.5 million in the year 2000. The $11-million signing bonus must be prorated over the course of the contract, so that an additional $1.375 million is paid each year over the 8-year contract period. With the salary paid at the beginning of each season, the net annual payment schedule looks like the following:

<hr>

[1] *The New York Times*, December 24, 1993. (Reprinted with permission.)

Beginning of Season	Contract Salary	Prorated Signing Bonus	Actual Annual Payment
1993	$2,500,000	$1,375,000	$3,875,000
1994	1,750,000	1,375,000	3,125,000
1995	4,150,000	1,375,000	5,525,000
1996	4,900,000	1,375,000	6,275,000
1997	5,250,000	1,375,000	6,625,000
1998	6,200,000	1,375,000	7,575,000
1999	6,750,000	1,375,000	8,125,000
2000	7,500,000	1,375,000	8,875,000

(a) How much is Troy's contract actually worth at the time of signing?

(b) For the signing bonus portion, suppose that the Dallas Cowboys allow Troy to take either the prorated payment option as described above or a lump sum payment option in the amount of $8 million at the time of contract. Should Troy take the lump sum option instead of the prorated one?

Assume that Troy can invest his money at 6% interest.

Solution
Given: Payment series given in Figure 2.16, i = 6% per year
Find: P

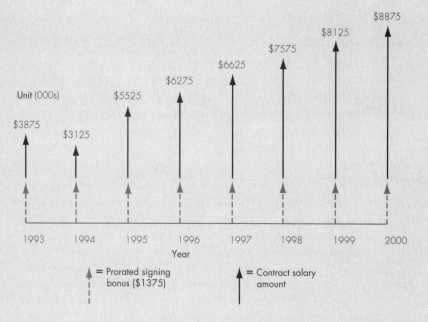

Figure 2.16   Troy Aikman's $50 million Dallas Cowboys contract (Example 2.12)

(a) Actual worth of the contract at the time of signing:

$$
\begin{aligned}
P_{\text{Contract}} &= \$3{,}875{,}000 + \$3{,}125{,}000(P/F, 6\%, 1) \\
&\quad + \$5{,}525{,}000(P/F, 6\%, 2) + \$6{,}275{,}000(P/F, 6\%, 3) \\
&\quad + \dots + \$8{,}875{,}000(P/F, 6\%, 7) \\
&= \$39{,}547{,}242.
\end{aligned}
$$

(b) Choice between the prorated payment option and the lump-sum payment: The equivalent present worth of the prorated payment option is

$$
\begin{aligned}
P_{\text{Bonus}} &= \$1{,}375{,}000 + \$1{,}375{,}000(P/F, 6\%, 1) \\
&\quad + \dots + \$1{,}375{,}000(P/F, 6\%, 7) \\
&= \$9{,}050{,}775,
\end{aligned}
$$

which is greater than $8 million. Therefore, Troy would be better off taking the prorated option if his money could be invested at 6% interest.

Comments: Note that the actual contract is worth less than $50 million as published. This "brute force" approach of breaking cash flows into single amounts will always work, but it is slow and subject to error because of the many factors that must be included in the calculation. We will develop more efficient methods in the next sections for cash flows with certain patterns.

2.3.4 Equal Cash Flow Series

As we learned in Example 2.12, the present worth of a stream of future cash flows can always be found by summing the present worth of each individual cash flow. However, if cash flow regularities are present within the stream (such as we just saw in the prorated bonus payment series in Example 2.12), the use of short-cuts such as finding the present worth of a uniform series may be possible. We often encounter transactions in which a uniform series of payments exists. Rental payments, bond interest payments, and commercial installment plans are based on uniform payment series.

Compound Amount Factor—Find F, Given A, i, N

Suppose we are interested in the future amount, F, of a fund to which we contribute A dollars each period and on which we earn interest at a rate of i per period. The contributions are made at the end of each of the following N periods. These transactions are graphically illustrated in Figure 2.17. Looking at this diagram, we see that if an amount, A, is invested at the end of each period, for N periods, the total amount, F, that can be withdrawn at the end of N periods will be the sum of the compound amounts of the individual deposits.

As shown in Figure 2.18, the A dollars we put into the fund at the end of the first period will be worth $A(1 + i)^{N-1}$ at the end of N periods. The A dollars we put

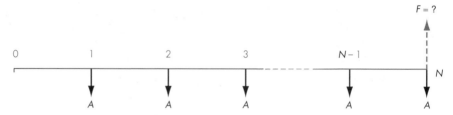

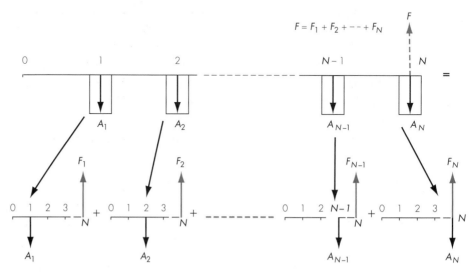

Figure 2.17 Cash flow diagram of the relationship between A and F

Figure 2.18 The future worth of a cash flow series obtained by summing the future worth figures of each individual flow

into the fund at the end of the second period will be worth $A(1 + i)^{N-2}$, and so forth. Finally, the last A dollars that we contribute at the end of the Nth period will be worth exactly A dollars at that time. This means there exists a series in the form:

$$F = A(1 + i)^{N-1} + A(1 + i)^{N-2} + \dots + A(1 + i) + A,$$

or expressed alternatively,

$$F = A + A(1 + i) + A(1 + i)^2 + \dots + A(1 + i)^{N-1}. \tag{2.8}$$

Multiplying Eq. (2.8) by $(1 + i)$ results in

$$(1 + i)F = A(1 + i) + A(1 + i)^2 + \dots + A(1 + i)^N. \tag{2.9}$$

Subtracting Eq. (2.8) from Eq. (2.9) to eliminate common terms gives us:

$$F(1 + i) - F = -A + A(1 + i)^N.$$

Solving for F yields

$$F = A\left[\frac{(1 + i)^N - 1}{i}\right] = A(F/A, i, N). \tag{2.10}$$

The bracketed term in Eq. (2.10) is called the **equal cash flow series compound amount factor**, or the **uniform series compound amount factor**; its factor notation is $(F/A, i, N)$. This interest factor has been calculated for various combinations of i and N in the interest tables.

EXAMPLE
2.13
$

Uniform series: Find *F*, given *i*, *A*, *N*

Suppose you make an annual contribution of $3000 to your savings account at the end of each year for 10 years. If your savings account earns 7% interest annually, how much can be withdrawn at the end of 10 years (Figure 2.19)?

Solution

Given: $A = \$3000$, $N = 10$ years, and $i = 7\%$ per year
Find: F

$$
\begin{aligned}
F &= \$3000(F/A, 7\%, 10) \\
&= \$3000(13.8164) \\
&= \$41{,}449.20.
\end{aligned}
$$

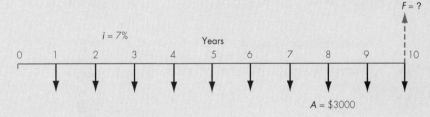

Figure 2.19 Cash flow diagram (Example 2.13)

EXAMPLE
2.14
$

Handling time shifts in a uniform series

In Example 2.13, the first deposit of the 10-deposit series was made at the end of period 1 and the remaining nine deposits were made at the end of each following period. Suppose that all deposits were made at the *beginning* of each period instead. How would you compute the balance at the end of period 10?

Solution

Given: Cash flow as shown in Figure 2.20, $i = 7\%$ per year
Find: F_{10}

Compare Fig. 2.20 to Fig. 2.19: Each payment has been shifted to 1 year earlier; thus each payment would be compounded for 1 extra year. Note that with the end-of-year deposit, the ending balance (F) was $41,449.20. With the beginning-of-year deposit, the same balance accumulates by the end of period 9. This balance can earn interest for one additional year. Therefore, we can easily calculate the resulting balance by

$$F_{10} = \$41,449.20(1.07) = \$44,350.64.$$

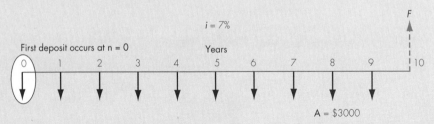

Figure 2.20   Cash flow diagram (Example 2.14)

Comments: Another way to determine the ending balance is to compare the two cash flow patterns. By adding the $3000 deposit at period 0 to the original cash flow and subtracting the $3000 deposit at the end of period 10, we obtain the second cash flow. Therefore, the ending balance can be found by making adjustment to the $41,449.20:

$$F_{10} = \$41,449.20 + 3000(F/P, 7\%, 10) - 3000 = \$44,350.64.$$

Sinking-Fund Factor—Find *A*, Given *F*, *i*, *N*

If we solve Eq. (2.10) for *A*, we obtain

$$A = F\left[\frac{i}{(1 + i)^N - 1}\right] = F\,(A/F,\ i,\ N). \tag{2.11}$$

The term within the brackets is called the **equal cash flow series sinking-fund factor**, or **sinking-fund factor**, and is referred to by the notation (*A/F, i, N*). A sinking fund is an interest-bearing account into which a fixed sum is deposited each interest period; it is commonly established for the purpose of replacing fixed assets.

EXAMPLE 2.15

$

Combination of a uniform series and a single present and future amount

To help you reach a $5000 goal 5 years from now, your father offers to give you $500 now. You plan to get a part-time job and make five additional deposits at the end of each year. (The first deposit is made at the end of the first year.) If all your money is deposited in a bank that pays 7% interest, how large must your annual deposit be?

Discussion: If your father is unable to contribute the $500, the calculation of the required annual deposit is easy because your five deposits fit the standard end-of-period pattern for a uniform series. All you need to evaluate is

$$A = \$5000(A/F, 7\%, 5) = \$5000(0.1739) = \$869.50.$$

If you do receive the $500 contribution from your father at $n = 0$, you may divide the deposit series into two parts: One contributed by your father at $n = 0$ and five equal annual deposit series contributed by yourself. Then you can use the F/P factor to find how much your father's contribution will be worth at the end of year 5 at a 7% interest rate. Let's call this amount F_c. The future value of your five annual deposits must then make up the difference, $\$5000 - F_c$.

Solution

Given: cash flow as shown in Figure 2.21, $i = 7\%$ per year, and $N = 5$ years
Find: A

$$
\begin{aligned}
A &= (\$5000 - F_c)(A/F, 7\%, 5) \\
&= [\$5000 - \$500(F/P, 7\%, 5)](A/F, 7\%, 5) \\
&= [\$5000 - \$500(1.4026)](0.1739) \\
&= \$747.55.
\end{aligned}
$$

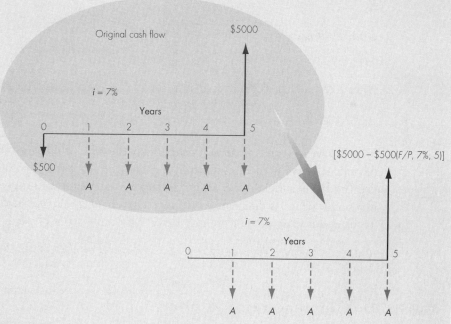

Figure 2.21   Equivalent cash flow diagram (Example 2.15)

Capital Recovery Factor (Annuity Factor)—Find *A*, Given *P*, *i*, *N*

We can determine the amount of a periodic payment, *A*, if we know *P*, *i*, and *N*. Figure 2.22 illustrates this situation. To relate *P* to *A*, recall the relationship between *P* and *F* in Eq. (2.3), $F = P(1 + i)^N$. By replacing *F* in Eq. (2.11) by $P(1 + i)^N$ we get

$$A = P (1 + i)^N \left[\frac{i}{(1 + i)^N - 1} \right],$$

or

$$A = P \left[\frac{i(1 + i)^N}{(1 + i)^N - 1} \right] = P (A/P, i, N). \tag{2.12}$$

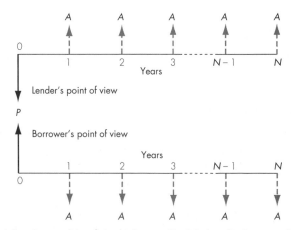

Figure 2.22   Cash flow diagram of the relationship between *P* and *A* where *P* is the amount borrowed and *A* is a series of payments of a fixed amount for *N* periods

Now we have an equation for determining the value of the series of end-of-period payments, *A*, when the present sum, *P*, is known. The portion within the brackets is called the **equal cash flow series capital recovery factor**, or simply **capital recovery factor**, which is designated (*A/P*, *i*, *N*). In finance, this *A/P* factor is referred to as the **annuity factor**. The annuity factor indicates a series of payments of a fixed, or constant, amount for a specified number of periods.

EXAMPLE

2.16

Uniform series: Find *A*, given *P*, *i*, *N*

BioGen Company, a small biotechnology firm, has borrowed $250,000 to purchase laboratory equipment for gene splicing. The loan carries an interest rate of 8% per year and is to be repaid in equal installments over the next 6 years. Compute the amount of this annual installment (Figure 2.23).

Solution

Given: $P = \$250,000$, $i = 8\%$ per year, $N = 6$ years
Find: A

$$
\begin{aligned}
A &= \$250,000(A/P, 8\%, 6) \\
&= \$250,000(0.2163) \\
&= \$54,075.
\end{aligned}
$$

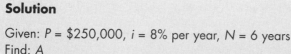

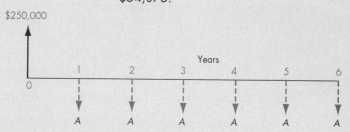

Figure 2.23 A loan cash flow diagram from BioGen's point of view (Example 2.16)

EXAMPLE 2.17 Deferred loan repayment

In Example 2.16, suppose that BioGen wants to negotiate with the bank to defer the first loan repayment until the end of year 2 (but still desires to make six equal installments at 8% interest). If the bank wishes to earn the same profit as in Example 2.16, what should be the annual installment (Figure 2.24)?

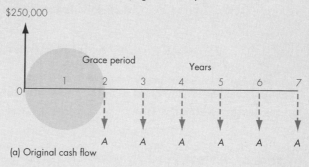

(a) Original cash flow

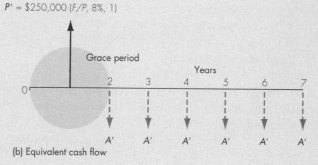

(b) Equivalent cash flow

Figure 2.24 A deferred loan cash flow diagram from BioGen's point of view (Example 2.17)

Solution

Given: $P = \$250{,}000$, $i = 8\%$ per year, $N = 6$ years, but the first payment occurs at the end of year 2

Find: A

By deferring 1 year, the bank will add the interest during the first year to the principal. In other words, we need to find the equivalent worth of $\$250{,}000$ at the end of year 1, P':

$$
\begin{aligned}
P' &= \$250{,}000(F/P, 8\%, 1)\\
&= \$270{,}000.
\end{aligned}
$$

In fact, BioGen is borrowing $\$270{,}000$ for 6 years. To retire the loan with six equal installments, the deferred equal annual payment, P', will be

$$
\begin{aligned}
A' &= \$270{,}000(A/P, 8\%, 6)\\
&= \$58{,}401.
\end{aligned}
$$

By deferring the first payment for one year, BioGen needs to make additional payments of $\$4326$ in each year.

Present Worth Factor—Find *P*, Given *A, i, N*

What would you have to invest now in order to withdraw A dollars at the end of each of the next N periods? We face just the opposite of the equal payment capital recovery factor situation—A is known, but P has to be determined. With the capital recovery factor given in Eq. (2.12), solving for P gives us

$$
P = A\left[\frac{(1 + i)^N - 1}{i(1 + i)^N}\right] = A\,(P/A, i, N). \tag{2.13}
$$

The bracketed term is referred to as the **equal cash flow series present worth factor** and is designated $(P/A, i, N)$.

EXAMPLE
2.18
$

Uniform series: Find *P*, given *A, i, N*

Frequently, people share the cost of lottery tickets. In one instance, 21 factory workers had agreed to pool $\$21$ to play the New York lottery, and split any winnings.[2] Their winning ticket was worth $\$13{,}667{,}667$, which would be distributed in 21 annual payments of $\$650{,}793$. According to their lawyer, this meant each member of the pool would receive 21 annual payments of about $\$24{,}000$ after taxes. John Brown, one of the lucky workers, wanted to quit the factory and start his own business, which required him to secure a $\$250{,}000$ bank loan. John offered to put up his future lottery earnings (as collateral) to secure the loan. If the bank's interest rate is 10% per year, how much can John borrow against his future lottery earnings?

---

[2] *The New York Times*, August 23, 1985. (Reprinted with permission.)

Discussion: We need to identify the critical data in this problem, because some numbers ultimately have nothing to do with the solution method. Basically John wants to borrow $250,000 from a bank, but the bank will not assure that he will get the full amount. (Normally a lending officer determines the maximum amount that one can borrow based on the borrower's capability of repaying the loan.) If the bank views John's lottery earnings as his only source of future income for repaying the loan, it must find the equivalent present worth of his 21 annual receipts of $24,000 in order to set the maximum loan amount (Figure 2.25).

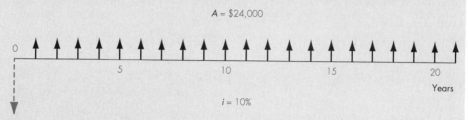

Figure 2.25   Cash flow diagram (Example 2.18)

Solution

Given: i = 10% per year, A = $24,000, and N = 21 years
Find: P

$$P = \$24,000(P/A, 10\%, 21) = \$24,000(8.6487) = \$207,569.$$

The bank may lend John a maximum of $207,569. John will have to borrow the remaining balance from other sources.

Comments: Note that the critical data included the actual cash flow over time rather than the total sum of the winnings because of the different time values of the payments. Of course, depending upon John's creditworthiness, the actual amount that the bank is willing to lend could deviate from the $207,569 amount.

EXAMPLE 2.19

$

Composite series that requires both (P/F, i, N) and (P/A, i, N) factors

Reconsider the 1995 Publishers' Clearing House, Ten Million Dollar SuperPrize payment options in Figure 2.8, page 58. At an interest rate of 8%, which option is preferred?

- Option A: $1,000,000 now, $200,000 yearly, plus a $3,400,000 final payment at the end of year 29.

- Option B: $500,000 now, $250,000 a year thereafter, plus $2,500,000 final payment at the end of year 29.

Note that 30 payments are required in either option.

Solution

Given: Cash flows given as in Fig. 2.8, $i = 8\%$ per year
Find: P for each option

Both payment options consist of a uniform series and two lump-sum payments during the first and the last year. Therefore, we can compute the equivalent present worth in two steps: (1) Find the equivalent present worth of the two lump-sum payment using the $(P/F, i, n)$ factor, and (2) find the equivalent present worth of the uniform series by using the $(P/A, i, n)$ factor. Note also that the yearly payment begins at the beginning of each year. Then, the equivalent worth calculation would look like the following:

$$
\begin{aligned}
P_{\text{Option A}} &= \$1{,}000{,}000 + \$200{,}000(P/A, 8\%, 28) \\
&\quad + \$3{,}400{,}000(P/F, 8\%, 29) \\
&= \$3{,}575{,}129. \\
P_{\text{Option B}} &= \$500{,}000 + \$250{,}000(P/A, 8\%, 28) \\
&\quad + \$2{,}500{,}000(P/F, 8\%, 29) \\
&= \$3{,}531{,}089.
\end{aligned}
$$

Comments: The difference between the two options is $44,040, favoring option A. The result indicates that having an additional $500,000 during the first year and $900,000 less during the last year will allow more wealth accumulation than receiving an additional $50,000 each year for 28 years. As you can verify, the result will be the reverse whenever your interest rate is lower than 6.18%.

2.3.5 Linear Gradient Series

Engineers frequently encounter situations involving periodic payments that increase or decrease by a constant amount (G) from period to period. This situation occurs often enough to warrant the use of special equivalence factors that relate the arithmetic gradient to other cash flows. Figure 2.26 illustrates a **strict gradient series**, $A_n = (n - 1)G$ for $n \geq 1$. Note that the origin of the strict gradient series is at the end of the first period with a zero value. The gradient G can be either positive or negative. If $G > 0$, the series is referred to as an *increasing* gradient series. If $G < 0$, it is a *decreasing* gradient series.

Unfortunately, the strict form of the increasing or decreasing gradient series does not correspond with the form that most engineering economic problems take. A typical problem involving a linear gradient includes an initial cash flow during period 1 that increases by G during some number of interest periods, a situation illustrated in Figure 2.27. This contrasts with the strict form illustrated in Fig. 2.26 a) in which no payment is made during period 1, and the gradient is added to the previous payment beginning in period 2.

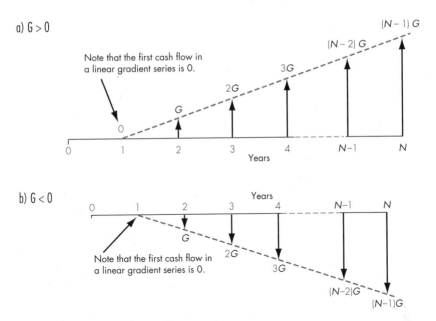

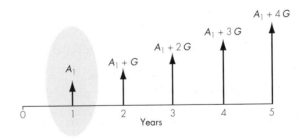

Figure 2.26 Cash flow diagram of a strict gradient series

Figure 2.27 Cash flow diagram for a typical problem involving linear gradient series. Note that a nonzero cash flow occurs at the end of the first period

Gradient Series as Composite Series

In order to utilize the strict gradient series to solve typical problems we must view cash flows as shown in Fig. 2.27 as a **composite series**, or a set of two cash flows, each corresponding to a form that we can recognize and easily solve. Figure 2.28 illustrates that the form in which we find a typical cash flow can be separated into two components: A uniform series of N payments of amount, A_1, and a gradient series of increments of constant amount, G. The need to view cash flows, which involve linear gradient series as composites of two series, is very important for the solution of problems, as we shall now see.

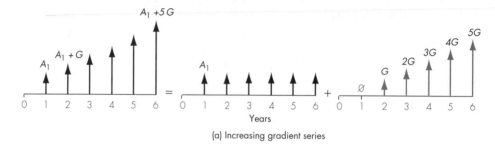

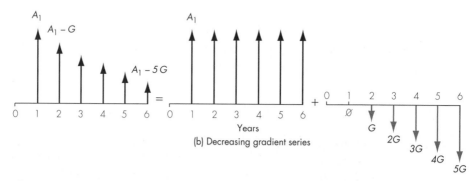

Figure 2.28   Two types of linear gradient series as composites of a uniform series of N payments of A_1 and the gradient series of increments of constant amount G

Present Worth Factor—Linear Gradient: Find P, Given G, N, i

How much would you have to deposit now to withdraw the gradient amounts specified in Fig. 2.26(a)? To find an expression for the present amount P, we apply the single-payment present worth factor to each term of the series and obtain

$$P = 0 + G/(1 + i)^2 + 2G/(1 + i)^3 + \dots + (N-1)G/(1 + i)^N,$$

or

$$P = \sum_{n=1}^{N} (n-1)G(1 + i)^{-n}. \tag{2.14}$$

Letting $G = a$ and $1/(1 + i) = x$ yields

$$\begin{aligned} P &= 0 + ax^2 + 2ax^3 + \dots + (N-1)ax^N \\ &= ax\,[0 + x + 2x^2 + \dots + (N-1)x^{N-1}]. \end{aligned} \tag{2.15}$$

Since an arithmetic-geometric series $\{0, x, 2x^2, \dots, (N-1)x^{N-1}\}$ has the finite sum of

$$0 + x + 2x^2 + \dots + (N-1)x^{N-1} = x\left[\frac{1 - Nx^{N-1} + (N-1)x^N}{(1-x)^2}\right],$$

we can rewrite Eq. (2.15) as

$$P = ax^2 \left[\frac{1 - Nx^{N-1} + (N - 1)x^N}{(1 - x)^2} \right].$$

(2.16)

Replacing the original values for A and x, we obtain

$$P = G \left[\frac{(1 + i)^N - iN - 1}{i^2(1 + i)^N} \right] = G(P/G, i, N).$$

(2.17)

The resulting factor in brackets above is called the **gradient series present worth factor**, for which we use the notation $(P/G, i, N)$.

EXAMPLE 2.20

Linear gradient: Find P, given A_1, G, i, N

A textile mill has just purchased a lift truck that has a useful life of 5 years. The engineer estimates that the maintenance costs for the truck during the first year will be $1000. Maintenance costs are expected to increase as the truck ages at a rate of $250 per year over the remaining life. Assume that the maintenance costs occur at the end of each year. The firm wants to set up a maintenance account that earns 12% annual interest. All future maintenance expenses will be paid out of this account. How much does the firm have to deposit in the account now?

Solution

Given: A_1 = $1000, G = $250, i = 12% per year, and N = 5 years
Find: P

This is equivalent to asking the equivalent present worth for this maintenance expenditure if 12% interest is used. The cash flow may be broken into its two components as shown in Figure 2.29.

The first component is an equal-payment series (A_1), and the second is a linear gradient series (G).

$$
\begin{aligned}
P &= P_1 + P_2 \\
P &= A_1(P/A, 12\%, 5) + G(P/G, 12\%, 5) \\
&= \$1000(3.6048) + \$250(6.397) \\
&= \$5204.
\end{aligned}
$$

Note that the value of N in the gradient factor is 5, not 4. This occurs because, by definition of the series, the first gradient value begins at period 2.

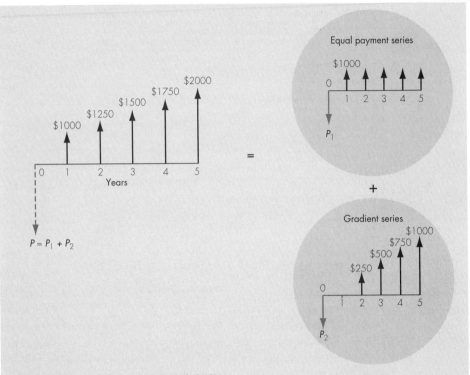

Figure 2.29   Cash flow diagram (Example 2.20)

Comments: As a check, we can compute the present worth of the cash flow by using the $(P/F, 12\%, n)$ factors:

Period (n)	Cash Flow	(P/F, 12%, n)	Present Worth
1	$1000	0.8929	$892.90
2	1250	0.7972	996.50
3	1500	0.7118	1067.70
4	1750	0.6355	1112.13
5	2000	0.5674	1134.80
			Total $5204.03

The slight difference is due to a rounding error.

Gradient-to-Equal-Cash Flow Series Conversion Factor—Find *A*, Given *G*, *i*, *N*

We can obtain an equal cash flow series equivalent to the gradient series, as depicted in Figure 2.30 by substituting Eq. (2.17) into Eq. (2.12) for *P* to obtain

$$A = G\left[\frac{(1 + i)^N - iN - 1}{i[(1 + i)^N - 1]}\right] = G(A/G, i, N), \qquad (2.18)$$

where the resulting factor in brackets is referred to as the **gradient-to-equal cash flow series conversion factor** and is designated $(A/G, i, N)$.

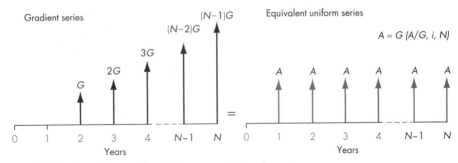

Figure 2.30 Converting a gradient series into an equivalent uniform series

EXAMPLE 2.21
$

Linear gradient: Find *A*, given *A₁*, *G*, *i*, *N*

John and Barbara have just opened two savings accounts at their credit union. The accounts earn 10% annual interest. John wants to deposit $1000 in his account at the end of the first year and increase this amount by $300 for each of the following 5 years. Barbara wants to deposit an equal amount each year for next 6 years. What should be the size of the Barbara's annual deposit so that the two accounts would have equal balances at the end of 6 years (Figure 2.31)?

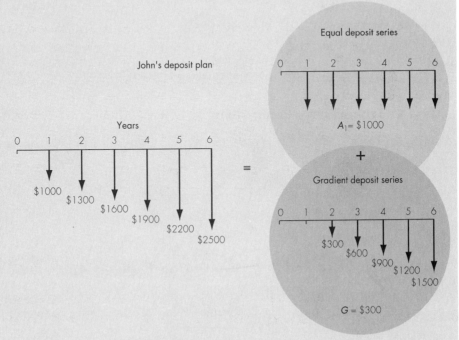

Figure 2.31 John's deposit series viewed as a combination of uniform and gradient series (Example 2.21)

Solution

Given: $A_1 = \$1000$, $G = \$300$, $i = 10\%$, and $N = 6$

Find: A

Since we use the end-of-period convention unless otherwise stated, this series begins at the end of the first year, and the last contribution occurs at the end of the sixth year. We can separate the constant portion of $1000 from the series, leaving the gradient series of 0, 0, 300, 600, ... , 1500.

To find the equal payment series beginning at the end of year 1 and ending at year 6 that would have the same present worth as that of the gradient series, we may proceed as follows:

$$
\begin{aligned}
A &= \$1000 + \$300(A/G, 10\%, 6) \\
&= \$1000 + \$300(2.2236) \\
&= \$1667.08.
\end{aligned}
$$

Barbara's annual contribution should be $1667.08.

Comments: Alternatively, we can compute Barbara's annual deposit by first computing the equivalent present worth of John's deposits and then finding the equivalent uniform annual amount. The present worth of this combined series is

$$
\begin{aligned}
P &= \$1000(P/A, 10\%, 6) + \$300(P/G, 10\%, 6) \\
&= \$1000(4.3553) + \$300(9.6842) \\
&= \$7260.56.
\end{aligned}
$$

The equivalent uniform deposit is

$$
A = \$7260.56(A/P, 10\%, 6) = \$1667.02.
$$

(The slight difference in cents is due to a rounding error.)

Future Worth Factor—Find F, Given G, i, N

To obtain the future worth equivalent of a gradient series, we substitute Eq. (2.18) into Eq. (2.10) for A:

$$
F = \frac{G}{i}\left[\frac{(1 + i)^N - 1}{i} - N\right] = G(F/G, i, N). \tag{2.19}
$$

EXAMPLE 2.22 $ Declining linear gradient: Find F, given A_1, G, i, N

Suppose that you make a series of annual deposits into a bank account that pays 10% interest. The initial deposit at the end of the first year is $1200. The deposit amounts decline by $200 in each of the next 4 years. How much would you have immediately after the fifth deposit?

Solution

Given: Cash flow shown in Figure 2.32, $i = 10\%$ per year, $N = 5$ years
Find: F

The cash flow includes a decreasing gradient series. Recall that we derived the linear gradient factors for an increasing gradient series. For a decreasing gradient series, the solution is most easily obtained by separating the flow into two components: A uniform series and an increasing gradient which is *subtracted* from the uniform series (Figure 2.32). The future value is

$$
\begin{aligned}
F &= F_1 - F_2 \\
&= A_1(F/A,10\%,5) - \$200(F/G,10\%,5) \\
&= A_1(F/A,10\%,5) - \$200(P/G,10\%,5)(F/P,10\%,5) \\
&= \$1200(6.105) - \$200(6.862)(1.611) \\
&= \$5115.
\end{aligned}
$$

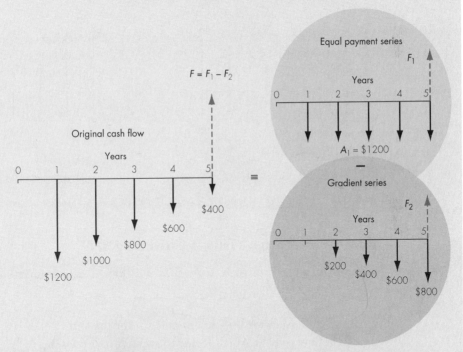

Figure 2.32   A series of decreasing gradient deposits viewed as a combination of uniform and gradient series (Example 2.22)

2.3.6   Geometric Gradient Series

Many engineering economic problems, particularly those relating to construction costs, involve cash flows that increase or decrease over time, not by a constant amount (linear gradient), but rather by a constant percentage (**geometric**), which is called **compound growth**. Price changes due to inflation are a good example of such a geometric series. If we use g to designate the percentage change in a payment from one period to the next, the magnitude of the nth payment, A_n, is related to the first payment A_1 as expressed by

$$A_n = A_1(1 + g)^{n-1}, \ n = 1, 2, \ldots, N. \tag{2.20}$$

The g can take either a positive or a negative sign depending on the type of cash flow. If $g > 0$, the series will increase, and if $g < 0$, the series will decrease. Figure 2.33 illustrates the cash flow diagram for this situation.

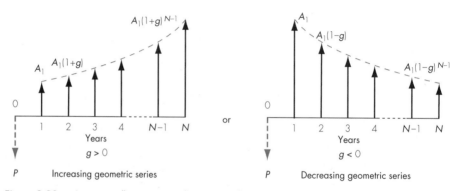

Figure 2.33   A geometrically increasing or decreasing gradient series at a constant rate g

Present Worth Factor—Find P, Given A₁, g, i, N

Notice that the present worth, P_n, of any cash flow A_n at interest rate i is

$$P_n = A_n(1 + i)^{-n} = A_1(1 + g)^{n-1}(1 + i)^{-n}.$$

To find an expression for the present amount for the entire series, P, we apply the **single cash flow present worth factor** to each term of the series:

$$P = \sum_{n=1}^{N} A_1(1 + g)^{n-1}(1 + i)^{-n}. \tag{2.21}$$

Bringing the constant term $A_1(1 + g)^{-1}$ outside the summation yields

$$P = \frac{A_1}{(1 + g)} \sum_{n=1}^{N} \left[\frac{1 + g}{1 + i}\right]^n. \tag{2.22}$$

Let $a = \dfrac{A_1}{1 + g}$ and $x = \dfrac{1 + g}{1 + i}$. Then, rewrite Eq. (2.22) as

$$P = a\,(x + x^2 + x^3 + \dots + x^N). \qquad (2.23)$$

Since the summation in Eq. (2.23) represents the first N terms of a geometric series, we may obtain the closed-form expression as follows: First, multiply Eq. (2.23) by x:

$$xP = a\,(x^2 + x^3 + x^4 + \dots + x^{N+1}). \qquad (2.24)$$

Then, subtract Eq. (2.24) from Eq. (2.23):

$$
\begin{aligned}
P - xP &= a(x - x^{N+1}) \\
P(1 - x) &= a(x - x^{N+1}) \\
P &= \frac{a(x - x^{N+1})}{1 - x}, \text{ where } x \neq 1. \qquad (2.25)
\end{aligned}
$$

If we replace the original values for a and x, we obtain

$$
P = \begin{cases}
A_1\left[\dfrac{1 - (1 + g)^N (1 + i)^{-N}}{i - g}\right], & \text{if } i \neq g \\[2ex]
\dfrac{NA_1}{(1+i)}, & \text{if } i = g
\end{cases} \qquad (2.26)
$$

or

$$P = A_1(P/A_1, g, i, N).$$

The factor within brackets is called the **geometric-gradient-series present worth factor** and designated $(P/A_1, g, i, N)$. In the special case where $i = g$, Eq. (2.22) becomes $P = [A_1/(1 + i)]N$.

EXAMPLE 2.23

Geometric gradient: Find *P*, given *A₁, g, i, N*

Ansell Inc., a medical device manufacturer, uses compressed air in solenoids and pressure switches in its machines to control various mechanical movements. Over the years, the manufacturing floor has changed layouts numerous times. With each new layout, more piping was added to the compressed air delivery system to accommodate new locations of manufacturing machines. None of the extra, unused old pipe was capped or removed; thus the current compressed air delivery system is inefficient and fraught with leaks. Because of the leaks in the current system, the compressor is expected to run 70% of the time that the plant will be in operation during the upcoming year. This will require 260 kWh of electricity at a rate of $0.05/kWh. (The plant runs 250 days a year, 24 hours per day.) If Ansell continues to operate the current air delivery system, the compressor run time will increase by 7% per year for the next 5 years due to ever-worsening leaks. (After 5 years, the current system will not be able to meet the plant's compressed air requirement, so it will have to be replaced.)

If Ansell decides to replace all of the old piping now, it will cost $28,570. The compressor will still run the same number of days; however, it will run 23% less (or 70%(1 − 0.23) = 53.9% usage during the day) because of the reduced air pressure loss. If Ansell's interest rate is 12%, is it worth fixing now?

Solution

Given: Current power consumption, $g = 7\%$, $i = 12\%$, $N = 5$ years
Find: A_1 and P

Step 1: We need to calculate the cost of power consumption of the current piping system during the first year. The power consumption is equal to the following:

$$
\begin{aligned}
\text{Power cost} &= \text{\% of day operating} \\
&\quad \times \text{days operating per year} \\
&\quad \times \text{hours per day} \\
&\quad \times \text{kWh} \times \text{\$/kWh} \\
&= (70\%) \times (250 \text{ days/year}) \times (24 \text{ hours/day}) \\
&\quad \times (260 \text{ kWh}) \times (\$0.05/\text{kWh}) \\
&= \$54{,}440.
\end{aligned}
$$

Step 2: Each year the annual power cost will increase at the rate of 7% over the previous year's cost. The anticipated power cost over the 5-year period is summarized in Figure 2.34. The equivalent present lump-sum cost at 12% for this geometric gradient series is

$$
\begin{aligned}
P_{\text{Old}} &= -\$54{,}440(P/A_1, 7\%, 12\%, 5) \\
&= -\$54{,}440 \left[\frac{1 - (1 + 0.07)^5(1 + 0.12)^{-5}}{0.12 - 0.07} \right] \\
&= -\$222{,}283.
\end{aligned}
$$

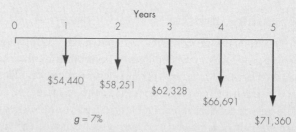

Figure 2.34   If repair is not performed, expected power expenditure over next 5 years will increase at a rate of 7% per year (Example 2.23)

Step 3: If Ansell replaces the current compressed air system with the new one, the annual power cost will be 23% less during the first year and will remain at that level over the next 5 years. The equivalent present lump sum cost at 12% is

$$P_{New} = -\$54,440(1 - 0.23)(P/A, 12\%, 5)$$
$$= -\$41,918.80(3.6048)$$
$$= -\$151,109.$$

Step 4: The net cost for not replacing the old system now is −$71,174 (= −$222,283 + $151,109). Conversely, installation of the new system avoids this net cost and the $71,174 can be viewed as the net savings or revenues attributable to the new system. Since the new system costs only $28,570, the replacement should be made now.

Comments: In this example, we assumed that the cost of removing the old system was included in the cost of installing the new system. If the removed system has some salvage value, replacing it will result in even greater savings. We will consider many types of replacement issues in Chapter 6.

Future Worth Factor—Find F, Given A_1, g, N, i

The future worth equivalent of the geometric series can be obtained by multiplying Eq. (2.26) by the F/P factor, $(1 + i)^N$.

$$F = \begin{cases} A_1\left[\dfrac{(1 + i)^N - (1 + g)^N}{i - g} \right], & \text{if } i \neq g \\ NA_1(1 + i)^{N-1}, & \text{if } i = g \end{cases} \qquad (2.27)$$

or

$$F = A_1(F/A_1, g, i, N).$$

EXAMPLE 2.24
$

Geometric gradient: Find A_1, given F, g, i, N

A self-employed individual, Jimmy Carpenter, is opening a retirement account at a bank. His goal is to accumulate $1,000,000 in the account by the time he retires from work in 20 years' time. A local bank is willing to open a retirement account that pays 8% interest, compounded annually, throughout the 20 years. Jimmy expects his annual income will increase at a 6% annual rate during his working career. He wishes to start with a deposit at the end of year 1 (A_1) and increase the deposit at a rate of 6% each year thereafter. What should be the size of his first deposit (A_1)? The first deposit will occur at the end of year 1, and subsequent deposits will be made at the end of each year. The last deposit will be made at the end of year 20.

Solution

Given: $F = \$1,000,000$, $g = 6\%$ per year, $i = 8\%$ per year, and $N = 20$ years
Find: A_1 as in Figure 2.35.

$$F = A_1(F/A_1, g, i, N)$$
$$= A_1(F/A_1, 6\%, 8\%, 20)$$
$$= A_1(72.6911).$$

Solving for A_1 yields

$$A_1 = \$1,000,000/72.6911 = \$13,757$$

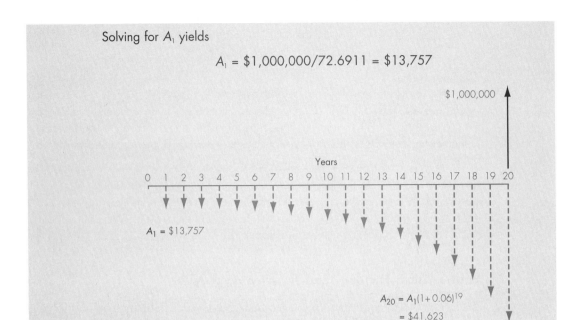

Figure 2.35 Jimmy Carpenter's retirement plan (Example 2.24)

EXAMPLE 2.25

Composite geometric gradient: Find *P*, given *A₁*, *g* (composite), *i*, *N*

A Labrador mining company expects to produce 400,000 tonnes of ore concentrate this year. The current selling price of the concentrate is $300/tonne. The company plans to continue its operations for another ten years. Price projections for concentrate indicate that the selling price is expected to increase at a rate of 5% per year. However, limitations in processing capacity and decreases in ore grade will reduce the production level from its present value. The rate of decrease in production is estimated to be 2% per year. If the mining company's interest rate is 10%, determine the present worth of the future revenues.

Solution

Given: Current production level and selling price, g (price) = 5%, g (production) = –2%, i = 10%, N = 10
Find: P

The revenue to be realized from concentrate sales this year, A_1, is the product of the total tonnes and the selling price.

$$
\begin{aligned}
A_1 &= 400{,}000 \text{ tonnes} \times \$300 \text{ per tonne} \\
&= \$120{,}000{,}000
\end{aligned}
$$

In each subsequent year, the tonnage will decrease by 2% and selling price will increase by 5%. Therefore, the revenue in year 2 will be

$$A_2 = 400{,}000(1 - 0.02) \text{ tonnes} \times \$300(1 + 0.05) \text{ per tonne}$$
$$= \$123{,}480{,}000$$

The year 3 revenue will be

$$A_3 = 400{,}000(1 - 0.02)^2 \text{ tonnes} \times \$300(1 + 0.05)^2 \text{ per tonne}$$
$$= \$127{,}060{,}920$$

For any year n, the revenue can be calculated as

$$A_n = 400{,}000(1 - 0.02)^{n-1} \text{ tonnes} \times \$300(1 + 0.05)^{n-1} \text{ per tonne}$$
$$= 400{,}000 \times \$300(1 - 0.02)^{n-1}(1 + 0.05)^{n-1}$$
$$= A_1(1 - 0.02)^{n-1}(1 + 0.05)^{n-1}$$

- Method 1: We can calculate A_n for each of the remaining years 4 through 10, and use a $(P/F, i, n)$ factor to discount each annual revenue value to time zero.

$$P = \$120{,}000{,}000(P/F, 10\%, 1)$$
$$+ \$123{,}480{,}000(P/F, 10\%, 2)$$
$$+ \$127{,}060{,}920(P/F, 10\%, 3)$$
$$+ \ldots + \$155{,}209{,}971(P/F, 10\%, 10)$$
$$= \$822{,}870{,}891$$

- Method 2: The revenue stream is the product of two quantities, each of which is represented by a geometric series. It follows that the revenue stream itself must be represented by a geometric series having some composite rate of change value $\hat{g}$. We can determine this composite value by going back to the geometric series definition in Eq. 2.20.

$$A_n = A_1(1 + \hat{g})^{n-1}$$
$$= A_1(1 - 0.02)^{n-1}(1 + 0.05)^{n-1}$$

Therefore, the composite value, $\hat{g}$, which incorporates the effects of the decrease in production and the increase in price is

$$\hat{g} = (1 - 0.02)(1 + 0.05) - 1$$
$$= 2.9\%$$

and the present worth of the revenues can be calculated as

$$P = \$120{,}000{,}000(P/A_1, 2.9\%, 10\%, 10)$$
$$= \$120{,}000{,}000(6.8543)$$
$$= \$822{,}876{,}000$$

The difference in the answers between Method 1 and 2 is due to rounding errors. The actual present worth as computed with Excel is \$822,879,887. Method 2 affords considerable efficiency compared to Method 1.

Note: The composite geometric rate of change of revenue does <u>not equal the difference</u> between the rate of price increase and production decrease, or 3%.

Table 2.3 summarizes the interest formulas developed in this section and the cash flow situations in which they should be used. Recall that all the interest formulas developed in this section are only applicable to situations where the interest (compounding) period is the same as the payment period (e.g., annual compounding with annual payment). Also, in this table we present some useful interest factor relationships.

Table 2.3
Summary of Discrete Compounding Formulas with Discrete Payments[3]

Flow Type	Factor Notation	Formula	Cash Flow Diagram	Factor Relationship
S I N G L E	Compound amount $(F/P, i, N)$	$F = P(1 + i)^N$		$(F/P,i,N) = i(F/A,i,N) + 1$
	Present worth $(P/F, i, N)$	$P = F(1 + i)^{-N}$		$(P/F,i,N) = 1 - (P/A,i,N)i$
E Q U A L P A Y M E N T S E R I E S	Compound amount $(F/A,i,N)$	$F = A\left[\dfrac{(1 + i)^N - 1}{i}\right]$		$(A/F,i,N) = (A/P,i,N) - i$
	Sinking fund $(A/F,i,N)$	$A = F\left[\dfrac{i}{(1 + i)^N - 1}\right]$		
	Present worth $(P/A,i,N)$	$P = A\left[\dfrac{(1 + i)^N - 1}{i(1 + i)^N}\right]$		$(A/P, i, N) = \dfrac{i}{1 - (P/F,i,N)}$
	Capital recovery $(A/P,i,N)$	$A = P\left[\dfrac{i(1 + i)^N}{(1 + i)^N - 1}\right]$		
G R A D I E N T S E R I E S	Linear gradient Present worth $(P/G,i,N)$	$P = G\left[\dfrac{(1 + i)^N - iN - 1}{i^2(1 + i)^N}\right]$		$(F/G,i,N) = (P/G,i,N)(F/P,i,N)$ $(A/G,i,N) = (P/G,i,N)(A/P,i,N)$
	Geometric gradient Present worth $(P/A_1,g,i,N)$	$P = \begin{bmatrix} A_1\left[\dfrac{1 - (1 + g)^N(1 + i)^{-N}}{i - g}\right] \\ \dfrac{NA_1}{1 + i} \quad (\text{if } i = g) \end{bmatrix}$		$(F/A_1,g,i,N) = (P/A_1,g,i,N)(F/P,i,N)$

[3]Park CS, Sharp-Bette GP. *Advanced Engineering Economics.* New York: John Wiley & Sons, 1990. (Reprinted by permission of John Wiley & Sons Inc.)

2.3.7 Limiting Forms of Interest Formulas

When the interest rate, i, is very small or the number of interest periods, N, is very large, the interest formulas in Table 2.3 reduce to the limiting forms summarized in Table 2.4. These limits are readily established by taking the limit as $i \to 0$ or as $N \to \infty$ of the formulas in Table 2.3. When indeterminate forms arise in the analysis, the usual calculus methods are used to establish the correct limit.

As we will see in later chapters, there are situations which satisfy the interest rate condition or the number of periods condition. If the interest factors required in such an analysis have finite limits, use of the limiting forms in Table 2.4 can save considerable effort.

Table 2.4

Limiting Forms of Discrete Compounding Formulas with Discrete Payments

	Limit as $N \to \infty$ (i is specified)		Limit as $i \to 0$ (N is specified)	
$(F/P, i, N)$	∞		1	
$(P/F, i, N)$	0		1	
$(F/A, i, N)$	∞		N	
$(A/F, i, N)$	0		$\dfrac{1}{N}$	
$(P/A, i, N)$	$\dfrac{1}{i}$		N	
$(A/P, i, N)$	i		$\dfrac{1}{N}$	
$(P/G, i, N)$	$\dfrac{1}{i^2}$		∞	
$(P/A_1, g, i, N)$	∞	$g > i$	$\dfrac{(1+g)-1}{g}$	$i \neq g$
	$\dfrac{1}{i-g}$	$g < i$		
	∞	$i = g$	N	$i = g$

2.4 Unconventional Equivalence Calculations

Throughout the preceding section, we occasionally presented two or more methods of attacking example problems even though we had standard interest factor equations by which to solve them. It is important that you become adept at examining problems from unusual angles and that you seek out unconventional solution methods, because not all cash flow problems conform to the neat patterns for which we have discovered and developed equations. Two categories of problems that demand unconventional treatment are composite (mixed) cash flows and problems in which we must determine the interest rate implicit in a financial contract. We will begin this section by examining instances of composite cash flows.

2.4.1 Composite Cash Flows

Although many financial decisions do involve constant or systematic changes in cash flows, many investment projects contain several cash flow components that do not exhibit an overall pattern. Consequently, it is necessary to expand our analysis to deal with these mixed types of cash flows.

To illustrate, consider the cash flow stream shown in Figure 2.36. We want to compute the equivalent present worth for this mixed payment series at an interest rate of 15%. Three different methods are presented.

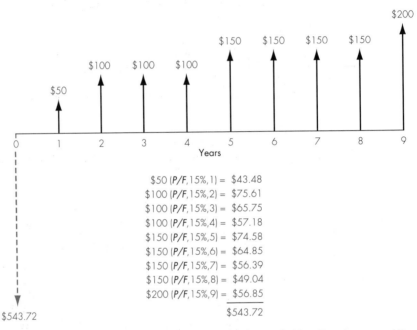

$50 (P/F, 15\%, 1) = \$43.48$
$100 (P/F, 15\%, 2) = \$75.61$
$100 (P/F, 15\%, 3) = \$65.75$
$100 (P/F, 15\%, 4) = \$57.18$
$150 (P/F, 15\%, 5) = \$74.58$
$150 (P/F, 15\%, 6) = \$64.85$
$150 (P/F, 15\%, 7) = \$56.39$
$150 (P/F, 15\%, 8) = \$49.04$
$200 (P/F, 15\%, 9) = \$56.85$

$543.72

Figure 2.36   Equivalent present worth calculation using P/F factors (Method 1 — "Brute force approach")

Method 1: A "brute force" approach is to multiply each payment by the appropriate (P/F, 10%, n) factors and then to sum these products to obtain the present worth of the cash flows, $543.72. Recall that this is exactly the same procedure we used to solve the category of problems called the uneven payment series, which were described in Section 2.3.3. Fig. 2.36 illustrates this computational method.

Method 2: We may group the cash flow components according to the type of cash flow pattern that they fit, such as the single payment, equal payment series and so forth, as shown in Figure 2.37. Then, the solution procedure involves the following steps:

* Group 1: Find the present worth of $50 due in year 1:

$$\$50(P/F, 15\%, 1) = \$43.48.$$

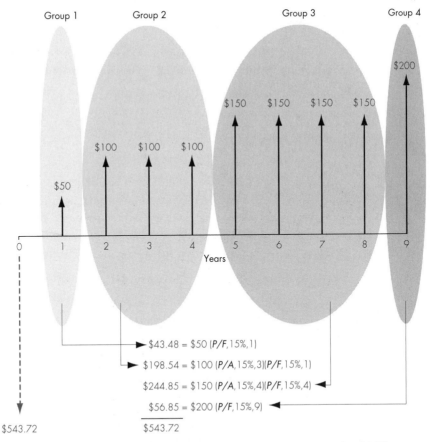

Figure 2.37 Equivalent present worth calculation for an uneven payment series using P/F and P/A factors (Method 2 — "Grouping Approach")

- Group 2: Find the equivalent worth of a $100 equal-payment series at year 1 (V_1) and then bring this equivalent worth to year 0.

$$\underbrace{\$100(P/A, 15\%, 3)}_{V_1} (P/F, 15\%, 1) = \$198.54.$$

- Group 3: Find the equivalent worth of a $150 equal-payment series at year 4 (V_4) and then bring this equivalent worth to year 0.

$$\underbrace{\$150(P/A, 15\%, 4)}_{V_4} (P/F, 15\%, 4) = \$244.85.$$

- Group 4: Find the equivalent present worth of the $200 due in year 9:

$$\$200(P/F, 15\%, 9) = \$56.85.$$

- Group total—sum the components:

$$P = \$43.48 + \$198.54 + \$244.85 + \$56.85 = \$543.72.$$

A pictorial view of this computational process is given in Fig. 2.37.

Method 3: In computing the present worth of the equal-payment series components, we may use an alternative method.

- Group 1: Same as in Method 2

- Group 2: Recognize that a $100 equal payment series will be received during years 2 through 4. Thus, we could determine the value of a 4-year annuity, subtract from it the value of a 1-year annuity, and have remaining the value of a 4-year annuity, whose first payment is due in year 2. This result is achieved by subtracting the $(P/A, 15\%, 1)$ for a 1-year, 15% annuity from that for a 4-year annuity, and then multiplying the difference by $100:

$$\begin{aligned} \$100[(P/A, 15\%, 4) - (P/A, 15\%, 1)] &= \$100(2.8550 - 0.8696) \\ &= \$198.54. \end{aligned}$$

Thus, the equivalent present worth of the annuity component of the uneven stream is $198.54.

- Group 3: We have another equal-payment series which starts in year 5 and ends in year 8.

$$\begin{aligned} \$150[(P/A, 15\%, 8) - (P/A, 15\%, 4)] &= \$150(4.4873 - 2.8550) \\ &= \$244.85. \end{aligned}$$

- Group 4: Same as Method 2

- Group total—Sum the components:

$$P = \$43.48 + \$198.54 + \$244.85 + \$56.85 = \$543.72.$$

Either the "brute force" method in Fig. 2.36 or the method utilizing both $(P/A, i, n)$ and $(P/F, i, n)$ factors can be used to solve problems of this type. Either Method 2 or Method 3 is much easier if the annuity component runs for many years, however. For example, the alternative solution would be clearly superior for finding the equivalent present worth of a stream consisting of $50 in year 1, $200 in years 2 through 19, and $500 in year 20.

Also, note that in some instances we may want to find the equivalent value of a stream of payments at some point other than the present (year 0). In this situation, we proceed as before, but compound and discount to some other points in time, say year 2, rather than year 0. Example 2.26 illustrates the situation.

EXAMPLE 2.26 Cash flows with subpatterns

The two cash flows in Figure 2.38 are equivalent at an interest rate of 12%, compounded annually. Determine the unknown value, C.

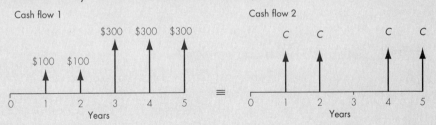

Figure 2.38   Equivalence calculation (Example 2.26)

Solution

Given: Cash flows as in Fig. 2.38, i = 12% per year
Find: C

- Method 1: Compute the present worth of each cash flow at time 0.

$$
\begin{aligned}
P_1 &= \$100(P/A, 12\%, 2) + \$300(P/A, 12\%, 3)(P/F, 12\%, 2) \\
&= \$743.42 \\
P_2 &= C(P/A, 12\%, 5) - C(P/F, 12\%, 3) \\
&= 2.8930C.
\end{aligned}
$$

Since the two flows are equivalent, $P_1 = P_2$.

$$743.42 = 2.8930C.$$

Solving for C, we obtain C = $256.97.

- Method 2: We may select a time point other than 0 for comparison. The best choice of a base period is largely determined by the cash flow patterns. Obviously, we want to select a base period that requires the minimum number of interest factors for the equivalence calculation. Cash flow 1 represents a combined series of two equal payment cash flows, whereas cash flow 2 can be viewed as an equal payment series with the third payment missing. For cash flow 1, computing the equivalent worth at period 5 will require only two interest factors:

$$
\begin{aligned}
V_{5,1} &= \$100(F/A, 12\%, 5) + \$200(F/A, 12\%, 3) \\
&= \$1310.16.
\end{aligned}
$$

For cash flow 2, computing the equivalent worth of the equal payment series at period 5 will also require two interest factors:

$$
\begin{aligned}
V_{5,2} &= C(F/A, 12\%, 5) - C(F/P, 12\%, 2) \\
&= 5.0984C.
\end{aligned}
$$

Therefore, the equivalence would be obtained by letting $V_{5,1} = V_{5,2}$:

$$\$1310.16 = 5.0984C.$$

Solving for C yields $C = \$256.97$, which is the same result obtained from Method 1. The alternative solution of shifting the time point of comparison will require only four interest factors, whereas Method 1 would require five interest factors.

2.27
$

Establishing a university fund

A couple with a newborn daughter wants to save for their child's university expenses in advance. The couple can establish a university fund that pays 7% annual interest. Assuming that the child enters university at age 18, the parents estimate that an amount of \$40,000 per year will be required to support the child's university expenses for 4 years. Determine the equal annual amounts the couple must save until they send their child to university. (Assume that the first deposit will be made on the child's first birthday and the last deposit on the child's 18th birthday. The first withdrawal will also be made at the beginning of the freshman year, which is the child's 18th birthday.)

Solution

Given: Deposit and withdrawal series shown in Figure 2.39, $i = 7\%$ per year
Find: Unknown annual deposit amount (X)

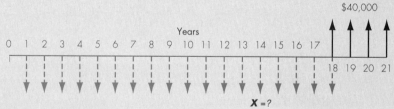

Figure 2.39   Establishing a college fund (Example 2.27)

- Method 1: Establish economic equivalence at period 0:

Step 1: Find the equivalent single lump sum deposit now:

$$
\begin{aligned}
P_{\text{Deposit}} &= -X(P/A, 7\%, 18) \\
&= -10.0591X.
\end{aligned}
$$

Step 2: Find the equivalent single lump sum withdrawal now:

$$
\begin{aligned}
P_{\text{Withdrawal}} &= \$40,000(P/A, 7\%, 4)(P/F, 7\%, 17) \\
&= \$42,892.
\end{aligned}
$$

Step 3: Since the fund is depleted after year 21, $P_{\text{Deposit}} + P_{\text{Withdrawal}} = 0$, we obtain X:

$$
\begin{aligned}
-10.0591X + \$42,892 &= 0 \\
X &= \$4264.
\end{aligned}
$$

- Method 2: Establish the economic equivalence at the child's 18th birthday.

Step 1: Find the accumulated deposit balance at the child's 18th birthday:

$$V_{18} = X(F/A, 7\%, 18)$$
$$= 33.9990X.$$

Step 2: Find the equivalent lump sum withdrawal at the child's 18th birthday:

$$V_{18} = \$40,000 + \$40,000(P/A, 7\%, 3)$$
$$= \$144,972.$$

Step 3: Since the two amounts must be the same, we obtain

$$33.9990X = \$144,972$$
$$X = \$4264.$$

The computational steps are also summarized in Figure 2.40. In general, the second method is the more efficient way to obtain an equivalence solution to this type of decision problem.

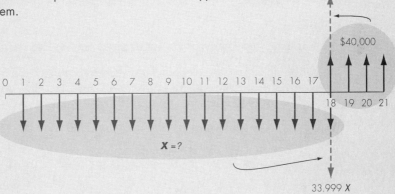

Figure 2.40   An alternative equivalence calculation (Example 2.27)

Comments: To verify if the annual deposits of \$4264 over 18 years would be sufficient to meet the child's university expenses, we can calculate the actual year-by-year balances: With the 18 annual deposits of \$4264, the balance at the child's 18th birthday is

$$\$4264(F/A, 7\%, 18) = \$144,972.$$

From this balance, the couple will make four annual tuition payments:

Year n	Beginning Balance	Interest Earned	Tuition Payment	Ending Balance
Freshman	\$144,972	\$0	\$40,000	\$104,972
Sophomore	104,972	7348	40,000	72,320
Junior	72,320	5062	40,000	37,382
Senior	37,382	2618	40,000	0

2.4.2 Determining an Interest Rate to Establish Economic Equivalence

Thus far, we have assumed that, in equivalence calculations, a typical interest rate is given. Now we can use the same interest formulas that we developed earlier to determine interest rates explicit in equivalence problems. For most commercial loans, interest rates are already specified in the contract. However, when making some investments in financial assets, such as stocks, you may want to know at what rate of growth (or rate of return) your asset is appreciating over the years. (This kind of calculation is the basis of rate-of-return analysis, which is covered in Chapter 4.) Although we can use interest tables to find the rate implicit in single payments and annuities, it is more difficult to find the rate implicit in an uneven series of payments. In such cases, a trial-and-error procedure or computer software may be used. To illustrate, consider Example 2.28.

EXAMPLE 2.28
$

Calculating an unknown interest rate with multiple factors

Consider again the sweepstake problem introduced at the beginning of this chapter. Suppose, that instead of receiving one lump sum of $1 million, you decide to accept the 20 annual installments of $100,000. If you are like most jackpot winners, you will be tempted to spend your winnings to improve your lifestyle during the first several years. Only after you get this type of spending "out of your system" will you save later sums for investment purposes. Suppose that you are considering the following two options:

Option 1: You save your winnings for the first 7 years and then spend every cent of the winnings in the remaining 13 years.

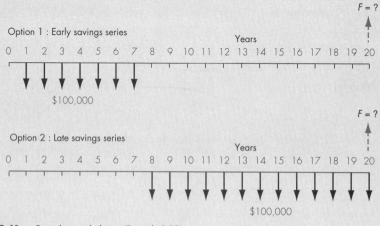

Figure 2.41   Equivalence calculation (Example 2.28)

Option 2: You do the reverse and spend for 7 years and then save for 13 years.

If you can save winnings at 7% interest, how much would you have at the end of 20 years? What interest rate on your savings will make these two options equivalent? (Cash flows into savings for the two options are shown in Figure 2.41.)

Solution

Given: Cash flows in Fig. 2.41
Find: (a) F and (b) i at which the two flows are equivalent

(a) In Option 1, the net balance at the end of year 20 can be calculated in two steps: Find the accumulated balance at the end of year 7 (V_7) first, and second, find the equivalent worth of V_7 at the end of year 20. For Option 2, find the equivalent worth of the 13 equal annual deposits at the end of year 20:

$$
\begin{aligned}
F_{\text{Option 1}} &= \$100,000(F/A, 7\%, 7)(F/P, 7\%, 13) \\
&= \$2,085,485 \\
F_{\text{Option 2}} &= \$100,000(F/A, 7\%, 13) \\
&= \$2,014,064.
\end{aligned}
$$

Option 1 accumulates $71,421 more than the Option 2.

(b) To compare the alternatives, we may compute the present worth for each option at period 0. By selecting time period 7, however, we can establish the same economic equivalence with fewer interest factors. As shown in Figure 2.42, we calculate the equivalent value for each option at the end of period 7, (V_7), remembering that the end of period 7 is also the beginning of period 8. (Recall from Example 2.4 that the choice of the point in time at which to compare two cash flows for equivalence is arbitrary.)

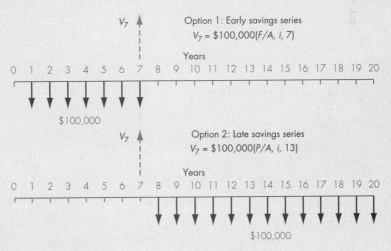

Figure 2.42   Establishing an economic equivalence at period 7 (Example 2.28)

For Option 1:

$$V_7 = \$100,000(F/A, i, 7).$$

For Option 2:

$$V_7 = \$100,000(P/A, i, 13).$$

To be equivalent, these values must be the same.

$$\$100,000(F/A, i, 7) = \$100,000(P/A, i, 13)$$

$$\frac{(F/A, i, 7)}{(P/A, i, 13)} = 1.$$

Here, we are looking for an interest rate that makes the ratio 1. When using the interest tables we need to resort to a trial-and-error method. Suppose that we guess the interest rate to be 6%. Then

$$\frac{(F/A, 6\%, 7)}{(P/A, 6\%, 13)} = \frac{8.3938}{8.8527} = 0.9482.$$

This is less than 1. To increase the ratio, we need to use an i value such that it increases the $(F/A, i, 7)$ factor value, but decreases the $(P/A, i, 13)$ value. This will happen if we use a larger interest rate. Let's try $i = 7\%$.

$$\frac{(F/A, 7\%, 7)}{(P/A, 7\%, 13)} = \frac{8.6540}{8.3577} = 1.0355.$$

Now the ratio is greater than 1.

Interest Rate	$(F/A, i, 7)/(P/A, i, 13)$
6%	0.9482
?	1.0000
7%	1.0355

As a result, we find that the interest rate is between 6% and 7% and may be approximated by linear interpolation as shown in Figure 2.43:

$$i = 6\% + (7\% - 6\%)\left[\frac{1 - 0.9482}{1.0355 - 0.9482}\right]$$

$$= 6\% + 1\%\left[\frac{0.0518}{0.0873}\right]$$

$$= 6.5934\%.$$

At 6.5934% interest, the options are equivalent, and you may decide to indulge your desire to spend like crazy for the first 7 years. However, if you could obtain a higher interest rate, you would be wiser to save for 7 years and spend for the next 13.

Comments: This example demonstrates that finding an interest rate is an iterative process, which is more complicated and generally less precise than the problem of finding an equivalent worth at a known interest rate. Since computers and financial calculators can speed the process of finding unknown interest rates, these tools are highly recommended for these types of problem solving. With a computer, a more precise break-even value of 6.60219% is found.

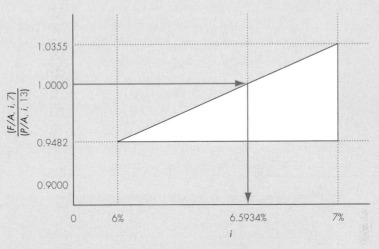

Figure 2.43 Linear interpolation to find unknown interest rate (Example 2.28)

2.5 Computer Notes

With the advent of personal computers (and workstations), we can easily access a great deal of computing power at a fractional cost. In particular, electronic spreadsheets such as Excel, Lotus 1-2-3, and Quattro Pro provide many useful financial functions that can expedite equivalence calculations.[3]

In electronic spreadsheets, most of the equivalence equations we've studied in this chapter are treated as built-in functions. However, despite their power, one drawback of these spreadsheets is the macro programming required to perform some of the higher economic functions.

Appendix B summarizes the built-in equivalence functions for Lotus 1-2-3, Excel, and Quattro Pro. As you will quickly see, the keystrokes and choices are similar for all three. Thus, although we begin with an Excel example, you should easily be able to translate the procedure described if you use a different spreadsheet package.

---

[3] Using the **Equivalence** command in **EzCash** enables you to calculate various economic equivalence values upon specifying proper input parameters. For example, with an equal-payment series with interest rate, constant amount, and present worth specified, you can solve for an unknown period. Four major cash flow patterns are provided: (1) Single-cash flow transactions, (2) equal-cash flow series, (3) linear gradient series, and (4) geometric gradient series.

We will use Excel to demonstrate some of the basic equivalence calculations. For the sweepstake problem in Example 2.28, we want to compute the value of each savings option at the end of period 7. One of the most useful features of any spreadsheet program is its ability to allow you to do "what if" analyses relatively quickly. For example: What if the winner in Example 2.28 can earn 10% instead of 6% interest on his/her savings? By varying the savings interest rate from 0% to 15%, we will be able to answer this type of "what-if" question and even locate the interest rate that makes the two options equivalent.

Figure 2.44 shows a reasonable format for the spreadsheet. We enter the known cash flows (columns B and C), and interest rate as shown in column E. In columns F and G, the equivalent values at $n = 7$ are computed by varying the savings interest rate. In column H, the difference in economic worth between the two options is listed. It is clear from column H that Option 1 is a better choice at a savings rate higher than 6% (more precisely, 6.6021%).

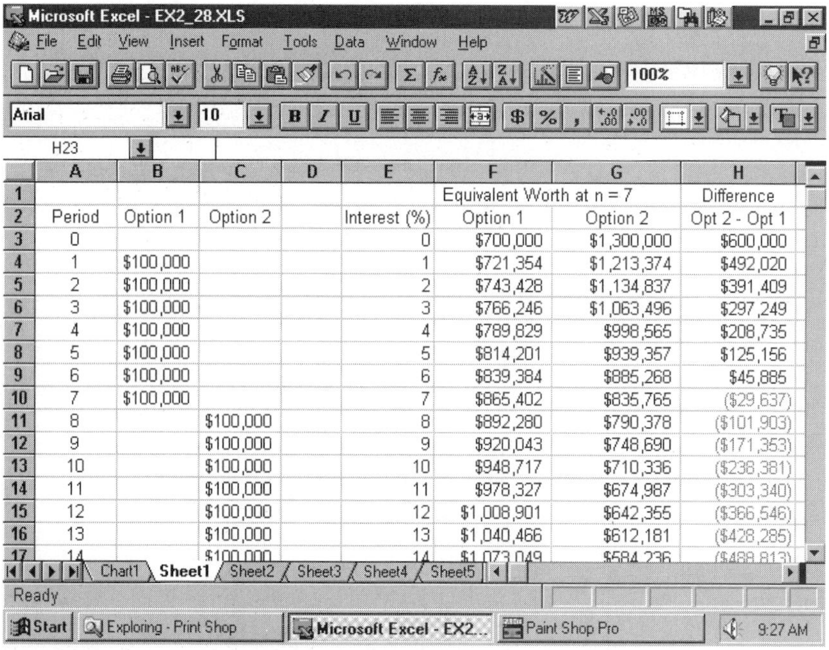

	A	B	C	D	E	F	G	H
1						Equivalent Worth at n = 7		Difference
2	Period	Option 1	Option 2		Interest (%)	Option 1	Option 2	Opt 2 - Opt 1
3	0				0	$700,000	$1,300,000	$600,000
4	1	$100,000			1	$721,354	$1,213,374	$492,020
5	2	$100,000			2	$743,428	$1,134,837	$391,409
6	3	$100,000			3	$766,246	$1,063,496	$297,249
7	4	$100,000			4	$789,829	$998,565	$208,735
8	5	$100,000			5	$814,201	$939,357	$125,156
9	6	$100,000			6	$839,384	$885,268	$45,885
10	7	$100,000			7	$865,402	$835,765	($29,637)
11	8		$100,000		8	$892,280	$790,378	($101,903)
12	9		$100,000		9	$920,043	$748,690	($171,353)
13	10		$100,000		10	$948,717	$710,336	($238,381)
14	11		$100,000		11	$978,327	$674,987	($303,340)
15	12		$100,000		12	$1,008,901	$642,355	($366,546)
16	13		$100,000		13	$1,040,466	$612,181	($428,285)
17	14		$100,000		14	$1,073,049	$584,236	($488,813)

Figure 2.44 Equivalence calculation using Excel (Example 2.28)

To obtain the equivalent worth at $n = 7$, we can use the following Excel financial functions:

- FV(*rate,nper,pmt,pv,type*) where *rate* is the interest rate per period, *nper* is the total number of payment periods in an annuity, and *pmt* is the payment made each period. *Pv* is the lump-sum amount included at the start of a cash flow series (i.e., at time zero). If *pv* is omitted, it is assumed to be 0. *Type* indicates when payments are due. If payments are due at the end of the period, you set *type* equal to 0 or omit it. If payments are due at the beginning of the period, you set *type* equal to 1. In our example, to compute the equivalent worth at $n = 7$ for Option 1 with interest rate of 6%, the Excel statement would be

$$= FV(0.06,7,100000,0).$$

- **PV(*rate,nper,pmt,fv,type*)** returns the present value of an investment. The present value is the total amount that a series of future payments is worth now. *Rate* is the interest rate per period. *Nper* is the total number of payment periods in an annuity and *pmt* is the payment made each period. *Fv* is the lump-sum amount included at the end of a cash flow series (i.e., at time *Nper*). If *fv* is omitted, it is assumed to be 0. *Type* is the number 0 or 1 and indicates when payments are due. If payments are due at the end of the period, you set *type* equal to 0 or omit it. If payments are due at the beginning of the period, you set *type* equal

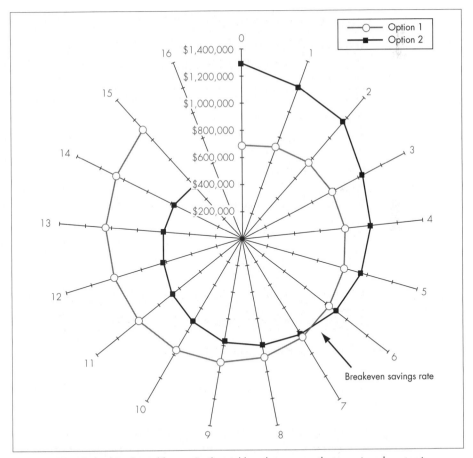

Figure 2.45 Radar chart obtained from an Excel spreadsheet that compares the two savings plans at various interest rates

to 1. In our example, to compute the equivalent worth at $n = 7$ for Option 2 with an interest rate of 6%, the Excel statement would be

$$= PV(0.06,13,100000,0).$$

Figure 2.45 shows a radar chart (built in spread sheet feature) that compares the two savings plans at various interest rates. Note that, as the interest rate increases, the smaller the economic difference between the two options up to the point at $i = 6.60\%$. It then increases again.

2.6 Summary

- Money has a time value because it can earn more money over time. A number of terms involving the time value of money were introduced in this chapter as follows:

  Interest is the cost of money. More specifically, it is a cost to the borrower and an earning to the lender, above and beyond the initial sum borrowed or loaned.

  Interest rate is a percentage periodically applied to a sum of money to determine the amount of interest to be added to that sum.

  Simple interest is the practice of charging an interest rate only to an initial sum.

  Compound interest is the practice of charging an interest rate to an initial sum *and* to any previously accumulated interest that has not been withdrawn from the initial sum. Compound interest is by far the most commonly used system in the real world.

  Economic equivalence exists between individual cash flows and/or patterns of cash flows that have the same value. Even though the amounts and timing of the cash flows may differ, the appropriate interest rate makes them equal.

- The compound interest formula is perhaps the single most important equation in this text:

$$F = P(1 + i)^N,$$

  where P is a present sum, i is the interest rate, N is the number of periods for which interest is compounded, and F is the resulting future sum. All other important interest formulas are derived from this one.

- **Cash flow diagrams** are visual representations of cash inflows and outflows along a timeline. They are particularly useful for helping us detect which of the five patterns of cash flows is represented by a particular problem.

- The five patterns of cash flows are as follows:
  1. Single payment: A single present or future cash flow.
  2. Uniform series: A series of flows of equal amounts at regular intervals.

3. Linear gradient series: A series of flows increasing or decreasing by a fixed amount at regular intervals.
4. Geometric gradient series: A series of flows increasing or decreasing by a fixed percentage at regular intervals.
5. Irregular series: A series of flows exhibiting no overall pattern. However, patterns might be detected for portions of the series.

- **Cash flow patterns** are significant because they allow us to develop **interest formulas**, which streamline the solution of equivalence problems. Table 2.3 summarizes the important interest formulas that form the foundation for all other analyses you will conduct in engineering economics.

Problems

Level 1

\$2.1* If you make the following series of deposits at an interest rate of 10%, compounded annually, what would be the total balance at the end of 10 years?

End of Period	Amount of Deposit
0	\$800
1 – 9	\$1500
10	0

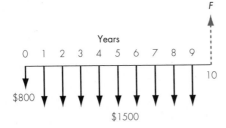

2.2* In computing the equivalent present worth of the following cash flow series at period 0, which of the following expressions is incorrect?

End of Period	Payment
0	
1	
2	
3	
4 – 7	\$100

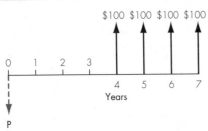

(a) $P = \$100(P/A, i, 4)(P/F, i, 4)$
(b) $P = \$100(F/A, i, 4)(P/F, i, 7)$
(c) $P = \$100(P/A, i, 7) - \$100(P/A, i, 3)$
(d) $P = \$100[(P/F, i, 4) + (P/F, i, 5) + (P/F, i, 6) + (P/F, i, 7)]$.

\$2.3* To withdraw the following \$1000 payment series, determine the minimum deposit (P) you should make now if your deposits earn an interest rate of 10%, compounded annually. Note that you are making another deposit at the end of year 7 in the amount of \$500. With the minimum deposit P, your balance at the end of year 10 should be zero.

End of Period	Deposit	Withdrawal
0	P	
1 – 6		\$1000
7	\$500	
8 –10		1000

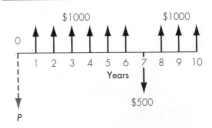

\$2.4* A couple is planning to finance their 5-year-old daughter's university education. They established a university fund that earns 10%, compounded annually. What annual deposit must be made from the daughter's 5th birthday (now) to her 16th birthday to meet the future university expenses shown in the following table. Assume that today is her 5th birthday.

Birthday	Deposit	Withdrawal
5 – 16	*A*	
17		
18		$25,000
19		27,000
20		29,000
21		31,000

$2.5* John and Susan just opened individual savings accounts at two different banks. They each deposited $1000. John's bank pays a simple interest at an annual rate of 10%, whereas Susan's bank pays compound interest at an annual rate of 9.5%. No interest would be taken out for a period of 3 years. At the end of 3 years, whose balance would be greater and by how much (in nearest dollars)?

$2.6* If you invest $2000 today in a savings account at an interest rate of 12%, compounded annually, how much principal and interest would you accumulate in 7 years?

$2.7* How much do you need to invest in equal annual amounts for the next 10 years if you want to withdraw $5000 at the end of the eleventh year and increase the annual withdrawal by $1000 each year thereafter until year 25? The interest rate is 6%, compounded annually.

2.8* What is the present worth of the following income stream at an interest rate of 10%? (All cash flows occur at year end.)

n	Net Cash Flow
1	$11,000
2	12,100
3	0
4	14,641

2.9* You want to find the equivalent present worth for the following cash flow series at an interest rate of 15%. Which of the following statements is *incorrect*?

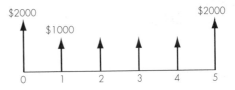

(a) $1000(P/A, 15\%, 4) + $2000 + $2000(P/F, 15\%, 5)

(b) $1000(P/F, 15\%, 5) + $1000(P/A, 15\%, 5) + $2000

(c) [$1000(F/A, 15\%, 5) + $1000] × (P/F, 15\%, 5) + $2000

(d) [$1000(F/A, 15\%, 4) + $2000] × (P/F, 15\%, 4) + $2000.

$2.10 You deposit $2000 in a savings account that earns 9% simple interest per year. To double your balance, you wait at least (?) years. But if you deposit the $2000 in another savings account that earns 8% interest, compounded yearly, it will take (?) years to double your balance.

2.11 Compare the interest earned by $1000 for 10 years at 7% simple interest with that earned by the same amount for 10 years at 7%, compounded annually.

$2.12* You are considering investing $1000 at interest rate of 6%, compounded annually, for 3 years or investing the $1000 at 7% per year simple interest rate for 3 years. Which option is better?

$2.13 Suppose you have the alternative of receiving either $10,000 at the end of 3 years or *P* dollars today. Currently you have no need for money, so you would deposit the *P* dollars in a bank that pays 6% interest. What value of *P* would make you indifferent in your choice between *P* dollars today and the promise of $10,000 at the end of 3 years?

$2.14 Suppose that you are obtaining a personal loan from your uncle in the amount of $10,000 (now) to be repaid in 2 years to cover some of your university expenses. If your uncle always earns 10% interest (annually) on his money invested in various sources, what minimum lump-sum payment 2 years from now would make your uncle happy?

$2.15 What will be the amount accumulated by each of these present investments?

(a) $8000 in 8 years at 9%, compounded annually

(b) $1250 in 11 years at 4%, compounded annually

(c) $3000 in 31 years at 7%, compounded annually

(d) $20,000 in 7 years at 8%, compounded annually.

2.16* What is the present worth of these future payments?

(a) $5300—6 years from now at 7%, compounded annually

(b) $7800—15 years from now at 8%, compounded annually

(c) $20,000—5 years from now at 9%, compounded annually

(d) $12,000—10 years from now at 10%, compounded annually.

$2.17 For an interest rate of 8%, compounded annually, find

(a) How much can be loaned now if $5000 will be repaid at the end of 5 years?

(b) How much will be required in 4 years to repay a $12,000 loan now?

$2.18 You bought 250 shares of Alberta Energy stock at $7800 on December 31, 1995. Your intention is to keep the stock until it doubles in value. If you expect 15% annual growth for Alberta Energy stock, how many years do you expect to hold onto the stock? Compare the solution obtained by the Rule of 72 (discussed in Example 2.8).

2.19 From the interest tables in the text, determine the following value of the factors by interpolation. Then compare these answers with those obtained by evaluating the *F/P* factor or the *P/F* factor.

(a) The single-payment compound amount factor for 38 periods at 6.5% interest.

(b) The single-payment present worth factor for 57 periods at 8% interest.

$2.20* If you desire to withdraw the following amounts over the next 5 years from the savings account which earns a 7% interest compounded annually, how much do you need to deposit now?

n	Amount
2	$1500
3	3000
4	3000
5	2000

$2.21 If $1000 is invested now, $1500 two years from now, and $2000 four years from now at an interest rate of 6% compounded annually, what will be the total amount in 10 years?

$2.22 A local newspaper headline blared: "Jim Smith signed for $10 Million." A reading of the article revealed that on April 1, 1997, Jim Smith, a junior hockey scoring sensation, signed a $10 million package with the Toronto Maple Leafs. The terms of the contract were $1 million immediately, $800,000 per year for the first 5 years (first payment after 1 year) and $1 million per year for the next 5 years (first payment at year 6). If Jim's interest rate is 8% per year, how much is his contract worth at the time of the signing?

$2.23 How much invested now at 6% would be just sufficient to provide three payments with the first payment in the amount of $2000 occurring 2 years hence, $3000 five years hence, and $4000 seven years hence?

2.24 What is the future worth of the following series of payments?

(a) $3000 at the end of each year for 5 years at 7%, compounded annually

(b) $2000 at the end of each year for 10 years at 8.25%, compounded annually

(c) $1500 at the end of each year for 30 years at 9%, compounded annually

(d) $4300 at the end of each year for 22 years at 10.75%, compounded annually.

$2.25 What equal annual series of payments must be paid into a sinking fund to accumulate the following amount?

(a) $10,000 in 13 years at 5%, compounded annually

(b) $25,000 in 10 years at 9%, compounded annually

(c) $15,000 in 25 years at 7%, compounded annually

(d) $8000 in 8 years at 12%, compounded annually.

2.26 Part of the income that a machine generates is put into a sinking fund to replace the machine when it wears out. If $1500 is deposited annually at 7% interest, how many years must the machine be kept before a new machine costing $25,000 can be purchased?

$2.27 A no-load (commission free) mutual fund has grown at a rate of 12%, compounded annually, since its beginning. If it is anticipated that it will continue to grow at this rate, how much must be invested every year so that $10,000 will be accumulated at the end of 5 years?

2.28 What equal annual payment series is required to repay the following present amounts?

(a) $11,000 in 5 years at 8% interest, compounded annually

(b) $5500 in 4 years at 13% interest, compounded annually

(c) $7000 in 3 years at 15% interest, compounded annually

(d) $60,000 in 25 years at 9% interest, compounded annually.

$2.29 You have borrowed $15,000 at an interest rate of 11%. Equal payments will be made over a 3-year period (first payment at the end of the first year). The annual payment will be (?) and the interest payment for the second year will be (?).

2.30* What is the present worth of the following series of payments?

(a) $1200 at the end of each year for 12 years at 6%, compounded annually

(b) $2000 at the end of each year for 10 years at 9%, compounded annually

(c) $500 at the end of each year for 5 years at 7.25%, compounded annually

(d) $4000 at the end of each year for 8 years at 8.75%, compounded annually.

2.31 From the interest tables in Appendix D determine the following value of the factors by interpolation. Then compare the results with those obtained from evaluating the *A/P* and *P/A* interest formulas.

(a) The capital recovery factor for 36 periods at 6.25% interest.

(b) The equal-payment series present worth factor for 125 periods at 9.25% interest.

$2.32 An individual deposits an annual bonus into a savings account that pays 7% interest, compounded annually. The size of the bonus increases by $1000 each year, and the initial bonus amount was $3000. Determine how much will be in the account immediately after the 5th deposit.

$2.33* Five annual deposits in the amounts of ($1200, $1000, $800, $600, and $400) are made into a fund that pays interest at a rate of 9%, compounded annually. Determine the amount in the fund immediately after the 5th deposit.

2.34 Compute the present worth for the cash flow diagram below. Assume $i = 10\%$.

2.35 What single amount at the end of the fifth year is equivalent to a uniform annual series of $3000 per year for 10 years, if the interest rate is 6%, compounded annually?

2.36 In computing either the equivalent present worth (P) or future worth (F) for the cash flow, at $i = 10\%$, identify all the correct equations from the list below to compute them.

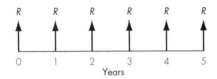

(1) $P = R(P/A, 10\%, 6)$

(2) $P = R + R(P/A, 10\%, 5)$

(3) $P = R(P/F, 10\%, 5) + R(P/A, 10\%, 5)$

(4) $F = R(F/A, 10\%, 5) + R(F/P, 10\%, 5)$

(5) $F = R + R(F/A, 10\%, 5)$

(6) $F = R(F/A, 10\%, 6)$

(7) $F = R(F/A, 10\%, 6) - R.$

$2.37 At what rate of interest, compounded annually, will an investment double itself in 10 years?

$2.38* You have $10,000 available for investment in stock. You are looking for a growth stock whose value can grow to $30,000 over 5 years. What kind of growth rate are you looking for?

2.39 Kersey Manufacturing Co., a small plastic fabricator, needs to purchase an extrusion molding machine for $120,000. Kersey will borrow money from a bank at an interest rate of 9% over 5 years. Kersey expects its product sales to be slow during the first year, but to increase subsequently at an annual rate of 10%. Kersey therefore arranges with the bank to pay off the loan on a "balloon scale," which results in the lowest payment at the end of first year, each subsequent payment to be 10% more than the preceding one. Determine these five annual payments.

Level 2

2.40* Consider the following cash flow series.

End of Period	Deposit	Withdrawal
0	$1000	
1	800	
2	600	
3	400	
4	200	
5		
6		C
7		$2C$
8		$3C$
9		$4C$
10		$5C$

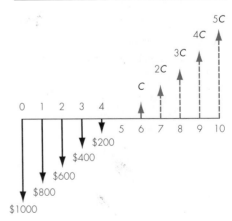

What value of C makes the deposit series equivalent to the withdrawal series at an interest rate of 12%, compounded annually?

$2.41 You are about to borrow $3000 from a bank at an interest rate of 12%, compounded annually. You are required to make three equal annual repayments in the amount of $1249.05 per year, with the first repayment occurring at the end of year 1. In each year, show the interest payment and principal payment.

2.42 How many years will it take an investment to triple itself if the interest rate is 11%, compounded annually?

$2.43* What is the future worth of a series of equal year-end deposits of $1200 for 10 years in a savings account that earns 9%, annual interest, if

(a) All deposits were made at the *end* of each year?

(b) All deposits were made at the *beginning* of each year?

2.44 What is the equal payment series for 10 years that is equivalent to a payment series of $12,000 at the end of the first year, decreasing by $1000 each year over 10 years. Interest is 8%, compounded annually.

$2.45 What is the amount of 10 equal annual deposits that can provide five annual withdrawals, when a first withdrawal of $1000 is made at the end of year 11, and subsequent withdrawals increase at the rate of 6% per year over the previous year's, if

(a) The interest rate is 8%, compounded annually?

(b) The interest rate is 6%, compounded annually?

2.46 By using only those factors given in the interest tables, find the values of the following factors, which are not given in your tables. Show the relationship between the factors by using the factor notation and calculate the value of the factor. Then compare the solution obtained by using the factor formulas to calculate the factor values directly.

Example: $(F/P, 8\%, 10) = (F/P, 8\%, 4)(F/P, 8\%, 6) = 2.159$

(a) $(P/F, 8\%, 67)$
(b) $(A/P, 8\%, 42)$
(c) $(P/A, 8\%, 135)$.

2.47 Prove the following relationships among interest factors:

(a) $(F/P, i, N) = i(F/A, i, N) + 1$
(b) $(P/F, i, N) = 1 - (P/A, i, N)i$
(c) $(A/F, i, N) = (A/P, i, N) - i$
(d) $(A/P, i, N) = i/[1 - (P/F, i, N)]$.

2.48 Find the present worth of the cash receipts where $i = 10\%$, compounded annually, with only four interest factors.

$2.49 Find the equivalent present worth of the cash receipts where $i = 10\%$. In other words, how much do you have to deposit now (with the second deposit in the amount of $100 at the end of the first year) so that you will be able to withdraw $100 at the end of second year, $60 at the end of third year, and so forth, where the bank pays you a 10% annual interest on your balance?

2.50 What value of A makes the two annual cash flows equivalent at 10% interest, compounded annually?

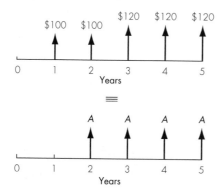

2.51 The two cash flow transactions shown in the cash flow diagram are said to be equivalent at 10% interest, compounded annually. Find the unknown X value which satisfies the equivalence.

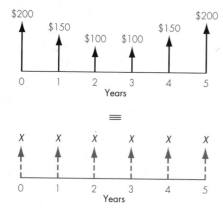

2.52 From the cash flow diagram, find the value of C that will establish the economic equivalence between the deposit series and the withdrawal series at an interest rate of 8%, compounded annually.

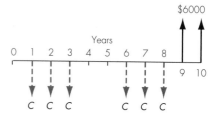

2.53 The following equation describes the conversion of a cash flow into an equivalent equal payment series with $N = 10$. Given the equation, reconstruct the original cash flow diagram:

$$A = [800 + 20(A/G, 6\%, 7)] \times$$
$$(P/A, 6\%, 7)(A/P, 6\%, 10)$$
$$+ [300(F/A, 6\%, 3) - 500](A/F, 6\%, 10).$$

2.54 Consider the following cash flow. What value of C makes the inflow series equivalent to the outflow series at an interest rate of 12%?

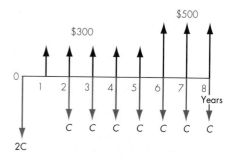

2.55 Find the value of X so that the two cash flows in the figure are equivalent for an interest rate of 10%.

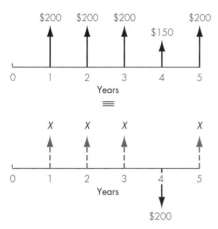

$2.56 On the day his baby was born, a father decided to establish a savings account for the child's university education. Any money that is put into the account will earn an interest rate of 8%, compounded annually. He will make a series of annual deposits in equal amounts on each of his child's birthdays from the first through the child's 18th, so that his child can make four annual withdrawals from the account in the amount of $20,000 on each of his birthdays. Assuming that the first withdrawal will be made on the child's 18th birthday, which of the following statements are correct to calculate the required annual deposit?

(1) $A = (\$20{,}000 \times 4)/18$

(2) $A = \$20{,}000(F/A, 8\%, 4) \times (P/F, 8\%, 21)(A/P, 8\%, 18)$

(3) $A = \$20{,}000(P/A, 8\%, 18) \times (F/P, 8\%, 21)(A/F, 8\%, 4)$

(4) $A = [\$20{,}000(P/A, 8\%, 3) + \$20{,}000](A/F, 8\%, 18)$

(5) $A = \$20{,}000[(P/F, 8\%, 18) + (P/F, 8\%, 19) + (P/F, 8\%, 20) + (P/F, 8\%, 21)](A/P, 8\%, 18)$.

2.57* Find the equivalent equal-payment series (A) using an A/G factor, such that the two cash flows are equivalent at 10%, compounded annually.

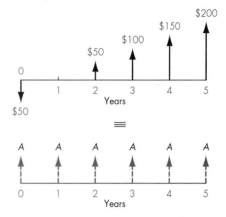

2.58 Consider the following cash flow:

Year End	Payment
0	$500
1 – 5	1000

In computing F at the end of year 5 at an interest rate of 12%, which of the following statements is incorrect?

(a) $F = \$1000(F/A, 12\%, 5) - \$500(F/P, 12\%, 5)$

(b) $F = \$500(F/A, 12\%, 6) + \$500(F/A, 12\%, 5)$

(c) $F = [\$500 + \$1000(P/A, 12\%, 5)] \times (F/P, 12\%, 5)$

(d) $F = [\$500(A/P, 12\%, 5) + \$1000] \times (F/A, 12\%, 5)$.

2.59 Consider the cash flow series given. In computing the equivalent worth at $n = 4$, which of the following statements is incorrect?

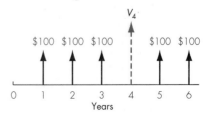

(a) $V_4 = [\$100(P/A, i, 6) - \$100(P/F, i, 4)](F/P, i, 4)$

(b) $V_4 = \$100(F/A, i, 3) + \$100(P/A, i, 2)$

(c) $V_4 = \$100(F/A, i, 4) - \$100 + \$100(P/A, i, 2)$

(d) $V_4 = [\$100(F/A, i, 6) - \$100(F/P, i, 2)](P/F, i, 2)$.

$2.60 Henry Cisco is planning to make two deposits, $25,000 now and $30,000 at the end of year 6. He wants to withdraw C each year for the first 6 years and $(C + \$1000)$ each year for the subsequent 6 years. Determine the value of C if the deposits earn 10% interest, compounded annually.

2.61 Determine the interest rate (i) that makes the pairs of cash flows economically equivalent.

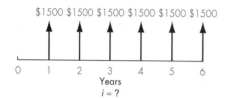

$2.62 Read the following letter from a magazine publisher:

Dear Parent:

Currently your Growing Child/Growing Parent subscription will expire with your 24-month issue. To renew on an annual basis until your child reaches 72 months would cost you a total of $63.84 ($15.96 per year). We feel it is so important for you to continue receiving this material until the 72nd month that we offer you an opportunity to renew now for $57.12. Not only is this a savings of 10% over the regular rate, it is an excellent inflation hedge for you against increasing rates in the future. Please act now by sending $57.12.

(a) If your money is worth 6% per year, determine whether this offer can be of any value.

(b) At what rate of interest would you be indifferent between the two renewal options?

Level 3

2.63 Suppose that an oil well is expected to produce 10,000 barrels of oil during its first production year. However, its subsequent production (yield) is expected to decrease by 10% over the previous year's production. The oil well has a proven reserve of 100,000 barrels.

(a) Suppose that the price of oil is expected to be $18 per barrel for next several years. What would be the present worth of the anticipated revenue stream at an interest rate of 15%, compounded annually, over the next 7 years?

(b) Suppose that the price of oil is expected to start at $18 per barrel during the first year, but to increase at the rate of 5% over the previous year's price. What would be the present worth of the anticipated revenue stream at an interest rate of 15%, compounded annually, over next 7 years?

(c) Reconsider (b) above. After 3 years' production, you decide to sell the oil well. What would be the fair price for the oil well if its production life is unlimited?

2.64 An engineer has estimated the annual toll revenues from a proposed toll highway over 20 years as follows:

$$A_n = (\$2,000,000)(n)(1.06)^{n-1},$$
$$n = 1, 2, \ldots, 20.$$

During an assessment of this project, the engineer was asked to present the estimated total present value of toll revenue at an interest rate of 6%. Assuming annual compounding, find the present value of the estimated toll revenue.

2.65 Solve for the present worth of this cash flow using at most three interest factors at 10% interest, compounded annually.

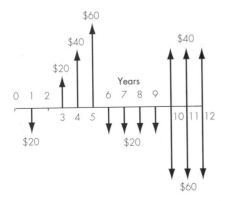

2.66 A major lottery corporation sold a total of 36.1 million lottery tickets at $1 each during the first week of January 1996. As prize money, a total of $41 million will be distributed ($1,952,381 at the *beginning* of each year) over the next 21 years. The distribution of the first-year prize money occurs now, and the remaining lottery proceeds will be put into the province's reserve funds, which earns an interest at the rate of 6%, compounded annually. After making the last prize distribution (at the beginning of Year 21), how much would be left over in the reserve account?

$2.67 *The Sporting News* carried the following story on June 19, 1989:

> Dallas Cowboys quarterback Troy Aikman, the number one pick in the National Football League draft, will earn either $11,406,000 over 12 years or $8,600,000 over 6 years. Aikman, represented by Leigh Steinberg, must declare which plan he prefers. The $11 million package is deferred through the year 2000, while the nondeferred arrangement ends after the 1994 season. Regardless of which plan is chosen, Aikman will be playing through the 1994 season.[4]

Deferred Plan		Nondeferred Plan	
1989	$2,000,000	1989	$2,000,000
1990	566,000	1990	900,000
1991	920,000	1991	1,000,000
1992	930,000	1992	1,225,000
1993	740,000	1993	1,500,000
1994	740,000	1994	1,975,000
1995	740,000		
1996	790,000		
1997	540,000		
1998	1,040,000		
1999	1,140,000		
2000	1,260,000		
Total	$11,406,000	Total	$8,600,000

(a) As it happens, Troy was under the non-deferred plan prior to his much publicised $50 million contract in 1993. In retrospect, if Aikman's interest rate were 6%, did he make a wise decision in 1989?

(b) At what interest rate would the two plans be economically equivalent?

2.68 Fairmont Textile has a plant in which employees have been having trouble with carpal tunnel syndrome (CTS) (inflammation of the nerves that pass through the carpal tunnel, a tight space at the base of the palm), resulting from long-term repetitive activities, such as years of sewing operation. It seems as if 15 of the employees working in this facility developed signs of CTS over the last 5 years. Avon Mutual, the company's insurance firm, has been increasing Fairmont's liability insurance steadily due to this CTS problem. Avon Mutual is willing to lower the insurance premiums to $16,000 a year (from the current $30,000 a year) for the next 5 years if Fairmont implements an acceptable CTS-prevention program that includes making the employees aware of CTS and how to reduce chances of it developing. What would be the max-

[4] *The Sporting News.* (Reprinted with permission.)

imum amount that Fairmont should invest in the CTS prevention program to make this program worthwhile? The firm's interest rate is 12%, compounded annually.

2.69 The Research and Development (R&D) department of Boswell Electronics Company (BEC) has developed a voice-recognition system that could lead to greater acceptance of personal computers among the Japanese. Currently, the complicated and voluminous set of Japanese characters required for a keyboard make it unwieldy and bulky in comparison to the trim, easily portable keyboards, to which Westerners are accustomed. Boswell has used voice-recognition systems in more traditional settings, such as medical reporting in the English language. However, adapting this technology to the Japanese language could result in a breakthrough new level of acceptance of personal computers in Japan. The investment required to develop a full-scale commercial version would cost BEC $10 million, which will be financed at an interest rate of 12%. The system will sell for about $4000 (or net cash profit of $2000) and run on high-powered PCs. The product will have a 5-year market life. Assuming that the annual demand for the product remains constant over the market life, how many units does BEC have to sell each year to pay off the initial investment and interest?

CHAPTER 3

Extending Equivalence to Real-World Transactions

Ford Credit Red Carpet Lease Advance Payment Program: Make One Payment, Save a Bundle. This could be the least expensive way to lease a new vehicle. It's the Ford Credit's innovative Red Carpet Lease Advance Payment Program.[1] Now you can combine the savings of paying cash with all the conveniences of a Red Carpet Lease. Find out how to drive the model you want most, for less.

With the Ford Credit Advance Payment Program you save $1163 when compared to a conventional 24-month lease. (Each lease payment is due at the beginning of each month.)

Now put yourself in the position of leasing an automobile from Ford. Would you go ahead and lease the car under the terms of the advance payment program? Are you really saving $1163 as claimed by Ford? Under what circumstances would you prefer to go with the advance payment program?

---

[1]Source: Red Carpet Lease—Vehicle Leasing Plan, Ford Motor Company, 1994. (Courtesy of the Ford Motor Company.)

	1994 Mustang GT Convertible	
	Conventional 24-month Red Carpet Lease	Advance Payment Program 24-month Red Carpet Lease
MSRP*	$24,204	$24,204
First month's payment	513	0
Refundable security deposit	525	475
Cash due at signing	1,038	11,624
Total cash outlay	$12,312†	$11,149‡

*Manufacturer's suggested retail price; †$513 3 24 = $12,312; ‡ $11,624 − $475 = $11,149.

In this chapter, several concepts crucial to managing money are considered. In Chapter 2, we examined how time affects the value of money, and we developed various interest formulas for this purpose. Using these basic formulas, we will now extend the concept of equivalence to determine interest rates implicit in many financial contracts. To this end, several examples in the area of loan transactions are introduced. For example, many commercial loans require that interest compound more frequently than once a year—for instance, monthly or quarterly. To consider the effect of more frequent compounding, we must begin with an understanding of the concepts of nominal and effective interest.

3.1 Nominal and Effective Interest Rates

In all the examples in Chapter 2, the implicit assumption was that payments are received once a year, or annually. However, some of the most familiar financial transactions in both personal financial matters and, engineering economic analysis involve payments not based on one annual payment; for example, monthly mortgage payments and quarterly earnings on savings accounts. Thus, if we are to compare different cash flows with different compounding periods, these cash flows need to be addressed on a common basis. The need to do this has led to the development of the concepts of **nominal interest rate** and **effective interest rate**.

3.1.1   Nominal Interest Rates

Even if a financial institution uses a unit of time other than a year—a month or a quarter (e.g., when calculating interest payments), the institution usually quotes the interest rate on an annual basis. Commonly this rate is stated as

$$r\% \text{ compounded } M\text{–ly,}$$

where

r	=	The nominal interest rate per year,
M	=	The compounding frequency or the number of compounding periods per year,
r/M	=	The interest rate per compounding period.

Many banks, for example, state the interest arrangement for credit cards in this way:

"18% compounded monthly."

We say 18% is the **nominal interest rate** or **annual percentage rate** (APR), and the compounding frequency is monthly (M = 12). To obtain the interest rate per compounding period, 18% is divided by 12, i.e., 1.5% per month (Figure 3.1). Therefore, the credit card statement above means that each month the bank will charge 1.5% interest on an unpaid balance.

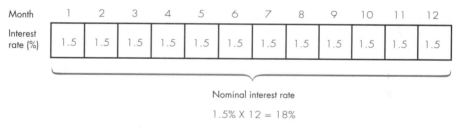

Figure 3.1   The nominal interest rate is determined by summing the individual interest rates per period

Although the annual percentage rate, or APR, is commonly used by financial institutions, and is familiar to customers, when compounding takes place more frequently than annually, the APR does not explain accurately the amount of interest that will accumulate in a year. To explain the true effect of more frequent compounding on annual interest amounts, the term effective interest rate needs to be introduced.

3.1.2   Effective Interest Rates

The **effective interest rate** is the one rate that truly represents the interest earned in a year or some other time period. For instance, in our credit card example, the bank will charge 1.5% interest on unpaid balance at the end of each month. Therefore, the 1.5% rate represents the effective interest rate per month. On a monthly basis, it is that rate which determines the actual interest charges on your outstanding credit card balance.

Suppose you obtain a loan from a bank at an interest rate of 12%, compounded monthly. Here, 12% represents the nominal interest rate, and the interest rate per month is 1% ($r/M = 12\%/12$). The total annual interest payment (assuming no withdrawals) for a principal amount of $1 may be computed using the formula given in Eq. (2.3). If $P = \$1$, $i = 12\%/12$, and $N = 12$, we obtain

$$
\begin{aligned}
F &= P(1 + i)^N \\
&= \$1(1 + 0.01)^{12} \\
&= \$1.1268.
\end{aligned}
$$

The implication is that, for each dollar borrowed for 1 year, you owe $1.1268 at the end of that year, including principal and interest. The annual interest payment can easily be obtained by subtracting the principal amount from the F value as follows:

$$
\begin{aligned}
I &= F - P \\
&= \$1.1268 - \$1 \\
&= 12.68 \text{ cents.}
\end{aligned}
$$

For each dollar borrowed, an equivalent annual interest of 12.68 cents is paid. In terms of an effective annual interest rate (i_a), the interest payment can be rewritten as a percentage of the principal amount:

$$
i_a = (1 + 0.01)^{12} - 1 = 0.1268, \text{ or } 12.68\%.
$$

Thus, the effective annual interest rate is 12.68%. The nominal rate may be related to the annual effective rate as shown in Figure 3.2.

For the same loan made at an interest rate of 12%, compounded quarterly, the interest rate per compounding period would be 3% (12%/4) for each 3-month period in the year. Since there are four quarters in a year, we find

$$
i_a = (1 + 0.03)^4 - 1 = 0.1255, \text{ or } 12.55\%.
$$

Table 3.1 shows effective interest rates at various compounding intervals for 12% APR and several other frequently encountered nominal rates. As you can see, depending on the frequency of compounding, the effective interest earned or paid can differ significantly from the APR. Therefore, truth-in-lending laws require that financial institutions quote both nominal interest rate and compounding frequency, i.e., effective interest, when you deposit or borrow money.

Clearly, compounding more frequently increases the amount of interest paid over a year at the same nominal interest rate. We can generalize the result for any arbitrary compounding scheme by providing a formula for computing the effective rate. Assuming that the nominal interest rate is r, and M compounding periods occur during the year, then the effective annual interest rate (i_a) can be calculated as follows:

$$
i_a = (1 + r/M)^M - 1. \tag{3.1}
$$

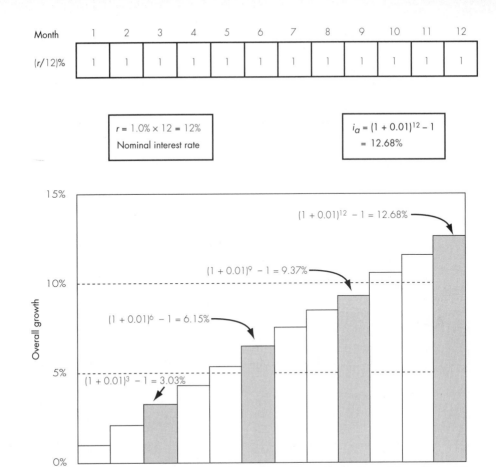

Figure 3.2 Relationship between nominal interest rate and annual effective interest rate based on monthly compounding

Table 3.1
Nominal and
Effective Interest
Rates with
Different
Compounding
Periods

	Annual Effective Rates				
Nominal Rate	**Compounding Annually**	**Compounding Semi-annually**	**Compounding Quarterly**	**Compounding Monthly**	**Compounding Daily**
4 %	4.00%	4.04%	4.06%	4.07%	4.08%
5	5.00	5.06	5.09	5.12	5.13
6	6.00	6.09	6.14	6.17	6.18
7	7.00	7.12	7.19	7.23	7.25
8	8.00	8.16	8.24	8.30	8.33
9	9.00	9.20	9.31	9.38	9.42
10	10.00	10.25	10.38	10.47	10.52
11	11.00	11.30	11.46	11.57	11.62
12	12.00	12.36	12.55	12.68	12.74

When $M = 1$, we have the special case of annual compounding. Substituting $M = 1$ into Eq. (3.1) reduces it to $i_a = r$. That is, when compounding takes place once annually, effective interest is equal to nominal interest. Thus, in the examples in Chapter 2, where only annual interest was considered, we were, by definition, using effective interest rates.

EXAMPLE 3.1

$

Determining a compounding period

Consider the following bank advertisement that appeared in a local newspaper: "Open a Term Deposit (TD) account at CIBC and get a guaranteed rate of return on as little as $1000. It's a smart way to manage your money for months."

TD Type	6-month	1-year	2-year	5-year
Rate (nominal)	8.00%	8.50%	9.20%	9.60%
Yield (effective)	8.16%	8.87%	9.64%	10.07%
Minimum deposit	$1000	$1000	$10,000	$20,000

In this advertisement, no mention is made of specific interest compounding frequencies. Find the compounding frequency for each TD.

Solution

Given: r and i_a
Find: M

First we will consider the 6-month TD. The nominal interest rate is 8% per year, and the effective annual interest rate (yield) is 8.16%. Using Eq. (3.1), we obtain the expression

$$0.0816 = (1 + 0.08/M)^M - 1$$

or

$$1.0816 = (1 + 0.08/M)^M.$$

By trial and error we find $M = 2$, which indicates semi-annual compounding. Thus, the 6-month TD earns 8% interest, compounded semi-annually. If the TD is not cashed at maturity, normally it will be renewed automatically at the prevailing interest rate. For example, if you leave the TD in the bank for another 6 months, and the prevailing interest rate is the same as the original interest rate, your TD will earn 4% interest on $1040:

$$\text{TD value after 1-year deposit} = \$1040(1 + 0.04)$$
$$= \$1081.60.$$

Note: This is equivalent to earning 8.16% interest on $1000 for 1 year.

Similarly, we can find the interest periods for the other TDs: For the 1-year TD, we note that the difference between nominal and effective interest is greater than in the case of the 6-month TD, where the compounding period was semi-annual. A greater difference between nominal and effective rates suggests a greater number of compounding periods, so let us guess at the other end of the spectrum, i.e., daily compounding. This guess proves to be correct, as $i_a = (1 + 0.085/365)^{365} - 1 = 8.87\%$. The same reasonable guess will prove true for the 2-year and 5-year TD—both are compounded daily.

EXAMPLE 3.2 $ Future value of a term deposit

Suppose you purchase the 5-year TD described in Example 3.1. How much would the TD be worth when it matures at the end of 5 years?

Solution

Given: $P = \$20,000$, $r = 9.6\%$ per year, $M = 365$ periods per year
Find: F

If you purchase the 5-year TD in Example 3.1, it will earn 9.60% interest, compounded daily. This means that your TD earns an effective annual interest of 10.07%:

$$
\begin{aligned}
F &= P(1 + i_a)^N \\
&= \$20,000(1 + 0.1007)^5 \\
&= \$20,000(F/P, 10.07\%, 5) \\
&= \$32,313.
\end{aligned}
$$

3.1.3 Effective Interest Rates per Payment Period

We can generalize the result of Eq. (3.1) to compute the effective interest rate for *any time duration*. As you will see later, the effective interest rate is usually computed based on the payment (transaction) period. For example, if cash flow transactions occur quarterly, but interest is compounded monthly, we may wish to calculate the effective interest rate on a quarterly basis. To consider this, we may redefine Eq. (3.1) as

$$
\begin{aligned}
i &= (1 + r/M)^C - 1 \\
&= \left(1 + \frac{r}{CK}\right)^C - 1,
\end{aligned}
\tag{3.2}
$$

where

M = The number of compounding periods per year,
C = The number of compounding periods per payment period,
K = The number of payment periods per year.

Note that $M = CK$ in Eq. (3.2).

EXAMPLE
3.3
$

Effective rate per payment period

Suppose that you make quarterly deposits in a savings account which earns 9% interest, compounded monthly. Compute the effective interest rate per quarter.

Solution

Given: r = 9%, C = three compounding periods per quarter, K = four quarterly payments per year, and M = 12 compounding periods per year
Find: i

Using Eq. (3.2), we compute the effective interest rate per quarter as

$$i = (1 + 0.09/12)^3 - 1$$
$$= 2.27\%.$$

Comments: For the special case of annual payments with annual compounding, we obtain $i = i_a$ with $C = M$ and $K = 1$. Figure 3.3 illustrates the relationship between the nominal and effective interest rates per payment period.

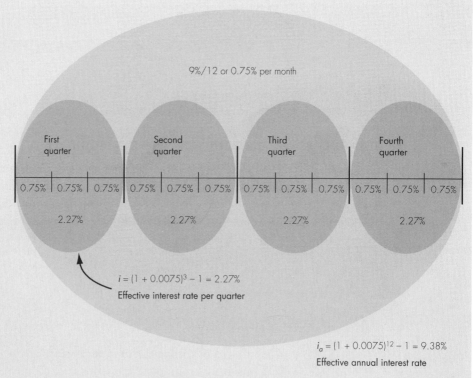

Figure 3.3 Functional relationships among r, i, and i_a where interest is calculated based on 9%, compounded monthly, and payments occur quarterly (Example 3.3)

3.1.4 Continuous Compounding

To be competitive on the financial market, or to entice potential depositors, some financial institutions offer frequent compounding. As the number of compounding periods (M) becomes very large, the interest rate per compounding period (r/M) becomes very small. As M approaches infinity and r/M approaches zero, we approximate the situation of **continuous compounding**.

By taking limits on both sides of Eq. (3.2), we obtain the effective interest rate per payment period as

$$
\begin{aligned}
i &= \lim_{CK \to \infty} [(1 + \tfrac{r}{CK})^C - 1] \\
&= \lim_{CK \to \infty} (1 + \tfrac{r}{CK})^C - 1 \\
&= \lim_{CK \to \infty} [(1 + \tfrac{r}{CK})^{CK}]^{1/K} - 1 \\
&= (e^r)^{1/K} - 1.
\end{aligned}
$$

Therefore,

$$
i = e^{r/K} - 1. \tag{3.3}
$$

To calculate the effective annual interest rate for continuous compounding, we set K equal to 1, resulting in

$$
i_a = e^r - 1. \tag{3.4}
$$

As an example, the effective annual interest rate for a nominal interest rate of 12% compounded continuously is $i_a = e^{0.12} - 1 = 12.7497\%$.

EXAMPLE 3.4
$

Calculating an effective interest rate

Assuming a $1000 initial deposit, find the effective interest rate per quarter, at a nominal rate of 8%, compounded (a) weekly, (b) daily, and (c) continuously. Also find the final balance at the end of 3 years under each compounding scheme as shown in Figure 3.4.

Solution

Given: $r = 8\%$, M, C, $K = 4$ payment periods per year, $P = \$1000$, $N = 12$ quarters
Find: i, F

Please note that there are no quarterly payments at all in this problem. Since we need to find the effective interest rate per quarter, we need to treat each quarter as a payment period in order to use the formulas we have developed. As a result, we have $K = 4$.

(a) Weekly compounding:

$r = 8\%$, $M = 52$, $C = 13$ compounding periods per quarter, $K = 4$ payment periods per year;

$$
i = (1 + 0.08/52)^{13} - 1 = 2.0186\%.
$$

With P = \$1000, i = 2.0187%, and N = 12 quarters in Eq.(2.3),

$$F = \$1000(F/P, 2.0187\%, 12) = \$1271.03.$$

(b) Daily compounding:
r = 8%, M = 365, C = 91.25 days per quarter, K = 4;

$$i = (1 + 0.08/365)^{91.25} - 1 = 2.0199,$$

$$F = \$1000(F/P, 2.0199\%, 12) = \$1271.21.$$

(c) Continuous compounding:
r = 8%, $M \rightarrow \infty$, $C \rightarrow \infty$, K = 4; using Eq. (3.3)

$$i = e^{0.08/4} - 1 = 2.0201\%,$$

$$F = \$1000(F/P, 2.0201\%, 12) = \$1271.23.$$

Comments: Note that the difference between daily compounding and continuous compounding is often negligible. Many banks offer continuous compounding to entice deposit customers, but the extra benefits are small.

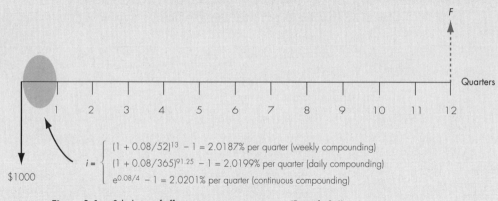

$$i = \begin{cases} (1 + 0.08/52)^{13} - 1 = 2.0187\% \text{ per quarter (weekly compounding)} \\ (1 + 0.08/365)^{91.25} - 1 = 2.0199\% \text{ per quarter (daily compounding)} \\ e^{0.08/4} - 1 = 2.0201\% \text{ per quarter (continuous compounding)} \end{cases}$$

Figure 3.4   Calculation of effective interest rate per quarter (Example 3.4)

3.2 Equivalence Calculations When Payment Periods and Compounding Periods Coincide

All the examples in Chapter 2 assumed annual payments and annual compounding. Whenever a situation occurs where the compounding and payment periods are equal ($M = K$), whether compounded annually or at some other interval, this solution method can be used:

1. Identify the number of compounding periods (M) per year.

2. Compute the effective interest rate per payment period, i.e., using Eq. (3.2), and with $C = 1$ and $K = M$, we have

$$i = r/M.$$

3. Determine the total number of compounding periods:

$$N = M \times \text{(number of years)}.$$

EXAMPLE 3.5 $

Calculating auto loan payments

Suppose you want to buy a car. You have surveyed the dealers' newspaper advertisements, and the following one has caught your attention:

8.5% Annual Percentage Rate! 48-month financing on all Mustangs in stock. 60 to choose from.
ALL READY FOR DELIVERY! Prices Starting as Low as $21,599.
You just add GST and 1% for dealer's freight. We will pay the tag, title, and license.

Add 5% PST = $1079.95
Add 7% GST = $1511.93
Add 1% dealer's freight = $215.99
Total purchase price = $24,406.87.

You can afford to make a down payment of $4406.87, so the net amount to be financed is $20,000. What would the monthly payment be? (Figure 3.5).

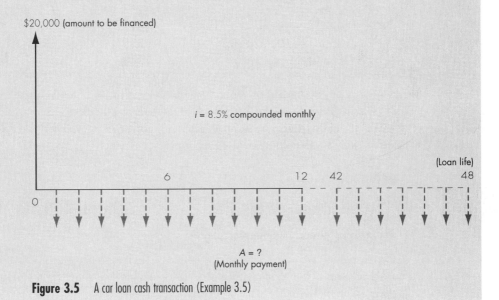

Figure 3.5 A car loan cash transaction (Example 3.5)

Solution

The advertisement does not specify a compounding period, but in automobile financing, the interest and the payment periods are almost always monthly. Thus, the 8.5% APR means 8.5%, compounded monthly.

Given: $P = \$20{,}000$, $r = 8.5\%$ per year, $K = 12$ payments per year, $N = 48$ months, $M = 12$ compounding periods per year
Find: A

In this situation, we can easily compute the monthly payment using Eq. (2.12):

$$
\begin{aligned}
i &= 8.5\%/12 = 0.7083\% \text{ per month,} \\
N &= (12)(4) = 48 \text{ months,} \\
A &= \$20{,}000(A/P, 0.7083\%, 48) = \$492.97.
\end{aligned}
$$

EXAMPLE 3.6 $ Ford's Red-Carpet auto-lease plan

Consider once more the Ford Motor Company's 24-month auto-lease plans that were introduced at the beginning of this chapter.

	1994 Mustang GT Convertible	
	Conventional	**Advance Payment**
Retail price	$24,204	$24,204
First month's payment	513	0
Refundable security deposit	525	475
Cash due at signing	1038	11,624

The lease payments should be made at the *beginning* of each month, and the refundable security deposits will be paid back at the end of 24 months. Suppose that you are able to invest your money in an account that pays 6% interest, compounded monthly; which lease plan would be the more economical?

Solution

Given: Lease payment series shown in Figure 3.6, $r = 6\%$, payment period = monthly, compounding period = monthly
Find: The most economical lease option

For each option, we will calculate the net equivalent lease cost at $n = 0$. Since the lease payments occur monthly, we need to determine the effective interest rate per month, which is 0.5%.

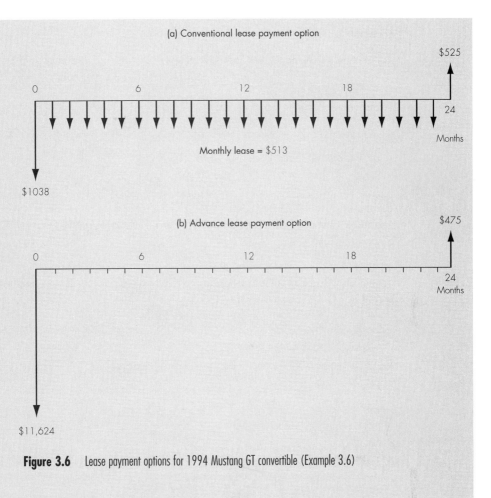

Figure 3.6 Lease payment options for 1994 Mustang GT convertible (Example 3.6)

- Conventional 24-month Red Carpet lease:

  Equivalent present cost of the total lease payments:

  $$
  \begin{aligned}
  P_1 &= \$1038 + \$513(P/A, 0.5\%, 23) \\
  &= \$1038 + \$11,119.62 \\
  &= \$12,157.62.
  \end{aligned}
  $$

  Equivalent present worth of the security deposit refund:

  $$
  P_2 = \$525(P/F, 0.5\%, 24) = \$465.77.
  $$

  Equivalent net lease cost:

  $$
  \begin{aligned}
  P &= \$12,157.62 - \$465.77 \\
  &= \$11,691.84.
  \end{aligned}
  $$

- Advance payment program 24-month Red-Carpet lease

Equivalent present cost of the total lease payments:

$$P_1 = \$11,624.$$

Equivalent present worth of the security deposit refund:

$$P_2 = \$475(P/F, 0.5\%, 24) = \$421.41.$$

Equivalent net lease cost:

$$P = \$11,624 - \$421.41$$
$$= \$11,202.59.$$

It appears that the advance payment program is more economical at 6% interest compounded monthly.

Comments: By varying the interest rate, the interest range that makes the conventional Red-Carpet lease option more attractive can be determined. At an interest rate of 10.80%, compounded monthly (or 0.90% per month), both lease options would be equivalent. At nominal rates greater than 10.80%, the conventional lease plan would be the more economical choice. In other words, if you can invest your money somewhere else, at a rate higher than 10.80%, compounded monthly, you would be better off selecting the conventional lease option.

3.3 Equivalence Calculations When Payment Periods and Compounding Periods Differ

A number of situations involve cash flows that occur at intervals not the same as the compounding intervals. Whenever payment and compounding periods differ from each other, *one or the other must be transformed so that both conform to the same unit of time.* For example, if payments occur quarterly and compounding occurs monthly, the most logical procedure is to calculate the effective interest rate per quarter. On the other hand, if payments occur monthly and compounding occurs quarterly, we may be able to find the equivalent monthly interest rate. The bottom line is that to proceed with equivalency analysis, the compounding and payment periods must be in the same order.

3.3.1 Compounding More Frequent Than Payments

The computational procedure is as follows:

1. Identify the number of compounding periods per year (M), the number of payment periods per year (K), and the number of compounding periods per payment period (C);

2. Compute the effective interest rate per payment period:

   - For discrete compounding, compute

   $$i = [(1 + r/M]^C - 1,$$

   - For continuous compounding, compute

   $$i = e^{r/K} - 1.$$

3. Find the total number of payment periods:

   $$N = K \times (\text{number of years}).$$

4. Use i and N in the appropriate formulas in Table 2.3.

EXAMPLE 3.7

$

Compounding more frequent than payments (discrete compounding case)

Suppose you make equal quarterly deposits of $1000 into a fund that pays interest at a rate of 12% compounded monthly. Find the balance at the end of year two (Figure 3.7).

Solution

Given: $A = \$1000$ per quarter, $r = 12\%$ per year, $M = 12$ compounding periods per year, $N = 8$ quarters
Find: F

We follow the procedure as described above.

1. Identify the parameter values for M, K, and C, where

$$M = 12 \text{ compounding periods per year,}$$
$$K = \text{four payment periods per year,}$$
$$C = \text{three compounding periods per payment period.}$$

2. Use Eq. (3.2) to compute effective interest for a payment period:

$$i = (1 + 0.12/12)^3 - 1$$
$$= 3.030\% \text{ per quarter.}$$

3. Find the total number of payment periods, N, where

$$N = K(\text{number of years}) = 4(2) = 8 \text{ quarters.}$$

4. Use i and N in the appropriate equivalence formulas:

$$F = \$1000(F/A, 3.030\%, 8) = \$8901.81.$$

Comment: No 3.030% interest table appears in Appendix D. You may use the interest formula to calculate $(F/A, 3.030\%, 8)$. Another way to calculate F is $F = \$1000(A/F, 1\%, 3)(F/A, 1\%, 24)$, where the first interest factor finds its equivalent monthly payment and the second interest factor converts the monthly payment series to an equivalent lump-sum future payment.

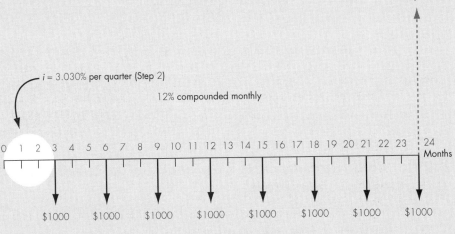

Figure 3.7 Quarterly deposits with monthly compounding (Example 3.7)

EXAMPLE 3.8
$

Compounding more frequent than payments (continuous compounding case)

A series of equal quarterly receipts of $500 extends over a period of 5 years. What is the present worth of this quarterly payment series at 8% interest compounded continuously? (Figure 3.8).

Discussion: The equivalent problem statement is "How much do you need to deposit now in a savings account that earns 8% interest, compounded continuously, so that you can withdraw $500 at the end of each quarter for 5 years?" Since the payments are quarterly, we need to compute i per quarter for the equivalence calculations.

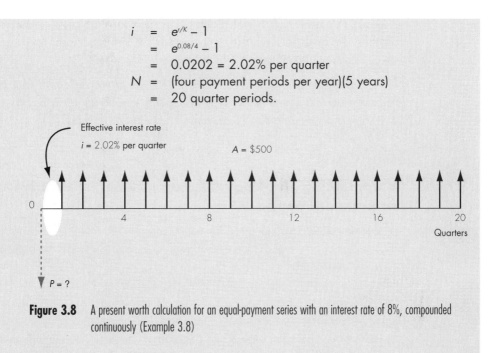

$$
\begin{aligned}
i &= e^{r/K} - 1 \\
&= e^{0.08/4} - 1 \\
&= 0.0202 = 2.02\% \text{ per quarter} \\
N &= (\text{four payment periods per year})(5 \text{ years}) \\
&= 20 \text{ quarter periods.}
\end{aligned}
$$

Effective interest rate
$i = 2.02\%$ per quarter

$A = \$500$

0

4

8

12

16

20

Quarters

$P = ?$

Figure 3.8 A present worth calculation for an equal-payment series with an interest rate of 8%, compounded continuously (Example 3.8)

Solution

Given: $i = 2.02\%$ per quarter, $N = 20$ quarters, and $A = \$500$ per quarter
Find: P

Using the $(P/A, i, N)$ factor with $i = 2.02\%$ and $N = 20$, we find that

$$
\begin{aligned}
P &= A(P/A, 2.02\%, 20) \\
&= \$500(16.3199) \\
&= \$8159.96.
\end{aligned}
$$

3.3.2 Compounding Less Frequent Than Payments

The two following examples contain identical parameters for savings situations in which compounding occurs less frequently than payments. However, two different underlying assumptions govern how interest is calculated. In Example 3.9, the assumption is that, whenever a deposit is made, it starts to earn interest. In Example 3.10, the assumption is that the deposits made within a quarter do not earn interest until the end of that quarter. As a result, in Example 3.9, we transform the compounding period to conform to the payment period. In Example 3.10, we lump several payments together to match the compounding period. In the real world, which assumption is applicable depends on the transactions and the financial institutions involved. The accounting methods used by many firms record cash transactions that occur within a compounding period as if they had occurred at the end of that period. For example, when cash flows occur daily, but the compounding period is monthly, the cash flows within each month are summed (ignoring interest) and treated as a single payment on which interest is calculated.

Note: *In this textbook, we assume that whenever the timing of a cash flow is specified, one cannot move it to another time point without considering the time value of money, i.e., the practice demonstrated in Example 3.9 should be followed.*

EXAMPLE 3.9

$

Compounding less frequent than payments: Effective rate per payment period

Suppose you make $500 monthly deposits to a registered retirement savings plan that pays interest at a rate of 10%, compounded quarterly. Compute the balance at the end of 10 years.

Solution

Given: r = 10% per year, M = four quarterly compounding periods per year, K = 12 payment periods per year, A = $500 per month, and N = 120 months, and interest is accrued on flow during the compounding period
Find: i, F

As in the case of Example 3.7, we follow the same procedure:

1. The parameter values for M, K, and C are

$$M = \text{four compounding periods per year,}$$
$$K = 12 \text{ payment periods per year,}$$
$$C = 1/3 \text{ compounding periods per payment period.}$$

2. As shown in Figure 3.9, the effective interest rate per payment period is calculated using Eq. (3.2):

$$i = (1 + 0.10/4)^{1/3} - 1$$
$$= 0.826\% \text{ per month.}$$

3. Find N

$$N = (12)(10) = 120 \text{ payment periods.}$$

4. Use i and N in the appropriate equivalence formulas (Figure 3.10):

$$F = \$500(F/A, 0.826\%, 120)$$
$$= \$101,907.89.$$

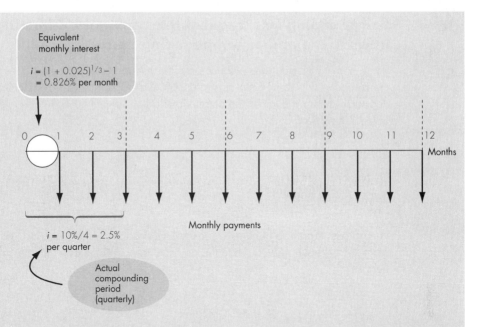

Figure 3.9 Calculation of equivalent monthly interest when the quarterly interest rate is specified (Example 3.9)

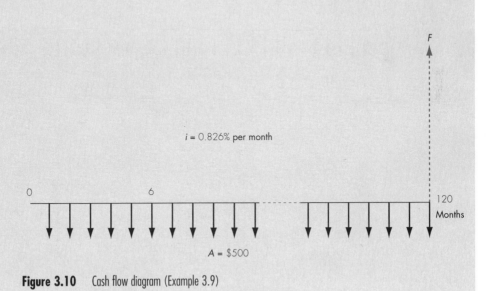

Figure 3.10 Cash flow diagram (Example 3.9)

EXAMPLE
3.10

$

Compounding less frequent than payment: Summing cash flows to the end of compounding period

Some financial institutions will not pay interest on funds deposited after the start of the compounding period. To illustrate, reconsider Example 3.9. Assume that money deposited during a quarter (the compounding period) will not earn any interest (Figure 3.11). Compute F at the end of 10 years.

Solution

Given: Same as those for Example 3.9; however, no interest on flow during the compounding period
Find: F

In this case, the three monthly deposits during each quarterly period will be placed at the end of each quarter. Then the payment period coincides with the compounding period:

$$
\begin{aligned}
i &= 10\%/4 = 2.5\% \text{ per quarter,} \\
A &= 3(\$500) = \$1500 \text{ per quarter,} \\
N &= 4(10) = 40 \text{ payment periods,} \\
F &= \$1500(F/A, 2.5\%, 40) = \$101{,}103.83.
\end{aligned}
$$

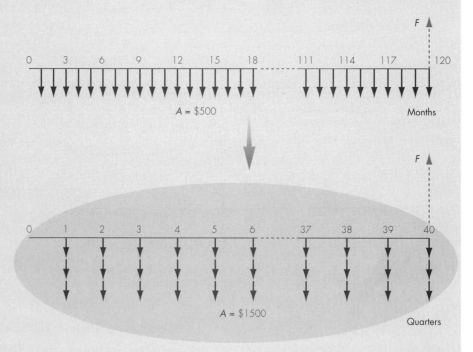

Figure 3.11 Transformed cash flow diagram created by summing monthly cash flows to the end of the quarterly compounding period (Example 3.10)

Comments: In Example 3.10, the balance will be $804.06 less than with the compounding situation in Example 3.9, a fact that is consistent with our understanding that increasing the frequency of compounding increases future value. Some financial institutions follow the practice illustrated in Example 3.9. As an investor, you should reasonably ask yourself whether it makes sense to make deposits in an interest-bearing account more frequently than interest is paid. In the interim between interest compounding, you may be tying up your funds prematurely and foregoing other opportunities to earn interest.

Figure 3.12 illustrates a decision chart that allows you to sum up how you can proceed to find the effective interest rate per payment period given the various possible compounding/interest arrangements.

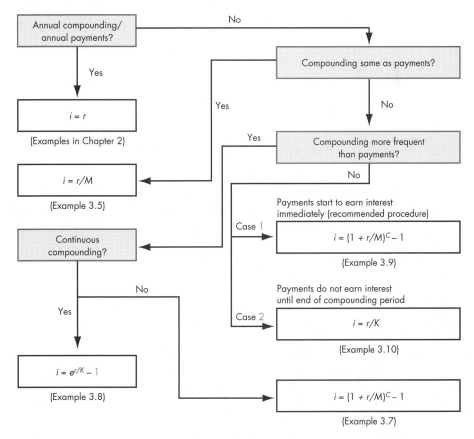

Figure 3.12   A decision flow chart demonstrating how to compute effective interest rate per payment period (i)

3.4 Equivalence Calculations with Continuous Payments

As we have seen so far, interest can be compounded annually, semi-annually, monthly, or even continuously. Discrete compounding is appropriate for many financial transactions: Mortgages, bonds, and installment loans, which require payments or receipts at discrete times, are good examples. In most businesses, however, transactions occur continuously throughout the year. In these circumstances, we may describe the financial transactions as having a continuous flow of money, for which continuous compounding and discounting are more realistic. This section illustrates how one establishes the economic equivalence between cash flows under continuous compounding.

Continuous cash flows represent situations where money flows continuously and at a known rate throughout a given time period. In business, many daily cash flow transactions can be viewed as continuous cash flows. An advantage of the continuous flow approach is that it more closely models the realities of business transactions. Costs for labor, carrying inventory, and operating and maintaining equipment are typical examples. Others include capital improvement projects that conserve energy, water, or process steam, whose savings can occur continuously.

3.4.1 Single Cash Flow Transactions

First we will illustrate how single cash flow formulas for continuous compounding and discounting are derived. Suppose that you invested P dollars at a nominal rate of $r\%$ interest for N years. If interest is compounded continuously, the effective annual interest is $i = e^r - 1$. The future value of the investment at the end of N years is obtained with the F/P factor by substituting $e^r - 1$ for i:

$$
\begin{aligned}
F &= P(1 + i)^N \\
&= P(1 + e^r - 1)^N \\
&= Pe^{rN}.
\end{aligned}
$$

This implies that $1 invested now at an interest rate of $r\%$, compounded continuously, accumulates to e^{rN} dollars at the end of N years.

Correspondingly, the present value of F due N years from now and discounted continuously at an interest rate of $r\%$ is equal to

$$P = Fe^{-rN}.$$

We can say that the present value of $1 due N years from now, discounted continuously at an annual interest rate of $r\%$, is equal to e^{-rN} dollars.

3.4.2   Continuous Funds-Flow

Suppose that an investment's future cash flow per unit of time (e.g., per year) can be expressed by a continuous function ($f(t)$), which can take any shape. Assume also that the investment promises to generate cash of $f(t)\Delta t$ dollars between t and $t + \Delta t$, where t is a point in the time interval $0 \leq t \leq N$ (Figure 3.13). If the nominal interest rate is constant r during this time interval, the present value of this cash stream is given approximately by the expression

$$\Sigma (f(t)\Delta t)e^{-rt},$$

where e^{-rt} is the discounting factor that converts future dollars into present dollars. With the project's life extending from 0 to N, we take the summation over all subperiods (compounding periods) in the interval from 0 to N. As the division of the interval becomes smaller and smaller, that is, as Δt approaches zero, we obtain the present value expression by the integral

$$P = \int_0^N f(t)e^{-rt}dt. \tag{3.5}$$

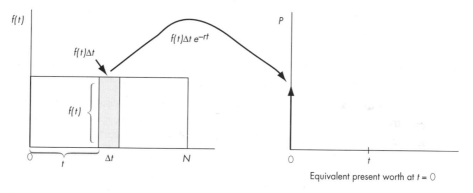

Figure 3.13   Finding an equivalent present worth of a continuous flow payment function $f(t)$ at a nominal rate of $r\%$

Similarly, the future value expression of the cash flow stream is given by the expression

$$F = Pe^{rN} = \int_0^N f(t)e^{r(N-t)}dt, \tag{3.6}$$

where $e^{r(N-t)}$ is the compounding factor that converts present dollars into future dollars. It is important to observe that the time unit is *year*, because the effective interest rate is expressed in terms of a year. Therefore, all time units in equivalence calculations must be converted into years. Table 3.2[3] summarizes some typical continuous cash functions that can facilitate equivalence calculations.

Table 3.2
Summary of Interest Factors for Typical Continuous Cash Flows with Continuous Compounding

Type of Cash Flow	Cash Flow Function	Parameters Find	Parameters Given	Algebraic Notation	Factor Notation
Uniform (step)	$f(t) = \bar{A}$	P	$\bar{A}$	$\bar{A}\left[\dfrac{e^{rN} - 1}{re^{rN}}\right]$	$(P/\bar{A}, r, N)$
		$\bar{A}$	P	$P\left[\dfrac{re^{rN}}{e^{rN} - 1}\right]$	$(\bar{A}/P, r, N)$
		F	$\bar{A}$	$\bar{A}\left[\dfrac{e^{rN} - 1}{r}\right]$	$(F/\bar{A}, r, N)$
		$\bar{A}$	F	$F\left[\dfrac{r}{e^{rN} - 1}\right]$	$(\bar{A}/F, r, N)$
Gradient (ramp)	$f(t) = Gt$	P	G	$\dfrac{G}{r^2}(1 - e^{-rN}) - \dfrac{G}{r}(Ne^{-rN})$	
Decay	$f(t) = ce^{-jt}$ $j' = $ decay rate with time	P	c, j	$\dfrac{c}{r + j}(1 - e^{-(r+j)N})$	

[3] Park CS, Sharp-Bette GP. *Advanced Engineering Economics*. New York: John Wiley & Sons, 1990. (Reprinted by permission of John Wiley & Sons Inc.)

EXAMPLE 3.11

Comparison of daily flows and daily compounding with continuous flows and continuous compounding

Consider a situation where money flows daily. Suppose you own a retail shop and generate $200 cash each day. You establish a special business account and deposit these daily cash flows in an account for 15 months. The account earns an interest rate of 6%. Compare the accumulated cash values at the end of 15 months, assuming

(a) Daily compounding, and

(b) Continuous compounding, respectively.

Solution

(a) With daily compounding:

Given: A = $200 per day, r = 6% per year, M = 365 compounding periods per year, N = 455 days
Find: F

Assuming there are 455 days in the 15-month period, we find

$$\begin{aligned} i &= 6\%/365 \\ &= 0.01644\% \text{ per day,} \\ N &= 455 \text{ days.} \end{aligned}$$

The balance at the end of 15 months will be

$$\begin{aligned} F &= \$200(F/A, 0.01644\%, 455) \\ &= \$200(472.4095) \\ &= \$94,482. \end{aligned}$$

(b) With continuous compounding:

Now we approximate this discrete cash flow series by a uniform continuous cash flow function as shown in Figure 3.14. Here we have the situation where an amount flows at the rate of $\bar{A}$ per year for N years.

Note: *Our time unit is a year.* Thus, a 15-month period is 1.25 years. Then, the cash flow function is expressed as

$$\begin{aligned} f(t) &= \bar{A}, 0 \le t \le 1.25 \\ &= \$200(365) \\ &= \$73,000 \text{ per year.} \end{aligned}$$

Given: $\bar{A}$ = $73,000 per year, r = 6% per year, compounded continuously, N = 1.25 years
Find: F

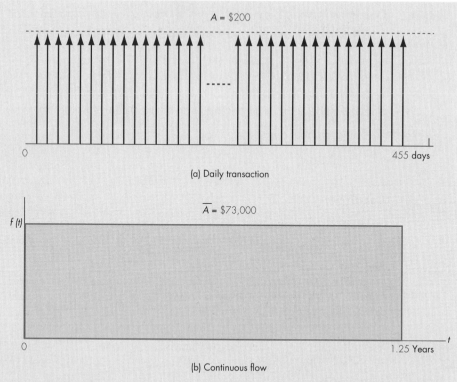

Figure 3.14 Comparison between daily transaction and continuous fund flow transaction (Example 3.11)

Substituting these values back into Eq. (3.6) yields

$$F = \int_{0}^{1.25} 73{,}000 \, e^{0.06(1.25 - t)} dt$$

$$= \$73{,}000 \left[\frac{e^{0.075} - 1}{0.06} \right]$$

$$= \$94{,}759.$$

The factor in the bracket is known as the **funds flow compound amount factor** and is designated $(F/\overline{A}, r, N)$ as shown in Table 3.2. Also notice that the difference between the two methods is only $277 (less than 0.3%).

Comments: As shown in this example, the differences between discrete (daily) and continuous compounding have no practical significance in most cases. Consequently, as a mathematical convenience, instead of assuming that money flows in discrete increments at the end of each day, we could assume that money flows continuously during the time period at a uniform rate. This type of cash flow assumption is a common practice in the chemical industry.

3.5 Changing Interest Rates

Up to this point, we have assumed a constant interest rate in our equivalence calculations. When an equivalence calculation extends over several years, more than one interest rate may be applicable to properly account for the time value of money. This is to say, over time, interest rates available in the financial marketplace fluctuate, and a financial institution, committed to a long-term loan, may find itself in the position of losing the opportunity to earn higher interest because some of its holdings are tied up in a lower interest loan. A financial institution may attempt to protect itself from such lost earning opportunities by building gradually increasing interest rates into a long-term loan at the outset. Variable home mortgage loans are perhaps the most common examples of variable interest rates. In this section, we will consider variable interest rates in both single-payment and a series of cash flows.

3.5.1 Single Sums of Money

To illustrate the mathematical operations involved in computing equivalence under changing interest rates, first consider the investment of a single sum of money, P, in a savings account for N interest periods. If i_n denotes the interest rate applicable during time period n, the future equivalent for a single sum of money can be expressed as

$$F = P(1 + i_1)(1 + i_2)...(1 + i_{N-1})(1 + i_N), \tag{3.7}$$

and solving for P yields the inverse relation

$$P = F[(1 + i_1)(1 + i_2)...(1 + i_{N-1})(1 + i_N)]^{-1}. \tag{3.8}$$

EXAMPLE 3.12 $

Changing interest rates with lump-sum amount

You deposit $2000 in a registered retirement savings plan (RRSP) that pays interest at 12%, compounded quarterly, for the first 2 years and 9%, compounded quarterly, for the next 3 years. Determine the balance at the end of 5 years (Figure 3.15).

Solution

Given: P = $2000, r = 12% per year for first 2 years, 9% per year for last 3 years, M = 4 compounding periods per year, N = 20 quarters
Find: F

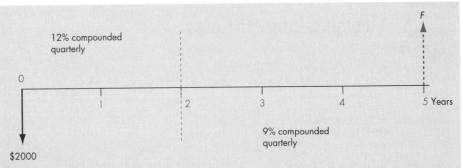

Figure 3.15 Changing interest rates (Example 3.12)

We will compute the value of F in two steps. First we will compute the balance at the end of 2 years, B_2. With 12%, compounded quarterly

$$
\begin{aligned}
i &= 12\%/4 = 3\%, \\
N &= 4(2) = 8 \text{ (quarters)}, \\
B_2 &= \$2000(F/P, 3\%, 8) \\
&= \$2000(1.2668) \\
&= \$2533.60.
\end{aligned}
$$

Since the fund is not withdrawn, but reinvested at 9%, compounded quarterly, as a second step, we compute the final balance as follows:

$$
\begin{aligned}
i &= 9\%/4 = 2.25\%, \\
N &= 4(3) = 12 \text{ (quarters)}, \\
F &= B_2(F/P, 2.25\%, 12) \\
&= \$2533.60(1.3060) \\
&= \$3309.
\end{aligned}
$$

3.5.2   Series of Cash Flows

This consideration of changing interest rates can be easily extended to a series of cash flows. In this case, the present worth of a series of cash flows can be represented as

$$
\begin{aligned}
P = {} & A_1(1 + i_1)^{-1} + A_2\,[(1 + i_1)^{-1}(1 + i_2)^{-1}] + \ldots \\
& + A_N\,[(1 + i_1)^{-1}(1 + i_2)^{-1}\ldots(1 + i_N)^{-1}].
\end{aligned} \tag{3.9}
$$

The future worth of a series of cash flows is given by the inverse of Eq. (3.9):

$$
\begin{aligned}
F = {} & A_1\,[(1 + i_2)(1 + i_3)\ldots(1 + i_N)] \\
& + A_2\,[(1 + i_3)(1 + i_4)\ldots(1 + i_N)] + \ldots + A_N.
\end{aligned} \tag{3.10}
$$

The uniform series equivalent is obtained in two steps. First, the present worth equivalent of the series is found using Eq. (3.9). Then, A is solved for after establishing the following equivalence equation:

$$P = A(1 + i_1)^{-1} + A\left[(1 + i_1)^{-1}(1 + i_2)^{-1}\right] + \ldots$$
$$+ A\left[(1 + i_1)^{-1}(1 + i_2)^{-1}\ldots(1 + i_N)^{-1}\right]. \qquad (3.11)$$

EXAMPLE 3.13

Changing interest rates with uneven cash flow series

Consider the cash flow in Figure 3.16 with the interest rates indicated and determine the uniform series equivalent for the cash flow series.

Discussion: In this problem and many others, the easiest approach involves collapsing the original flow into a single equivalent amount, for example, at time zero and then converting the single amount into the final desired form.

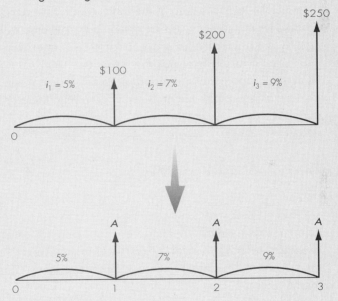

Figure 3.16 Equivalence calculation with changing interest rates (Example 3.13)

Solution

Given: Cash flows and interest rates as shown in Fig. 3.16, $N = 3$
Find: A

Using Eq. (3.9), we find the present worth:

$$P = \$100(P/F, 5\%, 1) + \$200(P/F, 5\%, 1)(P/F, 7\%, 1)$$
$$+ \$250(P/F, 5\%, 1)(P/F, 7\%, 1)(P/F, 9\%, 1)$$
$$= \$477.41.$$

Then, we obtain the uniform-series equivalent as follows:

$$\$477.41 = A(P/F, 5\%, 1) + A(P/F, 5\%, 1)(P/F, 7\%, 1)$$
$$+ A(P/F, 5\%, 1)(P/F, 7\%, 1)(P/F, 9\%, 1)$$
$$= 2.6591A$$
$$A = \$179.54.$$

3.6 Commercial Loan Transactions

Commercial loans are among the most significant financial transactions involving interest. Many types of loans are available, but here we will focus on those most frequently used by individuals and in business—**amortized loans**. Add-on loans are also covered to illustrate possible variations from conventional amortized loans. Moreover, since it is often important to know how much interest is represented in loan payments, several methods for determining interest and principal amounts are examined.

3.6.1 Amortized Loans

One of the most important applications of compound interest involves loans that are paid off in **installments** over time. If the loan is to be repaid in equal periodic amounts (weekly, monthly, quarterly, or annually), it is said to be an amortized loan. Examples include automobile loans, loans for appliances, home mortgage loans, and most business debts other than very short-term loans. Most commercial loans have interest that is compounded monthly.

So far, we have considered many instances of amortized loans in which we calculated present or future values of the loans or the amounts of the installment payments. An additional aspect of amortized loans, which will be of great interest to us, is calculating the amount of interest versus the portion of the principal that is paid off in each installment. As we shall explore more fully in Chapter 9, the interest paid on a loan is an important element in calculating taxable income and has repercussions for both personal and business loan transactions. For now, we will focus on several methods of calculating interest and principal paid at any point in the life of a loan.

In a typical amortized loan, the amount of interest owed for a specified period is calculated based on the remaining balance of the loan at the beginning of the period. It is possible to develop a set of formulas to compute the remaining loan balance, interest payment, and principal payment for a specified period. Suppose we borrow an amount, P, at an interest rate, i, and agree to repay this principal sum, P, including interest, in equal payments, A over N periods. The payment size (A) is $A = P(A/P, i, N)$. Each payment A is divided into an amount that is interest and a remaining amount for the principal payment. Let

$$B_n = \text{Remaining balance at the end of period } n, \text{ with } B_0 = P,$$
$$I_n = \text{Interest payment in period } n, \text{ where } I_n = B_{n-1}i,$$
$$PP_n = \text{Principal payment in period } n.$$

Then, each payment can be defined as

$$A = PP_n + I_n. \tag{3.12}$$

The interest and principal payments for an amortized loan can be determined in several ways; three methods are presented here. No clear-cut reason is available to prefer one method over another. Method 1, however, may be easier to adopt when the computational process is automated through a spreadsheet application, whereas methods 2 and 3 may be more suitable for obtaining a quick solution when a time period is specified. You should become comfortable with at least one of these methods for use in future problem solving. Pick the one that comes most naturally to you.

Method 1—Tabular Method

The first method is tabular. The interest charge for a given period is computed progressively based on the remaining balance at the beginning of that period. The interest due at the end of the first period will be

$$I_1 = B_0 i = Pi.$$

Thus, the principal payment at that time will be
$$PP_1 = A - Pi,$$
and the balance remaining after the first payment will be

$$B_1 = B_0 - PP_1 = P - PP_1.$$

At the end of the second period, we will have

$$I_2 = B_1 i = (P - PP_1)i,$$
$$P_2 = A - (P - PP_1)i = (A - Pi) + PP_1 i = PP_1(1 + i),$$
$$B_2 = B_1 - PP_2 = P - (PP_1 + PP_2).$$

If we continue, we can show that, at the nth payment,

$$
\begin{aligned}
B_n &= P - (PP_1 + PP_2 + \ldots + PP_n) \\
&= P - [PP_1 + PP_1(1 + i) + \ldots + PP_1(1 + i)^{n-1}] \\
&= P - PP_1(F/A, i, n) \\
&= P - (A - Pi)(F/A, i, n).
\end{aligned}
$$

Then, the interest payment during the nth payment period is expressed as

$$
I_n = (B_{n-1})i. \tag{3.13}
$$

The portion of payment A at period n that can be used to reduce the remaining balance is

$$
PP_n = A - I_n. \tag{3.14}
$$

EXAMPLE 3.14
$

Loan balance, principal, and interest—tabular method

Suppose you secure a home improvement loan in the amount of $5000 from a local bank. The loan officer computes your monthly payment as follows:

Contract amount	=	$5000,
Contract period	=	24 months,
Annual percentage rate	=	12%,
Monthly installments	=	$235.37.

Figure 3.17 is the cash flow diagram. Construct the loan payment schedule by showing the remaining balance, interest payment, and principal payment at the end of each period over the life of the loan.

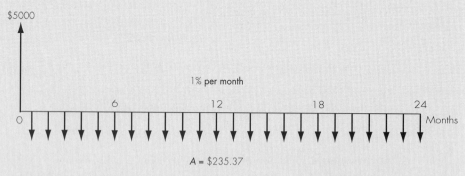

Figure 3.17 Cash flow diagram of home improvement loan with an APR of 12% (Example 3.14)

Solution

Given: $P = \$5000$, $A = \$235.37$ per month, $r = 12\%$ per year, $M = 12$ compounding periods per year, $N = 24$ months
Find: B_n, I_n, PP_n for $n = 1$ to 24

We can easily see how the bank calculated the monthly payment of $235.37. Since the effective interest rate per payment period on this loan transaction is 1% per month, we establish the following equivalence relationship:

$$\$235.37(P/A, 1\%, 24) = \$235.37(21.2431) = \$5000.$$

The loan payment schedule can be constructed as in Table 3.3. The interest due at $n = 1$ is $50.00, 1% of the $5000 outstanding during the first month. The $185.37 left over is applied to the principal, reducing the amount outstanding in the second month to $4814.63. The interest due in the second month is 1% of $4814.63, or $48.15, leaving $187.22 for repayment of the principal. At $n = 24$, the last $235.37 payment is just sufficient to pay the interest on the unpaid loan principal and to repay the remaining principal. Figure 3.18 illustrates the ratios between the interest and principal payments over the life of the loan.

Comments:　Certainly, generation of a loan repayment schedule such as that in Table 3.3 can be a tedious and time consuming process unless a computer is used. As you will see in Section 3.8, an electronic spreadsheet can solve this type of problem more effectively.

Table 3.3
Creating a Loan Repayment Schedule Using Excel (Example 3.14)

Payment No.	Payment Size	Principal Payment	Interest Payment	Loan Balance
1	$235.37	$185.37	$50.00	$4814.63
2	235.37	187.22	48.15	4627.41
3	235.37	189.09	46.27	4438.32
4	235.37	190.98	44.38	4247.33
5	235.37	192.89	42.47	4054.44
6	235.37	194.82	40.54	3859.62
7	235.37	196.77	38.60	3662.85
8	235.37	198.74	36.63	3464.11
9	235.37	200.73	34.64	3263.38
10	235.37	202.73	32.63	3060.65
11	235.37	204.76	30.61	2855.89
12	235.37	206.81	28.56	2649.08
13	235.37	208.88	26.49	2440.20
14	235.37	210.97	24.40	2229.24
15	235.37	213.08	22.29	2016.16
16	235.37	215.21	20.16	1800.96
17	235.37	217.36	18.01	1583.60
18	235.37	219.53	15.84	1364.07
19	235.37	221.73	13.64	1142.34
20	235.37	223.94	11.42	918.40
21	235.37	226.18	9.18	692.21
22	235.37	228.45	6.92	463.77
23	235.37	230.73	4.64	233.04
24	235.37	233.04	2.33	0.00

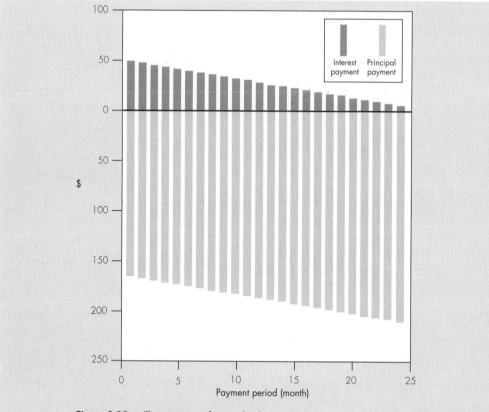

Figure 3.18 The proportions of principal and interest payments over the life of the loan (monthly payment = $235.37) (Example 3.14)

Method 2—Remaining Balance Method

Alternatively, we can derive B_n by computing the equivalent payments remaining after the nth payment. Thus, the balance with $N - n$ payments remaining is

$$B_n = A(P/A, i, N - n), \tag{3.15}$$

and the interest payment during period n is

$$I_n = (B_{n-1})i = A(P/A, i, N - n + 1)i, \tag{3.16}$$

where $A(P/A, i, N - n + 1)$ is the balance remaining at the end of period $n - 1$, and

$$
\begin{aligned}
PP_n = A - I_n &= A - A(P/A, i, N - n + 1)i \\
&= A[1 - (P/A, i, N - n + 1)i].
\end{aligned}
$$

Knowing the interest factor relationship from Table 2.3, $(P/F, i, n) = 1 - (P/A, i, n)i$, we obtain

$$PP_n = A(P/F, i, N - n + 1). \tag{3.17}$$

As we can see in Figure 3.19, this latter method provides more concise expressions for computing the loan balance.

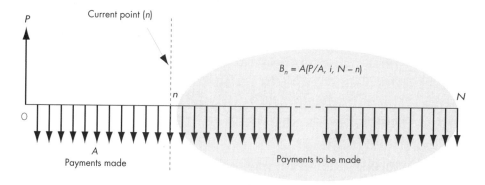

Figure 3.19 Calculating the remaining loan balance based on method 2

Method 3—Computing Equivalent Worth at nth Payment

The third method is to compute the remaining balance by the following equivalence relationship:

$$B_n = P(F/P, i, n) - A(F/A, i, n). \tag{3.18}$$

The first term $P(F/P, i, n)$ indicates the equivalent amount of P at the end of period n, whereas the second term indicates the equivalent lump sum amount at the end of period n of the n payments that have been made. The difference between the sums is the remaining balance of the loan. Calculation of the interest payment is the same as in Eq. (3.16).

EXAMPLE 3.15 $

Loan balances, principal, and interest—remaining balance method (method 2) and the method of equivalent worth of n payments (method 3)

Consider the home improvement loan in Example 3.14 and

(a) For the sixth payment, compute both the interest and principal payments.

(b) Immediately after making the sixth monthly payment, you would like to pay off the remainder of the loan in a lump sum. What is the required amount of this lump sum?

Solution

(a) Interest and principal payments for the sixth payment: Whether you use method 2 or method 3, the equations for calculating interest and principal payments are the same.

Given: (as for Example 3.14)
Find: I_6 and PP_6

Using Eqs. (3.16) and (3.17), we compute I_6 and P_6 as follows:

$$I_6 = \$235.37(P/A, 1\%, 19)(0.01)$$
$$= (\$4054.44)(0.01)$$
$$= \$40.54.$$
$$PP_6 = \$235.37(P/F, 1\%, 19) = \$194.82,$$

or simply subtract the interest payment from the monthly payment as follows:

$$PP_6 = \$235.37 - \$40.54 = \$194.83.$$

(b) Remaining balance after sixth payment: Figure 3.20 shows the cash flow diagram that applies to this part of the problem. Using method 2, we can compute the amount you owe after you make the sixth payment by calculating the equivalent worth of the remaining 18 payments at the end of the sixth month, with the time scale shifted by 6 (equation (3.15)). Using method 3, we can use the future worth of the borrowed amount at the end of the sixth payment subtracted by the future equivalent of the six payments at the end of the sixth payment (equation (3.18)).

Given: $A = \$235.37$, $i = 1\%$ per month, $n = 18$ months for method 2
$n = 6$ months for method 3
Find: Remaining balance after 6 months (B_6)

Method 2: $B_6 = \$235.37(P/A, 1\%, 18) = \$3859.62.$
Method 3: $B_6 = \$5000(F/P, 1\%, 6) - \$235.37(F/A, 1\%, 6)$
$\$3859.60$

The difference between the two B_6 values calculated is due to round-off error. If you desire to pay off the remainder of the loan at the end of the sixth payment, you must come up with \$3859.62. To verify our results, compare the answers to the value given in Table 3.3.

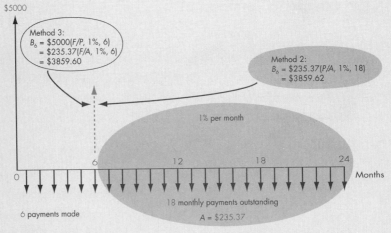

Figure 3.20 Computing the outstanding loan balance after making the sixth payment for the home improvement loan (Example 3.15)

3.6.2 Add-On Loans

The add-on loan is totally different from the popular amortized loan. In this type of loan, the total interest to be paid is precalculated and added to the principal. The principal plus this precalculated interest amount is then paid in equal installments. In such a case, the interest rate quoted is not the effective interest rate, but what is known as **add-on interest**. If you borrow P for N years at an add-on rate of i, with equal payments due at the end of each month, a typical financial institution might compute the monthly installment payments as follows:

$$
\begin{aligned}
\text{Total add-on interest} &= P(i)(N), \\
\text{Principal plus add-on interest} &= P + P(i)(N) = P(1 + iN) \\
\text{Monthly installments} &= P(1 + iN)/(12 \times N). \qquad (3.19)
\end{aligned}
$$

Notice that the add-on interest is *simple interest*. Once the monthly payment is determined, the financial institution computes the APR based on this payment, and you will be told what this value will be. Even though the add-on interest is specified along with the APR value, many ill-informed borrowers think that they are actually paying the add-on rate quoted for this installment loan. To see how much you actually pay interest under a typical add-on loan arrangement, consider Example 3.16.

EXAMPLE
3.16
$

Effective interest rate for an add-on loan

Reconsider the home improvement loan problem in Example 3.14. Suppose that you borrow $5000 with an add-on rate of 12% for 2 years. You will make 24 equal monthly payments.

(a) Determine the amount of the monthly installment.

(b) Compute the nominal and the effective annual interest rate on the loan.

Solution

Given: Add-on rate = 12% per year, loan amount (P) = $5000, N = 2 years
Find: (a) A and (b) i_a and i

(a) First we determine the amount of add-on interest:

$$iPN = (0.12)(\$5000)(2) = \$1200.$$

Then we add this simple interest amount to the principal and divide the total amount by 24 months to obtain A:

$$A = (\$5000 + \$1200)/24 = \$258.33.$$

(b) Putting yourself in the lender's position, compute the APR value of the loan just described. Since you are making monthly payments, with monthly compounding, you need to find the effective interest rate that makes the present $5000 sum equivalent to 24 future monthly payments of $258.33. In this situation, we are solving for i in the equation

$$\$258.33 = \$5000(A/P, i, 24)$$

or

$$(A/P, i, 24) = 0.0517.$$

You know the value of the A/P factor, but you do not know the interest rate, i. As a result, you need to look through several interest tables and determine i by interpolation as shown in Figure 3.21. The interest rate falls between 1.75% and 2.0% and may be computed by a **linear interpolation**:

$(A/P, i, 24)$	i
0.0514	1.75%
0.0517	?
0.0529	2.00%

$$i = 1.75\% + (0.25\%)\left[\frac{0.0517 - 0.0514}{0.0529 - 0.0514}\right] = 1.8\% \text{ per month.}$$

The nominal interest rate for this add-on loan is $1.8 \times 12 = 21.60\%$, and the effective annual interest rate is $(1 + 0.018)^{12} - 1 = 23.87\%$, rather than the 12% quoted add-on interest. When you take out a loan, you should not confuse the add-on interest rate stated by the lender with the actual interest cost of the loan.

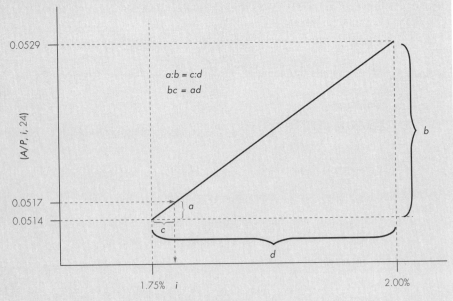

Figure 3.21 Linear interpolation to find unknown interest rate (Example 3.16)

As an alternative to linear interpolation based on interest tables, you can also calculate the effective interest rate for the add-on loan using one of the many financial calculators or spreadsheet programs currently available. If you solved the equation on either a financial calculator or a computer, the exact value would be 1.79749%. In practice, the small error introduced by this type of interpolation is rarely of significance.

Comments: In the real world, truth-in-lending laws require that APR information always be provided in mortgage and other loan situations—you would not have to calculate nominal interest as a prospective borrower (although you might be interested in calculating actual or effective interest). However, in later engineering economic analyses, you will discover that solving for implicit interest rates, or rates of return on investment, is regularly part of the solution procedure. Our purpose is to periodically give you some practice with this type of problem, even though the example/problem scenario does not exactly model the real-world information you would be given.

3.7 Mortgages

A mortgage is generally a long-term amortized loan used primarily for the purpose of purchasing a piece of property such as a home. The mortgage itself is a legal document in which the borrower agrees to give the lender certain rights to the property being purchased as security for the loan. The borrower is referred to as the *mortgagor* and the lender as the *mortgagee*. The mortgage document specifies the rights that the lender has to the property in the event of default by the borrower on the terms of the mortgage. In the sections that follow we will explain some concepts related to mortgages and give examples on how payment schedule, regular payment amount, and the interest charges are calculated. The amount of the loan—the cash that you actually borrow—is called the **principal**. The difference between the price of the property and the amount that one owes on the mortgage is called the purchaser's **equity**.

3.7.1 Types of Mortgages

Canadian banks, credit unions, trust companies, mortgage companies, private lenders, and others offer four main types of mortgages. These four types of mortgages are briefly described below.

1. **National Housing Act (NHA) mortgage:** These mortgages are loans granted under the provisions of the National Housing Act of 1954. Lenders are insured against loss by the Canada Mortgage and Housing Corporation (CMHC). Borrowers must pay an application fee which usually includes the property appraisal fee to CMHC and an insurance fee. The insurance fee is usually added to the principal amount of the mortgage, although it may be paid in cash.

2. **Conventional mortgage:** The conventional mortgage is the most common type of financing for principal residence or residential investment property. In this type of mortgage, the loan amount generally does not exceed 75% of the appraised value or the purchase price of the property, whichever is lower. The purchaser is responsible for raising the other 25% as a down payment.

3. **High-ratio mortgage:** If a potential buyer is unable to raise the necessary 25% funding to complete the purchase of the property, then he or she may obtain a high-ratio mortgage. Essentially, these are conventional mortgages that exceed the 75% referred to above. These mortgages must, by law, be insured. High-ratio mortgages are available for up to 95% of the purchase price or of the appraisal, whichever is lower.

4. **Collateral mortgage:** A collateral mortgage provides backup protection of a loan that is filed against a property. It is secondary to a main form of security taken by the lender for the loan, for example, a promissory note. When the promissory note is paid off, the collateral mortgage is automatically discharged. The money borrowed may be used for the purchase of the property itself or for other purposes, such as home improvements and other investments.

3.7.2   Terms and Conditions of Mortgages

To make the best mortgage decision, one has to consider many factors. The key factors are amortization, term of the mortgage, whether the mortgage is open or closed, interest rate, payment schedule, prepayment privilege, and portability. A brief explanation of each of these concepts is provided below:

1. **Amortization:** Amortization refers to the number of years it would take to repay a mortgage loan in full for a given interest rate and payment schedule. The usual amortization period is 25 years, although there is a wide range of choices. The longer the amortization period, the smaller the regular payment (usually monthly) and the larger the total interest payments.

2. **Term of the mortgage:** Term refers to the number of months or years which the mortgage—the legal document—covers. Terms may vary from 6 months to 10 years. At the end of a term, the unpaid principal is due and payable. One has the option to renew the mortgage with the same bank or refinance it through a different lending institution.

3. **Interest rate:** By law, mortgages must contain a statement showing, among other things, the rate of interest calculated annually or semi-annually. Mortgage interest has traditionally been quoted as a nominal annual rate based on semi-annual compounding. A *fixed-rate mortgage* is one where the rate of interest is set for the whole term. A *variable-rate mortgage* is one where the rate of interest varies according to the premium interest rate set by the lender every month. For

both a fixed-rate mortgage and a variable-rate mortgage, the required regular payment amount does not change within the term. However, the interest portion of the regular payment amount for a payment period is dependent on the outstanding balance at the beginning of the period and the interest rate applicable for the period.

4. **Open or closed mortgage:** An *open mortgage* provides the borrower with the flexibility to repay the loan more quickly. One can pay off the mortgage in full at any time before the term is over without any penalty or extra charges. Because of this flexibility, the interest rate for an open mortgage is higher than a closed mortgage, when other conditions remain the same. *A closed mortgage* does not allow the borrower to repay the loan more quickly than agreed. Payments must be made as specified in the agreement. If the borrower wants to pay off the mortgage before the term is over, a penalty charge is applied. The penalty charge is often equal to the greater of (1) three months' interest on the amount of the prepayment; and (2) the interest rate differential. The "interest rate differential" refers to the amount, if any, by which the existing interest rate exceeds the interest rate at which the lender would lend to the same borrower for a term commencing on the prepayment date and expiring at the existing term date. However, even for a closed mortgage, there may be some prepayment privileges such that no penalty will be levied.

5. **Payment schedule:** Most mortgage loans are amortized. Constant regular payments are made to pay off the principal. Monthly payments are the most common, although some mortgages may be paid weekly, bi-weekly, semi-monthly, quarterly, semi-annually, or annually.

6. **Prepayment privileges:** Many financial institutions offer closed mortgages with some prepayment privileges. For example, the borrower taking the closed mortgage may be allowed to make a prepayment of up to 10% of the original principal every calendar year or every anniversary. Another privilege allows the borrower to increase the regular payment by up to 100% once per year. If you are able to take advantage of these prepayment privileges, you may dramatically decrease the actual amortization period.

7. **Portability:** Some lenders offer mortgages with a feature called portability. This feature means that a borrower can sell one home and buy another during the term of the mortgage. The mortgage can be transferred from one property to the other without penalty.

3.7.3   An Example of Mortgage Calculations

Throughout this book, we will assume that the quoted mortgage interest rates are based on semi-annual compounding.

EXAMPLE
3.17
$

Closed mortgage with prepayment privileges

John Montgomery is considering buying a $125,000 home with a $25,000 down payment. He can get a conventional mortgage in the amount of $100,000 with a 3-year term at 8% per annum from the Toronto Dominion Bank. He has selected an amortization period of 25 years. The mortgage is a closed mortgage with the following prepayment privileges:

- Once each calendar year, on any regular payment date, John can prepay on account of principal a sum not more than 10% of the original borrowed amount, without notice or charge. If this privilege is not exercised in a certain year, it cannot be carried forward to the following years.

- Once each calendar year, on any regular payment date, on written notice, John, without charge, can increase the amount of the regular installment of principal and interest. The total of such increases cannot exceed 100% of the installment of principal and interest set out in the mortgage. If the regular installment has been increased, the mortgagor may decrease the installments to an amount not less than the installment of principal and interest set out in the mortgage, on written notice, without charge.

- On any regular payment date, John can prepay the whole or any part of the principal amount then outstanding on payment of an amount equal to the greater of

  (1) 3 months' interest, at the rate specified in the mortgage, on the amount prepaid or

  (2) the amount, if any, by which interest at the rate specified in the mortgage exceeds interest at the prevailing rate, calculated on the amount of principal prepayment, for a term commencing on the date of prepayment and expiring on the maturity date of the mortgage.

Answer the following questions regarding the mortgage in consideration:

(a) What is the amount of his regular payment if he chooses to pay weekly, semimonthly, or monthly?

(b) What would the balance be at the end of the term for each of the three payment frequencies if the calculated regular payment amount is followed exactly?

(c) Assume that John has selected the option of monthly payment because he receives only one salary payment per month. In year 2, he increases his monthly payment by 50%. In year 3, he doubles his calculated monthly payment in (a). What is the balance of the mortgage at the time of renewal?

(d) In addition to (c), if he makes lump sum payments of $8000 and $10,000 at the first and the second anniversaries, respectively, what is the balance of the mortgage at the time of renewal?

(e) After John has made only the calculated monthly payments for 1 year, the interest rate for a 2-year term has dropped to 6%. What would be the total penalty charge if he chooses to pay off his mortgage completely? What if the prevailing rate for a 2-year mortgage has increased to 9%?

Solution

Given: $P = \$100{,}000$, $r = 8\%$ per year, $M = 2$ compounding periods per year, amortization = 25 years, term = 3 years
Find:

(a) The regular payment amounts on the following payment schedules: weekly, semi-monthly, and monthly.

Figure 3.22(a) illustrates the cash flow diagram for the computation of regular payment.

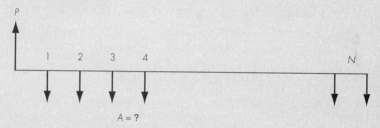

Figure 3.22(a)   Cash flow diagram for the term loan (Example 3.17(a))

where

N is the number of payment periods in the amortization period.

- For weekly payment: $N = (52)(25) = 1300$ weeks

$$i_{wk} = (1 + r/M)^c - 1 = (1 + 0.08/2)^{1/26} - 1 = 0.1510\%$$
$$A_{wk} = \$100{,}000 \, (A/P, 0.1510\%, 1300) = \$175.68$$

- For semi-monthly payment: $N = (25)(24) = 600$ half-months

$$i_{1/2\,mon} = (1 + 0.08/2)^{1/12} - 1 = 0.3274\%$$
$$A_{1/2\,mon} = \$100{,}000 \, (A/P, 0.3274\%, 600) = \$380.98$$

- For monthly payment: $N = (25)(12) = 300$ months

$$i_{mon} = (1 + 0.08/2)^{1/6} - 1 = 0.6558\%$$
$$A_{mon} = \$100{,}000 \, (A/P, 0.6558\%, 300) = \$763.20$$

Comments: The more frequent the payments, the smaller the total amount paid per month (i.e., $A_{mon} > 2 \times A_{1/2\,mon} > 4.33 \times A_{wk}$).

(b) What are the end-of-term balances for weekly, semi-monthly, and monthly payments? Figure 3.22(b) illustrates the cash flow diagram for the end-of-term balance calculation.

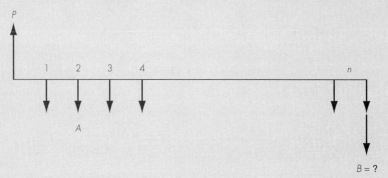

Figure 3.22(b)   Equivalent cash flow diagram (Example 3.17(b))

where

n is the number of payment periods in a term,

A is the calculated regular payment amount, and

B is the end-of-term balance to be computed.

- For weekly payment: $n = (3)(52) = 156$ weeks

$$
\begin{aligned}
i_{wk} &= 0.1510\% \\
B_{wk} &= P(F/P, i_{wk}, n) - A_{wk}(F/A, i_{wk}, n) \\
&= \$100{,}000(F/P, 0.1510\%, 156) \\
&\quad - \$175.68(F/A, 0.1510\%, 156) \\
&= \$95{,}655.93
\end{aligned}
$$

- For semi-monthly payment: $n = (3)(24) = 72$ half-months

$$
\begin{aligned}
i_{1/2\,mon} &= 0.3274\% \\
B_{1/2\,mon} &= \$100{,}000(F/P, 0.3274\%, 72) \\
&\quad - \$380.98(F/A, 0.3274\%, 72) \\
&= \$95{,}655.54
\end{aligned}
$$

- For monthly payment: $n = (3)(12) = 36$ months

$$
\begin{aligned}
i_{mon} &= 0.6558\% \\
B_{mon} &= \$100{,}000(F/P, 0.6558\%, 36) \\
&\quad - \$763.20(F/A, 0.6558\%, 36) \\
&= \$95{,}655.54
\end{aligned}
$$

Comments: Note that the end-of-term balances are the same (ignoring the round-off errors) for all three payment options. This is because all three options will pay off the mortgage loan in exactly 25 years, if the calculated regular payment amounts are made for 25 years. The difference between the three payment schedules then is the total amount of interest paid. As shown in part (a), more money is paid out over the whole term when the frequency of payments decreases. Since the same amount of principal is paid off over the three-year term, the excess amount corresponds to additional interest paid.

(c) What is the end-of-term balance, when monthly payments and some prepayment privileges are used?

$$
\begin{aligned}
B_{(c)} &= \$100{,}000(F/P, 0.6558\%, 36) \\
&\quad - \$763.20(F/A, 0.6558\%, 36) \\
&\quad - \$381.60(F/A, 0.6558\%, 24) \\
&\quad - \$381.60(F/A, 0.6558\%, 12) \\
&= \$81{,}023.51
\end{aligned}
$$

Figure 3.22(c) shows the payments for the whole term.

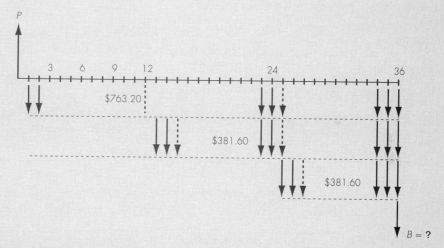

Figure 3.22(c) Equivalent cash flow diagram (Example 3.17(c))

(d) What is the end-of-term balance with some additional lump sum payments?

Notice that the lump sum payments are within the prepayment privilege limits. We can utilize the result calculated in (c), as shown in Figure 3.22(d).

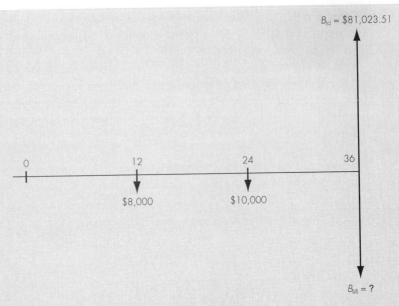

Figure 3.22(d) Equivalent cash flow diagram (Example 3.17(d))

$$B_{(d)} = \$81{,}023.51 - \$8000(F/P, 0.6558\%, 24)$$
$$- \$10{,}000(F/P, 0.6558\%, 12)$$
$$= \$60{,}848.71$$

(e) What are the prepayment penalties?

To find the prepayment penalty, we need to calculate the total prepayment amount first (i.e., the balance of the loan after monthly payments have been made for a year).

$$B = \$100{,}000(F/P, 0.6558\%, 12)$$
$$- \$763.20(F/A, 0.6558\%, 12)$$
$$= \$98{,}663.79$$

Three months' simple interest = $\$98{,}663.79 \times 0.6558\% \times 3 = \$1941.11$.

When the prevailing rate has dropped to 6% and there are two more years left in the term, the interest rate differential can be estimated as

$$\$98{,}663.79 \times (8\% - 6\%) \times 2 = \$3946.55.$$

When the prevailing rate has increased to 9%, the penalty charged based on the interest rate differential is zero.

As a result, the prepayment penalty would be $3946.55 (i.e., the larger of $1941.11 and $3946.55) when the prevailing rate is 6% and $1941.11 (the larger of $1941.11 and zero) when the prevailing rate is 9%.

> **Comment:** The penalty charge calculation illustrated in this case is only an approximate method. However, it does provide an indication of the magnitude of the required penalty payment.

3.8 Investment in Bonds

Bonds are a specialized form of a loan in which the creditor—usually a business or the federal, provincial, or local government—promises to pay a stated amount of interest at specified intervals for a defined period and then to repay the principal at a specific date known as the maturity date of the bond. Bonds are an important financial instrument by which the business world may raise funds to finance engineering projects.

In the case of bonds the lenders are investors (known as bondholders) who may be individuals or other businesses. Bonds can be a significant investment opportunity. In addition to the interest they earn, once purchased by the initial bondholder, they may be sold again for amounts other than their stated face value and thus enhance the bondholder's opportunity to increase the return on his or her initial investment. Given these complications, the concept of economic equivalence can be important in determining the worth of bonds. The following sections will illustrate some typical bond investment problems in the context of economic equivalence.

3.8.1   Bond Terminology

We will first look at how a typical bond may be issued on the financial market. We will consider a government bond issued by Ontario Hydro. The example in Figure 3.23(a) and Figure 3.23(b) is a reproduction of a $1000 bond issued by Ontario Hydro and known as "Ontario Hydro 9 1/4s of 2004." The certificate is the document one receives as evidence of an investment. Before explaining how bond values are determined, some of the terms associated with a typical bond will be defined.

Debentures and mortgage bonds: Traditionally, a debenture is a debt security or an unsecured promise to pay. That is, no property or assets are pledged as security for the loan. Most government bonds fall in this category. The Ontario Hydro bond is precisely this. You will note that this bond is guaranteed as to principal and interest by the province of Ontario. Hence the issue is known as a *Provincial Guarantee.* If a company or government agency issues bonds and backs them with specific pieces of property such as buildings, the bonds are called *mortgage bonds.*

Par value (or face value): Individual bonds are normally issued in even denominations, such as $1000 or multiples of $1000. The stated face value on the individual bond is termed the *par value*. For this type of bond, it is usually set at $1000, which is the case for the Ontario Hydro bond.

Maturity date: Bonds generally have a specified *maturity date* on which the par value is to be repaid. The Ontario Hydro bonds, which were issued on March 10, 1977, will mature on January 6, 2004; thus, they had a 26.17-year term at time of issue. Depending on the time remaining to maturity, bonds can be classified into the following categories: short term bonds (maturing within three years), medium term bonds (maturing from three to ten years), and long term bonds (maturing in more than 10 years). The term status of a bond or debenture depends on the time remaining to maturity, and the bond will change its term status as the maturity date approaches.

Coupon rate: The interest paid on the par value of a bond is called the annual *coupon rate*. For example, the Ontario Hydro bonds have a $1000 par value and pay $92.50 in interest (9.25%) each year ($46.25 every 6 months). The bond's coupon interest is $92.50 so its annual coupon rate is 9.25%. Even though the coupon rate is stated as an annual rate, Ontario Hydro will pay 4.625% interest semi-annually on the face value (i.e., $46.25) to the bond holders. Shown in Figure 3.23(c) is a sample of one of the 54 coupons originally attached to the Ontario Hydro bond. Each coupon has a date on which, or after which, it may be cashed.

Discount or premium bond: A bond that sells below its par value is called a *discount bond*. When a bond sells above its par value, it is called a *premium bond*. A bond that is purchased for its par value is said to have been bought *at par*. The Ontario Hydro bonds were offered *at par*. The price one has to pay to purchase a bond is called the bond's *market value*.

A call or redemption feature: Issuers of bonds sometimes reserve the right to pay off the bonds before the maturity date. This privilege is known as the *call feature* or *redemption feature* and a bond bearing this clause is known as a *callable bond* or a *redeemable bond*. As a rule, the issuer needs to give prior notice to the bondholders to call the bond.

Bearer bonds and registered bonds: A bond in bearer form does not have the name of the owner printed on the bond. It has interest coupons attached. The holder of the bond is presumed to be the owner of the bond. A fully registered bond shows the name of the current owner and interest is paid by cheque or direct deposit in the owner's bank account.

Figure 3.23(a) The face side of the Ontario Hydro bond certificate (Source: *The Canadian Securities Course.* Prepared and published by the Canadian Securities Institute, Toronto, Ontario. Copyright 1990.)

Figure 3.23(b) The back side of the Ontario Hydro bond certificate (Source: *The Canadian Securities Course.* Prepared and published by the Canadian Securities Institute, Toronto, Ontario. Copyright 1990.)

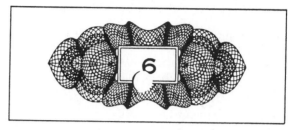

Sample of one of the 54 coupons originally attached to the $1000 Ontario Hydro 9¼ s of January 6, 2004. Top illustration: face of the sixth coupon which paid $46.25 on January 6, 1980. Bottom illustration: back of same coupon is numbered "6". Other coupons originally attached to the certificate were numbered up to 54 and payable up to January 6, 2004.

Figure 3.23(c)   Sample of one of the 54 coupons originally attached to the Ontario Hydro bond certificate (Source: *The Canadian Securities Course*. Prepared and published by the Canadian Securities Institute, Toronto, Ontario. Copyright 1990.)

3.8.2   Types of Bonds

Government Bonds

There are three main issuers of government bonds: federal government, provinces, and municipalities. The federal government has very broad taxation power and thus has guaranteed income. Securities issued by the federal government, therefore, have the highest credit rating of any bonds issued in Canada. The federal government is also the largest issuer in the Canadian bond market.

Government of Canada Treasury Bills: Treasury bills are short-term debts. Depending on the amount of money invested, they may have terms to maturity of 30 days, 60 days, 90 days, 180 days, and one year. Treasury Bills are sold at a discount (below the par value) and mature at the par value. The difference between the purchase price and par at maturity gives the investor the yield to maturity. There are no interest payments. Large financial institutions buy treasury bills at the bi-weekly auctions by the government. They then offer these treasury bills to individual investors. Individual investors cannot buy treasury bills directly from the government. The prices of treasury bills vary from day to day. Treasury bills may be bought or sold at market prices. Their denominations begin as low as $1000.

Marketable bonds: Compared to treasury bills, marketable bonds have a longer term to maturity that can range from 1 year to 30 years. They have a fixed semi-annual interest payment. Marketable bonds are issued and offered to large financial institutions at auctions on a scheduled basis by the government. These financial institutions will then offer these bonds to individual investors. Marketable bonds can be bought or sold at market prices. Their market prices may change from day to day. Their denominations also begin as low as $1000.

Canada premium bonds: Canada premium bonds have a fixed interest rate. The government guarantees both the principal and the interest rate. You may buy them at any bank, credit union, or most investment dealers. Their prices are fixed. You can only redeem your bonds each year on the anniversary date and during the 30 days thereafter. Canada premium bonds have two forms. You may choose the regular interest bonds (R-bonds) wherein interest can be deposited directly into a banking account or paid by cheque. You can also purchase the compound interest bonds (C-bonds). In this case, your interest will be automatically reinvested to purchase more C-bonds until your bonds reach maturity. Compound interest bonds are available in denominations as small as $100 while regular interest bonds are available starting at $300.

Canada savings bonds: Canada savings bonds offer minimum guaranteed interest rates which may be increased depending on the market conditions. The government guarantees both the principal and the minimum interest rate. Canada savings bonds can be redeemed at any time. The prices of the bonds are fixed. If you cash your bonds within 3 months of the issue date you will receive the full face value of the bond only with no interest. If you cash them after that date, you'll receive the full face value plus all the interest earned for each full month elapsed since the issue date. Like the Canada premium bonds, there are also two forms of Canada savings bonds. They are available in similar denominations.

Provincial bonds: Similar to the federal government, the provinces also issue debt directly and may guarantee them through agencies under their jurisdictions. Provincial bonds or debentures are issued to raise funds for major public capital projects. All provinces have statutes covering the use of funds raised through the issue of bonds. The bonds may be available in a wide range of denominations from a few hundred dollars to thousands of dollars. The terms, interest rates, and other conditions depend on the debt issues.

Municipal debentures: Municipalities are responsible for construction and maintenance of streets, sewers, waterworks, schools, and other services for individual communities. To provide these essential services and facilities, a municipality needs to take on capital projects from time to time. One method to raise the required capital for such projects is to issue bonds or debentures. Investors need to check with the issuers for further details such as denominations, interest rates, maturity dates, and guarantees.

Corporate Bonds and Debentures

Corporations issue bonds and debentures to borrow money. These securities have various terms, interest rates, and payment arrangements at the discretion of the issuer. Bonds and debentures for larger corporations are publicly traded.

Mortgage bonds: A mortgage is a legal document specifying that the borrower's fixed assets such as land, buildings, and equipment are pledged as security for a loan and the lender is entitled to ownership or partial ownership of the assets if the borrower fails to pay interest or repay principal when due. The lender holds the mortgage document until the loan is paid off. The mortgage bond is very similar to a home mortgage. Both are issued to protect the lender if the borrower fails to repay the loan. Mortgage bonds are issued so that the company can borrow money from many individuals who purchase the bonds. Instead of creating a mortgage document between each individual buyer and the company issuing the bonds, the mortgage document is deposited with a trust company on the behalf of all individual buyers. This mechanism is a mortgage bond.

First mortgage bonds, second mortgage bonds, and general mortgage bonds: A first mortgage bond has the first priority on charging the company's assets and earnings. It may be closed or open. If it is closed, no additional first mortgage bonds may be issued. If it is open, additional first mortgage bonds may be issued under certain conditions, usually when new assets are acquired. A second mortgage bond has the right to the assets and earnings of the company after the claims of the first mortgage bonds have been satisfied. A general mortgage bond ranks after the first mortgage bonds and other prior mortgage bonds.

Corporate debentures and notes: These are debt securities that are not backed by fixed assets or other property. They rank after all mortgage bonds. Their only security is the owners' investment and the retained earnings of the company.

Convertible bonds and debentures: Convertible bonds or debentures may be exchanged for the company's stocks at the holder's option. These bonds or debentures have a fixed interest rate and maturity date. Their values may also increase because of the holder's right to convert them into common stocks at stated prices over stated periods.

3.8.3　Bond Valuation

Certain types of bonds can be traded just like stocks on the market. Once purchased, a bond may be kept to maturity or for a variable number of interest periods before being sold. Bonds may be purchased or sold at prices other than face value, depending on the economic environment. Furthermore, bond prices change over time as a result of various factors—the risk of nonpayment of interest or the par value at maturity, supply and demand, market interest rates, and the future outlook for economic conditions. These factors affect the **yield to maturity** (or **return on investment**). The yield to maturity represents the actual interest earned from a bond over the holding period. We will explain these values with numerical examples in the following section.

Yield to Maturity

The yield to maturity on a bond is the interest rate that establishes the equivalence between all future interest/face value receipts and the market price of the bond. To illustrate the point, let us consider buying a $1000 denomination of the Ontario Hydro bond on January 7, 1994. Recall that the par value is $1000, but assume the market price was $1088. Since the interest will be paid semi-annually, the interest rate per payment period will be simply 4.625%, and there will be 20 interest payments until the bond matures on January 6, 2004. The resulting cash flow to the investor is shown in Fig. 3.24.

The yield to maturity is found by determining the interest rate that makes the present worth of the receipts equal to the market price of the bond:

$$\$1088 = \$46.25(P/A, i, 20) + \$1000(P/F, i, 20).$$

The value of i that makes the present worth of the receipts equal to $1088 lies between 3% and 4%. Solving for i by interpolation yields $i = 3.98\%$.

Present Worth of Receipts	i
$1084.94	4%
1088.00	?
1241.76	3%

$$i = 3\% + 1\% \left[\frac{\$1241.76 - \$1088}{\$1241.76 - \$1084.94} \right] = 3.98\%$$

Note that this is a 3.98% yield on maturity per semiannual period. The nominal (annual) yield is $2 \times 3.98\% = 7.96\%$ compounded semiannually. When compared with the coupon rate of 9¼% (or 9.25%), purchasing the bond at a price higher than the face value brings about a lower yield. The effective annual interest rate is then

$$i_a = (1 + 0.0398)^2 - 1 = 8.12\%.$$

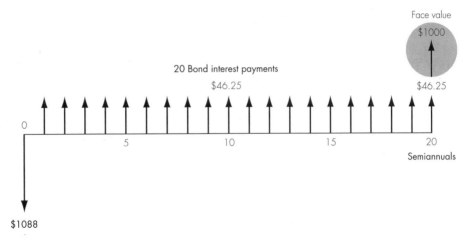

Figure 3.24 A typical cash flow transaction associated with an investment in the Ontario Hydro bond

This 8.12% represents the **effective annual yield** to maturity on the bond. Notice that when a bond is purchased at a par value, the yield to maturity will be exactly the *same* as the coupon rate of the bond, provided that they are both expressed as effective rates for the same length of time.

Until now, we have observed differences in nominal and effective interest rates as a result of frequency of compounding. In the case of bonds, the reason is very different: The stated (par) value of the bond and the actual price paid for it are different. The nominal interest is stated as a percentage of par value, but when the bond is purchased at a premium, you earn the same nominal interest on a larger initial investment; hence, your effective interest earnings are smaller than the stated nominal rate.

Current Yield

The **current yield** of a bond is the annual interest earned as a percentage of the current market price. This current yield provides an indication of the annual return realized from the bond investment. For our example of the Ontario Hydro Bond, the current yield is computed as follows:

$$
\begin{aligned}
\$46.25/1088 &= 4.25\% \text{ semiannually} \\
2 \times 4.25\% &= 8.5\% \text{ per year (nominal current yield)} \\
i_a &= (1 + 0.0425)^2 - 1 = 8.68\%.
\end{aligned}
$$

This effective current yield is 0.56% larger than the yield to maturity computed above (8.12%). The current yield is larger than the yield to maturity because the bond is sold at premium. If the bond were sold at a discount, the current yield would be smaller than the yield to maturity. There can be a significant difference between the yield to maturity and the current yield of a bond, because the market price of a bond may be more or less than its face value. Moreover, both the current yield and the yield to maturity may differ considerably from the stated coupon value of the bond.

EXAMPLE
3.18
$

Bond yields

John Brewer purchased a new corporate bond for $1000. The issuing corporation promised to pay the bondholder $45 interest on the $1000 face (par) value of the bond every 6 months, and to repay the $1000 at the end of 10 years. After 2 years John sold the bond to Kimberly Crane for $900.

(a) What was the yield on John's investment?

(b) If Kimberly keeps the bond for its remaining 8-year life, what is the yield to maturity on her bond investment?

(c) What was the current yield when Kimberly purchased the bond?

Solution

Given: Initial purchase price = par value = $1000, coupon rate = 9% per year paid semiannually, 10 years maturity, sold after 2 years for $900
Find: (a) Yield to initial owner, (b) yield to maturity for second owner, (c) current yield at 2 years

(a) Since John has received $45 every 6 months for 2 years, the cash flow diagram for this situation would look like Fig. 3.25. John should find the yield on bond investment by determining the interest rate that makes the expenditure of $1000 equivalent to the present worth of the semiannual receipts and selling price:

$$\$1000 = \$45(P/A, i, 4) + \$900(P/F, i, 4).$$

Try i = 3%:

$$\$45(P/A, 3\%, 4) + \$900(P/F, 3\%, 4) = \$966.91.$$

The present worth of the annual receipts is too low. Try a lower interest rate, say, i = 2%:

$$\$45(P/A, 2\%, 4) + \$900(P/F, 2\%, 4) = \$1002.81.$$

Figure 3.25   John Brewer's cash flow transaction associated with the bond investment (Example 3.18)

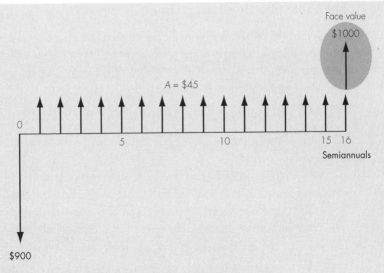

Face value
$1000

A = $45

0

5   10   15 16

Semiannuals

$900

Figure 3.26   Kimberly Crane's bond investment (Example 3.18)

Solving by interpolation, the yield on the bond investment is 2.08% over 6 months. This means the annual nominal yield is 2 × 2.08% = 4.16% based on semiannual compounding. (The effective annual yield is $(1.0208)^2 - 1 = 4.20\%$.)

(b) Kimberly's bond investment would look like Fig. 3.26.

$$\$900 = \$45(P/A, i, 16) + \$1000(P/F, i, 16)$$

Try $i = 5.4\%$:

$$\$45(P/A, 5.4\%, 16) + \$1000(P/F, 5.4\%, 16) = \$905.17.$$

Try $i = 5.5\%$:

$$\$45(P/A, 5.5\%, 16)$$
$$+ \$1000(P/F, 5.5\%, 16) = \$895.38$$
$$i = 5.4\% + (0.10\%)\left[\frac{\$905.17 - \$900}{\$905.17 - \$895.38}\right]$$
$$= 5.453\%.$$

The nominal (annual) yield is 2 × 5.453% = 10.906%, and the effective annual yield is $(1 + 0.05453)^2 - 1 = 11.20\%$. Compare this interest with the coupon rate of 9%.

(c) The current yield when Kimberly purchased the bond is

$$\$45/\$900 = 5\% \text{ per semiannual}$$
$$2 \times 5\% = 10\% \text{ per nominal}$$
$$i_a = (1 + 0.05)^2 - 1 = 10.25\%.$$

The Value of a Bond over Time

Reconsider the Ontario Hydro bond investment introduced earlier. If the nominal (annual) yield to maturity remains constant at 7.96%, what will the value of the bond be 1 year after it was purchased? We can find this value using the same valuation procedure, but now the term to maturity is only 9 years.

$$\$46.25(P/A, 3.98\%, 18) + \$1000(P/F, 3.98\%, 18) = \$1081.79$$

The value of the bond has decreased because there is a smaller number of interest payments to be received.

Now suppose interest rates in the economy rose after the Ontario Hydro bonds were issued, and as a result, the prevailing rate of interest was 10% one year later. Both the coupon interest payments and the maturity value would remain constant, but an interest rate of 10% would be used in calculating the value of the bond. The value of the bond at the end of the first year would be

$$\$46.25(P/A, 5\%, 18) + \$1000(P/F, 5\%, 18) = \$956.16.$$

Thus, the bond would sell at a discount relative to its par value.

The arithmetic of the bond price decrease should be clear, but what is the logic behind it? We can explain the reason for the decrease as follows: The fact that interest rates in the bond market rose to 10% means that, if we had had $1000 to invest, we would have bought new bonds rather than Ontario Hydro bonds, since the new bonds would pay $100 of interest every year instead of $92.50. If we were interested in buying Ontario Hydro Bonds, we would be willing to pay less than face value because they would have lower interest payments. All investors would recognize these facts, and as a result, Ontario Hydro bonds would be discounted to a lower price of $956.16, at which point, they would provide the same yield to maturity (rate of return) to a potential investor as the new bonds, 10%.

3.9 Computer Notes

As mentioned in Section 3.6.1, generating a loan payment schedule can be tedious. The power of an electronic spreadsheet can greatly facilitate the generation of such schedules. To illustrate how we might generate a loan repayment schedule with Excel, we will use the financial data in Example 3.14.

Using Figure 3.27 as a model, set up your Excel screen and specify the contract amount (C5), contract period (C6), and the interest rate (C7). Excel calculates the monthly payment (C8), with the principal (column E) as well as the interest (column F) itemized for each period. By separating the interest payment from the periodic payment, the remaining loan balance for each period (column G) can be generated.

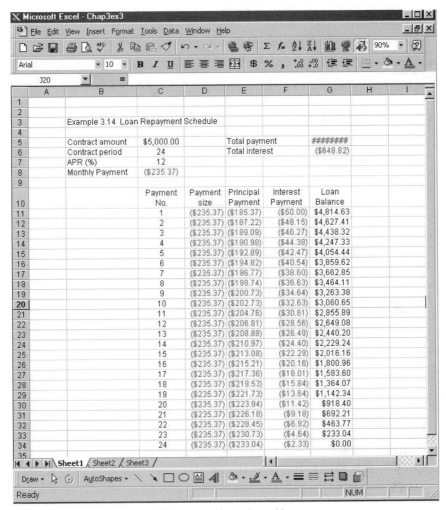

Figure 3.27   Loan repayment schedule generated by Excel spreadsheet

The built-in loan functions available for Excel are as follows:

- **PMT**(*rate,nper,pv,fv,type*) returns the periodic payment for an annuity based on constant payments and a constant interest rate;

- **IMPT**(*rate,per,nper,pv,fv,type*) returns the interest payment for an investment for a given period; and

- **PPMT**(*rate,per,nper,pv,fv,type*) returns the principal payment for an investment for a given period, where

  rate is the interest rate per period.

  per specifies the period and must be in the range 1 to *nper*.

nper is the total number of payment periods in an annuity.

pv is the present value, or the lump-sum amount that a series of future payments is currently worth.

fv is the future value, or a cash balance, you want to attain after the last payment is made. If *fv* is omitted, it is assumed to be 0 (the future value of a loan at the end of its term, for example, is 0).

type indicates when payments are due. If payments are due at the end of the period, set *type* equal to 0 or omit it. If payments are due at the beginning of the period, set *type* equal to 1.

Before showing the cell formula and values used in Figure 3.27, note the difference between C7 and *$C$7*. The latter notation indicates the absolute cell location. The dollar signs prevent the "C" and the "7" from changing whenever a cell formula containing this absolute cell reference is moved or copied to another cell. **Note:** *The given APR interest rate must be divided by 12 to obtain effective interest rate per month.* To convert the interest rate to a decimal, you also need to divide the effective interest rate per payment period by 100. For example:

- In cell C8: = **PMT(*$C$7*/1200, *$C$6*, *$C$5*, 0)** to calculate the monthly payment.

- In cell E11: = **PPMT(*$C$7*/1200, *$C$11*, *$C$6*, *$C$5*, 0)** to calculate the principal payment for the first payment period.

- In cell F11: = **IPMT(*$C$7*/1200, *$C$11*, *$C$6*, *$C$5*, 0)** to calculate the interest payment for the first payment period.

- In cell G6: = **SUM(*F11:F34*)** to calculate the total interest payment over the life of the loan.

As mentioned in Section 2.5, one of the most useful features of any spreadsheet is to allow you to do "what if" analyses relatively quickly. An example follows: What if you can borrow an amount at 9%, compounded monthly, instead of at 12%? Enter this new APR in cell C7, and you will see that Excel automatically recalculates (or updates) the entire worksheet.[3]

---

[3] Even when using the Copy function of Excel, whereby a piece of data (or cell formulas) can be copied into many cells simultaneously, you can see that to generate the Excel schedule, a great deal of manipulation was required. An alternative is to go to the Web HomePage described in Appendix B and download and use **EzCash**, which generates a loan repayment schedule in a much more efficient way.

3.10 Summary

- Interest rate is most frequently quoted by financial institutions as an **APR** or **annual percentage rate**. However, compounding commonly occurs more often than annually, and the APR does not account for the effect of this more frequent compounding. The situation leads to the distinction between nominal and effective interest:

- **Nominal interest rate** is a stated rate of interest for a given period (usually a year).

- **Annual effective interest rate** is the actual rate of interest, which accounts for the interest amount accumulated over a given year. The **annual effective rate** is related to the APR by the following equation:

$$i_a = (1 + r/M)^M - 1,$$

  where r = APR, M = number of compounding periods per year, and i_a = annual effective interest rate.

  In any equivalence problem, the interest rate to use is the effective interest rate per payment period:

$$i = (1 + \frac{r}{M})^C - 1,$$

  where C = number of compounding periods per payment period. Fig. 3.12 outlines the possible relationships between compounding and payment periods and indicates which version of the effective interest formula to use.

- The equation for determining the effective interest rate per payment period of **continuous compounding** is as follows:

$$i = e^{r/K} - 1,$$

  where K is the number of payment periods per year.

  The difference in accumulated interest between continuous compounding and very frequent compounding (M > 50) is minimal.

- Cash flows, as well as compounding, can be continuous. Table 3.2 shows the interest factors to use for continuous cash flows with continuous compounding.

- Nominal (and hence effective) interest rates may fluctuate over the life of a cash flow series. Some forms of home mortgages and bond yields are typical examples.

- An **amortized loan** is repaid in equal periodic amounts (weekly, monthly, quarterly, or annually). Examples of an amortized loan include automobile loans, loans of major appliances, home mortgage loans, and most intermediate and long term business debts.

- For amortized loans, three methods are introduced for calculation of remaining balances, principal payments, and interest payments. They are the tabular method, the remaining balance method, and the equivalent future worth method.

- The **add-on loan** is totally different from the popular amortized loan. In an add-on loan, the total interest to be paid is precalculated and added to the principal. The principal plus this precalculated interest amount is then divided by the total number of payment periods to find the constant periodic payment.

- A **mortgage loan** is a long-term amortized loan used primarily for the purpose of purchasing a piece of property such as a home. The covered mortgage terminology includes amortization, open or closed mortgage, term of the mortgage, prepayment privilege, and portability.

- **Bonds** are a specialized form of a loan in which the creditor promises to pay a stated amount of interest at specified intervals for a defined period of time and then to repay the principal at a specific date called the bond maturity date. Covered bond terminology includes debentures and mortgage bonds, par value or face value, maturity date, coupon rate, discount and premium bonds, the redemption feature, and bearer bonds and registered bonds.

- The types of **Government of Canada bonds** include treasury bills, marketable bonds, premium bonds, and savings bonds. Types of **corporate bonds** include mortgage bonds, debentures and notes, and convertible debentures and notes.

- The **yield to maturity** on a bond is the interest rate that establishes the equivalence between all future interest/face value receipts and the purchase price of the bond. The current yield of a bond is the annual interest earned as a percentage of the current market price.

Problems

Level 1

$3.1* You have just received credit card applications from two banks, A and B. The interest terms on any unpaid balance are stated as follows:

(1) Bank A: 15%, compounded quarterly
(2) Bank B: 14.8%, compounded daily.

Which of the following statements is incorrect?

(a) The effective annual interest rate for Bank A is 15.865%.
(b) The nominal interest rate for Bank B is 14.8%.
(c) Bank B's term is a better deal because you will pay less interest on your unpaid balance.
(d) Bank A's term is a better deal because you will pay less interest on your unpaid balance.

3.2* What is the future equivalent of an equal quarterly payment series of $2500 for 10 years, if the interest rate is 9%, compounded monthly?

$3.3* John secured a home improvement loan from a local bank in the amount of $10,000 at an interest rate of 9%, compounded monthly. He agreed to pay back the loan in 60 equal monthly installments. Immediately after the 24th payment, John decides to pay off the remainder of the loan in a lump sum. What will be the size of this payment?

$3.4* You want to open a savings plan for your future retirement. You are considering two options as follows:

• Option 1: You deposit $1000 at the end of each quarter for the first 10 years. At the end of 10 years, you make no further deposits, but you leave the amount accumulated at the end of 10 years for the next 15 years.
• Option 2: You do nothing for the first 10 years. Then you deposit $6000 at the end of each year, for the next 15 years.

If your deposits or investments earn an interest rate of 6%, compounded quarterly, and you choose Option 2 over Option 1, which of the following statements is correct? At the end of 25 years from now, I will accumulate

(a) $7067 more   (b) $8523 more
(c) $14,757 less   (d) $13,302 less.

$3.5* Vi Wilson is interested in buying an automobile priced at $18,000. From her personal savings, she can come up with a down payment in the amount of $3000. The remaining balance will be financed by the dealer over a period of 36 months at an interest rate of 6.25%, compounded monthly. Which of the following statements is correct?

(a) The dealer's annual percentage rate (APR) is 6.432%.
(b) The monthly payment can be calculated by using A = $15,000 (A/P, 6.25%, 3)/12.
(c) The monthly payment can be calculated by using A = $15,000 (A/P, 0.5208%, 36).
(d) The monthly payment can be calculated by using A = $15,000 (A/P, 6.432%/12, 3).

3.6* A series of equal semi-annual payments of $1000 for 3 years is equivalent to what present amount at an interest rate of 12%, compounded annually?

3.7* At what rate of interest, compounded quarterly, will an investment double itself in 5 years?

3.8* A series of equal quarterly deposits of $1000 extends over a period of 3 years. What is the future equivalent of this quarterly deposit series at 9% interest, compounded monthly?

$3.9* You have been offered a credit card by an oil company that charges interest at 1.8% per month, compounded monthly. What is the effective annual interest rate that this oil company charges?

3.10* A series of equal quarterly receipts of $1000 extends over a period of 5 years. What is the present equivalent of this quarterly payment series at 8% interest, compounded *continuously*?

3.11* You are considering purchasing a piece of industrial equipment that costs $30,000. You decide to make a down payment in the amount of $5000 and to borrow the remainder from a local bank at an interest rate of 9%, compounded monthly. The loan is to be paid off in 36 monthly installments. What is the amount of the monthly payment?

3.12* A building is priced at $100,000. If a down payment of $30,000 is made, and a payment of $1000 every month thereafter is required, how many months will it take to pay for the building? Interest is charged at a rate of 12%, compounded monthly.

$3.13* You obtained a loan of $20,000 to finance the purchase of an automobile. Based on monthly compounding for 24 months, the end-of-the-month equal payment was figured out to be $922.90. Immediately after making the 12th payment, you decide you want to pay off the loan in lump sum. What is this lump sum payment amount?

3.14* To raise money for your business, you need to borrow $20,000 from a local bank. If the bank asks you to repay the loan in 5 equal annual installments of $5548.19, determine the bank's effective annual interest rate on this loan transaction.

3.15* How many years will it take for an investment to double itself if the interest rate is 9%, compounded quarterly?

$3.16* Consider the following bank advertisement appearing in a local newspaper: "A Bank of Montreal Guaranteed Investment Certificate (GIC) pays a guaranteed rate of return (effective annual yield) of 8.87%." If there are 365 compounding periods per year, what is the nominal interest rate (annual percentage rate) for this GIC?

$3.17* You borrowed $1000 at 8%, compounded annually. The loan was repaid according to the following schedule.

n	Repayment Amount
1	$100
2	$300
3	$500
4	X

Find X, the amount that is required to pay off the loan at the end of year 4.

$3.18 Suppose you deposit $C at the end of each month for 10 years at an interest rate of 12%, compounded continuously. What equal end-of-year deposit over 10 years would accumulate the same amount at the end of 10 years under the same interest compounding?

(a) $A = [\$12C(F/A, 12.75\%, 10)] \times (A/F, 12.75\%, 10)$

(b) $A = \$C(F/A, 1.005\%, 12)$

(c) $A = [\$C(F/A, 1\%, 120)] \times (A/F, 12\%, 10)$

(d) $A = \$C(F/A, 1.005\%, 120) \times (A/F, 12.68\%, 10)$

(e) None of the above.

3.19 A loan company offers money at 1.8% per month, compounded monthly.

    (a) What is the nominal interest rate?

    (b) What is the effective annual interest rate?

    (c) How many years will it take an investment to triple itself if interest is compounded monthly?

    (d) How many years will it take an investment to triple itself if the nominal rate is compounded continuously?

$3.20 A department store has offered you a credit card that charges interest at 1.25% per month, compounded monthly. What is the nominal interest (annual percentage) rate for this credit card? What is the effective annual interest rate?

$3.21 The Toronto Dominion Bank advertised the following information: Interest rate 7.55%—effective annual yield 7.842%. No mention is made of the compounding frequency in the advertisement. Can you figure out the compounding scheme used by the bank?

Level 2

$3.22 A financial institution is willing to lend you $40. However, $45 is repaid at the end of one week.

    (a) What is the nominal interest rate?

    (b) What is the effective annual interest rate?

$3.23 War Eagle Financial Sources, which makes small loans to college students, offers to lend $400. The borrower is required to pay $26.61 at the end of each week for 16 weeks. Find the interest rate per week. What is the nominal interest rate per year? What is the effective interest rate per year?

$3.24* The Cadillac Motor Car Company is advertising a 24-month lease of a Cadillac Deville for $470 payable at the end of each month. The lease requires a $2200 down payment, plus a $500 refundable security deposit. As an alternative, the company offers a 24-month lease with a single up-front payment of $11,970, plus a $500 refundable security deposit. The security deposit will be refunded at the end of the 24-month lease. Assuming an interest rate of 6%, compounded monthly, which lease is the preferred one?

$3.25 As a typical middle class consumer, you are making monthly payments on your home mortgage (9% annual interest rate), car loan (12%), home improvement loan (14%), and past due charge accounts (18%). Immediately after getting a $100 monthly raise, your friendly mutual fund broker tries to sell you some investment funds, with a guaranteed return of 10% per year. Assuming that your only other investment alternative is a savings account, should you buy?

3.26 What will be the amount accumulated by each of these present investments?

    (a) $2455 in 10 years at 6%, compounded semiannually,

    (b) $5500 in 15 years at 8%, compounded quarterly,

    (c) $21,000 in 7 years at 9%, compounded monthly.

3.27 How many years will it take an investment to triple itself if the interest rate is 9% compounded

    (a) Quarterly,

    (b) Monthly,

    (c) Continuously?

3.28* A series of equal quarterly payments of $4000 each for 12 years is equivalent to what present amount at an interest rate of 9%, compounded as follows:

    (a) Quarterly,

    (b) Monthly,

    (c) Continuously?

3.29 What is the future equivalent of an equal payment series of $5000 each year for 5 years if the interest rate is 6%, compounded continuously?

$3.30 Suppose that $500 is placed in a bank account at the end of each quarter over the next 20 years. What is the future worth at the end of 20 years when the interest rate is 8%, compounded as follows:

(a) Quarterly,
(b) Monthly,
(c) Continuously?

3.31 A series of equal quarterly deposits of $1000 extends over a period of 3 years. It is desired to compute the future worth of this quarterly deposit series at 12%, compounded monthly. Which of the following equations is correct?

(a) $F = 4(\$1000)(F/A, 12\%, 3)$,
(b) $F = \$1000(F/A, 3\%, 12)$,
(c) $F = \$1000(F/A, 1\%, 12)$,
(d) $F = \$1000(F/A, 3.03\%, 12)$.

3.32 If the interest rate is 7.25%, compounded continuously, what is the required quarterly payment to repay a loan of $10,000 in 4 years?

3.33 What is the future equivalent of a series of equal monthly payments of $2000, if the series extends over a period of 6 years at 12% interest, compounded as follows:

(a) Quarterly,
(b) Monthly,
(c) Continuously?

3.34 What will be the required quarterly payment to repay a loan of $50,000 in 5 years, if the interest rate is 8%, compounded continuously?

3.35 A series of equal quarterly payments of $500 extends over a period of 5 years. What is the present equivalent of this quarterly payment series at 9.75% interest, compounded continuously?

$3.36 Suppose you deposit $1000 at the end of each quarter for 5 years at an interest rate of 8%, compounded continuously. What equal end-of-year deposit over 5 years would accumulate the same amount at the end of 5 years under the same interest compounding? Choose one of the following answers.

(a) $A = [\$1000(F/A, 2\%, 20)] \times (A/F, 8\%, 5)$,
(b) $A = \$1000(F/A, e^{0.02} - 1, 4)$,
(c) $A = \$1000(F/A, e^{0.02} - 1, 20) \times (A/F, e^{0.08} - 1, 5)$
(d) None of the above.

3.37 A series of equal quarterly payments of $3000 for 15 years is equivalent to what future lump-sum amount at the end of 10 years at an interest rate of 8%, compounded continuously?

3.38* What is the future equivalent of the following series of payments?

(a) $1500 at the end of each 6-month period for 10 years at 6%, compounded semi-annually,
(b) $2500 at the end of each quarter for 6 years at 8%, compounded quarterly,
(c) $3000 at the end of each month for 14 years at 9%, compounded monthly.

3.39 What equal series of payments must be paid into a sinking fund to accumulate the following amount?

(a) $12,000 in 10 years at 6%, compounded semi-annually, when payments are semi-annual,
(b) $7000 in 15 years at 9%, compounded quarterly, when payments are quarterly,
(c) $34,000 in 5 years at 7.55%, compounded monthly, when payments are monthly?

$3.40 James Hogan is purchasing a $24,000 automobile, which is to be paid for in 48 monthly installments of $583.66. What effective annual interest is being paid for this financing arrangement?

$3.41 A loan of $12,000 is to be financed to assist in buying an automobile. Based upon monthly compounding for 30 months, the end-of-the-month equal payment is quoted as $465.25. What nominal interest rate is being charged?

$3.42 You are purchasing a $9000 used automobile, which is to be paid for in 36 monthly installments of $288.72. What nominal interest rate is being paid on this financing arrangement?

$3.43 Suppose a newlywed couple is planning to buy a home 2 years from now. To save the down payment required at the time of purchasing a home worth $220,000 (let's assume the down payment is 10% of the sales price, $22,000), they have decided to set aside some money from their salaries at the end of each month. If they can earn 6% interest (compounded monthly) on their savings, determine the equal amount this couple must deposit each month so that they can buy the home in 2 years.

3.44 What is the present equivalent of the following series of payments?

    (a) $300 at the end of each 6-month period for 10 years at 8%, compounded semi-annually,

    (b) $1500 at the end of each quarter for 5 years at 8%, compounded quarterly,

    (c) $2500 at the end of each month for 8 years at 9%, compounded monthly?

$3.45* What is the amount of the quarterly deposits, *A*, such that you will be able to withdraw the amounts shown in the cash flow diagram, if the interest rate is 8%, compounded quarterly?

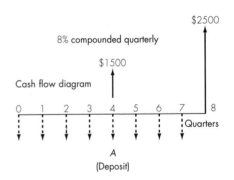

$3.46 Georgi Rostov deposits $3000 in a savings account that pays 6% interest, compounded monthly. Three years later he deposits $3500. Two years after the $3500 deposit, he makes another deposit in the amount of $2500. Four years after the $2500 deposit, half of the accumulated funds is transferred to a fund that pays 8% interest, compounded quarterly. How much money will be in each account 6 years after the transfer?

$3.47 A man is planning to retire in 25 years. He wishes to deposit a regular amount every 3 months until he retires so that, beginning 1 year following his retirement, he will receive annual payments of $45,000 for the next 10 years. How much must he deposit if the interest rate is 8%, compounded quarterly?

3.48 A building is priced at $125,000. If a down payment of $25,000 is made, and a payment of $1000 every month thereafter is required, how many months will it take to pay for the building? Interest is charged at a rate of 9%, compounded monthly.

$3.49 You obtained a loan of $20,000 to finance an automobile. Based on monthly compounding over 60 months, the end-of-the-month equal payment was figured to be $422.48. What is the APR used for this loan?

$3.50 *The Engineering Economist* (a professional journal) offers three types of subscriptions payable in advance: 1 year at $20, 2 years at $38, and 3 years at $56. If money can earn 6%, compounded monthly, which subscription should you take? (Assume that you plan to subscribe to the journal over the next 3 years.)

$3.51 A couple is planning to finance their 3-year-old son's college education. Money can be deposited at 6%, compounded quarterly. What quarterly deposit must be made after the son's 3rd birthday to his 18th birthday to provide $50,000 on each birthday from the 18th to the 21st? (Note that the last deposit is made on the date of the first withdrawal.)

$3.52 Sam Salvetti is planning to retire in 15 years. Money can be deposited earning 8%, compounded quarterly. What quarterly deposit must be made at the end of each quarter until he retires so that he can make a withdrawal of $25,000 semiannually over the 5 years after his retirement? Assume that his first withdrawal occurs at the end of 6 months after his retirement.

$3.53 Emily Lacy received $250,000 from an insurance company after her husband's death. Emily wants to deposit this amount in a savings account which earns interest at a rate of 6%, compounded monthly. Then, she would like to make 60 equal monthly withdrawals over the 5-year deposit period, such that, when she makes the last withdrawal, the savings accounts will have a balance of zero. How much can she withdraw each month?

3.54 Anita Tahani, who owns a travel agency, bought an old house to use as her business office. She found that the ceiling was poorly insulated and that the heat loss could be cut significantly if 6 inches of foam insulation were installed. She estimated that with the insulation she could cut the heating bill by $40 per month and the air conditioning cost by $25 per month. Assuming that the summer season is 3 months (June, July, August) of the year and that the winter season is another 3 months (December, January, and February) of the year, how much can she spend on insulation if she expects to keep the property for 3 years? Assume that neither heating nor air conditioning would be required during the fall and spring seasons. If she decides to install the insulation, it will be done at the beginning of May. Anita's interest rate is 9%, compounded monthly.

3.55* A new chemical production facility, which is under construction, is expected to be in full commercial operation 1 year from now. Once in full operation, the facility will generate $55,000 cash profit daily over the plant service life of 12 years. Determine the present equivalent of the future cash flows generated by the facility at the beginning of commercial operation, assuming

(a) 12% interest, compounded daily, with the daily flows,
(b) 12% interest, compounded continuously, with the daily flow series approximated by a uniform continuous cash flow function.

Also compare the difference between (a) discrete (daily) and (b) continuous compounding above.

3.56 Income from a project is projected to decline at a constant rate from an initial value of $500,000 at time 0 to a final value of $40,000 at the end of year 3. If interest is compounded continuously at a nominal annual rate of 11%, determine the present equivalent of this continuous cash flow.

3.57 A sum of $100,000 will be received uniformly over a 5-year period beginning 2 years from today. What is the present equivalent of this deferred funds flow if interest is compounded continuously at a nominal rate of 9%?

3.58* Consider the cash flow diagram, which represents three different interest rates applicable over the 5-year time span.

(a) Calculate the equivalent amount P at the present.
(b) Calculate the single-payment equivalent to F at $n = 5$.
(c) Calculate the equal-payment-series cash flow A that runs from $n = 1$ to $n = 5$.

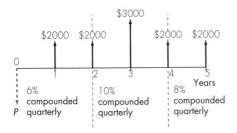

3.59 Consider the cash flow transactions depicted in the cash flow diagram, with the changing interest rates specified.

(a) What is the present equivalent? (In other words, how much do you have to deposit now, so that you can withdraw $300 at the end of year 1, $300 at the end of year 2, $500 at the end of year 3, and $500 at the end of year 4)?
(b) What is the single effective annual interest rate over 4 years?

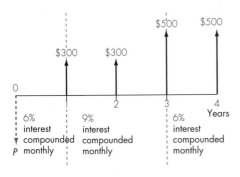

3.60 Compute the future equivalent for the cash flows with the different interest rates specified.

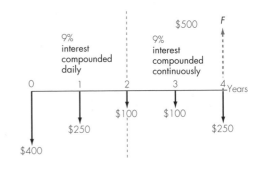

3.61 An automobile loan of $15,000 at a nominal rate of 9%, compounded monthly for 48 months, requires equal end-of-month payments of $373.28. Complete the following table for the first six payments, as you would expect a bank to calculate the values:

End of Month (n)	Interest Payment	Repayment of Principal	Remaining Loan Balance
1			$14,739.22
2			
3		$264.70	
4	$106.59		
5	104.59		
6			13,405.71

$3.62 Mr. Smith wants to buy a new car which will cost $12,500. He will make a down payment in the amount of $2500. He would like to borrow the remainder from a bank at an interest rate of 9%, compounded monthly. He agrees to pay the loan payment monthly for a period of 2 years. Select the correct answer for the following questions.

(a) What is the amount of the monthly payment (A)?

(i) $A = \$10,000(A/P, 0.75\%, 24)$,
(ii) $A = \$10,000(A/P, 9\%, 2)/12$,
(iii) $A = \$10,000(A/F, 0.75\%, 24)$,
(iv) $A = \$12,500(A/F, 9\%, 2)/12$?

(b) Mr. Smith has made 12 payments and wants to figure out the remaining balance immediately after 12th payment. What is the remaining balance?

(i) $B_{12} = 12A$,
(ii) $B_{12} = A \, (P/A, 9\%, 1)/12$,
(iii) $B_{12} = A \, (P/A, 0.75\%, 12)$,
(iv) $B_{12} = 10,000 - 12A$?

$3.63 Talhi Hafid is considering the purchase of a used automobile. The price including the title and taxes is $8260. Talhi is able to make a $2260 down payment. The balance, $6000, will be borrowed from his bank at an interest rate of 9.25%, compounded daily. The loan should be paid in 48 equal monthly payments. Compute the monthly payment. What is the total amount of interest he has to pay over the life of the loan?

$3.64 Suppose you are in the market for a new car worth $18,000. You are offered a deal to make a $1800 down payment now and to pay the balance in equal end-of-month payments, $421.85, over a 48-month period. Consider the following situations:

(a) Instead of going through the dealer's financing, you want to make a down payment of $1800 and take out an auto loan from a bank at 11.75% compounded monthly. What would be your monthly payment to pay off the loan in 4 years?

(b) If you were to accept the dealer's offer, what would be the effective rate of interest per month the dealer charges on your financing?

$3.65 Bob Pearson borrowed $12,000 from a bank at an interest rate of 12%, compounded monthly. This loan will be repaid in 36 equal monthly installments over 3 years. Immediately after his 20th payment, Bob desires to pay the remainder of the loan in a single payment. Compute the amount he must pay.

$3.66 David Kapamagian borrowed money from a bank to finance a small fishing boat. The bank's loan terms allowed him to defer payments (including interest) for 6 months and to make 36 equal end-of-month payments thereafter. The original bank note was for $4800 with an interest rate of 12%, compounded monthly. After 16 monthly payments, David found himself in a financial bind and went to a loan company for assistance in lowering his monthly payments. Fortunately, the loan company has offered to pay his debts in one lump sum, if he will pay the company $104 per month for the next 36 months. What monthly rate of interest is the loan company charging on this transaction?

$3.67* You are buying a home for $175,000. If you make a down payment of $30,000 and take out a mortgage on the rest at 8.5%, compounded semi-annually, what will be your monthly payment to retire the mortgage in 15 years?

$3.68 With a $230,000 home mortgage loan with a 20-year term at 9% APR, compounded semi-annually, compute the total payments on principal and interest over first 5 years of ownership. Assume monthly payment.

$3.69 A lender requires that monthly mortgage payments be no more than 25% of gross monthly income with a maximum term of 30 years. If you can make only a 15% down payment, what is the minimum monthly income needed to purchase a $200,000 house when the interest rate is 9%, compounded semi-annually?

$3.70 You buy a $150,000 house and you take out a 9% (APR) mortgage for $120,000. Five years later you sell the house for

$185,000 (after all other selling expenses). What equity (the amount that you can keep before any tax) would you realize with a 30-year amortization? Assume semi-annual compounding and monthly payments.

$3.71   Just before their 15th payment

- Family *A* had a balance of $80,000 on a 9%, 30-year mortgage;
- Family *B* had a balance of $80,000 on a 9%, 15-year mortgage; and
- Family *C* had a balance of $80,000 on a 9%, 20-year mortgage.

How much interest did each family pay on their 15th monthly payment?

3.72   Home mortgage lenders in the United States usually charge points on a loan to avoid exceeding a legal limit on interest rates or to be competitive with other lenders. As an example for a 2-point loan, the lender would loan only $98 for each $100 borrowed. The borrower would receive only $98, but would have to make payments just as if he or she had received $100. Suppose that you receive a loan of $130,000 payable at the end each month for 30 years with an interest rate of 9%, compounded monthly (not semi-annually as in Canada), but you have been charged 3 points. What is the effective interest rate on this home mortgage loan?

3.73   A restaurant is considering purchasing a lot adjacent to its business to provide adequate parking space for its customers. The restaurant needs to borrow $35,000 to secure the lot. A deal has been made between a local bank and the restaurant, so that the restaurant would pay the loan back over a 5-year period with the following payment terms: 15%, 20%, 25%, 30%, and 35% of the initial loan at the end of first, second, third, fourth, and fifth years, respectively.

(a) What rate of interest is the bank earning from this loan transaction?
(b) What would be the total interest paid by the restaurant over the 5-year period?

$3.74   Don Harrison's current salary is $40,000 per year, and he is planning retirement 25 years from now. He anticipates that his annual salary will increase by $2500 each year (first year, $40,000, second year, $42,500, third year, $45,000, and so forth), and he plans to make an annual deposit in the amount of 5% of his yearly salary into a retirement fund that earns 7% interest, compounded daily. What will be the amount accumulated at the time of his retirement?

$3.75*   Katerina Unger wants to purchase a set of furniture worth $3000. She plans to finance the furniture for 2 years. The furniture store tells Katerina that the interest rate is only 1% per month, and her monthly payment is computed as follows:

- Installment period = 24 months,
- Interest = 24(0.01)($3000) = $720,
- Loan processing fee = $25,
- Total amount owed = $3000 + $720 + $25 = $3745,
- Monthly payment = $3745/24 = $156.04 per month.

(a) What is the annual effective interest rate, i_a, Katerina is paying for her loan transaction? What is the nominal interest (annual percentage rate) for this loan, based on monthly compounding?
(b) Katerina bought the furniture and made 12 monthly payments. Now she wants to pay off the remaining installments in one lump sum payment (at the end of 12 months). How much does she owe the furniture store?

$3.76 You are to purchase a piece of furniture worth $5000 on credit through a local furniture store. You are told that your monthly payment will be $146.35 at a 10% add-on interest rate over 48 months. After making 15 payments, you have decided to pay off the balance. Compute the remaining balance based on the conventional amortized loan.

$3.77* Paula Hunt bought a new car for $18,400. A dealer's financing was available through a local bank at an interest rate of 13.5%, compounded monthly. Dealer financing required a 10% down payment and 48 equal monthly payments. Because the interest rate was rather high, she checked her credit union for possible financing. The loan officer at the credit union quoted on 10.5% interest for a new car loan and 12.25% for a used car. But to be eligible for the loan, Paula has to be a member of the union at least for 6 months. Since she joined the union 2 months ago, she had to wait 4 more months to apply for the loan. Now she decided to go ahead with the dealer's financing, and 4 months later refinance the balance through the credit union for 44 months at an interest rate of 12.25%.

(a) Compute the monthly payment to the dealer.
(b) Compute the monthly payment to the union.
(c) What is the total interest payment for each loan transaction?

$3.78 A house can be purchased for $85,000, and you have $17,000 cash for a down payment. You are considering the following two financing options:

- Option 1: Getting a new mortgage with a 10% (APR) interest, and a 30-year term.
- Option 2: Assuming the seller's old mortgage, which has an interest rate of 8.5% (APR), a remaining term of 25 years (original term of 30 years), a remaining balance of $35,394, and payments of $281.51 per month. You can obtain a second mortgage for the remaining balance ($32,606) from your bank at 12% (APR), with a 10-year repayment period.

(a) What is the effective interest rate of the combined mortgage in option 2?
(b) Compute the monthly payments for each option over the mortgage life.
(c) Compute the total interest payment for each option.
(d) What homeowner's interest rate makes the two financing options equivalent?

3.79* A loan of $10,000 is to be financed over a period of 24 months. The agency quotes a nominal rate of 8% for the first 12 months and a nominal rate of 9% for any remaining unpaid balance after 12 months, compounded monthly. Based on these rates, what equal end-of-the-month payment for 24 months would be required to repay the loan with interest?

$3.80 Consider a short-term loan that Robert Carre got recently from the Toronto Dominion Bank. The amount was $12,000, interest rate was 12% compounded monthly, and the term of the loan was for five years. He has been making monthly payments based on the payment schedule specified by the bank. Now he has learned that an Internet-based financial organization is offering loans with interest rates as low as 7% compounded monthly. If Robert is to switch to the Internet-based company right after his 14th monthly payment, what is the amount that he has to borrow from the Internet-based company? If his monthly payment to the Internet-based company is the same as the amount he paid the TD Bank, how long will it take him to pay off the loan?

$3.81 If you borrow $120,000 with a 30-year term, 9% (APR) variable rate compounded monthly, and the interest rate can be changed every 5 years,

(a) What is the initial monthly payment?
(b) If at the end of 5 years, the lender's interest rate is 9.75% (APR), what will the new monthly payments be?

$3.82 The Jimmy Corporation issued a new series of bonds on January 1, 1991. The bonds were sold at par ($1000), have a 12% coupon rate, and mature in 30 years, on December 31, 2020. Coupon interest payments are made semiannually (on June 30 and December 31). An investor has been interested in these bonds.

(a) What was the yield-to-maturity (YTM) of the bond if the investor bought it on January 1, 1991?
(b) Assume that the level of interest rate fell to 9% on January 1, 1998. This meant that the coupon rate of newly issued bonds with quality similar to the Jimmy Corporation's bond was set at 9% with interest paid semiannually. Assume such newly issued bonds could be bought at par ($1000) and had the same maturity as the Jimmy Corporation's bond. What was the YTM of these newly issued bonds? What was the maximum amount that the investor was willing to pay for the Jimmy Corporation's bond on that date?
(c) On July 1, 1998, the investor bought the bonds for $922.38. What was the YTM at that date? What was the current yield at that date?

$3.83 A $1000, 9.50% semiannual bond is purchased for $1010. If the bond is sold at the end of 3 years and six interest payments, what should the selling price be to yield a 10% return on your investment?

$3.84 Mr. Gonzalez wishes to sell a bond that has a face value of $1000. The bond bears an interest rate of 8% with bond interests payable semiannually. Four years ago, $920 was paid for the bond. At least a 9% return (yield) in investment is desired. What must be the minimum selling price?

$3.85 Candi Yamaguchi is considering purchasing a 6% bond with a face value of $1000 with interest paid semi-annually. She desires to earn a 9% annual return on her investment. Assume that the bond will mature to its face value 5 years hence. What is the required purchasing price of the bond?

(a) $P = \$60 \ (P/A, 6\%, 5)$
    $+ \$1000 \ (P/F, 6\%, 5)$
(b) $P = \$90 \ (P/A, 9\%, 5)$
    $+ \$1000 (P/F, 9\%, 5)$
(c) $P = \$30 \ (P/A, 4.5\%, 10)$
    $+ \$1000(P/F, 4.5\%, 10)$
(d) $P = \$30 \ (P/A, 3\%, 10)$
    $+ \$1000(P/F, 3\%, 10)$.

$3.86 Suppose you have the choice of investing in (1) a zero coupon bond which costs $513.60 today, pays nothing during its life, and then pays $1000 after 5 years; or (2) a bond which costs $1000 today, pays $113 in interest semiannually, and matures at par ($1000) at the end of 5 years. Which bond would provide the higher yield?

$3.87 Suppose you were offered a 12-year, 15% coupon, $1000 par value bond at a price of $1298.68. What rate of interest (yield-to-maturity) would you earn if you bought the bond and held it to maturity (semi-annual interest)?

3.88 The Diversified Products Company has two bond issues outstanding. Both bonds pay $100 semiannual interest plus $1000 at maturity. Bond A has a remaining maturity of 15 years, and bond B a maturity of 1 year. What will be the value of each of these bonds now, when the going rate of interest is 9%?

3.89 The AirJet Service Company's bonds have 4 years remaining to maturity. Interest is paid annually; the bonds have a $1000 par value; and the coupon interest rate is 8.75%.

(a) What is the yield-to-maturity at a current market price of $1108?

(b) Would you pay $935 for one of these bonds if you thought that the market rate of interest was 9.5%?

3.90 Suppose Ford sold at par an issue of bonds with a 15-year maturity, a $1000 par value, a 12% coupon rate, and semiannual interest payments.

(a) Two years after the bonds were issued, the going rate of interest on bonds such as these fell to 9%. At what price would an investor buy the Ford bonds?

(b) Suppose that, 2 years after the issue, the going interest rate had risen to 13%. At what price would an investor buy the Ford bonds?

(c) Today, the closing price of this bond is $783.58. What is the current yield?

$3.91 Jim Norton, an engineering student, has received two credit card applications in the mail from two different banks. Each bank offers a different annual fee and finance charge.

Terms	Bank A	Bank B
Annual fee	$20	$30
Finance charge	1.55% monthly interest rate	16.5% compounded monthly

Jim expects his average monthly balance after payment to be $300 and plans to keep the card he chooses for only 24 months (after graduation he will apply for a new card). Jim's interest rate (on his savings account) is 6%, compounded daily.

(a) Compute the effective annual interest rate for each card.

(b) Which bank's credit card should Jim choose?

3.92 A small chemical company, a producer of an epoxy resin, expects its production volume to decay exponentially according to the relationship

$$y_t = 5e^{-0.25t},$$

where y_t is the production rate at time t. Simultaneously, the unit price is expected to increase linearly over time at the rate of

$$u_t = \$55(1 + 0.09t).$$

What is the expression for the present worth of sales revenues from $t = 0$ to $t = 20$ at 12% interest, compounded continuously?

Level 3

$3.93 You are considering buying a new car worth $15,000. You can finance the car by either withdrawing cash from your savings account, which earns 8% interest, or by borrowing $15,000 from your dealer for 4 years at 11%. You could earn $5635 in interest from your savings account for 4 years if you left the money in the account. If you borrow $15,000 from your dealer, you only pay $3609 in interest over 4 years, so it makes sense to borrow for your new car and keep your cash in your savings account. Do you agree or disagree with the statement above? Justify your reasoning with a numerical calculation.

$3.94 The following is the actual promotional pamphlet prepared by Trust Company:

"Lower your monthly car payments as much as 48%." Now you can buy the car you want and keep the monthly payments as much as 48% lower than they would be if you financed with a conventional auto loan. Trust Company's *Alternative Auto Loan* (AAL)ᔆᔐ makes the difference. It combines the lower monthly payment advantages of leasing with tax and ownership of a conventional loan. And if you have your monthly payment deducted automatically from your Trust Company checking account, you will save 1/2% on your loan interest rate. Your monthly payments can be spread over 24, 36 or 48 months.

Amount Financed	Financing Period (months)	Monthly Payment	
		Alternative Auto Loan	Conventional Auto Loan
	24	$249	$477
$10,000	36	211	339
	48	191	270
	24	498	955
$20,000	36	422	678
	48	382	541

The amount of the final payment will be based on the residual value of the car at the end of the loan. Your monthly payments are kept low because you make principal payments on only a portion of the loan and not on the residual value of the car. Interest is computed on the full amount of the loan. At the end of the loan period you may:

1. Make the final payment and keep the car.
2. Sell the car yourself, repay the note (remaining balance), and keep any profit you make.
3. Refinance the car.
4. Return the car to Trust Company in good working condition and pay only a return fee.

So, if you've been wanting a special car, but not the high monthly payments that could go with it, consider the *Alternative Auto Loan*. For details, ask at any Trust Company branch.

Note 1: The chart above is based on the following assumptions: Conventional auto loan 13.4% annual percentage rate; *Alternative Auto Loan* 13.4% annual percentage rate.

Note 2: The residual value is assumed to be 50% of sticker price for 24 months; 45% for 36 months. The amount financed is 80% of sticker price.

Note 3: Monthly payments are based on principal payments equal to the depreciation amount on the car and interest in the amount of the loan.

Note 4: The residual value of the automobile is determined by a published residual value guide in effect at the time your Trust Company's *Alternative Auto Loan* is originated.

Note 5: The minimum loan amount is $10,000 (Trust Company will lend up to 80% of the sticker price). Annual household income requirement is $50,000.

Note 6: Trust Company reserves the right of final approval based on customer's credit history. Offer may vary at other Trust Company banks in Georgia.

(a) Show how the monthly payments were computed for the *Alternative Auto Loan*.

(b) Suppose that you have decided to finance a new car for 36 months from Trust Company. Assume also that you are interested in owning the car (not leasing it). If you decide to go with the *Alternative Auto Loan*, you will make the final payment and keep the car at the end of 36 months. Assume that your opportunity cost rate (personal interest rate) is an interest rate of 8%, compounded monthly. (You may view this opportunity cost rate as an interest rate at which you can invest your money in some financial instruments such as a savings account.) Compare this alternative option with the conventional option and determine your choice.

$3.95 Suppose you are going to buy a home worth $110,000 and you make a down payment in the amount of $50,000. The balance will be borrowed from the Capital Savings and Loan Bank. The loan officer offers the following two loan financing plans rather than a mortgage:

• Option 1: A conventional fixed loan at an interest rate of 13% compounded monthly over 30 years with 360 equal monthly payments.

• Option 2: A graduated payment schedule loan at 11.5% interest compounded monthly with the following monthly payment schedule.

Year (n)	Monthly Payment	Monthly Insurance
1	$497.76	$25.19
2	522.65	25.56
3	548.78	25.84
4	576.22	26.01
5	605.03	26.06
6 – 30	635.28	25.96

For the graduated payment loan, the additional insurance charges must be paid.

(a) Compute the monthly payment for option 1.

(b) What is the effective annual interest rate you are paying for option 2?

(c) Compute the outstanding balance for each option at the end of 5 years.

(d) Compute the total interest payment for each option.

(e) Assuming that your only investment alternative is a savings account that earns an interest rate of 6%, compounded monthly, which option is a better deal?

$3.96 Ms. Kennedy borrowed $4909 from a bank to finance a car at an add-on interest rate of 6.105%. The bank calculated the monthly payments as follows:

- Contract amount = $4909,
- Contract period = 42 months,
- Add-on interest at 6.105% = $4909(0.06105)(3.5) = $1048.90,
- Acquisition fee = $25,
- Loan charge = $1048.90 + $25 = $1073.90,
- Total of payments = $4909 + 1073.90 = $5982.90,
- Monthly installment = $5982.90/42 = $142.45.

After making the 7th payment, Ms. Kennedy wants to pay off the remaining balance. The following is the letter from the bank explaining the net balance Ms Kennedy owes:

Dear Ms. Kennedy

The following is an explanation of how we arrived at the payoff amount on your loan account.

Original note amount	$5982.90
Less 7 payments @ $142.45 each	997.15
	4985.75
Loan charge (interest)	1073.90
Less acquisition fee	25.00
	$1048.90

Rebate factor from Rule of 78ths chart is 0.6589 (loan ran 8 months on a 42 month term).

$1048.90 multiplied by 0.6589 = $691.12.

$691.12 represents the unearned interest rebate.

Therefore:

Balance	$4985.75
Less unearned interest rebate	691.12
Payoff amount	$4294.63

If you have any further questions concerning these matters, please contact us.

Sincerely,

S. Govia
Vice President

(a) Compute the effective annual interest rate for this loan.

(b) Compute the annual percentage rate (APR) for this loan.

(c) Show how would you derive the rebate factor (0.6589).

(d) Verify the payoff amount using the Rule of 78ths formula.

(e) Compute the payoff amount using the interest factor $(P/A, i, N)$.

Hint: The Rule of 78ths is used by some financial institutions to determine the outstanding loan balance. According to the Rule of 78ths, the interest charged during a given month is figured out by applying a changing fraction to the total interest over the loan period. For example, in the case of a 1-year loan, the fraction used in figuring the interest charge for the first month would be 12/78, 12 being the number of remaining months of the loan, and 78 being the sum of $1 + 2 + ... + 11 + 12$. For the second month, the fraction would be 11/78, and so on. In the case of 2-year loan, the fraction during the first month is 24/300, because there are 24 remaining payment periods, and the sum of the loan periods is $300 = 1 + 2 + ... + 24$.

CHAPTER 4

Analysis of Independent Investments

Five hundred miles from nowhere it'll give you a cold drink or a warm burger — Home refrigeration has come a long way since the days of the ice box and the block of ice, but, when we travel, we go back to the sloppy ice cooler with its soggy and sometimes spoiled food. No more! Now for the price of a good cooler and one or two seasons of buying ice (about the cost of five family restaurant meals), all the advantages of home cooling are available to you electronically and conveniently.[1] On motor trips, plug your Koolatron into your cigarette lighter; it uses less power than a tail light. If you decide to carry it to a picnic place or a fishing hole, the Koolatron will hold its cooling capacity for 24 hours. If you leave it plugged into your battery with the engine turned off, the Koolatron draws only 3 amps of current.

---

[1] Source: *USA Today*. May 11, 1994. (A featured promotional advertisement by Comtrad Industries).

Because Comtrad markets this product directly to consumers, you save the cost of distributors and retail mark-ups. This advanced, portable Koolatron refrigerator is available at the introductory price of $99.

Thermoelectric temperature control has now been proven after more than 25 years of use in some of the most rigorous space and laboratory applications. And Comtrad is the first manufacturer to make this technology available to families, fishermen, boaters, campers, and hunters—in fact to anyone on the move. On what basis did Comtrad decide to market the product? Certainly, we have no idea of how much Comtrad invested in developing and manufacturing the Koolatron. However, Comtrad management should have considered the following important issues: (1) How much additional investment in plant and equipment will be required to manufacture the product? (2) How long does Comtrad have to wait to recover its initial investment? and (3) Will Comtrad ever make a profit by selling the product at $99?

In Chapters 2 and 3, we presented the concept of the time value of money and developed techniques for establishing cash flow equivalence with compound interest factors. This background provides a foundation for accepting or rejecting a capital investment—the economic evaluation of a project's desirability.

There are four common measures of investment attractiveness or profitability. Three of these are based directly on cash flow equivalence: (1) equivalent present worth, (2) equivalent future worth, and (3) equivalent annual worth. Present worth represents a measure of future cash flows relative to the time point "now" with provisions that account for earning opportunities. Future worth is a measure of cash flows at some future planning horizon, which offers a consideration of the earning opportunities of intermediate cash flows. Annual worth is a measure of cash flows in terms of equivalent and equal cash flows occurring on an annual basis.

The fourth measure is rate of return which characterizes profitability or attractiveness as a calculated yield, interest rate, or rate of return generated by the investment. The apparent simplicity of this concept makes it the preferred choice in many organizations. However, the calculation can involve certain complications which cannot be ignored. If applied properly all four methods *must* and do provide identical conclusions with respect to the attractiveness of any given investment.

Our treatment of measures of investment attractiveness extends over three chapters. In Chapter 4, we define independent investments and distinguish these from mutually exclusive investments. Payback period is introduced as a useful project screening tool which provides no true measure of profitability. The remainder of this chapter is devoted to the development of four rigorous measures of investment attractiveness and the application of these to independent investments. Chapter 5 applies these same measures in the selection of the best project from a group of mutually exclusive projects. Chapter 6 discusses a number of special applications of the evaluation techniques developed in the two preceding chapters.

We must also recognize that one of the most important parts of the capital budgeting process is the estimation of relevant cash flows. For all examples in this chapter, and those in Chapters 5 and 6, however, net cash flows can be viewed as before-tax values or after-tax values for which tax effects have been incorporated. Since some organizations (e.g., governments and nonprofit organizations) are not subject to tax, the before-tax situation provides a valid base for this type of economic evaluation. Taking this view will allow us to focus on our main area of concern, the economic evaluation of investment projects. The procedures for determining after-tax net cash flows in taxable situations are developed in Chapter 9.

4.1 Describing Project Cash Flows

In Section 1.3, we described many engineering economic decision problems, but we did not provide suggestions on how to actually solve them. What do Examples 1.1 through 1.6 have in common? Note that all these problems involve two dissimilar types of amounts. First, there is the investment, which is usually made in a lump sum at the beginning of the project. Although not literally made "today," the investment is made at a specific point in time that, for analytical purposes, is called today, or time 0. Second, there is a stream of cash benefits that are expected to result from this investment over a period of future years.

An investment made in a fixed asset is similar to an investment made by a bank when it lends money. The essential characteristic of both transactions is that funds are committed today in the expectation of their earning a return in the future. In the case of the bank loan, the future return takes the form of interest plus repayment of the principal. This is known as the **loan cash flow**. In the case of the fixed asset, the future return takes the form of profits generated by productive use of the asset. As shown in Figure 4.1, the representation of these future earnings, along with the capital expenditures, and annual expenses (such as wages, raw materials, operating costs, maintenance costs, and income taxes), is the **project cash flow**. This similarity between the loan cash flow and the project cash flow brings us to an important conclusion—i.e., we can use the same equivalence techniques developed in Chapter 2 to measure economic worth. Example 4.1 illustrates a typical procedure for obtaining a project's cash flows.

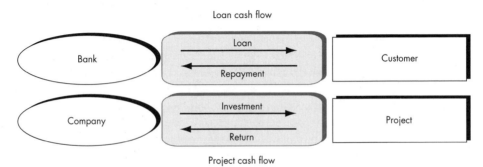

Figure 4.1 A bank loan versus an investment project

EXAMPLE 4.1 Identifying project cash flows

XL Chemicals is considering the installation of a computer process control system in one of its process plants. This plant is used about 40% of the time or 3500 operating hours per year to produce a proprietary demulsification chemical, and during the remaining 60% of the time it is used to produce other specialty chemicals. The annual production of the demulsification chemical amounts to 30,000 kilograms per year and it sells for $15 per kilogram. The proposed computer process control system will cost $650,000 and is expected to provide specific benefits in the production of the demulsification chemical. First, the selling price of the product could be increased by $2 per kilogram because the product will be of higher purity, which translates into better demulsification performance. Secondly, production volumes will increase by 4000 kilograms per year as a result of higher reaction yields without any increase in raw material quantities or production time. Finally, the number of process operators can be reduced by one per shift, which represents a savings of $25 per hour. The new control system would result in additional maintenance costs of $53,000 per year and has an expected useful life of 8 years. While the system is likely to provide similar benefits in the production of the other specialty chemicals manufactured in the process plant, these have not been quantified as yet.

Discussion: As in real-world engineering economic analysis, this problem contains a great deal of data from which we must extract and interpret critical cash flows. You would be wise to begin organizing your solution into a table consisting of the following categories.

Year (n)	Cash Inflows (Benefits)	Cash Outflows (Costs)	Net Cash Flows
0			
1			
⋮			
8			

Solution

Given: Cost and benefit information as stated above
Find: Net cash flow in each year over the life of the new system

Although we could assume similar benefits are derivable from the production of the other specialty chemicals, let's restrict our consideration to the demulsification chemical and allocate the full initial cost of the control system and the annual maintenance costs to this chemical product. (You could logically argue that only 40% of these costs belong to this production activity.) The gross benefits are the additional revenues realized from the increased selling price and the extra production, as well as the cost savings resulting from having one less operator. Revenues from the price increase are 30,000 kilograms per year × $2/kilogram or $60,000 per year. The added production volume at the new pricing adds revenues of 4000 kilograms per year × $17 per kilogram or $68,000 per year. The elimination of one operator results in annual savings of 3,500 operating hours per year × $25 per hour or $87,500 per year. The net benefits in each of the 8 years are the gross benefits less the maintenance costs ($60,000 + $68,000 + $87,500 − $53,000) = $162,500 per year. Now we are ready to summarize a cash flow table as follows:

Year (n)	Cash Inflows (Benefits)	Cash Outflows (Costs)	Net Cash Flows
0	0	$650,000	−$650,000
1	215,500	53,000	162,500
2	215,500	53,000	162,500
⋮	⋮	⋮	⋮
8	215,500	53,000	162,500

Comments: In Example 4.1, if the company purchases the computer process control system for $650,000 now, it could expect an annual net cash flow of $162,500 for 8 years. (Note that these savings occur in discrete lumps at the ends of years.) We also considered only the benefits associated with the production of the demulsification chemical. We could quantify some benefits attributable to the production of the other chemicals from this plant. Suppose that the demulsification chemical benefits alone justify the acquisition. Then, it is obvious that, had benefits deriving from the other chemicals been considered, acquisition would have been even more clearly justified.

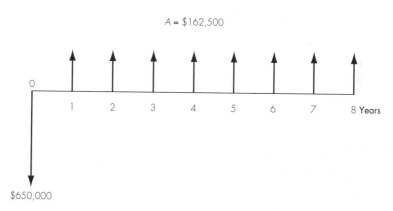

Figure 4.2   Cash flow diagram for the computer process control project described in Example 4.1

We can draw a cash flow diagram for this situation as shown in Figure 4.2. Assuming these cost savings and cash flow estimates are correct, should management give the go-ahead for installation of the system? If management decides not to purchase the computer process control system, what should it do with the $650,000 (assuming they have this amount in the first place)? The company could buy $650,000 of Treasury bills, or it could invest the amount in other revenue generating and/or cost saving projects. These are the types of questions this chapter is designed to help you answer.

4.1.1   Independent Investment Projects

Most firms will have a number of unrelated investment opportunities available. For example, in the case of XL Chemicals other projects being considered in addition to the computer process control project in Example 4.1 are a new waste heat recovery boiler, a CAD system for the engineering department, and a new warehouse. The economic attractiveness of each of these projects can be measured and an accept or reject decision made without reference to any of the other projects. The decision on any one project has no effect on the decision made on another project. Such projects are said to be **independent**.

In Chapter 5, we will see that in many engineering situations we are faced with selecting the most economically attractive project from a number of alternative projects which all resolve the same problem or meet the same need. It is unnecessary to choose more than one project in this situation and acceptance of one automatically rejects all of the others. Such projects are said to be **mutually exclusive**.

As long as the total cost of all the independent projects found to be economically attractive is less than the investment funds available to a firm, then all of these projects could proceed. However, this is rarely the case. The selection of those projects which should proceed when investment funds are limited is the subject of capital budgeting.

Apart from Chapter 10, which deals with capital budgeting, funds availability will not be a consideration in accepting or rejecting projects. Our discussion of economic evaluation methods in Chapter 4 is restricted to independent projects.

4.2 Initial Project Screening Method

Let's suppose that you are in the market for a new punch press for your company's machine shop, and you visit an equipment dealer. As you give a serious look at one of the punch press models in the display room, an observant equipment sales person approaches you and says, "That press you are looking at is state-of-the-art in its category. If you buy that top of the line model, it will cost a little bit more, but it will pay for itself in less than 2 years."

Before studying the four measures of investment attractiveness, we will review a simple, but nonrigorous, method commonly used to screen capital investments. One of the primary concerns of most business people is whether, and when, the money invested in a project can be recovered. The **payback method** screens projects on the basis of how long it takes for net receipts to equal investment outlays. This calculation can take one of two forms by either including time value of money considerations or ignoring them. The former case is usually designated as the **discounted payback method**.

A common standard used to determine whether or not to pursue a project is that a project does not merit consideration unless its payback period is shorter than some specified period of time. (This time limit is largely determined by management policy. For example, a high-tech firm, such as a computer manufacturer, would set a short time limit for any new investment because high-tech products rapidly become obsolete.) If the payback period is within the acceptable range, a formal project evaluation (such as a present worth analysis) may begin. It is important to remember that **payback screening** is not an *end* in itself, but rather a method of screening out certain obvious unacceptable investment alternatives before progressing to an analysis of potentially acceptable ones.

4.2.1 Payback Period—The Time It Takes to Pay Back

Determining the relative worth of new production machinery by calculating the time it will take to pay back what it cost is the single most popular method of project screening. If a company makes investment decisions based solely on the payback period, it considers only those projects with a payback period *shorter* than the maximum acceptable payback period. (However, due to shortcomings of the payback screening method, which we will discuss later, it is rarely used as the only decision criterion.)

What does the payback period tell us? One consequence of insisting that each proposed investment has a short payback period is that investors can assure themselves of being restored to their initial position within a short span of time. By restoring their initial position, investors can take advantage of additional, perhaps better, investment possibilities that may come along.

EXAMPLE 4.2

Payback period for the computer process control system project

Consider the cash flows given in Example 4.1. Determine the payback period for this computer process control system project.

Solution

Given: Initial cost = $650,000, annual net benefits = $162,500
Find: Payback period

Given a uniform stream of receipts, we can easily calculate the payback period by dividing the initial cash outlay by the annual receipts:

$$\text{Payback period} = \frac{\text{Initial cost}}{\text{Uniform annual benefit}} = \frac{\$650,000}{\$162,500}$$
$$= 4 \text{ years.}$$

If the company's policy is to consider only projects with a payback period of 5 years or less, this computer process control system project passes the initial screening.

In Example 4.2, dividing the initial payment by annual receipts to determine payback period is a simplification we can make because the annual receipts are uniform. Wherever the expected cash flows vary from year to year, however, the payback period must be determined by adding the expected cash flows for each year until the sum is equal to, or greater than, zero. The significance of this procedure can be easily explained. The cumulative cash flow equals zero at the point where cash inflows exactly match or pay back the cash outflows; thus, the project has reached the payback point. Similarly, if the cumulative cash flows exceed zero, the cash inflows exceed the cash outflows, and the project has begun to generate a profit, thus exceeding its payback point. To illustrate, consider Example 4.3.

EXAMPLE 4.3 Payback period with salvage value

Autonumerics Company has just bought a new spindle machine at a cost of $105,000 to replace one that had a salvage value of $20,000. The projected annual after-tax savings via improved efficiency, which will exceed the investment cost, are as follows:

Period	Cash Flow	Cumulative Cash Flow
0	−$105,000 + $20,000	−$85,000
1	15,000	−70,000
2	25,000	−45,000
3	35,000	−10,000
4	45,000	35,000
5	45,000	80,000
6	35,000	115,000

Solution

Given: Cash flow series as shown in Figure 4.3(a)
Find: Payback period

The salvage value of retired equipment becomes a major consideration in most justification analysis. (In this example, the salvage value of the old machine should be taken into account, as the company already decided to replace the old machine.) When used, the salvage value of the retired equipment is subtracted from the purchase price of new equipment, revealing a closer true cost of the investment. As we see from the cumulative cash flow in Fig.4.3(b), the total investment is recovered during year 4. If the firm's stated maximum payback period is 4 years, the project would pass the initial screening stage.

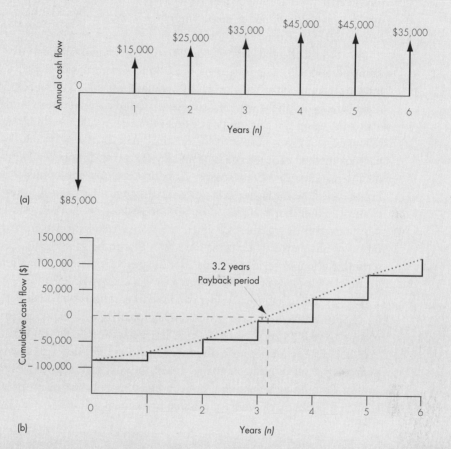

Figure 4.3   Illustration of conventional payback period (Example 4.3)

Comments: In Example 4.2, we assumed that cash flows only occur in discrete lumps at the ends of years. If cash flows occur continuously throughout the year, the payback period calculation needs adjustment. A negative balance of $10,000 remains at the start of year 4. If the $45,000 is expected to be received as a more or less continuous flow during year 3, the total investment will be recovered about two tenths ($10,000/$45,000) of the way through the 4th year. In this situation, the payback period is thus 3.2 years.

4.2.2   Benefits and Flaws of Payback Screening

The simplicity of the payback method is one of its most appealing qualities. Initial project screening by the payback method reduces the information search by focusing on that time at which the firm expects to recover the initial investment. The method may also eliminate some alternatives, thus reducing a firm's need to make further analysis efforts.

But the much-used payback method of equipment screening has a number of serious drawbacks. The principal objection to the payback method is that it fails to measure profitability, i.e., no "profit" is made during the payback period. Simply measuring how long it will take to recover the initial investment outlay contributes little to gauging the earning power of a project. (In other words, you already know that the money you borrowed for the drill press is costing you 12% per year. The payback method can't tell you how much your invested money is contributing toward the interest expense.) Because payback period analysis ignores differences in the timing of cash flows, it fails to recognize the difference between the present and future value of money. For example, although the payback for both investments can be the same in terms of numbers of years, the front-loaded investment is better because money available today is worth more than that to be gained later. Because payback screening also ignores all proceeds after the payback period, it does not allow for the possible advantages of a project with a longer economic life.

By way of illustration, consider two investment projects in Table 4.1. Each requires an initial investment outlay of $90,000. Project 1, with expected annual cash proceeds of $30,000 for the first 3 years, has a payback period of 3 years. Project 2 is expected to generate annual cash proceeds of $25,000 for 6 years; hence, its payback period is 3.6 years. If the company's maximum payback period is set to 3 years, then project 1 would pass the initial project screening, whereas project 2 would fail even though it is clearly the more profitable investment.

Table 4.1

Investment Cash Flows for Two Competing Projects

n	Project 1	Project 2
0	–$90,000	–$90,000
1	30,000	25,000
2	30,000	25,000
3	30,000	25,000
4	1000	25,000
5	1000	25,000
6	1000	25,000
	$3000	$60,000

4.2.3   Discounted Payback Period

To remedy one of the shortcomings of the payback period, we may modify the procedure to consider the time value of money, i.e., the cost of funds (interest) used to support the project. This modified payback period is often referred to as the **discounted payback period**. In other words, we may define the discounted payback period as the number of years required to recover the investment from *discounted* cash flows.

For the project in Example 4.3, suppose the company requires a rate of return of 15%. To determine the period necessary to recover both the capital investment and the cost of funds required to support the investment, we may construct Table 4.2 showing cash flows and costs of funds to be recovered over the project life. To illustrate, let's consider the cost of funds during the first year: With $85,000 committed at the beginning of the year, the interest in year 1 would be $12,750 ($85,000 × 0.15). Therefore, the total commitment grows to $97,750 but the $15,000 cash flow in year 1 leaves a net commitment of $82,750. The cost of funds during the second year would be $12,413 ($82,750 × 0.15). But with the $25,000 receipt from the project, the net commitment reduces to $70,163. When this process repeats for the remaining project years, we find that the net commitment to project ends during year 5. Depending on the cash flow assumption, the project must remain in use about 4.2 years (continuous cash flows) or 5 years (year end cash flows) in order for the company to cover its cost of capital and also recover the funds invested in the project. Figure 4.4 illustrates this relationship.

Inclusion of time value of money effects has increased the payback period calculated for this example by a year. Certainly, this modified measure is an improved one, but it does not show the complete picture of the project profitability, either.

Table 4.2
Payback Period Calculation Considering the Cost of Funds (Example 4.3)

Period	Cash Flow	Cost of Funds (15%)*	Cumulative Cash Flow
0	–$85,000	0	–$85,000
1	15,000	–$85,000(0.15) = –$12,750	–82,750
2	25,000	–$82,750(0.15) = –12,413	–70,163
3	35,000	–$70,163(0.15) = –10,524	–45,687
4	45,000	–$45,687(0.15) = –6853	–7540
5	45,000	–$ 7540(0.15) = –1131	36,329
6	35,000	$36,329(0.15) = 5449	76,778

*Cost of funds = Unrecovered beginning balance × interest rate

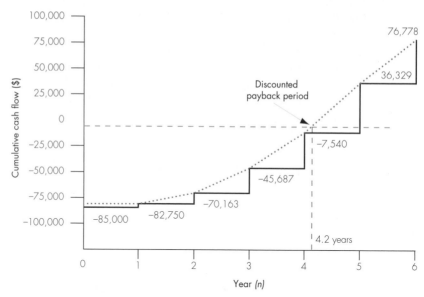

Figure 4.4 Illustration of discounted payback period

4.2.4 Where Do We Go from Here?

Should we abandon the payback methods? Certainly not. But, if you use payback screening exclusively for analysis of capital investments, look again. You may be missing something that another method can help you spot. Therefore, it is illogical to claim that payback is either a good or bad method of justification. Clearly, it is not a measure of profitability. But when it is used to supplement other methods of analysis, it can provide useful information. For example, payback can be useful when a company needs a measure of speed of cash recovery, when the company has a cash flow problem, when a product is built to last only for a short period of time, and when the machine itself is known to have a short market life.

4.3 Present Worth Analysis

Until the 1950s, the payback method was widely used as a means of making investment decisions. As flaws in this method were recognized, however, business people began to search for methods to improve project evaluations. The result was the development of **discounted cash flow techniques (DCF)**, which take into account the time value of money. One of the DCFs is the **net present worth (NPW)** or **present equivalent (PE)** method. A capital investment problem is essentially one of determining whether the anticipated cash inflows from a proposed project are sufficiently attractive to invest funds in the project. In developing the PE criterion, we will use the concept of cash flow equivalence discussed in Chapter 2. As we observed, the most convenient point at which to calculate the equivalent values is often at time 0. Under the PE criterion, the present worth of all cash inflows is compared against the present worth of all cash outflows associated with an investment project. The difference between the present worth of these cash flows, referred to as the **net present worth (NPW)** or **present equivalent (PE)**, determines whether or not the project is an acceptable investment.

4.3.1 Present Equivalent (Net Present Worth) Criterion

We will first summarize the basic procedure for applying the present worth criterion to a typical investment project.

- Determine the interest rate that the firm wishes to earn on its investments. The interest rate you determine represents the rate at which the firm can always invest the money in its **investment pool**. This interest rate is often referred to as either a **required rate of return** or a **minimum attractive rate of return** (MARR). Usually this selection is a policy decision made by top management. It is possible for the MARR to change over the life of a project, as we saw in Section 3.5, but for now we will use a single rate of interest when calculating PE.

- Estimate the service life of the project.

- Estimate the cash inflow for each period over the service life.

- Estimate the cash outflow over each service period.

- Determine the net cash flows for each period (net cash flow = cash inflow − cash outflow).

- Find the present worth of each net cash flow at the MARR. Add up these present worth figures; their sum is defined as the project's PE:

$$
\begin{aligned}
PE(i) &= \frac{A_0}{(1 + i)^0} + \frac{A_1}{(1 + i)^1} + \frac{A_2}{(1 + i)^2} + \ldots + \frac{A_N}{(1 + i)^N} \\
&= \sum_{n=0}^{N} \frac{A_n}{(1 + i)^n} \\
&= \sum_{n=0}^{N} A_n(P/F, i, n),
\end{aligned}
\tag{4.1}
$$

$$
\begin{aligned}
\text{where } PE(i) &= \text{PE calculated at } i, \\
A_n &= \text{Net cash flow at end of period } n, \\
i &= \text{MARR}, \\
N &= \text{Service life of the project.}
\end{aligned}
$$

A_n will be positive if the corresponding period has a net cash inflow or negative if there is a net cash outflow.

- In this context, a positive PE means the equivalent worth of the inflows is greater than the equivalent worth of outflows, so the project earns a rate of return greater than MARR. Therefore, if the PE(i) is positive for a single project when i is equal to MARR, the project should be accepted; if negative, it should be rejected. The decision rule is

$$\text{If } PE(\text{MARR}) > 0, \text{ accept the investment.}$$
$$\text{If } PE(\text{MARR}) = 0, \text{ remain indifferent to the investment.}$$
$$\text{If } PE(\text{MARR}) < 0, \text{ reject the investment.}$$

If MARR is the minimum attractive rate of return, why are we indifferent to a project for which $PE(\text{MARR}) = 0$? There is a number of things to consider in this regard. First, most organizations have more investment opportunities than available funds so the projects which are approved have large positive PE values. Secondly, cost and benefit estimates are always subject to uncertainty. When the PE value is zero, there is no room to offset the effect if actual costs are slightly higher or actual benefits are slightly lower than estimated. In general, an organization would be satisfied if on average, across the totality of all projects implemented, it earned MARR. The reality is that some projects will ultimately provide a return equal to or greater than that originally promised, whereas others will do considerably more poorly than promised and may well have an actual return well below MARR. Therefore, in the overall picture, a project which is at the margin $PE(\text{MARR}) = 0$, may not be particularly appealing.

Note that the decision rule is for the evaluation of independent projects where you can estimate the revenues as well as the costs associated with a project. There may be projects that cannot be avoided, e.g., the installation of pollution control equipment. In a case such as this, the project would be accepted even though its PE < 0. This type of project will be discussed in Chapter 5.

EXAMPLE 4.4 **Present equivalent—uniform flows**

Consider the investment cash flows associated with the computer process control project in Example 4.1. If the firm's MARR is 15%, compute the PE of this project. Is this project acceptable?

Solution

Given: Cash flows in Fig. 4.2, MARR = 15% per year
Find: PE

Since the computer process control project requires an initial investment of $650,000 at $n = 0$ followed by the eight equal annual cash inflows of $162,500, we can easily determine the PE as follows:

$$PE(15\%)_{\text{Outflow}} = \$650,000$$
$$PE(15\%)_{\text{Inflow}} = \$162,500(P/A, 15\%, 8)$$
$$= \$729,190.$$

Then, the PE of the project is

$$PE(15\%) = PE(15\%)_{Inflow} - PE(15\%)_{Outflow}$$
$$= \$729{,}190 - \$650{,}000$$
$$= \$79{,}190,$$

or, using Eq. (4.1)

$$PE(15\%) = -\$650{,}000 + \$162{,}500(P/A, 15\%, 8)$$
$$= \$79{,}190.$$

Since $PE(15\%) > 0$, the project would be acceptable.

Now let's consider an example where the investment cash flows are not uniform over the service life of the project.

EXAMPLE 4.5 Present equivalent—uneven flows

Tiger Machine Tool Company is considering the acquisition of a new metal cutting machine. The required initial investment of $75,000 and the projected cash benefits[2] over the 3-year's project life are as follows.

End of Year	Net Cash Flow
0	-$75,000
1	24,400
2	27,340
3	55,760

You have been asked by the president of the company to evaluate the economic merit of the acquisition. The firm's MARR is known to be 15%.

Solution

Given: Cash flows as tabulated, MARR = 15% per year
Find: PE

If we bring each flow to its equivalent at time zero we find

$$PE(15\%) = -\$75{,}000 + \$24{,}400(P/F, 15\%, 1) + \$27{,}340(P/F, 15\%, 2)$$
$$+ \$55{,}760(P/F, 15\%, 3)$$
$$= \$3553.$$

Since the project results in a positive PE of $3553, the project is acceptable.

In Example 4.5, we computed the PE of the project at a fixed interest rate of 15%. If we compute the PE at varying interest rates, we obtain the data in Table 4.3. Plotting the PE as a function of interest rate gives Figure 4.5, the present worth profile. When $i = 0$, the PE simply equals the sum of the cash flows. In the limit as the

---

[2] As we stated at the beginning of this chapter, we treat net cash flows as before tax values or as having their tax effects precalculated. Explaining the process of obtaining cash flows requires an understanding of income taxes and the role of depreciation, which are discussed in Chapters 7, 8 and 9.

Table 4.3	$i(\%)$	$PE(i)$	$i(\%)$	$PE(i)$
Present Worth Amounts at Varying Interest Rates (Example 4.5)	0	$32,500	20	−$3412
	2	27,743	22	−5924
	4	23,309	24	−8296
	6	19,169	26	−10,539
	8	15,296	28	−12,662
	10	11,670	30	−14,673
	12	8270	40	−23,302
	14	5077	100	−48,995
	16	2076	500	−69,915
	17.45*	0	1000	−72,514
	18	−750	2000	−73,770
			∞	−75,000

*Break-even interest rate (also known as the rate of return)

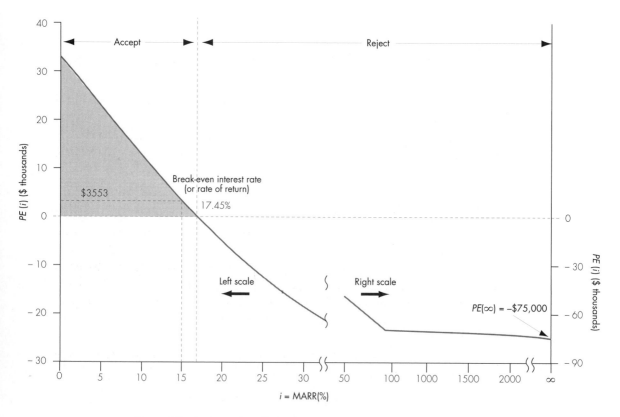

Figure 4.5   Present worth profile described in Example 4.5

interest rate becomes very large, the present value of each of the positive cash flows in years 1, 2, and 3 approach zero. So the present worth profile becomes asymptotic to a PE (∞) value equal to the initial investment. (As you will see in Section 4.8, you may use a spreadsheet program such as Excel to generate such results and the corresponding plot.)

Fig. 4.5 indicates that the investment project has a positive PE, if the interest rate is below 17.45%, and a negative PE if the interest rate is above 17.45%. As we will see in Section 4.6, this **break-even interest rate** is known as the **internal rate of return**. If the firm's MARR is 15%, the project has a PE of $3553 and so may be accepted. The figure of $3553 measures the equivalent immediate gain in present worth to the firm following the acceptance of the project. On the other hand, at i = 20%, $PE(20\%) = -\$3412$, the firm should reject the project. (Note that either accepting or rejecting an investment is influenced by the choice of a MARR. So it is crucial to estimate the MARR correctly. We will defer this important issue until Chapter 10. For now we will assume that the firm has an accurate MARR estimate available for use in investment analysis.)

4.3.2   Meaning of Present Equivalent

In present worth analysis, we assume that all the funds in a firm's treasury can be placed in investments that yield a return equal to the MARR. We may view these funds as an **investment pool**. Alternatively, if no funds are available for investment, we assume the firm can borrow them at the MARR from the capital market. In this section, we will examine these two views when explaining the meaning of MARR in PE calculations.

Investment Pool Concept

An investment pool is equivalent to a firm's treasury where all fund transactions are administered and managed by the firm's comptroller. The firm may withdraw funds from this investment pool for other investment purposes, but if left in the pool, these funds will earn at the MARR. Thus, in investment analysis, net cash flows will be net cash flows relative to this investment pool. To illustrate the investment pool concept, we consider again the project in Example 4.5 that required an investment of $75,000.

If the firm did not invest in the project and left $75,000 in the investment pool for 3 years, these funds would grow as follows:

$$\$75,000(F/P, 15\%, 3) = \$114,066.$$

Suppose the company decided instead to invest $75,000 in the project described in Example 4.5. Then the firm would receive a stream of cash inflows during its project life of 3 years in the amounts

Period (n)	Net Cash Flow (A_n)
1	$24,400
2	27,340
3	55,760

Since the funds that return to the investment pool earn interest at a rate of 15%, it would be of interest to see how much the firm would benefit from this investment. For this alternative, the returns after reinvestment are

$$\$24,400(F/P, 15\%, 2) = \$32,269$$
$$\$27,340(F/P, 15\%, 1) = \$31,441$$
$$\$55,760(F/P, 15\%, 0) = \$55,760$$
$$\text{Total} \qquad \$119,470.$$

These returns total $119,470. The additional cash accumulation at the end of 3 years from investing in the project is

$$\$119,470 - \$114,066 = \$5404.$$

If we compute the equivalent present worth (at time 0) of this net cash surplus after year 3, we obtain

$$\$5404(P/F, 15\%, 3) = \$3553,$$

which is exactly the same as the PE of the project computed by Eq. (4.1). Clearly, on the basis of its positive PE, the alternative of purchasing a new machine should

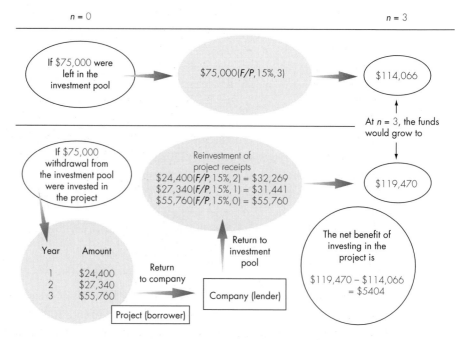

Figure 4.6 The concept of an investment pool with the company as a lender and the project as a borrower

be preferred to that of simply leaving the funds in the investment pool at the MARR. Thus, in PE analysis, any investment is assumed to be returned at the MARR. If a surplus exists at the end of the project, then *PE*(MARR) > 0. Figure 4.6 illustrates the reinvestment concept related to the firm's investment pool.

Borrowed Funds Concept

Suppose that the firm does not have $75,000 at the outset. In fact, the firm doesn't have to have an investment pool at all. Let's further assume that the firm borrows all its capital from a bank at an interest rate of 15%, invests in the project, and uses the proceeds from the investment to pay off the principal and interest on the bank loan. How much is left over for the firm at the end of the project period?

At the end of first year, the interest on the project's use of the bank loan would be $75,000(0.15) = $11,250. Therefore, the total loan balance grows to $75,000 × (1 + 0.15) = $86,250. Then, the firm receives $24,400 from the project and applies the entire amount to repay the loan portion. This repayment leaves a balance due of

$$\$75,000(1 + 0.15) - \$24,400 = \$61,850.$$

This amount becomes the net amount the project is borrowing at the beginning of year 2, which is also known as **project balance**. At the end of period 2, the bank debt grows to $61,850(1.15) = $71,128, but with the receipt of $27,340, the project balance reduces to

$$\$61,850(1.15) - \$27,340 = \$43,788.$$

Similarly, at the end of year 3, the loan balance becomes

$$\$43,788(1.15) = \$50,356.$$

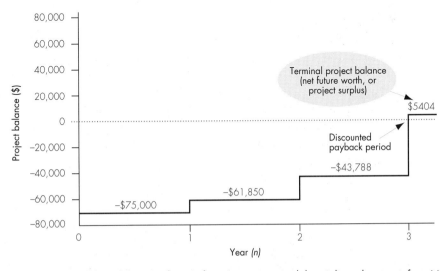

Figure 4.7 Project balance diagram as a function of time (a negative project balance indicates the amount of remaining loan to be paid off)

But, with the receipt of $55,760 from the project, the firm should be able to pay off the remaining balance and come out with a surplus in the amount of $5404. This terminal project balance is also known as the **net future worth** or **future equivalent** of the project. In other words, the firm fully repays its initial bank loan and interest at the end of period 3, with a resulting profit of $5404. If we compute the equivalent present worth of this net profit at time 0, we obtain

$$PE(15\%) = \$5404(P/F, 15\%, 3) = \$3553.$$

The result is identical to the case where we directly computed the PE of the project at $i = 15\%$, shown in Example 4.5. Figure 4.7 illustrates the project balance as a function of time. Note that the sign of the project balance changes from negative to positive during year 3. The time at which the project balance becomes zero is the **discounted payback period** as discussed in Section 4.2.3.

4.4 Future Worth Analysis

A second evaluation method which derives from discounted cash flow techniques is future worth analysis. Net present worth measures the surplus in an investment project at time 0. **Net future worth (NFW)** or **future equivalent (FE)** measures this surplus at a time period other than 0. Net future worth analysis is particularly useful in an investment situation where we need to compute the equivalent worth of a project at the end of its investment period, rather than at its beginning. For example, it may take 7 to 10 years to build a nuclear power plant because of the complexities of engineering design and the many time-consuming regulatory procedures that must be satisfied to ensure public safety. In this situation, it is more common to measure the worth of the investment at the time of the project's commercialization, i.e., to conduct an FE analysis at the end of the investment period.

4.4.1 Future Equivalent (Net Future Worth) Criterion

When A_n represents the cash flow at time n for $n = 0, 1, 2, ..., N$ for a typical investment project that extends over N periods, the net future worth or future equivalent expression at the end of period N is

$$
\begin{aligned}
FE(i) &= A_0(1 + i)^N + A_1(1 + i)^{N-1} + A_2(1 + i)^{N-2} + ... + A_N \\
&= \sum_{n=0}^{N} A_n(1 + i)^{N-n} \\
&= \sum_{n=0}^{N} A_n(F/P, i, N-n). \tag{4.2}
\end{aligned}
$$

FE is the equivalent project cash flow measured at the end of period N. When Equation (4.2) is compared with Equation (4.1) which is the expression for PE, the equivalent project cash flow at time zero, it is obvious that FE and PE for any project must be related in the following way.

$$FE(i) = PE(i)(F/P, i, N).\qquad(4.3)$$

Since the $(F/P, i, N)$ factor in Equation (4.3) is positive for $-100\% < i < \infty$, FE and PE will both have the same sign. As a result, the independent project evaluation decision rules for FE must be identical to those for the PE criterion. For a single project evaluation,

If $FE(\text{MARR}) > 0$, accept the investment.
If $FE(\text{MARR}) = 0$, remain indifferent to the investment.
If $FE(\text{MARR}) < 0$, reject the investment.

EXAMPLE 4.6 Future equivalent—at the end of project

Consider the project cash flows in Example 4.5. Compute the FE at the end of year 3 at $i = 15\%$.

Solution

Given: Cash flows in Example 4.5, MARR = 15% per year
Find: FE

As seen in Figure 4.8, the FE of this project at an interest rate of 15% would be

$$
\begin{aligned}
FE(15\%) &= -\$75,000\,(F/P, 15\%, 3) + \$24,400(F/P, 15\%, 2) \\
&\quad + \$27,340(F/P, 15\%, 1) + \$55,760 \\
&= \$5404.
\end{aligned}
$$

Note that the net future worth of this project is equivalent to the terminal project balance as calculated in Section 4.3.2. Since $FE(15\%) > 0$, the project is acceptable. We reached this same conclusion under present worth analysis.

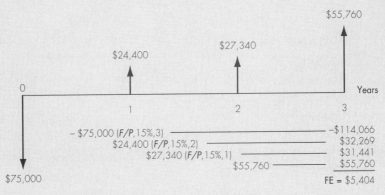

Figure 4.8 Future worth calculation (Example 4.6)

EXAMPLE
4.7

Future equivalent—at an intermediate time

Higgins Corporation (HC), a Montreal robot-manufacturing company, has developed a new advanced-technology robot called Helpmate, which incorporates advanced technology such as vision systems, tactile sensing, and voice recognition. These features allow the robot to roam the corridors of a hospital or office building without following a predetermined track or bumping into objects. HC's marketing department plans to target sales of the robot toward major hospitals to ease nurses' workloads by performing low-level duties such as delivering medicines and meals to patients.

The firm would need a new plant to manufacture the Helpmates; this plant could be built and made ready for production in 2 years. It would require a 12-hectare site, which can be purchased for $1.5 million in year 0. Building construction would begin early in year 1 and continue throughout year 2. The building would cost an estimated $10 million, would involve a $4 million payment due to the contractor at the end of year 1, and another $6 million would be payable at the end of year 2. The necessary manufacturing equipment would be installed late in year 2 and would be paid for at the end of year 2. The equipment would cost $13 million, including transportation and installation. When the project terminates, the land is expected to have an after-tax market value of $2 million, the building an after-tax value of $3 million, and the equipment an after-tax value of $3 million.

For capital budgeting purposes, assume that the cash flows occur at the end of each year. Because the plant would begin operations at the beginning of year 3, the first operating cash flows would occur at the end of year 3. The Helpmate plant's estimated economic life is 6 years after completion, with the following expected after-tax operating cash flows in millions:

Calendar Year	00	01	02	03	04	05	06	07	08
End of Year	0	1	2	3	4	5	6	7	8
After-tax cash flows									
A. Operating revenue				$6	$8	$13	$18	$14	$8
B. Investment									
Land	−1.5								+2
Building		−4	−6						+3
Equipment			−13						+3
Net cash flow	−$1.5	−$4	−$19	$6	$8	$13	$18	$14	$16

Compute the equivalent worth of this investment at the start of operation. Assume that HC's MARR is 15%.

Solution

Given: Cash flows above, MARR = 15% per year
Find: FE at the end of calendar year 2

One easily understood method involves calculating the present worth, then transforming this to the equivalent worth at the end of year 2. First, we can compute PE(15%) at time 0 of this project.

$$
\begin{aligned}
PE(15\%) &= -\$1.5 - \$4(P/F, 15\%, 1) - \$19(P/F, 15\%, 2) \\
&\quad +\$6(P/F, 15\%, 3) + \$8(P/F, 15\%, 4) + \$13(P/F, 15\%, 5) \\
&\quad +\$18(P/F, 15\%, 6) + \$14(P/F, 15\%, 7) + \$16(P/F, 15\%, 8) \\
&= \$13.91 \text{ million.}
\end{aligned}
$$

Then, the equivalent project worth at the start of operation is

$$
\begin{aligned}
FE_2(15\%) &= PE(15\%)(F/P, 15\%, 2) \\
&= \$18.40 \text{ million.}
\end{aligned}
$$

A second method brings all flows prior to year 2 up to that point and discounts future flows back to year 2. The equivalent worth of the earlier investment, when the plant begins full operation, is

$$
-\$1.5(F/P, 15\%, 2) - \$4(F/P, 15\%, 1) - \$19 = -\$25.58 \text{ million,}
$$

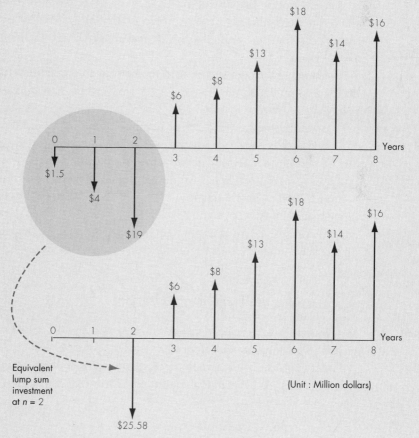

Figure 4.9 Cash flow diagram for the Helpmate project (Example 4.7)

which produces an equivalent flow as shown in Figure 4.9. If we discount the future flows to the start of operation, we obtain

$$
\begin{aligned}
FE_2(15\%) &= -\$25.58 + \$6(P/F, 15\%, 1) + \$8(P/F, 15\%, 2) + \ldots \\
&\quad + \$16(P/F, 15\%, 6) \\
&= \$18.40 \text{ million.}
\end{aligned}
$$

Comments: If another company is willing to purchase the plant and the right to manufacture the robots immediately after completion of the plant (year 2), HC would set the price of the plant at $43.98 million ($18.40 + $25.58) at a minimum.

4.5 Annual Equivalent Worth Analysis

Annual equivalent worth analysis is the third equivalence method for evaluating the attractiveness or profitability of independent investment projects. While present worth and future worth analysis measure project profitability referenced to time zero or some future point in time, **annual equivalent worth (AE)** represents it as an equivalent annual value.

Referencing the equivalence measurement to an annual timeframe is convenient in situations in which the desired or preferred answer is an annual or unit value. For example, you are considering buying a new car and expect to drive 24,000 kilometres per year on business. What rate of reimbursement on a per kilometre basis would your employer need to provide to cover your annual equivalent costs of owning the car? Or consider a real estate developer who is planning to build a shopping centre of 50,000 square metres. What would be the minimum annual rental fee per square metre required to justify this investment? It should also be noted that since corporations issue annual reports and develop yearly budgets, annual measurements of costs and benefits may be more useful than overall values. Hence, management may insist on a consistent and annual reporting basis for investment projects.

4.5.1 Annual Equivalent Criterion

In Chapter 2, we developed relationships which could be used to convert any single cash flow into an equivalent uniform series of cash flows. If we now consider a typical investment project where A_n represents the cash flow at time n over N periods ($n = 0, 1, 2\ldots, N$), the annual equivalent worth (AE) expression is

$$
\begin{aligned}
AE(i) \ &= \ A_0(A/P, i, N) + A_1(P/F, i, 1)(A/P, i, N) \\
&\quad + A_2(P/F, i, 2)(A/P, i, N) + \ldots + A_N(P/F, i, N)(A/P, i, N) \\
&= \ \left[\sum_{n=0}^{N} A_n(P/F, i, n), \right] (A/P, i, N) \\
&= \ PE(i)(A/P, i, N) \hspace{5.5cm} (4.4)
\end{aligned}
$$

or

$$
\begin{aligned}
AE(i) \ &= \ A_0(F/P, i, N)(A/F, i, N) + A_1(F/P, i, N-1)(A/F, i, N) \\
&\quad + A_2(F/P, i, N-2)(A/F, i, N) + \ldots + A_N(A/F, i, N) \\
&= \ \left[\sum_{n=0}^{N} A_n(F/P, i, N-n) \right] (A/F, i, N) \\
&= \ FE(i)(A/F, i, N) \hspace{5.5cm} (4.5)
\end{aligned}
$$

Since the factors $(A/P, i, N)$ and $(A/F, i, N)$ are positive for $-100\% < i < \infty$, the AE value will be positive if PE and FE are positive, and negative if PE and FE are negative. It then follows that the independent project investment decision rules for AE are identical to those for the PE and FE criteria.

If $AE(\text{MARR})$ $>$ 0, accept the investment.

If $AE(\text{MARR})$ $=$ 0, remain indifferent to the investment.

If $AE(\text{MARR})$ $<$ 0, reject the investment.

EXAMPLE 4.8

Present, future, and annual equivalence criteria relationships

Example 4.5 proved to be an attractive investment because its PE value at MARR = 15% was greater than zero. Compute future and annual equivalent values from the known PE value.

Given: PE(15%) = \$3553 and $N = 3$
Find: FE(15%) and AE(15%)

The equivalence criteria are related as indicated in Equations (4.3), (4.4) and (4.5). Therefore,

$$
\begin{aligned}
FE(15\%) \ &= \ PE(15\%)(F/P, 15\%, 3) \\
&= \ \$3553 \times 1.5209 \\
&= \ \$5404
\end{aligned}
$$

which is the same result calculated in Example 4.6 and

$$
\begin{aligned}
AE(15\%) \ &= \ PE(15\%)(A/P, 15\%, 3) \\
&= \ \$3553 \times 0.4380 \\
&= \ \$1556
\end{aligned}
$$

or

$$
\begin{aligned}
AE(15\%) &= FE(15\%)(A/F, 15\%, 3) \\
&= \$5404 \times 0.2880 \\
&= \$1556
\end{aligned}
$$

Since each of the equivalence criteria has a value greater than zero, each criterion leads to the same conclusion with respect to investment profitability, i.e., this is an attractive investment. The choice of a particular criterion is dependent on the type of problem and the preference of the analyst.

EXAMPLE 4.9 Annual equivalent by conversion from present equivalent (PE)

The Skyward Communications Company (SCC) is planning to develop satellite-based systems that will enable airline passengers to make phone calls or send facsimiles from virtually anywhere in the world. The systems, which use a network of satellites to bounce signals from a plane to ground stations, are linked with conventional telephone networks. The systems are based upon digital technology, and calls will be crisper than those made on currently used airborne telephones, which are based on conventional radio technology. One stumbling block remains: The new systems will not allow people on the ground to call airborne passengers. The airlines have yet to install the necessary digital cockpit electronics. Five international airlines have agreed to offer the in-flight telephone service on a total of 120 airplanes if SCC develops a satellite-based telephone system. SCC has estimated the projected cash flows (in million dollars) to install these 120 systems as follows:

n	A_n
0	−$15.0
1	−3.5
2	5.0
3	9.0
4	12.0
5	10.0
6	8.0

First, SCC wants to determine whether this project can be justified at MARR = 15%. Then it wants to know the annual benefit (or loss) that could be generated after installation of these systems.

Discussion: When a cash flow has no special pattern, it is easiest to find AE in two steps: (1) Find the PE (or FE) of the flow; and (2) find the AE of the PE (or FE). This method is presented below. You might want to try another method with this type of cash flow to demonstrate how difficult this can be.

Solution

Given: Cash flow in Figure 4.10, $i = 15\%$
Find: AE

We first compute the PE at $i = 15\%$:

$$PE(15\%) = -\$15 - \$3.5(P/F, 15\%, 1) + \$5(P/F, 15\%, 2) + \ldots$$
$$+ \$10(P/F, 15\%, 5) + \$8(P/F, 15\%, 6)$$
$$= \$6.946 \text{ million.}$$

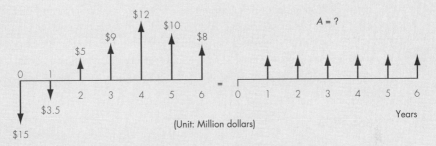

Figure 4.10 Cash flow diagram (Example 4.9)

Since $PE(15\%) > 0$, the project would be acceptable under the PE analysis. Now, spreading the PE over the project life gives

$$AE(15\%) = \$6.946(A/P, 15\%, 6) = \$1.835 \text{ million.}$$

Since $AE(15\%) > 0$, the project is worth undertaking. The positive AE value indicates that the project is expected to bring in a net annual benefit of $1.835 million over the project life.

Comments: How critical is the accuracy of SCC's estimated time zero, $15 million, investment cost to this project's viability? If this cost had been $22.5 million or 50% greater, the project PE value at 15% would be negative (viz., the previously calculated PE of $6.946 million minus the additional $7.5 million of costs at time zero) so on a PE (or AE) basis, the project becomes unacceptable. Is a relative change of 50% in project cost surprising or unusual? The answer to this question depends on a number of factors including project complexity, the stage the project has reached in terms of detailed engineering, experience with similar or identical projects in the past, and the newness of the technology involved. For projects which are complex, the final project cost is frequently considerably greater than the first estimated costs at the conceptual or initial stage of a project.

In some situations a **cyclic cash flow pattern** may be observed over the project life. Unlike the situation in Example 4.9, where we first computed the PE of the entire cash flow and then calculated the AE value from this PE, we can compute the AE by examining the first cash flow cycle. By computing the PE for the first cash flow cycle the AE over the first cash flow cycle can be derived. This short-cut method provides the same solution as calculating the PE over the entire project, and then computing AE from this PE.

EXAMPLE 4.10   Annual equivalent worth—repeating cash flow cycles

SOLEX Company is producing electricity directly from a solar source by using a large array of solar cells and selling the power to the local utility company. SOLEX decided to use amorphous silicon cells because of their low initial cost, but these cells degrade over time, thereby resulting in lower conversion efficiency and power output. The cells must be replaced every 4 years, which results in a particular cash flow pattern that repeats itself as shown in Figure 4.11. Determine the annual equivalent cash flows at $i = 12\%$.

Solution

Given: Cash flows in Fig. 4.11, $i = 12\%$
Find: Annual equivalent benefit

To calculate the AE, we need only consider one cycle over its 4-year period. For $i = 12\%$, we first obtain the PE for the first cycle as follows:

$$
\begin{aligned}
PE(12\%) &= -\$1,000,000 \\
&\quad + [(\$800,000 - \$100,000(A/G, 12\%, 4)](P/A, 12\%, 4) \\
&= -\$1,000,000 + \$2,017,150 \\
&= \$1,017,150.
\end{aligned}
$$

Then, we calculate the AE value over the 4-year life cycle:

$$
\begin{aligned}
AE(12\%) &= \$1,017,150(A/P, 12\%, 4) \\
&= \$334,880.
\end{aligned}
$$

We can now say that the two cash flow series are equivalent:

Original Cash Flows			Annual Equivalent Flows	
n	A_n	≡	n	A_n
0	-$1,000,000		0	0
1	800,000		1	$334,880
2	700,000		2	334,880
3	600,000		3	334,880
4	500,000		4	334,880

We can extend this cash flow equivalency over the remaining cycles of the cash flow. The reasoning is that each similar set of five values (one disbursement and four receipts) is equivalent to four annual receipts of $334,880 each.

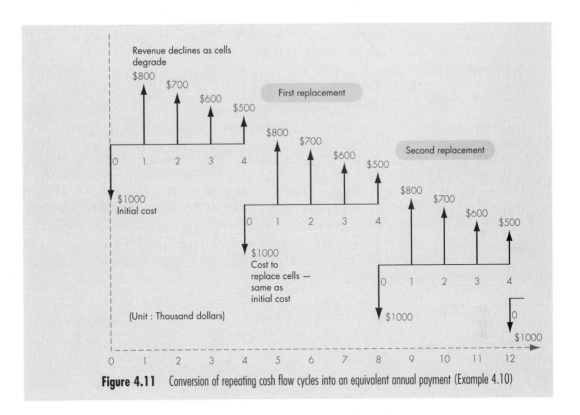

Figure 4.11 Conversion of repeating cash flow cycles into an equivalent annual payment (Example 4.10)

4.5.2 Capital Costs versus Operating Costs

When only costs are involved, the AE method is sometimes called the **annual equivalent cost** method. In this case, revenues must cover two kinds of costs: **operating costs** and **capital costs**. Operating costs are incurred by the operation of physical plant or equipment needed to provide service—examples include items such as labor and raw materials. Capital costs are incurred by purchasing assets to be used in production and service. Normally, capital costs are nonrecurring (i.e., one-time costs), whereas operating costs recur for as long as an asset is owned.

Because operating costs recur over the life of a project, they tend to be estimated on an annual basis anyway, so for the purposes of annual equivalent cost analysis, no special calculation is required. However, because capital costs tend to be one-time costs, in conducting an annual equivalent cost analysis, we must translate this one-time cost into its annual equivalent over the life of the project. The annual equivalent of a capital cost is given a special name: **capital recovery cost**, designated $CR(i)$.

Two general monetary transactions are associated with the purchase and eventual retirement of a capital asset: its initial cost (P) and its salvage value (S). Taking into account these sums, we calculate the capital recovery cost as follows:

$$CR(i) = P(A/P, i, N) - S(A/F, i, N). \tag{4.6}$$

Recall algebraic relationships between factors in Table 2.3, and notice that the $(A/F, i, N)$ factor can be expressed as

$$(A/F, i, N) = (A/P, i, N) - i.$$

Then, we may rewrite the $CR(i)$ as

$$\begin{aligned} CR(i) &= P(A/P, i, N) - S[(A/P, i, N) - i] \\ &= (P - S)(A/P, i, N) + iS. \end{aligned} \quad (4.7)$$

We may interpret this situation thus: To obtain the machine, one borrows a total of P dollars, S dollars of which are returned at the end of the Nth year. The first term $(P - S)(A/P, i, N)$ implies that the balance $(P - S)$ will be paid back in equal installments over the N-year period at a rate of i, and the second term iS implies that simple interest in the amount iS is paid on S until it is repaid (Figure 4.12). Thus, the amount to be financed is $P - S(P/F, i, N)$, and the installments of this loan over the N-periods are

$$\begin{aligned} AE(i) &= -[P - S(P/F, i, N)](A/P, i, N) \\ &= -P(A/P, i, N) + S(P/F, i, N)(A/P, i, N) \\ &= -[P(A/P, i, N) - S(A/F, i, N)] \\ &= -CR(i). \end{aligned} \quad (4.8)$$

Therefore, the $CR(i)$ tells us what the bank would charge each year. Many auto leases are based on this arrangement in that most require a guarantee of S dollars in salvage. From an industry viewpoint, $CR(i)$ is the annual cost to the firm of owning the asset.

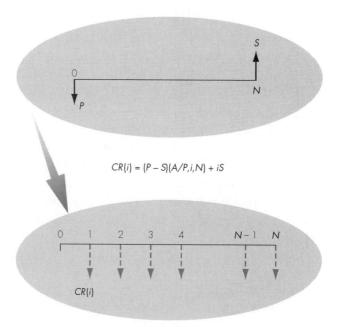

$$CR(i) = (P - S)(A/P, i, N) + iS$$

Figure 4.12 Capital recovery (ownership) cost calculation

With this information, the amount of annual savings required to recover the capital and operating costs associated with a project can be determined. As an illustration, consider Example 4.11.

EXAMPLE
4.11

Annual equivalent—capital recovery cost

Consider a machine that costs $20,000 and has a 5-year useful life. At the end of the 5 years, it can be sold for $4000 after tax adjustment. If the firm could earn an after-tax revenue of $4400 per year with this machine, should it be purchased at an interest rate of 10%? (All benefits and costs associated with the machine are accounted for in these figures.)

Solution

Given: $P = \$20,000$, $S = \$4000$, $A = \$4400$, $N = 5$ years, $i = 10\%$ per year
Find: AE, and determine whether to purchase the machine or not

We will compute the capital costs in two different ways:

Method 1: First compute the PE of the cash flows and then compute the AE from the calculated PE:

$$
\begin{aligned}
PE(10\%) &= -\$20,000 + \$4400(P/A, 10\%, 5) \\
&\quad + \$4000(P/F, 10\%, 5) \\
&= -\$20,000 + \$4400(3.7908) + \$4000(0.6209) \\
&= -\$836.88 \\
AE(10\%) &= -\$836.88(A/P, 10\%, 5) = -\$220.76.
\end{aligned}
$$

This negative AE value indicates that the machine does not generate sufficient revenue to recover the original investment so we may reject the project. In fact, there will be an equivalent loss of $220.76 per year over the machine's life (Figure 4.13a).

Method 2: The second method is to separate cash flows associated with the asset acquisition and disposal from the normal operating cash flows. Since the operating cash flows—the $4400 yearly income—are already given in equivalent annual flows $(AE(i)_1)$, we only need to convert the cash flows associated with asset acquisition and disposal into equivalent annual flows $(AE(i)_2)$ (Fig. 4.13b). Using Eq. (4.7),

$$
\begin{aligned}
CR(i) &= (P - S)(A/P, i, N) + iS \\
AE(i)_1 &= \$4400 \\
AE(i)_2 &= - CR(i) \\
&\quad -[(\$20,000 - \$4000)(A/P, 10\%, 5) + (0.10)\$4000] \\
&= -\$4620.76 \\
AE(10\%) &= AE(i)_1 + AE(i)_2 \\
&= \$4400 - \$4620.76 \\
&= -\$220.76.
\end{aligned}
$$

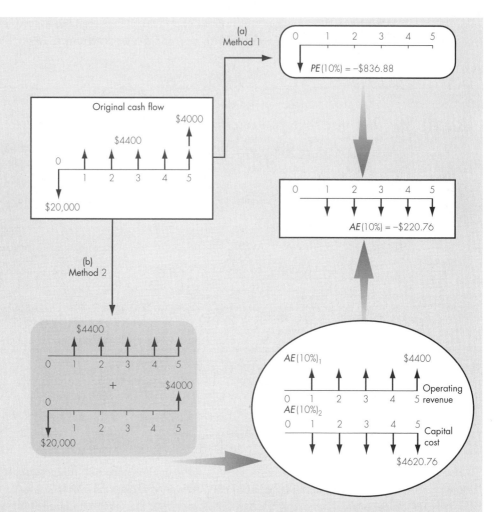

Figure 4.13 Alternative ways of computing capital recovery cost for an investment (Example 4.11)

Comments: Obviously, Method 2 saves a calculation step, so we may prefer it over Method 1. We may interpret Method 2 as determining that the annual operating benefits must be at least $4620.76 to recover the asset cost. However, the annual operating benefits actually amount to only $4400, resulting in a loss of $220.76 per year. Therefore, the project is not worth undertaking.

4.6 Rate of Return Analysis

The fourth method for evaluating independent investment projects measures profitability in terms of a return expressed as an interest rate value. This measurement derives from rate of return analysis and may be referred to as **rate of return**, **yield**, (i.e., the yield to maturity commonly used in bond valuation), **internal rate of return**, or **marginal efficiency of capital**.

Although the equivalence based methods for measuring profitability are easy to calculate and apply, as explained in Sections 4.3, 4.4, and 4.5, many engineers and financial managers prefer rate of return analysis to these other methods because they find it intuitively more appealing to analyze investments in terms of percentage rates of return rather than dollars of PE, FE or AE. Consider the following statements regarding an investment's profitability:

- This project will bring in a 15% rate of return on the investment.
- This project will result in a net surplus of $10,000 in the PE.

Neither statement describes the nature of an investment project in any complete sense. However, the rate of return figure is somewhat easier to understand because many of us are so familiar with savings and loan interest rates, which are in fact rates of return.

Unfortunately, rate of return analysis is complicated by a number of considerations, including the definition and calculation of rate of return. We will first review three common definitions and select the definition of internal rate of return as the profitability measure to be used throughout the text.

4.6.1   Return on Investment: Loan Transaction Basis

There are several ways of defining the concept of rate of return on investment. The first is based on a typical loan transaction.

Definition 1
Rate of return is the interest rate earned on the unpaid balance of an amortized loan.

Suppose that a bank lends $10,000, and it is repaid with payments of $4021 at the end of each year for 3 years. How would you determine the interest rate that the bank charges on this transaction? As we learned in Chapter 2, you would set up the following equivalence equation:

$$\$10,000 = \$4021(P/A, \, i, \, 3)$$

and solve for i. It turns out that $i = 10\%$. In this situation, the bank will earn a return of 10% on its investment of $10,000. The bank calculates the loan balances over the life of the loan as follows:

Year	Unpaid Balance at Beginning Year	Return on Unpaid Balance (10%)	Payment Received	Unpaid Balance at End of Year
0	-$10,000	$0	$0	-$10,000
1	-10,000	-1000	4021	-6979
2	-6979	-698	4021	-3656
3	-3656	-366	4021	0

A negative balance indicates an unpaid balance.

Observe that for the repayment schedule shown above, the 10% interest is calculated only for each year's outstanding balance. In this situation, only part of the $4021 annual payment represents interest; the remainder goes toward repaying the principal. Namely, the three annual payments repay the loan itself and additionally provide a return of 10% on the *amount still outstanding each year*.

Note that, when the last payment is made, the outstanding principal is eventually reduced to zero.[3] If we calculate the PE of the loan transaction at its rate of return (10%), we see

$$PE(10\%) = -\$10,000 + \$4021(P/A, 10\%, 3) = 0,$$

which indicates that the bank can break even at a 10% rate of interest.

4.6.2   Return on Investment: PE = 0 Basis

The break-even rate of interest for a loan transaction can be viewed as the interest rate that equates the present value of future cash repayments to the amount of the loan. This observation prompts the second definition of rate of return.

Definition 2

Rate of return is the break-even interest rate, i^*, which equates the present worth of a project's cash outflows to the present worth of its cash inflows, or

$$PE(i^*) = PE_{\text{Cash inflows}} - PE_{\text{Cash outflows}}$$
$$= 0.$$

Note that the PE expression is equivalent to

$$PE(i^*) = \frac{A_0}{(1 + i^*)^0} + \frac{A_1}{(1 + i^*)^1} + \ldots + \frac{A_N}{(1 + i^*)^N} = 0. \qquad (4.9)$$

Here we know the value of A_n for each period, but not the value of i^*. Since it is the only unknown, we can solve for i^*. (Inevitably, there will be a total of N real and

---

[3] As we learned in Section 4.3.2, this terminal balance is equivalent to the net future worth of the investment. If the net future worth of the investment is zero, its PE should also be zero.

imaginary values of i^* that satisfy this equation. Imaginary roots are not meaningful in our context. Negative real roots must be greater than –100% because Eq. (4.9) becomes discontinuous at this point. In any case, negative real roots are of little interest as they indicate the initial investment can never be recovered. In most project cash flows, you would be able to find a unique positive i^* that satisfies Eq. (4.9). However, you may encounter some cash flows that cannot be solved for a single rate of return greater than –100%. By the nature of the PE function in Eq. (4.9), it is certainly possible to have more than one real positive break even rate for certain types of cash flows. For some cash flows, we may not find a single positive rate of return at all.)

Note that the i^* formula in Eq. (4.9) is simply the PE formula, Eq. (4.1), solved for the particular interest rate (i^*) at which $PE(i)$ is equal to zero. By multiplying both sides of Eq. (4.9) by $(1+i^*)^N$, we obtain

$$PE(i^*)(1 + i^*)^N = FE(i^*) = 0.$$

If we multiply both sides of Eq. (4.9) by the capital recovery factor, $(A/P, i^*, N)$, we obtain the relationship $AE(i^*) = 0$. Therefore, the i^* of a project may be defined as the rate of interest that equates the present worth, future worth, and annual equivalent worth of the entire series of cash flows to zero.

4.6.3   Return on Invested Capital

Investment projects can be viewed as analogous to bank loans. We will now introduce the concept of rate of return based on the return on invested capital in terms of a project investment. A project's return is referred to as the internal rate of return (IRR) or the **yield** promised by an **investment project** over its **service life**.

Definition 3
Internal rate of return is the interest rate charged on the unrecovered project balance of the investment such that, when the project terminates, the unrecovered project balance will be zero.

Suppose a company invests $10,000 in a computer with a 3-year useful life and equivalent annual labor savings of $4021. Here, we may view the investing firm as the lender and the project as the borrower. The cash flow transaction between them would be identical to the amortized loan transaction described under Definition 1.

n	Beginning Project Balance	Return on Invested Capital	Ending Cash Payment	Project Balance
0	$0	$0	–$10,000	–$10,000
1	–10,000	–1000	4021	–6979
2	–6979	–698	4021	–3656
3	–3656	–366	4021	0

In our project balance calculation, we see that 10% is earned (or charged) on $10,000 during year 1, 10% is earned on $6979 during year 2, and 10% is earned on $3656 during year 3. This indicates that the firm earns a 10% rate of return on funds that remain *internally* invested in the project. Since it is a return *internal* to the project, we refer to it as the **internal rate of return**, or IRR. This means that the computer project under consideration brings in enough cash to pay for itself in 3 years and also to provide the firm with a return of 10% on its invested capital. To put it differently, if the computer is financed with funds costing 10% annually, the cash generated by the investment will be exactly sufficient to repay the principal and the annual interest charge on the fund in 3 years.

Notice also that only one cash outflow occurs at time 0, and the present worth of this outflow is simply $10,000. There are three equal receipts, and the present worth of these inflows is $4021(P/A, 10%, 3) = $10,000. Since the PE = PE_{Inflow} − PE_{Outflow} = $10,000 − $10,000 = 0, 10% also satisfies Definition 2 for rate of return. Even though the above simple example implies that i^* coincides with IRR, only Definitions 1 and 3 correctly describe the true meaning of internal rate of return. As we will see later, if the cash expenditures of an investment are not restricted to the initial period, several real, positive, break-even interest rates (i^*s) may exist that satisfy Eq. (4.9). However, none of these may be the rate of return *internal* to the project.

4.6.4   Simple versus Nonsimple Investments

We can classify an investment project by counting the number of sign changes in its net cash flow sequence. A change from either "+" to "−" or "−" to "+" is counted as one sign change. (We ignore a zero cash flow.) Then,

- a **simple (or conventional) investment** is one in which the initial cash flows are negative, and only one sign change occurs in the net cash flow series. If the initial flows are positive and only one sign change occurs in the subsequent net cash flows, they are referred to as **simple borrowing** cash flows.

- a **nonsimple (or nonconventional) investment** is one in which more than one sign change occurs in the cash flow series.

Multiple real positive i^*s, as we will see later, occur only in nonsimple investments. The different types of investment may have the following possible individual cash flow signs:

Investment Type	Cash Flow Sign at Period					
	0	1	2	3	4	5
Simple	−	+	+	+	+	+
Simple	−	−	+	+	0	+
Nonsimple	−	+	−	+	+	−
Nonsimple	−	+	+	−	0	+

EXAMPLE
4.12

Investment classification

Consider the three cash flow series shown below and classify them into either simple or nonsimple investments.

| Period | | Net Cash Flow | |
n	Project A	Project B	Project C
0	–$1000	–$1000	+$1000
1	–500	3900	–450
2	800	–5030	–450
3	1500	2145	–450
4	2000		

Solution

Given: Cash flow sequences above
Find: Classify into either simple or nonsimple investments

- Project A represents a simple investment. This common type of investment has a PE profile as shown in Figure 4.14(a). The curve only crosses the *i*-axis once.

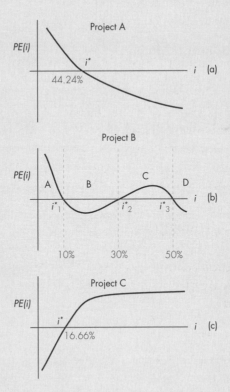

Figure 4.14 Present worth profiles: (a) Simple investment, (b) nonsimple investment with multiple positive *i**s, and (c) simple borrowing cash flows

- Project B represents a nonsimple investment. The PE profile for this investment has the shape shown in Fig. 4.14(b). The *i*-axis is crossed at 10%, 30%, and 50%.

- Project C represents neither a simple nor a nonsimple investment, even though only one sign change occurs in the cash flow sequence. Since the first cash flow is positive, this is a **simple borrowing** cash flow, not an investment flow. The PE profile for this type of investment looks like the one in Fig. 4.14(c).

Comments: Not all PE profiles for nonsimple investments have multiple crossings of the *i*-axis. In Appendix 4A, we illustrate when to expect such multiple crossings by examining types of cash flows.

4.6.5 Methods of Finding *i**

Once we identify the type of an investment cash flow, several ways to determine a break even rate, or *i**, are available. We will discuss some of the most practical methods here. They are as follows:

- Direct solution method,
- Trial-and-error method, and
- Computer solution method.

Direct Solution Method

For the very special case of a project with only a two-flow transaction (an investment followed by a single future payment) or a project with a service life of 2 years of return, we can seek a direct mathematical solution for determining the rate of return. These two cases will be examined in Example 4.13.

EXAMPLE 4.13

Finding *i by direct solution: Two flows and two periods**

Consider two independent investment projects with the following cash flow transactions. Compute the rate of return for each project.

n	Project 1	Project 2
0	–$1000	–$2000
1	0	1300
2	0	1500
3	0	
4	1500	

Solution

Given: Cash flows for two projects
Find: *i** for each

Project 1: Solving for i^* in $PE(i^*) = 0$ is identical to solving $FE(i^*) = 0$ because FE equals PE times a constant. We could do either here, but we will set $FE(i^*) = 0$ to demonstrate the latter. Using the single-payment future worth relationship, we obtain

$$\begin{aligned} FE(i^*) &= -\$1000(F/P, i^*, 4) + \$1500 = 0 \\ \$1500 &= \$1000(F/P, i^*, 4) = \$1000(1 + i^*)^4 \\ 1.5 &= (1 + i^*)^4. \end{aligned}$$

Solving for i^* yields

$$\begin{aligned} i^* &= \sqrt[4]{1.5} - 1 \\ &= 0.1067 \text{ or } 10.67\%. \end{aligned}$$

Project 2: We may write the PE expression for this project as follows:

$$PE(i^*) = -\$2000 + \frac{\$1300}{(1 + i^*)} + \frac{\$1500}{(1 + i^*)^2} = 0.$$

Let $X = \dfrac{1}{(1 + i^*)}$. We may then rewrite the $PE(i^*)$ as a function of X as follows:

$$PE(i^*) = -\$2000 + \$1300X + \$1500X^2 = 0.$$

This is a quadratic equation that has the following solution:[4]

$$\begin{aligned} X &= \frac{-1300 \pm \sqrt{1300^2 - 4(1500)(-2000)}}{2(1500)} \\ &= \frac{-1300 \pm 3700}{3000} \\ &= 0.8 \text{ or } -1.667 \end{aligned}$$

Replacing X values and solving for i^* gives us

$$0.8 = \frac{1}{(1 + i^*)} \rightarrow i^* = 25\%$$

$$-1.667 = \frac{1}{(1 + i^*)} \rightarrow i^* = -160\%.$$

Since an interest rate less than -100% has no economic significance, we find that the project's i^* is 25%.

Comments: In both projects, one sign change occurred in the net cash flow series, so we expected a unique real i^*. Also these projects had very simple cash flows. When cash flows are more complex, generally we must use a trial-and-error method or a computer to find i^*.

---

[4] Given $aX^2 + bX + c = 0$, the solution of the quadratic equation is

$$X = \frac{-b \pm \sqrt{b^2 - 4ac}}{2a}$$

Trial-and-Error Method

The first step in the trial-and-error method is to make a reasonable **guess**[5] at the value of i^*. For a simple investment, we compute the present worth of net cash flows using the "guessed" interest rate and observe whether it is positive, negative, or zero. Suppose the $PE(i)$ is negative. Since we are aiming for a value of i that makes $PE(i) = 0$, we must raise the present worth of the cash flow. To do this we lower the interest rate and repeat the process. If $PE(i)$ is positive, however, we raise the interest rate in order to lower $PE(i)$. The process is continued until $PE(i)$ is approximately equal to zero. Whenever we reach the point where $PE(i)$ is bounded by one negative and one positive value, we use **linear interpolation** to approximate the i^*. This process is somewhat tedious and inefficient. (The trial-and-error method does not work for nonsimple investments in which the PE function is not, in general, a monotonically decreasing function of interest rate.)

EXAMPLE 4.14 **Finding i^* by trial and error**

Agdist Corporation distributes agricultural equipment. The board of directors is considering a proposal to establish a facility to manufacture an electronically controlled "intelligent" crop sprayer invented by a professor at a local university. This crop sprayer project would require an investment of $10 million in assets and would produce an annual after-tax net benefit of $1.8 million over a service life of 8 years. All costs and benefits are included in these figures. When the project terminates, the net proceeds from the sale of the assets would be $1 million (Figure 4.15). Compute the rate of return of this project.

Figure 4.15 Cash flow diagram for a simple investment (Example 4.14)

[5] As we see later in this chapter, the ultimate objective of finding the i^* is to compare it to the MARR. Therefore, it is a good idea to use the MARR as the initial guess value.

Solution

Given: Initial investment $(P) = \$10$ million, $A = \$1.8$ million, $S = 1$ million, $N = 8$ years

Find: i^*

We start with a guessed interest rate of 8%. The present worth of the cash flows in millions of dollars is

$$PE(8\%) = -\$10 + \$1.8(P/A, 8\%, 8) + \$1(P/F, 8\%, 8) = \$0.88.$$

Since this present worth is positive, we must raise the interest rate to bring this value toward zero. When we use an interest rate of 12%, we find that

$$PE(12\%) = -\$10 + \$1.8(P/A, 12\%, 8) + \$1(P/F, 12\%, 8) = -\$0.65.$$

We have bracketed the solution: $PE(i)$ will be zero at i somewhere between 8% and 12%. Using straight-line interpolation, we approximate

$$
\begin{aligned}
i^* &\cong 8\% + (12\% - 8\%)\left[\frac{0.88 - 0}{0.88 - (-0.65)}\right] \\
&= 8\% + 4\%(0.5752) \\
&= 10.30\%.
\end{aligned}
$$

Now we will check to see how close this value is to the precise value of i^*. If we compute the present worth at this interpolated value, we obtain

$$
\begin{aligned}
PE(10.30\%) &= -\$10 + \$1.8(P/A, 10.30\%, 8) + \$1(P/F, 10.30\%, 8) \\
&= -\$0.045.
\end{aligned}
$$

As this is not zero, we may recompute the i^* at a lower interest rate, say 10%:

$$PE(10\%) = -\$10 + \$1.8(P/A, 10\%, 8) + \$1(P/F, 10\%, 8) = \$0.069.$$

With another round of linear interpolation, we approximate

$$
\begin{aligned}
i^* &\cong 10\% + (10.30\% - 10\%)\left[\frac{0.069 - 0}{0.069 - (-0.045)}\right] \\
&= 10\% + 0.30\%(0.6053) \\
&= 10.18\%.
\end{aligned}
$$

At this interest rate,

$$
\begin{aligned}
PE(10.18\%) &= -\$10 + \$1.8(P/A, 10.18\%, 8) + \$1(P/F, 10.18\%, 8) \\
&= \$0.0007,
\end{aligned}
$$

which is practically zero, so we may stop here. In fact, there is no need to be more precise about these interpolations because the final result can be no more accurate than the basic data, which ordinarily are only rough estimates. Computing the i^* for this problem on computer, incidentally, gives us 10.1819%.

Computer Solution Method

We don't need to do laborious manual calculations to find i^*. Many financial calculators have built-in functions (root finding algorithms) for calculating i^*. It is worth noting that spreadsheet packages also have i^* functions, which solve Eq. (4.9) very rapidly.[6] This is usually done by entering the cash flows via a keyboard or by reading a cash flow data file. For example, Microsoft Excel has an IRR financial function that analyzes investment cash flows, as will be illustrated in Section 4.8.

The most easily generated and understandable graphic method of solving for i^* is to create the **net present worth** or **PE profile** on computer. In the graph, the horizontal axis indicates interest rate, and the vertical axis indicates PE. For a given project's cash flows, the PE is calculated at an interest rate of zero (which gives the vertical axis intercept) and several other interest rates. Points are plotted, and a curve sketched. Since i^* is defined as the interest rate at which $PE(i^*) = 0$, the point at which the curve crosses the horizontal axis closely approximates the i^*. The graphical approach works for both simple and nonsimple investments.

EXAMPLE 4.15 **Graphical approach to estimate i^***

Consider the cash flow series shown in Figure 4.16(a). Estimate the rate of return by generating the PE profile on a computer.

Solution

Given: Cash flow series in Fig. 4.16(a)
Find: i^* by plotting the PE profile

The present worth function for the project cash flow series is

$$PE(i) = -\$10,000 + \$20,000(P/A, i, 2) - \$25,000(P/F, i, 3)$$

We first use $i = 0$ in this equation, to obtain PE = \$5000, which is the vertical axis intercept. Substitute several other interest rates—10%, 20%, ..., 140%—and plot these values of $PE(i)$ as well. The result is Fig. 4.16(b), which shows the curve crossing the horizontal axis at roughly 140%. This value can be verified by other methods. Note that, in addition to establishing the interest rate that makes PE = 0, the PE profile indicates where positive and negative PE values fall, thus giving us a broad picture of those interest rates for which the project is acceptable or unacceptable. (Note that a trial-and-error method would lead to some confusion—as you increase the interest rate from 0% to 20%, the PE value also keeps increasing, instead of decreasing.) Even though the project is a nonsimple investment, the curve crosses the horizontal axis only once. As mentioned in the previous section, however, most nonsimple projects have more than one value of i^*, which makes PE = 0; i.e., more than one real

---

[6] An alternative method of solving i^* problems is to use a computer-aided economic analysis program. **EzCash** finds i^* visually by specifying the lower and upper bounds of the interest search limit and generates PE profiles when given a cash flow series. In addition to the savings of calculation time, the advantage of computer-generated results is their precision.

positive i^* for a project. In such a case, the PE profile would cross the horizontal axis more than once.[7]

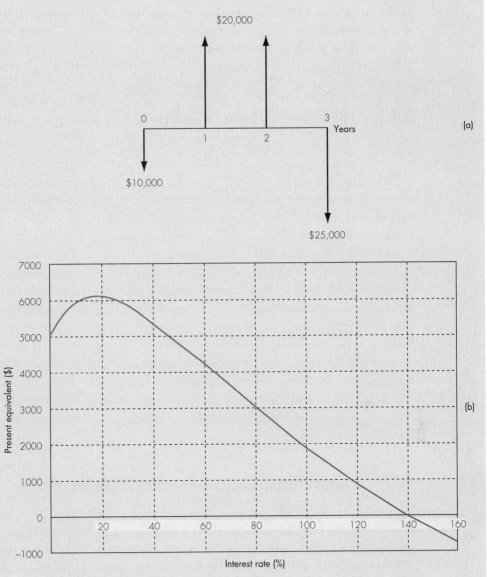

Figure 4.16 Graphical solution to rate of return for a typical nonsimple investment (Example 4.15)

---

[7] In Appendix 4A, we discuss methods of predicting the number of i^* values by looking at cash flows. However, generating a PE profile to discover multiple i^*s is as practical and informative as any other method.

4.7 Internal Rate of Return Criterion

Now that we have classified investment projects and learned methods to determine the i^* value for a given project's cash flows, our objective is to develop an accept/reject decision rule that gives results consistent with those obtained from NPW analysis.

4.7.1 Relationship to the PE Analysis

As we already observed in Example 4.5, page 213, the decision to accept or reject a project based on PE analysis is dependent on the rate of interest used for the PE computation. A different rate may change a project from being considered acceptable to being unacceptable. Consider again the PE profile as drawn for the simple project in Fig. 4.14(a). For interest rates below i^*, this project should be accepted as PE > 0; for interest rates above i^* it should be rejected.

On the other hand, for certain nonsimple projects, the PE may look like the one shown in Fig. 4.14(b). Use of PE analysis would lead you to accept the projects in regions **A** and **C**, but reject those in regions **B** and **D**. Of course, this result goes against intuition, since a higher interest rate would change an unacceptable project into an acceptable one. The situation graphed in Fig. 4.14(b) is one of the cases of multiple i^*s mentioned in Definition 2. Therefore, for the simple investment situation in Fig. 4.14(a), the i^* can serve as an appropriate index for either accepting or rejecting the investment. However, for the nonsimple investment in Fig. 4.14(b), it is not clear which i^* to use to make an accept/reject decision. Therefore, the i^* value fails to provide an appropriate measure of profitability for an investment project with multiple rates of return.

4.7.2 Accept/Reject Decision Rule for Simple Investments

Suppose we have the simple investment in Example 4.5, page 213. Why are we interested in finding the particular interest rate that equates a project's cost with the present worth of its receipts? Again, we may easily answer this by re-examining Fig. 4.5. In this figure, we notice two important characteristics of the PE profile. First, as we compute the project's $PE(i)$ at a varying interest rate (i), we see that the PE is positive for $i < i^* = 17.45\%$, indicating that the project would be acceptable under the PE analysis for those values of i. Second, the PE is negative for $i > i^* = 17.45\%$, indicating that the project is unacceptable for those values of i. Therefore, the i^* serves as a **benchmark** interest rate. By knowing this bench mark rate, we will be able to make an accept/reject decision consistent with the PE analysis.

Note that, for any simple investment, i^* is indeed the IRR of the investment (see Section 4.6.3). So IRR is 17.45% for Example 4.5. Merely knowing the i^* does not by itself indicate anything about the attractiveness of the investment. It is the interest rate for which PE = 0 and therefore we are indifferent to the project. However, firms typically wish to do better than break even and require projects to meet a minimum attractive rate of return (MARR) which is determined by company policy, management, or the project decision maker. If the IRR exceeds this MARR, we are assured that the company will more than break even. Thus, the IRR becomes a useful gauge to judge project acceptability, and the decision rule for a simple investment project is as follows:

> If IRR > MARR, accept the project.
> If IRR = MARR, remain indifferent.
> If IRR < MARR, reject the project.

Note that this decision rule is designed to be applied for a single project evaluation. When we have to compare mutually exclusive investment projects, we need to apply the **incremental analysis approach**, as we shall see in Chapter 5.

EXAMPLE 4.16

Investment decision for a simple investment

Merco Inc., a machinery builder in Toronto, is considering making an investment of $1,250,000 in a complete, structural beam-fabrication system. The increased productivity resulting from the installation of the system is central to the project's justification. Merco estimates the following figures as a basis for calculating productivity:

- Increased fabricated steel production: 2000 tonnes/year
- Average sales price/tonne fabricated steel: $2566/tonne
- Labor rate : $10.50/hour
- Cost of steel per tonne: $1950/tonne
- Additional maintenance cost: $128,500/year

The cost of producing a tonne of fabricated steel with this new system is estimated at $2170. With a selling price of $2566 per tonne, the resulting contribution to overhead and profit becomes $396 per tonne. Assuming that Merco will be able to sustain an increased production of 2000 tonnes per year by purchasing the system, the projected additional contribution has been estimated to be 2000 tonnes × $396 = $792,000.

Since the system has the capacity to fabricate the full range of structural steel, two workers can run it, one on the saw and the other on the drill. A third operator is required to operate a crane for loading and unloading materials. Merco's current conventional manufacturing system employs 17 people for center punching, hole making with a radial or magnetic drill, and material handling compared to three workers with the new system. This translates into a labor savings in the amount of $294,000 per year (14 × $10.50 × 40 hours/week × 50 weeks/year). The system can last for 15 years with an estimated after-tax salvage value of $80,000. However, after an annual deduction of $226,000 in corporate income taxes, the net investment cost as well as savings are as follows:

- Project investment cost: $1,250,000
- Projected annual net savings:
  ($792,000 + $294,000) – $128,500 – $226,000 = $731,500
- Projected after-tax salvage value at the end of year 15: $80,000
  (a) What is the projected IRR on this fabrication investment?
  (b) If Merco's MARR is known to be 18%, is this investment justifiable?

Solution

Given: Projected cash flows as shown in Figure 4.17, MARR = 18%

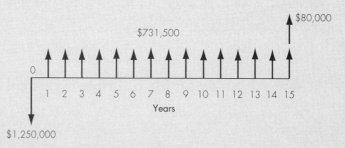

Figure 4.17 Cash flow diagram (Example 4.16)

Find: (a) IRR and (b) whether to accept or reject the investment

(a) Since only one sign change occurs in the net cash flow series, the fabrication project is a simple investment. This indicates that there will be a unique rate of return that is internal to the project:

$$PE(i) = -\$1,250,000 + \$731,500(P/A, i, 15)$$
$$+ \$80,000(P/F, i, 15)$$
$$= 0.$$

Using the trial-and-error approach outlined in Section 4.6.5, let's calculate the present values at two interest rates:

At $i = 50\%$:

$$PE(50\%) = -\$1,250,000 + \$731,500(P/A, 50\%, 15)$$
$$+ \$80,000(P/F, 50\%, 15)$$
$$= \$209,842.$$

At $i = 60\%$:

$$PE(60\%) = -\$1,250,000 + \$731,500(P/A, 60\%, 15)$$
$$+ \$80,000(P/F, 60\%, 15)$$
$$= -\$31,822.$$

After making further iterative calculations, you will find that the IRR is about 58.71% for the net investment of $1,250,000.

(b) The IRR figure far exceeds Merco's MARR, indicating that the fabrication system project is an economically attractive one. Merco's management believes that, over a broad base of structural products, there is no doubt that the installation of the system would result in a significant savings, even after considering some potential deviations from the estimates used in the above analysis.

4.7.3 Accept/Reject Decision Rule for Nonsimple Investments

When applied to independent simple investment projects, the i^* provides an unambiguous criterion for measuring profitability. However, when multiple rates of return occur, none of them is an accurate portrayal of project acceptability or profitability. Clearly, then, we should place a high priority on discovering this situation early in our analysis of a project's cash flows. The most reliable way to predict multiple i^*s is to generate a PE profile on computer and to check if it crosses the horizontal axis more than once.

In addition to the PE profile, there are good—although somewhat more complex—analytical methods for predicting multiple i^*s. Perhaps more importantly, there is a good method, which uses an **external interest rate**, for refining our analysis when we do discover multiple i^*s. An external rate of return allows us to calculate a single valid rate of return; it is discussed in Appendix 4A.

If you choose to avoid these more complex applications of rate of return techniques, you must be able to predict multiple i^*s via the PE profile and, when they occur, select one of the equivalent worth methods (PE, FE or AE) in your analysis for determining project acceptability. Figure 4.18 summarizes the decision logic that should be followed in making investment decisions for independent projects based on the IRR criterion.

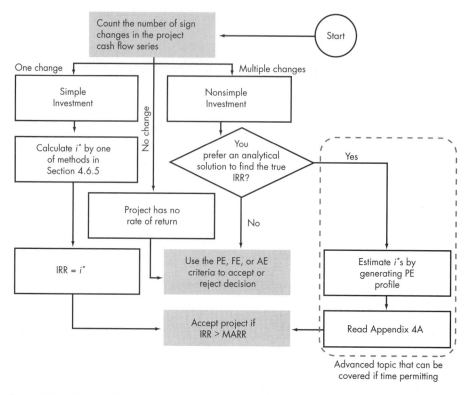

Figure 4.18　Summary of IRR criterion—a flow chart that summarizes how to proceed when applying the net cash flow sign rule

---

**EXAMPLE
4.17**　**Investment decision for a nonsimple project**

By outbidding its competitors Trane Image Processing (TIP) has received a government contract worth $7,300,000 to build flight simulators for pilot training over 2 years. For some contracts, the government makes an advance payment when the contract is signed, but in this case, the government will make two progressive payments; $4,300,000 at the end of the first year, and the $3,000,000 balance at the end of the second year. The expected cash outflows required to produce these simulators are estimated to be $1,000,000 now, $2,000,000 during the first year, and $4,320,000 during the second year. The expected net cash flows from this project are summarized as follows:

Year	Cash Inflow	Cash Outflow	Net Cash Flow
0		$1,000,000	–$1,000,000
1	$4,300,000	2,000,000	2,300,000
2	3,000,000	4,320,000	–1,320,000

In normal situations, TIP would not even consider a marginal project such as this one in the first place. However, hoping that it can establish itself as a technology leader in the field, management felt that it was worth outbidding its competitors by providing the lowest bid. Financially, what is the economic worth of outbidding the competitors for this project?

(a) Compute the values of i^* for this project.
(b) Make an accept/reject decision based on the results in part (a). Assume that the contractor's MARR is 15%.

Solution

Given: Cash flow shown above, MARR = 15%
Find: (a) i^* and (b) determine whether to accept the project

(a) Since this project has a 2-year life, we may solve the net present worth equation directly via the quadratic formula method:

$$PE(i^*) = -\$1,000,000 + \$2,300,000/(1 + i^*) - \$1,320,000/(1 + i^*)^2 = 0$$

If we let $X = 1/(1 + i^*)$, we can rewrite the expression

$$-1,000,000 + 2,300,000X - 1,320,000X^2 = 0.$$

Solving for X gives $X = 10/11$ and $10/12$, or $i^* = 10\%$ and 20%. As shown in Figure 4.19, the PE profile intersects the horizontal axis twice, once at 10% and again at 20%. The investment is obviously not a simple one, and thus neither 10% nor 20% represents the true internal rate of return of this government project.

(b) Since the project is a nonsimple project, we may abandon the IRR criterion for practical purposes and use the PE criterion. If we use the present worth method at MARR = 15%, we obtain

$$
\begin{aligned}
PE(15\%) \quad = \quad & -\$1,000,000 + \$2,300,000(P/F, 15\%, 1) \\
& -\$1,320,000(P/F, 15\%, 2) \\
= \quad & \$1890 > 0,
\end{aligned}
$$

which verifies that the project is marginally acceptable, and it is not as bad as initially believed.

Comments: Example 4A.4 in Appendix 4A illustrates how you go about finding the true rate of return for this nonsimple investment project.

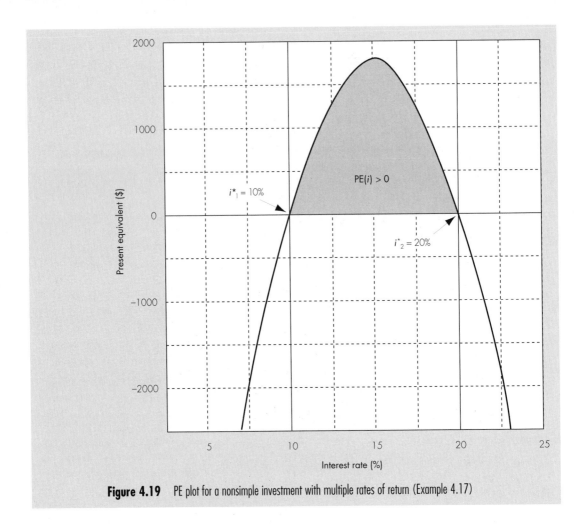

Figure 4.19 PE plot for a nonsimple investment with multiple rates of return (Example 4.17)

4.8 Computer Notes

In this section, we will demonstrate how to create a net present worth or present equivalent table and plot as a function of interest rate by using Excel.[8] We will use the cash flow data described in Example 4.9.

---

[8] **EzCash** has a built-in command to generate a net present worth or present equivalent table and plot when the interest range is specified.

4.8.1   Creating an NPW Table

Using Figure 4.20 as a model, enter the known time periods and their cash flows (the figures in parentheses represent negative amounts). The interest rate (MARR) is specified in cell C1. To calculate the PE of the cash flows at a specified interest rate, you can use the **NPV(rate, range)** function. The NPV(rate, range) function assumes that cash flows occur at the ends of periods. To find the PE of an investment where you make an initial cash outflow immediately, or at period 0, and follow it by a series of future cash flows, you must factor the initial flow separately because it is not affected by the interest.

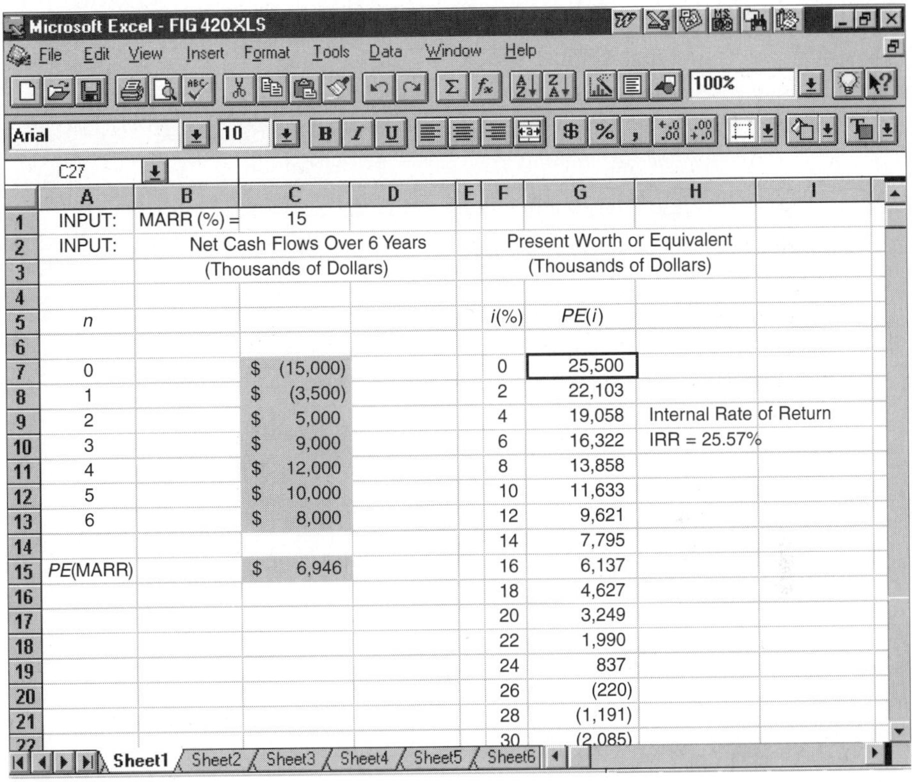

Figure 4.20   Excel's output: Example 4.9

Cell C7 is your initial flow (at period 0), cells from C8 to C13 contain a range of future cash flows, and 15%, which is stored in cell C1, is the annual interest rate. The total PE is calculated by

$$= NPV(C1\%, C8:C13) + C7.$$

The result in the amount of $6946 thousands is displayed in cell C15.

To create a present worth profile, you may enter the range of interest rates in column F (say 0% through 30%, in increments of 2%). The $ sign in front of either component of a cell reference (i.e., the column letter or the row number) provides an absolute anchor by preventing that component of the reference from changing when copying or moving the formula to another cell. For example, to obtain the net present values at varying interest rates, you may follow these two steps:

- Step 1: Enter the PE cell formula at 0% interest in cell G7:

$$= NPV(F7\%, C\$8{:}C\$13) + C\$7$$

- Step 2: To copy to adjacent cells, make a selection at cell G7 and then position the mouse pointer over the fill handle. The pointer changes to a black cross. Drag the fill handle in the direction you want to copy (downward to cell G22). When you release the mouse button, the cell formula is copied into each cell in the range (Figure 4.21).

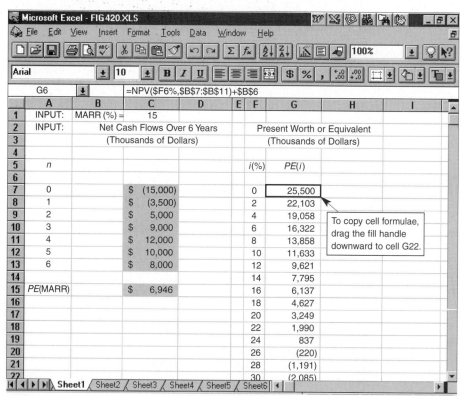

Figure 4.21   Excel's output: A copy function provided in Excel

4.8.2   Creating a Present Worth Chart

You can now easily build a PE graph using the **ChartWizard** command in Excel, which appears in the main menu. **ChartWizard** is a series of dialog boxes that simplifies creating a chart. Additionally, **ChartWizard** guides you through the process step by step: You verify your data selection, select a chart type, and decide whether to add items such as titles and a legend. You can create an embedded chart as an object on a worksheet when you want to display a chart along with its associated data, or you can create a chart sheet as a separate sheet in a workbook when you want to display a chart apart from its associated data. You might do this when you want to show overhead projections of your charts as part of a presentation. Whether you create an embedded chart or a chart sheet, your chart data is automatically linked to the worksheet from which you created it. When you change the data on your worksheet, the chart is updated to reflect these changes.

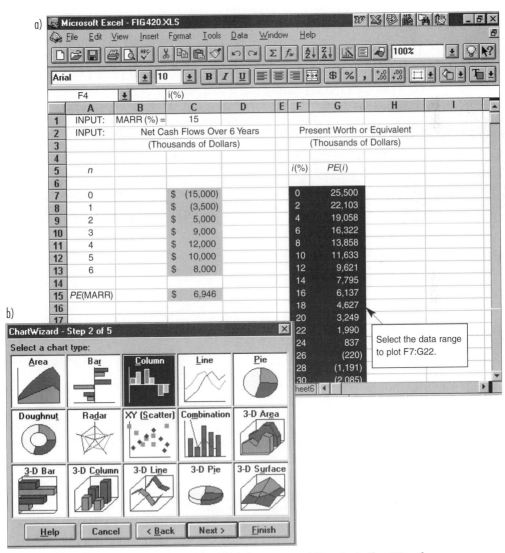

Figure 4.22 Excel's output: Types of graphical chart options available in Excel's **ChartWizard**

For example, to create an XY graph of this present worth data, you may follow these steps:

- Step1: Select the worksheet data you want to display in the chart as shown in Figure 4.22(a).
- Step 2: Click the **ChartWizard** button and proceed as follows:

1. Select an XY chart (chart type) with data points connected by a smooth line (subchart type), Figure 4.22(b).
2. Press the **Next** button to display the data per the chart selection in the previous step.
3. Press the **Next** button again, add required titles, etc., and choose options on grid lines, etc., to further customize the chart.
4. Press the **Next** button again and choose between embedding the chart in the worksheet or leaving it as a separate sheet.
5. Press the **FINISH** button and the line chart is added to the worksheet as shown in Figure 4.23.

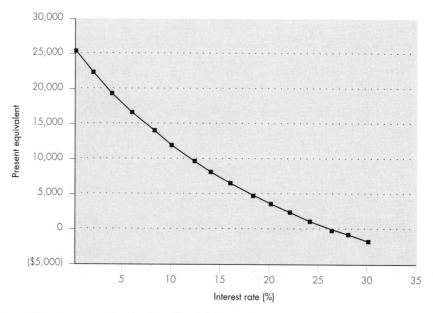

Figure 4.23 Present equivalent plot obtained by Excel for Example 4.9

4.8.3 Printing the Worksheet and Chart

Before you can print, you must select a printer. From the **File** menu, choose **Print**. Click the Printer button, and then select the printer you want to use. Once you select the printer, you can control the appearance of the printed sheets by changing the options, such as margins, page orientation, headers, and footnotes in the **Page Setup** dialog box. By previewing your sheet, you can see each page exactly as it will print, with the correct margins, page breaks, and headers and footers in place.

4.8.4 IRR Financial Functions with Excel

For rate of return analysis, we can utilize the IRR function of Excel. The IRR function requires two arguments: A guess concerning the resulting i^* and the range of cells containing the cash flows. Excel uses the guess as a first approximation of the i^*, then refines this guess until it converges into the correct value. (If you enter a guess value that deviates too much from the true value, you may see an error message. Either increase or decrease the guess value.)

= **IRR (values, guess)**: Returns a rate of return or an i^* value for a series of cash flows.

Values is an array or a reference to cells that contain numbers for which you want to calculate the IRR. Values must contain at least one positive value and one negative value to calculate the IRR.

Guess is a number that you guess is close to the result of IRR. Microsoft Excel uses an iterative technique for calculating the IRR. Starting with guess, the IRR cycles through the calculation until the result is accurate within 0.00001 percent. If the IRR can't find a result that works after 20 tries, the "#NUM!" error value is returned. In most cases you do not need to provide a value for guess in the IRR calculation. If guess is omitted, it is assumed to be 0.1 (10%). If IRR gives the "#NUM!" error value, or if the result is not close to what you expected, you need to try again with a different value for guess.

To calculate IRR for Example 4.9, we use the cash flows in Figure 4.20. The IRR is calculated as

$$= \text{IRR}(C7\text{: } C13, 0.15)$$

We have used an initial guess for IRR equal to MARR, i.e., 15%. The IRR for this series of cash flows is 25.57% and appears in cell H10 in Figure 4.20.

4.9 Summary

In this chapter, we have presented four methods for measuring the economic attractiveness or profitability of independent projects. We observed the following important results:

- Present, future, and annual worth are three equivalence methods of analysis in which a project's cash flows are converted to a single present (PE), future (FE), or annual (AE) equivalent value.
- The interest rate used in these equivalence calculations is MARR or the minimum attractive rate of return. It is generally dictated by management and is the interest rate at which a firm can always earn or borrow money.
- The three equivalent worth criteria are related in the following way

$$FE(i) = PE(i)(F/P, i, N)$$
$$AE(i) = PE(i)(A/P, i, N) = FE(i)(A/F, i, N)$$

  when FE is referenced to end of the project, N.

- The capital recovery cost factor, or $CR(i)$, is an important application of AE analysis in that it allows managers to calculate an annual equivalent cost of capital for ease of itemization with annual operating costs. The equation for $CR(i)$ is

$$CR(i) = (P - S)(A/P, i, N) + iS$$

  where P = initial cost and S = salvage value.

- Rate of return analysis measures project profitability in terms of a calculated interest rate or yield for the project.
- Mathematically we can determine the rate of return for a given project cash flow series by locating a real positive interest rate that equates the net present worth of its cash flows to zero. This break-even interest rate is denoted by the symbol i^*.
- **Rate of return (ROR)** is the break-even interest rate, i^*, earned on unrecovered project balances such that an investment's cash receipts make the terminal project balance equal to zero.
- **Internal rate of return (IRR)** is a specific form of ROR that stresses the fact that we are concerned with the interest earned on the portion of the project that is internally invested, not those portions that are released by (borrowed from) the project.
- To apply rate of return analysis correctly, we need to classify an investment into either a simple or a nonsimple investment. A **simple investment** is defined as one in which the initial cash flows are negative and only one sign change in the net cash flow occurs, whereas a **nonsimple investment** is one for which more than one sign change in the cash flow series occurs. Multiple i^*s occur only in nonsimple investments. However, not all nonsimple investments will have multiple i^*s.
- For simple investments, ROR = IRR = i^*.
- For a nonsimple investment, because of the possibility of having multiple rates of return, it is recommended that the IRR analysis be abandoned and one of the

Table 4.4
Summary of Project Analysis Methods for Independent Projects

Analysis Method/ Equation Form(s)	Description	Comments/Decision Rule
Payback period	A method for determining *when* in a project's history it breaks even.	Should be used as a screening method only—reflects liquidity, not profitability of project.
Present equivalent $$PE(i) = \sum_{n=0}^{N} A_n(P/F, i, n)$$	An equivalence method that translates a project's cash flows into a net present worth value.	If $PE(MARR) > 0$, accept. If $PE(MARR) = 0$, remain indifferent. If $PE(MARR) < 0$, reject
Future equivalent $$FE(i) = \sum_{n=0}^{N} A_n(F/P, i, N-n)$$	An equivalence method that translates a project's cash flows into a net future worth value.	If $FE(MARR) > 0$, accept. If $FE(MARR) = 0$, remain indifferent. If $FE(MARR) < 0$, reject
Annual equivalent $$AE(i) = \left[\sum_{n=0}^{N} A_n(P/F, i, n) \right](A/P, i, N)$$	An equivalence method that translates a project's cash flows into a net annual worth value.	If $AE(MARR) > 0$, accept. If $AE(MARR) = 0$, remain indifferent. If $AE(MARR) < 0$, reject
Rate of return $PE(i^*) = 0$ where $i^* = IRR$ (For simple investments only; see Appendix 4A for nonsimple investment case)	A method which calculates the interest rate earned on funds internal to the project.	For both simple and nonsimple investments, If $IRR > MARR$, accept. If $IRR = MARR$, remain indifferent. If $IRR < MARR$, reject

equivalent worth methods be used to make an accept/reject decision. Procedures are outlined in Appendix 4A for determining the rate of return internal to non-simple investments. Once you find the IRR (or return on invested capital), you can use the same decision rule for simple investments.

- The three equivalent worth methods and the rate of return method result in identical conclusions with respect to the attractiveness or profitability of a project. Table 4.4 summarizes the decision rules for each of the methods.

- The choice of one method over another is dependent on the specifics of the project being evaluated and analyst preference. Because rate of return is an intuitively familiar and understandable measure of project profitability, many managers prefer it to the equivalent worth methods.

Problems

Note: *Unless otherwise stated, all cash flows represent after-tax cash flows. The interest rate (MARR) is also given on an after-tax basis.*

Note: *The symbol i^* represents the interest rate that makes the net present value of the project equal to zero. The symbol IRR represents the* **internal** *rate of return of the investment. For a simple investment, $IRR = i^*$. For a nonsimple investment, generally i^* is not equal to IRR.*

Level 1

4.1* An investment project costs P. It is expected to have an annual net cash flow of $0.125P$ for 20 years. What is the project's payback period?

4.2* You are given the following financial data:

- Investment cost at $n = 0$: $10,000
- Investment cost at $n = 1$: $15,000
- Useful life: 10 years (after year 1)
- Salvage value (at the end of 11 years): $5000
- Annual revenues: $12,000 per year
- Annual expenses: $4000 per year
- MARR: 10%.

(**Note:** *The first annual revenue and expenses will occur at the end of year 2.*)

Determine the projects payback and discounted payback periods.

4.3* Find the net present worth of the following cash flow series at an interest rate of 9%.

End of Period	Cash Flow
0	−$100
1	−200
2	−300
3	−400
4	−500
5 – 8	900

4.4* Find the net present worth of the following cash flow series at an interest rate of 9%.

End of Period	Cash Flow	End of Period	Cash Flow
0	−$100	5	−$300
1	−150	6	−250
2	−200	7	−200
3	−250	8	−150
4	−300	9	−100

4.5 Camptown Togs, Inc., a children's clothing manufacturer, has always found payroll processing to be costly because it must be done by a clerk so that the number of piece-goods coupons for each employee can be collected and the types of tasks performed by each employee can be calculated. Recently an industrial engineer has designed a system that partially automates the process by means of a scanner that reads the piece-goods coupons. Management is enthusiastic about this system because it utilizes some personal computer systems that were purchased recently. It is expected that this new automated system will save $30,000 per year in labor. The new system will cost about $25,000 to build and test prior to operation. It is expected that operating costs, including income taxes, will be about $5000 per year.

The system will have a 5-year useful life. The expected net salvage value of the system is estimated to be $3000.

(a) Identify the cash inflows over the life of the project.
(b) Identify the cash outflows over the life of the project.
(c) Determine the net cash flows over the life of the project.

4.6 Refer to Problem 4.5:

(a) How long does it take to recover the investment?
(b) If the firm's interest rate is 15% after tax, what would be the discounted payback period for this project?

4.7 For each of the following cash flows:

	Project's Cash Flow			
n	A	B	C	D
0	–$1500	–$4000	–$4500	–$3000
1	300	2000	2000	5000
2	300	1500	2000	3000
3	300	1500	2000	–2000
4	300	500	5000	1000
5	300	500	5000	1000
6	300	1500		2000
7	300			3000
8	300			

(a) Calculate the payback period for each project.
(b) Determine whether it is meaningful to calculate a payback period for project D.
(c) Assuming $i = 10\%$, calculate the discounted payback period for each project.

4.8 Consider the following independent investment projects. All projects have a 3-year service life:

	Project's Cash Flow			
n	A	B	C	D
0	–$1000	–$1000	–$1000	–$1000
1	0	600	1200	900
2	0	800	800	900
3	3000	1500	1500	1800

(a) Compute the net present worth of each project at $i = 10\%$.
(b) Plot the present worth as function of interest rate (from 0% to 30%) for project B.

4.9 Consider the following independent investment projects. All projects have a 3-year service life.

Period	Project's Cash Flow			
(n)	A	B	C	D
0	–$2500	–$1000	$2500	–$3000
1	5400	–3000	–7000	1500
2	14400	1000	2000	5500
3	7200	3000	4000	6500

(a) Compute the net present worth of each project at $i = 13\%$.
(b) Compute the net future worth of each project at $i = 13\%$.

Which project(s) are acceptable?

4.10* Consider the following independent investment projects:

	Project's Cash Flow				
n	A	B	C	D	E
0	–$1000	–$5000	–$1000	–$3000	–$5000
1	500	2000	0	500	1000
2	900	–3000	0	2000	3000
3	1000	5000	3000	3000	2000
4	2000	5000	7000	4000	
5	–500	3500	13,000	1250	

(a) Compute the future worth at the end of life for each project, at $i = 15\%$.
(b) Determine the acceptability of each project.

4.11 Refer to Problem 4.7 and use future worth and rate of return analysis to determine the viability of projects A, B and C for MARR = 10%.

4.12 Refer to Problem 4.8 and compute the future worth for each project at $i = 12\%$.

4.13* Compute the annual equivalent of an income stream of $1000 per year, to be received at the end of each of the next 3 years, at an interest rate of 12%.

4.14* Your firm has purchased an injection molding machine at a cost of $100,000. The machine's useful life is estimated at 8 years. Your accounting department has estimated the capital recovery cost for this machine at about $25,455 per year. If your firm's MARR is 20%, how much salvage value do you think the accounting department assumed at the end of 8 years?

4.15* You purchased a cutting machine for $18,000. It is expected to have a useful life of 10 years and a salvage value of $3000. At $i = 15\%$, what is the capital recovery cost of this machine?

4.16 Consider the following cash flows and compute the equivalent annual worth at $i = 12\%$.

	A_n	
n	Investment	Revenue
0	−$10,000	
1		$2000
2		2000
3		3000
4		3000
5		1000
6	+ 2000	500

4.17* Consider the cash flow diagram. Compute the equivalent annual worth at $i = 12\%$.

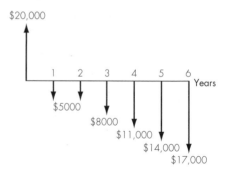

4.18 Consider the cash flow diagram. Compute the equivalent annual worth at $i = 10\%$.

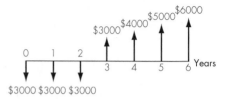

4.19 Consider the cash flow diagram. Compute the equivalent annual worth at $i = 13\%$.

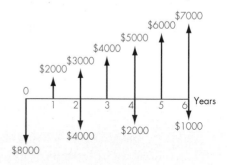

4.20 Consider the cash flow diagram. Compute the equivalent annual worth at $i = 8\%$.

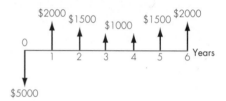

4.21 Consider the following independent investment projects:

Project's Cash Flow				
n	A	B	C	D
0	−$2000	−$4000	−$3000	−$9000
1	400	3000	−2000	2000
2	500	2000	4000	4000
3	600	1000	2000	8000
4	700	500	4000	8000
5	800	500	2000	4000

Compute the equivalent annual worth of each project at $i = 10\%$ and determine the acceptability of each project.

4.22 In Problem 4.21, plot the equivalent annual worth of each project as a function of interest rate (0% – 40%).

4.23 Consider the following independent investment projects:

Period (n)	Project's Cash Flow			
	A	B	C	D
0	−$3000	−$3000	−$3000	−$3000
1	0	1500	3000	1800
2	0	1800	2000	1800
3	5500	2100	1000	1800

Compute the equivalent annual worth of each project at $i = 13\%$ and determine the acceptability of each project.

4.24 Consider the following project's cash flows:

	Net Cash Flow	
n	Investment	Operating Income
0	−$800	
1		$500
2		400
3	−800	300
4		500
5		400
6	−800	300
7		500
8		400
9		300

Find the equivalent annual worth for this project at $i = 10\%$ and determine the acceptability of the project.

4.25* Consider the cash flows for the following independent investment projects:

	Project's Cash Flow	
n	A	B
0	−$4000	$5500
1	1000	−1400
2	X	−1400
3	1000	−1400
4	1000	−1400

(a) For project A, find the value of X that makes the equivalent annual receipts equal the equivalent annual disbursement at $i = 13\%$.

(b) Would you accept B at $i = 15\%$ based on AE criterion?

4.26* You are considering an investment that costs $2000. It is expected to have a useful life of 3 years. You are very confident about the revenues during the first and the third year, but you are unsure about the revenue in year 2. If you hope to make at least a 10% rate of return on your investment ($2000), what should be the minimum revenue in year 2?

Year	Cash Flow
0	–$2000
1	1000
2	X
3	1200

4.27* You are considering a cutting machine that costs $150,000. This machine will have an estimated service life of 10 years with a net after-tax salvage value of $15,000. Its annual after-tax operating and maintenance costs are estimated to be $50,000. To expect an 18% rate of return on investment after-tax, what would be the required minimum annual after-tax revenues?

$4.28 Assume that you are going to buy a new car worth $22,000. You will be able to make a down payment of $7000. The remaining $15,000 will be financed by the dealer. The dealer computes your monthly payment to be $512 for 48 months' financing. What is the dealer's rate of return on this loan transaction?

$4.29* Mr. Smith wishes to sell a bond that has a face value of $1000. The bond bears a coupon rate of 8% with bond interest payable semi-annually. Four years ago, the bond was purchased at $900. At least 9% annual return on investment is desired. What must be the minimum selling price for the bond today in order to make the desired return on the investment?

$4.30 The Vincent Van Gogh painting, "Irises," was purchased by John Whitney Payson for $80,000 in 1947. He sold it for $53.9 million in 1988. If Mr. Payson invested his $80,000 in another investment vehicle (such as stock), how much interest would he need to earn to accumulate the same wealth as from the painting investment? Assume for simplicity that the investment period is 40 years, and the interest is compounded annually.

4.31 Consider four investments with the following sequences of cash flows:

	Net Cash Flow			
n	Project A	Project B	Project C	Project D
0	–$18,000	–$20,000	$34,578	–$56,500
1	10,000	32,000	–18,000	–2500
2	20,000	32,000	–18,000	–6459
3	30,000	–22,000	–18,000	–78,345

(a) Identify all the simple investments.
(b) Identify all the nonsimple investments.
(c) Compute i^* for each investment.
(d) Which project(s) has no rate of return?

$4.32 An investor bought 100 shares of stock at a cost of $10 per share. He held the stock for 15 years and then sold it for a total of $4000. For the first 3 years, he received no dividends. For each of the next 7 years, he received total dividends of $50 per year. For the remaining period he received total dividends of $100 per year. What rate of return did he make on the investment?

4.33 Consider the following independent investment projects:

n	A	B	C	D	E
0	–$100	–$100		–$200	–$50
1	60	70	$20	120	–100
2	900	70	10	40	–50
3		40	5	40	0
4		40	–180	–20	150
5			60	40	150
6			50	30	100
7			40		100
8			30		
9			20		
10			10		

(a) Classify each project as either simple or nonsimple.

(b) Compute the i^* for A using the quadratic equation.

(c) Obtain the rate(s) of return for each project by plotting the PE as a function of interest rate.

4.34 Consider the following independent projects:

n	A	B	C	D
		Net Cash Flow		
0	−$1000	−$1000	−$1700	−$1000
1	500	800	5600	360
2	100	600	4900	4675
3	100	500	−3500	2288
4	1000	700	−7000	
5			−1400	
6			2100	
7			900	

(a) Classify each project as either simple or nonsimple.

(b) Identify all positive i^*s for each project.

(c) Plot the present worth as a function of interest rate (i) for each project.

4.35* Consider an investment project with the following cash flows:

n	Cash Flow
0	−$5000
1	0
2	4840
3	1331

Compute the IRR for this investment. Is this project acceptable at MARR = 10 %?

4.36 Consider the following project's cash flow:

n	Net Cash Flow
0	−$2000
1	800
2	900
3	X

If the project's IRR is 10%,

(a) Find the value of X.

(b) Is this project acceptable at MARR = 8%?

Level 2

4.37* Which of the following statements is incorrect for i) the payback method and ii) the discounted payback method?

(a) If two investors are considering the same project, the payback period will be longer for the investor with the higher MARR.

(b) If you were to consider the cost of funds in a payback period calculation, you would have to wait longer as you increase the interest rate.

(c) Considering the cost of funds in a payback calculation is equivalent to finding the time period when the project balance becomes zero.

(d) The simplicity of the payback period method is one of its most appealing qualities even though it fails to measure project profitability.

$4.38* You are considering buying an old house that you will convert into an office building for rental. Assuming that you will own the property for 10 years, how much would you be willing to pay for the old house now given the following financial data?

- Remodelling cost at period 0 = $20,000
- Annual rental income = $25,000
- Annual upkeep costs (including taxes) = $5000
- Estimated net property value (after taxes) at the end of 10 years = $225,000
- The time value of your money (interest rate) = 8% per year.

4.39* Your R&D group has developed and tested a computer software package that helps engineers control the proper chemical mix for the various process-manufacturing industries. If you decide to market the software, your first year operating net cash flow is estimated to be $1,000,000. Because of market competition, product life will be about 4 years, and the product's market share will decrease by 25% each year over the previous year's share. You are approached by a big software house which wants to purchase the right to manufacture and distribute the product. Assuming that your interest rate is 15%, for what minimum price would you be willing to sell the software?

4.40* The following table contains a summary of how a project's balance is expected to change over its 5-year service life at 10% interest.

End of Period	Project Balance
0	−$1000
1	−1500
2	600
3	900
4	1500
5	2000

Which of the following statements is incorrect?

(a) The required additional investment at the end of period 1 is $500.
(b) The net present worth of the project at 10% interest is $1242.
(c) The net future of the project at 10% interest is $2000.
(d) Within 2 years, the company will recover all its investments and the cost of funds (interest) from the project.

4.41* NasTech Corporation purchased a vibratory finishing machine for $20,000 in year 0. The useful life of the machine is 10 years, at the end of which the machine is estimated to have a zero salvage value. The machine generates net annual revenues of $6000. The annual operating and maintenance expenses are estimated to be $1000. If NasTech's MARR is 15%, how many years does it take before this machine becomes profitable?

4.42* You need to know if the building of a new warehouse is justified under the following conditions:

The proposal is for a warehouse costing $100,000. The warehouse has an expected useful life of 35 years and a net salvage value (net proceeds from sale after tax adjustments) of $25,000. Annual receipts of $17,000 are expected, annual maintenance and administrative costs will be $4000 per year, and annual income taxes are $2000.

Given these data, which of the following statements are correct?

(a) The proposal is justified for a MARR of 9%.
(b) The proposal has a net present worth of $62,730.50, when 6% is used as the interest rate.
(c) The proposal is acceptable as long as MARR ≤ 10.77%.
(d) All of the above are correct.

4.43 A large food-processing corporation is considering using laser technology to speed up and eliminate waste in the potato-peeling process. To implement the system, the company anticipates needing $3 million to purchase the industrial-strength lasers. The system will save $1,200,000 per year in labor and materials. However, it will require an additional operating and maintenance cost of $250,000. Annual income taxes will also increase by $150,000. The system is expected to have a 10-year service life and will have a salvage value of about $200,000. If the company's MARR is 18%, justify the economics of the project based on:

(a) PE method   (c) AE method
(b) FE method   (d) Rate of return

4.44 Consider the following project balances for a typical investment project with a service life of 4 years:

n	A_n	Project Balance
0	−$1000	−$1000
1	( )	−1100
2	( )	−800
3	460	−500
4	( )	0

(a) Construct the original cash flows of the project.
(b) Determine the interest rate used in computing the project balance.
(c) At i = 15%, would this project be acceptable?

4.45* Consider the project balance diagram for a typical investment project with a service life of 5 years. The numbers in the figure indicate the beginning project balances.

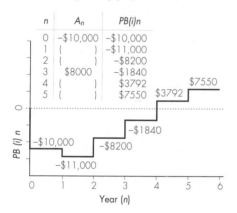

(a) From the project balance diagram, construct the project's original cash flows.
(b) What is the project's conventional payback period (without interest)?

4.46 Consider the following cash flows and present worth profile:

	Net Cash Flows ($)	
Year	Project 1	Project 2
0	−$100	−$100
1	40	30
2	80	Y
3	X	80

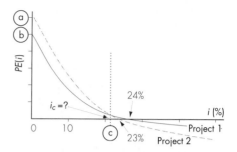

(a) Determine the values for X and Y.
(b) Calculate the terminal project balance of project 1 at MARR=24%.
(c) Find the values for ⓐ, ⓑ and ⓒ in the PE plot.

4.47 Consider the project balances for a typical investment project with a service life of 5 years:

n	A_n	Project Balance
0	−$1000	−$1000
1	( )	−900
2	490	−500
3	( )	0
4	( )	−100
5	200	( )

(a) Construct the original cash flows of the project and the terminal balance and fill in the blanks above.
(b) Determine the interest rate used in the project balance calculation, and compute the present worth of this project at the computed interest rate.

4.48 Refer to Problem 4.7:

(a) Graph the project balances (at i = 10%) of each project as a function of n.
(b) By examining the graphical results in part (a), determine which project appears to be the safest to undertake if there is some possibility of premature termination of the projects at the end of year 2.

4.49 Refer to Problem 4.10.

(a) Plot the future worth for each project as a function of interest rate (0% – 50%).

(b) Compute the project balance of each project at $i = 15\%$.

(c) Compare the terminal project balances calculated in (b) with the results obtained in Problem 4.10(a). Without using the interest factor tables, compute the future worth based on the project balance concept.

4.50 Consider the following independent investment projects:

	Project Cash Flows		
n	A	B	C
0	−$100	−$100	$100
1	50	40	−40
2	50	40	−40
3	50	40	−40
4	−100	10	
5	400	10	
6	400		

Assume MARR=10% for the following questions:

(a) Compute the net present worth for each project and determine the acceptability of each project.

(b) Compute the net future worth of each project and determine the acceptability of each project.

(c) Compute the future worth of each project at the end of 6 years with variable MARRs as follows: 10% for $n = 0$ to $n = 3$ and 15% for $n = 4$ to $n = 6$.

4.51 Consider the following project balance profiles for proposed independent investment projects:

	Project Balances		
n	A	B	C
0	−$1000	−$1000	−$1000
1	−1000	−650	−1200
2	−900	−348	−1440
3	−690	−100	−1328
4	−359	85	−1194
5	105	198	−1000
Interest rate used	10%	?	20%
PE	?	$79.57	?

Project balance figures are rounded to nearest dollars.

(a) Compute the net present worth of projects A and C, respectively.

(b) Determine the cash flows for project A.

(c) Identify the net future worth of project C.

(d) What interest rate would be used in the project balance calculations for project B?

4.52 Consider the following project balance profiles for proposed independent investment projects:

	Project Balances		
n	A	B	C
0	−$1000	−$1000	−$1000
1	−800	−680	−530
2	−600	−302	X
3	−400	−57	−211
4	−200	233	−89
5	0	575	0
Interest rate used	0%	18%	12%

Project balance figures are rounded to nearest dollars.

(a) Compute the net present worth of each investment.

(b) Determine the project balance at the end of period 2 for project C, if $A_2 = \$500$.

(c) Determine the cash flows for each project.

(d) Identify the net future worth of each project.

4.53 The owner of a business is considering investing $55,000 in new equipment. He estimates that the net cash flows during the first year will be $5000, but these will increase by $2500 per year the next year, and each year thereafter. The equipment is estimated to have a 10-year service life and a net salvage value at that time of $6000. The firm's interest rate is 12%.

(a) Determine the annual capital recovery cost (ownership cost) for the equipment.
(b) Determine the equivalent annual savings (revenues).
(c) Determine if this is a wise investment using annual equivalent.

4.54 Nelson Electronics Company just purchased a soldering machine to be used in its assembly cell for flexible disk drives. This soldering machine cost $250,000. Because of the specialized function it performs, its useful life is estimated to be 5 years. At that time its salvage value is also estimated to be $40,000. What is the equivalent annual capital cost for this investment if the firm's interest rate is 18%?

4.55 The present price (year 0) of kerosene is $0.26 per litre, and its cost is expected to increase by $0.03 per year. (Kerosene at the end of year 1 will cost $0.29 per litre.) Mr. Garcia uses about 4000 litres of kerosene during a winter season for space heating. He has an opportunity to buy a storage tank for $400, and at the end of 4 years, he can sell the storage tank for $100. The tank has a capacity to supply 4 years of Mr. Garcia's heating needs, so he would buy 4 years of kerosene at its present price ($0.26). He can invest his money elsewhere at 8%. Use AE analysis to determine if he should purchase the storage tank. Assume that kerosene purchased on a pay-as-you-go basis is paid for at the end of the year. (However, kerosene purchased for the storage tank is purchased now.) Would this decision change if kerosene increased $0.06 per litre per year rather than $0.03?

4.56 Consider the following advertisement that appeared in a local paper.

Pools-Spas-Hot Tubs—Pure Water without Toxic Chemicals: The IONETICS water purification system has proven highly effective in killing algae and bacteria in pools and spas. Here is how it works: The "Ion Chamber" installed in the return water line contains copper/silver electrodes. A safe, low-voltage current is sent through those electrodes from a "Computerized Controller." Copper and silver ions enter the water stream and the pool or spa where they attack and kill the algae and bacteria. These charged, dead microorganisms mutually attract, forming larger particles easily removed by the existing filtration system. The IONETICS system can make your pool water pure enough to drink without the use of chlorine or other toxic chemicals. The ion level need only be tested about once per week and ion output is easily adjusted. The comparative costs between the conventional chemical system (chlorine) and the IONETICS systems are as follows:

Item	Conventional System	IONETICS System
Annual costs		
Chemical	$471	
IONETICS		$85
Pump	$576	$100
($0.667/kWh)		
Capital investment		$1200

Note that the IONETICS system pays for itself in less than 2 years.

Assume that the IONETICS system has a 12-year service life, and the annual interest rate is 6%. What is the equivalent monthly cost of operating the IONETICS system?

4.57 A construction firm is considering establishing an engineering computing center. This center will be equipped with three engineering workstations that would cost

$25,000 each, and each has a service life of 5 years. The expected salvage value of each workstation is $2000. The annual operating and maintenance cost would be $15,000 for each workstation. At an MARR of 15%, determine the equivalent annual cost for operating the engineering center.

4.58 An industrial firm is considering purchasing several programmable controllers and automating their manufacturing operations. It is estimated that the equipment will initially cost $100,000 and the labor to install it will cost $35,000. A service contract to maintain the equipment will cost $5000 per year. Trained service personnel will have to be hired at annual salary of $30,000. Also estimated is an approximate $10,000 annual income-tax savings (cash inflow). How much will this investment in equipment and services have to increase the annual revenues after taxes to break even? The equipment is estimated to have an operating life of 10 years with no salvage value because of obsolescence. The firm's MARR is 10%.

$4.59* Consider the following two investment situations:

- In 1970, when Wal-Mart Stores, Inc. went public, an investment of 100 shares cost $1650. That investment would have been worth $2,991,080 after 25 years. The Wal-Mart investors' rate of return would be around 35%.
- In 1980, if you bought 100 shares of Fidelity Mutual Funds, it would have cost $5245. That investment would have been worth $80,810 after 15 years.

Which of the following statements is correct?

(a) If you bought only 50 shares of Wal-Mart stocks in 1970 and kept them for 25 years, your rate of return would be 0.5 times 35%.

(b) The investors in Fidelity Mutual Funds would have made profit at the annual rate of 30% on the funds remaining invested.

(c) If you bought 100 shares of Wal-Mart in 1970 but sold them after 10 years (assume that the Wal-Mart stocks grew at the annual rate of 35% for the first 10 years) then immediately put all the proceeds into Fidelity Mutual Funds, after 15 years the total worth of your investment would be around $511,140.

(d) None of the above.

4.60 Consider the following financial data for a project:

Initial investment	$10,000
Project life	8 years
Salvage value	$0
Annual revenue	$5,229
Annual expenses (including income taxes)	$3,000

(a) What is the i^* for this project?

(b) If annual expense increases at a 7% rate over the previous year's expenses but annual income is unchanged, what is the new i^*?

(c) In part (b), at what annual rate will annual income have to increase to maintain the same i^* obtained in part (a)?

4.61 The InterCell Company wants to participate in the World Fair to be held in Mexico. To participate, the firm needs to spend $1 million at year 0 to develop a showcase. The showcase will produce a cash flow of $2.5 million at the end of year 1. Then, at the end of year 2, $1.54 million must be expended to restore the land to its original condition. Therefore, the project's expected net cash flows are as follows (in thousands of dollars):

n	Net Cash Flow
0	−$1000
1	2500
2	−1540

(a) Plot the present worth of this investment as a function of i.

(b) Compute the i^*s for this investment.

(c) Would you accept this investment at MARR = 14%?

4.62 Recent technology has made possible a computerized vending machine that can grind coffee beans and brew fresh coffee on demand. The computer also makes possible such complicated functions as changing $5 and $10 bills and tracking the age of an item, moving the oldest stock to the front of the line, thus cutting down on spoilage. With a price tag of $4500 for each unit, Easy Snack has estimated the cash flows in millions of dollars over the product's 6-year useful life, including the initial investment, as follows:

n	Net Cash Flow
0	−$20
1	8
2	17
3	19
4	18
5	10
6	3

(a) If the firm's MARR is 18%, is this product worth marketing based on the IRR criterion?

(b) If the required investment remains unchanged, but the future cash flows are expected to be 10% higher than the original estimates, how much increase in IRR do you expect?

(c) If the required investment has increased from $20 million to $22 million, but the expected future cash flows are projected to be 10% smaller than the original estimates, how much decrease in IRR do you expect?

Level 3

4.63 Your firm is considering the purchase of an old office building with an estimated remaining service life of 25 years. The tenants have recently signed long-term leases, which leads you to believe that the current rental income of $150,000 per year will remain constant for the first 5 years. Then the rental income will increase by 10% for every 5-year interval over the remaining asset life. For example, the annual rental income would be $165,000 for year 6 through 10, $181,500 for year 11 through 15, $199,650 for year 16 through 20, and $219,615 for year 21 through 25. You estimate that operating expenses, including income taxes, will be $45,000 for the first year, and that they will increase by $3000 each year thereafter. You estimate that razing the building and selling the lot on which it stands will realize a net amount of $50,000 at the end of the 25-year period. If you had the opportunity to invest your money elsewhere, and thereby earn interest at the rate of 12% per annum, what would be the maximum amount you would be willing to pay for the building and lot at the present time?

4.64 Consider the following investment project:

n	A_n	i
0	−$2000	10%
1	2400	12
2	3400	14
3	2500	15
4	2500	13
5	3000	10

Suppose the company's reinvestment opportunities change over the life of the project as shown above (i.e., the firm's MARR changes over the life of the project). For example, the company can invest funds available now at 10% for the first year, 12% for the second year, and so forth. Calculate the net present worth of this investment and determine the acceptability of the investment.

4.65 Cable television companies and their equipment suppliers are on the verge of installing new technology that will pack many more channels into cable networks, thereby creating a potential programming revolution with implications for broadcasters, telephone companies, and the consumer electronics industry.

Digital compression uses computer techniques to squeeze 3 to 10 programs into a single channel. A cable system fully using digital compression technology would be able to offer a few hundred channels, compared with about 100 for the average cable television system now used. If the new technology is combined with the increased use of optical fibers the number of available channels would become even greater.

A cable company is considering installing this new technology to increase subscription sales and save on satellite time. The company estimates that the installation will take place over two years. The system is expected to have an 8-year service life and have the following savings and expenditures:

Digital Compression

Investment	
Now	$500,000
First year	$3,200,000
Second year	$4,000,000
Annual savings in satellite time	$2,000,000
Incremental annual revenues due to new subscriptions	$4,000,000
Incremental annual expenses	$1,500,000
Incremental annual income taxes	$1,300,000
Economic service life 8 years, net salvage value	$1,200,000

Note that the project has a 2-year investment period which is followed by an 8-year service life (a total 10-year life project). This implies that the first annual savings will occur at the end of year 3 and the last savings will occur at the end of year 10. If the firm's MARR is 15%, justify the economic worth of the project based on the PE method.

4.66 The Engineering department of a large firm is overly crowded. In many cases, several engineers share one office. It is evident that the distraction caused by crowded conditions considerably reduces the productive capacity of the engineers. Management is considering the possibility of new facilities for the department, which could result in fewer engineers per office and a private office for some. In the new building, three engineers would be assigned to offices measuring 4×8 metres. This same amount of space is occupied by five engineers in the existing facility. What productivity improvement is required of these engineers to justify such a change? The following data apply:

- Average annual salary of $60,000 per engineer.
- Cost of building per square metre is $650.
- Estimated life of building of 25 years.
- Estimated salvage value of the building is 10% of the first cost.
- Annual taxes, insurance, and maintenance are 6% of the first cost.
- Cost of janitor service, heating and illumination, etc. per square metre per year is $30.00.
- Interest rate is 12%.

Assume that engineers reassigned to other office space will maintain their present productive capability as a minimum.

4.67 Champion Chemical Corporation is planning to expand one of its propylene manufacturing facilities. At $n = 0$, a piece of property costing $1.5 million must be purchased to build a plant. The building, which needs to be constructed during the first year, costs $3 million. At the end of the first year, the company needs to spend about $4 million on equipment and other start-up costs. Once the plant becomes operational, it will generate revenue in the amount of $3.5 million during the first operating year. This will increase at the annual rate of 5% over the previous year's revenue, for 10 years. After 10 years, the sales revenue will stay constant for another 3 years before the operation is phased out. (It will have a project life of 13 years after construction.) The expected salvage value of the land at the end of the project life would be about $2 million, the building about $1.4 million, and the equipment about $500,000. The annual operating and maintenance costs are estimated to be about 40% of the sales revenue each year. (Assume that all figures incorporate the effect of the income tax.)

a) What is the IRR for this investment? If the company's MARR is 12%, determine if this is a good investment.

b) When the operation was discontinued, the salvage value actually realized on the building and equipment were zero. All other dollar values had been correctly estimated. In hindsight, was this a good investment?

4.68 Critics have charged that the commercial nuclear power industry does not consider the cost of "decommissioning," or "moth-balling," a nuclear power plant when doing an economic analysis, and that the analysis is therefore unduly optimistic. As an example, consider the nuclear generating facility currently under construction: The first cost is $1.5 billion (present worth at start of operations), the estimated life is 40 years, the annual operating and maintenance costs the first year are assumed to be 4.6% of the first cost and are expected to increase at the fixed rate of 0.05% of the first cost each year, and annual revenues have been estimated to be three times the annual operating and maintenance costs throughout plant life.

(a) The criticism of over-optimism in the economic analysis caused by omitting "mothballing" costs is not justified since the addition of a cost to "mothball" the plant equal to 50% of the first cost only decreases the 10% rate of return to approximately 9.9%.

(b) If the estimated life of the plants is more realistically taken as 25 years instead of 40 years, then the criticism is justified. By reducing the life to 25 years, the rate of return of approximately 9% without a "mothballing" cost drops to approximately 7.7% when a cost to "mothball" the plant equal to 50% of the first cost is added to the analysis.

Verify these calculations and comment on these statements.

4.69 The estimates of investment costs, revenues (and savings), expenses, and salvage value for the digital compression project in Problem 4.65 are subject to considerable uncertainty.

a) Would a 25% increase in investment costs now and in the first and second year change the project's acceptability?

b) If annual expenses were underestimated by 40%, what effect would this have on the original project attractiveness?

4.70 Repeat Problem 4.67 using PE analysis.

Computing IRR for Nonsimple Investments

To comprehend the nature of multiple i^*s, we need to understand the investment situation represented by any cash flow. The net investment test will indicate whether the i^* computed represents the true rate of return earned on the money invested in a project while it is actually in the project. As we shall see, the phenomenon of multiple i^*s occurs only when the net investment test fails. When multiple positive rates of return for a cash flow are found, in general none is suitable as a measure of project profitability, and we must proceed to the next analysis step: introducing an external rate of return.

4A.1 Predicting Multiple i^*s

As hinted at in Section 4.6.2, for certain series of project cash flows, we may uncover the complication of multiple i^* values that satisfy Eq. (4.9). By analyzing and classifying cash flows, we may anticipate this difficulty and adjust our analysis approach. Here we will focus on the initial problem of whether we can predict a unique i^* for a project by examining its cash flow pattern. Two useful rules allow us to focus on sign changes (1) in net cash flows and (2) in accounting net profit (accumulated net cash flows).

4A.1.1   Net Cash Flow Rule of Signs

One useful method for predicting an upper limit on the number of positive i^*s of a cash flow stream is to apply the rule of signs: *The number of real i^*s that are greater than -100% for a project with N periods is never greater than the number of sign changes in the sequence of the A_n. A zero cash flow is ignored.*

An example would be

Period	A_n	Sign Change
0	−$100	
1	−20	
2	50	1
3	0	
4	60	
5	−30	1
6	100	1

Three sign changes occur in the cash flow sequence, so three or fewer real positive i^*s exist.

It must be emphasized that the rule of signs provides an indication only of the possibility of multiple rates of return: The rule only predicts the *maximum* number of possible i^*s. Many projects have multiple sign changes in their cash flow sequence, but still possess a unique real i^* in the $(-100\%, +\infty)$ range.

4A.1.2   Cumulative Cash Flow Sign Test

The accumulated cash flow is the sum of the net cash flows up to, and including, a given time. If the rule of cash flow signs indicates the possibility of multiple i^*s, we should proceed to the **cumulative cash flow sign test** to eliminate some possibility of multiple rates of return.

If we let A_n represent the net cash flow in period n and S_n represent the accumulated cash flow up to period n, we have the following:

Period (n)	Cash Flow (A_n)	Accumulated Cash Flow (S_n)
0	A_0	$S_0 = A_0$
1	A_1	$S_1 = S_0 + A_1$
2	A_2	$S_2 = S_1 + A_2$
⋮	⋮	⋮
N	A_N	$S_N = S_{N-1} + A_N$

We then examine the sequence of accumulated cash flows (S_0, S_1, S_2, S_3, ... , S_N) to determine the number of sign changes. *If the series S_n starts negatively and changes sign only once, a unique positive i^* exists.* This cumulative cash flow sign rule is a more discriminating test for identifying the uniqueness of i^* than the previously described method.

EXAMPLE 4A.1 Predicting the number of i^*s

Predict the number of real positive rate(s) of return for each cash flow series:

Period	A	B	C	D
0	−$100	−$100	$ 0	−$100
1	−200	+ 50	−50	+50
2	+200	−100	+115	0
3	+200	+ 60	−66	+200
4	+200	−100		−50

Solution

Given: Four cash flow series and cumulative flow series
Find: The upper limit on number of i^*s for each series

The cash flow rule of signs indicates the following possibilities for the positive values of i^*:

Project	Number of Sign Changes in Net Cash Flows	Possible Number of Positive Values of i^*
A	1	1 or 0
B	4	4, 3, 2, 1 or 0
C	2	2, 1, or 0
D	2	2, 1, or 0

For cash flows B, C, and D, we would like to apply the more discriminating cumulative cash flow test to see if we can specify a smaller number of possible values of i^*.

Project B		Project C		Project D	
A_n	S_n	A_n	S_n	A_n	S_n
–$100	–$100	$ 0	$ 0	–$100	–$100
+50	–50	–50	–50	+50	–50
–100	–150	+115	+65	0	–50
+60	–90	–66	–1	+200	+150
–100	–190			–50	+100

Recall the test: If the series starts *negatively* and changes sign only once, a unique positive i^* exists. Only project D begins negatively and passes the test; we may predict a unique i^* value, rather than 2, 1, or 0 as predicted by the cash flow rule of signs. Project B, with no sign change in the cumulative cash flow series, has no rate of return. Project C fails the test, and we cannot eliminate the possibility of multiple i^*s. (If projects do not begin negatively, they are borrowing projects rather than investment projects.)

4A.2 Net Investment Test

A project is said to be a **net investment** when the project balances computed at the project's i^* values, $PB(i^*)_n$, are either less than or equal to zero throughout the life of the investment with $A_0 < 0$. The investment is *net* in the sense that the firm does not overdraw on its return at any point and hence is *not indebted* to the project. This type of project is called a **pure investment**. [On the other hand, **pure borrowing** is defined as the situation where $PB(i^*)_n$ values are positive or zero throughout the life of the loan with $A_0 > 0$.] *Simple investments will always be pure investments.* If a nonsimple project passes the net investment test (a pure investment), then the accept/reject decision rule will be the same as in the simple investment case given in Section 4.7.2.

If any of the project balances calculated at the project's i^* are positive, the project is not a pure investment. A positive project balance indicates that, at some time during the project life, the firm acts as a borrower [$PB(i^*)_n > 0$] rather than an investor in the project [$PB(i^*)_n < 0$]. This type of investment is called a **mixed** investment.

Pure versus mixed investments

Consider the following four independent investment projects with known i^* values. Determine which projects are pure investments.

n	A	B	C	D
0	–$1000	–$1000	–$1000	–$1000
1	–1000	1600	500	3900
2	2000	–300	–500	–5030
3	1500	–200	2000	2145
i^*	33.64%	21.95%	29.95%	10%, 30%, 50%

Solution

Given: Four projects with cash flows and i^*s as shown
Find: Which projects are pure investments

We will first compute the project balances at the projects' respective i^*s. If multiple i^*s exist, we may use the largest value of i^* greater than zero.[1]

Project A:

$PB(33.64\%)_0 = -\$1000$.
$PB(33.64\%)_1 = -\$1000(1 + 0.3364) + (-\$1000) = -\$2336.40$.
$PB(33.64\%)_2 = -\$2336.40 (1 + 0.3364) + \$2000 = -\$1122.36$.
$PB(33.64\%)_3 = -\$1122.36 (1 + 0.3364) + \$1500 = 0$.

(–, –, –, 0): Passes the net investment test (pure investment).

Project B:

$PB(21.95\%)_0 = -\$1000$.
$PB(21.95\%)_1 = -\$1000(1 + 0.2195) + \$1600 = \$380.50$.
$PB(21.95\%)_2 = +\$380.50(1 + 0.2195) - \$300 = \$164.02$.
$PB(21.95\%)_3 = +\$164.02(1 + 0.2195) - \$200 = 0$.

(–, +, +, 0): Fails the net investment test (mixed investment).

Project C:

$PB(29.95\%)_0 = -\$1000$.
$PB(29.95\%)_1 = -\$1000(1 + 0.2995) + \$500 = -\$799.50$.
$PB(29.95\%)_2 = -\$799.50(1 + 0.2995) - \$500 = -\$1538.95$.
$PB(29.95\%)_3 = -\$1538.95(1 + 0.2995) + \$2000 = 0$.

(–, –, –, 0): Passes the net investment test (pure investment).

---

[1] In fact, it does not matter which rate we use in applying the net investment test. If one value passes the net investment test, they will all pass. If one value fails, they all fail.

Project D: (There are three rates of return. We can use any of them for the net investment test.)

$PB(50\%)_0 = -\$1000.$
$PB(50\%)_1 = -\$1000(1 + 0.50) + \$3900 = \$2400.$
$PB(50\%)_2 = +\$2400(1 + 0.50) - \$5030 = -\$1430.$
$PB(50\%)_3 = -\$1430(1 + 0.50) + \$2145 = 0.$

$(-, +, -, 0)$: Fails the net investment test (mixed investment).

Comments: As shown in Figure 4A.1, projects A and C are the only pure investments. Project B demonstrates that the existence of a unique i^* is a necessary but not sufficient condition for a pure investment.

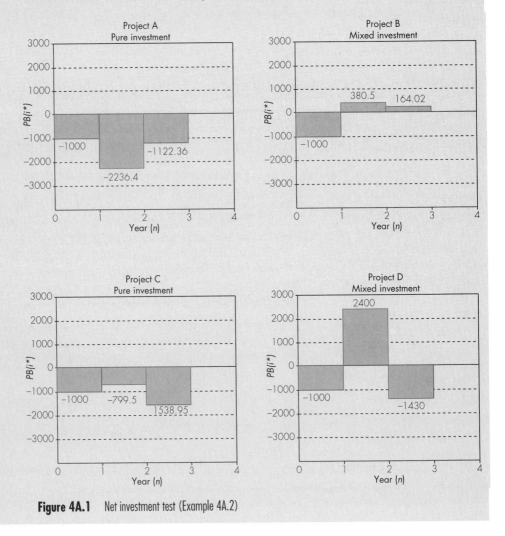

Figure 4A.1 Net investment test (Example 4A.2)

IRR for nonsimple project: Pure investment

Consider project C in Example 4A.2. Assume that all costs and benefits are stated explicitly for this independent project. Apply the net investment test and determine the acceptability of the project using the IRR criterion.

Solution

Given: Cash flow, i^* = 29.95%, MARR = 15%
Find: IRR, and determine whether the project is acceptable

The net investment test was already applied in Example 4A.2, and it indicated that project C is a pure investment. In other words, the project balances were all less than or equal to zero, and the final balance was zero, as it must be if the interest rate used is an i^*. This proves that 29.95% is the true internal rate of return for the cash flow. At MARR = 15%, the IRR > MARR, thus, the project is acceptable. If we compute the PE of this project at i = 15%, we obtain

$$\begin{aligned} PE(15\%) &= -\$1000 + \$500(P/F, 15\%, 1) - \$500(P/F, 15\%, 2) \\ &\quad + \$2000(P/F, 15\%, 3) \\ &= \$371.74. \end{aligned}$$

Since $PE(15\%) > 0$, the project is also acceptable under the PE criterion. The IRR and PE criteria will always produce the same decision if the criteria are applied correctly.

Comments: In the case of a pure investment, the firm has funds committed to the project over the life of the project and at no time withdraws money from the project. The IRR is the return earned on the funds that remain internally invested in the project.

4A.3 External Interest Rate for Mixed Investments

Even for a nonsimple investment, in which there is only one positive rate of return, the project may fail the net investment test, as demonstrated by project B in Example 4A.2. In this case, the unique i still may not be a true indicator of the project's profitability. That is, when we calculate the project balance at an i^* for mixed investments, we notice an important point—cash borrowed (released) from the project is assumed to earn the same interest rate through external investment as money that remains internally invested. In other words, in solving a cash flow for an unknown interest rate it is assumed that money released from a project can be reinvested to yield a rate of return equal to that received from the project. In fact, we have been making this assumption whether or not a cash flow produces a unique positive i^*. Note that money is borrowed from the project only when $PB(i^*) > 0$, and the magnitude of the borrowed amount is the project balance. When $PB(i^*) < 0$, no money is borrowed from the project, even though the cash flow may be positive at that time.

In reality, it is not always possible for cash borrowed (released) from a project to be reinvested to yield a rate of return equal to that received from the project. Instead, it is likely that the rate of return available on a capital investment in the business is much different—usually higher—from the rate of return available on other external investments. Thus, it may be necessary to compute the project balances for a project's cash flow at two rates of interest—one on the internal investment and one on the external investments. As we will see later, by separating the interest rates, we can measure the **true rate of return** of any internal portion of investment project.

Because the net investment test is the only way to accurately predict project borrowing (i.e., external investment), its significance now becomes clear: In order to calculate accurately a project's true IRR, we should always test a solution by the net investment test and, when the test fails, take the further analytical step of introducing an external interest rate. Even the presence of a unique positive i^* is a necessary but not sufficient condition to predict net investment, so if we find a unique value for a nonsimple investment we should still subject it to the net investment test.

4A.4 Calculation of Return on Invested Capital for Mixed Investments

A failed net investment test indicates a combination of internal and external investment. When this combination exists, we must calculate a rate of return on the portion of capital that remains invested internally. This rate is defined as the **true IRR** for the mixed investment or commonly known as **return on invested capital** (**RIC**).

How do we determine the IRR of this investment? Insofar as a project is not a net investment, one or more periods when the project has a net outflow of money (positive project balance) must later be returned to the project. This money can be put into the firm's investment pool until such time as it is needed in the project. The interest rate of this investment pool is the interest rate at which the money can in fact be invested outside the project.

Recall that the PE method assumed that the interest rate charged to any funds withdrawn from a firm's investment pool would be equal to the MARR. In this book, we will use the MARR as an established external interest rate (i.e., the rate earned by money invested outside of the project). We can then compute IRR, or RIC, as a function of MARR by finding the value of IRR that will make the terminal project balance equal to zero. (This implies that the firm wants to fully recover any investment made in the project and pays off any borrowed funds at the end of the project life.) This way of computing rate of return is an accurate measure of the profitability of the project represented by the cash flow. The following procedure outlines the steps for determining the IRR for a mixed investment:

Step 1: Identify the MARR (or external interest rate).

Step 2: Calculate $PB(i, MARR)_n$ (or simply PB_n) according to the rule

$$PB(i, MARR)_0 = A_0.$$

$$PB(i, MARR)_1 = \begin{cases} PB_0(1 + i) + A_1 \\ \quad \text{if } PB_0 < 0. \\ PB_0(1 + MARR) + A_1 \\ \quad \text{if } PB_0 > 0. \end{cases}$$

$$\vdots$$

$$PB(i, MARR)_n = \begin{cases} PB_{n-1}(1 + i) + A_n \\ \quad \text{if } PB_{n-1} < 0. \\ PB_{n-1}(1 + MARR) + A_n \\ \quad \text{if } PB_{n-1} > 0. \end{cases}$$

(As defined in the text, A_n stands for the net cash flow at the end of period n. Note also that the terminal project balance must be zero.)

Step 3: Determine the value of i by solving the terminal project balance equation

$$PB(i, MARR)_N = 0.$$

That interest rate is the IRR for the mixed investment.

Using the MARR as an external interest rate, we may accept a project if the IRR exceeds MARR, and should reject the project otherwise. Figure 4A.2 summarizes the IRR computation for a mixed investment.

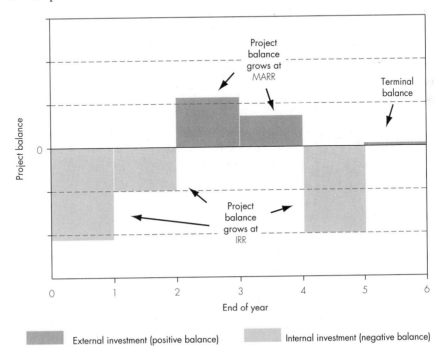

Figure 4A.2 Computational logic for IRR (mixed investment)

EXAMPLE
4A.4

IRR for nonsimple project: Mixed investment

Reconsider the contractor's flight-simulator project in Example 4.17. The project was a nonsimple and mixed investment. To simplify the decision-making process, we abandoned the IRR criterion and used PE to make an accept/reject decision. Apply the procedures outlined to find the true IRR, or return on invested capital, for this mixed investment.

(a) Compute the IRR (RIC) for this project, assuming the MARR = 15%.

(b) Make an accept/reject decision based on the results in part (a).

Solution

Given: Cash flow shown in Example 4.17, MARR = 15%
Find: (a) IRR and (b) determine whether to accept the project

(a) As calculated in Example 4.17, the project has multiple rates of return. This is obviously not a net investment, as shown below. Because the net investment test indicates external as well as internal investment, neither 10% nor 20% represents the true internal rate of return of this government project. Since the project is a mixed investment, we need to find the IRR by applying the steps shown previously.

Net Investment Test	Using $i^* = 10\%$			Using $i^* = 20\%$		
n	0	1	2	0	1	2
Beginning balance	$0	–$1000	$1200	$0	–$1000	$1100
Return on investment	0	–100	120	0	–200	220
Payment	–1000	2300	–1320	–1000	2300	–1320
Ending balance	–$1000	$1200	0	–$1000	$1100	0

(Unit: $1000)

At $n = 0$, there is a net investment to the firm so that the project balance expression becomes

$$PB(i, 15\%)_0 = -\$1,000,000.$$

The net investment of $1,000,000 that remains invested internally grows at i for the next period. With the receipt of $2,300,000 in year 1, the project balance becomes

$$
\begin{aligned}
PB(i, 15\%)_1 &= -\$1,000,000(1 + i) + \$2,300,000 \\
&= \$1,300,000 - \$1,000,000i \\
&= \$1,000,000(1.3 - i).
\end{aligned}
$$

At this point, we do not know whether $PB(i, 15\%)_1$ is positive or negative: It depends on the value of i, which we want to determine. Therefore, we need to consider two situations: (1) $i < 1.3$ and (2) $i > 1.3$.

- Case 1: $i < 1.3 \rightarrow PB(i, 15\%)_1 > 0$.

Since this would be a positive balance, the cash released from the project would be returned to the firm's investment pool to grow at the MARR until it is required back in the project. By the end of year 2, the cash placed in the investment pool would have grown at the rate of 15% [to $\$1,000,000(1.3 - i)(1 + 0.15)$], and must equal the investment into the project of $\$1,320,000$ required at that time. Then, the terminal balance must be

$$
\begin{aligned}
PB(i, 15\%)_2 &= \$1,000,000(1.3 - i)(1 + 0.15) - \$1,320,000 \\
&= \$175,000 - \$1,150,000i \\
&= 0.
\end{aligned}
$$

Solving for i yields

$$\text{IRR (or RIC)} = 0.1522 \text{ or } 15.22\%.$$

The computational process is shown graphically in Figure 4A.3.

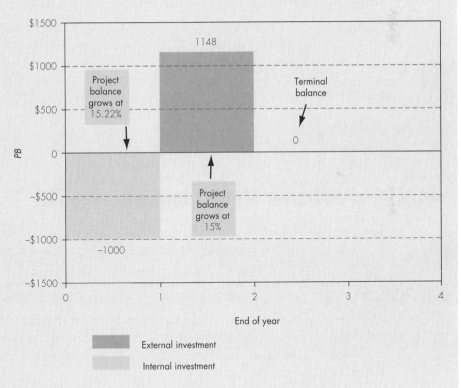

Figure 4A.3 Calculation of the IRR for a mixed investment (Example 4A.4)

- Case 2: $i > 1.3 \rightarrow PB(i, 15\%)_1 < 0$.

The firm is still in an investment mode. Therefore, the balance at the end of year 1 that remains invested will grow at the i for the next period. With the investment of $1,320,000 required in year 2, and the fact that the net investment must be zero at the end of project life, the balance at the end of year 2 should be

$$PB(i, 15\%)_2 = \$1,000,000(1.3 - i)(1 + i) - \$1,320,000$$
$$= -\$20,000 + \$300,000i - \$1,000,000i^2$$
$$= 0.$$

Solving for i gives

$$\text{IRR} = 0.1 \text{ or } 0.2 < 1.3,$$

which violates the initial assumption $(i > 1.3)$. Therefore, Case 1 is the correct situation.

(b) Case 1 indicates IRR > MARR, so the project would be acceptable, resulting in the same decision as obtained in Example 4.17 by applying the PE criterion.

Comments: In this example we could have seen by inspection that Case 1 was correct. Since the project required an investment as the final cash flow, the project balance at the end of the previous period (year 1) had to be positive in order for the final balance to equal zero. Inspection does not generally work for more complex cash flows.

4A.5 Trial-and-Error Method for Computing IRR

The trial-and-error approach for finding IRR (RIC) for a mixed investment is similar to the trial-and-error approach for finding i^*. We begin with a given MARR and a guess for IRR and solve for the project balance. (A value of IRR close to the MARR is a good starting point for most problems.) Since we desire the project balance to approach zero, we can adjust the value of IRR as needed after seeing the result of the initial guess. For example, for a given pair of interest rates (IRR, MARR), if the terminal project balance is positive, the IRR value is too low, so we raise it and recalculate. We can continue adjusting our IRR guesses in this way until we obtain a project balance equal or close to zero.

EXAMPLE 4A.5 IRR for mixed investment by trial and error

Consider project D in Example 4A.2, which has the following cash flow. We know from an earlier calculation that this is a mixed investment.

n	A_n
0	–$1000
1	3900
2	–5030
3	2145

Compute the IRR for this project. Assume that MARR = 6%.

Solution

Given: Cash flow as stated for mixed investment, MARR = 6%
Find: IRR

For MARR = 6%, we must compute i by trial and error. Suppose we guess $i = 8\%$:

$PB(8\%, 6\%)_0 = -\$1000.$
$PB(8\%, 6\%)_1 = -\$1000(1 + 0.08) + \$3900 = \$2820.$
$PB(8\%, 6\%)_2 = +\$2820(1 + 0.06) - \$5030 = -\$2040.80.$
$PB(8\%, 6\%)_3 = -\$2040.80(1 + 0.08) + \$2145 = -\$59.06.$

The net investment is negative at the end of the project, indicating that our trial $i = 8\%$ is in error. After several trials, we conclude that for MARR = 6%, IRR is approximately at 6.13%. To verify the results,

$PB(6.13\%, 6\%)_0 = -\$1000.$
$PB(6.13\%, 6\%)_1 = -\$1000.00(1 + 0.0613) + \$3900 = \$2838.66.$
$PB(6.13\%, 6\%)_2 = +\$2838.66(1 + 0.0600) - \$5030 = -\$2021.02.$
$PB(6.13\%, 6\%)_3 = -\$2021.02(1 + 0.0613) + \$2145 = 0.$

The positive balance at the end of year 1 indicates the need to borrow from the project during year 2. However, note that the net investment becomes zero at the end of project life, confirming that 6.13% is the IRR for the cash flow. Since IRR > MARR, the investment is acceptable. Figure 4A.4 is a visual representation of the occurrence of internal and external interest rates for the project.

Comments: At the end of year 1, the project releases $2838.66 that must be invested outside of the project at an interest rate of 6%. The money invested externally must be returned to the project at the end of year 2 to provide another needed disbursement of $5030. At the end of year 2, there is no external investment of money and, hence, no need for an external interest rate. Instead, the amount of $2021.02 that remains invested during year 3 has to bring in a 6.13% of return. Finally, with the receipt of $2145, the net investment becomes zero at the end of project life. We conclude that IRR (or RIC)= 6.13% is a correct measure of the profitability of the project.

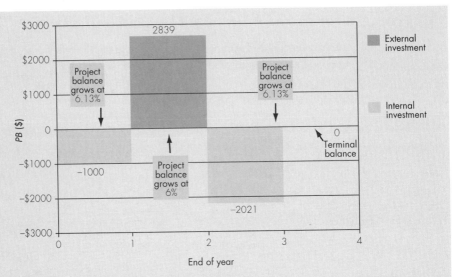

FIGURE 4A.4 Computing the IRR for a mixed investment (Example 4A.5)

Using the PE criterion, the investment would be acceptable if the MARR was between zero and 10% or between 30% and 50%. The rejection region is 10% < *i* < 30% and *i* > 50%. This can be verified in Figure 4.14(b). Note that the project also would be marginally accepted using present worth analysis at MARR = *i* = 6%:

$$
\begin{aligned}
PE(6\%) \quad &= \quad -\$1000 + 3900(P/F,\ 6\%,\ 1) \\
&\quad -\$5030(P/F,\ 6\%,\ 2) + 2145(P/F,\ 6\%,\ 3) \\
&= \quad \$3.55 > 0.
\end{aligned}
$$

   The flow chart in Figure 4A.5 summarizes how you should proceed to apply the net cash flow sign test, cumulative cash flow sign test, and net investment test to calculate an IRR, and make an accept/reject decision for a single project.

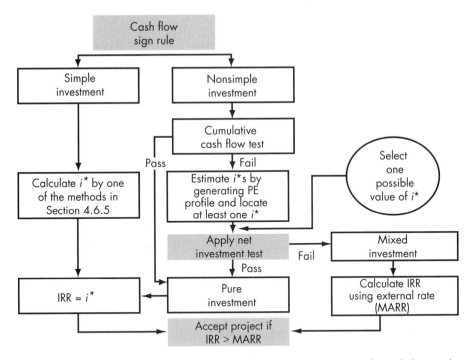

Figure 4A.5 Summary of IRR criteron: A flow chart that summarizes how you may proceed to apply the net cash flow sign rule and net investment test to calculate IRR for a pure, as well as a mixed, investment

4A.6 Summary

1. The possible presence of multiple i^*s (rates of return) can be predicted by

   - The net cash flow sign test
   - The cumulative cash flow sign test

   When multiple rates of return cannot be ruled out by the two methods, it is useful to generate a PE profile to approximate the value of i^*.

2. For all non-simple investments, an i^* value should be exposed to the **net investment test**. Passing the net investment test indicates that the i^* is an internal rate of return and is therefore a suitable measure of project profitability. Failure to pass the test indicates project borrowing, a situation that requires further analysis by use of an **external interest rate**.

3. **Return on invested capital** analysis uses one rate (the firm's MARR) on externally invested balances and solves for another rate (IRR) on internally invested balances.

Problems

Level 1

4A.1 Consider the following independent investment projects:

			Project Cash Flows			
n	A	B	C	D	E	F
0	−$100	−$100	−$100	−$100	−$100	−$100
1	200	470	200	300	300	300
2	−300	−720	200	−300	−250	100
3	400	360	−250	50	40	−400

(a) Apply the sign rule to predict the number of possible i^*s for each project.
(b) Plot the present worth profile as a function of i between 0 and 200% for each project.
(c) Compute the value(s) of i^* for each project.

4A.2 Consider an investment project with the following cash flows:

n	Net Cash Flow
0	−$20,000
1	94,000
2	−144,000
3	72,000

(a) Find the IRR for this investment.
(b) Plot the present worth of the cash flow as a function of i.
(c) Using the IRR criterion, should the project be accepted at MARR = 15%?

4A.3 Consider the following independent investment projects:

	Net Cash Flow		
n	Project 1	Project 2	Project 3
0	−$1000	−$2000	−$1000
1	500	1560	1400
2	840	944	−100
IRR	?	?	?

Assume that MARR = 12% in the following questions:

(a) Compute the i^* for each investment. If the problem has more than one i^*, identify all of them.
(b) Compute the IRR for each project.
(c) Determine the acceptability of each investment.

Level 2

4A.4 Consider the following independent investment projects:

		Project Cash Flow			
n	A	B	C	D	E
0	−$100	−$100	−$5	−$100	$200
1	100	30	10	30	100
2	24	30	30	30	−500
3		70	−40	30	−500
4		70		30	200
5				30	600

(a) Compute the i^* for A using the quadratic equation.
(b) Classify each project into either simple or nonsimple.
(c) Apply the cash flow sign rules to each project and determine the number of possible positive i^*s. Identify all projects having unique i^*.
(d) Apply the net investment test to each project.

(e) Compute the IRRs for projects B through E.

(f) With MARR = 10%, which projects will be acceptable using the IRR criterion?

4A.5 Consider the following independent investment projects:

	Net Cash Flow		
n	Project 1	Project 2	Project 3
0	-$1,600	-$5,000	-$1000
1	10,000	10,000	4000
2	10,000	30,000	-4000
3		-40,000	

Assume that MARR = 12% in the following questions:

(a) Identify the $i^*(s)$ for each investment. If the problem has more than one i^*, identify all of them.

(b) Which project(s) is (are) a mixed investment?

(c) Compute the IRR for each project.

(d) Determine the acceptability of each project.

4A.6 Reconsider Problem 4.33.

(a) Compute the i^*s for projects B through E.

(b) Identify the projects that fail the net investment test.

(c) Determine the IRR for each investment.

(d) With MARR = 12%, which projects will be acceptable using the IRR criterion?

4A.7* Consider the following independent investment projects:

	Net Cash Flow		
n	Project A	Project B	Project C
0	-$100	-$150	-$100
1	30	50	410
2	50	50	-558
3	80	50	252
4		100	
IRR	23.24%	21.11%	20%,40%,50%

Assume the MARR = 12% for the following questions:

(a) Identify the pure investment(s).

(b) Identify the mixed investment(s).

(c) Determine the IRR for each investment.

(d) Which project would be acceptable?

4A.8 Spar Canada Ltd. has received a NASA contract worth $460 million to make design modifications to the Canadarm to be used in future space missions. NASA will pay $50 million when the contract is signed, another $360 million at the end of the first year, and the $50 million balance at the end of second year. The expected cash outflows required to make these modifications are estimated to be $150 million now, $100 million during the first year, and $218 million during the second year. The firm's MARR is 12%.

			Net Cash
n	Outflow	Inflow	Flow
0	$150	$50	-$100
1	100	360	260
2	218	50	-168

(a) Show whether or not this project is a mixed investment.

(b) Compute the IRR for this investment.

(c) Should Spar accept the project?

4A.9 Consider the following independent investment projects:

	Net Cash Flow		
n	Project A	Project B	Project C
0	-$100		-$100
1	-216	-150	50
2	116	100	-50
3		50	200
4		40	
i^*	?	15.51%	29.95%

(a) Compute the i^* for project A. If there is more than one i^*, identify all of them.

(b) Identify the mixed investment(s).

(c) Assuming that MARR = 10%, determine the acceptability of each project based on the IRR criterion.

4A.10 Consider the following independent investment projects:

			Net Cash Flow		
n	A	B	C	D	E
0	–$1000	–$5000	–$2000	–$2000	–$1000
1	3100	20,000	1560	2800	3600
2	–2200	–12,000	944	–200	–5700
3		–3000			3600
i^*	?	?	18%	32.45%	35.39%

Assume that MARR = 12% in the following questions:

(a) Compute the i^* for projects A and B. If the project has more than one i^*, identify all of them.

(b) Classify each project as either a pure or mixed investment.

(c) Compute the IRR for each investment.

(d) Determine the acceptability of each project.

4A.11 Consider an investment project whose cash flows are given as follows:

n	Net Cash Flow
0	–$5000
1	10,000
2	30,000
3	–40,000

(a) Plot the present worth curve by varying i from 0% to 250%.

(b) Is this a mixed investment?

(c) Should the investment be accepted at MARR = 18%?

4A.12 Consider the following project's cash flows:

n	Net Cash Flow
0	–$100,000
1	310,000
2	–220,000

The project's i^*s are computed as 10% and 100%, respectively. The firm's MARR is 8%.

(a) Show why this investment project fails the net investment test.

(b) Compute the IRR, and determine the acceptability of this project.

4A.13* Consider the following independent investment projects:

	Net Cash Flow		
n	Project 1	Project 2	Project 3
0	–$1000	–$1000	–$1000
1	–1000	1600	1500
2	2000	–300	–500
3	3000	–200	2000

Which of the following statements is correct?

(a) All projects are nonsimple investments.

(b) Project 3 should have three real rates of return.

(c) All projects will have a unique positive real rate of return.

(d) None of the above.

CHAPTER 5

Comparing Mutually Exclusive Alternatives

T

he proliferation of computers into all aspects of business has created an ever-increasing need for data capture systems that are fast, reliable and cost-effective. One technology which has been adopted by many manufacturers, distributors and retailers is a bar-coding system. Hermes Electronics, a leading manufacturer of underwater surveillance equipment, evaluated the economic benefits of installing an automated data acquisition system into their Halifax plant. They could use the system on a limited scale, such as for tracking parts and assemblies for inventory management, or opt for a broader implementation by recording information useful for quality control, operator efficiency, attendance, and other functions. All of these aspects are currently monitored, but although computers are used to manage the information, the recording is primarily conducted manually. The advantages of an automated data collection system, which include faster and more accurate data capture, quicker analysis and response to production changes, and savings due to tighter control over operations, could easily outweigh the cost of the new system.

Two alternative systems from competing suppliers are under consideration. System 1 relies on hand-held bar-code scanners which transmit their data to the host server on the local area network (LAN) using radio frequencies (RF). The hub of this wireless network can then be connected to their existing LAN, and integrated with their current MRP II system and other management software. System 2 consists primarily of specialized data terminals installed at every collection point, with connected bar-code scanners where required. This system is configured in such a way to facilitate either phasing in the components over two stages or installing the system all at once. The former would allow Hermes to defer some of the capital investment, while becoming thoroughly familiar with the functions introduced in the first stage.[1]

Either of these systems would satisfy Hermes' data collection needs. They each have some unique elements, and the company needs to compare the relative benefits of the features offered by each system. From the point of view of engineering economics, the two systems have different capital costs, and their operating and maintenance costs are not identical. One system may be also rated to last longer than the other system before replacement is required, particularly if the option of acquiring System 2 in phases is selected. There are many issues to be considered before making the best choice.

Until now, we have considered situations involving only a single project, or projects that were independent of each other. In both these cases, we made the decision to reject or accept each project individually based on whether it met the MARR requirements, evaluated using the PE, FE, AE, or IRR criteria.

In the real world of engineering practice, however, it is more typical for us to have two or more choices of projects for accomplishing a business objective. (As we shall see, even when it appears that we have only one project to consider, the implicit "do nothing" alternative must be factored into the decision-making process.) In this chapter, we extend our evaluation techniques to multiple projects which are mutually exclusive. Other dependencies between projects will be considered in Chapter 10.

Often, various projects or investments under consideration do not have the same duration, or do not match the desired study period. Adjustments must be made when comparing multiple options to properly account for such differences. In this chapter, the concept of an analysis period, and the process of accommodating for different lifetimes, are explained as important considerations for selecting among several alternatives. In the first few sections of the chapter, all available options in a decision problem are assumed to have equal lifetimes. In Section 5.5, this restriction is relaxed.

---

[1] Source: Hermes Electronics, Halifax, Nova Scotia.

5.1 Meaning of Mutually Exclusive and "Do-Nothing"

Mutually exclusive means that any one of several alternatives will fulfill the same need and that the selection of one alternative implies that the others will be excluded. Take, for example, buying versus leasing an automobile for business use—when one alternative is accepted, the other is excluded. We will use the terms **alternative** and **project** interchangeably to mean **decision option**.

When considering an investment, we are in one of two situations. The project is either aimed at replacing an existing asset or system, or it is a new endeavor. In either case, a **do-nothing** alternative may exist. If a process or system already in place to accomplish our business objectives is still adequate, then we must determine which, if any, new proposals are economical replacements. If none are feasible, then we do nothing. On the other hand, if the existing system has terminally failed, the choice amongst proposed alternatives is **mandatory** (i.e., do-nothing is not an option).

New endeavors occur as alternatives to the "green fields" do-nothing situation which has zero revenues and zero costs (i.e., nothing currently exists). For most new endeavors, do-nothing is generally an alternative, as we won't proceed unless at least one of the proposed alternatives is economically sound. In fact, undertaking even a single project entails making a decision between two alternatives when it is optional, because the do-nothing alternative is implicitly included. Occasionally, a new initiative must be undertaken, cost notwithstanding, and in this case the goal is to choose the most economical alternative since do-nothing is not an option.

When the option of retaining an existing asset or system is available, there are two ways to incorporate it into the evaluation of the new proposals. One way is to treat the do-nothing option as a distinct alternative, but this approach will be primarily covered in Chapter 6, where methodologies specific to replacement analysis are presented. The second approach, mostly used in this chapter, is to generate the cash flows of the new proposals relative to that of the do-nothing alternative. That is, for each new alternative the **incremental costs** (and **incremental savings** or **revenues** if applicable) relative to do-nothing are used for the economic evaluation. For a replacement type problem, this is calculated by subtracting the do-nothing cash flows from those of each new alternative. For new endeavors, the incremental cash flows are the same as the absolute amounts associated with each alternative, since the do-nothing values are all zero.

Since the main purpose of this chapter is to illustrate how to choose among mutually exclusive alternatives, most of the problems are structured such that one of the options presented *must* be selected. Therefore, unless otherwise stated, is is assumed that do-nothing is not an option, and costs and revenues can be viewed as incremental to do-nothing.

5.2 Revenue Projects versus Service Projects

When comparing mutually exclusive alternatives, we need to classify investment projects into either service or revenue projects. **Service projects** are those whose revenues do not depend on the choice of project. For example, suppose an electric utility is considering building a new power plant to meet the peak load demand during either hot summer or cold winter days. Two alternative service projects could meet this peak-load demand: A combustion turbine plant or a fuel-cell power plant. No matter which type of plant is selected, the firm will generate the same amount of revenue from its customers. The only difference is how much it will cost to generate electricity from each plant. If we were to compare these service projects, we would be interested in knowing which plant could provide the cheaper power (lower production cost). Further, if we were to use the PE criterion to compare these alternatives to minimize expenditures, we would choose the alternative with the **lower present value** production cost over the service life.

On the other hand, **revenue projects** are those whose revenues depend on the choice of alternative. For example, a TV manufacturer is considering marketing two types of high-resolution monitor. With its present production capacity, the firm can market only one of them. Distinct production processes for the two models could incur very different manufacturing costs, so the revenues from each model would be expected to differ due to divergent market prices, and potentially different sales volumes. In this situation, if we were to use the PE criterion, we would select the model that promises to bring in the higher net present worth.

5.3 Total Investment Approach

Applying an evaluation criterion to each mutually exclusive alternative individually, and then comparing the results to make a decision, is referred to as the **total investment approach**. Unfortunately, this approach only guarantees valid results when using PE, AE and FE criteria, as demonstrated in the following sections.

5.3.1 Present Equivalent Comparison

We compute the PE for each individual alternative, as we did in Chapter 4, and select the one with the highest PE.

EXAMPLE 5.1

Present equivalent worth—three alternatives

Bullard Company (BC) is considering expanding its range of industrial machinery products by manufacturing machine tables, saddles, machine bases, and other similar parts. Several combinations of new equipment and manpower could serve to fulfill this new function:

- Method 1 (M1): new machining center, with 3 operators

- Method 2 (M2): new machining center with an automatic pallet changer, and with 3 operators

- Method 3 (M3): new machining center with an automatic pallet changer, and with 2 task-sharing operators

Each of these arrangements incurs different costs and revenues. The time for loading and unloading parts is reduced in the pallet-changer cases. It costs more to acquire, install and tool-fit a pallet-changer, but because it is more efficient and versatile, it can generate larger annual revenues. Although saving on labor costs, task-sharing operators take longer to train and are more inefficient initially. As operators become more experienced at their tasks and used to collaborating, it is expected that the annual benefits will increase by 13% per year over the 5-year study period. BC has estimated the investment costs and additional revenues as follows:

| | Machining Center Methods | | |
	M1	M2	M3
Investment			
Machine tool purchase	$121,000	$121,000	$121,000
Automatic pallet changer		$66,600	$66,600
Installation	$30,000	$42,000	$42,000
Tooling expense	$58,000	$65,000	$65,000
Total investment	$209,000	$294,600	$294,600
Annual benefits: Year 1			
Additional revenues	$65,000	$69,300	$36,000
Direct labor savings			$17,300
Setup savings		$4700	$4700
Year 1: Net revenues	$65,000	$74,000	$58,000
Years 2–5: Net revenues	constant	constant	g = +13%/year
Salvage value in year 5	$90,000	$200,000	$200,000

All cash flows include any tax effects. Do-nothing is obviously an option, since BC will not undertake this expansion if none of the proposed methods are economically viable. If a method is chosen, BC expects to operate the machining center over the next 5 years. Which option would be selected based on the use of the PE measure at $i = 12\%$?

Solution

Given: Cash flows for three projects, $i = 12\%$ per year
Find: PE for each project, which to select

For these revenue projects, present equivalent figures at $i = 12\%$ would be as follows:

- For Option M1:

$$\begin{aligned} PE(12\%)_{M1} &= -\$209,000 + \$65,000(P/A, 12\%, 5) + \\ &\quad \$90,000(P/F, 12\%, 5) \\ &= \$76,379. \end{aligned}$$

- For Option M2:

$$\begin{aligned} PE(12\%)_{M2} &= -\$294,600 + \$74,000(P/A, 12\%, 5) + \\ &\quad \$200,000(P/F, 12\%, 5) \\ &= \$85,639. \end{aligned}$$

- For Option M3:

$$\begin{aligned} PE(12\%)_{M3} &= -\$294,600 + \$58,000(P/A_1, 13\%, 12\%, 5) + \\ &\quad \$200,000(P/F, 12\%, 5) \\ &= \$82,479. \end{aligned}$$

Clearly, option M2 is the most profitable. Given the nature of BC parts and shop orders, management has decided that the best way to expand would be with an automatic pallet changer but without task-sharing.

Comments: Of course, calculating the **future equivalent** of these mutually exclusive options will give the same optimal choice as the present equivalent analysis.

$$\begin{aligned} FE(12\%)_{M1} &= \$76,379(F/P, 12\%, 5) &= \$134,606. \\ FE(12\%)_{M2} &= \$85,639(F/P, 12\%, 5) &= \$150,925. \\ FE(12\%)_{M3} &= \$82,479(F/P, 12\%, 5) &= \$145,356. \end{aligned}$$

Since the alternatives are compared over a fixed 5-year period, the ratio of the future equivalent to the present equivalent is the same for all options:

$$\frac{FE_{M1}}{PE_{M1}} = \frac{FE_{M2}}{PE_{M2}} = \frac{FE_{M3}}{PE_{M3}} = (F/P, i, N)$$

This also implies that the ratio of future equivalents of any two options with identical life times is equal to the ratio of their present equivalents, as you can easily check:

$$\frac{FE_1}{FE_2} = \frac{PE_1}{PE_2}$$

5.3.2 Annual Equivalent Comparison

Mutually exclusive projects can also be compared by calculating the annual equivalent for each option, as shown in the following continuation of Example 5.1. We choose the project with the largest AE value.

EXAMPLE 5.2 **Annual equivalent worth—three alternatives**

Recall the cash flow estimates for the Bullard Company from Example 5.1:

| | Machining Center Methods | | |
	M1	M2	M3
Net investment cost	$209,000	$294,600	$294,600
Year 1: Net revenues	$65,000	$74,000	$58,000
Years 2–5: Net revenues	constant	constant	g = +13%/year
Salvage value in year 5	$90,000	$200,000	$200,000

Solution

Given: Cash flows for three projects, i = 12% per year
Find: AE of each project, which to select

- For Option M1:

$$AE(12\%)_{M1} = -\$209,000(A/P, 12\%, 5) + \$65,000 + \$90,000(A/F, 12\%, 5)$$
$$= \$21,188.$$

- For Option M2:

$$AE(12\%)_{M2} = -\$294,600(A/P, 12\%, 5) + \$74,000 + \$200,000(A/F, 12\%, 5)$$
$$= \$23,757.$$

- For Option M3:

$$AE(12\%)_{M3} = -\$294,600(A/P, 12\%, 5) + \$58,000(P/A_1, 13\%, 12\%, 5)(A/P, 12\%, 5) + \$200,000(A/F, 12\%, 5)$$
$$= \$22,881.$$

The best choice is option M2 because it has the highest AE value. As we expect, this result is consistent with the PE analysis. In fact, the annual equivalents calculated directly from the PE values in Example 5.1 would have yielded the same results as above. When two options' lifetimes are equal, the ratios of their AEs and PEs are equal, as follows:

$$\frac{AE_1}{AE_2} = \frac{PE_1(A/P, i, N)}{PE_2(A/P, i, N)} = \frac{PE_1}{PE_2}$$

So, the option with the higher AE always has a higher PE as well. This congruity of assessment criteria was noted in Chapter 4 for independent projects. Although the required consistency of analysis methods seems self-evident, particular care must be taken when applying the IRR criterion to mutually exclusive options, as shown in the following section.

5.4 Incremental Investment Analysis

In this section, we will present a method known as **incremental analysis**, a decision procedure that must be used when comparing two or more mutually exclusive projects based on the rate of return measure. This method may also be used when assessing alternatives using the PE, FE or AE criteria.

5.4.1 Flaws in Project Ranking by IRR

Under PE, FE or AE analysis, the mutually exclusive project with the highest worth figure is preferred. Unfortunately, the analogy does not carry over to IRR analysis. The project with the highest IRR may *not* be the preferred alternative. To illustrate the flaws of comparing IRRs to choose from mutually exclusive projects, suppose you have two mutually exclusive alternatives, each with a 1-year service life: One requires an investment of $1000 with a return of $2000, and the other requires $5000 with a return of $7000. You already obtained the IRRs and PEs at MARR = 10% as follows:

n	A1	A2
0	–$1000	–$5000
1	2000	7000
IRR	100%	40%
PE(10%)	$818	$1364

Assuming that you have enough money in your investment pool to select either one, would you prefer the first project simply because you expect a higher rate of return?

We can see that A2 is preferred over A1, by the PE measure. On the other hand, the IRR measure gives a numerically higher rating for A1. This inconsistency in ranking occurs because the PE, FE, and AE are **absolute (dollar)** measures of investment worth, whereas the IRR is a **relative (percentage)** measure and cannot be applied in the same way. That is, the IRR measure ignores the **scale** of the investment. Therefore, the answer is no; instead, you would prefer the second project with the lower rate of return, but higher PE. Either the PE, FE or the AE measure would lead to that choice, but comparison of IRRs would rank the smaller project higher. Another approach, referred to as **incremental analysis**, is needed.

5.4.2   Incremental Analysis Method

In our previous ranking example, the more costly option requires an incremental investment of $4000 for an incremental return of $5000.

- If you decide to invest in option A1, you will need to withdraw only $1000 from your investment pool. The remaining $4000 will continue to earn 10% interest. One year later, you will have $2000 from the outside investment and $4400 from the investment pool. With an investment of $5000, in 1 year you will have $6400. The equivalent present worth of this wealth change is $PE(10\%) = -\$5000 + \$6400(P/F, 10\%, 1) = \$818$.

- If you decide to invest in option A2, you will need to withdraw $5000 from your investment pool, leaving no money in the pool, but you will have $7000 from your outside investment. Your total wealth changes from $5000 to $7000 in a year. The equivalent present worth of this wealth change is $PE(10\%) = -\$5000 + \$7000(P/F, 10\%, 1) = \$1364$.

In other words, if you decide to take the more costly option, certainly you would be interested in knowing whether this additional investment is justified at the MARR. The 10% MARR value implies that you can always earn that rate from other investment sources (i.e., $4400 at the end of 1 year for a $4000 investment). However, in the second option, by investing the additional $4000, you would make an additional $5000, which is equivalent to earning at the rate of 25%. Therefore, the incremental investment can be justified.

Now we can generalize the decision rule for comparing mutually exclusive projects. For a pair of mutually exclusive projects (A and B, with B defined as a more costly option), we may rewrite B as

$$B = A + (B - A).$$

In other words, B has two cash flow components: (1) The same cash flow as A and (2) the incremental component $(B - A)$. Therefore, the only situation in which B is preferred to A is when the rate of return on the incremental component $(B - A)$ exceeds the MARR. Therefore, for two mutually exclusive projects, rate of return analysis is done by computing the *internal rate of return on incremental investment* (IRR_Δ) between the projects. Since we want to consider increments of investment, we compute the cash flow for the difference between the projects by subtracting the cash flow for the lower investment-cost project (A) from that of the higher investment-cost project (B). Then, the decision rule is

If IRR_{B-A} > MARR, select B,

If IRR_{B-A} = MARR, select either one,

If IRR_{B-A} < MARR, select A,

where B – A is an investment increment (negative cash flow). If the "do-nothing" option is allowed, but not explicitly included in the evaluation, then the smaller cost option A must be profitable to be initially retained as a feasible alternative (i.e., its IRR must be greater than MARR).

Since the incremental analysis method may also be used for PE, FE and AE analyses, comparable decision rules apply:

	> 0	= 0	< 0
PE_{B-A} (MARR)	select B	select either one	select A
FE_{B-A} (MARR)	select B	select either one	select A
AE_{B-A} (MARR)	select B	select either one	select A

It may seem odd to you how this simple rule allows us to select the right project. Example 5.3 will illustrate the incremental investment decision rule for you.

EXAMPLE 5.3

IRR on incremental investment: Two alternatives

John Covington, a college student, wants to start a small-scale painting business during his off-school hours. To economize the start-up business, he decides to purchase some used painting equipment. He has two mutually exclusive options: Do most of the painting by himself by limiting his business to only residential painting jobs (B1) or purchase more painting equipment and hire some helpers to do both residential and commercial painting jobs that he expects will have a higher equipment cost, but provide higher revenues as well (B2). In either case, he expects to fold the business in 3 years when he graduates from college.

The cash flows and equivalence measures for the two mutually exclusive alternatives are as follows:

n	B1	B2	B2 – B1
0	–$3000	–$12,000	–$9000
1	1350	4200	2850
2	1800	6225	4425
3	1500	6330	4830
IRR	25.00%	17.43%	15.00%
PE (10%)	$841	$1718	$877
AE (10%)	$338	$691	$353
FE (10%)	$1120	$2287	$1167

Which project would he select at MARR=10%?

Solution

Given: Incremental cash flow between two alternatives, MARR=10%
Find which alternative is preferable using: (a) IRR criterion, (b) PE criterion, (c) AE criterion and (d) FE criterion

Given that $IRR_{B1} > MARR$, $PE_{B1} > 0$, $AE_{B1} > 0$, and $FE_{B1} > 0$, by any criterion the cheapest option B1 is economically acceptable. The question is whether to make the additional investment, and opt for B2 instead. To choose the best project, we first calculate the incremental cash flow B2 − B1.

a) We compute the IRR on this increment of investment by solving

$$-\$9000 + \$2850(P/F, i, 1) + \$4425(P/F, i, 2) + \$4830(P/F, i, 3) = 0.$$

We obtain $i^*_{B2-B1} = 15\%$ as shown in the table above. By inspection of the incremental cash flows, we know it is a simple investment, so $IRR_{B2-B1} = i^*_{B2-B1}$. Since $IRR_{B2-B1} > MARR$, we select B2, which is consistent with choosing the option with the largest PE. This result is illustrated in Figure 5.1. Notice that the IRR_{B2-B1} for the incremental cash flow curve B2 − B1 occurs at the same interest rate at which the B1 and B2 curves intersect each other. This makes sense, as $PE_{B2-B1} = PE_{B2} - PE_{B1} = 0$ at this interest rate of 15%. Note however that $IRR_{B2-B1} \neq IRR_{B2} - IRR_{B1}$, which is why we cannot compare the IRRs for the individual options to arrive at a valid conclusion.

(b) $PE_{B2-B1}(10\%) = \$877$, which is greater than zero, so the incremental investment is worthwhile, and we choose B2. Since $PE_{B2-B1} = PE_{B2} - PE_{B1}$, stating that $PE_{B2-B1} > 0$ is the same as saying that $PE_{B2} - PE_{B1} > 0$, or $PE_{B2} > PE_{B1}$. This is why we can directly compare the PEs of all the mutually exclusive options, and simply pick the best, rather than performing an incremental PE analysis which takes longer.

(c) $AE_{B2-B1}(10\%) = \$353 > 0$ so choose B2, which is consistent with the PE evaluation. Again, since $AE_{B2-B1} = AE_{B2} - AE_{B1}$, the AE of each option may be compared directly, which is faster than applying this criterion to the incremental cash flow.

(d) $FE_{B2-B1}(10\%) = \$1167 > 0$, which leads to the same optimal choice of B2, and this result could have been obtained directly from the difference of the future equivalents of the two options $FE_{B2} - FE_{B1}$.

Comments: Why did we choose to look at the increment B2 − B1 instead of B1 − B2? We want the first flow of the incremental cash flow series to be negative (investment flow) so that we can calculate an IRR. By subtracting the lower initial investment project from the higher, we guarantee that the first increment will be an investment flow. If we ignore the investment ranking, we might end up with an increment that involves borrowing cash flow and has no internal rate of return. This is the case for B1 − B2. (i^*_{B1-B2} is also 15%, not −15%.) If, erroneously, we had compared this

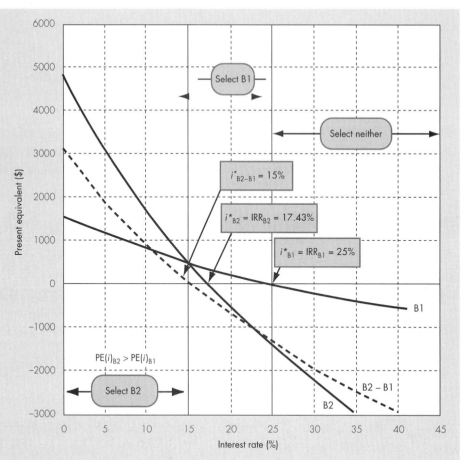

Figure 5.1 PE profiles for B1 and B2 (Example 5.3)

i^* with the MARR, we might have accepted project B1 over B2. This undoubtedly would have damaged our credibility with management! We will revisit this problem in Example 5.6.

When you have more than two mutually exclusive alternatives, they can be compared in pairs by successive examination. Example 5.4 will illustrate how to compare three alternative projects using incremental analysis. This example will indicate that the ranking inconsistency between PE and IRR can also occur when differences in the timing of a project's future cash flows exist, even if their initial investments are the same.

EXAMPLE 5.4

IRR on incremental investment: Three alternatives

Reconsider the data from Example 5.1:

	Machining Center Methods		
n	M1	M2	M3
0	−$209,000	−$294,600	−$294,600
1	65,000	74,000	58,000
2	65,000	74,000	65,540
3	65,000	74,000	74,060
4	65,000	74,000	83,688
5	$155,000	$274,000	$294,567
IRR	24.03%	20.88%	20.10%

Which project would you select based on rate of return on incremental investment, assuming that MARR = 12%?

Solution

Given: Cash flows given above, MARR = 12%
Find: IRR on incremental investment and which alternative is preferable

Step 1: Examine the IRR for each alternative. When "do-nothing" is an option, we can eliminate any alternative that fails to meet the MARR[1]. In this example, all three alternatives exceed the MARR of 12%.

Step 2: Compare M1 and M2 in pairs[2]. M1, being the cheapest option, becomes the default choice, and is termed the defender. To see whether the challenger, M2, is worth the extra investment, calculate the IRR on the incremental cash flow:

n	M2 – M1
0	−$85,600
1	9000
2	9000
3	9000
4	9000
5	119,000
IRR_{M2-M1}	14.76%

The incremental cash flow represents a simple investment, so the break-even interest rate calculated in the table is the valid internal rate of return. Since IRR_{M2-M1} exceeds the MARR of 12%, M2 is preferred over M1, and M2 becomes the defender.

---

[1] When it is mandatory to pick one of the "new options" (i.e. if "do-nothing" is not allowed), then individual options cannot be rejected based on their IRR. Rather, their rate of return on incremental investment compared to any cheaper option must then be used as a screening criterion.

[2] When faced with many alternatives, you may arrange them in order of increasing initial cost. This is not a required step, but it makes the comparison more tractable.

Step 3: Compare M3 and M2. When initial investments are equal, we progress through the cash flows until we find the first difference, then set up the increment so that this first non-zero flow is negative (i.e., an investment). In this problem, the existing order of M2 and M3 satisfies this requirement as shown in the following incremental investment table:

n	M3 – M2
0	0
1	–$16,000
2	–8460
3	60
4	9688
5	20,567
IRR$_{M3 - M2}$	6.66%

The incremental cash flows show a simple investment pattern. The (M3 – M2) increment has an unsatisfactory 6.66% rate of return; therefore, M3 is not preferred over M2. In summary, we conclude that M2 is the best alternative, consistent with our prior PE, AE and FE analyses of this problem.

Comments: It may be emphasized again that the IRR for each individual option, shown in the data table at the start of this problem, can be used for screening the alternatives for acceptability, but cannot be used to select the best option. Figure 5.2 demonstrates again that the option with the highest IRR is not necessarily the one with the highest PE at MARR.

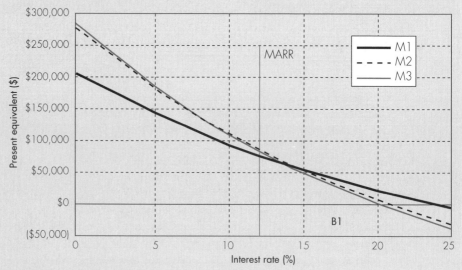

Figure 5.2 PE profiles for the three machining center options (Example 5.4)

EXAMPLE 5.5 Incremental analysis for cost-only projects

Falk Corporation is considering two types of manufacturing systems to produce its shaft couplings over 6 years: (1) A cellular manufacturing system (CMS) and (2) a flexible manufacturing system (FMS). The average number of pieces to be produced on either system would be 544,000 per year. Operating cost, initial investment, and salvage value for each alternative are estimated as follows:

Items	CMS Option	FMS Option
Annual operating and maintenance costs		
Annual labor cost	$1,169,600	$707,200
Annual material cost	832,320	598,400
Annual overhead cost	3,150,000	1,950,000
Annual tooling cost	470,000	300,000
Annual inventory cost	141,000	31,500
Annual income taxes	1,650,000	1,917,000
Total annual costs	$7,412,920	$5,504,100
Investment	$4,500,000	$12,500,000
Net salvage value	$500,000	$1,000,000

Figure 5.3 illustrates the cash flows associated with each alternative. The firm's MARR is 15%. Which alternative would be a better choice based on the IRR criterion?

Discussion: Since we can assume that both manufacturing systems would provide the same level of revenues over the analysis period, we can compare these alternatives based on cost only. (These are service projects.) Although we cannot compute the IRR for each option without knowing the revenue figures, we can still calculate the IRR on incremental cash flows. Since the FMS option requires a higher initial investment than that for the CMS, the incremental cash flow is the difference (FMS – CMS).

n	CMS Option	FMS Option	Incremental (FMS – CMS)
0	–$4,500,000	–$12,500,000	–$8,000,000
1	–7,412,920	–5,504,100	1,908,820
2	–7,412,920	–5,504,100	1,908,820
3	–7,412,920	–5,504,100	1,908,820
4	–7,412,920	–5,504,100	1,908,820
5	–7,412,920	–5,504,100	1,908,820
6	–7,412,920	–5,504,100	
	+ $500,000	+ $1,000,000	$2,408,820

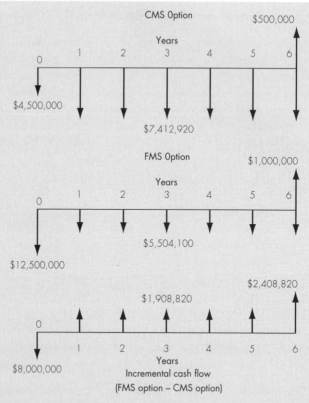

Figure 5.3 Comparison of mutually exclusive alternatives with equal revenues (cost only) (Example 5.5)

Solution

Given: Cash flows shown in Fig. 5.3, $i = 15\%$ per year
Find: Incremental cash flows, and select the better alternative based on the IRR criterion

$$PE(i^*)_{FMS-CMS} = -\$8,000,000 + \$1,908,820(P/A, i, 5)$$
$$+ \$2,408,820(P/F, i, 6)$$
$$= 0.$$

Solving for i^* by trial and error yields 12.43%. Since $IRR_{FMS-CMS} = 12.43\% < 15\%$, we would select CMS. Although the FMS would provide an incremental annual savings of $1,908,820 in operating costs, the savings do not justify the incremental investment of $8,000,000.

Comments: Note that the CMS option was marginally preferred to the FMS option. However, there are dangers in relying solely on the easily quantified savings in input factors—such as labor, energy, and materials—from FMS and in not considering gains from improved manufacturing performance that are more difficult and subjective to quantify. Factors such as improved product quality, increased manufacturing flexibility (rapid response to customer demand), reduced inventory levels, and increased capacity for product innovation are frequently ignored because we have inadequate means for quantifying their benefits. If these intangible benefits were considered, however, the FMS option could come out better than the CMS option.

5.4.3   Incremental Borrowing Approach

Subtracting the less costly alternative from the more costly one is not absolutely necessary in incremental analysis. In fact, we can examine the difference between two projects A and B as either an (A – B) increment or a (B – A) increment. If the difference in flow (B – A) represents an increment of **investment**, then (A – B) is an increment of **borrowing**. When looking at increments of investment, we accepted the increment when its rate of return exceeded the MARR. When considering an increment of borrowing, however, the rate we calculated (that is, i^* for borrowing) was essentially the rate we paid to borrow money from the increment. We will call this the **borrowing rate of return (BRR)**. Conceptually, we would prefer to get a loan from the increment rather than from our initial investment pool, if the loan rate is less than the MARR. In other words, it is cheaper to borrow than to use your own money. Therefore, the decision rule is reversed:

$$\text{If BRR}_{A-B} \quad < \quad \text{MARR, select A,}$$
$$\text{If BRR}_{A-B} \quad = \quad \text{MARR, select either one,}$$
$$\text{If BRR}_{A-B} \quad > \quad \text{MARR, select B,}$$

where A – B is a borrowing increment (positive first cash flow).

EXAMPLE 5.6   Borrowing rate of return on incremental projects

Consider Example 5.3 again, but this time compute the rate of return on the increment B1 – B2.

n	B1	B2	B1 – B2
0	–$3000	–$12,000	+$9000
1	1350	4200	–2850
2	1800	6225	–4425
3	1500	6330	–4830

Note that the first incremental cash flow is positive, and all others are negative, which indicates that the cash flow difference is an increment of borrowing. What is the rate of return on this increment of borrowing, and which project should we prefer?

Solution

Given: Incremental borrowing cash flow, MARR = 10%
Find: BRR on increment, and determine whether it is acceptable

Note that the cash flow signs are simply reversed from Example 5.3. The rate of return on this increment of borrowing will be the same as that on the increment of investment, $i^*_{B1-B2} = i^*_{B2-B1} = 15\%$. However, because we are *borrowing* $9000, the 15% interest rate is what we are losing, or paying, by not investing the $9000 in B2. If we invest in B1, we are saving $9000 now, but we are giving up the opportunity of making $2850 at the end of first year, $4425 at the end of second year, and $4830 at the end of third year. This is equivalent to a situation where we are borrowing $9000, with the repayment series of ($2850, $4425, $4830). In effect, we are paying 15% interest on a borrowed sum. Is this an acceptable rate for a loan, considering the cost of using our own money is only 10%? Since the firm's MARR is 10%, we can assume that our maximum interest rate for borrowing should also be 10%. Since 15% > 10%, this borrowing situation is not desirable, and we should reject the lower cost alternative—B1—which produced the increment of borrowing (Figure 5.4). We choose instead B2, the same result as in Example 5.3.

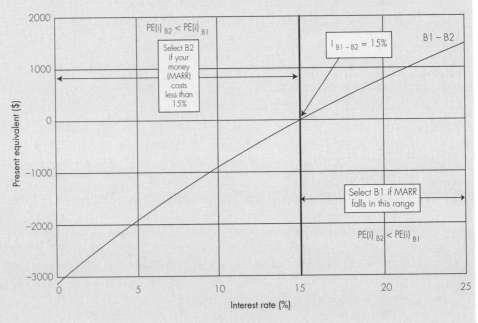

Figure 5.4   Incremental PE profile for B1 − B2 (Example 5.6)

Since the incremental investment and incremental borrowing methods yield the same results, if you find one intuitively easier to understand, you should feel free to set up project comparisons consistently by that method. Remember that a negative increment indicates investment and a positive increment indicates borrowing. However, we recommend that alternatives be compared on the basis of incremental *investment* rather than *borrowing*. This strategy allows you to avoid having to remember two decision rules. It also circumvents the ranking problem of deciding which rule to apply to mixed increments. (The first rule always applies to nonsimple investments: If the IRR on the increment is greater than the MARR, the increment is acceptable.)

5.5 Analysis Period

The **analysis period** is the time span over which the economic effects of an investment will be evaluated. The analysis period may also be called the **study period** or **planning horizon**. The length of the analysis period may be determined in several ways: It may be a predetermined amount of time set by company policy, or it may be either implied or explicit in the need the company is trying to fulfill—for example, a diaper manufacturer sees the need to dramatically increase production over a 10-year period in response to an anticipated "baby boom." In either of these situations, we consider the analysis period to be the **required service period**. When the required service period is not stated at the outset, the analyst must choose an appropriate analysis period over which to study the alternative investment projects. In such a case, one convenient choice of analysis period is the period of the useful life of the investment project.

When the useful life of an investment project does not match the analysis or required service period, we must make adjustments in our analysis. A further complication, in a consideration of two or more mutually exclusive projects, is that the investments themselves may have differing useful lives. We must compare projects with different useful lives over an **equal time span**, which may require further adjustments in our analysis. (Figure 5.5 is a flow chart showing the possible combinations of the analysis period and the useful life of an investment.) In the preceding examples (5.1 to 5.6), the project lives were all assumed to be equal, and the same as the study period. In the sections that follow, we will explore in more detail how to handle situations in which project lives differ from the analysis period and from each other.

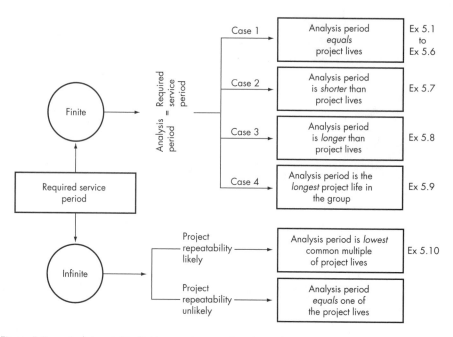

Figure 5.5   Analysis period implied in comparing mutually exclusive alternatives

5.5.1   Analysis Period Differs from Project Lives

In Examples 5.1 to 5.6, we assumed the simplest scenario possible when analyzing mutually exclusive projects: The projects had useful lives equal to each other and to the required service period. In practice, this is seldom the case. Often project lives do not match the required analysis period and/or do not match each other. For example, two machines may perform exactly the same function, but one lasts longer than the other, and both of them last longer than the analysis period for which they are being considered. In the following sections and examples, we will develop some techniques for dealing with these complications. Only the PE criterion will be applied to the examples in this section, so that we can focus on the issue of differing lifetimes.

Project's Life Is Longer than Analysis Period

Project lives frequently do not coincide with a firm's predetermined required analysis period; they are often too long or too short. The case of project lives that are too long is the easier one to address.

Consider the case of a firm that undertakes a 5-year production project when all of the alternative equipment choices have useful lives of 7 years. In such a case, we analyze each project for only as long as the required service period (in this case, 5 years). We are then left with some unused portion of the equipment (in this case, 2 years' worth), which we include as salvage value in our analysis. **Salvage value** is the amount of money for which the equipment could be sold after its service to the project has been rendered, or the dollar measure of its remaining usefulness.

A common instance of project lives that are longer than the analysis period occurs in the construction industry, where a building project may have a relatively short completion time, but the equipment purchased has a much longer useful life.

EXAMPLE 5.7

Present equivalent worth comparison—project lives longer than the analysis period

Waste Management Company (WMC) has won a contract that requires the firm to remove radioactive material from government-owned property and transport it to a designated dumping site. This task requires a specially made ripper-bulldozer to dig and load the material onto a transportation vehicle. Approximately 400,000 tons of waste must be moved in a period of 2 years. Model A costs $150,000 and has a life of 6000 hours before it will require any major overhaul. Two units of model A would be required to remove the material within 2 years, and the operating cost for each unit would run to $40,000/year for 2000 hours of operation. At this operational rate, the model would be operable for 3 years, and at the end of that time, it is estimated that the salvage value will be $25,000 for each machine.

A more efficient model B costs $240,000 each, has a life of 12,000 hours without any major overhaul, and costs $22,500 to operate for 2000 hours per year to complete the job within 2 years. The estimated salvage value of model B if it were used for 6 years is $30,000. Once again, two units of model B would be required.

Since the lifetime of either model exceeds the required service period of 2 years (Figure 5.6), WMC has to assume some things about the used equipment at the end of that time. Therefore, the engineers at WMC estimate that, after 2 years, the model A units could be sold for $45,000 each and the model B units for $125,000 each. After considering all tax effects, WMC summarized the resulting cash flows (in thousand dollars) for each project in the following table:

Period	Model A		Model B	
0	–$300		–$480	
1	–80		–45	
2	–80	+90	–45	+250
3	–80	+ 50	–45	
4			–45	
5			–45	
6			–45	+ 60

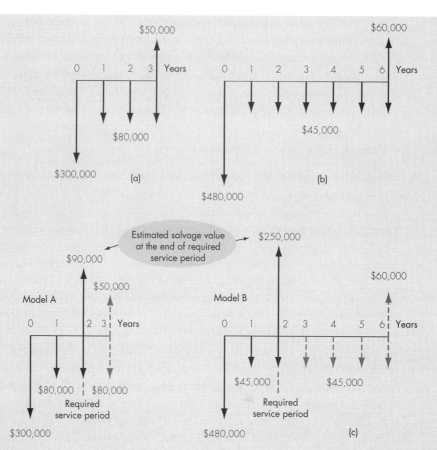

Figure 5.6 (a) Cash flow for model A; (b) cash flow for model B; (c) comparison of unequal-lived projects when the required service period is shorter than the individual project life (Example 5.7)

Here, the figures in the boxes represent the estimated salvage values at the end of the analysis period (end of year 2). Assuming that the firm's MARR is 15%, which option would be acceptable?

Solution

Given: Cash flows for two alternatives as shown in Fig. 5.6, i = 15% per year
Find: PE for each alternative, and which is preferred

First, note that these are service projects and we can assume the same revenues for both configurations. Since the firm explicitly estimated the market values of the assets at the end of the analysis period (2 years), we can compare the two models directly. Since the benefits (removal of the wastes) are equal, we can concentrate on the costs:

$$PE(15\%)_A = -\$300 - \$80(P/A, 15\%, 2) + \$90(P/F, 15\%, 2)$$
$$= -\$362.$$
$$PE(15\%)_B = -\$480 - \$45(P/A, 15\%, 2) + \$250(P/F, 15\%, 2)$$
$$= -\$364.$$

Model A has the less negative or larger PE and thus would be preferred.

Project's Life Is Shorter than Analysis Period

When project lives are shorter than the required service period, we must consider how, at the end of the project lives, we will satisfy the rest of the required service period. Replacement projects—additional projects to be implemented when the initial project has reached the limits of its useful life—are needed in such a case. Sufficient replacement projects must be analyzed to match or exceed the required service period.

To simplify our analysis, we could assume that the replacement project will be exactly the same as the initial project, with the same corresponding costs and benefits. However, this assumption is not necessary. For example, depending on our forecasting skills, we may decide that a different kind of technology—in the form of equipment, materials, or processes—is a preferable replacement. Whether we select exactly the same alternative, or a new technology as the replacement project, we are ultimately likely to have some unused portion of the equipment to consider as salvage value, just as in the case when project lives are longer than the analysis period. On the other hand, we may decide to lease the necessary equipment or subcontract the remaining work for the duration of the analysis period. In this case, we can probably exactly match our analysis period and not worry about salvage values.

In any event, we must make some initial guess concerning the method of completing the analysis period at its outset. Later, when the initial project life is closer to its expiration, we may revise our analysis with a different replacement project. This is only reasonable, since economic analysis is an ongoing activity in the life of a company and an investment project, and we should always use the most reliable, up-to-date data we can reasonably acquire.

EXAMPLE 5.8

Present equivalent worth comparison—project lives shorter than analysis period

The Smith Novelty Company, a mail-order firm, wants to install an automatic mailing system to handle product announcements and invoices. The firm has a choice between two different types of machines. The two machines are designed differently but have identical capacities and do exactly the same job. The $12,500 semi-automatic model A will last 3 years, while the fully automatic model B will cost $15,000 and last 4 years. The expected cash flows for the two machines including maintenance, salvage value, and tax effects are as follows:

n	Model A	Model B
0	-$12,500	-$15,000
1	-5,000	-4,000
2	-5,000	-4,000
3	-5,000 + 2,000	-4,000
4		-4,000 + 1,500
5		

As business grows to a certain level, neither of the models may be able to handle the expanded volume after year 5. If that happens, a fully computerized mail-order system will need to be installed to handle the increased business volume. With this scenario, which model should the firm select at MARR = 15%?

Solution

Given: Cash flows for two alternatives as shown in Figure 5.7, analysis period of 5 years, $i = 15\%$
Find: PE of each alternative and which to select

Since both models have a shorter life than the required service period (5 years), we need to make an explicit assumption of how the service requirement is to be met. Suppose that the company considers leasing comparable equipment that has an annual lease payment of $6000 (after taxes) for the remaining required service period. In this case, the cash flow would look like Fig. 5.7.

n	Model A	Model B
0	-$12,500	-$15,000
1	-5,000	-4,000
2	-5,000	-4,000
3	-5,000 + 2,000	-4,000
4	-5,000 - 6,000	-4,000 + 1,500
5	-5,000 - 6,000	-5,000 - 6,000

Here, the boxed figures represent the annual lease payments. (It costs $6000 to lease the equipment and $5000 to operate it annually. Other maintenance costs will be paid by the leasing company.) Note that both alternatives now have the same required service period of 5 years. Therefore, we can use PE analysis:

$$PE(15\%)_A = -\$12,500 - \$5000(P/A, 15\%, 2) - \$3000(P/F, 15\%, 3)$$
$$-\$11,000(P/A, 15\%, 2)(P/F, 15\%, 3)$$
$$= -\$34,359.$$
$$PE(15\%)_B = -\$15,000 - \$4000(P/A, 15\%, 3) - \$2500(P/F, 15\%, 4)$$
$$-\$11,000(P/F, 15\%, 5)$$
$$-\$31,031.$$

Since these are service projects, model B is the better choice.

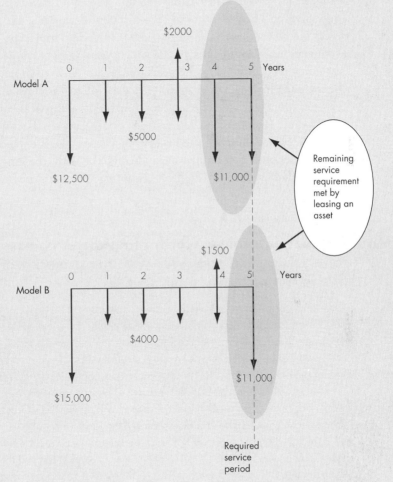

Figure 5.7 Comparison for projects with unequal lives when the required service period is longer than the individual project life (Example 5.8)

Analysis Period Coincides with Longest Project Life

As seen in the preceding pages, equal future time periods are generally necessary to achieve comparability of alternatives. In some situations, however, revenue projects with different lives can be compared if they require only a one-time investment because the task or need within the firm is a one-time task or need. We may find an example of this situation in the extraction of a fixed quantity of a natural resource such as oil or coal. Consider two mutually exclusive processes: One that requires 10 years to recover the available coal and a second that can accomplish the task in only 8 years. There is no need to continue the project if the short-lived process is used and all the coal has been retrieved. In this example, the two processes can be compared over an analysis period of 10 years (longest project life in the group), assuming no cash flows after 8 years for the shorter-lived project. The revenues must be included in the analysis even if the price of coal is constant because of the time value of money. Even if the total revenue (undiscounted) is equal for either process, that for the faster process has a larger present worth. Therefore, the two projects could be compared using the PE of each over its own life. Note that in this case the analysis period is determined by, and coincides with, the longest project life in the group. (Here we are still, in effect, assuming an analysis period of 10 years.)

EXAMPLE 5.9

Present equivalent worth comparison—a case where the analysis period coincides with the project with the longest life in the mutually exclusive set

The family-operated Foothills Ranching Company (FRC) owns the mineral rights for land used for growing grain and grazing cattle. Recently, oil was discovered on this property. The family has decided to recover the oil, sell the land, and retire. The company can either lease the necessary equipment and recover and sell the oil itself, or it can lease the land to an oil production company. If the company chooses the former, it will require $300,000 leasing expenses up front, but the net annual cash flow after taxes from production operations will be $600,000 at the end of each year for the next 5 years. The company can sell the land for a net cash flow of $1,000,000 in 5 years when the oil is depleted. If the company chooses the latter, the production company can extract all the oil in only 3 years, and it can sell the land for a net cash flow of $800,000 in 3 years. (The difference in resale value of the land is due to the increasing rate of land appreciation anticipated for this property.) The net cash flow from the lease payments to FRC will be $630,000 at the *beginning* of each of the next 3 years. All benefits and costs associated with the two alternatives have been accounted for in the figures listed above. Which option should the firm select at $i = 15\%$?

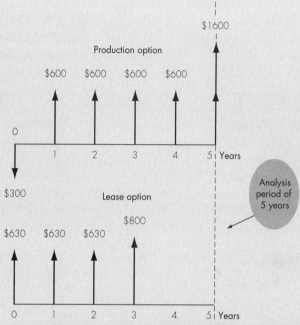

Figure 5.8 Comparison of projects with unequal lives where the analysis period coincides with the project with the longest life in the mutually exclusive set (Example 5.9). In our example, the analysis period is 5 years, assuming zero cash flows in years 4 and 5 for the lease option

Solution

Given: Cash flows shown in Figure 5.8, $i = 15\%$ per year
Find: PE of each alternative and which to select

As illustrated in Fig. 5.8, the cash flows associated with each option look like this:

n	Drill	Lease
0	−$300,000	$630,000
1	600,000	630,000
2	600,000	630,000
3	600,000	800,000
4	600,000	
5	1,600,000	

After depletion of the oil, the project will terminate.

$$PE(15\%)_{Production} = -\$300,000 + \$600,000(P/A, 15\%, 4)$$
$$+ \$1,600,000(P/F, 15\%, 5)$$
$$= \$2,208,470.$$

$$PE(15\%)_{Lease} = \$630,000 + \$630,000(P/A, 15\%, 2)$$
$$+ \$800,000(P/F, 15\%, 3)$$
$$= \$2,180,210.$$

Note that these are revenue projects. Therefore, the production option appears to be the marginally better option.

Comments: The relatively small difference between the two PE amounts ($28,260) suggests that the actual decision between producing and leasing might be decided on noneconomic issues. Even if the production option were slightly better, the company might prefer to forego the small amount of additional income and select the lease option rather than undertake an entirely new business venture to produce the oil. A variable that might also have a critical effect on this decision is the sales value of the land in each alternative. The value of land is often difficult to forecast over any long period of time, and the firm may feel some uncertainty about the accuracy of its guesses. In Chapter 13, we will discuss sensitivity analysis, which is a method by which we can factor uncertainty about the accuracy of project cash flows into our analysis.

5.5.2 Analysis Period Not Specified

Our coverage so far has focused on situations in which an analysis period is known. When an analysis period is not specified, either explicitly by company policy or practice or implicitly by the projected life of the investment project, it is up to the analyst to choose an appropriate one. In such a case, the most convenient procedure is to choose one based on the useful lives of the alternatives. When the alternatives have equal lives, this is an easy selection. When the lives of alternatives differ, we must select an analysis period that allows us to compare different-lived projects on an equal time basis, i.e., a **common service period**.

Lowest Common Multiple of Project Lives

A required service period of infinity may be assumed if we anticipate an investment project will be ongoing at roughly the same level of activity for some indefinite period. It is certainly possible to do so mathematically, though the analysis is likely to be complicated and tedious. Therefore, in the case of an indefinitely ongoing investment project, we typically select a finite analysis period by using the **lowest common multiple** of project lives. For example, if alternative A has a 3-year useful life and alternative B has a 4-year life, we may select 12 years as the analysis or common service period. We would consider alternative A through 4 life cycles and alternative B through 3 life cycles; in each case, we would use the alternatives completely. We then accept the finite model's results as a good prediction of what will be the economically wisest course of action for the foreseeable future. The following example is a case in which we conveniently use the lowest common multiple of project lives as our analysis period.

EXAMPLE
5.10

Present equivalent worth comparison—unequal lives—lowest common multiple method

Consider Example 5.8. Suppose that both models A and B can handle the increased future volume and that the system is not going to be phased out at the end of 5 years. Instead, the current mode of operation is expected to continue for an indefinite period of time. We also assume that these two models will be available in the future without significant changes in price and operating costs. At MARR = 15%, which model should the firm select?

Solution

Given: Cash flows for two alternatives as shown in Figure 5.9, i = 15% per year, indefinite period of need
Find: PE of each alternative and which to select

Recall that the two mutually exclusive alternatives have different lives, but provide identical annual benefits. In such a case, we ignore the common benefits and can make the decision based solely on costs, as long as a common analysis period is used for both alternatives.

To make the two projects comparable, let's assume that, after either the 3- or 4-year period, the system would be reinstalled repeatedly using the same model and that the same costs would apply. The lowest common multiple of 3 and 4 is 12, so we will use 12 years as the common analysis period. Note that any cash flow difference between the alternatives will be revealed during the first 12 years. After that, the same cash flow pattern repeats every 12 years for an indefinite period. The replacement cycles and cash flows are shown in Fig. 5.9.

- Model A: Four replacements occur in a 12-year period. The PE for the first investment cycle is

$$PE(15\%) = -\$12,500 - \$5000(P/A, 15\%, 2)$$
$$-\$3000(P/F, 15\%, 3)$$
$$= -\$22,601.$$

With four replacement cycles, the total PE is

$$PE(15\%) = -\$22,601 [1 + (P/F, 15\%, 3)$$
$$+ (P/F, 15\%, 6) + (P/F, 15\%, 9)]$$
$$= -\$53,657.$$

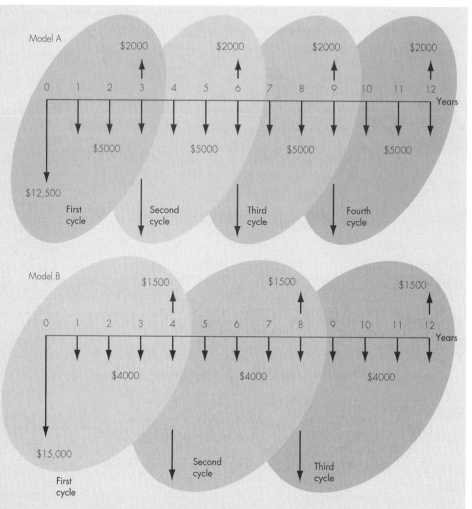

Figure 5.9 Comparison of projects with unequal lives when the required service period is infinite and project repeatability is likely with the same investment and O&M costs in the future replacement (Example 5.10)

- Model B: Three replacements occur in a 12-year period. The PE for the first investment cycle is

$$PE(15\%) = -\$15{,}000 - \$4000(P/A,\ 15\%,\ 3)$$
$$- \$2500(P/F,\ 15\%,\ 4)$$
$$= -\$25{,}562.$$

With three replacement cycles in 12 years, the total PE is

$$PE(15\%) = -\$25{,}562\ [1 + (P/F,\ 15\%,\ 4) + (P/F,\ 15\%,\ 8)]$$
$$= -\$48{,}534.$$

Comments: In Example 5.10, an analysis period of 12 years seems reasonable. The number of actual reinvestment cycles needed with each type of system will depend on the technology of the future system, so we may or may not actually need the four (model A) or three (model B) reinvestment cycles we used in our analysis. The validity of the analysis also depends on the costs of system and labor remaining constant. If we assume **constant dollar prices** (see Chapter 12), this analysis would provide us with a reasonable result. (As you will see in Example 5.11, the annual equivalent approach will make it mathematically easier to solve this type of comparison.) If we cannot assume the constant dollar prices in future replacements, we need to estimate the costs for each replacement over the analysis period. This will certainly complicate the problem significantly.

AE Analysis for Unequal Project Lives

Annual worth analysis also requires establishing common analysis periods, but AE offers some computational advantages as opposed to present worth analysis, provided the following criteria are met:

1. The service of the selected alternative is required on a continuous basis.

2. Each alternative will be replaced by an *identical* asset which has the same performance and associated cash flows of equivalent magnitude and timing as the original asset.

When these two criteria are in effect, we may solve for the AE of each project based on its initial life span, rather than on the lowest common multiple of the projects' lives.

EXAMPLE 5.11

Annual equivalent cost comparison—unequal project lives

Continuing Example 5.10, we apply the annual equivalent approach to select the most economical equipment.

Solution

Given: Cost cash flows shown in Figure 5.10, $i = 15\%$ per year
Find: AE cost, and which is the preferred alternative

An alternative procedure for solving Example 5.10 is to compute the annual equivalent cost of an outlay of $12,500 for model A every 3 years and the annual equivalent cost of an outlay of $15,000 for model B every 4 years. Notice that the AE of each 12-year cash flow is the same as that of the corresponding 3- or 4-year cash flow (Fig. 5.10). From Example 5.10, we calculate

- Model A:

  For a 3-year life:

  $$PE(15\%) = -\$22,601$$
  $$AE(15\%) = -22,601(A/P, 15\%, 3)$$
  $$= -\$9899.$$

  For 12-year period (computed for the entire analysis period):

  $$PE(15\%) = -\$53,657$$
  $$AE(15\%) = -53,657(A/P, 15\%, 12)$$
  $$= -\$9899.$$

- Model B:

  For a 4-year life:

  $$PE(15\%) = -\$25,562$$
  $$AE(15\%) = -\$25,562(A/P, 15\%, 4)$$
  $$= -\$8954.$$

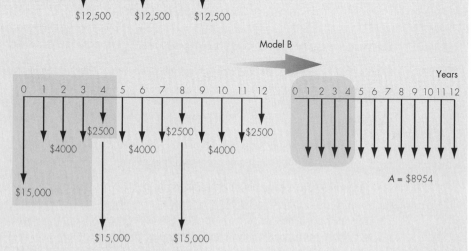

Figure 5.10 Comparison of projects with unequal lives and an indefinite analysis period using the equivalent annual worth criterion (Example 5.11)

For a 12-year period (computed for the entire analysis period):

$$PE(15\%) = -\$48,534$$
$$AE(15\%) = -\$48,534(A/P, 15\%, 12)$$
$$= -\$8954.$$

Notice that the annual equivalent values that were calculated based on the common service period are the same as those that were obtained over their initial life spans. Thus, for alternatives with unequal lives, we will obtain the same selection by comparing PE over a common service period using repeated projects or comparing AE for initial lives.

IRR Analysis for Unequal Project Lives

Above, we discussed the use of the PE, FE and AE criteria as bases for comparing projects with unequal lives. The IRR measure can also be used to compare projects with unequal lives, as long as we can establish a common analysis period. The decision procedure is then exactly the same as for projects with equal lives. It is likely, however, that we will have a multiple-root problem, which creates a substantial computational burden. For example, suppose we apply the IRR measure to a case in which one project has a 5-year life and the other project has an 8-year life, resulting in a least common multiple of 40 years. When we determine the incremental cash flows over the analysis period, we are bound to observe many sign changes. This leads to the possibility of having many i^*s. The following two examples use i^* to compare mutually exclusive projects, one with only one sign change in the incremental cash flows, and one with several sign changes. (Our purpose is not to encourage you to use the IRR approach to compare projects with unequal lives. Rather, it is to show the correct way to compare them if the IRR approach must be used.)

EXAMPLE 5.12

IRR analysis for projects with unequal lives in which the increment is a simple investment

Consider the following mutually exclusive investment projects (E1, E2):

n	E1	E2
0	-$2000	-$3000
1	1000	4000
2	1000	
3	1000	

Project E1 has a service life of 3 years, whereas project E2 has only 1 year of service life. Assume that project E2 can be repeated with the same investment costs and benefits over the analysis period of 3 years. Assume that the firm's MARR is 10% and that "do nothing" is not an option. Determine which project should be selected.

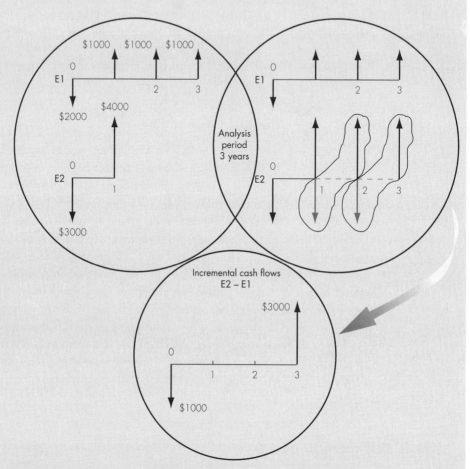

Figure 5.11 Comparison of projects with unequal lives when the required service period is indefinite (Example 5.12)

Solution

Given: Two alternatives with unequal lives, cash flows as shown, MARR = 10%
Find: IRR on incremental investment, and determine which alternative is preferable

By assuming three repetitions of project E2 over the analysis period (as shown in Figure 5.11) and taking the incremental cash flows (E2 – E1), we obtain

n	E1	E2			E2 – E1
0	–$2000		–$3000	= –$3000	–$1000
1	1000	4000 – 3000	=	1000	0
2	1000	4000 – 3000	=	1000	0
3	1000		4000	= 4000	3000

By inspection, the increment in this case is an investment. To compute i^*_{E2-E1}, we evaluate:

$$PE = -\$1000 + \frac{\$3000}{(1+i)^3} = 0.$$

Solving for i yields

$$IRR_{E2-E1} = 44.22\% > 10\%.$$

Therefore, we select project E2.

Now we revisit the mail-order firm in Example 5.10, and show how the IRR decision criterion can become intractable for repeated life comparisons.

EXAMPLE 5.13   IRR analysis for projects with different lives in which the increment is a nonsimple investment

Consider Example 5.10, in which a mail-order firm wants to install an automatic mailing system to handle product announcements and invoices. Using the IRR as a decision criterion, select the best machine. Assume a MARR of 15% as before.

Solution

Given: Cash flows for two projects with unequal lives, as shown in Figure 5.12, MARR = 15%
Find: The alternative that is preferable

Since the analysis period is equal to the least common multiple of 12 years, we may compute the incremental cash flow over this 12-year period. As shown in Fig. 5.12, we subtract cash flows of model A from those of model B to form the increment of investment. (We want the first cash flow difference to be a negative value.)

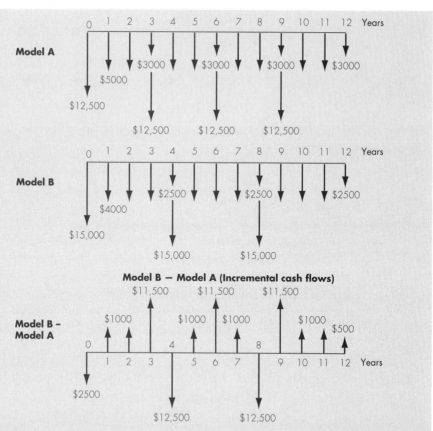

Figure 5.12 Comparison of projects with unequal lives (Example 5.13)

n	Model A		Model B		Model B – Model A
0	–$12,500		–$15,000		–$2,500
1		–5,000		–4,000	1,000
2		–5,000		–4,000	1,000
3	–12,500	–3,000		–4,000	11,500
4		–5,000	–15,000	–2,500	–12,500
5		–5,000		–4,000	1,000
6	–12,500	–3,000		–4,000	11,500
7		–5,000		–4,000	1,000
8		–5,000	–15,000	–2,500	–12,500
9	–12,500	–3,000		–4,000	11,500
10		–5,000		–4,000	1,000
11		–5,000		–4,000	1,000
12		–3,000		–2,500	500

Five sign changes occur in the incremental cash flows, indicating a nonsimple incremental investment. As suggested in Section 4.7.3, we may abandon the rate of return analysis and use the PE criterion.

$$
\begin{aligned}
PE(15\%)_{B-A} &= -\$2500 + \$1000(P/F, 15\%, 1) \\
&\quad + \ldots + \$500(P/F, 15\%, 12) \\
&= \$5123 > 0.
\end{aligned}
$$

This implies $PE(15\%)_B > PE(15\%)_A$, indicating that project B is preferable.

If you insist on applying the IRR analysis to comparing the nonsimple investments with unequal lives, you need to find the rate of return on this incremental investment. Even though there are five sign changes in the cash flow, there is only one positive i^* for this problem, 63.12%. Unfortunately, however, this is not a pure investment. We need to employ an external rate to compute the IRR to make a proper accept/reject decision. Assuming that the firm's MARR is 15%, we will use a trial-and-error approach (as described in Section 4.A.5). Try $i = 20\%$:

$$
\begin{aligned}
PB(20\%, 15\%)_0 &= -\$2500. \\
PB(20\%, 15\%)_1 &= -\$2500(1.20) + \$1000 = -\$2000. \\
PB(20\%, 15\%)_2 &= -\$2000(1.20) + \$1000 = -\$1400. \\
PB(20\%, 15\%)_3 &= -\$1400(1.20) + \$11{,}500 = \$9820. \\
PB(20\%, 15\%)_4 &= \$9820(1.15) - \$12{,}500 = -\$1207. \\
PB(20\%, 15\%)_5 &= -\$1207(1.20) + \$1000 = -\$448.40. \\
PB(20\%, 15\%)_6 &= -\$448.40(1.20) + \$11{,}500 = \$10{,}961.92. \\
PB(20\%, 15\%)_7 &= \$10{,}961.92(1.15) + \$1000 = \$13{,}606.21. \\
PB(20\%, 15\%)_8 &= \$13{,}606.21(1.15) - \$12{,}500 = \$3147.14. \\
PB(20\%, 15\%)_9 &= \$3147.14(1.15) + \$11{,}500 = \$15{,}119.21. \\
PB(20\%, 15\%)_{10} &= \$15{,}119.21(1.15) + \$1000 = \$18{,}387.09. \\
PB(20\%, 15\%)_{11} &= \$18{,}387.09(1.15) + \$1000 = \$22{,}145.16. \\
PB(20\%, 15\%)_{12} &= \$22{,}145.16(1.15) + \$500 = \$25{,}966.93.
\end{aligned}
$$

Since $PB(20\%, 15\%)_{12} > 0$, the guessed 20% is not the IRR. We may increase the value of i and repeat the calculations. After several trials, we find that the IRR is 50.68%.[3] Since $IRR_{B-A} > MARR$, model B would be selected, which is consistent with the PE analysis. In other words, the additional investment over the years to obtain model B (−$2500 at $n = 0$, −$12{,}500 at $n = 4$, −$12{,}500 at $n = 8$) yields a satisfactory rate of return: Model B, therefore, is the preferred project. (Note that model B was picked when PE analysis was used in Example 5.10.)

---

[3] It will be tedious to solve this type of problem by a trial-and-error method on your calculator. The problem can be solved quickly by using the **EzCash** software.

Given the complications involved in using IRR analysis to compare alternative projects, it is usually more desirable to use one of the other equivalence techniques for this purpose. As an engineering manager, you should keep in mind the intuitive appeal of the rate of return measure. Once you have selected a project on the basis of PE, FE or AE analysis, you may also wish to express its worth as a rate of return, for the benefit of your associates.

Other Common Analysis Periods

In some cases, the lowest common multiple of project lives is an unwieldy analysis period to consider. Suppose, for example, that you were considering alternatives with lives of 7 and 12 years, respectively. Besides making for tedious calculations, an 84-year analysis period may lead to inaccuracies, since over a long period of time we can be less and less confident about the ability to install identical replacement projects with identical costs and benefits. In a case like this, it would be reasonable to use the useful life of one of the alternatives by either factoring in a replacement project or salvaging the remaining useful life, as the case may be. The important rule is to compare both projects on the same time basis.

5.6 Computer Notes

Mutually exclusive project comparisons can be made easily using a spreadsheet. The built-in functions and formula copying capabilities can be used to advantage to very quickly enter and solve a problem. Sensitivity analyses can be performed effectively from two perspectives: changing the value in a cell to see how it affects the outcomes in other cells; and using spreadsheet analysis tools to determine what cell values must be changed to achieve a desired outcome in a target cell.

5.6.1   Mutually Exclusive Project Comparisons

The data from the machining center decision problem, Example 5.1, is reproduced in Figure 5.13. To speed up the task and help avoid data entry errors, formulae and copying cells are used as much as possible. In Figure 5.13, the only "numbers" that are entered are in the shaded cells A10, B3, B4, B5, B6, B10, B11, C10, C11 and D11. In this problem statement, for example, the initial cost of options M2 and M3 are identical, so cell D10 contains a cell reference to C10. This helps avoid errors, because if any of the acquisition costs for M2 and M3 change, only the value in cell C10 must be updated, and consistency is automatically ensured in cell D10. Similarly, any change in the data for options M1, or M2 or M3 will automatically be updated in the incremental cash flow columns.

Analogous to the example in Section 4.8, sensitivity analysis can be conducted on the discount rate by generating a range of *i* values in a column of the spreadsheet, and then in the three adjacent columns entering the PE formula for M1, M2 and M3 respectively as functions of the *i* column values. The resulting PE profile presented earlier in the chapter (Figure 5.2) illustrates the sensitivity of each option to changes in *i*, and provides an effective comparison between these alternatives. Naturally, a PE profile for the incremental cash flows can be similarly plotted.

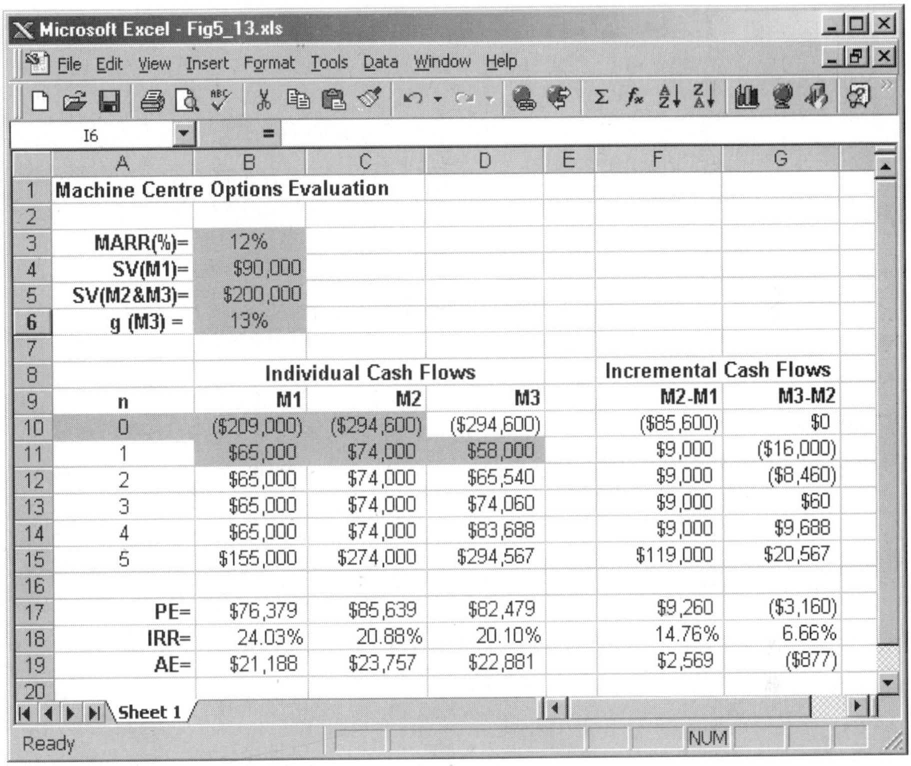

	A	B	C	D	E	F	G
1	**Machine Centre Options Evaluation**						
2							
3	MARR(%)=	12%					
4	SV(M1)=	$90,000					
5	SV(M2&M3)=	$200,000					
6	g (M3) =	13%					
7							
8			Individual Cash Flows			Incremental Cash Flows	
9	n	M1	M2	M3		M2-M1	M3-M2
10	0	($209,000)	($294,600)	($294,600)		($85,600)	$0
11	1	$65,000	$74,000	$58,000		$9,000	($16,000)
12	2	$65,000	$74,000	$65,540		$9,000	($8,460)
13	3	$65,000	$74,000	$74,060		$9,000	$60
14	4	$65,000	$74,000	$83,688		$9,000	$9,688
15	5	$155,000	$274,000	$294,567		$119,000	$20,567
16							
17	PE=	$76,379	$85,639	$82,479		$9,260	($3,160)
18	IRR=	24.03%	20.88%	20.10%		14.76%	6.66%
19	AE=	$21,188	$23,757	$22,881		$2,569	($877)
20							

Figure 5.13 Spreadsheet set-up for Machining Center Method evaluation

5.6.2 Incremental Analysis with Excel

To solve Example 5.5, enter the cash flows for both models over the analysis period (6 years) as shown in Figure 5.14(a). The incremental cash flows in column E are obtained by subtracting the cash flows of CMS from those of FMS. Using the IRR function at a guessed interest rate of 10%, you obtain the i^* value of 12.43%, which indicates that the CMS option is preferable.

With **Microsoft Excel Solver**, you can conduct a break-even analysis by creating a worksheet model with multiple changing cells. You can set constraints on your problem that must be satisfied before a solution is reached. For example, to determine

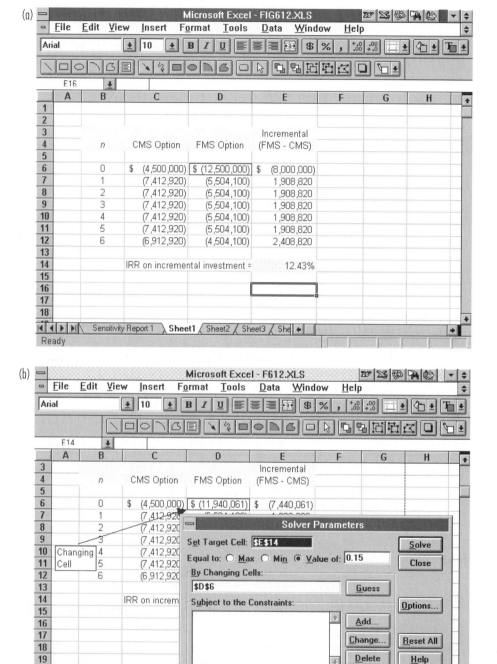

Figure 5.14 Excel's output: (a) An incremental analysis for data in Example 5.5 and (b) an example of using Excel's Solver to find the break-even investment amount for the FMS option

the level of the investment required for the FMS option to break even with the CMS option, you may perform the following steps:

Step 1: Select the "Solver" command from the "Tool" menu. Excel will display the Solver Parameters dialog box as shown in Fig. 5.14(b).

Step 2: Specify the target cell that you want to attain a specified value, or that you want to maximize or minimize. In our example, the target cell is $E$14, the IRR on incremental investment.

Step 3: Set "Equal To" to the desired target value. In our example, to break even, the target cell value must be 15%, which is the value of MARR.

Step 4: Specify the adjustable cells. An adjustable cell is one that can be changed by Solver until the constraints in the problem are satisfied and the cell in the "set target cell" box reaches its target. In our example, the adjustable cell is $D$6, where we entered the investment amount for the FMS option.

Step 5: Specify other constraints if necessary. In our example, no additional constraints are needed. Press the "Solve" button to start the problem-solving process.

Solver will display the search result in cell $D$6, which indicates that, to break even with the CMS option, the required investment for the FMS option would be $11,940,061. At any amount less than this value, the FMS becomes a more economically attractive option. You can perform other break-even analyses in a similar fashion.

5.7 Summary

- **Revenue projects** are those for which the income generated depends on the choice of project. **Service projects** are those for which income remains the same, regardless of which project is selected.

- The term **mutually exclusive** means that, when one of several alternatives that meet the same need is selected, the others will be rejected.

- In many cases, a **do-nothing alternative** is available, either because the need is being met by an existing system which could be replaced, or because we are assessing the viability of a new endeavor. When do-nothing is not an option, one of the mutually exclusive alternatives under consideration must be chosen.

- In **total investment analysis**, the PE (or FE or AE) of each mutually exclusive alternative is calculated separately, and the best choice is the one with the highest PE (or FE or AE).

- In **incremental investment analysis**, the differences in cash flows between pairs of mutually exclusive alternatives are used as the basis for analysis. When using the IRR criterion, incremental investment analysis must be used to arrive at the correct optimal choice. This approach may also be used in conjunction with the PE, FE or AE criteria. Given that B − A is an investment increment, the decision rules are as follows:

$$\text{If } IRR_{B-A} \quad > \quad \text{MARR, select B,}$$
$$\text{If } IRR_{B-A} \quad = \quad \text{MARR, select either one,}$$
$$\text{If } IRR_{B-A} \quad < \quad \text{MARR, select A,}$$

or

	> **0**	= **0**	< **0**
PE_{B-A} (MARR)	select B	select either one	select A
FE_{B-A} (MARR)	select B	select either one	select A
AE_{B-A} (MARR)	select B	select either one	select A

- Mutually exclusive alternatives must be compared over the same **analysis period**.

- When not specified by management or company policy, the analysis period to use in a comparison of mutually exclusive projects may be chosen by an individual analyst. Several efficiencies can be applied when selecting an analysis period. In general, the analysis period should be chosen to cover the required service period as highlighted in Fig. 5.5.

- AE analysis is recommended for many situations involving the comparison of options with differing lives. If we can use a **least common multiple** duration, or **infinite planning horizon** with project repeatability, then the AE criterion involves less computation than other methods.

- When comparing options with different lifetimes, the incremental cash flows often exhibit multiple sign changes, hence the IRR method becomes impractical to use.

Problems

Note: *Assume that "do-nothing" is not an option, and that one of the mutually exclusive alternatives must be selected, unless the problem statement specifies otherwise.*

Level 1

5.1* Consider the following two mutually exclusive projects, where doing nothing is an option:

End of Year	Net Cash Flow Project A	Project B
0	−$1000	−$2000
1	475	915
2	475	915
3	475	915

At an interest rate of 12%, what would you recommend using PE analysis?

5.2* Consider the following two mutually exclusive investment alternatives:

End of Year	Net Cash Flow Machine A	Machine B
0	−$1000	−$2000
1	900	2500
2	800	800 + 200
3	700	

Suppose that your firm needs either machine for only 2 years. The net proceeds from the sale of machine B is estimated to be $200. What should be the required net proceeds from the sale of machine A so that both machines could be considered economically indifferent at an interest rate of 10%?

5.3 Consider the following cash flow data for two competing investment projects.

	Cash Flow Data (Unit: $ thousand)	
n	Project A	Project B
0	−$500	−$2520
1	−1500	−565
2	−435	820
3	775	820
4	775	1080
5	1275	1880
6	1275	1500
7	975	980
8	675	580
9	375	380
10	660	840

Using PE analysis, which of the two projects would be a better choice given $i = 12\%$?

5.4* Consider the following two mutually exclusive investment projects. Assume that the MARR=12%.

	Project's Cash Flow	
n	A	B
0	−$4200	−$2500
1	2610	1210
2	2930	1720
3	2300	1500

(a) Which alternative would be selected using the PE criterion?
(b) Which alternative would be selected using the FE criterion?

5.5* Jones Construction Company needs a temporary office building at a construction site. Two types of heating schemes are being considered. The first method is to use "bottled gas" for floor-type furnaces. The second is to install electric radiant panels in the walls and ceiling. The temporary building will be used for 5 years before being dismantled.

	Bottled Gas	Electric Panels
Investment cost	$6000	$8500
Service life	5 years	5 years
Salvage value	0	$1000
Annual O&M cost	$2000	$1000
Extra expenses for income taxes		$ 220

Compare the alternatives based on the present equivalent criterion at $i = 10\%$.

Adapted from Grant EL, Ireson W, Leavenworth WG. *Engineering Economy*. 8th Ed. New York: John Wiley and Sons, 1990.

5.6 Two alternative machines are being considered for a manufacturing process. Machine A has a first cost of $75,200, and its salvage value at the end of 6 years of estimated service life is $21,000. The operating costs of this machine are estimated to be $6800 per year. Extra income taxes are estimated at $2400 per year. Machine B has a first cost of $44,000, and its estimated salvage value at the end of 6 years' service is estimated to be negligible. The annual operating costs will be $11,500. Compare these two alternatives by the present equivalent method at $i = 13\%$.

5.7* Two options are available for painting your house: (1) Oil-based painting, which costs $5000 and (2) water-based painting, which costs $3000. The estimated lives are 10 years and 5 years respectively. For either option, no salvage value will remain at the end of

their respective service lives. Assume that you will keep and maintain the house for 10 years. If your personal interest rate is 10% per year, which of the following statements is correct?

(a) On an annual basis, Option 1 will cost about $850.
(b) On an annual basis, Option 2 is about $22 cheaper than Option 1.
(c) On an annual basis, both options cost about the same.
(d) On an annual basis, Option 2 will cost about $820.

5.8* An industrial firm can purchase a special machine for $40,000. A down payment of $4000 is required, and the balance can be paid in five equal year-end installments at 7% interest on the unpaid balance. As an alternative, the machine can be purchased for $36,000 in cash. If the firm's MARR is 10%, determine which alternative should be accepted, based on the annual equivalent method.

5.9* A manufacturing company must choose between two types of industrial projects that require the same level of initial investment but provide different levels of operating cash flows over the project life. The in-house engineer has compiled the following financial data related to both projects, including the rate of return figures.

	Net Cash Flow		
			Incremental
n	Project A	Project B	(B – A)
0	–$18,000	–$18,000	0
1	960	11,600	$10,640
2	7,400	6,500	–900
3	13,100	4,000	–9,100
4	7,560	3,122	–4,438
i^*	18%	20%	14.72%

The company wants to use the internal rate

of return as a project justification tool. The firm's minimum required rate of return is known to be 12%. Which project should you select and on what basis?

(a) Select A, because the increment of borrowing for B exceeds 12%.
(b) Select B, because its increment of investment exceeds 12%.
(c) Select B, because its increment of borrowing exceeds 12%.
(d) Select B, because it can generate 20% profit (as opposed to 18% for project A) for every dollar invested.

5.10 The following information on two potential revenue projects is given:

- IRR of project A = 17%.
- IRR of project B = 16%.
- Both the projects have a service life of 6 years and need the same initial investment of $23,000.
- IRR on incremental cash flows (A – B) = 10%.

If your MARR is 20%, which of the following statements is correct?

(a) Select A.
(b) Select B.
(c) Select either one of the projects.
(d) Select neither project.
(e) Information is insufficient to make a decision.

5.11* The following information on three mutually exclusive projects is given:

Project	Investment in Year 0	IRR or BRR(%)
A	–$1250	43%
B	–$1000	57
C	–$1200	48
B – C		27
B – A		23
A – C		15

All three projects have the same service life and require investment in year 0 only. Which project, if any, would you choose based on the IRR criterion at a MARR of 25%?

5.12 A manufacturing firm is considering the following mutually exclusive alternatives:

	Net Cash Flow	
n	Project A	Project B
0	–$2000	–$3000
1	1400	2400
2	1640	2000

Determine which project is a better choice at a MARR = 15%, based on the IRR criterion.

5.13 Consider the following two mutually exclusive alternatives:

	Net Cash Flow	
n	Project A1	Project A2
0	–$10,000	–$12,000
1	5,000	6,100
2	5,000	6,100
3	5,000	6,100

(a) Determine the IRR on the incremental investment of $2000.
(b) If the firm's MARR is 10%, which alternative is the better choice?

5.14* The E. F. Fedele Company may upgrade to an automatic screwing machine for its assembly operation of a personal computer. Three different models with varying automatic features are under consideration. The required investments are $360,000 for model A, $380,000 for model B, and $405,000 for model C, respectively. All three models are expected to have the same service life of 8 years. The following financial information is available. In the following, model (B – A) represents the incremental cash flow determined by subtracting model A's cash flow from model B's.

Model	IRR (%)
A	30%
B	15
C	25

Model	Incremental IRR (%)
(B – A)	5%
(C – B)	40
(C – A)	15

If the firm's MARR is known to be 12%, which model, if any, should be selected?

5.15 The GeoStar Company, a leading wireless communication device manufacturer, is considering three cost-reduction proposals in its batch job shop manufacturing operations. The company already calculated rates of return for the three projects along with some incremental rates of return. A_0 denotes the do-nothing alternative. The required investments are $420,000 for A_1, $550,000 for A_2, and $720,000 for A_3. If the MARR is 12%, what system, if any, should be selected?

Incremental Investment	Incremental Rate of Return (%)
$A_1 - A_0$	20%
$A_2 - A_0$	17
$A_3 - A_0$	16
$A_2 - A_1$	11
$A_3 - A_1$	13
$A_3 - A_2$	14

Level 2

5.16* Two 150-horsepower (HP) motors are being considered for installation at a municipal sewage-treatment plant (1 HP = 0.7457 kW). The first motor costs $4500 and has an operating efficiency of 83% at full load. The second motor costs $3600 and has an efficiency rating of 80% at full load. Both motors are projected to have zero salvage value after a life of 10 years. The annual operating and maintenance cost (excepting power cost) amounts to 15% of the original cost of each motor. The power costs are a flat 5 cents per kilowatt-hour. Find the minimum number of hours of full-load operation per year necessary to justify the purchase of the more expensive motor at $i = 6\%$.

5.17* Consider the following two mutually exclusive service projects with project lives of 3 years and 2 years, respectively. (The mutually exclusive service projects will have identical revenues for each year of service.) The interest rate is known to be 12%.

	Net Cash Flow	
End of Year	Project A	Project B
0	–$1000	–$800
1	–400	–200
2	–400	–200 + 0
3	–400 + 200	(salvage)
	(salvage)	

If the required service period is 6 years, and both projects can be repeated with the given costs and better service projects are unavailable in the future, which of the following statements is correct?

(a) Select project B because it will save you $344 in present worth over the required service period.

(b) Select project A because it will cost $1818 in PE each cycle, with only one replacement, whereas project B will cost \$1138 in PE each cycle, with two replacements.

(c) Select project B because its PE exceeds that of project A by \$680.

(d) None of the above.

5.18 Consider the following two mutually exclusive investment projects. Assume MARR = 15%.

	Project's Cash Flow	
n	A	B
0	−\$3,000	−\$8,000
1	400	11,500
2	7,000	400

(a) Using the PE criterion, which project would be selected?

(b) Sketch the PE(i) function for each alternative on the same chart between 0% and 50%. For what range of i would you prefer project B?

5.19* An electric motor is rated at 10 horsepower (HP) and costs \$800. Its full load efficiency is specified to be 85%. A newly designed, high-efficiency motor of the same size has an efficiency of 90%, but costs \$1200. It is estimated that the motors will operate at a rated 10 HP output for 1500 hours a year, and the cost of energy will be \$0.07 per kilowatt-hour. Each motor is expected to have a 15-year life. At the end of 15 years, the first motor will have a salvage value of \$50, and the second motor will have a salvage value of \$100. Consider the MARR to be 8%. (Note: 1 HP = 0.7457 kW.)

(a) Determine which motor should be installed based on the PE criterion.

(b) In (a), what if the motors operated 2500 hours a year instead of 1500 hours a year? Would the same motor in (a) be the choice?

5.20* Consider the following two mutually exclusive investment projects.

	Project's Cash Flow	
n	A	B
0	−\$10,000	−\$22,000
1	7,500	15,500
2	7,000	18,000
3	5,000	

Which project would be selected if you use the infinite planning horizon with project repeatability likely (same costs and benefits) based on the PE criterion? Assume that i = 12%.

5.21 Consider the following two mutually exclusive investment projects, which have unequal service lives.

	Project's Cash Flow	
n	A1	A2
0	−\$900	−\$1800
1	−400	−300
2	−400	−300
3	−400 + 200	−300
4		−300
5		−300
6		−300
7		−300
8		−300 + 500

(a) What assumption(s) do you need to compare a set of mutually exclusive investments with unequal service lives?

(b) With the assumption(s) defined in (a) and using i = 10%, determine which project should be selected based on PE analysis.

(c) If your analysis period (study period) is just 3 years, what should be the salvage value of project A2 at the end of year 3 to make the two alternatives economically indifferent?

5.22* An electric utility is taking bids on the purchase, installation, and operation of microwave towers.

	Cost per Tower Bid A	Bid B
Equipment cost	$65,000	$58,000
Installation cost	$15,000	$20,000
Annual maintenance and inspection fee	$1,000	$1,250
Annual extra income taxes		$500
Life	40 years	35 years
Salvage value	$0	$0

Which is the most economical bid, if the interest rate is considered to be 11%? Either tower will have no salvage value after 20 years of use.

5.23 Refer to Problem 2.67. Consider the two different payment plans for Troy Aikman's contract. Compare these two mutually exclusive plans based on the PE method.

5.24 Consider the following two investment alternatives:

	Project's Cash Flow	
n	A1	A2
0	−$15,000	−$25,000
1	9,500	0
2	12,500	X
3	7,500	X
PE(15%)	?	9300

The firm's MARR (minimum attractive rate of return) is known to be 15%.

(a) Compute the PE(15%) for A1.
(b) Compute the unknown cash flow X in years 2 and 3 for A2.
(c) Compute the project balance (at 15%) of A1 at the end of period 3.
(d) If these two projects are mutually exclusive alternatives, which project would you select?

5.25 A bi-level mall is under construction. It is planned to install only nine escalators at the start, although the ultimate design calls for 16. The question arises in the design whether to provide necessary facilities (stair supports, wiring conduits, motor foundations, etc.), which would permit the installation of the additional escalators at the mere cost of their purchase and installation, or to defer investment in these facilities until the escalators need to be installed.

Option 1: Provide these facilities now for all seven future escalators at $200,000.

Option 2: Defer the facility investment as needed. It is planned to install two more escalators in 2 years, three more in 5 years, and the last two in 8 years. The installation of these facilities at the time they are required is estimated to cost $100,000 in year 2, $160,000 in year 5, and $140,000 in 8 years.

Additional annual expenses are estimated at $3000 for each escalator facility installed. At an interest rate of 12%, compare the net present equivalent of each option over 8 years.

5.26 A large refinery-petrochemical complex is planning to manufacture caustic soda, which will use feed water of 10,000 gallons per day. Two types of feeder-water storage installation are being considered over 40 years of useful life.

Option 1: Build a 20,000-gallon tank on a tower. The cost of installing the tank and tower is estimated to be $164,000. The salvage value is estimated to be negligible.

Option 2: Place a tank of equal capacity on a hill, which is 150 yards away from the refinery. The cost of installing the tank on the hill, including the extra length of service lines, is estimated to be $120,000 with negligible salvage value. Because of its hill location, an additional investment of $12,000 in pumping equipment is required. The pumping equipment is expected to have a service life of 20 years with a salvage value of $1000 at the end of that time. The annual operating and maintenance cost (including any income tax effects) for the pumping operation is estimated at $1000.

If the firm's MARR is known to be 12%, which option is better on the basis of present equivalent criterion?

5.27 A certain factory building has an old lighting system, and lighting this building costs, on average, $20,000 a year. A lighting consultant tells the factory supervisor that the lighting bill can be reduced to $8000 a year if $50,000 were invested in relighting the factory building. If the new lighting system is installed, an incremental maintenance cost of $3000 per year must be considered. If the old lighting system has zero salvage value, and the new lighting system is estimated to have a life of 20 years, what is the net annual benefit for this investment in new lighting? Consider the MARR to be 12%. Also consider that the new lighting system has zero salvage value at the end of its life.

$5.28* Travis Wenzel has $2000 to invest. Normally, he would deposit the money in his savings account, which earns 6% interest, compounded monthly. However, he is considering three alternative investment opportunities:

Option 1: Purchasing a bond for $2000. The bond has a face value of $2000 and pays $100 every 6 months for 3 years. The bond matures in 3 years.

Option 2: Buying and holding a growth stock that grows 11% per year for 3 years.

Option 3: Making a personal loan of $2000 to a friend and receiving $250 per year for 3 years.

Determine the equivalent annual cash flows for each option, and select the best option.

5.29* A chemical company is considering two types of incinerators to burn solid waste generated by a chemical operation. Both incinerators have a burning capacity of 20 tons per day. The following data have been compiled for comparison:

	Incinerator A	Incinerator B
Installed cost	$1,200,000	$750,000
Annual O&M costs	$50,000	$80,000
Service life	20 years	10 years
Salvage value	$60,000	$30,000
Income taxes	$40,000	$30,000

If the firm's MARR is known to be 13%, determine the processing cost per ton of solid waste by each incinerator. Assume that incinerator B will be available in the future at the same cost.

5.30 Consider the cash flows for the following investment projects (assume MARR = 15%):

	Project's Cash Flow		
n	A	B	C
0	−$3000	−$4000	−$5000
1	1000	1600	1800
2	1800	1500	1800
3	1000	1500	2000
4	400	1500	2000

(a) Suppose that projects A and B are mutually exclusive. Which of these three projects would you select based on the AE criterion?

(b) Assume that projects B and C are mutually exclusive. Which of these three projects would you select based on the AE criterion?

5.31 Norton Auto-Parts, Inc., is considering one of two forklift trucks for their assembly plant. Truck A costs $15,000 and requires $3000 annually in operating expenses. It will have a $5000 salvage value at the end of its 3-year service life. Truck B costs $20,000, but requires only $2000 annually in operating expenses; its service life is 4 years, at which time truck B's expected salvage value will be $8000. The firm's MARR is 12%. Assuming that the trucks are needed for 12 years and that no significant changes are expected in the future price and functional capacity of both trucks, select the most economical truck based on AE analysis.

5.32 Consider two investments, A and B, with the following sequences of cash flows:

	Net Cash Flow	
n	**Project A**	**Project B**
0	−$25,000	−$25,000
1	2,000	10,000
2	6,000	10,000
3	12,000	10,000
4	24,000	10,000
5	28,000	5,000

(a) Compute the i^* for each investment.
(b) Plot the present worth curve for each project on the same chart, and find the interest rate that makes the two projects equivalent.

5.33* Consider two investments A and B with the following sequences of cash flows:

	Net Cash Flow	
n	**Project A**	**Project B**
0	−$120,000	−$100,000
1	20,000	15,000
2	20,000	15,000
3	120,000	130,000

(a) Compute the IRR for each investment.

(b) At a MARR = 15%, determine the acceptability of each project.
(c) If A and B are mutually exclusive projects, which project would you select based on the rate of return on incremental investment?

$5.34 With $10,000 available, you have two investment options. The first option is to buy a guaranteed investment certificate (GIC) from a bank at an interest rate of 10% annually for 5 years. The second choice is to purchase a bond for $10,000 and invest the bond's interests in the bank at an interest rate of 9%. The bond pays 10% interest annually and will mature to its face value of $10,000 in 5 years. Based on the IRR method, determine which option is better. Assume your MARR is 9% per year.

$5.35 You are considering two types of automobiles. Model A costs $18,000 and model B costs $15,624. Although the two models are essentially the same, model A can be sold for $9000 while model B can be sold for $6500 after 4 years of use. Model A commands a better resale value because its styling is popular among university students. Determine the rate of return on the incremental investment of $2376. For what range of values of your MARR is model A preferable?

5.36* A plant engineer must choose between two types of solar water heating system:

	Model A	Model B
Initial cost	$7000	$10,000
Annual savings	$700	$1,000
Annual maintenance	$100	$50
Expected life	20 years	20 years
Salvage value	$400	$500

The firm's MARR is 12%. Based on the IRR criterion, which system is the better choice?

5.37 Fulton National Hospital is reviewing ways of cutting the stocking costs of medical supplies. Two new stockless systems are being considered, to lower the hospital's holding and handling costs. The hospital's industrial engineer has compiled the relevant financial data for each system as follows. Dollar values are in millions.

	Current Practice	Just-in-Time System	Stockless Supply System
Start-up cost	$0	$2.5	$5
Annual stock holding cost	$3	$1.4	$0.2
Annual operating cost	$2	$1.5	$1.2
System life	8 years	8 years	8 years

The system life of 8 years represents the contract period with the medical suppliers. If the hospital's MARR is 10%, use the IRR method to determine which system is more economical.

5.38 Consider the following investment projects:

	Net Cash Flow		
n	Project 1	Project 2	Project 3
0	−$1000	−$5000	−$2000
1	500	7500	1500
2	2500	600	2000

Assume the MARR = 15%.

(a) Compute the IRR for each project.
(b) If the three projects are mutually exclusive investments, which project should be selected?

5.39* Consider the following two investment alternatives:

	Net Cash Flow	
n	Project A	Project B
0	−$10,000	−$20,000
1	5,500	0
2	5,500	0
3	5,500	40,000
IRR	30%	?
PE(15%)	?	6300

The firm's MARR is known to be 15%.

(a) Compute the IRR of project B.
(b) Compute the PE of project A.
(c) Suppose that projects A and B are mutually exclusive. Using the IRR, which project would you select?

5.40 An electronic circuit board manufacturer is considering six mutually exclusive cost reduction projects for its PC-board manufacturing plant. All have lives of 10 years and zero salvage values. The required investment and the estimated after-tax reduction in annual disbursements are given for each alternative. Along with these gross rates of return, rates of return on incremental investments are also computed.

Proposal A_j	Required Investment	After-Tax Savings	Rate of Return (%)
A_1	$60,000	$22,000	34.2 %
A_2	100,000	28,200	25.2
A_3	110,000	32,600	26.9
A_4	120,000	33,600	25.0
A_5	140,000	38,400	24.3
A_6	150,000	42,200	25.1

Incremental Investment	Incremental Rate of Return (%)
$A_2 - A_1$	8.9%
$A_3 - A_2$	42.8
$A_4 - A_3$	0.0
$A_5 - A_4$	20.2
$A_6 - A_5$	36.3

Which project, if any, would you select based on the rate of return on incremental investment if it is stated that the MARR is 15%?

5.41 Consider the following two mutually exclusive investment projects:

	Net Cash Flow	
n	Project A1	Project A2
0	−$10,000	−$15,000
1	5,000	20,000
2	5,000	
3	5,000	

(a) To use the IRR criterion, what assumption must be made to compare a set of mutually exclusive investments with unequal service lives?
(b) With the assumption defined above, determine the range of MARR that will indicate the selection of project A1.

Level 3

5.42 Consider the cash flows for the following investment projects.

Assume MARR=15%.

	Project's Cash Flow				
n	A	B	C	D	E
0	−$1500	−$1500	−$3000	1500	−$1800
1	1350	1000	1000	−450	600
2	800	800	X	−450	600
3	200	800	1500	−450	600
4	100	150	X	−450	600

(a) Suppose Projects A and B are mutually exclusive. Which project would be selected based on the PE criterion?
(b) Repeat (a) using the FE criterion.
(c) Find the minimum value of X that makes project C acceptable.
(d) Would you accept D at $i = 18\%$?
(e) Assume that projects D and E are mutually exclusive. Which project would you select based on the PE criterion (at MARR = 15%)?

5.43 Consider the following two mutually exclusive projects B1 and B2.

	B1		B2	
n	Cash Flow	Salvage Value	Cash Flow	Salvage Value
0	−$12,000		−$10,000	
1	−2,000	6,000	−2,100	6,000
2	−2,000	4,000	−2,100	3,000
3	−2,000	3,000	−2,100	1,000
4	−2,000	2,000		
5	−2,000	2,000		

Salvage values represent the net proceeds (after tax) from disposal of the assets if they are sold at the end of a given year. Both B1 and B2 will be available (or can be repeated) with the same costs and salvage values for an indefinite period.

(a) With the infinite planning horizon assumption, which project is a better choice at MARR = 12%?
(b) With a 10-year planning horizon, which project is a better choice at MARR = 12%?

5.44 For each of the following after-tax cash flows:

n	A	B	C	D
	Project's Cash Flow			
0	−$2500	−$7000	−$5000	−$5000
1	650	−2500	−2000	−500
2	650	−2000	−2000	−500
3	650	−1500	−2000	4000
4	600	−1500	−2000	3000
5	600	−1500	−2000	3000
6	600	−1500	−2000	2000
7	300		−2000	3000
8	300			

(a) Compute the project balances for projects A and D as a function of project year at $i = 10\%$.

(b) Compute the future equivalent values for projects A and D at the end of their respective lives ($i = 10\%$).

(c) Suppose that projects B and C are mutually exclusive. Assume also that the required service period is 8 years, and the company is considering leasing comparable equipment that has an annual lease expense of $3000 for the remaining years of the required service period. Determine the best project using a PE analysis.

5.45 An electrical utility is experiencing a sharp power demand, which continues to grow at a high rate in a certain local area.

Two alternatives are under consideration. Each alternative is designed to provide enough capacity during the next 25 years. Both alternatives will consume the same amount of fuel, so fuel cost is not considered in the analysis.

Alternative A: Increase the generating capacity now so that the ultimate demand can be met without additional expenditures later. An initial investment of $30 million would be required, and it is estimated that this plant facility would be in service for 25 years

and have a salvage value of $0.85 million. The annual operating and maintenance costs (including income taxes) would be $0.4 million.

Alternative B: Spend $10 million now and follow this expenditure with future additions during the 10th year and the 15th year. These additions would cost $18 million and $12 million, respectively. The facility would be sold 25 years from now with a salvage value of $1.5 million. The annual operating and maintenance costs (including income taxes) initially will be $250,000, increasing to $0.35 million after the second addition (from 11th year to 15th year) and to $0.45 million during the final 10 years. (Assume that these costs begin 1 year subsequent to the actual addition.)

If the firm uses 15% as a MARR, which alternative should be undertaken based on the present equivalent criterion?

5.46 Apex Corporation requires a chemical finishing process for a product under contract for a period of 6 years. Three options are available. Neither Option 1 nor Option 2 can be repeated after its process life. However, Option 3 will always be available from H&H Chemical Corporation at the same cost during the contract period.

- Option 1: Process device A, which costs $100,000, has annual operating and labor costs of $60,000, and a useful service life of 4 years with an estimated salvage value of $10,000.
- Option 2: Process device B, which costs $150,000, has annual operating and labor costs of $50,000, and a useful service life of 6 years with estimated salvage value of $30,000.
- Option 3: Subcontract out the process at a cost of $100,000 per year.

According to present equivalent criterion, which option would you recommend at $i = 12\%$?

5.47 Provincial Electric was faced with providing electricity to a newly developed industrial park complex. The distribution engineering department needs to develop guidelines for design of the distribution circuit. The "main feeder," which is the backbone of each 13 kV distribution circuit, represents a substantial investment by the company.[4]

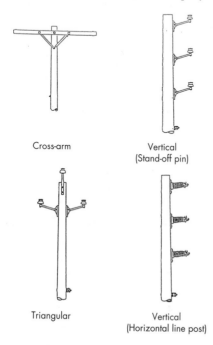

Cross-arm

Vertical
(Stand-off pin)

Triangular

Vertical
(Horizontal line post)

Provincial Electric has four approved main feeder construction configurations. They are: (1) cross-arm, (2) vertical (horizontal line post), (3) vertical (stand-off pin), and (4) triangular, as illustrated. The width of the easement sought depends on the planned construction configuration. If cross-arm construction is planned, a 5 metre easement is sought. A 3 metre wide easement is sought for vertical and triangular configurations. Once the required easements are obtained, the line clearance department clears any foliage that would impede the construction of the line. The clearance cost is dictated by the typical tree densities along road rights-of-way. The average cost to trim one tree is estimated at $20, and the average

tree density in the service area is estimated to be 75 trees per kilometre. The costs of each construction type are as follows:

| Factors | Design Configurations | | | |
	Cross Arm	Triangular	Horizontal Line	Stand Off
Easements	$487,000	$388,000	$388,000	$388,000
Line clearance	$613	$1,188	$1,188	$1,188
Line construction	$7,630	$7,625	$12,828	$8,812

Additional factors to consider in selecting the best main feeder configuration are as follows:

In certain sections of Provincial Electric's service territory, osprey often nest on transmission and distribution poles. These osprey nests reduce the structural and electrical integrity of the pole on which the nest is built. Cross-arm construction is most vulnerable to osprey nesting, since the cross arm and braces provide a secure area for nest construction. Vertical and triangular construction do not provide such spaces. In areas where osprey are known to nest, vertical and triangular configuration have added advantages. The insulation strength of a construction configuration may favorably or adversely affect the reliability of the line for which the configuration is used. A common measure of line insulation strength is the critical flashover (CFO) voltage. The greater the value of CFO, the less susceptible the line to suffer from nuisance flashovers from lightning and other electrical phenomenon. The existing inventory of cross arms is used primarily for main feeder construction and maintenance. Use of another configuration for main feeder construction would result in a substantial reduction of cross-arm inventory. The line crews complain that line spacing on vertical and triangular construction is too restrictive for safe live line work. Each accident would cost $65,000 in lost work and other medical expenses. The average cost of each flashover repair would be $3000.

[4]Example provided by Andrew Hanson.

Factors	Design Configurations			
	Cross Arm	Triangular	Horizontal Line	Stand Off
Nesting Insulation strength	Severe	None	None	None
CFO (kV)	387	474	476	362
Annual flashover occurrence (n)	1	2	2 1	
Annual inventory savings		$4521	$4521	$4521
Safety	OK	Problem	Problem	Problem

All configurations would last about 20 years with no salvage values. It appears that non-cross-arm designs are better, but engineers need to consider other design factors, such as safety, besides monetary factors when implementing the project. It is true that the line spacing on triangular construction is restrictive. However, with a better clearance design between phases for vertical construction, the safety issue would be minimized. In the utility industry, the typical opposition to new construction types is due to the confidence acquired from constructing lines in the cross-arm configuration for many years. As more vertical and triangular lines are built, the opposition to these configurations should decrease. Which of the four designs would you recommend to the management?

5.48 An airline is considering two types of engine systems for use in its planes. Each has the same life and the same maintenance and repair record.

- System A costs $100,000 and uses 100,000 litres per 1000 hours of operation at the average load encountered in passenger service.
- System B costs $200,000 and uses 80,000 litres per 1000 hours of operation at the same level.

Both engine systems have 3-year lives before any major overhaul. Based on the initial investment, the systems have 10% salvage values. If jet fuel costs $0.50 per litre currently, and fuel consumption is expected to increase at the rate of 6% due to degrading engine efficiency (each year), which engine system should the firm install? Assume 2000 hours of operation per year, and a MARR of 10%. Use the AE criterion. What is the equivalent operating cost per hour for each engine?

5.49 A small manufacturing firm is considering the purchase of a new machine to modernize one of its current production lines. Two types of machines are available on the market. The lives of machine A and machine B are 4 years and 6 years, respectively, but the firm does not expect to need the service of either machine for more than 5 years. The machines have the following expected receipts and disbursements.

Item	Machine A	Machine B
First cost	$6500	$8500
Service life	4 years	6 years
Estimated salvage value	$600	$1000
Annual O&M costs	$800	$520
Change oil filter after 2 years	$100	None
Engine overhaul	$200 (every 3 years)	$280 (every 4 years)

The firm always has another option: To lease a machine for $3000 paid at the start of the year, fully maintained by the leasing company. After 4 years of use, the salvage value for machine B will remain at $1000.

(a) How many decision alternatives are there?

(b) Which decision appears to be the best at $i = 10\%$?

5.50 A plastic manufacturing company owns and operates a polypropylene production facility that converts the propylene from one of its cracking facilities to polypropylene plastics for outside sale. The polypropylene production facility is currently forced to operate at less than capacity due to lack of propylene production capacity in its hydrocarbon cracking facility. The chemical engineers are considering alternatives for supplying additional propylene to the polypropylene production facility. Some of the feasible alternatives are as follows: Option 1: Build a pipeline to the nearest outside supply source; Option 2: Provide additional propylene by truck from an outside source. The engineers also gathered the following projected cost estimates:

- Future costs for purchased propylene excluding delivery: $0.215 per kg
- Cost of pipeline construction: $200,000 per pipeline km
- Estimated length of pipeline: 180 km
- Transportation costs by tank truck: $0.05 per kg, utilizing common carrier
- Pipeline operating costs: $0.005 per kg, excluding capital costs
- Projected additional propylene needs: 180 million kg per year
- Projected project life: 20 years
- Estimated salvage value of the pipeline: 8% of the installed costs

Determine the propylene cost per kilogram under each option, if the firm's MARR is 18%. Which option is more economical?

5.51 Consider the following two mutually exclusive investment alternatives:

	Net Cash Flow	
n	**Project A1**	**Project A2**
0	–$15,000	–$20,000
1	7,500	8,000
2	7,500	15,000
3	7,500	5,000
IRR	23.5%	20%

(a) Determine the IRR on the incremental investment of $5000. (Assume that MARR = 10%.)
(b) Which alternative is the better choice?

5.52 Consider the following sets of investment projects. Assume that MARR = 15%.

	Net Cash Flow					
n	A	B	C	D	E	F
0	–$100	–$200	–$4000	–$2000	–$2000	–$3000
1	60	120	2410	1400	3700	2500
2	50	150	2930	1720	1640	1500
3	50					
*i**	28.89%	21.65%	20.86%	34.12%	121.95%	23.74%

(a) Projects A and B are mutually exclusive projects. Assuming that both projects can be repeated for an indefinite period, which project would you select based on the IRR criterion?
(b) Suppose projects C and D are mutually exclusive. Using the IRR criterion, which project would be selected?
(c) Suppose projects E and F are mutually exclusive. Which project is better based on the IRR criterion?

5.53 Consider the cash flows for the following investment projects.

Assume that the MARR = 12%.

	Project Cash Flow				
n	**A**	**B**	**C**	**D**	**E**
0	–$1000	–$1000	–$2000	$1000	–$1200
1	900	600	900	–300	400
2	500	500	900	–300	400
3	100	500	900	–300	400
4	50	100	900	–300	400

(a) Suppose A, B, and C are mutually exclusive projects. Which project would be selected based on the IRR criterion?

(b) What is the BRR (borrowing rate of return) for D?

(c) Would you accept D at MARR = 20%?

(d) Assume that projects C and E are mutually exclusive. Using the IRR criterion, which project would you select?

5.54 A worker at the Baby Doll Shop makes wooden parts for doll houses. The worker is paid $8.10 an hour and, using a hand saw, can produce a year's required production (1600 parts) in just 8 weeks of 40 hours per week. That is, the worker averages five parts per hour when working by hand. The shop is considering the purchase of a power band saw with associated fixtures, to improve the productivity of this operation. Three models of power saw could be purchased: Model A (economy version), model B (high-powered version), and model C (deluxe high-end version). The major operating difference between these models is their speed of operation. The investment costs, including the required fixtures and other operating characteristics, are summarized as follows:

Category	By Hand	Model A	Model B	Model C
Production rate (parts/hour)	5	10	15	20
Labor hours required (hours/year)	320	160	107	80
Annual labor cost (@ $8.10/hour)	2592	1296	867	648
Annual power cost ($)		400	420	480
Initial investment ($)		4000	6000	7000
Salvage value ($)		400	600	700
Service life (years)		20	20	20

Assume that MARR = 10%. Are there enough savings to purchase any of the power band saws? Which model is most economical based on the rate of return principle? (Source: This problem is adapted with the permission of Professor Peter Jackson of Cornell University.)

5.55 Consider the following two mutually exclusive investment projects. Assume that the MARR = 15%.

	Net Cash Flow	
n	**Project A**	**Project B**
0	–$100	–$200
1	60	120
2	50	150
3	50	
IRR	28.89%	21.65%

Which project would be selected under infinite planning horizon with project repeatability likely, based on the IRR criterion?

5.56 Akadaka Toy Company is facing an uncertain but impending deadline to phase out polyvinyl chloride (PVC) from its infant toy products. Young children chewing on the material may be exposed to a phthalate called DINP, which is used to "soften" the vinyl. Experiments on animals have shown that intense DINP exposure can cause potential damage to the kidneys and liver, but the risks to infants from sucking on these products is still under investigation. Since a few European counties have already banned the use of DNIPs in such products, Akadaka is predicting that Health Canada may follow suit, when the results of some major studies are released in three years. Akadaka has been pursuing other means of creating similar plastics using safer compounds, and they are considering two options for modifying their manufacturing process:

- Option 1: Retrofitting the plant now to adapt the chemical process to be DNIP-free, and remain a market leader in infant toys. Because implementing the new processes on a large scale is untested, it may cost more to operate the facility for some time while learning the new system.

- Option 2: Deferring the retrofitting for 3 years until the anticipated federal regulations are introduced. With expected improvement in plastics processing technology and know-how, the retrofitting cost will be cheaper, but there will be tough market competition, and the revenue would be less than that of Option 1.

The financial data for the two options are as follows:

	Option 1	Option 2
Investment timing	Now	3 years from now
Initial investment	$6 million	$5 million
System life	8 years	8 years
Salvage value	$1 million	$2 million
Annual revenue	$15 million	11 million
Annual O&M costs	$6 million	$7 million

(a) What assumptions must be made to compare these two options?

(b) If Akadaka's MARR is 15%, which option is the better choice, based on the IRR criterion?

5.57 An oil company is considering changing the size of the downhole pump that is operational in wells in an oil field. If the existing pump is kept, it will extract 50% of the known crude oil reserve in the first year of operation and the remaining 50% in the second year. A pump larger than the existing pump will cost $1.6 million, but it will extract 100% of the known reserve in the first year. The total oil revenues over the 2 years is the same for both pumps, $20 million. The advantage of the large pump is that it allows 50% of the revenues to be realized a year earlier than with the small pump.

	Existing Pump	Larger Pump
Investment, year 0	0	$1.6 million
Revenue, year 1	$10 million	$20 million
Revenue, year 2	$10 million	0

If the firm's MARR is known to be 20%, what do you recommend, using an IRR analysis?

5.58 Consider the following two mutually exclusive investment projects. Assume that MARR = 15%.

	Net Cash Flow	
n	Project A	Project B
0	−$300	−$800
1	0	1150
2	690	40
i^*	51.66%	47.15%

(a) Using the IRR criterion, which project would be selected?

(b) Sketch the $PE(i)$ function on incremental investment (B − A).

5.59 You have been asked by the president of the company to evaluate the proposed acquisition of a new injection molding machine for the firm's manufacturing plant. Two types of injection molding machines have been identified with the following estimated cash flows:

	Net Cash Flow	
n	Project A	Project B
0	−$30,000	−$40,000
1	20,000	43,000
2	18,200	5,000
IRR	18.1%	18.1%

You return to your office and quickly retrieve your old engineering economics text, then begin to smile: Aha—this is a classic rate of return problem! Now, using a calculator, you find out that both projects have about the same rate of return, 18.1%. This rate of return figure seems to be high enough for project justification, but you recall that the ultimate justification should be done in reference to the firm's MARR. You call the accounting department to find out the current MARR the firm should use for project justification. "Oh boy, I wish I could tell you, but my boss will be back next week, and she can tell you what to use," said the accounting clerk.

A fellow engineer approaches you and says, "I couldn't help overhearing you talking to the clerk. I think I can help you. You see, both projects have the same IRR and on top of that, project 1 requires less investment but returns more cash flows (−$30,000 + $20,000 − $18,200 = $8200, and −$40,000 + 43,000 + $5000 = $8000), thus project 1 dominates project 2. For this type of decision problem, you don't need to know an MARR!"

(a) Comment on your fellow engineer's statement.
(b) At what range of MARR would you recommend the selection of project 2?

CHAPTER 6

Applications of Economic Evaluation Techniques

Suppose you are considering buying a car. If you expect to drive 12,000 kilometres per year, can you figure out how much it will cost you per kilometre? You would have a good reason to want to know this cost if you were to be reimbursed by your employer on a per kilometre basis for business use of your car. Or consider a real estate developer who is planning to build a shopping centre of 50,000 square metres. What would be the minimum annual rental fee per square metre to recover the initial investment?

Engineers are frequently involved in making design decisions that provide the required functional quality at the lowest cost. For example, General Motors engineers used the "design for manufacturability" principle to trim part counts for the front suspension design on the 1992 Cadillac Seville.

This redesign eliminates two parts and 68 seconds of assembly time. For the overall suspension, 50 parts were trimmed, and assembly time was cut from nearly 19 minutes to under 6 minutes! Redesigns in the suspension alone yielded annual savings of over $2 million. Engineering economy is one of the most important design considerations in today's competitive environment.

Companies or organizations providing taxi services, bus services, and trucking services have a large fleet of vehicles. They need to know the economic life of each type of vehicle. In other words, how long should a vehicle be operated so that it incurs the lowest total cost? If this economic life is known, the company can then design a schedule for replacing vehicles from year to year.

Consider some major capital equipment in a chemical company, a reformer furnace for producing hydrogen. The cost of such a furnace is in the order of $10 million. Suppose that the design life of the furnace is 11 years. However, because the actual operating condition such as temperature and pressure may not always be the same as the design condition, the actual economic life of the furnace will not be exactly 11 years. When the furnace is 10 years old, you need to make a decision whether to replace the furnace with a new one the following year. How do you conduct this kind of replacement analysis?

In Chapters 4 and 5, we presented methods for evaluating independent projects and comparing mutually exclusive projects. In this chapter, we will show you how to apply these fundamental methods in the following special decision-making situations: (1) the project life is very long, (2) unit cost/profit calculation, (3) break-even analysis, (4) design economics, (5) economic life calculation, and (6) replacement analysis.

6.1 Capitalized Equivalent Method

A special case of the PE criterion is useful when the life of a proposed project is **perpetual** or the planning horizon is extremely long (say, 40 years or more). Many public projects such as bridges, waterway constructions, irrigation systems, and hydro-electric dams are expected to generate benefits over an extended period of time (or forever). In this section, we will examine the **capitalized equivalent** ($CE(i)$) method for evaluating such projects.

Perpetual Service Life

Consider the cash flow series shown in Figure 6.1. How do we determine the PE for an infinite (or almost infinite) uniform series of cash flows or a repeated cycle of cash flows? The process of computing the PE cost for this infinite series is referred to as the **capitalization** of project cost. The cost is known as the **capitalized cost**. The capitalized cost represents the amount of money that must be invested today to generate a certain amount of income A at the end of each and every period forever, assuming an interest rate of i. Recall the limiting forms of the interest formulas developed in Section 2.3.7. As N approaches infinity:

$$\lim_{N\to\infty}(P/A, i, N) = \lim_{N\to\infty}\left[\frac{(1 + i)^N - 1}{i(1 + i)^N}\right] = \frac{1}{i}.$$

Thus, it follows that

$$PE(i) = A(P/A, i, N\to\infty) = \frac{A}{i}. \tag{6.1}$$

Another way of looking at this is to ask what constant income stream could be generated by $PE(i)$ dollars today in perpetuity. Clearly, the answer is $A = iPE(i)$. If withdrawals were greater than A, you would be eating into the principal, which would eventually reduce it to 0.

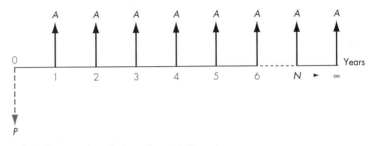

Figure 6.1 Equivalent present worth of an infinite cash flow series

EXAMPLE 6.1 **Capitalized equivalent cost**

A major university has just completed a new engineering complex worth $50 million dollars. A campaign, targeting alumni, is planned to raise funds for future maintenance costs, which are estimated at $2 million per year. Any unforeseen costs above $2 million per year would be obtained by raising tuition. Assuming that the university can create a trust fund that earns 8% interest annually, how much has to be raised now to cover the perpetual string of $2 million annual costs?

Solution

Given: A = $2 million, i = 8% per year, N = ∞
Find: $CE(8\%)$

The capitalized cost equation is

$$CE(i) = \frac{A}{i}$$
$$CE(8\%) = \$2,000,000/0.08$$
$$= \$25,000,000.$$

Comments: It is easy to see that this lump-sum amount should be sufficient to pay maintenance expenses on the building forever. Suppose the university deposited $25 million in a bank that paid 8% interest annually. At the end of the first year, the $25 million would earn 8%($25 million) = $2 million interest. If this interest were withdrawn, the $25 million would remain in the account. At the end of the second year, the $25 million balance would again earn 8%($25 million) = $2 million. The annual withdrawal could be continued forever, and the endowment (gift funds) would always remain at $25 million.

Project's Service Life is Extremely Long

The benefits of typical civil engineering projects, such as bridge and highway construction, although not perpetual, can last for many years. In this section, we will examine the use of the $CE(i)$ criterion to approximate the PE of engineering projects with long lives.

EXAMPLE 6.2 **Comparison of present equivalent for long life and infinite life**

Mr. Gaynor L. Bracewell amassed a small fortune developing real estate in British Columbia over the past 30 years. He sold more than 700 hectares of timber and farmland to raise $800,000 to build a small hydroelectric plant, known as Edgemont Hydro, which has been a decade in the making. The design for the plant, which the entrepreneur developed using his military training as a civil engineer, is relatively simple. A 7-metre-deep canal, blasted out of solid rock, just above the higher of two dams on his property, carries water 350 metres along the river to a "trash rack," where leaves and other debris are caught. A 2-metre-wide pipeline capable of holding 1.5 million kilograms of liquid then funnels the water into the powerhouse at 3.5 metres per second, creating the thrust to run the turbines.

Government regulations encourage private power development, and any electricity generated must be purchased by the power company that owns the provincial power grid. The plant can generate 6 million kilowatt hours per year. Suppose that after paying income taxes and operating expenses, the annual net income from the hydroelectric plant will be $120,000. With normal maintenance, the plant is expected to provide service for at least 50 years. Was the $800,000 investment a wise one? How long does this entrepreneur have to wait to recover his initial investment, and will he ever make a profit? Examine the situation by computing the project worth at varying interest rates.

(a) If Mr. Bracewell's interest rate is 8%, compute the PE (at time 0 in Figure 6.2) of this project with a 50-year service life and infinite service, respectively.

(b) Repeat part (a) assuming an interest rate of 12%.

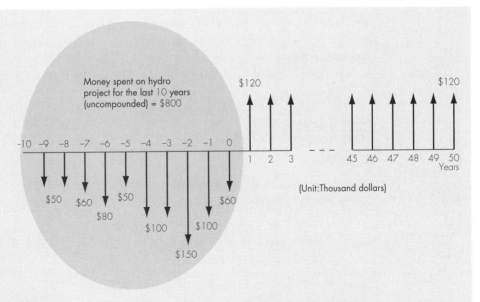

Money spent on hydro
project for the last 10 years
(uncompounded) = $800

(Unit: Thousand dollars)

Figure 6.2 Net cash flow diagram for Mr. Bracewell's hydroelectric project (Example 6.2)

Solution

Given: Cash flow in Figure 6.2 (to 50 years or ∞), $i = 8\%$ or 12%
Find: PE at time 0

One of the main questions was whether Mr. Bracewell's plant would be profitable.

Now we will compute the equivalent total investment and the equivalent worth of receiving future power revenues at the start of generation, i.e., at time 0.

(a) At $i = 8\%$

- With a plant service life of 50 years, we can make use of single payment compound amount factors in the invested cash flow to help us find the equivalent total investment at the start of power generation. Using "K" to indicate thousands,

$$
\begin{aligned}
V_1 &= -\$50K(F/P, 8\%, 9) - \$50K(F/P, 8\%, 8) \\
&\quad - \$60K(F/P, 8\%, 7) \dots -\$100(F/P, 8\%, 1) - \$60K \\
&= -\$1101K.
\end{aligned}
$$

The equivalent total benefit at the start of generation is

$$
V_2 = \$120K(P/A, 8\%, 50) = \$1468K.
$$

Summing, we find the net equivalent worth at the start of power generation:

$$
\begin{aligned}
V_1 + V_2 &= -\$1101K + \$1468K \\
&= \$367K.
\end{aligned}
$$

- With an infinite service life, the net equivalent worth is called the capitalized equivalent worth. The investment portion prior to time 0 is identical, so the capitalized equivalent worth is

$$CE(8\%) = -\$1101K + \$120K/(0.08)$$
$$= \$399K.$$

Note that the difference between the infinite situation and the planning horizon of 50 years is only $32,000.

(b) At $i = 12\%$

- With a service life of 50 years, proceeding as for (a), the equivalent total investment at the start of generation is

$$V_1 = -\$50K(F/P, 12\%, 9) - \$50K(F/P, 12\%, 8)$$
$$- \$60K(F/P, 12\%, 7) \ldots - \$100K(F/P, 12\%, 1) - 60K$$
$$= -\$1299K.$$

Equivalent total benefits at the start of generation:

$$V_2 = \$120K(P/A, 12\%, 50) = \$997K.$$

Net equivalent worth at the start of power generation:

$$V_1 + V_2 = -\$1299K + \$997K$$
$$= -\$302K.$$

- With infinite cash flows, the capitalized equivalent worth at time 0 is

$$CE(12\%) = -\$1299K + \$120K/(0.12)$$
$$= -\$299K.$$

Note that the difference between the infinite situation and a planning horizon of 50 years is merely $3000, which demonstrates that we may closely approximate the present worth of long cash flows (i.e., 50 years or more) by using the capitalized equivalent value. The accuracy of the approximation improves as the interest rate increases (or number of years is greater).

Comments: At $i = 12\%$, Mr. Bracewell's investment is not a profitable one, but at 8% it is. The outcome indicates the importance of using the appropriate i in investment analysis. The issue of selecting an appropriate i will be presented in Chapter 10.

6.2 Unit Profit/Cost Calculation

In many situations we need to know the unit profit (or cost) of operating an asset. To obtain a unit profit (or cost), we may proceed as follows:

- Determine the number of units to be produced (or serviced) each year over the life of the asset.

- Identify the cash flow series associated with production or service over the life of the asset.

- Calculate the net present worth of the project cash flow series at a given interest rate and then determine the equivalent annual worth.

- Divide the equivalent annual worth by the number of units to be produced or serviced during each year. When the number of units varies each year, you may need to convert them into equivalent annual units.

To illustrate the procedure, we will consider Example 6.3, where the annual equivalent concept is useful in estimating the savings per machine hour for a proposed machine acquisition.

EXAMPLE 6.3 **Unit profit per machine hour—where annual operating hours remain constant**

Consider the investment in the metal-cutting machine in Example 4.5, page 213. Recall that this 3-year investment was expected to generate a PE of $3553. Suppose that the machine will be operated for 2000 hours per year. Compute the equivalent savings per machine hour at $i = 15\%$.

Solution

Given: PE = $3553, N = 3 years, i = 15% per year, 2000 machine hours per year
Find: Equivalent savings per machine hour

We first compute the annual equivalent savings from the use of the machine. Since we already know the PE of the project, we obtain the AE by

$$AE(15\%) = \$3553(A/P, 15\%, 3) = \$1556.$$

With an annual usage of 2000 hours, the equivalent savings per machine hour would be

$$\text{Savings per machine hour} = \$1556/2000 \text{ hours} = \$0.78/\text{hour}.$$

Comments: Note that we cannot simply divide the PE amount ($3553) by the total number of machine hours over the 3-year period (6000 hours), or $0.59/hour. This $0.59 figure represents the instant savings in present worth for each hourly use of the equipment, but does not consider the time over which the savings occur. Once we have the annual equivalent worth, we can divide it by the desired time unit if the compounding period is 1 year. If the compounding period is shorter, then the equivalent worth should be calculated for the compounding period.

EXAMPLE 6.4

Unit profit per machine hour—where annual operating hours fluctuate

Reconsider Example 6.3 and suppose that the metal cutting machine will be operated according to varying hours: the first year—1500 hours; the second year—2500 hours; and third year—2000 hours. The total operating hours still remain at 6000 over 3 years. Compute the equivalent savings per machine hour at $i = 15\%$.

Solution

Given: PE = $3553, $N = 3$ years, $i = 15\%$ per year, operating hours—1500 hours first year; 2500 hours second year; and 2000 hours third year
Find: Equivalent savings per machine hour

As calculated in Example 6.3, the annual equivalent savings is $1556. Let C denote the equivalent annual savings per machine hour that needs to be determined. Now, with varying annual usages of the machine, we can represent the equivalent annual savings as a function of C:

$$\begin{aligned}
\text{Equivalent annual savings} &= [C(1500)(P/F, 15\%, 1) \\
&\quad + C(2500)(P/F, 15\%, 2) \\
&\quad + C(2000)(P/F, 15\%, 3)](A/P, 15\%, 3) \\
&= 1975.16C.
\end{aligned}$$

We can equate this amount to $1556 in Example 6.3 and solve for C. This gives us

$$C = \$1556/1975.16 = \$0.79/\text{hour},$$

which is about a penny more than the situation in Example 6.3.

6.3 Make-or-Buy Decision

Make-or-buy problems are among the most common of business decisions. At any given time, a firm may have the option of either buying an item or producing it. Unlike the make-or-buy situation we looked at in Section 1.7.2, *if either the "make" or the "buy" alternative requires the acquisition of machinery and/or equipment, then it becomes an investment decision.* Since the cost of an outside service (the "buy" alternative) is usually quoted in terms of dollars per unit, it is easier to compare the two alternatives if the differential costs of the "make" alternative are also given in dollars per unit. This unit cost comparison requires the use of annual worth analysis. The specific procedure is as follows:

Step 1: Determine the time span (planning horizon) for which the part (or product) will be needed.

Step 2: Determine the annual quantity of the part (or product).

Step 3: Obtain the unit cost of purchasing the part (or product) from the outside firm.

Step 4: Determine the equipment, manpower, and all other resources required to make the part (or product).

Step 5: Estimate the net cash flows associated with the "make" option over the planning horizon.

Step 6: Compute the annual equivalent cost of producing the part (or product).

Step 7: Compute the unit cost of making the part (or product) by dividing the annual equivalent cost by the required annual volume.

Step 8: Choose the option with the minimum unit cost.

EXAMPLE 6.5

Equivalent worth—make or buy

Ampex Corporation currently produces both videocassette cases and metal particle magnetic tape for commercial use. An increased demand for metal particle tapes is projected, and Ampex is deciding between continuing the internal production of empty cassette cases or purchasing empty cassette cases from an outside vendor. If Ampex purchases the cases from a vendor, the company must also buy specialized equipment to load the magnetic tapes, since its current loading machine is not compatible with the cassette cases produced by the vendor under consideration. The projected production rate of cassettes is 79,815 units per week for 48 weeks of operation per year. The planning horizon is 7 years. After considering the effects of income taxes, the accounting department has itemized the annual costs associated with each option as follows:

- Make option (annual costs):

Labor	$1,445,633
Materials	$2,048,511
Incremental overhead	$1,088,110
Total annual cost	$4,582,254

- Buy option:

Capital expenditure	
Acquisition of a new loading machine	$405,000
Salvage value at end of 7 years	$ 45,000

Annual operating costs

Labor	$ 251,956
Purchasing empty cassette ($0.85/unit)	$3,256,452
Incremental overhead	$ 822,719
Total annual operating costs	$4,331,127

(Note the conventional assumption that cash flows occur in discrete lumps at the ends of years, as shown in Figure 6.3.) Assuming that Ampex's MARR is 14%, calculate the unit cost under each option.

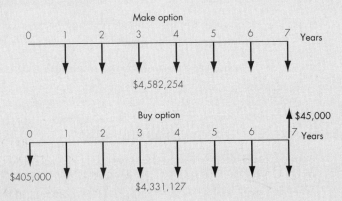

Figure 6.3 Make-or-buy analysis (Example 6.5)

Solution

Given: Cash flows for two options, $i = 14\%$
Find: Unit cost for each option, and which one is preferred

The required annual production volume is
$$79,815 \text{ units/week} \times 48 \text{ weeks} = 3,831,120 \text{ units per year.}$$

We now need to calculate the annual equivalent cost under each option.

- Make option: Since the "make option" is already given on an annual basis, the equivalent annual cost will be
  $$AE(14\%)_{Make} = -\$4,582,254.$$

- Buy option: The two cost components are capital cost and operating cost.

  Capital recovery cost [using equation (4.7)]:
  $$CR(14\%) = (\$405,000 - \$45,000)(A/P, 14\%, 7)$$
  $$+ (0.14)(\$45,000)$$
  $$= \$90,249$$
  $$AE(14\%)_1 = -CR(14\%) = -\$90,249.$$

  Operating cost:
  $$AE(14\%)_2 = -\$4,331,127.$$

Total annual equivalent cost:
$$AE(14\%)_{Buy} = AE(14\%)_1 + AE(14\%)_2 = -\$4,421,376.$$

Obviously, this annual equivalent calculation indicates that Ampex would be better off buying cassette cases from the outside vendor. However, Ampex wants to know the unit costs in order to set a price for the product. In this situation, we need to calculate the unit cost of producing the cassette tapes under each option. (The negative sign indicates that AE is a cash outflow or cost. When comparing the options, we are interested in the magnitude costs and can ignore the negative sign.) Do this by dividing the magnitude of the annual equivalent cost for each option by the annual quantity required:

- Make option:
  Unit cost = \$4,582,254/3,831,120 = \$1.20/unit.

- Buy option:
  Unit cost = \$4,421,376/3,831,120 = \$1.15/unit.

Buying the empty cassette cases from the outside vendor and loading the tape in-house will save Ampex 5 cents per cassette before any tax consideration.

Comments: Two important noneconomic factors should also be considered. The first is the question of whether the quality of the supplier's component is better than (or equal to) or worse than the component the firm is presently manufacturing. The second is the reliability of the supplier in terms of providing the needed quantities of the cassette cases on a timely basis. A reduction in quality or reliability should virtually always rule out a switch from making to buying.

6.4 Break-Even Point: Cost Reimbursement

Companies often need to calculate the cost of equipment that corresponds to a **unit of use** of that equipment. A familiar example is an employer's reimbursement of costs for the use of an employee's personal car for business purposes. If an employee's job is dependent on obtaining and using a personal vehicle on the employer's behalf, reimbursement on the basis of the employee's overall costs per kilometre seems fair. Although many car owners think of costs in terms of outlays for gasoline, oil, tires, and tolls, a careful examination shows that, in addition to these operating costs, which are directly related to the use of the car, there are also **ownership costs**, which occur whether or not the vehicle is driven.

Ownership costs include depreciation, insurance, finance charges, registration and titling fees, scheduled maintenance, accessory costs, and garaging (storage). Even if the vehicle is permanently garaged, a portion of each of these costs occurs. **Depreciation** is the loss in value of the vehicle during the time it is owned due to (1) the passage of time, (2) its mechanical and physical condition, and (3) the number of kilometres it is driven. This type of depreciation, known as **economic depreciation** (as opposed to **accounting depreciation**), is discussed in Chapter 7.

Operating costs include nonscheduled repairs and maintenance, gasoline, oil, tires, parking and tolls, and taxes on gasoline and oil. Certainly, the more a car is used the greater these costs become.

Once the cost of owning and operating a personal vehicle is determined, you may wonder what minimum reimbursement rate per kilometre allows you to break even. Using the reimbursement cost equation, we solve for the unknown reimbursement rate. The reimbursement rate that's exactly equal to the cost of owning and operating the vehicle is known as the **break-even point**. Example 6.6 illustrates how you obtain the cost of reimbursement for the use of an employee's vehicle for business purposes.

EXAMPLE 6.6

Break-even point—per unit of equipment use

Sam Tucker is a sales engineer at Buford Chemical Engineering Company. Sam owns two vehicles, and one of them is entirely dedicated to his business use. His business car is a 1998 subcompact automobile purchased with personal savings for $11,000. Based on his own records and an analysis published in the *National Post*, Sam has estimated the costs of owning and operating his business vehicle for the first three years as follows:

Year	First Year	Second Year	Third
Depreciation	2879	1776	1545
Scheduled maintenance	100	132	172
Insurance	635	635	635
Registration fee	78	78	78
Total ownership cost	$ 3692	$ 2621	$ 2450
Nonscheduled repairs	70	115	227
Accessories	15	13	12
Gasoline and taxes	688	650	522
Oil	80	100	100
Parking and tolls	135	125	110
Total operating costs	$ 988	$ 1003	$ 971
Total of all costs	$ 4680	$ 3624	$ 3421

Sam expects to drive 14,500, 13,000, and 11,500 business kilometres respectively for the next three years. If his interest rate is 6%, what should be the reimbursement rate per kilometre so that Sam can break even?

Discussion: You may wonder why the initial cost of the vehicle ($11,000) is not explicitly considered in the costs Sam has estimated. The answer is in the depreciation listings. In true economic analysis, capital costs are not considered all at once in the year in which they are incurred. Rather, the costs are spread out over the useful life of the asset, based on how much of the cost of the asset is used up each year. Thus, the depreciation amount of $2879 during the first year represents the fact that, if the car were bought for $11,000 and then sold at the end of the first year when it had been driven 14,500 kilometres, Sam would expect the sale price to be $2879 less than the original purchase price. In Chapter 7, we will explore depreciation in more detail and learn the conventions for its calculation. For this example, depreciation amounts are given to you to more closely represent a true economic analysis of the problem.

Solution

Given: Yearly costs and kilometres, i = 6% per year
Find: Equivalent cost per kilometre

Suppose Buford pays Sam $X per kilometre for his personal car. Assuming that Sam expects to travel 14,500 kilometres the first year, 13,000 kilometres the second year, and 11,500 kilometres the third year, his annual reimbursements would be

Year	Total Kilometres Driven	Reimbursement
1	14,500	(X) (14,500) = 14,500X
2	13,000	(X) (13,000) = 13,000X
3	11,500	(X) (11,500) = 11,500X

As depicted in Fig. 6.4, the annual equivalent reimbursement would be

$$[14,500X(P/F, 6\%, 1) + 13,000X(P/F, 6\%, 2)$$
$$+ 11,500X(P/F, 6\%, 3)](A/P, 6\%, 3)$$
$$= 13,058X.$$

The annual equivalent costs of owning and operating would be

$$[\$4680(P/F, 6\%, 1) + \$3624(P/F, 6\%, 2) + \$3421(P/F, 6\%, 3)](A/P, 6\%, 3)$$
$$= \$3933.$$

Then, the minimum reimbursement rate X should be

$$13,058X = \$3933$$
$$X = 30.12 \text{ cents per kilometre.}$$

If Buford pays him 30.12 cents per kilometre or more, Sam's decision to use his car for business makes sense economically.

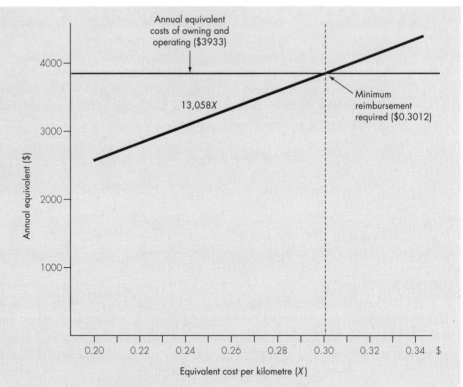

Figure 6.4 Annual equivalent reimbursement as a function of cost per kilometre (Example 6.6)

6.5 Design Economics

Another valuable extension of AE analysis is minimum-cost analysis. This method is useful when we have two or more cost components that are affected differently by the same design element. That is, for a single design variable, some costs increase while others decrease. When the equivalent annual total cost of a design variable is a function of increasing and decreasing cost components, we usually can find the optimal value that will minimize its cost.

$$AE(i) = a + bx + \frac{c}{x},\tag{6.2}$$

where x is a common design variable, and a, b, and c are constants.

To find the value of the common design variable that minimizes the $AE(i)$, we need to take the first derivative, equate the result to zero, and solve for x:

$$\frac{dAE(i)}{dx} = b - \frac{c}{x^2}$$
$$= 0$$
$$x^* = \sqrt{\frac{c}{b}}. \tag{6.3}$$

The value x^* is the minimum cost point for the design alternative.

The logic of the first-order requirement is that an activity should, if possible, be carried to a point where its **marginal yield** ($dAE(i)/dx$) is zero. However, to be sure whether we have found a maximum or a minimum when we have located a point whose marginal yield is zero, we must examine it in terms of what are called the **second-order conditions**. Second-order conditions are equivalent to the usual requirements that the second derivative be negative in the case of a maximum, and positive for a minimum. In our situation,

$$\frac{d^2 AE(i)}{dx^2} = \frac{2c}{x^3}. \tag{6.4}$$

As long as $c > 0$, the second derivative will be positive, indicating that the value x^* is the minimum cost point for the design alternative. To illustrate the optimization concept, two examples are provided. The first example is about designing the optimal cross-sectional area of a conductor, and the second example is about selecting an optimal size for a pipe.

EXAMPLE 6.7 **Optimal cross-sectional area**

A constant electric current of 5000 amperes is to be transmitted a distance of 300 metres from a power plant to a substation for 24 hours a day, 365 days a year. Copper conductor can be installed for $16.50 per kilogram and will have an estimated life of 25 years and a salvage value of $1.65 per kilogram. Power loss from conductors is inversely proportional to the cross-sectional area (A) of the conductor. The resistance of a conductor one metre long with a one square centimetre cross-sectional area is $1.7241 \times 10^{-4}\,\Omega$. The cost of energy is $0.0375 per kilowatt-hour, the interest rate is 9%, and the density of copper is 8894 kilograms per cubic metre. For the data given, calculate the optimum cross-sectional area (A) of the conductor.

Discussion: The resistance of transmission-line conductors is the most important cause of power loss in a transmission line. The resistance of an electrical conductor varies with its length and inversely with its cross-sectional area:

$$R = \rho(L/A), \tag{6.5}$$

where R = resistance of the conductor, L = length of the conductor, A = cross-sectional area of the conductor, ρ = resistivity of the conductor material.

A consistent set of units must be used. In SI units (the official designation for the Système International d'Unités system of units), L is in metres, A in square metres, and ρ in ohm-metres. In terms of the SI units, the copper conductor has a ρ value of 1.7241×10^{-8} Ω-metre, or 1.7241×10^{-4} Ω-cm^2/m. When current flows through a conductor, power is used to overcome the resistance. The unit of electrical work is kilowatt-hour (kWh), which is equal to the power in kilowatts multiplied by the hours during which work is performed. If the current (I) is steady, the energy loss through a conductor in time T can be expressed as

$$\text{Energy loss} = I^2\,RT/1000 \text{ kWh}, \tag{6.6}$$

where I = current in amperes, R = resistance in ohms (Ω), and T = time duration in hours.

Solution

Given: Cost components as a function of cross-sectional area (A), $N = 25$ years, $i = 9\%$
Find: Optimal A value

Step 1: This classical minimum-cost example, the design of the cross-sectional area of an electrical conductor, involves increasing and decreasing cost components. Since resistance is inversely proportional to the size of the conductor, the energy loss will decrease with the increased conductor size. More specifically, the annual energy loss in kilowatt-hours in a conductor due to resistance is equal to

$$
\begin{aligned}
\text{Energy loss in kilowatt-hours} \ &= \ \frac{I^2RT}{1000} = \frac{I^2T}{1000} \times \frac{\rho L}{A} \\[2mm]
&= \ \frac{5000^2 \times 24 \times 365}{1000} \times \frac{1.7241 \times 10^{-4} \times 300}{A} \\[2mm]
&= \ \frac{11{,}327{,}337}{A} \text{ kWh},
\end{aligned}
$$

where A = Cross sectional area in cm^2.

Step 2: The total cost of energy in dollars per year for the specified conductor material is

$$
\begin{aligned}
\text{annual energy loss cost} \ &= \ \frac{11{,}327{,}337}{A} \times \Phi \\[2mm]
&= \ \frac{11{,}327{,}337}{A} \times \$0.0375 \\[2mm]
&= \ \frac{\$424{,}775}{A},
\end{aligned}
$$

where Φ = lost energy in dollars per kilowatt-hour.

Step 3: As we increase the size of the conductor, however, it will cost more to build. First, we need to calculate the total amount of conductor material in kilograms. Since the cross-sectional area is given in square centimetres, we need to convert it to square metres before finding the material mass.

$$\text{material mass in kilograms} = \frac{(300)(8894)A}{100^2}$$
$$= 267A$$
$$\text{total material cost} = 267(A)(\$16.50)$$
$$= \$4406A$$

Here, we are looking for the trade-off between the cost of installation and the cost of energy loss.

Step 4: Since the copper material will be salvaged at the rate of $1.65 per kilogram at the end of 25 years, we can compute the capital recovery cost as follows:

$$CR(9\%) = [\$4406A - \$1.65(267A)](A/P, 9\%, 25) + \$1.65(267A)(0.09)$$
$$= 404A + 40A$$
$$= 444A.$$

Step 5: Using Eq. (6.2), we express the total annual equivalent cost (AEC) as a function of a design variable (A) as follows:

$$AEC(9\%) = + \overbrace{444A}^{\text{Capital cost}} + \underbrace{\frac{424{,}775}{A}}_{\text{Operating cost}}$$

To find the annual equivalent cost of minimum magnitude, we use the result of Eq. (6.3):

$$\frac{dAEC(9\%)}{dA} = + 444 - \frac{424{,}775}{A^2} = 0$$
$$A^* = \sqrt{\frac{424{,}775}{444}}$$
$$= 31 \text{ cm}^2.$$

The minimum annual equivalent total cost is

$$AEC(9\%) = 444 \times 31 + \frac{424{,}775}{31}$$
$$= \$27{,}466.$$

Figure 6.5 illustrates the nature of this design trade-off problem.

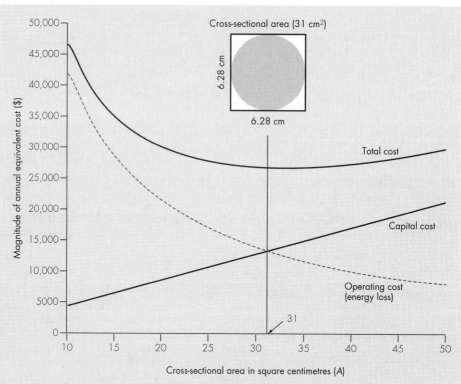

Figure 6.5 Optimal cross-sectional area for a copper conductor (Example 6.7). Note that the minimum point coincides with the crossing point of the capital cost and operating cost lines. In general, this is not always true. Since the cost components can have a variety of cost patterns, the minimum point usually does not occur at the crossing point.

EXAMPLE

6.8

Economical pipe size

As a result of the conflict in the Persian Gulf, Kuwait is studying the feasibility of running a steel pipeline across the Arabian Peninsula to the Red Sea. The pipeline will be designed to handle 3 million barrels of crude oil per day under optimum conditions. The length of the line will be 1000 kilometres. Calculate the optimum pipeline diameter that will be used for 20 years for the following data at $i = 10\%$:

Pumping power = $Q\Delta P/1000$ kW

Q = volume flow rate, cubic metres/second

$\Delta P = \dfrac{128Q\mu L}{\pi D^4}$, pressure drop, pascals

L = pipe length, metres

D = inside pipe diameter, metres

$t = 0.01\ D$, pipeline wall thickness, metres

$\mu = 3.5138$ kg/metre-second, oil viscosity

Power cost, $0.02 per kWh

Oil cost, $18 per barrel

Pipeline cost, $2.2 per kg of steel

Pump and motor costs, $261.50 per kW.

The salvage value of the steel after 20 years is assumed to be zero because removal costs exhaust scrap profits from steel. (See Figure 6.6 for the relationship between D and t.)

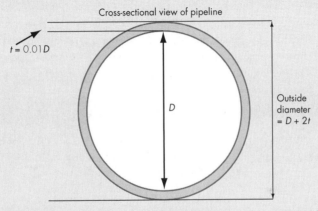

Cross-sectional view of pipeline

$t = 0.01D$

D

Outside
diameter
$= D + 2t$

Figure 6.6 Designing economical pipe size to handle 3 million barrels of crude oil per day (Example 6.8)

Discussion:　　In general, when a progressively larger size pipe is used to carry a given fluid at a given volume flow rate, the energy required to move the fluid will progressively decrease. However, as we increase the pipe size, the cost of its construction will increase. In practice, to obtain the best pipe size for a particular situation, you may choose a reasonable, but small, starting size. Compute the energy cost of pumping fluid through this size and the total construction cost. The process may be repeated with progressively larger pipe sizes until the added construction cost exceeds the savings in energy cost. As soon as this happens, the best pipe size to use in the particular application is identified. However, this search process can be simplified by using the minimum cost concept as represented by Eqs. (6.2) and (6.3).

Solution

Given: Cost components as a function of pipe diameter (D), N = 20 years, i = 10%
Find: Optimal pipe size (D)

Step 1: Several different units have been introduced; however, we need to work with common units. We will assume the following conversion units.

1000 km	=	10^6 metres
1 barrel	=	0.159 m³
Density of steel	=	7861 kg/m³

Step 2: Power required to pump oil:

For any given set of operating conditions involving the flow of incompressible fluid, such as oil, through a pipe of constant diameter, it is well known that a small-diameter pipe will have a high fluid velocity and a high fluid pressure drop through the pipe. This will require a pump that will deliver a high discharge pressure and a motor with large energy consumption. To determine the pumping power, we need to determine both the volume flow rate and the total pressure drop. Then, we can calculate the cost of the power required to pump the oil.

* Volume flow rate per hour:

$$
\begin{aligned}
Q &= 3{,}000{,}000 \text{ barrels/day} \times 0.159 \text{ m}^3\text{/barrel} \\
&= 477{,}000 \text{ m}^3\text{/day} \\
&= 19{,}875 \text{ m}^3\text{/hour} \\
&= 5.52 \text{ m}^3\text{/second.}
\end{aligned}
$$

* Pressure drop:

$$
\begin{aligned}
\Delta P &= \frac{128 Q \mu L}{\pi D^4} \\
&= \frac{128 \times 5.52 \times 3.5138 \times 10^6}{3.14159 \times D^4} \\
&= \frac{790{,}271{,}973}{D^4} \text{ Pascals.}
\end{aligned}
$$

* Pumping power required:

$$
\begin{aligned}
\text{Power} &= \frac{Q\Delta P}{1000} \\
&= \frac{1}{1000} \times 5.52 \times \frac{790{,}271{,}973}{D^4} \\
&= \frac{4{,}362{,}301}{D^4} \text{ kW.}
\end{aligned}
$$

Power cost to pump oil:

$$\text{Power cost} = \frac{4{,}362{,}301}{D^4} \text{ kW} \times \$0.02/\text{kWh}$$

$$\times \text{ 24 hours/day} \times \text{ 365 days/year}$$

$$= \frac{\$764{,}275{,}135}{D^4}/\text{year}.$$

Step 3: Pump and motor cost calculation:

Once we determine the required pumping power, we can determine the size of the pump and the motor costs. The size of the pump and motor is proportional to the required pumping power.

$$\text{Pump and motor cost} = \frac{4{,}362{,}301}{D^4} \times \$261.50/\text{kW}$$

$$= \frac{\$1{,}140{,}741{,}712}{D^4}.$$

Step 4: Required amount and cost of steel:

The pumping cost will be counterbalanced by the lower costs for the smaller pipe, valves, and fittings. If the pipe diameter is made larger, the fluid velocity drops markedly and the pumping costs become substantially lower. Conversely, the capital cost for the larger pipe, fittings, and valves becomes greater. For a given cross-sectional area of the pipe, we can determine the total volume of the pipe as well as the weight. Once the total weight of the pipe is determined, we can easily convert it into the required investment cost.

$$\begin{aligned}
\text{Cross-sectional area} &= 3.14159[(0.51D)^2 - (0.50D)^2] \\
&= 0.032D^2 \text{ m}^2 \\[4pt]
\text{Total volume of pipe} &= 0.032D^2 \text{ m}^2 \times 1{,}000{,}000 \text{ m} \\
&= 32{,}000D^2 \text{ m}^3 \\[4pt]
\text{Total weight of steel} &= 32{,}000D^2 \text{ m}^3 \times 7861 \text{ kg/m}^3 \\
&= 251{,}552{,}000D^2 \text{ kg} \\[4pt]
\text{Total pipeline cost} &= \$2.20/\text{kg} \times 251{,}552{,}000D^2 \text{ kg} \\
&= \$553{,}414{,}000D^2.
\end{aligned}$$

Step 5: Annual equivalent cost calculation:

Given the total cost of the pipeline and pumping equipment and its salvage value at the end of 20 years of service life, we can find the equivalent annual capital cost by using the capital cost recovery formula.

$$\text{Capital recovery cost} = (\$553{,}414{,}000D^2 + \frac{\$1{,}140{,}741{,}712}{D^4})(A/P, 10\%, 20)$$

$$= \$65{,}026{,}145D^2 + \frac{\$134{,}037{,}151}{D^4}$$

$$\text{Annual power cost} = \frac{\$764{,}275{,}135}{D^4}.$$

Step 6: Economical pipe size:

Now that we have determined the annual pumping and motor costs and the equipment capital recovery cost, we can express the total equivalent annual cost (AEC) as a function of the pipe diameter (D).

$$AEC(10\%) = 65{,}026{,}145D^2 + \frac{\$134{,}037{,}151}{D^4} + \frac{\$764{,}275{,}135}{D^4}.$$

To find the optimal pipe size (D) that results in the minimum annual equivalent cost, we take the first derivative of AEC(10%) with respect to D, equate the result to zero, and solve for D:

$$\frac{dAEC(10\%)}{dD} = 130{,}052{,}290D - \frac{3{,}593{,}249{,}144}{D^5}$$

$$= 0.$$

$$130{,}052{,}290D^6 = 3{,}593{,}249{,}144$$
$$D^6 = 27.6293$$
$$D^* = 1.7387 \text{ m.}$$

Note that the velocity in the pipeline should ideally be no more than approximately 3 m/sec. Higher velocities cause excessive wear rates which reduce the life of the pipeline. To check whether the answer is reasonable, we may compute

$$Q = \text{velocity} \times \text{pipe inner area:}$$

$$5.52 \text{ m}^3/\text{s} = V\frac{3.14159 \times 1.7387^2}{4}$$

$$V = 2.32 \text{ m/s,}$$

which is less than 3 m/sec. Therefore, the optimal answer as calculated is practical.

Step 7: Equivalent annual cost at optimal pipe size:

$$\text{Capital cost} = (\$553{,}414{,}000 \times 1.7387^2 + \frac{\$1{,}140{,}741{,}712}{1.7387^4})(A/P, 10\%, 20)$$

$$= \$65{,}026{,}145 \times 1.7387^2 + \frac{\$134{,}037{,}151}{1.7387^4}$$

$$= \$211{,}245{,}591.$$

$$\text{Annual power cost} \quad = \quad \frac{\$764,275,135}{1.7387^4}$$

$$= \quad \$83,627,885.$$

$$\text{Total annual cost} \quad = \quad \$211,245,591 + \$83,627,985$$
$$= \quad \$294,873,476.$$

Step 8: Total annual oil revenue:

$$\text{Annual oil revenue} \quad = \quad \$18/\text{bbl} \times 3,000,000 \text{ bbls/day}$$
$$\text{x } 365 \text{ days/year}$$
$$= \quad \$19,710,000,000 \text{ year.}$$

Comments: A variety of other criteria exists for choosing pipe size for a particular fluid transfer application. For example, low velocity may be required where erosion or corrosion concerns must be considered. Alternatively, higher velocities may be desirable for slurries where settling is a concern. Ease of construction may also weigh significantly when choosing pipe size. A small pipe size may not accommodate the head and flow requirements efficiently, whereas space limitations may prohibit the selection of large pipe sizes.

6.6 Replacement Analysis Fundamentals

In March, 1996, Ms. Karen Horseford, a Production Engineer at Hilton Furniture Company, was considering replacing a 1000-kilogram-capacity industrial forklift truck. The truck was being used to move assembled furniture from the dyeing department to the finishing department, and, then to the warehouse. Recently, the truck had not been dependable and was frequently out of service while awaiting repairs. Its maintenance expenses had been rising steadily, currently amounting to about $3000 per year. If the truck was not available, the company had to rent one. Additionally, the forklift truck was diesel-operated, and workers in the plant were complaining about the air pollution. The old truck was purchased 6 years ago at a cost of $15,000. The truck was originally expected to have a useful life of 8 years and an estimated salvage value of $2000 at the end of that period. The truck could be sold today for $4000. If retained, it would require an immediate $1500 overhaul to keep it in operable condition. The overhaul would neither extend the originally estimated service life nor would it increase the value of the truck.

Two types of forklift trucks were recommended as replacements. One was electric-operated and the other was gasoline-operated. The electric truck would eliminate the air pollution problem entirely, but would necessitate a battery change twice a day, which would significantly increase the operating cost. If the gasoline-operated truck were to be used, it would require more frequent maintenance.

Ms. Horseford was undecided about whether the company should buy the electric- or the gasoline-operated forklift truck at this point. She felt she should do some homework before approaching upper management for authorization of the replacement. Two questions came to her mind immediately:

1. Should the forklift truck be replaced now by one of the more advanced and fuel-efficient trucks?

2. If not now, when should the replacement occur?

To answer the first question, one has to know the annual cost of the current truck and the annual equivalent cost of the best truck available on the market. The cheaper option should then be selected. If the decision is to continue to use the current truck, then we also need to answer the second question. In this case, the current truck can be maintained in serviceable condition for some time—when will be the economically optimal time to make the replacement? Furthermore, when is the technologically optimal time to switch? By waiting for a more technologically advanced truck to evolve and gain commercial acceptance, the company may be able to obtain even better results in the future than by just switching to whatever lift-truck is available on the market today.

The engineering economic evaluation techniques that we presented in Chapters 4 and 5 can be used to make decisions on whether to buy new and more efficient equipment or to continue to use existing equipment. This class of decision problems is known as the **replacement problem**. In this section and the following two sections, we will examine three aspects of the replacement problem: (1) approaches for comparing defender and challenger, (2) determination of economic service life, and (3) replacement analysis when the required service period is long. The impact of income tax regulations will be ignored in these sections. In Chapter 11, we will revisit these replacement problems considering income taxes.

6.6.1 Basic Concepts and Terminology

Replacement projects are decision problems involving the replacement of existing obsolete or worn-out assets. The continuation of operations is dependent on these assets. Failure to make an appropriate decision will result in a slowdown or shutdown of the operations. The question is when existing equipment should be replaced with more efficient equipment. This situation has given rise to the use of the terms **defender** and **challenger**, terms commonly used in the boxing world. In every boxing class, the current defending champion is constantly faced with a new challenger who wants to take his position. In replacement analysis, the defender is the existing machine (or system), and the challenger is the best available replacement equipment.

An existing piece of equipment will eventually be removed at some future time, either when the task it performs is no longer necessary or when the task can be performed more efficiently by newer and better equipment. The question is *not* whether the existing piece of equipment will be removed, but *when* it will be removed. A variation of this question is why we should replace existing equipment at *this time* rather than postponing replacement of the equipment by repairing or overhauling it. Another aspect of the defender-challenger comparison concerns deciding exactly which equipment is the best challenger. If the defender is to be replaced by the challenger, we would generally want to install the very best of the possible alternatives.

Current Market Value

The most common problem encountered in considering the replacement of existing equipment is the determination of what financial information is actually relevant to the analysis. Often, a tendency to include irrelevant information in the analysis is apparent. To illustrate this type of decision problem, let us consider Example 6.9.

EXAMPLE 6.9 **Relevant information for replacement analysis**

Macintosh Printing, Inc., purchased a $20,000 printing machine 2 years ago. The company expected this machine to have a 5-year life and a salvage value of $5000. The company spent $5000 last year on repairs, and current operating costs are now running at the rate of $8000 per year. Furthermore, the anticipated salvage value has now been reduced to $2500 at the end of the printer's remaining useful life. In addition, the company has found that the current machine has a market value of $10,000 today. The equipment vendor will allow the company this full amount as a trade-in on a new machine. What value(s) for the defender are relevant in our analysis?

Solution

In this example, three different dollar amounts relating to the defender are presented:

1. Original cost: The printing machine was purchased for $20,000.

2. Market value: The company estimates the old machine's current market value at $10,000.

3. Trade–in allowance: It is the same as the market value. (This value, however, could be different from the market value.)

In this example, and in all defender analyses, the relevant cost is the current market value of the equipment. The original cost, repair cost, and trade-in value are irrelevant. A common misconception is held that the trade-in value is the same as the **current market value** of the equipment and thus could be used to assign a suitable current value to the equipment. This is not always true, however. For example, a car dealer typically offers a trade-in value on a customer's old car to reduce the price of

---

a new car. Would the dealer offer the same value on the old car if he were not also selling the new one? This is not generally the case. In many instances, the trade-in allowance is inflated to make the deal look good, and the price of the new car is also inflated to compensate for the dealer's trade-in cost. In this type of situation, the trade-in value does not represent the true value of the item, so we should not use it in economic analysis.[1]

Sunk Costs

A sunk cost is any past cost unaffected by any future investment decision. In Example 6.9, the company spent $20,000 to buy the machine two years ago. Last year, $5000 more was spent on this machine. The total accumulated expenditure on this machine is $25,000. If the machine is sold today, the company can only get $10,000 back (Figure 6.7). It is tempting to think that the company would lose $15,000 in addition to the cost of the new machine if the machine were to be sold and replaced with a new one. This is an *incorrect* way of doing economic analysis. In a proper engineering economic analysis, only future costs should be considered; past or sunk costs should be ignored. Thus, the value of the defender that should be used in a replacement analysis should be its current market value, not what it cost when it was originally purchased and not the cost of repairs that have already been made on the machine.

Sunk costs are the money that is gone and no present action can recover it. They represent past actions. They are the results of decisions made in the past. In making economic decisions at the present time, one should only consider the possible outcomes of different decisions and pick the one with the best possible future results. Using sunk costs in arguing one option over the other would only lead to more bad decisions.

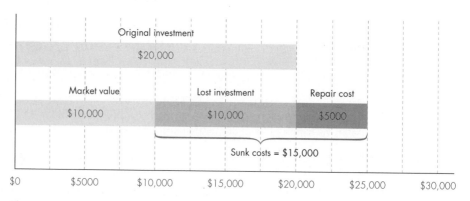

Figure 6.7 Sunk cost associated with an asset's disposal as described in Example 6.9.

---

[1] If we do make the trade, however, the actual *net* cash flow at this time, properly used, is certainly relevant.

Operating Costs

The driving force for replacing existing equipment is that it becomes more expensive to operate with time. The total cost of operating a piece of equipment may include repair and maintenance costs, wages for the operators, energy consumption costs, and costs of materials. Increases in any one or a combination of these cost items over a period of time may cause us to look for a replacement for the existing asset. The challenger is usually newer than the defender and often incorporates design improvements and newer technology. As a result, some or all of the cost items for the challenger are likely to be less than those for the defender.

We will call the sum of the various cost items related to the operation of an asset the operating costs. As will be illustrated in the following sections, keeping the defender involves a lower initial investment cost than purchasing the challenger but higher annual operating costs. These operating costs usually increase over time for both the defender and the challenger. In many instances, the labor costs, material costs and energy costs are the same for the defender and the challenger and do not change with time. It is the repair and maintenance costs that increase and cause the operating costs to increase each year as an asset ages. When repair and maintenance costs are the only cost items that differ between the defender and the challenger on a year by year basis, we only need to include repair and maintenance costs in the operating costs used in the analysis. Regardless of which cost items we choose to include in the operating costs, it is essential that the same items are included for both the defender and the challenger. For example, if energy costs are included in the operating costs of the defender, they should also be included in the operating costs of the challenger.

A more comprehensive discussion of the various types of costs incurred in a complex manufacturing facility is provided in Section 8.2.1.

6.6.2   Approaches for Comparing Defender and Challenger

Although replacement projects are a subcategory of the mutually exclusive project decisions we studied in Chapter 5, they do possess unique characteristics that allow us to use specialized concepts and analysis techniques in their evaluation. We will consider two basic approaches to analyzing replacement problems: the **cash flow approach** and the **opportunity cost approach**. We will start with a replacement decision problem where both the defender and the challenger have the *same useful life* which begins now.

Cash Flow Approach

The cash flow approach can be used as long as the *analysis period* is the *same* for all replacement alternatives. In this approach, we consider explicitly the actual cash flow consequences for each replacement alternative and compare them based on either PE or AE values.

EXAMPLE
6.10

Replacement analysis using the cash flow approach

Consider Example 6.9. The company has been offered a chance to purchase another printing machine for $15,000. Over its 3-year useful life, the machine will reduce labor and raw materials usage sufficiently to cut operating costs from $8000 to $6000. It is estimated that the new machine can be sold for $5500 at the end of year 3. If the new machine were purchased, the old machine would be sold to another company rather than traded in for the new machine. Suppose that the firm will need either machine (old or new) for only 3 years and that it does not expect a new, superior machine, to become available on the market during this required service period. Assuming that the firm's interest rate is 12%, decide whether replacement is justified now.

Solution

- Option 1: Keep the defender.

  If the old machine is kept, there is no additional cash expenditure today. It is in perfect operational condition. The annual operating cost for the next 3 years will be $8000 per year and its salvage value 3 years from today will be $2500. The cash flow diagram for this option is shown in Figure 6.8(a).

- Option 2: Replace the defender with the challenger.

  If this option is taken, the defender (designated D) can be sold for $10,000. The cost of the challenger (designated C) is $15,000. Thus, the initial combined cash flow for this option is a negative cash flow of $15,000 – $10,000 = $5000. The annual operating cost of the challenger is $6000. The salvage value of the challenger 3 years later will be $5500. The actual cash flow diagram for this option is shown in Figure 6.8(b).

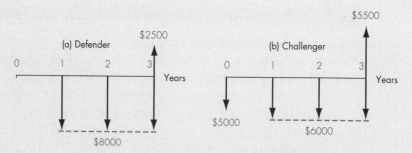

Figure 6.8 Comparison of defender and challenger based on the cash flow approach (Example 6.10)

$$PE(12\%)_D = \$2500(P/F, 12\%, 3) - \$8000(P/A, 12\%, 3)$$
$$= \$2500 \times 0.7118 - \$8000 \times 2.4018$$
$$= -\$17,434.90.$$

$$AE(12\%)_D = PE(12\%)_D/(P/A, 12\%, 3)$$
$$= -\$17,434.90/2.4018$$
$$= -\$7259.10.$$

$$PE(12\%)_C = \$5500(P/F, 12\%, 3) - \$5000 - \$6000(P/A, 12\%, 3)$$
$$= \$5500 \times 0.7118 - \$5000 - \$6000 \times 2.4018$$
$$= -\$15,495.90.$$

$$AE(12\%)_C = PE(12\%)_C/(P/A, 12\%, 3)$$
$$= -\$15,495.90/2.4018$$
$$= -\$6451.79.$$

Because of the annual difference of $807.31 in favor of the challenger, the replacement should be made now.

Comments: In this example, we did not address the question of whether or not the defender should be kept for 1 or 2 years before being replaced with the challenger. This is a valid question which requires more data on market values of the defender over time. We will address this situation later, in Section 6.8.

Opportunity Cost Approach

In the previous example, $10,000 in receipts from the sale of the old machine was foregone by not selling the defender. Another way to analyze such a problem is to charge the $10,000 as an **opportunity cost** of keeping the asset. That is, instead of deducting the current market value from the purchase cost of the challenger, we consider the current market value as a cash outflow for the defender (or investment required to keep the defender).

EXAMPLE 6.11 **Replacement analysis using the opportunity cost approach**

Rework Example 6.10 using the opportunity cost approach.

Solution

Recall that the cash flow approach in Example 6.10 credited proceeds in the amount of $10,000 from the sale of the defender toward the $15,000 purchase price of the challenger, and no initial outlay would have been required had the decision been to keep the defender. If the decision to keep the defender had been made, the opportunity cost approach treats the $10,000 current market value of the defender as an incurred cost. Figure 6.9 illustrates the cash flows related to these decision options.

Since the lifetimes are the same, we can use either PE or AE analysis as follows:

$$PE(12\%)_D = -\$10,000 - \$8000(P/A, 12\%, 3) + \$2500(P/F, 12\%, 3)$$
$$= -\$27,434.90.$$

$$AE(12\%)_D = PE(12\%)_D(A/P, 12\%, 3)$$
$$= -\$11,422.64.$$

$$PE(12\%)_C = -\$15,000 - \$6000(P/A, 12\%, 3) + \$5500(P/F, 12\%, 3)$$
$$-\$25,495.90.$$

$$AE(12\%)_C = PE(12\%)_C(A/P, 12\%, 3)$$
$$= -\$10,615.33.$$

The decision outcome is the same as in Example 6.10, that is, the replacement should be made. Since both the challenger and defender cash flows were adjusted by the same amount—$10,000—at time 0, this should not come as a surprise.

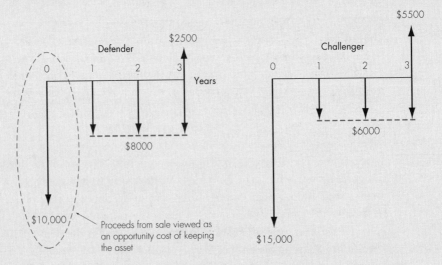

Figure 6.9 Comparison of defender and challenger based on the opportunity cost approach (Example 6.11)

Comments: Recall that we assumed the same service life for both the defender and the challenger in Examples 6.10 and 6.11. In general, however, old equipment has a relatively short remaining life compared to new equipment, so this assumption is overly simplistic. In the next section, we will discuss how to find the economic service life of equipment.

6.7 Economic Service Life

You have probably seen a 50-year-old automobile still in service. Provided it receives the proper repair and maintenance, almost anything can be kept operating for an extended period of time. If it's possible to keep a car operating for an almost indefinite period, why aren't more old cars spotted on the streets? There are several reasons.

Some people may get tired of driving the same old car. Others may want to keep a car as long as it will last, but they realize that repair and maintenance costs will become excessive.

In general, we need to consider explicitly how long an asset should be held, once it is placed in service. For instance, a truck-rental firm that frequently purchases fleets of identical trucks may wish to arrive at a policy on how long to keep each vehicle before replacing it. If an appropriate life span is computed, a firm could stagger a schedule of truck purchases and replacements to smooth out annual capital expenditures for overall truck purchases.

The costs of owning and operating an asset can be divided into two categories: capital costs and operating costs. Capital costs have two components: initial investment and the salvage value at the time of disposal. The initial investment for the challenger is simply its purchase price. For the defender, we should treat the opportunity cost as its initial investment. We will use N to represent the length of time in years the asset will be kept, P the initial investment, and S_N the salvage value at the end of the ownership period of N years.

The annual equivalent of capital costs, which is called capital recovery cost (refer to Section 4.5.2), over the period of N years can be calculated with the following equation:

$$CR = P(A/P, i, N) - S_N(A/F, i, N) \tag{6.7}$$

Generally speaking, as an asset becomes older, its salvage value becomes smaller. As long as the salvage value is less than the initial cost, the capital recovery cost is a decreasing function of N. In other words, the longer we keep an asset, the lower the capital recovery cost becomes. If the salvage value is equal to the initial cost no matter how long the asset is kept, the capital recovery cost is always constant.

As described earlier, the operating costs of an asset include repair and maintenance (R&M) costs, labor costs, material costs, and energy consumption costs. For the same equipment, labor costs, material costs, and energy costs are often constant from year to year if the usage of the equipment remains constant. However, R&M costs tend to increase as a function of the age of the asset. Because of the increasing trend of the R&M costs, the total operating costs of an asset usually increase as the asset ages. We will use OC_n to represent the total operating costs in year n of the ownership period and AEOC to represent the annual equivalent of the operating costs over a lifespan of N years. Then, AEOC can be expressed as

$$AEOC = \left(\sum_{n=1}^{N} OC_n(P/F, i, n) \right) (A/P, i, N). \tag{6.8}$$

As long as the annual operating costs increase with the age of the equipment, AEOC is an increasing function of the life of the asset. If the annual operating costs are the same from year to year, AEOC is constant and equal to the annual operating costs no matter how long the asset is kept.

The total annual equivalent cost of owning and operating an asset is a summation of the capital recovery costs and the annual equivalent of operating costs of the asset.

$$AEC = CR + AEOC$$

$$= P(A/P, i, N) - S_N (A/F, i, N) + \left(\sum_{n=1}^{N} OC_n(P/F, i, n) \right) (A/P, i, N). \quad (6.9)$$

The **economic service life** of an asset is defined to be the period of useful life that minimizes the annual equivalent costs of owning and operating the asset. Based on the foregoing discussions, we need to find the value of N that minimizes AEC as expressed in equation (6.9). If CR is a decreasing function of N and AEOC is an increasing function of N, as this is often the case, AEC will be a convex function of N with a unique minimum point (see Figure 6.10). In this book, **we will assume that AEC has a unique minimum point**. If the salvage value is constant and equal to the initial cost and the annual operating cost increases with time, AEC is an increasing function of N and attains its minimum at N = 1. In this case, we should try to replace the asset as soon as possible. If the annual operating cost is constant and the salvage value is less than the initial cost and decreases with time, AEC is a decreasing function of N. In this case, we would try to delay the replacement of the asset as much as possible. If the salvage value is constant and equal to the initial cost and the annual operating costs are constant, AEC will also be constant. In this case, when to replace the asset does not make any economic difference.

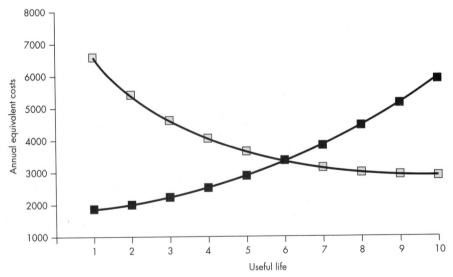

Figure 6.10　A schematic illustrating the trends of capital recovery cost and annual equivalent operating cost

If a new asset is purchased and operated for the length of its economic life, the annual equivalent cost is minimized. If we further assume that a new asset of identical price and features can be purchased repeatedly over an indefinite period, we would always replace this kind of asset at its economic life. By replacing perpetually according to an asset's economic life, we obtain the minimum AE cost stream over an indefinite period. However, if the identical replacement assumption cannot be made, we will have to use the methods to be covered in Section 6.8 to make replacement analysis. The next example explains the computational procedure for determining economic service life.

EXAMPLE 6.12 Economic service life for a lift truck

As a challenger to the forklift truck described at the beginning of Section 6.6, consider a new electric-lift truck that would cost $18,000, have operating costs of $4000 in the first year, and have a salvage value of $10,000 at the end of the first year. For the remaining years, operating costs increase each year by 40% over the previous year's operating costs. The salvage value declines each year by 25% from the previous year's salvage value. The lift truck has a maximum life of 7 years. An overhaul costing $5000 will be required at the end of the fifth year of service. The firm's required rate of return is 15%. Find the economic service life of this new machine.

Discussion: For an asset whose revenues are either unknown or irrelevant, we compute its economic life based on the costs for the asset and its year-by-year salvage values. To determine an asset's economic service life, we need to compare the options of keeping the asset for one year, two years, three years, and so forth. The option that results in the lowest annual equivalent cost (AEC) gives the economic service life of the asset.

- $N = 1$: One-year replacement cycle. In this case, the machine is bought, used for one year, and sold at the end of year 1. The cash flow diagram for this option is shown in Figure 6.11. The annual equivalent cost for this option is:

$$AEC(15\%) = \$18,000(A/P, 15\%, 1) + \$4000 - \$10,000$$
$$= \$14,700$$

Note that $(A/P, 15\%, 1) = (F/P, 15\%, 1)$ and the annual equivalent cost is the equivalent cost at the end of year 1 since $N = 1$. Because we are calculating the annual equivalent costs, we have treated cost items with a positive sign, while the salvage value has a negative sign in the above computation of $AEC(15\%)$.

- $N = 2$: Two-year replacement cycle. In this case, the truck will be used for two years and disposed of at the end of year 2. The operating cost in year 2 is 40% higher than that in year 1 and the salvage value at the end of year 2 is 25% lower than that at the end of year 1. The cash flow diagram for this option is also shown in Figure 6.11. The annual equivalent cost over the two year period is:

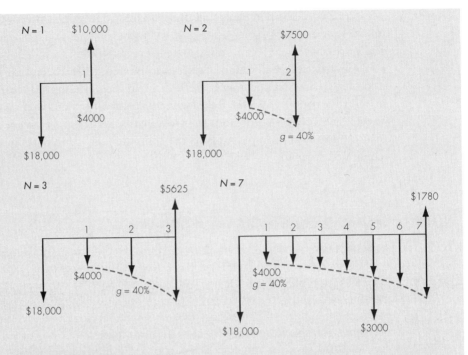

Figure 6.11 Cash flow diagrams for the options of keeping the asset for one year, two years, and three years (Example 6.12)

$$AEC(15\%) = [\$18,000 + \$4000(P/A_1, 40\%, 15\%, 2)](A/P, 15\%, 2)$$
$$- \$7500(A/F, 15\%, 2)$$
$$= \$12,328.$$

- $N = 3$: Three-year replacement cycle. In this case, the truck will be used for 3 years and sold at the end of year 3. The salvage value at the end of year 3 is 25% lower than that at the end of year 2, i.e., $\$7500 \times (1 - 25\%) = \$5625$. The operating cost per year increases at a rate of 40%. The cash flow diagram for this option is also shown in Figure 6.11.

$$AEC(15\%) = [\$18,000 + \$4000(P/A_1, 40\%, 15\%, 3)](A/P, 15\%, 3)$$
$$- \$5625(A/F, 15\%, 3)$$
$$= \$11,899.$$

Similarly, we can find the annual equivalent costs for the options of keeping the asset for 4 years, 5 years, 6 years, and 7 years. One has to note that there is an additional cost of overhaul in year 5. The cash flow diagram when $N = 7$ is shown in Figure 6.11. The computed annual equivalent costs for these options are:

$N = 4$, $AEC(15\%) = \$12,165.$
$N = 5$, $AEC(15\%) = \$13,632.$
$N = 6$, $AEC(15\%) = \$14,677.$
$N = 7$, $AEC(15\%) = \$16,158.$

From the above-calculated AEC values for $N = 1, 2, ..., 7$ we find that AEC(15%) is the smallest when $N = 3$. If the truck were to be sold after 3 years, it would have an annual equivalent cost of $11,899. If it were to be used for a period other than 3 years, the annual equivalent costs would be higher than $11,899. Thus, a life span of 3 years for this truck results in the lowest annual equivalent cost. We conclude that the economic service life of the truck is 3 years. By replacing the assets perpetually according to an economic life of 3 years, we obtain the minimum annual equivalent cost stream. Figure 6.12 illustrates this concept. Of course, we should envision a long period of required service for this kind of asset.

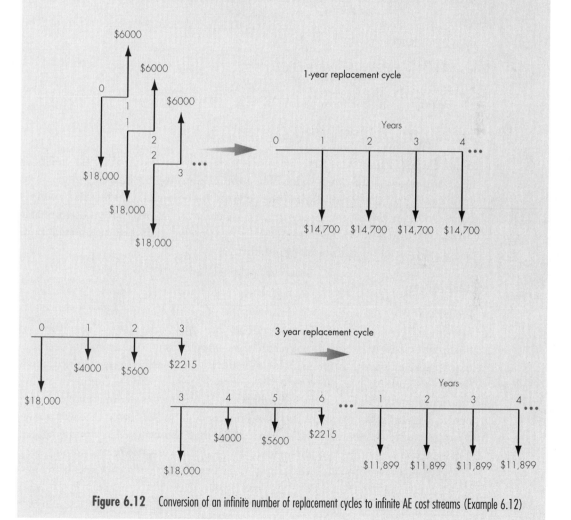

Figure 6.12 Conversion of an infinite number of replacement cycles to infinite AE cost streams (Example 6.12)

6.8 Replacement Analysis When Required Service Is Long

Now that we understand how the economic service life of an asset is determined, the next question is how to use these pieces of information to decide whether *now* is the time to replace the defender. If *now* is not the right time, *when* is the optimal time to replace the defender? Before presenting an analytical approach to answer this question, we will consider several important assumptions.

6.8.1   Required Assumptions and Decision Frameworks

In deciding whether *now* is the time to replace the defender, we need to consider the following three factors:

- Planning horizon (study period)
- Technology
- Relevant cash flow information

Planning Horizon (Study Period)

By planning horizon, we simply mean the service period required by the defender and a sequence of future challengers. The **infinite planning horizon** is used when we are simply unable to predict when the activity under consideration will be terminated. In other situations, it may be clear that the project will have a definite and predictable duration. In these cases, replacement policy should be formulated more realistically based on a **finite planning horizon.**

Technology

Predictions of technological patterns over the planning horizon refer to the development of types of challengers that may replace those under study. A number of possibilities exist in predicting purchase cost, salvage value, and operating cost dictated by the efficiency of the machine over the life of an asset. If we assume that all future machines will be the same as those now in service, we are implicitly saying no technological progress in the area will occur. In other cases, we may explicitly recognize the possibility of machines becoming available in the future that will be significantly more efficient, reliable, or productive than those currently on the market. (Personal computers are a good example.) This situation leads to recognition of technological change or obsolescence. Clearly, if the best available machine gets better and better over time, we should certainly investigate the possibility of delaying replacement for a couple of years, which contrasts with the situation where technological change is unlikely.

Revenue and Cost Patterns over Asset Life

Many varieties of predictions can be used to estimate the patterns of revenue, cost, and salvage value over the life of an asset. Sometimes revenue is constant, but costs increase, while salvage value decreases, over the life of a machine. In other situations, a decline in revenue over equipment life can be expected. The specific situation will determine whether replacement analysis is directed toward cost minimization (with constant revenue) or profit maximization (with varying revenue). We will formulate a replacement policy for an asset in which salvage values do not increase with age.

Decision Frameworks

To illustrate how a decision framework is developed, we will indicate a replacement sequence of assets by the notation (j_0, n_0), (j_1, n_1), (j_2, n_2), ... , (j_K, n_K). Each pair of numbers (j, n) indicates an asset type and the lifetime for which that asset will be retained. The defender, asset 0, is listed first; if the defender is replaced now, $n_0 = 0$. A sequence of pairs may cover a finite period or an infinite period. For example, the sequence $(j_0, 2)$, $(j_1, 5)$, $(j_2, 3)$ indicates retaining the defender for 2 years, replacing the defender with an asset of type j_1, using it for 5 years, replacing j_1 with an asset of type j_2, and using it for 3 years. In this situation, the total planning horizon covers 10 years $(2 + 5 + 3)$. The special case of keeping the defender for n_0 periods, followed by infinitely repeated purchases and the use of an asset of type j for n^* years, is represented by (j_0, n_0), $(j, n^*)_\infty$. This sequence covers an infinite period, and the relationship is illustrated in Figure 6.13.

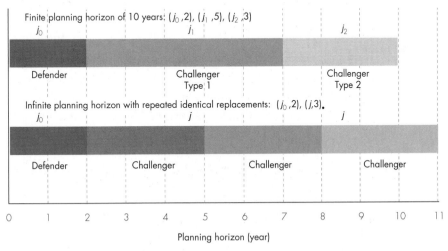

Figure 6.13 Types of typical replacement decision frameworks

Decision Criterion

While the economic life of the defender is defined as the additional number of years of service that minimizes the annual equivalent cost (or maximizes the annual equivalent revenue), this is *not* necessarily the *optimal time* to replace the defender. The correct replacement time depends on data for the challenger as well as on data for the defender.

As a decision criterion, the AE method provides a more direct solution when the planning horizon is infinite. When the planning horizon is finite, the PE method is more convenient to use. We will develop the replacement decision procedure for both situations: (1) the infinite planning horizon and (2) the finite planning horizon. We begin by analyzing an infinite planning horizon without technological change. While a simplified situation such as this is not likely to occur in real life, the analysis of this replacement situation introduces methods useful for analyzing infinite horizon replacement problems with technological change.

6.8.2   Replacement Strategies under the Infinite Planning Horizon

We will consider the situation where a firm has a machine in use in a process. The process is expected to continue for an indefinite period. Presently, a new machine will be on the market that is, in some ways, more effective for the application than the defender. The problem is when the defender should be replaced by the challenger.

Under the infinite planning horizon, the service is required for a very long time. Either we continue to use the defender to provide the service, or we replace the defender with the best available challenger for the same service requirement. In this case, the following procedure may be followed in replacement analysis:

1. Compute the economic lives of both defender and challenger. Let's use N_D^* and N_C^* to indicate the economic lives of the defender and the challenger, respectively. The annual equivalent costs for the defender and the challenger at their respective economic lives are indicated by AEC_D^* and AEC_C^*.

2. Compare AEC_D^* and AEC_C^*. If AEC_D^* is bigger than AEC_C^*, we know that it is more costly to keep the defender than to replace it with the challenger. Thus, the defender should be replaced by the challenger *now*. If AEC_D^* is smaller than AEC_C^*, it costs less to keep the defender than to replace it with the challenger now. Thus, the defender should not be replaced now. The defender should continue to be used at least for the duration of its economic life if there are no technological changes over the economic life of the defender.

3. If the defender should not be replaced now, when should it be replaced? First we need to continue to use it until its economic life is over. Then, we should calculate the cost of running the defender for one more year after its economic

life. If this cost is greater than AEC_C^*, the defender should be replaced at the end of its economic life. Otherwise, we should calculate the cost of running the defender for the second year after its economic life. If this cost is bigger than AEC_C^*, the defender should be replaced one year after its economic life. This process should be continued until you find the optimal replacement time. This approach is called marginal analysis, i.e., calculate the incremental cost of operating the defender for just one more year. In other words, we want to see whether the cost of extending the use of the defender for an additional year exceeds the savings resulting from delaying the purchase of the challenger. Here we have assumed the best available challenger does not change.

It should be noted that the above procedure may be applied dynamically. It may be performed annually for replacement analysis. Whenever there are updated data on the costs of the defender or new challengers available on the market, these new data should be used in the procedure. The following example illustrates the above procedure.

EXAMPLE 6.13 Replacement analysis under the infinite planning horizon

Advanced Electrical Insulator Company is considering replacing a broken inspection machine, which has been used to test the mechanical strength of electrical insulators, with a newer and more efficient one. If repaired, the old machine can be used for another 5 years. The firm does not expect to realize any salvage value from scrapping it in 5 years. But, the firm can sell it now to another firm in the industry for $5000. If the machine is kept, it will require an immediate $1200 overhaul to restore it to operable condition. The overhaul will neither extend the service life originally estimated nor increase the value of the inspection machine. The operating costs are estimated at $2000 during the first year, and these are expected to increase by $1500 per year thereafter. Future market values are expected to decline by $1000 per year.

The new machine costs $10,000 and will have operating costs of $2000 in the first year, increasing by $800 per year thereafter. The expected salvage value is $6000 after 1 year and will decline 15% each year. The company requires a rate of return of 15%. (1) Find the economic life of the defender and that of the challenger and (2) determine when the defender should be replaced.

Solution

1. Economic Service Life

- Defender

  If the company retains the inspection machine, it is in effect deciding to overhaul the machine and invest the machine's current market value in that alternative. The opportunity cost of the machine is $5000. Since an overhaul costing $1200 is also needed to make the machine operational, the total initial investment of the machine is considered to be $5000 + $1200 = $6200. Other data for the defender is summarized as follows:

n	Overhaul	Forecasted Operating Cost	Marked Value If Disposed of
0	$1200		$5000
1	0	$2000	$4000
2	0	$3500	$3000
3	0	$5000	$2000
4	0	$6500	$1000
5	0	$8000	0

We can calculate the annual equivalent costs if the defender is to be kept for 1 year, 2 years, 3 years, and so forth. For example, the cash flow diagram for $N = 4$ years is shown in Figure 6.14.

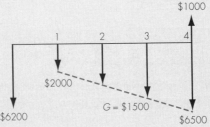

Figure 6.14 Cash flow diagram for defender when $N = 4$ years (Example 6.13)

$$N = 4 \text{ years: } AEC(15\%) = \$6200(A/P, 15\%, 4) + \$2000$$
$$+ \$1500(A/G, 15\%, 4)$$
$$- \$1000(A/F, 15\%, 4)$$
$$= \$5961.$$

The other AEC values can be calculated with the following equation:

$$AEC(15\%)_N = \$6200(A/P, 15\%, N) + \$2000 + \$1500(A/G, 15\%, N)$$
$$- \$1000(5 - N)(A/F, 15\%, N) \text{ for } N = 1, 2, 3, 4, 5$$

$$N = 1: AEC(15\%) = \$5130.$$
$$N = 2: AEC(15\%) = \$5116.$$
$$N = 3: AEC(15\%) = \$5500.$$
$$N = 4: AEC(15\%) = \$5961.$$
$$N = 5: AEC(15\%) = \$6434.$$

When $N = 2$ years, we get the lowest AEC value. Thus, the defender's economic life is 2 years. Using the notation we have defined in the procedure, we have

$$N_D^* = 2 \text{ years}$$
$$AEC_D^* = \$5116.$$

The AEC values as a function of N are plotted in Figure 6.15. Actually, after computing AEC for N = 1, 2, and 3, we can stop right there. There is no need to compute AEC for N = 4 and N = 5 because AEC is increasing when $N > 2$ and we have assumed that AEC has a unique minimum point.

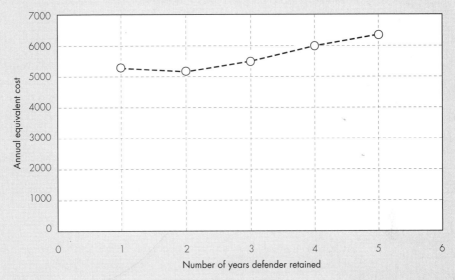

Figure 6.15   AEC as a function of the life of the defender (Example 6.13)

- Challenger

    The economic life of the challenger can be determined using the same procedure shown in this example for the defender and in Example 6.12. We summarize the general equation for AEC calculation for the challenger as follows. You don't have to summarize such an equation when you need to determine the economic life of an asset as long as you follow the procedure illustrated in Example 6.12.

$$AEC(15\%)_N = \$10,000(A/P, 15\%, N) + \$2000$$
$$+ \$800(A/G, 15\%, N)$$
$$- \$6000(1 - 15\%)^{N-1}(A/F, 15\%, N).$$

The obtained results are:

N = 1 year: $AEC(15\%)$ = \$7500.
N = 2 years: $AEC(15\%)$ = \$6151.
N = 3 years: $AEC(15\%)$ = \$5857.
N = 4 years: $AEC(15\%)$ = \$5826.
N = 5 years: $AEC(15\%)$ = \$5897.

The economic life of the challenger is 4 years, that is

$$N_c^* = 4 \text{ years}$$
$$AEC_c^* = \$5826.$$

2. Should the defender be replaced now?

   Since $AEC_D^* = \$5116 < AEC_C^* = \$5826$, the answer is not to replace the defender now. If there are no technological advances in the next few years, the defender should be used for at least $N_D^* = 2$ more years. It is not necessarily best to replace the defender right at its economic life.

3. When should the defender be replaced?

   If we need to find the answer to this question today, we have to calculate the cost of keeping and using the defender for the 3rd year from today. That is, what is the cost of not selling the defender at the end of year 2, using it for the 3rd year, and replacing it at the end of year 3? The following cash flows are related to this question:

   (a) Opportunity cost at the end of year 2: It is equal to the market value then, $3000

   (b) Operating cost for the 3rd year: $5000

   (c) Salvage value of the defender at the end of year 3: $2000

   The following cash flow diagram represents these cash flows.

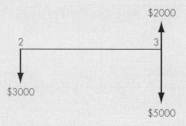

   The cost of using the defender for one more year from the end of its economic life is:

   $$\$3000 \times 1.15 + \$5000 - \$2000 = \$6450.$$

   Now compare this cost with the $AEC_C^* = \$5826$ of the challenger. It is greater than AEC_C^*. It is more expensive to keep the defender for the 3rd year than to replace it with the challenger at the end of year 2. Thus, the conclusion is to replace the defender at the end of year 2. If this one-year cost is still smaller than AEC_C^*, we need to calculate the cost of keeping the defender for the 4th year and then compare it with AEC_C^* of the challenger.

In replacement analysis, it is common for a defender and its challenger to have different economic service lives. The annual equivalent approach is frequently used in replacement analysis, but it is important to know that we use the AE method in replacement analysis, not because we have to deal with the unequal service life problem, but rather because the AE approach provides some computational advantage for a special class of replacement problem.

In Chapter 5, we discussed the general principle for comparing alternatives with unequal service lives. In particular, we pointed out that use of the AE method relies on the concept of **repeatability** of projects and one of two assumptions: an infinite planning horizon or a common service period. In defender-challenger situations, however, repeatability of the defender cannot be assumed. In fact, by virtue of our problem definition, we are not *repeating* the defender, but *replacing* it with its challenger, an asset that in some way constitutes an improvement over the current equipment. Thus, the assumptions we made for using an annual cash flow analysis with unequal service life alternatives are not valid in the usual defender-challenger situation.

The complication—the unequal life problem—can be resolved, however, if we recall that the replacement problem at hand is not whether to replace the defender, but when to do so. When the defender is replaced, it will always be *by the challenger* —the best available equipment. The challenger can then be replaced repeatedly by an identical challenger. In fact, we are comparing the following two options in replacement analysis:

1. Replace the defender now: The cash flows of the challenger will be used from today and will be repeated because an identical challenger will be used if replacement becomes necessary again in the future. This stream of cash flows is equivalent to a cash flow of AEC^*_C each year for an infinite number of years.

2. Replace the defender, say, x years later: The cash flows of the defender will be used in the first x years. Starting in year $x + 1$, the cash flows of the challenger will be used indefinitely.

The annual equivalent cash flows for the years beyond year x are the same for the above two options. We need only to compare the annual equivalent cash flows for the first x years to determine which option is better. This is why we can compare AEC^*_D with AEC^*_C to determine whether now is the time to replace the defender.

6.8.3 Replacement Strategies under the Finite Planning Horizon

If the planning period is **finite** (for example, 8 years), a comparison based on the AE method over a defender's economic service life does not generally apply. The procedure for solving such a problem with a finite planning horizon is to establish *all* "reasonable" replacement patterns and then use the PE value for the planning period to select the most economical pattern. To illustrate the procedure, let us consider Example 6.14.

EXAMPLE 6.14

Replacement analysis under the finite planning horizon (PE approach)

Reconsider the defender and the challenger in Example 6.13. Suppose that the firm has a contract to perform a given service, using the current defender or the challenger for the next 8 years. After the contract work, neither the defender nor the challenger will be retained. What is the best replacement strategy?

Solution

Recall again the annual equivalent costs for the defender and challenger under the assumed holding periods (a boxed number denotes the minimum AEC value at $N_D^* = 2$ and $N_c^* = 4$, respectively).

N	Annual Equivalent Cost ($) Defender	Challenger
1	5130	7500
2	5116	6151
3	5500	5857
4	5961	5826
5	6434	5897

Many ownership options would fulfill an 8-year planning horizon. Six options are shown in Figure 6.16 and the present equivalent cost (PEC) for each option is calculated as follows:

- Option 1: $(j_0, 0), (j, 4), (j, 4)$

    $PEC(15\%)_1$ = \$5826(P/A, 15\%, 8)

    = \$26,143.

- Option 2: $(j_0, 1), (j, 4), (j, 3)$

    $PEC(15\%)_2$ = \$5130(P/F, 15\%, 1)

    + \$5826(P/A, 15\%, 4)(P/F, 15\%, 1)

    + \$5857(P/A, 15\%, 3)(P/F, 15\%, 5)

    = \$25,573.

- Option 3: $(j_0, 2), (j, 4), (j, 2)$

    $PEC(15\%)_3$ = \$5116(P/A, 15\%, 2)

    + \$5826(P/A, 15\%, 4)(P/F, 15\%, 2)

    + \$6151(P/A, 15\%, 2)(P/F, 15\%, 6)

    = \$25,217 ← minimum cost.

- Option 4: $(j_0, 3), (j, 5)$

    $PEC(15\%)_3$ = \$5500(P/A, 15\%, 3)

    + \$5897(P/A, 15\%, 5)(P/F, 15\%, 3)

    = \$25,555.

- Option 5: $(j_0, 3), (j, 4), (j, 1)$

$$
\begin{aligned}
PEC(15\%)_4 &= \$5500(P/A, 15\%, 3) \\
&\quad + \$5826(P/A, 15\%, 4)(P/F, 15\%, 3) \\
&\quad + \$7500(P/F, 15\%, 8) \\
&= \$25,946.
\end{aligned}
$$

- Option 6: $(j_0, 4), (j, 4)$

$$
\begin{aligned}
PEC(15\%)_5 &= \$5961(P/A, 15\%, 4) \\
&\quad + \$5826(P/A, 15\%, 4)(P/F, 15\%, 4) \\
&= \$26,529.
\end{aligned}
$$

An examination of the present equivalent cost of a planning horizon of 8 years indicates that the least-cost solution appears to be Option 3: Retain the defender for 2 years, purchase the challenger and keep it for 4 years, and purchase another challenger and keep it for 2 years.

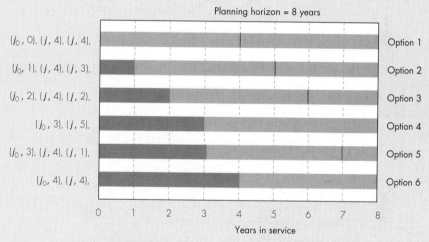

Figure 6.16   Some likely replacement patterns under a finite planning horizon of 8 years (Example 6.14)

Comments: In this example, we examined only six possible decision options likely to lead to the best solution, but it is important to note that several other possibilities have not been looked at. To explain, consider Figure 6.17, which shows a graphical representation of various replacement strategies under a finite planning horizon.

For example, the replacement strategy (shown as a solid line in Figure 6.17), $[(j_0, 2), (j, 3), (j, 3)]$ is certainly feasible, but we did not include it in the previous computation. Naturally, as we extend the planning horizon, the number of possible decision options can easily multiply. To make sure that we indeed find the optimal solution for such a problem, an optimization technique such as dynamic programming can be used.[2]

[2] Hillier FS, Lieberman GS. *Introduction to Operations Research.* Sixth Edition. New York: McGraw Hill, 1995: Chapter 10.

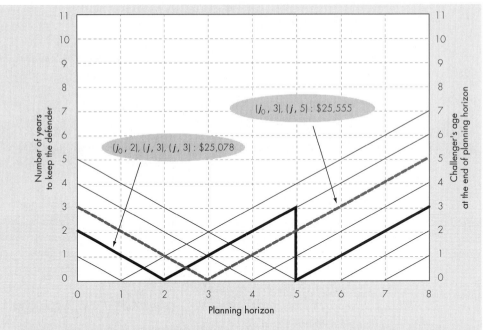

Figure 6.17   Graphical representation of replacement strategies under a finite planning horizon (Example 6.14)

6.8.4   Consideration of Technological Change

Thus far, we have defined the challenger simply as the best available replacement for the defender. It is more realistic to recognize that often the replacement decision involves an asset now in use versus a candidate for replacement that is, in some way, an improvement on the current asset. This, of course, reflects technological progress that is ongoing continually. Future models of a machine are likely to be more effective than a current model. In most areas, technological change appears as a combination of gradual advances in effectiveness; the occasional technological breakthrough, however, can revolutionize the character of a machine.

The prospect of improved future challengers makes a current challenger a less desirable alternative. By retaining the defender, we may later have an opportunity to acquire an improved challenger.[3] If this is the case, the prospect of improved future challengers may affect a current decision between a defender and its challenger. It is difficult to forecast future technological trends in any precise fashion. However, in developing a long-term replacement policy, we need to take technological change into consideration.

---

[3] This should not be interpreted as the faster technology is changing, the longer we should keep the defender. What we are saying is that the prospect of ever-improving future challengers would influence the timing of replacement decision.

6.9 Computer Notes

As indicated in Section 6.5, minimum cost analysis is useful in making an engineering design decision when two or more cost components are affected by a common design variable. This section illustrates how Excel can be used as a tool in this type of decision. The data given in Example 6.7 indicated that:

Electric current (amps) = 5000

Transmission distance (m) = 300

Operating hours per year (hours) = 8760

Service life (years) = 25

Material costs ($/kg) = 16.50

Scrap value ($/kg) = 1.65

Electrical resistance (ohms-cm^2/m) = 1.7241×10^{-4}

Cost of energy ($/kWh) = 0.0375

Material density (kg/m^3) = 8894

MARR (%) = 9

These data have been used to set up the spreadsheet shown in Figure 6.18. In addition, a cross-sectional area (A) is specified as a user input.

To illustrate the power of the spreadsheet approach, the cross-sectional area (A) is shown as a range of values from 10 to 50 cm^2, with an increment of 5 cm^2. Note that the minimum cost occurs with the A value in between 30 cm^2 and 35 cm^2. You can search the optimal value (minimum cost) by varying the value of A within these bounds. At $A = 31$, the annual equivalent cost is $27,466, the same number as previously calculated in this example.

Spreadsheet formulas are written as functions of the data in the lower section of the spreadsheet. To allow maximum flexibility in calculations, input data is expressed as absolute locations in all formulas. A plot of these total costs against design area (A) is easily obtained, as explained in Chapter 4. The plot is shown as Figure 6.19.

The spreadsheet can now be used to explore a wide variety of scenarios by merely changing input data values. This exploration is basic to the concept of sensitivity analysis, which will be discussed in detail in Chapter 13. Spreadsheet analysis makes the exploration considerably easier.

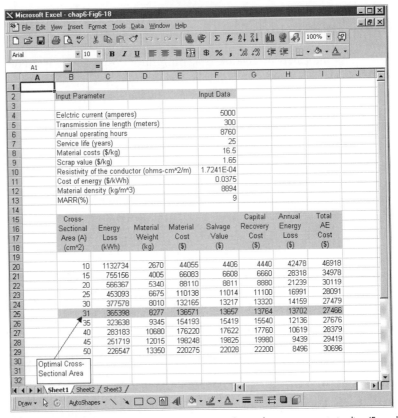

Figure 6.18   Excel output: Determining the optimal cross-sectional area of a copper transmission line (Example 6.7)

6.10 Summary

• The capitalized equivalent method is useful when the planning horizon is infinite or extremely long.

• AE analysis is recommended over PE analysis in many key real-world situations for the following reasons:

1. In many financial reports, an annual equivalent value is preferred to a present equivalent value.

2. Calculation of unit costs is often required to determine reasonable pricing for sale items.

3. Calculation of cost per unit of use is required to reimburse employees for business use of personal cars.

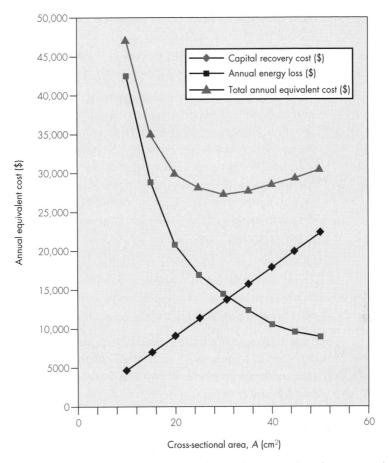

Figure 6.19   An Excel plot of annual equivalent cost as a function of cross-sectional area for a copper conductor

4. Make-or-buy decisions usually require the development of unit costs for the various alternatives.

5. Minimum cost analysis is easy to do when based on annual equivalent.

6. Determination of the economic life of a piece of equipment is required.

- Replacement analysis is an important application of engineering economic analysis techniques.

   1. In replacement analysis, the **defender** is an existing asset; the **challenger** is the best available replacement candidate.

   2. The **current market value** is the value to use in preparing a defender's economic analysis. **Sunk costs**—past costs that cannot be changed by any future invest-

ment decision—should not be considered in a defender's economic analysis. We call the sum of various cost items related to the operation of an asset the **operating costs**. Operating costs may include repair and maintenance costs, wages for the operators, energy consumption costs, and material costs.

3. Two basic approaches to analyzing replacement problems are the **cash flow approach** and the **opportunity cost approach**. The cash flow approach explicitly considers the actual cash flow consequences for each replacement alternative as they occur. Typically, the proceeds from sale of the defender are subtracted from the purchase price of the challenger. The opportunity cost approach views the current market value of the defender as an opportunity cost of keeping the defender. That is, instead of deducting the salvage value from the purchase cost of the challenger, it is considered as an investment required to keep the asset.

4. **Economic service life** is the remaining useful life of a defender, *or* a challenger, that results in the minimum equivalent annual cost or maximum annual equivalent revenue. We should use the respective economic service lives of the defender and the challenger when conducting a replacement analysis.

5. Ultimately, in replacement analysis, the question is not *whether* to replace the defender, but *when* to do so. The AE method is widely used in replacement analysis. It also provides a marginal basis on which to make a year-by-year decision about the best time to replace the defender.

6. The role of **technological change** in asset improvement should be weighed in making long-term replacement plans: If a particular item is undergoing rapid, substantial technological improvements, it may be prudent to delay replacement (to the extent where the loss in production does not exceed any savings from improvements in future challengers) until a desired future model is available.

Problems

Level 1

6.1* What is the capitalized equivalent amount, at 10% annual interest, for a series of annual receipts of $400 for the first 10 years, which will increase to $500 per year after 10 years, and which will remain constant thereafter?

6.2* Find the capitalized equivalent for the project cash flow series at an interest rate of 10%.

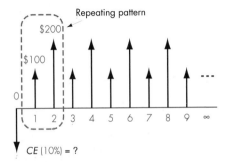

$CE (10\%) = ?$

6.3* Find the annual equivalent for the following infinite cash flow series at an interest rate of 10%:

n	Net Cash Flow
0	0
1 – 10	$1000
11 – ∞	$500

6.4* What is the annual equivalent at $i = 10\%$ for the following infinite series of annual receipts: $500 per year for the first 10 years and $1000 per year after that?

6.5* A machine now in use was bought 5 years ago for $4000. It can be sold today for $2500, but could be used for 3 more years (remaining useful life), at the end of which time it would have no salvage value. The

annual operating and maintenance costs for the old machine amount to $10,000. A new machine can be purchased at an invoice price of $14,000 to replace the present equipment. Because of the nature of the manufactured product, the new machine has an expected economic life of 3 years, and it will have no salvage value at the end of that time. The new machine's expected operating and maintenance costs amount to $2000 for the first year and $3000 for each of the next 2 years. The firm's interest rate is 15%.

(a) If you decide to retain the old machine for now, what will be the opportunity cost?

(b) If the old asset is to be sold now, what is its sunk cost?

6.6* A local delivery company has purchased a delivery truck for $15,000. The truck's market value (or selling price) is expected to be $2500 less each year. The O&M costs are expected to be $3000 per year. The firm's MARR is 15%. Compute the annual equivalent cost for retaining the truck for a 2-year period.

6.7* The annual equivalent costs of retaining a defender over its 3-year remaining life and the annual equivalent costs for its challenger over its 4-year physical life are as follows:

Holding Period	Annual Equivalent Cost	
	Defender	Challenger
1	$3000	$5000
2	2500	4000
3	2800	3000
4		4500

Assume a MARR of 12%. Would you recommend that the defender be replaced by the challenger now? Why? Without further calculations, when do you think that the defender should be replaced by the challenger?

Level 2

6.8 Maintenance money for a new building has been sought. Mr. Kendall would like to make a donation to cover all future expected maintenance costs for the building. These maintenance costs are expected to be $40,000 each year for the first 5 years, $50,000 for each year 6 through 10, and $60,000 each year after that. (The building has an indefinite service life.)

    (a) If the money is placed in an account that will pay 13%, compounded annually, how large should the donation be?

    (b) What is the equivalent annual maintenance cost over the infinite service life?

6.9 Consider an investment project, the cash flow pattern of which repeats itself every 5 years forever as shown. At an interest rate of 14%, compute the capitalized equivalent amount for this project.

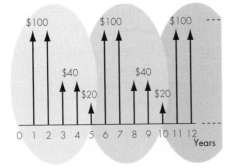

6.10* A group of concerned citizens has established a trust fund that pays 6% compounded quarterly to preserve a historical building by providing annual maintenance funds of $12,000 forever. Compute the capitalized equivalent amount for these

building maintenance expenses.

6.11 A newly constructed bridge costs $2,000,000. The same bridge is estimated to need renovation every 15 years at a cost of $500,000. Repairs and maintenance are estimated to be $50,000 per year.

    (a) If the interest rate is 5%, determine the capitalized cost of the bridge.

    (b) Suppose that, in (a), the bridge must be renovated at the same cost every 20 years, not every 15 years. What is the capitalized cost of the bridge?

    (c) Repeat (a) and (b) with an interest rate of 10%. What have you to say about the effect of interest on the results?

6.12 To decrease the costs of operating a lock in a large river, a new system of operation is proposed. The system will cost $650,000 to design and build. It is estimated that it will have to be reworked every 10 years at a cost of $100,000. In addition, an expenditure of $50,000 will have to be made at the end of the fifth year for a new type of gear that will not be available until then. Annual operating costs are expected to be $30,000 for the first 15 years and $35,000 per year thereafter. Compute the capitalized cost of perpetual service at $i = 8\%$.

6.13 Beginning next year, a foundation will support an annual seminar on campus by the earnings from a $100,000 gift it received this year. It is felt that 8% interest will be realized for the first 10 years, but that plans should be made to anticipate an interest rate of 6% after that time. What amount should be added to the foundation now to fund the seminar at the $10,000 level into infinity?

6.14* You just purchased a pin-inserting machine to relieve some bottleneck problems that have been created in manufacturing a PC board. The machine cost $56,000 and has an estimated service life of 5 years. At that time, the estimated salvage value would be

$5000. The machine is expected to operate 2500 hours per year. The expected annual operating and maintenance cost would be $6000. If your firm's interest rate is 15%, what would be the machine cost per hour?

6.15* The City of Greenville has decided to build a softball complex on land donated by one of the city residents. The city council has already voted to fund the project at a level of $800,000 (initial capital investment). The city engineer has collected the following financial information for the project.

- Annual upkeep costs: $120,000
- Annual utility costs: $13,000
- Renovation costs: $50,000 every 5 years
- Annual team user fees (revenues): $32,000
- Useful life: Infinite
- Interest rate: 5%.

If the city expects 40,000 visitors to the complex each year, what should be the minimum ticket price per person, so that the city can break even?

6.16* Consider manufacturing equipment that has an installed cost of $100K. The equipment is expected to generate $30K of annual energy savings during its first year of installation. The value of these annual savings is expected to increase by 3% per year because of increased fuel costs. Assume that the equipment has a service life of 5 years (with 3000 operating hours per year) with negligible salvage value. Determine the equivalent net savings per operating hour at $i = 14\%$.

6.17 Colgate Printing Co. (CPC) has the book binding contract for the Ralph Brown library. The library pays $25 per book to CPC. CPC binds 1000 books every year for the library. Ralph Brown library is considering the option of binding the books in-house in the basement of the library complex. In order to do this, the library would have to invest in a binding machine

and other printing equipment at a cost of $100,000. The useful life of the machine is 8 years, at the end of which time, the machine is estimated to have a salvage value of $12,000. The annual operating and maintenance costs of the machine are estimated to be $10,000. Assuming an interest rate of 12%, what is the cost of binding per book for the in-house option? What annual volume of books in need of binding would make both the options (in-house versus subcontracting) equivalent?

6.18* A small machine shop with an electrical load of 28 kW purchases its electricity at the following rates:

kWh/month	@ $/kWh
First 1500	0.025
Next 1250	0.015
Next 3000	0.009
All over 5750	0.008

The current monthly consumption of electric power averages 3200 kWh. A bid is to be made on some new business which would require an additional 3500 kWh per month. If the new business were to last 2 years, what would be the equivalent monthly power cost that should be allocated to the new business? Assume that the shop's interest rate is 9%, compounded monthly.

6.19 Two 150-horsepower (HP) motors are being considered for installation at a municipal sewage-treatment plant. The first costs $4500 and has an operating efficiency of 83%. The second costs $3600 and has an efficiency of 80%. Both motors are projected to have zero salvage value after a life of 10 years. If all the annual charges such as insurance, maintenance, etc., amount to a total of 15% of the original cost of each motor, and if power costs are a flat 5 cents per kilowatt-hour, how many minimum hours of full-load operation per year are necessary to justify purchase of the more expensive motor at $i = 6\%$? (A conversion factor you might find useful is 1 HP = 746 watts = 0.746 kilowatts.)

6.20* Danford Company, a farm-equipment manufacturer, currently produces 20,000 units of gas filters for use in its lawnmower production annually. The following costs are reported based on the previous year's production:

Item	Expense ($)
Direct materials	$60,000
Direct labor	180,000
Variable overhead (power and water)	135,000
Fixed overhead (light and heat)	70,000
Total cost	$445,000

It is anticipated that gas-filter production will last 5 years. If the company continues to produce the product in-house, the annual direct material costs will increase at the rate of 5%. (For example, the annual material costs during the first production year will be $63,000.) The direct labor will increase at the rate of 6% per year. The variable overhead costs would increase at the rate of 3%, but the fixed overhead would remain at the current level over the next 5 years. Tompkins Company has offered to sell Danford 20,000 units of gas filters for $25 per unit. If Danford accepts the offer, some of the manufacturing facilities currently used to manufacture the gas filter could be rented to a third party at annual rental of $35,000. Additionally, $3.5 per unit of the fixed overheard applied to gas-filter production would be eliminated. The firm's interest rate is known to be 15%. What is the unit cost of buying the gas filter from the outside source? Should Danford accept Tompkins' offer, and why?

6.21 Sentech Environmental Consulting (SEC), Inc., designs plans and specifications for asbestos abatement (removal) projects involving public, private, and governmental buildings. Currently, SEC must also conduct an air test before allowing the reoccupancy of a building from which asbestos has been removed. SEC sends the air-test sample to a subcontracted laboratory for analysis by a transmission electron microscope (TEM). In subcontracting the TEM analysis, SEC charges its client $100 above the subcontractor's analysis fee. The only expenses in this system are the costs of shipping the air-test samples to the subcontractor and the labor involved in this shipping. As business grows, SEC needs to consider either continuing to subcontract the TEM analysis to outside companies or developing its own TEM laboratory. With recent government regulations requiring the removal of asbestos from buildings, SEC expects about 1000 air-sample testings per year over 8 years. The firm's MARR is known to be 15%.

- **Subcontract option:** The client is charged $400 per sample, which is $100 above the subcontracting fee of $300. Labor expenses are $1500 per year, and the shipping expenses are estimated to be $0.50 per sample.
- **TEM purchase option:** The purchase and installation cost for the TEM is $415,000. The equipment would last for 8 years with no salvage value. The design and renovation cost is estimated to be $9500. The client is charged $300 per sample based on the current market price. One full-time manager and two part-time technicians are expected to be needed to operate the laboratory. Their combined annual salaries are $100,000. Material required to operate the lab includes carbon rods, copper grids, filter equipment, and acetone. These annual material costs are estimated at $6000. Utility costs, operating and maintenance costs, and indirect labor needed to maintain the lab are estimated to be $18,000

per year. The extra income tax expenses would be $20,000.

(a) Determine the cost per air-sample test by the TEM (in-house).
(b) What is the required number of air samples per year to make two options equivalent?

6.22* A company is currently paying an employee $0.25 per kilometre to drive a personal car for company business. The company is considering supplying the employee with a car, which would involve the following: purchase of the car for $15,000 with an estimated 3-year life, a net salvage value of $5000, taxes and insurance costing $500 per year, and operating and maintenance expenses of $0.10 per kilometre. If the interest rate is 10% and the employee's travels are anticipated to be 12,000 kilometres annually, what is the equivalent cost per kilometre (without considering any income tax)?

6.23 An electrical automobile can be purchased for $25,000. It is estimated to have a life of 12 years with annual travel of 20,000 kilometres. A new set of batteries will have to be purchased every 3 years at a cost of $3000. Annual maintenance to the vehicle is estimated to be $700 per year. The cost of recharging the batteries is estimated to be $0.015 per kilometre. The salvage value of the batteries and the vehicle at the end of 12 years is estimated to be $2000. Consider the MARR to be 7%. What is the cost per kilometre to own and operate this vehicle, based on the above estimates? The $3000 cost of the batteries is a net value with the old batteries traded in for the new ones.

6.24 A 40-kilowatt generator is estimated to cost $30,000 completely installed and ready to operate. The annual maintenance for this machine is estimated to be $500. If the value of the energy generated is considered to be $0.08 per kilowatt-hour, how many hours each year will it take to generate sufficient energy to equal the purchase price? Consider the MARR to be 9%. Also consider the salvage value of the machine to be $2000 at the end of its estimated life of 15 years. Assume that the annual energy generated is estimated to be 100,000 kilowatt-hours. What is the annual worth of this machine? How long will it take before this machine becomes profitable?

6.25 A large university facing severe parking problems on its campus is considering constructing parking decks off campus. Then, using a shuttle service, students could be picked up at the off-campus parking deck and quickly transported to various locations on campus. The university would charge a small fee for each shuttle ride, and the students could quickly and economically travel to their classes. The funds raised by the shuttle would be used to pay for the trolleys, which cost about $150,000 each. The trolley has a 12-year service life with an estimated salvage value of $3000. To operate each trolley, the following additional expenses must be considered:

Item	Annual Expenses
Driver	$25,000
Maintenance	7,000
Insurance	2,000

If students pay 10 cents for each ride, determine the annual ridership per trolley (number of shuttle rides per year) required to justify the shuttle project, assuming an interest rate of 6%.

6.26 The following cash flows represent the potential annual savings associated with two different types of production processes, each requiring an investment of $12,000.

n	Process A	Process B
0	−$12,000	−$12,000
1	9,120	6,350
2	6,840	6,350
3	4,560	6,350
4	2,280	6,350

Assuming an interest rate of 15%,

(a) Determine the equivalent annual savings for each process.
(b) Determine the hourly savings for each process if there were 2000 hours of operation each year.
(c) Which process should be selected?

6.27 Eradicator Food Prep, Inc., has invested $7 million to construct a food irradiation plant. This technology destroys organisms that cause spoilage and disease, thus extending the shelf life of fresh foods and the distances over which it can be shipped. The plant can handle about 200,000 kg of produce in an hour, and it will be operated for 3600 hours a year. The net expected operating and maintenance costs would be $4 million per year. The plant is expected to have a useful life of 15 years, with a net salvage value of $700,000. The firm's interest rate is 15%.

(a) If investors in the company want to recover the plant investment within 6 years of operation (rather than 15 years), what would be the equivalent after-tax annual revenues that must be generated?
(b) To generate annual revenues determined in part (a), what minimum processing fee per kilogram should the company charge to their producers?

6.28 The local government of Grand Manan Island is completing plans to build a desalination plant. A modern desalination plant could produce fresh water from seawater for $4 per cubic metre. On the island, the cost from natural sources is about the same as for desalting. The $3 million plant can produce 160m³ of fresh water a day (enough to supply 295 households daily), more than a quarter of the island's total needs. The desalination plant has an estimated service life of 20 years with negligible salvage value. The annual operating and maintenance costs would be about $250,000. Assuming an effective annual interest rate of 10% and monthly cash flows, what should be the minimum monthly water bill for each household?

6.29 A utility firm is considering building a 50-megawatt geothermal plant that generates electricity from naturally occurring underground heat. The binary geothermal system will cost $85 million to build and $6 million (including any income-tax effect) to operate per year. (Fuel costs are negligible compared to those from a conventional fossil fuel plant.) The geothermal plant is to last for 25 years. At that time, the expected salvage value will be about the same as the cost to remove the plant. The plant will be in operation for 70% (plant utilization factor) of the year (or 70% of 8760 hours per year). If the firm's MARR is 14% per year, determine the cost of generating electricity per kilowatt-hour.

6.30 A corporate executive jet with a seating capacity of 20 has the following cost data:

Item	Cost
Initial cost	$12,000,000
Service life	15 years
Salvage value	$2,000,000
Crew costs per year	$225,000
Fuel cost per kilometre	$0.67
Landing fee	$250
Maintenance per year	$237,500
Insurance cost per year	$166,000
Catering per passenger trip	$75

The company flies three round trips from Montreal to London per week, a distance of 5200 kilometres one way. How many round-trip passengers must be carried on an average trip in order to justify the use of the jet if the first-class round-trip fare is $4300? The firm's MARR is 15%.

6.31 A continuous electric current of 2000 amperes is to be transmitted from a power plant to a substation located 60 metres away. Copper conductors can be installed for $13.00 per kilogram, will have an estimated life of 25 years, and can be salvaged for $2.17 per kilogram. Power loss from each conductor will be inversely proportional to the cross-sectional area of the conductors and may be expressed as $\dfrac{41.378}{A}$ kilowatt, where A is given in square centimetres. The cost of energy is $0.0825 per kilowatt-hour, the interest rate is 11%, and the density of copper is 8894 kg/m³.

(a) Calculate the optimum cross-sectional area of the conductor.
(b) Calculate the annual equivalent total cost for the value obtained in part (a).
(c) Graph the two individual cost factors and the total cost as a function of cross-sectional area A and discuss the impact of increasing energy cost on the optimum obtained in part (a).

6.32* Inland Trucking Company is considering the replacement of a 500-kilogram-capacity forklift truck. The truck was purchased 3 years ago at a cost of $15,000. The diesel-operated forklift truck was originally expected to have a useful life of 8 years and a zero estimated salvage value at the end of that period. The truck has not been dependable and is frequently out of service while awaiting repairs. The maintenance expenses of the truck have been rising steadily and currently amount to about $3000 per year. The truck could be sold for $6000. If retained, the truck will require an immediate $1500 overhaul to keep it in operable condition. This overhaul will neither extend the originally estimated service life nor will

it increase the value of the truck. The updated annual operating costs, engine overhaul cost, and market values over the next 5 years are estimated as follows:

n	O&M	Depreciation	Engine Overhaul	Market Value
-3				
-2		$3000		
-1		4800		
0		2880	$1500	$6000
1	$3000	1728		4000
2	3500	1728		3000
3	3800	864		1500
4	4500	0		1000
5	4800	0	5000	0

A drastic increase in costs during the fifth year is expected due to another overhaul, which will be required to keep the truck in operating condition. The firm's MARR is 15%.

(a) If the truck is to be sold now, what will be its sunk cost?
(b) What is the opportunity cost of not replacing the truck now?
(c) What is the annual equivalent cost of owning and operating the truck for 2 more years?
(d) What is the annual equivalent cost of owning and operating the truck for 5 years?

6.33 Halifax Machine Tool Company is considering replacing one of its CNC machines with one that is newer and more efficient. The firm purchased the CNC machine 10 years ago at a cost of $135,000. It had an expected economic life of 12 years at the time of purchase and an expected salvage value of $12,000 at the end of the 12 years. The original salvage estimate is still good, and the machine has a remaining economic life of 2 years. The firm can sell this old machine now to another firm in the industry for $30,000. A new machine can be purchased for $165,000, including installation costs. It has an estimated useful (economic) life of 8 years. The new machine is expected to reduce cash operating expenses by $30,000 per year over its 8-year economic life. At the end of its useful life, the machine is estimated to be worth only $5000. The company has a MARR of 12%.

(a) If you decided to retain the old machine, what is the opportunity (investment) cost of retaining the old asset?

(b) Compute the cash flows associated with retaining the old machine in years 1 to 2.

(c) Compute the cash flows associated with purchasing the new machine in years 1 to 8 (use the opportunity cost concept).

(d) If the firm needs the service of these machines for an indefinite period and no technology improvement is expected in future machines, what will be your decision?

6.34 Air Links, a commuter airline company, is considering the replacement of one of its baggage loading/unloading machines with a newer and more efficient one.

- The old machine has a remaining economic life of 5 years. The salvage value expected from scrapping the old machine at the end of 5 years is zero, but the company can sell the machine now to another firm in the industry for $10,000.

- The new baggage handling machine has a purchase price of $120,000 and an estimated economic life of 7 years. It has an estimated salvage value of $30,000 and is expected to realize economic savings on electric power usage, labor, and repair costs and also to reduce the amount of damaged luggage. In total, an annual savings of $50,000 will be realized if the new machine is installed. The firm uses a 15% of MARR. Using the opportunity cost approach,

(a) What is the initial cash outlay required for the new machine?

(b) What are the cash flows for the defender in years 0 to 5?

(c) Should the airline purchase the new machine?

6.35 Winnipeg Medico purchased a digital image processing machine 3 years ago at a cost of $50,000. The machine had an expected life of 8 years at the time of purchase and an expected salvage value of $5000 at the end of the 8 years. The old machine has been slow at handling the increased business volume, so management is considering replacing the machine. A new machine can be purchased for $75,000, including installation costs. Over its 5-year life, the machine will reduce cash operating expenses by $30,000 per year. Sales are not expected to change. At the end of its useful life, the machine is estimated to be worthless. The old machine can be sold today for $10,000. The firm's interest rate for project justification is known to be 15%. The firm does not expect a better machine (other than the current challenger) to be available for the next 5 years. Assuming that the economic service life for the new machine and the remaining useful life for the old machine are 5 years,

(a) Determine the cash flows associated with each option (keeping the defender versus purchasing the challenger).

(b) Should the company replace the defender now?

6.36 The Northwest Manufacturing Company is currently manufacturing one of its products on a hydraulic stamping press machine. The unit cost of the product is $12, and in the past year, 3000 units were produced and sold for $19 each. It is expected that both the future demand of the product and the unit price will remain steady at 3000 units per year and $19 per unit. The old machine has a remaining useful life of

3 years. The old machine could be sold on the open market now for $5500. Three years from now the old machine is expected to have a salvage value of $1200. The new machine would cost $36,500, and the unit manufacturing cost on the new machine is projected to be $11. The new machine has an expected economic life of 5 years and an expected salvage of $6300. The appropriate MARR is 12%. The firm does not expect a significant improvement in technology, and it needs the service of either machine for an indefinite period of time.

(a) Compute the cash flows over the remaining useful life, if the firm decides to retain the old machine.
(b) Compute the cash flows over the economic service life, if the firm decides to purchase the machine.
(c) Should the new equipment be acquired now?

6.37* A firm is considering replacing a machine that has been used for making a certain kind of packaging material. The new improved machine will cost $31,000 installed and will have an estimated economic life of 10 years with a salvage value of $2500. Operating costs are expected to be $1000 per year throughout its service life. The old machine in use had an original cost of $25,000 four years ago, and at the time it was purchased, its service life (physical life) was estimated to be 7 years with a salvage value of $5000. The old machine has a current market value of $7700. If the firm retains the old machine further, its updated market values and operating costs for the next 4 years will be as follows:

Year End	Market Value	Operating Costs
0	$7700	
1	4300	$3200
2	3300	3700
3	1100	4800
4	0	5850

The firm's minimum attractive rate of return is 12%.

(a) Working with the updated estimates of market values and operating costs over the next 4 years, determine the remaining economic life of the old machine.
(b) Determine whether it is economical to make the replacement now.

6.38 A machine has a first cost of $10,000. The salvage values and annual operating costs are provided over its useful life as follows.

Year End	Salvage Value	Operating Costs
1	$5300	$1500
2	3900	2100
3	2800	2700
4	1800	3400
5	1400	4200
6	600	4900

(a) Determine the economic life if the MARR is 15%.
(b) Determine the economic life if the MARR is 28%.

6.39 Given the following data:

$$I = \$20,000$$
$$S_n = 12,000 - 2000n$$
$$O\&M_n = 3000 + 1000\,(n-1),$$

where
$$I = \text{Asset purchase price}$$
$$S_n = \text{Market value at the end of year } n$$
$$O\&M_n = \text{O\&M cost during year } n.$$

(a) Determine the economic service life if $i = 10\%$.
(b) Determine the economic service life if $i = 25\%$
(c) Assume $i = 0$: Determine the economic life mathematically (i.e., the calculus technique for finding the minimum point as described in this chapter).

6.40 The University Resume Service has just invested $8000 in a new desktop publishing system. From past experience, the owner of the company estimates its after-tax cash returns as follows:

$$A_n = \$8000 - \$4000(1 + 0.15)^{n-1}$$
$$S_n = \$6000(1 - 0.3)^n;$$

where A_n stands for net after-tax cash flows from operation during period n, and S_n stands for the after-tax salvage value at the end of period n.

(a) If the company's MARR is 12%, compute the economic service life of the desktop operating system.
(b) Explain how the economic service life varies with the interest rate.

6.41 A special purpose machine is to be purchased at a cost of $15,000. The following table shows the expected annual operating and maintenance cost and the salvage values for each year of service.

Year of Service	O&M Costs	Market Value
1	$2,500	$12,000
2	3,200	8,100
3	5,300	5,200
4	6,500	3,500
5	7,800	0

(a) If the interest rate is 10%, what is the economic service life for this machine?
(b) Repeat (a) above using $i = 15\%$.

6.42* Quintana Electronic Company is considering the purchase of new robot-welding equipment to perform operations currently being performed by less efficient equipment. The new machine's purchase price is $150,000, delivered and installed. A Quintana industrial engineer estimates that the new equipment will produce savings of $30,000 in labor and other direct costs annually, as compared with the present equipment. He estimates the proposed equipment's economic life at 10 years, with a zero salvage value. The present equipment is in good working order and will last, physically, for at least 10 more years. Quintana uses a 10% discount rate for analysis.

(a) Assuming that the present equipment has a zero current market value, should the company buy the proposed equipment?
(b) Assume that the new equipment will save only $15,000 a year, but that its economic life is expected to be 12 years. If other conditions are as described in (a) above, should the company buy the proposed equipment?

6.43 Quintana Company decided to purchase the equipment described in Problem 6.45 (hereafter called "Model A" equipment). Two years later, even better equipment (called "Model B") comes onto the market, which makes Model A completely obsolete, with no resale value. The Model B equipment costs $300,000 delivered and installed, but it is expected to result in annual savings of $75,000 over the cost of operating the Model A equipment. The economic life of Model B is estimated to be 10 years with a zero salvage value. The interest rate is 10%.

(a) What action should the company take?
(b) If the company decides to purchase the Model B equipment, a mistake must have been made, because good equipment, bought only 2 years previously, is being scrapped. How did this mistake come about?

6.44 A special-purpose turnkey stamping machine was purchased 4 years ago for $20,000. It was estimated at that time that this machine would have a life of 10 years and a salvage value of $3000, with a removal cost of $1500. These estimates are still good. This machine has annual operating costs of $2000. A new machine, which is more efficient, will reduce the operating costs to $1000, but it will require an investment of $20,000 plus $1000 for installation. The life of the new machine is estimated to be 12 years with a salvage of $2000 and a removal cost of $1500. An offer of $6000 has been made for the old machine, and the purchaser is willing to pay for removal of the machine. Find the economic advantage of replacement or of continuing with the present machine. State any assumptions that you make. (Assume the MARR = 8%.)

6.45 A 5-year-old defender has a current market value of $4000, and has costs of $3000 this year. These operating costs will increase by $1500 per year. Future market values are expected to decline by $1000 per year. Assume that the economic life of the defender is 3 years. The challenger costs $6000 and has O&M costs of $2000 per year, increasing by $1000 per year. The machine has an economic life of 3 years, and the salvage value at the end of 3 years is expected to be $2000. The MARR is 15%.

(a) Determine the annual cash flows for retaining the old machine for 3 years.
(b) Determine if now is the time to replace the old machine. First show the annual cash flows for the challenger.

6.46 Greenleaf Company is considering the purchase of a new set of air-electric quill units to replace an obsolete one. The machine being used for the operation has a market value of zero; however, it is in good working order, and it will last physically for at least an additional 5 years. The new quill units will perform the operation with so much more efficiency that the firm's engineers estimate that labor, material, and other direct costs will be reduced by $3000 a year if it is installed. The new set of quill units costs $10,000, delivered and installed, and its economic life is estimated to be 5 years with zero salvage value. The firm's MARR is 10%. The two options are to be compared over the five-year period.

(a) What is the investment required to keep the old machine?
(b) Compute the cash flow to use in the analysis for each option.
(c) If the firm uses the internal rate of return criterion, should the firm buy the new machine on that basis?

6.47 Wu Lighting Company is considering the replacement of an old, relatively inefficient vertical drill machine that was purchased 7 years ago at a cost of $10,000. The machine had an original expected life of 12 years and a zero estimated salvage value at the end of that period. Its current market value is $2000. The divisional manager reports that a new machine can be bought and installed for $12,000 which, over its 5-year life, will expand sales from $10,000 to $11,500 a year and, furthermore, will reduce labor and raw materials usage sufficiently to cut annual operating costs from $7000 to $5000. The new machine has an estimated salvage value of $2000 at the end of its 5-year life. The MARR is 15%. The two options are to be compared over a five-year period.

(a) Should the new machine be purchased now?
(b) What purchase price of the new machine would make the two options equivalent?

6.48 Advanced Robotics Company is faced with the prospect of replacing its old call-switching systems, which have been used in the company's headquarters for 10 years. This particular system was installed at a cost of $100,000, and it was assumed that it would have a 15-year life with no appreciable salvage value. The current annual operating costs are $20,000 for this old system, and these costs would be the same for the rest of its life. A sales representative from North Central Bell is trying to sell this company a computerized switching system. The new system would require an investment of $200,000 for installation. The economic life of this computerized system is estimated to be 10 years with a salvage value of $18,000, and the system will reduce annual operating costs to $5000. No detailed agreement has been made with the sales representative about the disposal of the old system. Determine the ranges of resale value associated with the old system that would justify installation of the new system at a MARR of 14%.

6.49 Five years ago a conveyor system was installed in a manufacturing plant at a cost of $35,000. It was estimated that the system, which is still in operating condition, would have a useful life of 8 years with a salvage value of $3000. If the firm continues to operate the system, its estimated market values and operating costs for the next 3 years are as follows.

Year End	Market Value	Operating Costs
0	$8000	
1	5200	$6000
2	3500	7000
3	1200	10000

A new system can be installed for $43,500; it would have an estimated economic life of 10 years with a salvage value of $3500. Operating costs are expected to be $1500 per year throughout its service life. This firm's MARR is 18%.

(a) Decide whether to replace the existing system now.

(b) If the decision is to not replace the existing system now, when should replacement occur?

6.50 A company is currently producing chem-icals in a process plant installed 10 years ago at a cost of $100,000. It was assumed that the process would have a 20-year life with a zero salvage value. The current market value of this equipment is $60,000, and the initial estimate of its economic life is still good. The annual operating costs associated with this process are $18,000. A sales representative from Northern Instrument Company is trying to sell a new chemical compound-making process to the company. This new process will cost $200,000, have an economic life of 10 years with a salvage value of $20,000, and reduce annual operating costs to $4000. Assuming the company desires a return of 12% on all investments, should it invest in the new process?

6.51 Eight years ago a lathe was purchased for $45,000. Its operating expenses were $8700 per year. An equipment vendor offers a new machine for $53,500. An allowance of $8500 would be made for the old machine on purchase of the new one. The old machine is expected to be scrapped at the end of 5 years. The new machine's economic service life is 5 years with a salvage value of $12,000. The new machine's O&M cost is estimated to be $4200 for the first year, increasing at an annual rate of $500 thereafter. The firm's MARR is 12%. What option would you recommend?

6.52 The Toronto Taxi Cab Company has just purchased a new fleet of 1999 models. Each brand-new cab cost $25,000. From past experience, the company estimates after-tax cash returns for each cab as follows.

$$A_n = \$32,900 - 15,125(1 + 15\%)^{n-1}$$
$$S_n = \$25,000(1 - 0.35)^n$$

where A_n stands for net after-tax cash flows from operation during period n, and S_n stands for the after-tax salvage value at the end of period n. The management views the replacement process as a constant and infinite chain.

(a) If the firm's MARR is 10%, and it expects no major technological and functional change in future models, what is the optimal time period (constant replacement cycle) to replace its cabs?

(b) What is the internal rate of return for a cab if it is retired at the end of its economic service life? What is the internal rate of return for a sequence of identical cabs if each cab in the sequence is replaced at the optimal time?

6.53 Four years ago an industrial batch oven was purchased for $23,000. If sold now, the machine will bring $2000. If sold at the end of the year, it will bring in $1500. Salvage value for subsequent years will decline by 20% each year. Annual operating costs are constant at $3800. However, the remaining physical life of the defender is 5 years. A new machine will cost $50,000 with a 12-year economic life and have a $3000 salvage value. The operating cost will be $3000 as of the end of each year with the $6000 per year savings due to better quality. If the firm's MARR is 10% should the machine be purchased now?

6.54 Johnson Ceramic Company has an automatic glaze sprayer that has been used for the past 10 years. The sprayer can be used for another 10 years and will have a zero salvage value at that time. The annual operating and maintenance costs for the sprayer amount to $15,000 per year. Due to an increase in business, a new sprayer must be purchased.

- Option 1: If the old sprayer is retained, a new smaller capacity sprayer will be purchased at a cost of $48,000, and it will have a $5000 salvage value in 10 years. This new sprayer will have annual operating and maintenance costs of $12,000.
- Option 2: If the old sprayer is sold, a new sprayer of larger capacity will be

p
h
a
n
sprayer has a current market value of $6000.

Which option should be selected at MARR = 12%?

6.55 A 6-year-old CNC machine that originally cost $8000 has a current market value of $1500. If the machine is kept in service for the next 5 years, its O&M costs and salvage value are estimated as follows:

	O&M Costs		
End of Year	Operation and Repairs	Delays Due to Breakdowns	Salvage Value
1	$1300	$600	$1200
2	1500	800	1000
3	1700	1000	500
4	1900	1200	0
5	2000	1400	0

It is suggested that the machine be replaced by a new CNC machine of improved design at a cost of $6000. It is believed that this purchase will completely eliminate breakdowns and the resulting cost of delays and that operation and repair costs will be reduced by $200 a year from those incurred with the old machine. Assume a 5-year economic life for the challenger and a $1000 terminal salvage value. The firm's MARR is 12%. Should the old machine be replaced now?

6.56* The annual equivalent after-tax costs of retaining a defender machine over 4 years (physical life), or operating its challenger over 6 years (physical life), are on the next page.

If you need the service of either machine for only the next 10 years, what is the best replacement strategy? Assume a MARR of

12% and no technology improvement in future challengers.

n	Defender	Challenger
1	−$3200	−$5800
2	−2500	−4230
3	−2650	−3200
4	−3300	−3500
5		−4000
6		−5500

6.57 The after-tax annual equivalent of retaining a defender over 4 years (physical life) or operating its challenger over 6 years (physical life) are as follows. The boxed numbers indicate the maximum AE, at their economic lives.

n	Defender	Challenger
1	$13,400	$12,300
2	13,500	13,000
3	13,800	13,600
4	13,200	13,400
5		13,000
6		12,500

If you need the service of either machine only for the next 8 years, what is the best replacement strategy? Assume a MARR of 12% and no technology improvement in future challengers.

6.58 An existing asset that cost $16,000 two years ago has a market value of $12,000 today, an expected salvage value of $2000 at the end of its remaining economic life of 6 more years, and annual operating costs of $4000. A new asset under consideration as a replacement has an initial cost of $10,000, an expected salvage value of $4000 at the end of its economic life of 5 years, and annual operating costs of $2000. It is assumed that this new asset could be replaced by another one identical in every

respect after 3 years at a salvage value of $5000, if desired.

(a) By using a MARR of 11%, a 6-year study period, and PW calculations, decide whether the existing asset should be replaced by the new one.

(b) Repeat (a) above based on the AE criterion.

Level 3

6.59* Automotive engineers at Ford are considering the laser blank welding (LBW) technique to produce a windshield frame rail blank (see below). The engineers believe that the LBW as compared with the conventional sheet metal blanks would result in a significant savings as follows:

1. Scrap reduction through more efficient blank nesting on coil.
2. Scrap reclamation (weld scrap offal into a larger usable blank).

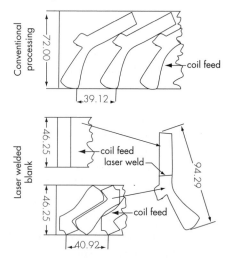

Use of a laser welded blank to provide a reduction in engineered scrap for production of a window frame rail blank. (Problem 6.59)

Based on an annual volume of 3000 blanks, Ford engineers have estimated the following financial data:

5 The example was provided by Mr. Joel M. Height of the Chevron Oil Company.

| Description | Blanking Method | |
	Conventional	Laser Blank Welding
Weight per blank (kgs/part)	63.764	34.870
Steel cost/part	$14.98	$8.19
Transportation/part	$0.67	$0.42
Blanking/part	$0.50	$0.40
Die investment	$106,480	$83,000

The LBW technique appears to achieve significant savings, so Ford's engineers are leaning toward adopting the LBW technique. Since Ford engineers have had no prior experience with the LBW, they are not sure if producing the windshield frames in-house at this time is a good strategy. For this windshield frame, it may be cheaper to use the services of a supplier, which has both the experience and the machinery for laser blanking. Ford's lack of skill in laser blanking may require that it take 6 months to get up to the required production volume. On the other hand, if Ford relies on a supplier, it can only assume that supplier labor problems will not halt production of Ford's parts. The make-or-buy decision depends on two factors: The amount of new investment that is required in laser welding, and whether additional machinery will be required for future products. Assuming an analysis of 10 years and an interest rate of 16%, recommend the best course of action. Assume also that the salvage value at the end of 10 years is estimated to be nonsignificant for either system. If Ford is considering the subcontracting option, what would be the acceptable range of contract bid (unit cost per part)?

6.60 Chevron Overseas Petroleum, Inc. entered into a 1993 joint venture agreement with the Republic of Kazakhstan, a former republic of the old Soviet Union, to develop the huge Tengiz oil field.[5] Unfortunately, the climate in the region is harsh, making it difficult to keep oil flowing. The untreated oil comes out of the ground at 46°C. Even though the pipelines are insulated, as the oil gets further from the well on its way to be processed, hydrate salts begin to precipitate out of the liquid phase as the oil cools. These hydrate salts create a dangerous condition as they form plugs in the line.

The method for preventing this trap pressure condition is to inject methanol (CH_3OH) into the oil stream. This keeps the oil flowing and prevents hydrate salts from precipitating out of the liquid phase. The present methanol loading and storage facility is a manually controlled system, with no fire protection and with a rapidly deteriorating tank that causes leaks. The scope of repairs and upgrades is extensive. The storage tanks are rusting and are leaking at the riveted joints. The manual level control system causes frequent tank overfills. There is no fire protection system, as water is not available at this site.

The present storage facility has been in service for 5 years. Due to permit requirements, upgrades are required to achieve minimum acceptable Kazakhstan standards. Upgrades, in the amount of $104,000, will extend the life of the current facility to about 10 years. However, upgrades will not completely stop the leaks. The expected spill and leak losses will amount to $5000 a year. The annual operating costs are expected to be $36,000.

As an alternative, a new methanol storage facility can be designed based on minimum acceptable international oil industry practice. The new facility, which would cost $325,000, would last about 12 years before a major upgrade would be required. With a lower risk of leaks, spills, and evaporation loss due to a more closely monitored system, the expected annual operating cost would be $12,000. However, it is believed that oil transfer technology will be such that methanol will not be necessary in 10 years. The pipeline heating and insulation systems will make methanol storage and use systems obsolete.

(a) Assume that the storage tanks (the new as well as the upgraded ones) will have no salvage values at the end of their useful lives (after considering the removal costs). If Chevron's interest rate is 20% for foreign projects, which option is a better choice?

(b) How would the decision change as you consider the risk of spills (clean-up costs) and evaporation of product related to environmental impact?

6.61 National Woodwork Company, a manufacturer of window frames, is considering replacing a conventional manufacturing system with a flexible manufacturing system (FMS). The company cannot produce rapidly enough to meet demand. Some manufacturing problems identified follow:

- The present system is expected to be useful for another 5 years, but will require an estimated $105,000 per year in operating costs including R&M, scrap, inventory space, and operators' salaries, which will increase by $10,000 each year as parts become more scarce. The current market value of the existing system is $140,000.

The proposed system will reduce, or entirely eliminate, the set-up times, and each window can be made as it is ordered by the customer. Customers phone their orders into the head office, where details are fed into the company's main computer. These manufacturing details are then dispatched to computers on the manufacturing floor, which are, in turn, connected to a computer that controls the proposed

FMS. This system eliminates the warehouse space and material handling time that are needed when using the conventional system.

Before installing the FMS, the old equipment will be removed from the job shop floor at an estimated cost of $100,000. This cost includes the needed electrical work for the new system. The proposed FMS will cost $1,200,000. The economic life of the machine is expected to be 10 years, and the salvage value is expected to be $120,000. The change in window styles has been minimal in the past few decades and is expected to continue to remain stable in the future. The total annual savings will be $664,243: $12,000 attributed to the reduction of defective windows, $511,043 from the elimination of 13 workers, $100,200 from the increase in productivity, and $41,000 from the near elimination of warehouse space and material handling. The repair and maintenance costs will be only $45,000, increasing by $2000 per year. The National Woodwork's MARR is about 15%.

(a) What assumptions are required to compare the conventional system with the FMS?

(b) With the assumptions defined in (a), should the FMS be installed now?

6.62 In 2 × 4 and 2 × 6 lumber production, significant amounts of wood are present in sideboards produced after the initial cutting of logs. Instead of processing the sideboards into wood chips for the paper mill, an "edger" is used to reclaim additional lumber, thus resulting in savings for the company. An edger is capable of reclaiming lumber by any of the following three methods: (1) removing rough edges, (2) splitting large sideboards, and (3) salvaging 2 × 4 lumber from low-quality 4 × 4 boards. Union Camp Company's engineers have discovered that a significant reduction in production costs could be achieved simply by replacing the original "edger" machine with a newer laser-controlled model.

Old Edger: The old edger was placed in service 12 years ago. Any machine scrap value would offset the removal cost of the equipment. No market exists for this obsolete equipment. The old edger needs two operators. During the cutting operation, the operator makes edger settings based on his/her own judgment. The operator has no means of determining exactly what dimension of lumber could be recovered from a given sideboard and must guess at the proper setting to recover the highest grade of lumber. Furthermore, the old edger is not capable of salvaging good-quality 2 × 4s from poor-quality 4 × 4s. The defender can continue in service for another 5 years with proper maintenance.

Current market value	$0
Current book value	0
Annual maintenance cost	$2,500 in year 1, increasing at a rate of 15% each year over the previous year's cost
Annual operating costs (labor and power)	$65,000

New Laser-Controlled Edger: The new edger has numerous advantages over its defender. These advantages include laser beams that indicate where cuts should be made to obtain the maximum yield by the edger. The new edger requires a single operator, and labor savings will be reflected in lower operating and maintenance costs of $35,000 a year. The data for the challenger is shown below.

Estimated Cost	
Equipment	$35,700
Equipment installation	21,500
Building	47,200
Conveyor modification	14,500
Electrical (wiring)	16,500
Subtotal	$135,400
Engineering	7,000
Construction management	20,000
Contingency	16,200
Total	$178,600

Useful life of new edger	10 years
Salvage value	
Building (tear down)	$0
Equipment	10% of the original cost
Annual O&M costs	$35,000

Twenty-five percent of total mill volume passed through the edger. A 12% yield improvement is expected to be realized on this production, which will result in an improvement of total mill volume of $(0.25)(0.12) = 3\%$, or an annual savings of $57,895. Should the defender be replaced now if the mill's MARR is 16%?

6.63 Rivera Industries, a manufacturer of home heating appliances, is considering the purchase of Amada Turret Punch Press, a more advanced piece of machinery, to replace its present system that uses four old presses. Currently, the four smaller presses are used (in varying sequences, depending on the product) to produce one component of a product until a scheduled time when all machines must retool to set up for a different component. Because setup cost is high, production runs of individual components are long. This results in large inventory buildups of one component, which are necessary to prevent extended backlogging while other products are being manufactured.

- The four presses in use now were purchased 6 years ago at a price of $100,000. The manufacturing engineer expects that these machines can be used for 8 more years, but they will have no market value after that. The present market value is estimated to be $40,000. The average setup cost, which is determined by the number of required labor hours times the labor rate for the old presses, is $80 per hour, and the number of setups per year expected by the production control department is 200. This yields a yearly setup cost of $16,000. The expected operating and maintenance cost for each year in the remaining life of this system is estimated as follows:

Year	Setup Costs	O&M Costs
1	$16,000	$15,986
2	16,000	16,785
3	16,000	17,663
4	16,000	18,630
5	16,000	19,692
6	16,000	20,861
7	16,000	22,147
8	16,000	23,562

These costs, which were estimated by the manufacturing engineer, with the aid of data provided by the vendor, represent a reduction in efficiency and an increase in needed service and repair over time.

- The price of the 2-year-old Amada turret punch press is $135,000 and would be paid for with cash from the company's capital fund. Additionally, the company would incur installation costs, which would total $1200. An expenditure of $12,000 would be required to recondition the press to its original condition. The reconditioning would extend the Amada's economic service life to 8 years. It would have no salvage value at that time. The cash savings of the Amada over the present system are due to the reduced setup time. The average setup cost of the Amada is $15, and the Amada would incur 1000 setups per year, yielding a yearly setup cost of $15,000. Savings due to the reduced setup time are realized because of the reduction in carrying costs associated with that level of inventory where the production run and ordering quantity are reduced. The Accounting Department has estimated that probably $36,000 per year could be saved by shortening production runs. The operating and maintenance costs of the Amada as estimated by the manufacturing engineer are similar, but somewhat less, than the O&M costs for the present system.

Year	Setup Costs	O&M Costs
1	$15,000	$11,500
2	15,000	11,950
3	15,000	12,445
4	15,000	12,990
5	15,000	13,590
6	15,000	14,245
7	15,000	14,950
8	15,000	15,745

The reduction in the O&M costs is due to the age difference of the machines and the reduced power requirements of the Amada.

- If Rivera Industries delays the replacement of the current four presses for another year, the second-hand Amada machine will no longer be available, and the company will have to buy a brand-new machine at an installed price of $200,450. The expected setup costs would be the same as those for the second-hand machine, but the annual operating and maintenance costs would be about 10% lower than the estimated

O&M costs for the second-hand machine. Similar inventory savings will be realized. The expected economic service life of the brand-new press would be 8 years with no salvage value. The market value of the defender a year later is estimated to be $30,000.

Rivera's MARR is 12%.

(a) Assuming that the company would need the service of either press for an indefinite period, what would you recommend?
(b) Assuming that the company would need the press for only 5 more years, what would you recommend?

CHAPTER 7

Depreciation

S uppose that you are a design engineer employed by a firm that manufactures die cast automobile parts. To enhance the firm's competitive position in the marketplace, management has decided to purchase a computer-aided design system that features 3-D solid modeling and full integration with sophisticated simulation and analysis capabilities. As part of the design team, you are excited at the prospect that the design of die cast molds, the testing of product variations, and the simulation of processing and service conditions can be made highly efficient by use of this state-of-the-art system. In fact, the more you think about it, the more you wonder why this purchase wasn't made earlier.

Now ask yourself, how does the cost of this system affect the financial position of the firm? In the long run, the system promises to create greater wealth for the organization by improving design productivity, increasing product quality, and cutting down design lead time. In the short run, however, the high cost of this system will have a negative impact on the organization's "bottom line," because it involves high initial costs that are only gradually recovered from the benefits of the system.

Another consideration should come to mind. This state-of-the-art equipment will inevitably wear out over time, and even if its productive service extends over many years, the cost of maintaining its high level of functioning will increase as the individual pieces of hardware wear out and need to be replaced. Of even greater concern is the question of how long this system will be "state-of-the-art." When will the competitive advantage the firm has just acquired become a competitive disadvantage through obsolescence?

One of the facts of life that organizations must deal with and account for is that fixed assets lose their value—even as they continue to function and contribute to the engineering projects that use them. This loss of value, called **depreciation**, can involve deterioration and obsolescence.

The main function of **depreciation accounting** is to account for the cost of fixed assets in a pattern that matches their decline in value over time. The cost of the CAD system we have just described, for example, will be allocated over several years in the firm's financial statements so that its pattern of costs roughly matches its pattern of service. In this way, as we shall see, depreciation accounting enables the firm to stabilize the statements of financial position that it distributes to stockholders and the outside world.

On a project level, engineers must be able to assess how the practice of depreciating fixed assets influences the investment value of a given project. To do this, they need to estimate the allocation of capital costs over the life of the project, which requires understanding the conventions and techniques that accountants use to depreciate assets. This chapter overviews the conventions and techniques of asset depreciation.

We begin by discussing the nature and significance of depreciation, distinguishing its general economic definition from the related but different accounting view of depreciation. We then focus our attention almost exclusively on the methods that accountants use to allocate depreciation expenses, and the rules and laws that govern asset depreciation under the Canadian income tax system. A knowledge of these rules will prepare you to apply them in assessing the depreciation of assets acquired in engineering projects.

Finally, we turn our attention to the subject of depletion, which utilizes similar ideas, but specialized techniques, to allocate the cost of the depletion of natural resource assets.

7.1 Asset Depreciation

Fixed assets such as equipment and real estate are economic resources that are acquired to provide future cash flows. Generally, **depreciation** can be defined as the gradual decrease in utility of fixed assets with use and time. While this general definition does not adequately capture the subtleties inherent in a more specific definition of depreciation, it does provide us with a starting point for examining the variety of underlying ideas and practices that are discussed in this chapter. Figure 7.1 will serve as a road map for understanding the distinctions inherent in the meaning of depreciation we will explore in this chapter.

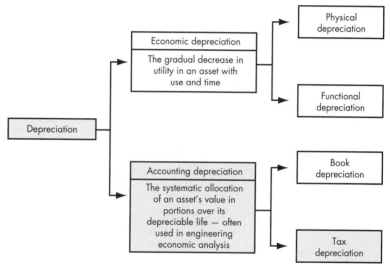

Figure 7.1   Classification of depreciation

7.1.1   Economic Depreciation

This chapter is primarily concerned with accounting depreciation, which is the form of depreciation that provides an organization with the information needed to assess its financial position and calculate its income taxes. It would also be useful, however, to discuss briefly the economic ideas upon which accounting depreciation is based. In the course of the discussion, we will develop a precise definition of economic depreciation that will help us distinguish between various conceptions of depreciation.

If you have ever owned a car, you are probably familiar with the term depreciation as it is used to describe the decreasing value of your vehicle (see Example 6.6). Because a car's reliability and appearance usually decline with age, the vehicle is worth less with each passing year. You can calculate the economic depreciation accumulated for your car by subtracting the current market value or "black book" value of the car from the price you originally paid for it. We can define **economic depreciation** as follows:

$$\text{Economic depreciation} = \text{Purchase price} - \text{market value}$$

Physical and functional depreciation are categories of economic depreciation.

Physical depreciation can be defined as a reduction in an asset's capacity to perform its intended service due to physical impairment. Physical depreciation can occur in any fixed asset in the form of (1) deterioration from interaction with the environment, including such agents as corrosion, rotting, and other chemical changes; and (2) wear and tear from use. Physical depreciation leads to a decline in performance and high maintenance costs.

Functional depreciation occurs as a result of changes in the organization or in technology that decrease or eliminate the need for an asset. Examples of functional depreciation include obsolescence attributable to advances in technology, a declining need for the services performed by an asset, or the inability to meet increased quantity and/or quality demands.

The measurement of economic depreciation does not require that an asset be sold: the market value of an asset can be closely estimated without actually testing its value in the marketplace. The need to have a precise scheme for recording the ongoing decline in the value of an asset as a part of the accounting process leads us to an exploration of how organizations account for depreciation.

7.1.2 Accounting Depreciation

The acquisition of fixed assets is an important activity for a business organization. This is true whether the organization is starting up or acquiring new assets to remain competitive. Like other disbursements, the cost of these fixed assets must be recorded as expenses on a firm's balance sheet and income statement. However, unlike costs such as maintenance, material, and labor, the costs of fixed assets are not treated simply as expenses to be accounted for in the year that they are acquired, rather, these assets are **capitalized**; that is, their costs are distributed by subtracting them as expenses from gross income—one part at a time over a number of periods. The systematic allocation of the initial cost of an asset in parts over a time, known as its **depreciable life**, is what we mean by **accounting depreciation**. Because accounting depreciation is the standard of the business world, we sometimes refer to it more generally as **asset depreciation.**

Accounting depreciation is based on the **matching concept**: A fraction of the cost of the asset is chargeable as an expense in each of the accounting periods in which the asset provides service to the firm, and each charge is meant to be a percentage of the whole cost which "matches" the percentage of the value utilized in the given period. The matching concept suggests that the accounting depreciation allowance generally reflects to some extent the actual economic depreciation of the asset.

In engineering economic analysis, we exclusively use the concept of accounting depreciation. This is because accounting depreciation provides a basis for determining the income taxes associated with any project undertaken. As we will see in Chapter 8, depreciation has a significant influence on the income and cash position of a firm. With a clear understanding of the concept of depreciation, we will be able to appreciate fully the importance of utilizing depreciation as a means to maximize the value both of engineering projects and of the organization as a whole.

7.2 Factors Inherent to Asset Depreciation

The process of depreciating an asset requires that we make several preliminary determinations: (1) what is the cost of the asset? (2) what is the asset's value at the end of its useful life? (3) what is the depreciable life of the asset? and, finally, (4) what method of depreciation do we choose? In this section, we will discuss each of these factors.

7.2.1 Depreciable Property

As a starting point, it is important to recognize what constitutes a **depreciable asset**, that is, a property for which a firm may take depreciation deductions against income. For the purposes of Canadian taxes, any depreciable property has the following characteristics:

1. It must be used in business or held for production of income.

2. It must have a definite useful life, and that life must be longer than 1 year.

3. It must be something that wears out, decays, gets used up, becomes obsolete, or loses value from natural causes.

Depreciable property includes buildings, machinery, equipment, and vehicles. Inventories are not depreciable property because these are held primarily for sale to customers in the ordinary course of business. If an asset has no definite useful life, the asset cannot be depreciated. For example, *you cannot depreciate land.*[1]

As long as assets meet the conditions listed above, depreciation may be claimed by any business enterprise, whether a sole proprietorship, partnership, or corporation. For example, an individual who periodically uses her car for business, or part of her house as a work space, may claim depreciation on a prorated portion of the capital cost of the property.

---

[1] This also means that you cannot depreciate the cost of clearing, grading, planting, and landscaping. All such expenses are considered part of the cost of the land. (An exception is land acquired as part of a timber limit; see Section 7.6.3.)

7.2.2   Cost Basis

The **cost basis** of an asset represents the total cost that is claimed as an expense over an asset's life, i.e., the sum of the annual depreciation expenses.

Cost basis generally includes the actual cost of an asset and all other incidental expenses, such as freight, site preparation, installation, and legal, accounting, and engineering fees. This total cost, rather than the cost of the asset only, must be the depreciation basis charged as an expense over an asset's life.

EXAMPLE 7.1   **Cost basis**

Lanier Corporation purchased an automatic hole-punching machine priced at $62,500. Lanier also paid the inbound transportation charges of $725 on the new machine as well as labor cost of $2150 to install the machine in the factory. Lanier also had to prepare the site at the cost of $3500 before installation. Determine the cost basis for the new machine for depreciation purposes.

Solution

Given: Invoice price = $62,500, freight = $725, installation cost = $2150, and site preparation = $3500
Find: The cost basis

The cost of machine that is applicable for depreciation is computed as follows:

Cost of new hole-punching machine	$62,500
Freight	725
Installation labor	2,150
Site preparation	3,500
Cost of machine (cost basis)	$68,875

Comments: Why do we include all the incidental charges relating to the acquisition of a machine in its cost? Why not treat these incidental charges as expenses of the period in which the machine is acquired? The matching of costs and revenue is the basic accounting principle. Consequently, the total costs of the machine should be viewed as an asset and allocated against the future revenue that the machine will generate. All costs incurred in acquiring the machine are costs of the services to be received from using the machine.

If the asset is purchased by trading in a similar asset, the difference between the **book value** (cost basis minus the total accumulated depreciation) and trade-in allowance must be considered in determining the cost basis for the new asset. If the trade-in allowance exceeds the book value, the difference (known as **unrecognized** gain) needs to be subtracted from the cost basis of the new asset. If the opposite is true (**unrecognized loss**), it should be added to the cost basis for the new asset.

EXAMPLE 7.2

Cost basis with trade-in allowance

In Example 7.1, suppose Lanier purchased the hole-punching press by trading in a similar machine and paying cash for the remainder. The trade-in allowance was $5000 and the book value of the hole-punching machine that was traded in was $4000.

Old hole-punching machine (book value)	$4,000
Less: Trade-in allowance	5,000
Unrecognized gains	$1,000
Cost of new hole-punching machine	$62,500
Less: Unrecognized gains	(1,000)
Freight	725
Installation labor	2,150
Site preparation	3,500
Cost of machine (cost basis)	$67,875

7.2.3 Useful Life and Salvage Value

As we discussed in Chapter 5, the useful life is an estimate of the duration over which the asset is expected to fulfill its intended service. Referring to Figure 7.1, the useful life may be based on the asset's anticipated physical longevity, which depends on many factors, such as its durability, severity of planned usage, and harshness of its environment. For other types of property, functionality may be the limiting factor, particularly for rapidly changing technologies such as computers, electronics, and communication equipment. Predicting the useful life of an asset is required for some depreciation computations and most economic analyses.

The salvage value is an asset's value at the end of its life; it is the amount eventually recovered through sale, trade-in, or salvage, net of all costs incurred for disposal and restoring the site to its original condition if required. Since there are many unknown factors which may affect this value, a rough estimate must be made in advance to perform an economic analysis. If the demolition or removal costs are expected to exceed receipts from selling the used asset, a negative salvage value is applied. For long-lived assets, a zero salvage value is often assumed for simplicity, as any small residual value may be negated by the disposal costs. Often, the estimated salvage value will not equal the projected book value at the end of the asset's useful life. Since accounting depreciation allocates the cost of the asset over time to match its anticipated usage, the book value is not intended to equal the salvage value. Adjustments must be made to account for differences between the salvage value (estimated, or actual at the time of disposal) and the book value, as shown in the following sections and Chapter 8.

7.2.4   Depreciation Methods: Book and Tax Depreciation

As mentioned earlier, most firms calculate depreciation in two different ways, depending on whether the calculation is (1) intended for financial reports (**book depreciation method**), such as for the balance sheet or income statement or (2) for the Canada Customs and Revenue Agency (CCRA: formerly called Revenue Canada), for the purpose of calculating taxes (**capital cost allowance**, or **CCA**). In Canada, this distinction is totally legitimate under Revenue Canada regulations, as it is in many other countries. Calculating depreciation differently for financial reports and for tax purposes allows for the following benefits:

- It enables firms to report depreciation to stockholders and other significant outsiders based on the matching concept. Therefore the actual loss in value of the assets is generally reflected.

- It allows firms to benefit from the tax advantages of depreciating assets more quickly than would be possible using the matching concept. In many cases, capital cost allowance permits firms to defer paying income taxes. This does not mean that they pay less tax overall, because the total depreciation expense accounted for over time is the same in either case. However, because CCA methods generally result in greater depreciation in earlier years than do common book depreciation methods, the tax benefit of depreciation is enjoyed earlier, and firms generally pay lower taxes in the initial years of an investment project. Typically this leads to a better cash position in early years, the added cash leading to greater future wealth because of the time value of the funds.

As we proceed through the chapter, we will make increasing use of the distinction between depreciation accounting for financial reporting and depreciation accounting used for income tax calculation. Now that we have established the context for our interest in both book depreciation and CCA for tax purposes, we can survey the different methods with an accurate perspective.

7.3 Book Depreciation Methods

Three different methods are commonly used to calculate the periodic depreciation allowances. These are the (1) straight-line method, (2) accelerated methods, and (3) the units-of-production method. In engineering economic analysis, we are primarily interested in depreciation in the context of income tax computation. Nonetheless, a number of reasons make the study of book depreciation methods useful. First, capital cost allowance is based largely on the same principles that are used in certain book depreciation methods. Second, firms continue to use book depreciation methods for financial reporting to stockholders and outside parties. Finally, some resource allowances discussed in Section 7.6 are based on some of these three book depreciation methods.

7.3.1 Straight-Line Method

The **straight-line method (SL)** of depreciation interprets a fixed asset as one that provides its services in a uniform fashion. The asset provides an equal amount of service in each year of its useful life. SL is the prevailing book depreciation method, due to its reasonable assumptions and simplicity.

The straight-line method charges, as an expense, an equal fraction of the net cost of the asset each year, as expressed by the relation

$$D_n = \frac{(P - S)}{N}, \quad (7.1)$$

where
$$
\begin{aligned}
D_n &= \text{Depreciation charge during year } n \\
P &= \text{Cost of the asset including installation expenses} \\
S &= \text{Salvage value at the end of useful life} \\
N &= \text{Useful life.}
\end{aligned}
$$

The book value of the asset at the end of n years is then defined as

Book value in a given year = Cost basis – total depreciation charges made

or

$$B_n = P - (D_1 + D_2 + D_3 + \ldots + D_n) = \frac{n(P - S)}{N} \quad (7.\_)$$

EXAMPLE 7.3 **Straight-line depreciation**

Consider the following automobile data:

$$
\begin{aligned}
\text{Cost basis of the asset, } P &= \$10{,}000 \\
\text{Useful life, } N &= 5 \text{ years} \\
\text{Estimated salvage value, } S &= \$2000
\end{aligned}
$$

Compute the annual depreciation allowances and the resulting book values using the straight-line depreciation method.

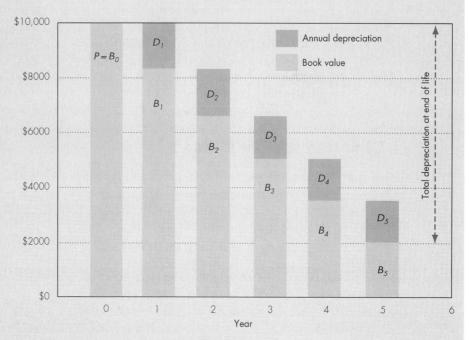

Figure 7.2 Straight-line depreciation method (Example 7.3)

Solution

Given: $P = \$10,000$, $S = \$2000$, $N = 5$ years
Find: D_n and B_n for $n = 1$ to 5

The straight-line depreciation rate is 1/5 or 20%. Therefore, the annual depreciation charge is

$$D_n = (0.20)(\$10,000 - \$2000) = \$1600.$$

Then, the asset would have the following book values during its useful life:

n	B_{n-1}	D_n	B_n
1	$10,000	$1600	$8400
2	8,400	1600	6800
3	6,800	1600	5200
4	5,200	1600	3600
5	3,600	1600	2000

where B_{n-1} represents the book value before the depreciation charge for year n. This situation is illustrated in Figure 7.2. The time zero book value, B_o, is the cost basis of the asset, P.

7.3.2 Accelerated Methods

The second depreciation concept recognizes that the stream of services provided by a fixed asset may decrease over time; in other words, the stream may be greatest in the first year of an asset's service life and least in its last year. This pattern may occur because the mechanical efficiency of an asset tends to decline with age, because maintenance costs tend to increase with age, or because of the increasing likelihood that better equipment will become available and make the original asset obsolete. This reasoning leads to a method that charges a larger fraction of the cost as an expense of the early years than of the later years. Any such method is called an **accelerated method**. The two most widely used accelerated methods are the **declining balance method** and the **sum-of-the-years'-digits method**.

Declining Balance Method (DB)

The **declining balance method** of calculating depreciation allocates a fixed fraction of the beginning book balance each year. The fraction, d, is obtained as follows:

$$d = (1/N)(\text{multiplier}). \tag{7.3}$$

The most prevalent multiplier in Canada is 1.0 (called DB), as it is the basis of most of the CCA rates for tax purposes. Other common mulipliers include 1.5 (called 150% DB) and 2.0 (called 200%, or double declining balance, DDB). As N increases, d decreases. This method results in a situation in which depreciation is highest in the first year and decreases over the asset's depreciable life.

The factor, d, can be utilized to determine depreciation charges for a given year, D_n, as follows:

$$D_1 = dP,$$
$$D_2 = d(P - D_1) = dP(1 - d),$$
$$D_3 = d(P - D_1 - D_2) = dP(1 - d)^2,$$

and thus for any year, n, we have a depreciation charge, D_n, of

$$D_n = dP(1 - d)^{n-1}, \; n \geq 1 \tag{7.4}$$

We can also compute the total declining balance depreciation (TDB) at the end of n years as follows:

$$
\begin{aligned}
TDB_n &= D_1 + D_2 + \cdots + D_n \\
&= dP + dP(1 - d) + dP(1 - d)^2 + \cdots + dP(1 - d)^{n-1} \\
&= dP[1 + (1 - d) + (1 - d)^2 + \cdots + (1 - d)^{n-1}]. \tag{7.5}
\end{aligned}
$$

Multiplying Eq. (7.5) by $(1 - d)$, we obtain

$$TDB_n(1 - d) = dP[(1 - d) + (1 - d)^2 + (1 - d)^3 + \cdots + (1 - d)^n]. \tag{7.6}$$

Subtracting Eq. (7.5) from Eq. (7.6) and dividing by d gives

$$TDB_n = P[1- (1 - d)^n]. \tag{7.7}$$

The book value, B_n, at the end of n years will be the cost of the asset P minus the total depreciation at the end of n years:

$$
\begin{aligned}
B_n &= P - TDB_n \\
&= P - P[1 - (1 - d)^n] \\
B_n &= P(1 - d)^n.
\end{aligned}
\tag{7.8}
$$

EXAMPLE 7.4 Declining balance depreciation

Consider the following accounting information for a photocopier:

$$
\begin{aligned}
\text{Cost basis of the asset, } P &= \$10{,}000 \\
\text{Useful life, } N &= 5 \text{ years} \\
\text{Estimated salvage value, } S &= \$3277.
\end{aligned}
$$

Compute the annual depreciation charges and the resulting book values using the declining balance depreciation method (Figure 7.3).

Solution

Given: $P = \$10{,}000$, $S = \$3277$, $N = 5$ years
Find: D_n and B_n for $n = 1$ to 5

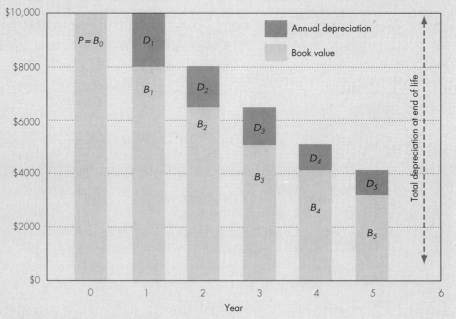

Figure 7.3 Declining balance method (Example 7.4)

The book value at the beginning of the first year is $10,000, and the declining balance rate (d) is (1/5) = 20%. Then, the depreciation deduction for the first year will be $2000 (20% $\times$ $10,000 = $2000). To figure the depreciation deduction in the second year, we must first adjust the book value for the amount of depreciation we deducted in the first year. The first year's depreciation is subtracted from the beginning book value ($10,000 – $2000 = $8000). This amount is multiplied by the rate of depreciation ($8000 $\times$ 20% = $1600). By continuing the process, we obtain

n	B_{n-1}	D_n	B_n
1	$10,000	$2,000	$8,000
2	$8,000	$1,600	$6,400
3	$6,400	$1,280	$5,120
4	$5,120	$1,024	$4,096
5	$4,096	$819	$3,277

Alternatively, we could use equations 7.4 and 7.8 to calculate D_n and B_n directly. The declining balance is illustrated in terms of the book value each year in Figure 7.3.

Salvage value (S) must be estimated independently of depreciation analysis. In Example 7.4, the final book value (B_N) conveniently equals the estimated salvage value of $3277, an occurrence that is rather unusual in the real world. When $B_N \neq S$, the following adjustments may be made in our calculations for book depreciation. The approaches below do not apply for tax purposes, however, as the CCRA has specific procedures to follow when the book and salvage values differ (see Section 7.4.5 and Chapter 8).

• **Case 1: $B_N > S$**

When $B_N > S$, we are faced with a situation in which we have not depreciated the entire cost of the asset. If you would prefer to reduce the book value of an asset to its salvage value as quickly as possible, it can be done by switching from DB to SL whenever SL depreciation would result in larger depreciation charges and therefore a more rapid reduction in the book value of the asset. The switch from DB to SL depreciation can take place in any of the N years, the objective being to identify the optimal year to switch. The switching rule is as follows: If depreciation by DB in any year is less than (or equal to) what it would be by SL, then we should switch to and remain with the SL method for the remaining duration of the project's depreciable life. Switch in year n when the following condition holds:

$$dB_{n-1} \leq \frac{B_{n-1} - S}{N - n + 1}. \tag{7.9}$$

EXAMPLE
7.5

Declining balance with conversion to straight line depreciation ($B_N > S$)

Suppose the asset given in Example 7.4 has a salvage value of $2700 instead of $3277. Determine the optimal time to switch from DB to SL depreciation and the resulting depreciation schedule.

Solution

Given: $P = \$10,000$, $S = \$2700$, $N = 5$ years, $d = 20\%$
Find: Optimal conversion time, D_n and B_n for $n = 1$ to 5

Computing the SL depreciation from the beginning of each year n, to the final year N, and comparing to the DB calculated in Example 7.4, we use the decision rule given in Equation (7.9) to determine when to switch:

If Switch to SL at Beginning of Year	SL Depreciation	DB Depreciation	Decision
2	($8000 – 2700)/4 = $1325	< $1600	Do not switch
3	($6400 – 2700)/3 = $1233	< $1280	Do not switch
☐4	($5120 – 2700)/2 = $1210	> $1024	Switch to SL

The optimal time (year 4) in this situation corresponds to n' in Figure 7.4(a). The resulting depreciation schedule is:

Year	DB with Switching to SL	End-of-Year Book Value
1	$2000	$8000
2	1600	6400
3	1280	5120
4	1210	3910
5	1210	2700
Total	$7300	

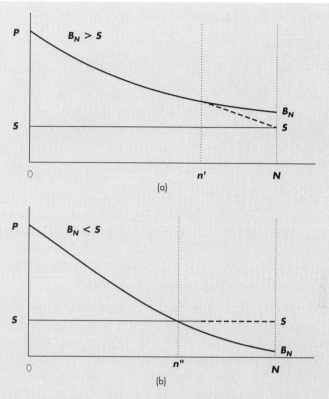

Figure 7.4 Adjustments to the declining balance method: (a) Switch from DB to the SL after n'; (b) no further depreciation allowances are available after n'' (Examples 7.5 and 7.6)

Case 2: $B_N < S$

With a relatively high salvage value, it is possible that the book value of the asset could decline below the estimated salvage value. To avoid deducting depreciation charges that would drop the book value below the salvage value, you simply stop depreciating the asset whenever you get down to $B_n = S$.

EXAMPLE 7.6 **Declining balance, $B_N < S$**

Suppose the asset described in Example 7.4 has a salvage value of $4500 instead of $3277. Determine the declining balance schedule which culminates in a final book value equal to the estimated salvage value.

Solution

Given: $P = \$10{,}000$, $S = \$4500$, $N = 5$ years, $d = 20\%$

Find: D_n and B_n for $n = 1$ to 5

End of Year	D_n			B_n		
1	0.2($10,000)	=	$2000	$10,000 − $2000	=	$8000
2	0.2(8000)	=	1600	$8000 − 1600	=	6400
3	0.2(6400)	=	1280	6400 − 1280	=	5120
4	0.2(5120) = 1024	>	620	5120 − 620	=	4500
5			0	4500 − 0	=	4500
	Total	=	$8000			

Note that B_4 would be less than $S = \$4500$, if the full deduction ($1024) had been taken. Therefore, we adjust D_4 to $620, making $B_4 = \$4500$. D_5 is zero and B_5 remains at $4500. Year 4 is equivalent to n'' in Figure 7.4(b).

Sum-of-Years'-Digits (SOYD) Method

Another accelerated method for allocating the cost of an asset is called **sum-of-years'-digits** (SOYD) depreciation. Compared with SL depreciation, SOYD results in larger depreciation charges during the early years of an asset's life and smaller charges as the asset reaches the end of its estimated useful life. Unlike the DB method, SOYD guarantees that $B_n = S$ automatically.

In the SOYD method, the numbers 1, 2, 3, ..., N are summed, where N is the estimated years of useful life. We find this sum by the equation[2]

$$SOYD = 1 + 2 + 3 + ... + N = \frac{N(N+1)}{2}. \tag{7.10}$$

The depreciation rate each year is a fraction in which the denominator is the SOYD and the numerator is, for the first year, N; for the second year, $N - 1$; for the third year, $N - 2$; and so on. Each year the depreciation charge is computed by dividing the remaining useful life by the SOYD and by multiplying this ratio by the total amount to be depreciated $(P - S)$.

$$D_n = \frac{N - n + 1}{SOYD}(P - S). \tag{7.11}$$

---

[2] You may derive this sum equation by writing the SOYD expression in two ways:

SOYD	=	$1 + 2 + 3 + ... + N$
SOYD	=	$N + (N - 1) + ... + 1.$

Then you add these two expressions and solve for SOYD

2SOYD	=	$(N + 1) + (N + 1) + ... + (N + 1)$
	=	$N(N + 1)$
SOYD	=	$N(N + 1)/2.$

EXAMPLE 7.7 SOYD depreciation

Compute the SOYD depreciation schedule for Example 7.3.

> Cost basis of the asset, P = $10,000
> Useful life, N = 5 years
> Salvage value, S = $2000.

Solution

Given: P = $10,000, S = $2000, N = 5 years
Find: D_n and B_n for n = 1 to 5

We first compute the sum-of-years' digits:

$$\text{SOYD} = 1 + 2 + 3 + 4 + 5 = 5(5 + 1)/2 = 15$$

Year	D_n			B_n
1	(5/15) ($10,000 – $2000)	=	$2667	$7333
2	(4/15) (10,000 – 2000)	=	2133	5200
3	(3/15) (10,000 – 2000)	=	1600	3600
4	(2/15) (10,000 – 2000)	=	1067	2533
5	(1/15) (10,000 – 2000)	=	533	2000

This situation is illustrated in Figure 7.5.

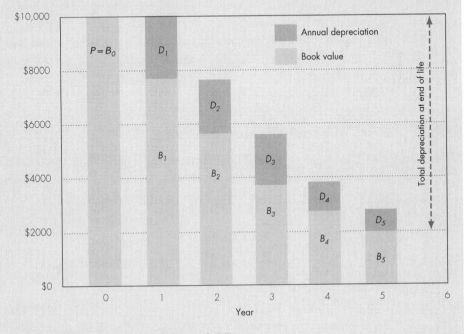

Figure 7.5 Sum-of-years' digits method (Example 7.7)

7.3.3 Units-of-Production (UP) Method

Straight-line depreciation is most applicable if the machine is used for the same amount of time each year. What happens when a punch press machine is run 1670 hours one year and 780 the next, or when some of its output is shifted to a new machining center? This leads us to a consideration of another depreciation concept that views the asset as consisting of a bundle of service units; unlike the SL and accelerated methods, however, this concept does not assume that the service units will be consumed in a time-phased pattern. The cost of each service unit is the net cost of the asset divided by the total number of such units. The depreciation charge for a period is then related to the number of service units consumed in that period. This leads to the **units-of-production method**. By this method, the depreciation in any year is given by

$$D_n = \frac{\text{Service units consumed for year}}{\text{Total service units}}(P - S) \tag{7.12}$$

When using the units-of-production method, depreciation charges are made proportional to the ratio of actual output to the total expected output, and usually this ratio is figured in machine hours. The advantage of using this method is that depreciation varies with production volume, and therefore the method gives a more accurate picture of machine usage. A disadvantage of the units-of-production method is that the collecting of data on machine use and the accounting methods are somewhat tedious. This method can be useful for depreciating equipment used to exploit natural resources, if the resources will be depleted before the equipment wears out. It is also the basis for depletion allowances for certain natural resources (see Section 7.6). It is not, however, considered a practical method for general use in depreciating industrial equipment.

EXAMPLE 7.8 **Units-of-production depreciation**

A truck for hauling coal has an estimated net initial cost of $55,000 and is expected to give service for 250,000 kilometres, resulting in a $5000 salvage value. Compute the allowed depreciation amount for the year in which the truck usage was 30,000 kilometres.

Solution

Given: P = $55,000, S = $5000, total service units = 250,000 kilometres, usage for this year = 30,000 kilometres
Find: Depreciation amount in this year

The depreciation expense in a year in which the truck traveled 30,000 kilometres would be

$$\frac{30{,}000 \text{ kilometres}}{250{,}000 \text{ kilometres}}(\$55{,}000 - \$5000) = (\frac{3}{25})(\$50{,}000)$$

$$= \$6000.$$

7.4 Capital Cost Allowance for Income Tax

The Federal and Provincial governments in Canada collect income taxes from businesses, as they do from individuals. Since income taxes often have a major impact on the economic viability of engineering projects, we will study this topic in some detail in Chapters 8, 9, and 11. The Income Tax Act, administered by the Canada Customs and Revenue Agency, also recognizes that depreciable assets lose value over time. Although the purchase cost of depreciable assets may not be deducted as an expense for tax purposes in the year of acquisition, a portion of the acquisition cost of such assets may be deducted from income each year, which may lower the taxes payable. This deduction is referred to as capital cost allowance (CCA).

Although many features of book depreciation also apply to capital cost allowance calculations, the Income Tax Act defines its own nomenclature specifically for the CCA context. The following table shows the correspondence between terms, although the interpretations are slightly different in some cases.

Book Depreciation Term	Tax Depreciation Term
asset	property
depreciation	capital cost allowance
cost basis	capital cost
book value	undepreciated capital cost
salvage value	proceeds of disposition

7.4.1 CCA System

The capital cost allowance system differs from book depreciation practices in two significant ways. First of all, the Income Tax Act specifies exactly how the CCA must be calculated, so there is no leeway as there is for financial reporting. This ensures fairness and consistency across all business taxpayers. The key provisions for calculating CCA are presented in the following sections.

Secondly, most **property** items are not depreciated individually, but are grouped into **classes** (also know as **asset pools**) as decreed by the tax regulations. The frequently used CCA classes are displayed in Table 7.1. The property in most classes must be depreciated using the declining-balance method, but a few classes follow more complex procedures, usually variants of the straight-line method. To limit the maximum allowance, the tax rules specify a percentage rate (**CCA rate**) for depreciation of each of the DB classes, and guidelines for determining the depreciable life for all property in the straight-line classes (Classes 13 and 14 in Table 7.1).

For example, the milling machines, saws, and drill presses owned by a machine shop would be combined into Class 43 (if they were all acquired after February 25, 1992). At the end of each year, the **undepreciated capital cost** (i.e., the income tax equivalent of current book value) of all this equipment may be depreciated by 30%, and the resulting capital cost allowance deducted as an expense before calculating income taxes. A company may claim up to the maximum allowance in each class in each year. This is the general practice, but for a variety of reasons, such as an unprofitable year, the full CCA is not always claimed. In the examples in this text, it is assumed that the maximum allowable CCA is claimed unless stated otherwise.

You may wonder what the rationale is for pooling property in this way. Three main principles apply. First, similar assets are generally grouped together, as they would have comparable usage characteristics and lifetimes, and the CCA rates are set accordingly. For example, buildings of sturdier materials (Classes 1 and 3) normally last longer than those in Class 6, and so are depreciated more slowly under the lower CCA rate. Machinery and vehicles wear out faster than buildings, and fall into classes with higher rates. Secondly, some combinations seem incompatible, such as the taxicabs and pinball machines in Class 16. However, the government has decided that they can be assigned to the same high CCA rate class to reflect the degree of wear-and-tear on such assets. Thirdly, since the capital cost allowance can act as an economic incentive for businesses, the CCA rates may reflect government policies or special circumstances. As an example of the former, for several years pollution abatement equipment could be depreciated at a very high rate to encourage companies to invest in such technology. As an example of the latter, notice that cash registers acquired up to 1993 used to record multiple sales tax fall into Class 12 (CCA rate = 100%), while other cash registers are in Class 8 (20%). This distinction was introduced to allow a faster depreciation which eases the tax burden on companies that had to replace or upgrade their registers to accommodate the Goods and Services Tax (GST) or Harmonized Sales Tax (HST).

Table 7.1 is included in the Income Tax Guides for Corporations and Businesses because it covers the majority of property owned by most businesses. Class numbers not listed here either relate to highly specialized assets, such as Class 42 for fibre-optic cable (CCA rate = 12%), or are applicable to property acquired during limited periods, such as Class 40, which includes specific assets purchased in 1988 or 1989. Several of the classes retained in Table 7.1 are also time-delimited. On the surface, it is not apparent why automobiles in Class 10.1 are not simply included in Class 10, since the same CCA rate applies in both cases. It is because certain depreciation details for more expensive cars (Class 10.1) are different, such as how to calculate the tax effect upon disposition. Note that systems software depreciates at 30%

Class Number	Description	CCA Rate
1	Most buildings made of brick, stone or cement acquired after 1987, including their component parts such as electric wiring, lighting fixtures, plumbing, heating and cooling equipment, elevators and escalators	4%
3	Most buildings made of brick, stone or cement acquired before 1988, including their component parts as listed in Class 1 above	5%
6	Buildings made of frame, log, stucco on frame, galvanized iron, or corrugated metal that are used in the business of farming or fishing, or that have no footings below ground; fences and most greenhouses	10%
7	Canoes, boats and most other vessels, including their furniture, fittings or equipment	15%
8	Property that is not included in any other class such as furniture, calculators and cash registers (that do not record multiple sales taxes), photocopy and fax machines, printers, display fixtures, refrigeration equipment, machinery, tools costing $200 or more, and outdoor advertising billboards and greenhouses with rigid frames and plastic covers acquired after 1987	20%
9	Aircraft, including furniture, fittings or equipment attached, and their spare parts	25%
10	Automobiles (except taxis and others used for lease or rent), vans, wagons, trucks, buses, tractors, trailers, drive-in theatres, general-purpose electronic data processing equipment (e.g., personal computers) and systems software, and timber cutting and removing equipment	30%
10.1	Passenger vehicles costing more than $27,000 if acquired after 1999 ($26,000 if acquired after 1997 and before 2000; $25,000 if acquired in 1997; $24,000 if acquired after August 31, 89, and before 1997; and $20,000 if acquired before September 1989)	30%
12	Chinaware, cutlery, linen, uniforms, dies, jigs, moulds or lasts, computer software (except systems software), cutting or shaping parts of a machine, certain property used for earning rental income such as apparel or costumes, and videotape cassettes; certain property costing less than $200 such as kitchen utensils, tools, and medical or dental equipment; certain property acquired after August 8, 1989, and before 1993 for use in a business of selling or providing services such as electronic bar-code scanners, and cash registers used to record multiple sales taxes	100%

Table 7.1
Capital Cost Allowance Rates and Classes

continued…

13	Property that is leasehold interest (the maximum CCA rate depends on the type of the leasehold and the terms of the lease)	N/A
14	Patents, franchises, concessions, and licenses for a limited period—the CCA is limited to the lesser of: • the capital cost of the property spread out over the life of the property; and • the undepreciated capital cost of the property at the end of the taxation year Class 14 also includes patents, and licences to use patents for a limited period, that you elect not to include in class 44	N/A
16	Automobiles for lease or rent, taxicabs, and coin-operated video games or pinball machines; certain tractors and large trucks acquired after December 6, 1991, that are used to haul freight and that weigh more than 11,788 kilograms	40%
17	Roads, sidewalks, parking-lot or storage areas, telephone, telegraph, or nonelectronic data communication switching equipment	8%
38	Most power-operated movable equipment acquired after 1987 used for moving, excavating, placing, or compacting earth, rock, concrete, or asphalt	30%
39	Machinery and equipment acquired after 1987 that is used in Canada primarily to manufacture and process goods for sale or lease	25%
43	Manufacturing and processing machinery and equipment acquired after February 25, 1992, described in Class 39 above	30%
44	Patents and licences to use patents for a limited or unlimited period, that the corporation acquired after April 26, 1993. However, you can elect not to include such property in Class 44 by attaching a letter to your return for the year the corporation acquired the property. In the letter, indicate the property you do not want to include in Class 44.	25%

Source: CCRA, 1998 T2 Corporation—Income Tax Guide

under Class 10, while all other software falls into Class 12, and the distinction may be difficult in some cases. Since tax laws and regulations are changed regularly, it is important to note that we must use whatever rates, classifications, and adjustments mandated *at the time an asset is acquired*.

These comments illustrate that the CCA rules can be very complicated, and recourse must often be made to accountants and other tax experts. However, sufficient information is provided in this text to deal with simple situations, or to develop after-tax cash flow estimates robust enough to make engineering economic decisions. The student interested in studying the CCA system or general tax rules in more detail can refer to the following sources (in increasing order of comprehensiveness):

- T2 Corporation—Income Tax Guide, Canada Customs and Revenue Agency
- Interpretation Bulletins (IT) and Circulars (IC) available from Revenue Canada offices, or on the WEB (www.ccra_adrc.gc.ca). In particular, bulletin *IT285R2 Capital Cost Allowance—General Comments* discusses many aspects of the CCA system.
- Income Tax Act and Regulations. Several private publishers compile this information, such as: *The Practitioner's Income Tax Act*, 14th Edition, David M. Sherman (ed.), Carswell, Thomson Canada Ltd., 1998; *Canadian Income Tax Regulations*, by CCH Canadian Ltd.; *Income tax act ... annotated*, published by R. De Boo, Toronto.

7.4.2   Available-for-Use Rule

The earliest taxation year in which capital cost allowance may be claimed for property is when it becomes **available-for-use**. This term has a specific meaning within the context of the Income Tax Act, and the restrictions differ for building versus non-building property, as described below.

A building is considered available for use on the earliest of several dates. The following are some examples of these dates:

- when the business uses all or substantially all of the building for its intended purpose;
- when the construction of the building is completed;
- the beginning of the first taxation year that starts at least 358 days after the taxation year during which the corporation acquired the property;
- immediately before disposal of the property.

Property other than a building is considered available for use on the earliest of several dates. The following are some examples of these dates:

- when the property is first used to produce income;
- when the business can use the property to either produce a saleable product or perform a saleable service;
- the beginning of the first taxation year that starts at least 358 days after the taxation year during which the corporation acquired the property;
- immediately before disposal of the property.

EXAMPLE
7.9

Available-for-use rule for CCA

Kasement Windows Ltd. decided to expand their production capacity and automate more of their operation in anticipation of a forecasted housing construction boom. On May 20, 1997 they purchased a vacant lot adjacent to their current plant and signed a contract to have a new facility built. On August 3rd, 1997 they received a new conveyor/assembly system which they temporarily housed in one corner of the new facility, which was not yet finished. The conveyor system required custom jigs to hold the window components, and these jigs had not yet arrived. On December 1st 1997, they acquired a used frame welder and an extruder at a good price, and stored them in the new facility. At the same time, they also purchased a forklift truck for the new production plant, but inventory was piling up so much at the existing plant that they started using the truck immediately. Construction of the new facility was completed on February 12, 1998. In the meantime, in January 1998 the company decided to invest in a state-of-the-art welder instead, and traded in the used one they had bought for a slight loss. They were ready for full production at the new facility on June 1, 1998. They initially had to rely on manual material handling and assembly, and by the end of 1998 they still hadn't used the conveyor system since the custom jigs never fit properly. Determine when each new piece of property became available for use. The company's taxation year matches the calendar year.

Solution

Given: Timing of acquisitions and usage of several pieces of property
Find: When each piece of property is considered available-for-use for CCA calculations

The building is available for use in 1998, because construction was completed then. The truck is available for use in 1997 because it was used immediately in December as part of their production (i.e., to generate income). Although the extruder was acquired in 1997, it was not installed or used until June 1998, which is the taxation year when it became available-for-use. The used frame welder was never installed, so it becomes available-for-use and added into the pool of depreciable property just before it was sold, hence in 1998. The state-of-the-art welder was first applied to production in the year 1998, which is when it became available-for-use. The conveyor still had not been used in the year 1998. Adding 358 days to the date of acquisition August 3rd 1997, leads to the date July 27th 1998. The first taxation year following this latter date is January 1st 1999, so this equipment is considered available-for-use in 1999, even if Kasement has not used the conveyor by then. (Note: the jigs are in a separate CCA class from the conveyor, as shown in following section).

Comments: The gist of this rule is that no CCA can be claimed for property in the year it is acquired, or in the following year, unless it is available-for-use. The 358-day criterion essentially places a maximum 2-year limit on this restriction.

7.4.3   Calculating the Capital Cost Allowance

To claim a capital cost allowance, a corporation must fill out a Schedule 8 form according to the T2 Corporation Income Tax Guide. Self-employed professionals, sole proprietors or business partnerships perform equivalent calculations on their respective tax forms T2032 or T2124. We will first present the CCA form in Figure 7.6, and accompanying guidelines for filling out each of the fields. As this is a summary only, and tax rules change periodically, you should refer to the current tax guides when performing calculations for an actual business. We will then proceed to provide examples to explain and illustrate most of the key elements in the CCA calculation. Each declining balance CCA class must be entered on a separate line of the form. Property belonging to one of the straight-line classes must have a separate calculation of CCA performed for each asset, which depends on many factors including the nature of the asset and its duration. The total allowable capital cost allowance for the firm is the sum of the CCAs across all classes.

- **Column 1—Class number** Each class of property is identified with the assigned class number. Generally, all depreciable property of the same class is grouped together. The CCA is then calculated based on the undepreciated capital cost of all the property in that class.

- **Column 2—Undepreciated capital cost (UCC) at the beginning of the year** This is the value of all the assets in the class at the start of the year. It is set equal to the UCC value of all the assets in the class at the end of the preceding taxation year.

- **Column 3—Cost of acquisitions during the year** The total cost (including shipping, installation, and related fees) of depreciable property acquired by the business and available for use in the taxation year is entered by class of assets. *Land is excluded*, as it is not a depreciable property.

- **Column 4—Net adjustments** Sometimes, amounts must be entered that either increase or decrease the capital cost of a property. A partial list of such quantities is included here. The capital cost of a property must be *reduced* by the following amounts:
  - any GST/HST input tax credit or rebate received in the year;
  - any federal **investment tax credit** (ITC) used to reduce taxes in the preceding taxation year (note: the government provides such incentives to encourage investment in certain activities or regions of the country; these transactions will be reviewed in Chapter 8).
  - any provincial tax credits received in the current year;
  - any government assistance received in the current year.

  The capital cost of a property must be *increased* by the following amounts:
  - any refund of GST/HST input tax credit previously deducted;
  - any depreciated property transferred from an amalgamated or wound-up subsidiary;

Revenue Canada / Revenu Canada

CAPITAL COST ALLOWANCE (CCA) (1998 and later taxation years)

SCHEDULE 8

Name of corporation

Business Number

Taxation year end
Year | Month | Day

For more information, see the section called "Capital Cost Allowance" in the *T2 Corporation Income Tax Guide*.

Is the corporation electing under regulation 1101(5q)? **101** 1 Yes ☐ 2 No ☐

1 Class number	2 Undepreciated capital cost at the beginning of the year (undepreciated capital cost at the end of the year from last year's CCA schedule)	3 Cost of acquisitions during the year (new property must be available for use) See note 1 below	4 Net adjustments (show negative amounts in brackets)	5 Proceeds of dispositions during the year (amount not to exceed the capital cost)	6 Undepreciated capital cost (column 2 **plus** column 3 **plus** or **minus** column 4 **minus** column 5)	7 50% rule (1/2 of the amount, if any, by which the net cost of acquisitions exceeds column 5) See note 2 below	8 Reduced undepreciated capital cost (column 6 **minus** column 7)	9 CCA rate %	10 Recapture of capital cost allowance	11 Terminal loss	12 Capital cost allowance (column 8 **multiplied** by column 9, or a lower amount) See note 3 below	13 Undepreciated capital cost at the end of the year (column 6 **minus** column 12)
200	**201**	**203**	**205**	**207**		**211**		**212**	**213**	**215**	**217**	**220**
1.												
2.												
3.												
4.												
5.												
6.												
7.												
8.												
9.												
10.												

Totals ☐ ☐ ☐

Note 1. Include any property acquired in previous years that has now become available for use. This property would have been previously excluded from column 3. List separately any acquisitions that are not subject to the 50% rule, see Regulation 1100(2) and (2.2).

Note 2. The net cost of acquisitions is the cost of acquisitions plus or minus certain adjustments from column 4.

Note 3. If the taxation year is shorter than 365 days, prorate the CCA claim. See the *T2 Corporation Income Tax Guide* for more information.

Enter the total of column 10 on line 107 of Schedule 1.
Enter the total of column 11 on line 404 of Schedule 1.
Enter the total of column 12 on line 403 of Schedule 1.

T2 SCH 8
Printed in Canada

1388

(Français au verso)

Canada

Figure 7.6 Capital cost allowance Form

– any government assistance the business repaid which was previously used to reduce the capital cost.

- **Column 5—Proceeds of dispositions during the year** The net proceeds of disposition (after deducting costs of removal of the asset or restoring the site) are entered for each class. If these proceeds exceed the capital cost of the property, enter the capital cost. (Note: the amount by which disposition proceeds exceed the original capital cost results in a **capital gain**—see Chapter 8—which is taxed at a different rate than other income.)

- **Column 6—Undepreciated capital cost** (UCC) The value in this cell is calculated by adding columns 2 and 3, subtracting or adding the amount in column 4 depending on the type of adjustment, and subtracting the amount in column 5. For each class, we must now distinguish between three possible states:

  (1) *the UCC in column 6 is positive, and property remains in the class at the end of the year:* Proceed with the calculations in column 7;
  (2) *the UCC in column 6 is negative:* This generally results when more CCA has been claimed than the actual loss of value on assets bought and sold in the class. Not only are you unable to claim CCA for the property in that class in the current year, but this **recapture of CCA** must be added to the firm's income, which is then taxed. The recapture of CCA is entered in column 10, and no further calculations are performed for that class.
  (3) *the UCC in column 6 is positive, but no property remains in the class at the end of the year:* This positive quantity results when the last asset(s) is sold for less than the undepreciated capital of all the property that was owned in the class. Since no assets remain in the class, this amount or **terminal loss** may be deducted directly from the business' income in the taxation year. The terminal loss is entered in column 11, and no further calculations are performed for that class.

  Note that recapture and terminal loss rules do not apply to a few specific classes, including passenger vehicles in Class 10.1.

- **Column 7—50% rule** Most new property that is available for use during the taxation year is only eligible for 50% of the normal maximum CCA for the year (see Section 7.4.4). A reduced UCC must be calculated accordingly before applying the CCA rate. The adjustment equals one-half of the net amount of additions to the class (the net cost of acquisitions minus the proceeds of dispositions). This amount must be entered into column 7. Sometimes, adjustments must be made to this total based on amounts in column 4.

- **Column 8—Reduced undepreciated capital cost** A reduced UCC is required because of the 50% rule, by subtracting the amount in column 7 from column 6.

- **Column 9—CCA rate** The prescribed CCA rate for the class is entered here. For a class where the rate is not specified in the Income Tax Act, enter NA.

- **Column 10—Recapture of capital cost allowance** The recapture amount calculated in column 6 is entered here (if applicable).

- **Column 11—Terminal loss** The terminal loss calculated in column 6 is entered here (if applicable).

- **Column 12—Capital cost allowance** The maximum allowable CCA for each class is calculated as the product of the reduced UCC from column 8 multiplied by the CCA rate in column 9. The business may claim any amount up to the maximum. If the taxation year is shorter than 365 days for a start-up business, the CCA for most property must be prorated (see tax guide for details).

- **Column 13—Undepreciated capital cost at the end of the year** The UCC from the end of the previous year, adjusted for acquisitions, dispositions, and government assistance (plus a few other adjustments as described in column 4) is reduced by the CCA claimed in the current year, resulting in the final UCC at the end of the current taxation year. This result is calculated by subtracting the amount in column 12 from the amount in column 6. For any class with a terminal loss or recapture of CCA, the UCC at the end of the year is always nil.

7.4.4   The 50% Rule

The **50% rule**, also known as the **half-year convention**, was introduced into the income tax system on November 12, 1981. For most classes, it limits the maximum capital cost allowance of new property acquired during the year to 50% of what it would otherwise be. This rule reflects an "average" amount of depreciation for the year of acquisition across all businesses' new assets, since property purchased early in the taxation year has been subject to almost an entire year of usage, while property acquired near the end is practically brand new. More precisely, the 50% rule is applied to the **net acquisitions** in each class, equal to the amount by which purchases exceed dispositions during the year.

For a single asset, let us compare the pattern of CCA and UCC over time with and without the 50%-rule.

Let: P =   the capital cost of the property
U_n =   the undepreciated capital cost at the end of year n
CCA_n =   the maximum claimable capital cost allowance in year n
d =   the prescribed CCA rate (declining-balance).

	Without 50% Rule		With 50% Rule	
Year	**CCA**	**UCC**	**CCA**	**UCC**
0		$U_0 = P$		$U_0 = P$
1	$CCA_1 = Pd$	$U_1 = P(1-d)$	$CCA_1 = Pd/2$	$U_1 = P(1-d/2)$
2	$CCA_2 = Pd(1-d)$	$U_2 = P(1-d)^2$	$CCA_2 = Pd(1-d/2)$	$U_2 = P(1-d/2)(1-d)$
$n\,(\geq 2)$	$CCA_n = Pd(1-d)^{n-1}$	$U_n = P(1-d)^n$	$CCA_n = Pd(1-d/2)(1-d)^{n-2}$	$U_n = P(1-d/2)(1-d)^{n-1}$

Most, but not all, depreciable property is subject to the 50% rule. Exemptions are listed below:

- Most Class 10 property must *comply* with the 50% rule, except for: television cable boxes or decoders; a Canadian film or video production.

- Most Class 12 property is *exempt* from the 50% rule *except for:* a die, jig, pattern, mould or last; the cutting or shaping part in a machine; the film or videotape comprising a television commercial message; a certified feature film or production; computer software (non-systems);

- All property in Classes 13, 14, 15, 23, 24, 27, 29 and 34 is *exempt* from the 50% rule (note: some provisions in classes 13, 24, 27, 29 and 34 require a reduction in their first-year CCA by 50%, but not using the same procedures as the 50% rule).

- Any property whose earliest available-for-use date is 358 days after acquisition (see section 7.4.2) is *exempt* from the 50% rule.

Take special note of the capital cost allowance for Class 12 items (d = 100%). Those assets which are exempt from the 50% rule may be fully depreciated in the year that they are available for use. The remaining items in Class 12 are fully depreciated over 2 years. Most Class 12 items that are of interest for engineering economic decisions (machine parts, software) fall under the 50% rule.

EXAMPLE 7.10 CCA calculations for multiple property classes

Reconsider Kasement Windows' situation from Example 7.9. Assume that their undepreciated capital costs at the end of 1996 by CCA class are given by:

CCA Class	UCC_{1996}
1	$1,766,419
3	194,980
8	63,971
10	0
12	17,518
43	483,602

At that time, Class 1 included their existing plant (which was constructed after 1987); the only asset in Class 3 was an old warehouse which they used for excess inventory; Class 8 contained various office equipment; Class 12 encompassed certain machine parts; and Class 43 included all of their manufacturing equipment. Aside from the property transactions described in Example 7.9, Kasement finally got the proper jigs installed and working on their conveyor/assembly system in the year 1999. In that year, they sold the forklift truck, deciding that the conveyor system satisfied their material handling needs. They also sold their old warehouse in the year 1999, since the new facility has ample storage capacity. Given the following depreciable property transaction data, calculate the maximum CCA that Kasement can claim for the years 1997, 1998 and 1999.

Depreciable Property Transactions 1997–1999					
Asset	CCA Class	Year Available for Use	Capital Cost	Disposition Year	Proceeds of Disposition
Forklift truck	10	1997	$18,000	1999	$5,000
New facility*	1	1998	$2,600,000		
Extruder	43	1998	$41,000		
Used welder	43	1998	$29,000	1998	$28,000
New welder	43	1998	$54,000		
Conveyor/assembler	43	1999	$112,000		
Conveyor jigs	12	1999	$13,000		
Warehouse	3			1999	$200,000

*This includes the new building costs only, excluding the land, as land is not depreciable.

Solution

Given: UCC$_{1996}$ for CCA classes, and depreciable property transactions for 1997–1999
Find: CCA for all classes for 1997, 1998 and 1999
By following the procedure for calculating CCA, a Schedule 8 form is completed for each of the three years, as shown in Table 7.2.

Comments on 1997 Schedule 8: The UCC at the start of 1997 (column 2) must equal the UCC at the end of 1996. Under the 50% rule, only half of the value of the truck may be claimed in the first year ($CCA_{1997} = 1/2Pd = 1/2*18,000*30\% = \$2,700$). Also, the remaining Class 12 property depreciates to $0 under the 100% rate.

Comments on 1998 Schedule 8: As with most CCA calculations, all of the new assets in Class 43 are pooled together. The 50% rule only applies to the *net* new acquisitions in each class (i.e., column 3 minus 5 in Class 43). Even though the used welder was never installed, it must still be recorded on Schedule 8 before it is sold. Therefore, the $1000 loss on this machine does not become a write-off, but must be depreciated annually. Although the used welder became available-for-use the year after it was acquired, it is still subject to the 50% rule. As opposed to accounting depreciation, where an explicit adjustment to the cost basis must be made when the trade-in value does not equal the book value (see Example 7.2), the CCA system takes care of this automatically.

Kasement Windows—Schedule 8 for 1997 (see Figure 7.6 for column headings)

1	2	3	4	5	6	7	8	9	10	11	12	13
1	$1,766,419				$1,766,419		$1,766,419	4%			$70,656	$1,695,762
3	$194,980				$194,980		$194,980	5%			$9,749	$185,231
8	$63,971				$63,971		$63,971	20%			$12,794	$51,176
10	$0	$18,000			$18,000	$9,000	$9,000	30%			$2,700	$15,300
12	$17,518				$17,518		$17,518	100%			$17,518	$0
43	$483,602				$483,602		$483,602	30%			$145,080	$338,521
										Total CCA	$258,497	

Kasement Windows—Schedule 8 for 1998 (see Figure 7.6 for column headings)

1	2	3	4	5	6	7	8	9	10	11	12	13
1	$1,695,762	$2,600,000			$4,294,762		$4,294,762	4%			$171,790	$4,122,972
3	$185,231				$185,231		$185,231	5%			$9,262	$175,969
8	$51,176				$51,176		$51,176	20%			$10,235	$40,941
10	$15,300				$15,300		$15,300	30%			$4,590	$10,710
12	$0				$0		$0	100%			$0	$0
43	$338,521	$124,000		$28,000	$434,521	$48,000	$386,521	30%			$115,956	$318,565
										Total CCA	$311,833	

Kasement Windows—Schedule 8 for 1999 (see Figure 7.6 for column headings)

1	2	3	4	5	6	7	8	9	10	11	12	13
1	$4,122,972				$4,122,972		$4,122,972	4%			$164,919	$3,958,053
3	$175,969		$200,000		-$24,031			5%	$24,031		$8,188	$32,753
8	$40,941				$40,941		$40,941	20%				
10	$10,710			$5,000	$5,710			30%		$5,710		
12	$0	$13,000			$13,000	$6,500	$6,500	100%			$6,500	$6,500
43	$318,565	$112,000			$430,565		$430,565	30%			$129,170	$301,396
										Total CCA	$308,777	

Table 7.2 Capital Cost Allowance for Kasement Windows (Example 7.10)

Comments on 1999 Schedule 8: In Class 3, because the sale price of the warehouse is higher than the UCC at the start of the year, the intermediate UCC result in column 6 is negative. Since Kasement has actually sold the property for more than the UCC of this class, Canada Customs and Revenue Agency taxes this difference by requiring the company to add this "recaptured CCA" to their income for 1999. This amount is transferred to column 10, and no further calculations are performed for this class. When the forklift truck in Class 10 is sold, it is the last property in the class. Since it is not fully depreciated, the balance in column 6 is copied to column 11, and this "terminal loss" is deducted as an expense in 1999. The jigs are one type of Class 12 assets governed by the 50% rule. The new conveyor system is *not* subject to the 50% rule, because it became available for use in the 2nd taxation year after acquisition (i.e., following the 358$^+$ days constraint).

Answer: $CCA_{1997} = \$258,497$ $CCA_{1998} = \$311,833$ $CCA_{1999} = \$308,777$

7.4.5   CCA for Individual Projects

Since a principal aim of engineering economic analysis is to evaluate the viability of one or more independent projects, the functioning of the CCA system presents a unique challenge. Because most depreciable property is pooled into classes, the tax implications of purchasing new assets for the project being evaluated depend on activity in the class (acquisitions and dispositions) unrelated to the current project. For example, we may apply the 50%-rule for the first-year CCA of a new Class 43 machine for a proposed project, but the effect would be incorrect if any Class 43 property from elsewhere in the company is disposed of in the same year (because the 50% rule applies to the *net* acquisitions).

Another difficulty arises when predicting the tax implications upon disposal of an asset in our project evaluation. If the asset is the only property in its class, we may foresee a terminal loss or recapture upon its disposal, depending on the assumed salvage value. However, in the much more common situation where:

(1) the asset for the proposed project is outlived by other property in the class, and
(2) the asset is sold for more, or less, than its undepreciated capital cost,

then the CCA effects of this asset continues indefinitely after its disposal. That is, although the UCC of a class accounts for the sale of property (column 5 in Figure 7.6), there are residual effects on the pool in future years if the salvage value S of an asset does not equal its undepreciated capital cost U. This complicates the calculation of equivalence measures for project evaluation, particularly when using the IRR method. The details of disposal tax effects are presented in Chapter 8.

For the sake of consistency and simplicity, the following assumptions are used in this text (unless noted otherwise) for calculating the CCA of new assets on a project basis:

- Ignore any other activity in the class from outside the project. Implications of this include applying the 50% rule (if applicable) to a new asset in the project, and allowing the asset to fully depreciate every year of its useful life (which otherwise may not happen if there is a recapture of CCA in the class due to assets outside of the project).

- The **disposal tax effect** is based solely on the difference $U - S$ in the year of disposal. That is, a gain relative to its UCC on disposal ($S > U$) gets taxed, while a relative loss from the sale ($S < U$) results in tax savings. This approximation is generally acceptable since the error compared with the actual CCA effect is often small, and the disposal often occurs years into the future with little impact on the project viability (see Chapter 8 examples).

- Property is disposed of at the end of the year. This allows the disposal tax effect to coincide with the salvage value transaction.

EXAMPLE 7.11 CCA for a project

Brigitte's Bakery in Port-aux-Basques Newfoundland is considering expanding to a second location in Corner Brook. They would take out a lease on January 1st, renewable every four years, in a vacant store. The lease costs $10,000 per year. They would install $13,000 worth of equipment for selling and storing the baked goods: a cash register, display counter, refrigeration unit for pastries, etc. If sales go as well as anticipated, after 2 years they will spend $6000 on capital improvements to the store, and install an oven for $3000 so that they can produce some goods locally, rather than ship everything from Port-aux-Basques. To evaluate this proposal over an eight-year time horizon, Brigitte assumes that the salvage value of all equipment combined in eight years would be $2000. What CCA amounts need to be included in this project evaluation?

Solution

Given: Capital costs for an assortment of depreciable properties, and project life
Find: CCA over project life

As a tenant, Brigitte's Bakery has a **leasehold interest** in the store property, so any capital improvements to the building are eligible for a CCA allowance under Class 13. The capital cost allowance for this straight-line class is calculated as follows:[3]

- A *prorated portion* of the capital cost equals the lesser of:

  (a) one fifth of the capital cost

  (b) the capital cost divided by the number of 12-month periods from the start of the year when the improvement was made to the end of the original lease plus first succeeding renewal (if applicable)

---

[3] See CCRA Interpretation Bulletin IT464R. Several other conditions can apply for this class of property, with respect to multiple concurrent leases, multiple renovations, etc.

- Only half of the prorated portion may be claimed in the first year, with the balance claimable in the year following the end of the eight-year amortization period.

There are 6 years from the time of the capital improvements until the end of the lease's first renewal, so the prorated portion equals $6000/6 = $1000 (half in the first year).

The equipment and oven fall into Class 8 (CCA rate = 20%).

End-of-Year	Equipment CCA	Oven CCA	Leasehold CCA	Total CCA
1	$1300			$1300
2	$2340			$2340
3	$1872	$300	$500	$2672
4	$1498	$540	$1000	$3038
5	$1198	$432	$1000	$2630
6	$958	$346	$1000	$2304
7	$767	$276	$1000	$2043
8	$614	$221	$1000	$1835

Comments: This example illustrates some features of evaluating a project in isolation, while also demonstrating calculations for a CCA class which is not based on declining balance. By treating the project independently of other assets in the company, we have avoided several complications. The CCA claimed for the equipment in year 1 is only accurate if no other equipment in that class was disposed of in that year. The same holds for the acquisition of the oven. The CCA in any year for the equipment or the leasehold improvement could be disallowed if the UCC were driven to zero by the disposition of property outside of the project. By using the disposal tax effect simplification described above, where U_8 = $3338 and S_8 = $2000, then a terminal loss on the equipment of $1338 is assumed at the end of the eight years, independently of any other equipment in that class owned by Brigitte's. A similar adjustment may be made for the leasehold which has a UCC of $500 after eight years, and an assumed salvage value of zero.

7.5 Additions or Alterations to Depreciable Assets

If any major alterations or repairs (engine overhaul) or additions (improvements) are made during the life of the asset, we need to determine whether these actions will extend the life of the asset or will increase the originally estimated salvage value. When either of these situations arises, a revised estimate of useful life should be made, and the periodic depreciation expense should be updated accordingly. We will examine how repairs, alterations or improvements affect both book and tax depreciations.

7.5.1   Revision of Book Depreciation

Recall that book depreciation rates are based on estimates of the useful life of assets. Estimates of useful life are seldom precise. Therefore, after a few years of use you may find that the asset could last for a considerably longer or shorter period than was originally estimated. If this happens, the annual depreciation expense, based on the estimated useful life, may be either excessive or inadequate. (If the repairs or improvements do not extend the life or increase the salvage value of the asset, these costs may be treated as maintenance expenses during that year.) The procedure for correcting the book depreciation schedule is to revise the current book value and to allocate this cost over the remaining years of useful life.

7.5.2   Revision of Tax Depreciation

Under the capital cost allowance system, capital expenditures (additions or alterations) on existing depreciable property are treated as new depreciable property acquisitions. The CCA class of the capital expenditure on additions or alterations is the class that would apply to the original property if it were made available-for-use at the same time as the addition or alteration.

Aside from the general prerequisites provided above regarding increased life or value of the property, Canada Customs and Revenue Agency publishes guidelines on how to distinguish between current expenditures on maintenance and repairs, versus capital upgrades (IT-128R bulletin). The principles include a test of **enduring benefit**. A part, added to a depreciable property, which requires regular replacement is not considered a capital expenditure. A distinction is also made between **maintenance** and **betterment**. Bringing a property back to its original condition is one indication that the repair should be treated as a current expenditure, unless it involves a material improvement beyond the property's original condition. For example, replacing a roof is normally a current expense, unless the new roof is clearly of better quality or durability than the original one. Other criteria involve the **relative value** of the addition or improvement compared with the capital cost of the existing property, and whether the upgrade is an **integral part** of the original asset, or a separate marketable asset. If the relative value of the addition is small, or the new part is an integral part of the original property, the cost is more likely treated as a current expense. If in doubt, these and related criteria may be reviewed with a qualified accountant or other tax expert.

EXAMPLE 7.12 Depreciation adjustment for an overhauled asset

In January 1995, Kendall Manufacturing Company purchased a new numerical control machine at a cost of $60,000. The machine had an expected life of 10 years at the time of purchase and a zero expected salvage value at the end of the 10 years. For book depreciation purposes, no major overhauls had been planned for that period and the machine was being depreciated using the straight-line method toward a zero salvage value, or $6000 per year. For tax purposes, the machine was Class 43 property (CCA rate = 30%). In January, 1998, however, the machine was thoroughly overhauled and rebuilt at a cost of $15,000. It was estimated that the overhaul would extend the machine's useful life by 5 years (see Figure 7.7).

(a) Calculate the book value and depreciation for the year 2000 on a straight-line basis.

(b) Calculate the CCA and UCC for the year 2000 for this machine.

Solution

Given: $P = \$60,000$, $S = \$0$, $N = 10$ years, machine overhaul = $15,000, extended life = 15 years from the original purchase date
Find: B_6 and D_6 for book depreciation, U_6 and CCA_6 for tax depreciation

(a) Since an improvement was made at the beginning of 1998, the book value of the asset at that time consists of the existing book value plus the cost added to the asset. First, the existing book value at the end of 1997 is calculated:

$$B_3 \text{ (before improvement)} = \$60,000 - 3(\$6000) = \$42,000.$$

After adding the improvement cost of $15,000, the revised book value is

$$B_3 \text{ (after improvement)} = \$42,000 + \$15,000 = \$57,000.$$

To calculate the book depreciation in the year 2000, which is 3 years after the improvement, we need to calculate the annual straight-line depreciation amount with the extended useful life. The remaining useful life before the improvement was made was 7 years. Therefore, the revised remaining useful life should be 12 years. The revised annual depreciation is then $57,000/12 = $4750. Using the straight-line depreciation method, we compute the book value and depreciation for year 2000 as follows:

$$B_6 = \$42,750 \quad D_6 = \$4750.$$

(b) For tax depreciation, overhaul expenditure is treated as new property. The amount is added to Class 43, because *if* the original machine had been bought in 1998, it would have been placed in Class 43 at that time. To determine the total CCA and UCC in year 6, we calculate them separately for the two properties (the original asset and the "improvement") and then add them.

	Original Machine		Overhaul	
Year	CCA (d = 30%)	UCC (year-end)	CCA (d = 30%)	UCC (year-end)
1995	$9,000*	$51,000		
1996	$15,300	$35,700		
1997	$10,710	$24,990		
1998	$7,497	$17,493	$2,250*	$12,750
1999	$5,248	$12,245	$3,825	$8,925
2000	$3,674	$8,571	$2,678	$6,247

* The 50% rule applies to both assets in their respective first years.

The total CCA for this machine and its overhaul in the year 2000 is $6352 (3674 + 2678) resulting in an end-of-year UCC of $14,818 (8571 + 6247).

These results could also be obtained directly by using the CCA formulas from section 7.4.4:

For the original machine, $n = 6$ years (from acquisition to the end of year 2000):

$$CCA_{2000} = Pd(1 - d/2)(1 - d)^{n-2} = \$60,000(0.30)(1 - 0.15)(1 - 0.30)^4 = \$3674$$
$$U_{2000} = P(1 - d/2)(1 - d)^{n-1} = \$60,000(1 - 0.15)(1 - 0.30)^5 = \$8571$$

For the overhaul, $n = 3$ years (from improvement completed to the end of year 2000):

$$CCA_{2000} = Pd(1 - d/2)(1 - d)^{n-2} = \$15,000(0.30)(1 - 0.15)(1 - 0.30)^1 = \$2678$$
$$U_{2000} = P(1 - d/2)(1 - d)^{n-1} = \$15,000(1 - 0.15)(1 - 0.30)^2 = \$6247$$

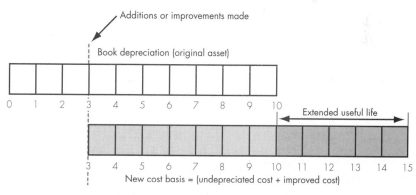

Figure 7.7 Revision of book depreciation as additions or improvements are made as described in Example 7.12

7.6 Natural Resource Allowances (Optional)

If you own a **resource property** such as oil, gas, mines, or timber, you may be able to claim a deduction as you deplete the resource. A capital investment in natural resources needs to be recovered as the natural resources are being removed and sold. The process of amortizing the cost of natural resources in the accounting periods is called **depletion**. The objective of depletion is the same as that for depreciation: to amortize the cost in a systematic manner over the asset's useful life.

Many of our industries in Canada are resource-related, and engineers play an integral role in all aspects of exploration, development, transportation and processing of these materials. It is important to be aware of the basic principles of depletion, as such allowances are a key aspect of financial accounting reports for resources. Since each province owns its own natural resources, the tax regime for these activities is very complex. The following sections illustrate some of the main deductions applicable under the *federal* corporate income tax regulations. Consultation with tax experts is essential to ensure proper treatment for a detailed after-tax economic analysis.

7.6.1 Depletion

The cost depletion method is based on the same concept as the units-of-production method. To determine the amount of cost depletion, the adjusted cost basis of the resource property is divided by the total number of recoverable units in the deposit and the resulting rate is multiplied by the number of units sold:

$$\text{Cost depletion} = \frac{\text{(adjusted basis of resource property)}}{\text{total number of recoverable units}} \times \text{(number of units sold)} \quad (7.13)$$

The adjusted basis represents all the depletion allowed (or allowable cost on the property). Estimating the number of recoverable units in a resource deposit is largely an engineering problem. Most capital equipment and structures associated with resource recovery fall into special CCA classes, in which case they do not form part of the capital cost base of the resource property itself.

EXAMPLE 7.13 **Cost depletion for an oil company**

Clampett Oil Inc. owns an oil well with estimated reserves of 1.8 million barrels (bbls). Capitalized costs come to $27 million, including acquisition costs, geological and geophysical expenditures, cost of drilling, and estimated future removal and site restoration costs. If they pump 200,000 bbls in the year 2000, determine the depletion allowance.

Solution

Given: Basis = $27 million, total recoverable volume = 1.8 million bbls, amount sold this year = 200,000 bbls

Find: The depletion allowance this year

Depletion allowance per bbl	= $27,000,000/1,800,000
	= $15.00/bbl.
Depletion allowance for the year	= 200,000 bbls x $15.00/bbl
	= $3,000,000.

A **depletion allowance** is one mechanism for the government to allow deductions with respect to resource properties. The amount of allowance is generally calculated based on an **earned depletion base**, and typically capped at a maximum value each year which depends on annual profits. In recent years the federal government has phased out depletion allowances in favour of other types of tax savings for resource properties. Since depletion that is earned, but unclaimed, can be carried forward indefinitely, allowances may still be deducted by some companies based on previous programs. Capital cost allowances also apply to certain resource properties, as shown below.

7.6.2   Allowance for Industrial Mineral Mines

Depletion allowances, as described above, are only applicable to **mineral resources**, but in the tax regulations, this is distinct from **industrial mineral mines**. The term industrial mineral means any non-metallic mineral capable of being used in industry, such as gravel, limestone, quartz or sulphur. More specific guidelines for distinguishing between these categories appear in CCRA Interpretation Bulletin IT-492.

Industrial mineral mines are eligible for a capital cost allowance deduction, calculated in a similar way to the units-of-production depreciation method. The capital cost of the mine minus its estimated residual value (once fully mined) is divided by the expected number of units of commercially mineable material. This provides an allowance rate in dollars per unit mined. In each tax year, the maximum CCA equals this rate times the number of units mined during the year. Of course, provisions are made in the income tax regulations for reassessing the depreciation rate if the estimated value of the property, number of mineable units, or other relevant quantities change significantly. Most capital mining equipment falls into CCA Class 10 ($d = 30\%$), while most mining structures and buildings are subject to a declining balance rate of 25% under CCA Class 41.

EXAMPLE 7.14 — CCA for an industrial mineral mine

Larmac Ltd. has just paid $733,000 for a mine in British Columbia, from which they will be extracting primarily mica for use in plastics and paints. The mine has been certified as an industrial mineral mine. The proven and probable ore reserves total about 250,000 tonnes, and at planned extraction rates the mine life is estimated to be 10 years, at the end of which the assessed residual value of the property is $50,000. Larmac mines 15,000 tonnes in the first year, 20,000 tonnes in the second year, and 25,000 tonnes each year thereafter once they are up to full capacity. After four years of operating the mine, exploratory drilling has resulted in estimates of additional reserves of 130,000 tonnes, with no change in the residual value once all the mineable material is removed. Calculate the CCA for this property over the first 7 years of operation.

Solution

Given: The ore reserve estimates, capital cost and estimated residual value
Find: The capital cost allowance each year

The allowance rate equals the capital cost less the residual value, divided by the number of mineable units:

CCA rate = (733,000 − 50,000)/250,000 = $2.732/tonne

Applying this rate to the actual extraction each year (units-of-production approach) gives:

End-of-Year	Reserves (tonnes)	CCA	UCC
0	250,000		$733,000
1	235,000	2.732 * 15,000 = $40,980	$692,020
2	215,000	2.732 * 20,000 = $54,640	$637,380
3	190,000	2.732 * 25,000 = $68,300	$569,080
4	165,000	2.732 * 25,000 = $68,300	$500,780

With the new discoveries, at the end of the 4th year the remaining reserves equal 295,000 tonnes. A revised CCA rate is calculated using the UCC at the beginning of year 5:

revised CCA rate = (500,780 − 50,000)/295,000 = $1.528/tonne

The CCA over the following three years is given by:

End-of-Year	Reserves (tonnes)	CCA	UCC
4 (after revision)	295,000		$500,780
5	270,000	1.528 * 25,000 = $38,200	$462,580
6	245,000	1.528 * 25,000 = $38,200	$424,380
7	222,000	1.528 * 25,000 = $38,200	$386,180

Comments: The exploration costs are considered current expenses in this case, which is why they were not added to the UCC at the end of year 4. Each mine that is eligible for CCA must be placed in a separate class. If the taxpayer does not claim the maximum allowance in a year, this does not give the right to adjust the CCA rate. Land acquired as part of the mine *is included in the depreciable property.*

7.6.3   Timber Allowances

It is first necessary to determine whether an acquired property is a **timber resource property** or a **timber limit** (or **cutting right**), because their respective income tax treatments differ significantly. A synopsis of the criteria and deductions allowed for each category is included here, but a more thorough discussion is available in the CCRA Interpretation bulletin IT-481.

Conditions for a property to be classified as a timber resource property include:

- a right or license to cut or remove timber from an area in Canada;
- the original right was acquired after May 6, 1974;
- the original right is extendable, renewable, or can be substituted for by another right.

If all relevant stipulations are met to qualify as a timber resource property, then a capital cost allowance is available under Class 33, at a CCA rate of 15%. All timber resource properties owned by the taxpayer are pooled into Class 33. If the net proceeds of disposition exceed the undepreciated capital cost of the class, the difference is added to income, as in recaptured CCA. Also, a timber resource property does not qualify for capital gains. (See Chapter 8.)

The following considerations apply to a timber limit or cutting right:

- a cutting right that does not satisfy the requirements for a timber resource property is a timber limit; this includes acquisition before May 7, 1974, or a nonrenewable right;
- if land is acquired with standing timber, it is classified as a timber limit.

The capital cost allowance calculations for timber limits are fairly detailed, but in essence they are calculated in a similar way to the units-of-production depreciation method. A rate is calculated as the capital cost of the limit minus its estimated residual value, divided by the quantity of timber in the limit (expressed in cords, board feet or cubic metres). In each tax year, the maximum CCA equals this rate times the number of units cut during the year, plus a percentage of eligible costs spent to survey and prepare to obtain the right. Each timber limit of the taxpayer must be placed in a separate class. Land acquired as part of a timber limit is *included in the depreciation base* for the timber property.

Certain assets used in resource industries are given preferential deductions, such as immediate write-offs, or higher CCA rates. For example, "immovable woods assets" such as camp buildings, railway track or wharves, acquired for cutting and removing on a timber limit may be depreciated in proportion to the rate of timber removal (i.e., units-of-production) according to Class 15 rules. In fact, such equipment may be written off as current expenses if their useful life is less than three years, due to all of the timber in the area being removed within that time frame. Other timber cutting and removing equipment belongs in CCA Class 10.

7.7 Computer Notes

Most depreciation calculations can easily be obtained using computers. Excel has several built-in functions for doing depreciation calculations.[4] Table 7.2 summarizes the built-in depreciation functions for Excel. In this section, we will briefly examine these features.

In Figure 7.8, the book and tax depreciation for the data in Example 7.3 are presented. The book depreciation schedules in columns B, C and D are calculated using built-in functions from Table 7.2. The VDB function even allows for switching from declining balance to straight-line if desired, as described in section 7.3.2. In Excel, no financial functions are equivalent to calculating capital cost allowances. Therefore, you need to provide your own programming when you wish to consider features such as the 50% rule and non-DB classes. Even the DB function available in Excel is not suitable for Canadian taxes, because the depreciation rate is automatically calculated based on the property's useful life, rather than externally specified as required by the CCA tables. Fortunately, it is easy to generate the appropriate CCA schedule using formulas as shown in columns E and F of Figure 7.8.

Table 7.2
Excel's Built-In Depreciation Functions

Function Description	Excel
Straight-line depreciation	SLN (cost, salvage, life)
Double-declining-balance depreciation	DDB (cost, salvage, life, period, factor)
Sum-of-years'-digits depreciation	SYD (cost, salvage, life, period)
Declining-balance depreciation (other than DDB; assumes no switching)	VDB (cost, salvage, life, life start, end period, factor, no switch)

---

[4] **EzCash** can easily generate depreciation schedules for DB and SL capital cost allowances.

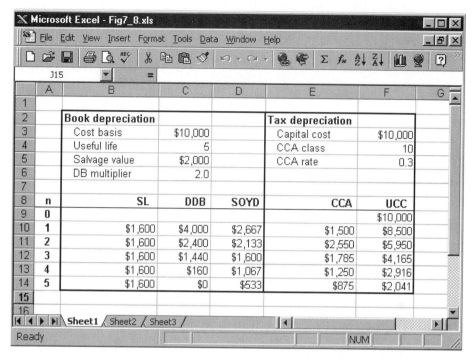

Figure 7.8 An Excel depreciation worksheet. Note that none of the electronic spreadsheets available in the market have built-in CCA calculations

7.8 Summary

- Machine tools and other manufacturing equipment, and even the factory buildings themselves, are subject to wear over time. This **economic depreciation** generally results in increased maintenance and lower reliability; as assets age, periodic replacement of capital property is required.

- The entire acquisition cost of a machine cannot be properly charged to any one year's production; rather, the cost should be spread (or capitalized) over the years in which the machine is in service, in accordance with the **matching principle**. The cost charged to operations during a particular year is called **depreciation**. Several different meanings and applications of depreciation have been presented in this chapter. From an engineering economics point of view, our primary concern is with **accounting depreciation**: The systematic allocation of an asset's value over its depreciable life.

- Accounting depreciation can be broken into two categories:
  1. **Book depreciation**—the method of depreciation used for financial reports and pricing products;

2. **Tax depreciation**—the method of depreciation used for calculating taxable income and income taxes; it is governed by tax legislation.

- The four components of information required to calculate depreciation are:
  1. The cost basis of the asset,
  2. The salvage value of the asset,
  3. The depreciable life of the asset, and
  4. The method of its depreciation.

Table 7.3 summarizes the differences in the way these components are treated for purposes of book and tax depreciation.

- Most firms in Canada select straight-line depreciation for book depreciation because of its relative ease of calculation.

- Many **capital cost allowance (CCA)** classes employ accelerated methods of depreciation, which gives taxpayers a break: It allows them to take earlier and faster advantage of the tax-deferring benefits of depreciation.

- Among the many regulations specific to tax depreciation, two rules are prevalent. The **available-for-use rule** dictates when property may be included in the capital cost base for its class. The **50% rule**, which applies to most CCA classes, specifies that only half of the amount of net acquisitions may be used for calculation of the first year's CCA.

- If the **undepreciated capital cost (UCC)** in a class becomes negative, a **recapture of CCA** is triggered. If the UCC is greater than zero in the year that the final asset in a class is disposed of, a **terminal loss** may be claimed.

- Natural resource property gets used up as the resource is extracted. Special **depletion** allowances or capital cost allowances may be applied to mining, oil and gas, and timber properties to amortize investment costs over the productive life of the resource.

- Given the frequently changing nature of depreciation and tax law, we must use whatever rates, classifications, and adjustments, mandated *at the time an asset is acquired.*

Table 7.3

Summary of Book Versus Tax Depreciation

Components of Depreciation	Book Depreciation	Tax Depreciation (CCA)
Cost basis (Capital cost)	Based on the actual cost of the asset, plus all incidental costs such as freight, site preparation, installation, etc.	Same as book depreciation
Salvage value (Proceeds of disposition)	Estimated at the outset of depreciation analysis. If the final book value does not equal the estimated salvage value, we may need to make adjustments in our depreciation calculations.	Estimated in order to calculate the terminal loss or recapture of CCA. Also needed for most SL classes, and some depletion calculations.
Depreciable life	Firms may select their own estimated useful lives, following generally acceptable accounting practices.	Not applicable to declining balance classes. Estimates required for other classes.
Method of depreciation	Firms may select from the following: • straight-line • accelerated methods (declining balance, double declining balance, and sum-of-years'-digits) • units-of-production	Depreciation method is specified by class; primarily declining balance and straight-line. Units-of-production is commonly applied to resource properties.

Problems

Assumptions:

- Tax year equals the calendar year (unless specified otherwise).
- Property is available-for-use when acquired (unless specified otherwise).
- 50% rule for CCA applies as described in the text.

Level 1

7.1* A machine, purchased for $45,000, had a depreciable life of 4 years. It will have an expected salvage value of $5000 at the end of the depreciable life. Using the straight-line method, what is the book value at the end of year 2?

7.2* Consider Problem 7.1. If the double declining balance (200% DB) method is used, what is the depreciation amount for year 2?

7.3* Consider Problem 7.1. If the sum-of-the-years'-digits (SOYD) method is used, what is the depreciation amount for year 2?

7.4* Your accounting records indicate that an asset in use has an undepreciated capital cost of $11,059. The asset cost $30,000 when it was purchased, and it has been depreciated under CCA Class 8 (rate = 20%). Based on the information available, determine how many years the asset has been in service.

7.5 A machine now in use was purchased 4 years ago at a cost of $5000. It has a book value of $1300. It can be sold for $2300, but could be used for 3 more years, at the end of which time it would have no salvage value. What is the amount of economic depreciation now for this asset?

7.6 To automate one of their production processes, Waterloo Corporation bought three flexible manufacturing cells at a price of $500,000 each. When they were delivered, Waterloo paid freight charges of $25,000 and handling fees of $12,000. Site preparation for these cells cost $35,000. Six foremen, each earning $15 an hour, worked five 40-hour weeks to set up and test the manufacturing cells. Special wiring and other materials applicable to the new manufacturing cells cost $1500. Determine the cost basis (amount to be capitalized) for these cells.

7.7* Leo Smith Inc. paid $495,000 in 1998 for a used office building. This purchase price represents $375,000 for the building and $120,000 for the land. Calculate the capital cost allowance over the first 5 years of ownership.

7.8* A new drill press was purchased for $95,000 by trading in a similar machine that had a book value of $25,000. Assuming that the trade-in allowance is $20,000 and that $75,000 cash is to be paid for the new asset, what is the cost basis of the new asset for book depreciation purposes?

7.9* A lift truck priced at $35,000 is acquired by trading in a similar lift truck and paying cash for the remaining balance. Assuming that the trade-in allowance is $10,000, and the book value of the asset traded in is $6000, what is the cost basis of the new asset for the computation of book depreciation?

7.10* A firm is trying to decide whether to keep an item of construction equipment which typically lasts for 8 years. The firm is using DDB for book depreciation purposes, and this is the fourth year of ownership. The item cost $150,000 new. What was the depreciation in year 3?

7.11 Compute the double declining balance (DDB) depreciation schedule for the following asset with switching to straight-line:

Cost of the asset, P	$60,000
Useful life, N	8 years
Salvage value, S	$5000

7.12* Compute the SOYD depreciation schedule for the following asset:

Cost of the asset, P	$12,000
Useful life, N	5 years
Salvage value, S	$2000

(a) What is the denominator of the depreciation fraction?

(b) What is the amount of depreciation for the first full year of use?

(c) What is the book value of the asset at the end of the fourth year?

7.13 A truck for hauling coal has an estimated net cost of $85,000 and is expected to give service for 250,000 kilometres. Its salvage value will be $5000 and depreciation will be charged at a rate of 32 cents per kilometre. Compute the allowed depreciation amount for the truck usage amounting to 55,000 kilometres per year.

7.14* A diesel-powered generator with a cost of $60,000 is expected to have a useful operating life of 50,000 hours. The expected salvage value of this generator is $8000. In its first operating year, the generator was operated 5000 hours. Determine the depreciation for the year.

7.15 The Harris Foundry Company purchased new casting equipment in 1999 at a cost of $180,000. Harris also paid $35,000 to have the equipment delivered and installed. The casting machine has an estimated useful life of 12 years, and it will be depreciated as a Class 43 asset (CCA rate = 30%).

(a) What is the capital cost of the casting equipment?

(b) What will be the CCA each year for the life of the casting equipment?

7.16* An item of equipment falls into CCA Class 8. Compute the undepreciated capital cost at the end of 3 years. The capital cost is $100,000.

7.17* A piece of machinery purchased at a cost of $68,000 has an estimated salvage value of $9000 and an estimated useful life of 5 years. It was placed in service on May 1 of the current fiscal year, which ends on December 31. The asset falls into CCA Class 43. Determine the capital cost allowances over the useful life.

7.18 Suppose that a taxpayer acquires new drafting software for $10,000. Calculate the CCA and UCC over the next 4 years.

Level 2

7.19 General Service Contractor Company paid $100,000 for a house and lot. The value of the land was appraised at $65,000, and the value of the house at $35,000. The house was then torn down at an additional cost of $5000 so that a warehouse could be built on the lot at a cost of $50,000. What is the total value of the property with the warehouse? For book depreciation purposes, what is the cost basis for the warehouse?

7.20 Consider the following data on an asset:

Cost of the asset, P	$100,000
Useful life, N	5 years
Salvage value, S	$10,000

Compute and graph the annual depreciation allowances and the resulting book values, using

(a) The straight-line depreciation method,

(b) Declining balance method (with switching to SL if required), and

(c) Sum-of-the-years' digits method.

7.21 Consider the following data on an asset:

Cost of the asset, P	$30,000
Useful life, N	7 years
Salvage value, S	$8000

Compute the annual depreciation allowances and the resulting book values using the DB with switching to SL if required.

7.22 The double declining balance method is to be used for an asset with a cost of $80,000, estimated salvage value of $22,000, and estimated useful life of 6 years.

(a) What is the depreciation for the first 3 fiscal years, assuming that the asset was placed in service at the beginning of the year?

(b) If switching to the straight-line method is allowed, when is the optimal time to switch?

7.23 Upjohn Company purchased new packaging equipment with an estimated useful life of 5 years. Cost of the equipment was $20,000 and the salvage value was estimated to be $3000 at the end of year 5. Compute the annual depreciation expenses through the 5-year life of the equipment under each of the following methods of book depreciation:

(a) Straight-line

(b) Double declining balance method (limit the depreciation expense in the fifth year to an amount that will cause the book value of the equipment at year end to equal the $3000 estimated salvage value).

(c) Sum-of-years'-digits method.

7.24* A second-hand bulldozer acquired at the beginning of the fiscal year at a cost of $58,000 has an estimated salvage value of $8000 and an estimated useful life of 12 years. Determine the following:

(a) The amount of annual depreciation by the straight-line method,

(b) The amount of depreciation for the third year computed by the declining balance method,

(c) The amount of depreciation for the second year computed by the sum-of-years'-digits method.

7.25 Ingot Land Company owned four trucks dedicated primarily for its landfill business. The company's accounting record indicates the following:

	Truck Type			
Description	A	B	C	D
Purchase cost ($)	50,000	25,000	18,500	35,600
Salvage value ($)	5,000	2,500	1,500	3,500
Useful life (kilometres)	200,000	120,000	100,000	200,000
Accumulated depreciation as year begins ($)	0	1,500	8,925	24,075
Kilometres driven during year	25,000	12,000	15,000	20,000

Determine the amount of depreciation for each truck during the year using units-of-production method.

7.26 Zerex Paving Company purchased a hauling truck on January 1, 2000, at a cost of $32,000. The truck has a useful life of 8 years with an estimated salvage value of $5000. The straight-line method is used for book purposes. For tax purposes the truck would be depreciated as Class 10 property with a declining balance rate of 30%. Determine the annual depreciation amount to be taken over the useful life of the hauling truck for both book and tax purposes.

7.27* On October 1, you purchased a residential home in which to locate your professional office for $150,000. The appraisal is divided into $30,000 for the land and $120,000 for the building.

(a) In your first year of ownership, how much CCA can you deduct? (Assume that the entire house is used for business.)

(b) Suppose that the property was sold at $187,000 at the end of the 4th year of ownership. What is the undepreciated capital cost of the property?

7.28 On July 9th 1996, Twovens Ltd. purchased a used spindle machine at a bankruptcy sale for $34,000. They finally installed it in their factory on January 12th 1999. Calculate the CCA and UCC for this Class 43 machine for 5 years, starting in the year when it first became available for use.

7.29 Armour Construction Company has just acquired a new Structural-Testing Sledge Hammer for $7500. By hitting a bridge or other structure, the feedback signals from the piezoelectric load cell embedded in the hammer head gives the engineer valuable information on structural integrity and resonant frequencies. They expect the equipment to last for 6 years, with a zero salvage value at that time. For book depreciation, Armour will use declining balance, with the rate determined by the useful life (no 50% rule). For tax purposes, the hammer falls into Class 8 property. Using the depreciation formulas presented in Section 7.4.4, calculate the book depreciation and capital cost allowance in the sixth year of ownership, and the UCC at the end of that time.

7.30 A manufacturing company has purchased four assets:

	Asset Type			
Item	Lathe	Truck	Building	Photocopier
Initial cost ($)	45,000	25,000	800,000	40,000
Book life	12 yr	200,000 mi	50 yr	5 yr
CCA class	43	10	1	8
Salvage value ($)	3,000	2,000	100,000	0
Book depreciation	DDB	UP	SL	SOYD

For book depreciation, the units-of-production (UP) method was used for the

truck. Usage of the truck was 22,000 kilometres and 25,000 kilometres during the first 2 years, respectively.

(a) Calculate the book depreciation for each asset in the first 2 years.

(b) Calculate the capital cost allowance for each asset in the first 2 years.

(c) If the lathe is to be depreciated over the early portion of its life using DDB and then by switching to SL for the remainder of the asset's life, when should the switch occur?

7.31 For each of four assets in the following table, determine the missing amounts for the year indicated. For asset type IV, annual usage is 15,000 kilometres.

Types of Asset	I	II	III	IV
Depreciating methods	SL	DDB	SOYD	UP
End of year	—	4	3	3
Initial cost ($)	10,000	18,000	—	30,000
Salvage value($)	2,000	2,000	7,000	0
Book value ($)	3,000	2,333	—	—
Depreciable life	8 yrs	5 yr	5 yr	90,000 km
Depreciating amount ($)	—	—	16,600	—
Accumulated depreciation ($)	—	15,680	66,400	—

7.32 Flint Metal Shop purchased a stamping machine for $147,000 on March 1, 2000. It is expected to have a useful life of 10 years, salvage value of $27,000, production of 250,000 units, and working hours of 30,000. During 2000 Flint used the stamping machine for 2450 hours to produce 23,450 units. From the information given, compute the book depreciation expense for 2000 under each of the following methods:

(a) Straight-line

(b) Units of production method

(c) Working hours

(d) Sum-of-years'-digits

(e) Declining balance

(f) Double declining balance.

7.33 Three assets were purchased in February, 1998 and placed in service according to the following table.

Property Type	Date Placed in Service	Capital Cost	CCA Class
Car	Feb. 12, 1998	$15,000	10
Arc Welder	Dec. 22, 1998	$12,000	43
Freezer	Jan. 6, 1999	$8,000	8

Compute the CCA by year for each asset up to, and including, the year 2002.

7.34 Otto-Rentals Ltd. is setting up a new car rental operation. In the first year they purchase twelve vehicles for a total of $148,000. In year 2, they buy two more vehicles for $27,000. In year 3, they sell four vehicles for $29,000, and buy five new cars for $71,000. In year 4, they sell three vehicles for $16,000. These vehicles belong in CCA Class 16. Using columns similar to Schedule 8 (Figure 7.6), calculate the CCA and UCC for the four years for this pool of vehicles.

7.35 Given the data below, identify the depreciation method used for each depreciation schedule as one of the following:

- Double declining balance (DDB) depreciation
- Sum-of-years'-digits depreciation
- DDB with conversion to straight-line, assuming a zero salvage value
- CCA with 50% rule.

First cost	$80,000
Book depreciation life	7 years
Salvage value	$24,000
CCA Class	8

Depreciation Schedule

n	A	B	C	D
1	$14,000	22,857	8,000	22,857
2	12,000	16,327	14,400	16,327
3	10,000	11,661	11,520	11,661
4	8,000	5,154	9,216	8,330
5	6,000	0	7,373	6,942
6	4,000	0	5,898	6,942
7	2,000	0	4,719	6,942
8	0	0	3,775	0

7.36* Having acquired a right to mine limestone on acreage near Kingston, Ontario, the capital cost associated with the property came to $335,000 in the year of acquisition for Bethel Construction. The estimated quantity of mineable limestone at the outset was 475,000 cubic metres. Bethel does not retain any residual rights to the property once the license has expired. In the first year of operation, Bethel extracts 45,000 cubic metres of limestone, which increases by 5% each year thereafter. Calculate the CCA of this industrial mineral mine for its first four years of operation.

7.37 Eldridge Inc., a large environmental consulting company, wishes to set up an office in Winnipeg. They sign a 7-year lease for commercial space, with no option for renewal. Occupancy begins next January 1st. In January they undertake major renovations, including new walls to partition the offices, an air conditioning system, and replacement of the drafty casement windows with new vinyl-clad sliding models. It is ready for the employees to move in on February 1st. The capital cost of these improvements totals $22,000, and since Eldridge has a leasehold interest, they may claim CCA under Class 13. Calculate the maximum CCA claimable in years 1 to 3.

Level 3

7.38* Perkins Construction Company bought a building for $1,200,000 25 years ago; it is to be used as a warehouse. A number of major structural repairs, completed at the beginning of the current year at a cost of $125,000, are expected to extend the life of the building 10 years beyond the original estimate. The building has been depreciated by the straight-line method. Salvage value is expected to be negligible and has been ignored. The book value of the building before the structural repairs is $400,000.

(a) What has the amount of annual depreciation been in past years?

(b) What is the book value of the building after the repairs have been recorded?

(c) What is the amount of depreciation for the current year, using the straight-line method?

7.39 The Dow Ceramic Company purchased a glass molding machine in 1996 for $140,000. The company has been depreciating the machine over an estimated useful life of 10 years, assuming no salvage value, by the straight-line method of depreciation. For tax purposes, the machine has been depreciated as a Class 43 property. At the beginning of 1999, Dow overhauled the machine at a cost of $25,000. As a result of the overhaul, Dow estimated that the useful life of the machine would extend 5 years beyond the original estimate.

(a) Calculate the book depreciation for year 2001.

(b) Calculate the CCA for year 2001.

7.40 Three recent Industrial Engineering graduates are considering setting up a new consulting company, ErgoTech Ltd., to provide services on ergonomic designs and evaluations. Initially, they expect to invest $7000 on furniture, photocopier and testing equipment (Class 8), $3500 on a computer including systems software (Class 10), $2400 on general and specialized software (Class 12), and $13,000 for a small car (Class 10) to visit clients. If the business grows as well as expected, they foresee the following property transactions over the coming years:

Year 1: purchase additional ergonomics software for $800

Year 2: purchase a second computer for $4000, and specialized software for $1000

Year 3: sell the car for $6000 and purchase a van for $20,000 for ease of transporting testing equipment; sell the old photocopier

for $1000, and buy a better model costing $2500

Year 4: sell the original computer for $500 (uninstalling all software except the systems software); spend $2000 to upgrade the motherboard, RAM and video card on the second computer which will significantly improve its performance

Fill out a table similar to Schedule 8 shown in Figure 7.6 to show the capital cost allowances for each of these classes, and an annual total, for ErgoTech's first four years of business.

7.41 On January 2, 1995, Hines Food Processing Company purchased a machine priced at $75,000 that dispenses a premeasured amount of tomato juice into a can. The estimated useful life of the machine is estimated at 12 years with a salvage value of $4500. At the time of purchase, Hines incurred the following additional expenses:

Freight-in	$800
Installation cost	2500
Testing costs prior to regular operation	1200

Book depreciation was calculated by the straight-line method but, for tax purposes, the machine belonged in Class 43. In January 1997, accessories costing $2000 were added to the machine to reduce its operating costs. These accessories need to be replaced every year as they wear out.

(a) Calculate the book depreciation expense for 1998.

(b) Calculate the tax depreciation expense for 1998.

7.42 On January 2, 1994, Allen Flour Company purchased a new machine at a cost of $63,000. Installation costs for the machine were $2000. The machine was expected to have a useful life of 10 years with a salvage value of $4000. The company uses straight-line depreciation for financial reporting. On January 3, 1996, the machine broke down, and an extraordinary repair had to be made to the machine at a cost of $6000. The repair resulted in extending the machine's life to 13 years, but left the salvage value unchanged. On January 2, 1997, an improvement was made to the machine in the amount of $3000, which increases the machine's productivity and increases the salvage value to $6000, but does not affect the remaining useful life. Determine book depreciation expenses for the years December 31, 1994, 1995, 1996, and 1997.

7.43 At the beginning of the fiscal year, Borland Company acquired new equipment at a cost of $65,000. The equipment has an estimated life of 5 years and an estimated salvage value of $5000.

(a) Determine the annual depreciation (for financial reporting) for each of the 5 years of estimated useful life of the equipment, the accumulated depreciation at the end of each year, and the book value of the equipment at the end of each year by (1) straight-line method, (2) the double declining balance method, and (3) the sum-of-years'-digits method.

(b) Determine the annual depreciation for tax purposes assuming that the equipment falls into CCA class 43.

(c) Assume that the equipment was depreciated under CCA class 43. In the first month of the fourth year, the equipment was traded in for similar equipment priced at $82,000. The trade-in allowance on the old equipment was $10,000, and cash was paid for the balance. Calculate the CCA for Class 43 in the fourth year, and the UCC at the end of that year.

7.44 A coal mine expected to contain 6.5 million tonnes of coal was purchased at a cost of $50 million. One million tonnes of coal are produced this year. The gross income for this coal is $600,000, and operating costs (excluding depletion expense) are $450,000. What is the depletion allowance for book purposes using the cost depletion method?

7.45 La Compagnie Concordier purchased a right to cut timber for 5 years in northern New Brunswick. The capital cost of acquiring the right comes to $730,000. The licence specifies that the lease is renewable every 5 years subject to adhering to certain environmental stipulations. This means that the acquisition cost is depreciated as a timber resource property, in Class 33 with a rate of 15%. Concordier also purchased $175,000 worth of logging equipment, which belongs in CCA Class 10. Calculate the CCA and UCC for each of these two classes of property for the first four years of operation.

7.46 Fitzgerald-Harker Ltd. has just purchased 125,000 hectares of land with standing timber for $8.2 million. The total estimated yield is 300,000 cords of wood, harvestable over 20 years. The company estimates that the value of the land once the timber is depleted would be about $1 million. The capital costs of "immovable woods assets" come to $1.3 million, for the installation of roads, culverts, and the logging camp. The range and useful life of these assets is assumed to apply to about one-quarter of the total timber limit. Fitzgerald-Harker also acquires logging equipment for $275,000 which belongs in Class 10, including a harvester, tractor, skidder, and barker. The production rates for the first three years of operation are 12,000, 13,000 and 16,000 cords respectively. Based on the information provided, calculate the total CCA for this operation for the first three years of production.

7.47 There are aspects of the capital cost allowance system that frequently arise during engineering economics evaluations, but are not described in detail in the T2 Tax Guide.

For each of the following, look up the associated Interpretation Bulletin on the Canada Customs and Revenue Agency web site *www.ccra-adrc.gc.ca*, and provide a short summary of the key points:

(a) IT128R Capital Cost Allowance— Depreciable Property
(b) IT422 Definition of Tools
(c) IT472 Capital Cost Allowance—Class 8 property

CHAPTER 8

Income Taxes

The total revenues and income taxes paid by five well known companies during 1998 are summarized below (dollars in millions).

Company	Revenues	Earnings Before Income Taxes*	Income Taxes	Net Income (1)	Effective (Average) Tax Rate (%)
Alcan**	$8,020	$653	$210	$443	32.15
Bell Canada	10,561	1613	781	832	48.42
Bombardier	11,500	827	273	554	33.01
Imperial Oil	9,145	747	193	554	25.84
Trans Canada Pipelines	17,228	547	121	426	22.12

*Before extraordinary items
**All amounts in U.S. dollars

What do these companies have in common? All are among the largest and most successful corporations in Canada. But how do we explain the apparent discrepancy in the rates of taxation which range from 22.12% in the case of Trans Canada Pipelines to 48.42% for Bell Canada. The fact is that although each company has a statutory combined federal/provincial tax rate, the unique nature of business activities in any given year can result in taxes being paid at effective rates which may be higher or lower than the statutory rate. Bell Canada's statutory rate is 42.4%; however, the nondeductibility of certain items raised 1998 income taxes by $96 million above the value based on 42.4%. On the other hand, almost two thirds of Trans Canada's net income in 1998 was exempt from taxes in 1998. Although Trans Canada's statutory rate is 44.6%, its effective rate was 22.12%.

In this chapter, we will focus on combined federal plus provincial/territorial income taxes. When you operate a business, any profits or losses you incur are subject to income tax consequences. Similarly, any profits or return from personal investments, such as stocks or bonds, are taxable. Therefore, we cannot ignore the impact of income taxes in project and personal investment evaluation.

Federal and provincial/territorial tax laws are exceptionally complex. While the overall conceptual framework for income tax calculation changes little from year to year, the specific details are subject to frequent changes. The approach presented in this chapter is a simplistic, but reasonably accurate, method of calculating corporate and personal income taxes which reflects the basic structure of the Canadian tax system. This treatment can be readily adapted to reflect future changes in the tax laws. The discussion and examples are based on 1999 regulations and rates.[1]

The chapter begins with a description of the general approach used in the calculation of both corporate and personal income taxes. This is followed by sections which describe the calculation of corporate and personal income taxes and, more specifically, tax rates to be used in the evaluation of a new project or personal investment.

It must be emphasized that corporations and individuals may have unique tax considerations which are well beyond the scope of the treatment presented. These situations should be referred to an accountant or a lawyer who specializes in tax matters.

8.1 Income Tax Fundamentals

There is a number of concepts which apply to both corporate and personal income tax calculations. In its most simple form, an income tax calculation can be represented as follows:

$$\text{Income taxes} = (\text{tax rate}) \times (\text{taxable income}).$$

The fundamental differences between the corporate and personal income tax calculations relate to the determination of applicable tax rates and taxable income. These are dealt with in the subsequent sections in which we focus specifically on corporate and then personal taxes.

---

[1] Check the web site for current tax rates and revised versions of Tables 8.1 through 8.5. The Federal Budget announced on February 28, 2000 introduced some significant changes for the year 2000 and beyond.

8.1.1. Tax Rates

Three different tax rates can be defined for corporate and personal tax situations.

The **average or effective tax rate** is that value which gives the total income taxes payable by a corporation or individual when it is multiplied by the total taxable income, i.e., income from **all** sources:

$$\text{Total income taxes} = (\text{average tax rate}) \times (\text{total taxable income}).$$

In general, the average tax rate is of little interest for our purposes. Furthermore, the determination of an average tax rate occurs only after the detailed tax calculations have provided values for the total income taxes and total taxable income. Then, the average tax rate can be calculated as the ratio of these two values.

The **marginal tax rate** represents the tax rate which is applicable to the next dollar of taxable income. Given the fact that a corporation or individual has an existing level of taxable income and associated income taxes, how much additional income tax is payable if the taxable income increases by one dollar? Thus:

$$\begin{array}{c}\text{Income tax on the next} \\ \text{dollar of taxable income}\end{array} = (\text{marginal tax rate}) \times (\$1).$$

As we will see in later sections, the marginal tax rate depends on a number of factors, including the current level of taxable income and the type of additional taxable income.

The **incremental tax rate** is the tax rate which applies to an increment of taxable income over and above the existing level of taxable income and results in a corresponding increment of income tax. When this increment of taxable income is the result of a new project or investment, we need the incremental tax rate, i.e., an average tax rate over the increment of new taxable income, to quantify the tax effects in an economic evaluation. Hence, the incremental tax rate is of most interest for our purposes:

$$\begin{array}{c}\text{Income taxes due to a new} \\ \text{project or investment}\end{array} = \begin{array}{c}(\text{incremental} \\ \text{tax rate})\end{array} \times \begin{array}{c}(\text{taxable income due to} \\ \text{a new project or investment}).\end{array}$$

To illustrate, let's consider a corporation or individual having an existing level of annual taxable income of $70,000 which is taxable at an average rate of 35%. A new investment is under consideration which will increase annual taxable income by $25,000. The first $10,000 is taxable at 40% and the next $15,000 is taxable at 45%.

	Before Investment	After Investment
Taxable income	$70,000	$95,000
Income taxes	$70,000 × 35% = $24,500	$70,000 × 35% = $24,500
		+
		$10,000 × 40% = $ 4,000
		+
		$15,000 × 45% = $ 6,750
		Total $35,250

Following the investment, the total taxes will be $35,250 on taxable income of $95,000 which corresponds to an average tax rate for the corporation or individual of 37.1% ($35,250/$95,000). The increase in taxes as a result of the investment is $10,750 ($35,250 – $24,500) from the additional $25,000 in taxable income. Therefore, the incremental tax rate for this investment is 43.0% ($10,750/$25,000). There are two marginal tax rates. In the interval of taxable income between $70,000 and $80,000, the marginal rate is 40% and between $80,000 and $95,000, it is 45%. With this knowledge of the marginal rates, the incremental rate for the investment could have been calculated by prorating these marginal rates in relation to the total increase in taxable income:

$$\text{Incremental tax rate for the investment} = \frac{\$10,000}{\$25,000} \times 40\% + \frac{\$15,000}{\$25,000} \times 45\%$$

$$= 43.0\%.$$

In certain situations, the average, marginal and incremental tax rates can be identical. We will demonstrate this in the sections on corporate and personal income taxes.

8.1.2 Capital Gains (Losses)

Capital assets, in the case of corporations, include all depreciable assets (buildings, equipment, etc.) and nondepreciable assets (land, stock, bonds, etc.). For individuals, capital assets refer to everything that is owned and used for personal purposes (house, car, furniture, boat, cottage, etc.) or investment (stocks, bonds, etc.).

When capital assets are sold for more than the purchase price a profit or **capital gain** may be realized. Conversely, if the selling price is less than the purchase price a **capital loss** may result. The determination of gain or loss is as follows:

Capital gain (loss) = selling price – cost base.

The selling price represents the sale proceeds minus any selling expense and the **cost base** usually includes the purchase price plus the cost of improvements and expenses incurred in acquiring the capital asset. For depreciable assets, the cost base equals the original capital cost.

Depreciable assets and personal use property do not incur capital losses. Furthermore, one house (an individual's principal residence) is exempt from capital gains considerations.

Capital gains (losses) must be considered in the calculation of taxable income. Any gains will increase taxable income while losses will decrease taxable income. However, losses can only be counted to the extent that there are offsetting capital gains. So, if there are a series of capital gains, CG_1, CG_2, CG_3..., and capital losses, CL_1, CL_2, CL_3..., the net capital gain, which must be greater than or equal to zero, is calculated as

$$\text{Net capital gain} = \sum_{n=1}^{X} CG_n + \sum_{n=1}^{Y} CL_n \geq 0$$

where X and Y represent the number of capital gains and losses respectively. Unused capital losses in any given year can be used to offset capital gains reported in previous years or retained and used to offset capital gains in future years.

For our purposes, whenever a project or investment incurs a capital loss, we will assume there are sufficient capital gains from other sources such that the capital loss can be used to reduce the taxable income from the project or investment in the same year.

The specific details on the inclusion of capital gains (losses) in corporate and personal tax calculations are described in Sections 8.3 and 8.4 respectively.

EXAMPLE 8.1 Capital gain on land transactions

Senstech is a Winnipeg-based manufacturer of electronic sensor and alarm systems. In 1995, it acquired land in three locations within the province to construct supply/distribution facilities. By 1999, only one of these facilities had been constructed and the company decided to sell the other two pieces of property. One of these was purchased for $65,000 and there were associated acquisition costs of $5000. When it was sold in 1999, the actual price was $95,000 and there were legal costs of $3500. The second property was purchased for $45,000 and sold for $35,000. The associated acquisition and selling costs were $3000 and $1500 respectively. Determine the net capital gain (loss) in 1999 for the land transactions.

Solution

Given: Land purchase and selling price and associated costs
Find: Net capital gain (loss) for 1999

First, we have to determine the selling price adjusted for associated costs.

Capital	Actual Selling Price	−	Associated Costs of Sale	=	Adjusted Selling Price for Gains Purposes
Property 1	$95,000		$3500		$91,500
Property 2	$35,000		$1500		$33,500

The cost base is determined as follows:

Capital	Purchase Price	+	Associated Costs of Acquisition	=	Cost Base
Property 1	$65,000		$5000		$70,000
Property 2	$45,000		$3000		$48,000

The capital gain (loss) is the difference between the adjusted selling price and the cost base.

$$\begin{aligned}
\text{Property 1 capital gain} &= \$91,500 - \$70,000 = \$21,500 \\
\text{Property 2 capital loss} &= \$33,500 - \$48,000 = (\$14,500) \\
\text{Net capital gain} &= \$7,000
\end{aligned}$$

So Senstech would have a net capital gain in 1999 of $7,000 as a result of the two land sales.

EXAMPLE 8.2 $ Capital gain (loss) on stock transactions

Ken Smith lives in Toronto and has a number of personal investments including a small stock portfolio. In 1993, he purchased 100 shares of Imperial Oil Limited at $45 per share. Brokerage fees amounted to $100. He sold these shares in 1999 for $75 per share and the brokerage charges on the sale were $150. He also sold 750 shares of Air Canada stock in 1999 at $7 per share and the brokerage charges were $100. These shares had been purchased in 1996 as a new stock issue at $12.00 per share so there were no brokerage costs. Determine Mr. Smith's net capital gain (loss) on these stock transactions.

Solution

Given: 1999 share transactions
Find: Net capital gain (loss) in 1999

As in Example 8.1, we calculate the adjusted selling prices and cost bases for the two stocks.

	Shares Sold	x	Selling Price Per Share	=	Selling Price	–	Brokerage Fees	=	Selling Price for Capital Gains Purposes
Imperial Oil	100		$75		$7500		$150		$7350
Air Canada	750		$ 7		$5250		$100		$5150

The cost base for each stock is determined as follows:

	Shares Purchased	x	Purchase Price Per Share	=	Purchase Price	+	Brokerage Fees	=	Cost Base
Imperial Oil	100		$45		$4500		$100		$4600
Air Canada	750		$12		$9000		—		$9000

The capital gain (loss) for each stock is the difference between the adjusted selling price and the cost base.

$$\text{Capital gain on Imperial Oil} = \$7350 - \$4600 = \$2750$$
$$\text{Capital loss on Air Canada} = \$5150 - \$9000 = (\$3850)$$
$$\text{Net capital loss} \quad (\$1100)$$

This net capital loss can only be used to reduce taxes in 1999 if there are capital gains from other investments of at least $1100.

8.2 Net Income

Firms invest in a project because they expect it to increase their wealth. If the project does this—if project revenues exceed project costs—we say it has generated a **profit**, or **income**. If the project reduces a firm's wealth—if project costs exceed project revenues—we say that the project has resulted in a **loss**. One of the most important roles of the accounting function within an organization is to measure the amount of profit or loss a project generates each year, or in any other relevant time period. Any profit generated will be taxed. The accounting measure of a project's after-tax profit during a particular time period is known as **net income**.

8.2.1 Calculation of Net Income

Accountants measure the net income of a specified operating period by subtracting expenses from revenues for that period. These terms can be defined as follows:

1. The **project revenue** is the income earned[2] by a business as a result of providing products or services to customers. Revenue comes from sales of merchandise to customers and from fees earned by services performed for clients.

2. The **project expenses** that are incurred[3] are the cost of doing business to generate the revenues of the specified operating period. Common expenses include labor, raw materials, supplies, supervision, sales, depreciation, and allocated portions of utilities, rent, insurance, property taxes and corporate functions such as marketing, engineering, administrative and management as well as interest on borrowed money and income taxes.

The business expenses listed above are accounted for in a straightforward fashion on a company's income statement and balance sheet. The amount paid by an organization for each item would translate dollar for dollar into expenses in financial reports for the period. For manufacturing organizations, accountants treat interest and income taxes separately but place the other expenses in two broad categories — **cost of goods sold** and **operating expenses**. The **operating expenses** include all the general, administrative and selling expenses. The **cost of goods sold** includes all other expenses, including depreciation, which are reported in the subcategories of direct labor, direct materials and manufacturing overhead.

The depreciation expense arises from the systematic allocation of the cost of an asset over time. In the following section, we will discuss how tax depreciation accounting is used in the calculation of taxable income and the resulting net income. It should be noted that this net income calculation may differ from that published in corporate financial reports. As noted in Section 7.2.4. these reports are based on book depreciation.

8.2.2 Treatment of Depreciation Expenses

Whether you are starting or maintaining a business, you will probably need to acquire assets (such as buildings and equipment). The cost of this property becomes part of your business expenses. The accounting treatment of capital expenditures differs from the treatment of manufacturing and operating expenses, such as cost of goods sold and business operating expenses. As you recall from Chapter 7, **capital expenditures must be capitalized**, i.e., they must be systematically allocated as expenses over their depreciable lives. Therefore, when you acquire a piece of property that has a productive life extending over several years, you cannot deduct the total costs from profits in the year the asset was purchased. Instead, a capital cost allowance[4] schedule is established over the life of the asset, and an appropriate allowance is included in the company's deductions from profit each year. Because it

---

[2] Note that the cash may be received in a different accounting period.

[3] Note that the *cash* may be paid in a different accounting period.

[4] This allowance is based on the total cost basis of the property.

plays a role in reducing taxable income, depreciation accounting is of special concern to a company. In the next section, we will investigate the relationship between tax depreciation and net income.

8.2.3 Taxable Income and Income Taxes

Corporate taxable income is defined as follows:

$$\text{Taxable income} = \text{gross income (revenues)} - \text{expenses.}$$

As discussed in Section 8.1, income taxes are calculated as:

$$\text{Income taxes} = \text{(tax rate)} \times \text{(taxable income).}$$

(We will discuss how we determine the applicable tax rate in Section 8.3.) We then calculate net income as follows:

$$\text{Net income} = \text{taxable income} - \text{income taxes.}$$

A more common format is to present the net income in the following tabular income statement:

Item
Gross income (revenues)
Expenses
Cost of goods sold[5]
Capital cost allowance (CCA)
Operating expenses
Interest[6]
Taxable income
Income taxes
Net income

Our next example illustrates this relationship using numerical values.

---

[5] The formal accounting definition of this item includes depreciation expenses. Here, depreciation in the form of CCA is represented as a separate item in the income statement and the definition of cost of goods sold is modified accordingly.

[6] Interest expense is included for completeness but is ignored elsewhere in the chapter. It is dealt with in Chapter 9.

EXAMPLE 8.3   Net income within a year

A company buys a numerically controlled (NC) machine for $40,000 (year 0) and uses it for 5 years, after which it is scrapped. This equipment falls into CCA class 43. Therefore, its declining balance CCA rate is 30% and the CCA claimable in the first year is $6000, or 15% of the initial cost. (The 50% rule applies in the first year.) The cost of the goods produced by this NC machine should include a charge for the depreciation of the machine. Suppose the company estimates the following revenues and expenses including the depreciation for the first operating year.

Sales revenue	=	$52,000
Cost of goods sold	=	$20,000
CCA on NC machine	=	$6,000
Operating expenses	=	$5,000

If the company pays taxes at an incremental rate of 40% on the taxable income from the project, what is the net project income during the first year?

Solution

Given: Gross income and expenses as stated, income tax rate = 40%
Find: Net income

At this point, we will defer the discussion of how the tax rate (40%) is determined and treat it as given. We consider the purchase of the machine to have been made at time 0, which is also the beginning of year 1. (Note that our example explicitly assumes that the only CCA claimed for year 1 is that for the NC machine, a situation that may not be typical.)

Item	Amount
Gross income (revenues)	$52,000
Expenses	
Cost of goods sold	20,000
CCA	6,000
Operating expenses	$5,000
Taxable income	$21,000
Taxes (40%)	$8,400
Net income	$12,600

Comments: In this example, the inclusion of a tax depreciation or CCA expense reflects the true cost of doing business. This expense is meant to match the amount of the $40,000 total cost of the machine that has been put to use or "used up" during the first year. This example also highlights some of the reasons why income tax laws govern the depreciation of assets for tax calculation purposes. If the company were allowed to claim the entire $40,000 as a year 1 expense, a discrepancy would exist between the one-time cash outlay for the machine's cost and the gradual benefits of its productive use. This discrepancy would lead to dramatic variations in the firm's income taxes and net income. Net income would become a less accurate measure of the organization's performance. On the other hand, failing to account for this cost would lead to increased reported profit during the accounting period. In this situation, the profit would be a "false profit" in that it would not accurately account for the usage of the machine. Depreciating the cost over time allows the company a logical distribution of costs that matches the utilization of the machine's value.

Identical considerations apply to net income values based on book depreciation which appear in financial reports. Unless the basis for book depreciation calculations is reasonable in terms of distributing an asset's cost over its life, the reported net income values will be misleading.

8.2.4   Net Income versus Cash Flow

Traditional accounting stresses net income as a means of measuring a firm's profitability, but we will explain why cash flows are the relevant data to be used in project evaluation. As seen in section 8.2.1, net income is an accounting measure based, in part, on the **matching concept**. Costs become expenses as they are matched against revenue. The actual timing of cash inflows and outflows is ignored.

Over the life of a firm, net incomes and net cash inflows will usually be the same. However, the timing of incomes and cash inflows can differ substantially. Given the time value of money, it is better to receive cash now rather than later, because cash can be invested to earn more cash. (You cannot invest net income.) For example, consider two firms and their income and cash flow schedules over 2 years as follows:

		Company A	Company B
Year 1	Net income	$1,000,000	$1,000,000
	Cash flow	1,000,000	
Year 2	Net income	1,000,000	1,000,000
	Cash flow	1,000,000	2,000,000

Both companies have the same amount of net income and cash sum over 2 years, but Company A returns $1 million cash yearly, while Company B returns $2 million at the end of the second year. If you received $1 million at the end of the first year from Company A, you could invest it at 10%, for example. While you would receive only $2 million in total from Company B at the end of the second year, you would receive in total $2.1 million from Company A.

Apart from the concept of the time value of money, certain expenses do not even require a cash outflow. Depreciation is the best example of this type of expense.

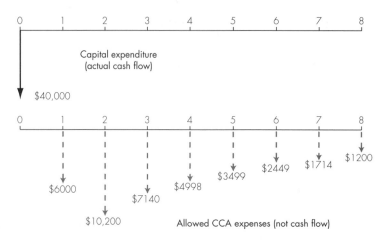

Figure 8.1 Cash flow versus CCA expenses for an asset with a cost basis of $40,000

Even though capital cost allowance is deducted from revenue, no cash is paid to anyone.

In Example 8.3, we have just seen that the annual capital cost allowance has an important impact on both taxable and net income. However, although capital cost allowance has a direct impact on net income, it is *not* a cash outlay; as such, it is important to distinguish between annual income in the presence of depreciation and annual cash flow.

The situation described in Example 8.3 serves as a good vehicle to demonstrate the difference between tax depreciation costs as expenses and the cash flow generated by the purchase of a fixed asset. In this example, cash in the amount of $40,000 was expended in year 0, but the $6000 of CCA charged against the income in year 1 is not a cash outlay. Figure 8.1 summarizes the difference.

Net income (**accounting profit**) is important for accounting purposes, but **cash flows** are more important for project evaluation purposes. As we will now demonstrate, net income can provide us with a starting point to estimate the cash flow of a project.

The procedure for calculating net income is identical to that used for obtaining net cash flow (after-tax) from operations, with the exception of CCA, which is excluded from the net cash flow computation (it is needed only for computing income taxes). **Assuming that revenues are received and expenses are paid in cash**, we can obtain the net cash flow by adding the **non-cash expense** (CCA) to net income, which cancels the operation of subtracting it from revenues.

$$\text{Cash flows} = \text{net income} + \text{non-cash expense (CCA)}.$$

Example 8.4 illustrates this relationship.

EXAMPLE 8.4 Cash flow versus net income

Using the situation described in Example 8.3, assume that (1) all sales were cash sales, and (2) all expenses except depreciation were paid during year 1. How much cash would be generated from operations?

Solution

Given: Net income components
Find: Cash flow

We can generate a cash flow statement by simply examining each item in the income statement and determining which items actually represent receipts or disbursements. Some of the assumptions listed in the problem statement make this process simpler.

Item	Income	Cash Flow
Gross income (revenues)	$52,000	$52,000
Expenses		
Cost of goods sold	20,000	–20,000
CCA	6,000	
Operating expenses	5,000	–5,000
Taxable income	21,000	
Taxes (40%)	8,400	–8,400
Net income	$12,600	
Net cash flow		$18,600

Column 2 shows the income statement, while Column 3 shows the statement on a cash flow basis. The sales of $52,000 are all cash sales. Costs other than depreciation were $25,000; these were paid in cash, leaving $27,000. CCA is not a cash flow—the firm did not pay out $6000 in CCA expenses. Taxes, however, are paid in cash, so the $8,400 for taxes must also be deducted from the $27,000, leaving a net cash flow from operations of $18,600.

Check: As shown in Figure 8.2, this $18,600 is exactly equal to net income plus CCA: $12,600 + $6000 = $18,600.

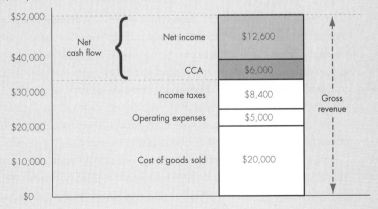

Figure 8.2 Net income versus net cash flow (Example 8.4)

As we've just seen, CCA has an important impact on annual cash flow in its role as an accounting expense that reduces taxable income and thus taxes. (Although capital cost allowance expenses are not actual cash flows, depreciation has a positive impact on the after-tax cash flow of the firm.) Of course, during the year in which an asset is actually acquired, the cash disbursed to purchase it creates a significant negative cash flow and, during the depreciable life of the asset, the depreciation charges will affect the taxes paid and, therefore, cash flows.

As shown in Example 8.4, we can see clearly that depreciation, through its influence on taxes, plays a critical role in project cash flow analysis, which we will explore further in Chapter 9.

8.3 Corporate Income Taxes

Now that we have learned what elements constitute taxable income, we turn our attention to the process of computing income taxes. The corporate tax rate is applied to the taxable income of a corporation, which is defined as its gross income minus allowable deductions. As we briefly discussed in Section 8.2, the allowable deductions include the cost of goods sold, CCA (and depletion), operating expenses and interest.

The cash expenses attributable to manufacturing a product or delivering a service are often referred to as operating and maintenance, or O and M, costs and may be shortened further to just operating costs as discussed in Section 6.6.1. Examples and problems throughout the book use both the more rigorous cost of goods sold and operating expense categories and the less well defined O and M or operating cost designations.

8.3.1 Combined Corporate Tax Rates on Operating Income

The combined corporate tax rate consists of a federal rate plus a provincial or territorial rate. A corporation's federal tax rate is determined by corporate size (as measured by taxable income), ownership (private versus public and Canadian versus non-Canadian), type of business (manufacturing versus nonmanufacturing), the types of income (operating versus investment), and the source of income (inside versus outside Canada). In general, provincial/territorial rates depend on the same considerations.

If we restrict the discussion of federal rates to operating income of Canadian controlled corporations with all such income earned in Canada, there are only three corporate tax rates—small business, nonmanufacturing, and manufacturing. The calculation of these rates is shown in Table 8.1. In all cases, the starting point is the basic federal tax rate of 38% which is reduced by the 10% federal tax abatement.

Table 8.1

Federal Corporate Tax Structure for 1999

	TYPE OF CORPORATION		
	Small Business	Manufacturing	Nonmanufacturing
Basic federal tax	38.00%	38.00%	38.00%
Less: federal tax abatement	10.00%	10.00%	10.00%
Plus: 4% federal surtax (4% of (38% – 10%))	1.12%	1.12%	1.12%
Less: SBD	16.00%	—	—
Less: MPPD	—	7.00%	—
Total federal tax rate	13.12%	22.12%	29.12%

Table 8.2

Provincial/ Territorial Corporate Tax Structure for 1999 (including surtax)[1]

Province/Territory	Small Business Tax Rate	Manufacturing Tax Rate	Nonmanufacturing Tax Rate
British Columbia	8.5%	16.5%	16.5%
Alberta	6.0%	14.5%	15.5%
Saskatchewan	8.0%	10.0%	17.0%
Manitoba	9.0%	17.0%	17.0%
Ontario	8.5%	13.5%	15.5%
Quebec	5.91%[2]	9.15%	9.15%
New Brunswick	6.0%	17.0%	17.0%
Nova Scotia	5.0%	16.0%	16.0%
Prince Edward Island	7.5%	7.5%	16.0%
Newfoundland	5.0%	5.0%	14.0%
Northwest Territories/ Nunavat	5.0%	14.0%	14.0%
Yukon	2.5/6.0%[3]	2.5%	15.0%

[1] From "Tax Facts 1999–2000," by KPMG, reprinted with permission.

[2] Increased to 9.15% effective July 1, 1999.

[3] The first number is for manufacturing and the second for nonmanufacturing businesses.

The 4% federal surtax also applies in each case. Private corporations whose taxable capital is less than $10 million qualify for the small business deduction (SBD) of 16% on the first $200,000 of taxable income. Corporations which derive at least 10% of their gross revenue from manufacturing and processing goods in Canada qualify for the 7% manufacturing and processing profits deduction (MPPD). The MPPD cannot be applied to income eligible for the small business deduction or SBD.

The **combined corporate tax** rate is obtained by adding the federal value for a given type of corporation from Table 8.1 to the corresponding provincial/territorial value from Table 8.2.

Since corporate tax rates are considered **nonprogressive**, i.e., independent of the level of taxable income, the average combined tax rate is the same as the marginal combined tax rate and any incremental combined tax rate. This statement must be qualified to the extent that the new project does not place the corporation in a different tax category. For example, a manufacturing firm whose entire taxable income qualifies for the small business deduction may become a manufacturing type of business as a result of a project which increases production and sales. These considerations are illustrated in Example 8.5.

8.3.2　Income Taxes on Operating Income

The corporate tax rates from Section 8.3.1 can be applied directly to determine income taxes on operating income as illustrated in the following example.

EXAMPLE 8.5

Corporate income taxes

A Canadian controlled and private mail-order computer company in Truro, Nova Scotia sells computer supplies and peripherals. The company leased showroom space and a warehouse for $20,000 a year and installed $100,000 worth of inventory checking and packaging equipment. At a 30% CCA rate, the allowed capital cost allowance (CCA) for this capital expenditure within the first year will amount to $15,000. The store was completed and operations began on January 1st. The company had a gross income of $1,250,000 for the calendar year. Supplies and all operating expenses other than the lease expense were itemized as follows:

Cost of merchandise sold in the year	$500,000
Employee salaries and benefits	$250,000
Other supplies and expenses	$ 90,000
Total expenses	$840,000

Compute the taxable income for the company. How much will the company pay in federal and provincial income taxes for the year?

Solution

Given: Income, cost information, and CCA
Find: Taxable income, federal and provincial income taxes

First we compute the taxable income as follows:

Gross revenues	$1,250,000
Expenses	$ 840,000
Lease expense	$ 20,000
CCA	$ 15,000
Taxable income	$ 375,000

Now we must calculate the federal and provincial income tax.

Federal Income Tax

Taxable income	$375,000

The first $200,000 is taxable at the rate of 13.12%
(Small business federal corporate tax rate)

Federal tax	= $200,000 × 13.12%	$ 26,240

The remaining $175,000 is taxable at the rate of 29.12%
(Nonmanufacturing corporate tax rate)

Federal tax	= $175,000 × 29.12%	$ 50,960
Total federal tax payable		$ 77,200

Provincial Income Tax
(Based upon Nova Scotia tax rates)

Taxable income	$375,000

The first $200,000 is taxable at the rate of 5%
(Small business provincial corporate tax rate)

Provincial tax	= $200,000 × 5%	$ 10,000

The remaining $175,000 is taxable at the rate of 16%
(Standard provincial corporate tax rate)

Provincial tax	= $175,000 × 16%	$ 28,000
Total provincial tax payable		$ 38,000
Total combined corporate income tax payable		$115,200

Comments: Because the mail order computer company is private and Canadian controlled, it qualifies for the small business deduction (SBD) on the first $200,000 of taxable income. Therefore, the combined corporate tax rate on the $200,000 is 18.12% (13.12% federal + 5.0% Nova Scotia). The combined rate on the taxable income above $200,000 is the nonmanufacturing rate of 45.12% (29.12% federal + 16.0% Nova Scotia). The computer company's combined average or effective tax rate is 30.72% (total income taxes of $115,200 divided by taxable income of $375,000).

Projects which this company undertakes to increase revenue or reduce expenses will keep taxable income well above the $200,000 small business maximum. Above this maximum all additional taxable income is taxed at a combined rate of 45.12%. Therefore, this is both the marginal rate and the incremental rate which should be used in the economic evaluation of any project which this company may be considering.

It should also be noted that if this company had been a public rather than a private company, the SBD would not apply. In such a case all the company's taxable income would have been taxed at a combined rate of 45.12%. The average, marginal and incremental rates would be identical and equal to 45.12%.

8.3.3 Corporate Operating Losses

Although corporations strive to be profitable, there may be years in which a corporation operates at a loss and pays no taxes. Ordinary corporate operating losses can be carried back to each of the preceding 3 years and forward for the following 7 years and can be used to offset taxable income in those years. The loss must be applied first to the earliest year, then to the next earliest year, and so forth. For example, an operating loss in 1999 can be used to reduce taxable income for 1996, 1997, or 1998, resulting in tax refunds or credits; any remaining losses can then be carried forward and used in 2000, 2001, and so forth, to the year 2006. If you are unable to deduct all of it as a noncapital loss within this allowed time frame, the unused part becomes a net capital loss, and you can carry it forward indefinitely to reduce taxable capital gains.

8.3.4 Income Taxes on Nonoperating Income

Nonoperating income is derived from activities outside a company's mainstream business. For example, a manufacturer of oil and gas processing equipment may have income which is not directly related to the sale of its equipment. Such income could include interest payments from investments, rent, and capital gains. The tax rates applicable to such income differ from those discussed in Section 8.3.1. We will only consider the tax considerations around capital gains.

Capital gains are taxable at three quarters of their value.[7] The applicable federal tax rate for 1999 is 35.8% and the provincial/territorial tax rate is, with the exception of Quebec, the same as the nonmanufacturing rates in Table 8.2. The Quebec rate for capital gains is 16.7%. The income tax on capital gains or the **capital gains tax** is calculated as follows:

$$\text{Tax on capital gain} = 3/4 \times \text{capital gain} \times (35.8\% + \text{provincial/territorial rate}).$$

Frequently, the 3/4 factor is combined with the tax rate term to give a capital gains combined tax rate, t_{CG}, and the tax calculation becomes

$$\text{Tax on capital gain} = t_{CG} \times \text{capital gain}.$$

You should recall from the discussion in Section 8.1.2 that capital losses must be offset against capital gains and we assumed that these were always available. Such a loss actually reduces the taxes payable. The tax saving resulting from a capital loss is:

$$\text{Tax saving from capital loss} = t_{CG} \times \text{capital loss}.$$

---

[7] Effective February 28, 2000 capital gains are taxed at two-thirds rather than three-quarters of their value.

8.3.5　Investment Tax Credits

The federal government may use the tax regulations to provide incentives which encourage investments in certain types of activities within specific areas of Canada. The **investment tax credit** is such an incentive. Income taxes can be reduced by a percentage of the expenditures which are eligible for an investment tax credit.

Eligible expenditures fall into three categories:

- Scientific Research and Experimental Development—Canadian-controlled private corporations earn investment tax credits at a rate of 35% on the first $2 million dollars of investment in scientific research and development activities.
- Certified Property—Expenditures on buildings, machinery and equipment in prescribed slow growth areas of Canada have earned investment tax credits at a rate of 30% in previous years. This category of investment tax credits is not available for 1999.
- Qualified Property—Expenditures on buildings, machinery and equipment for use in designated activities (manufacturing, exploration and development of natural resources, natural resource processing, logging, farming and fishing) within specific geographical areas (Newfoundland, Prince Edward Island, Nova Scotia, New Brunswick, Gaspé Peninsula, and certain off shore regions) earn investment tax credits at a rate of 10% for 1999.

The investment tax credit (ITC) is calculated as follows:

$$\text{investment tax credit} = \text{value of eligible property} \times \text{rate of ITC}.$$

This tax saving is equivalent to reducing the net cost of the capital investment by the amount of the investment tax credit. The investment tax credit is returned to the company as a tax credit at the end of the taxation year. The amount of the tax credit is subtracted from the value of the property prior to calculating the CCA.

Current information on available investment tax credits can be found on Canada Customs and Revenue Agency (CCRA) Web site *www.ccra-adrc.gc.ca*, Information Circular IC78-4R3.

8.4　Practical Issues Around Corporate Income Tax

The detailed material in Section 8.3 provides a basis for calculating the correct corporate tax rate. However, in practice (and most of the problems in this book are a case in point) the incremental corporate tax appropriate to a particular project has already been determined and this value is given. But do we know whether the company can actually use all the CCA available? How often are taxes paid? These questions need resolution before we can proceed to build income taxes into a project evaluation.

8.4.1   Incremental Corporate Tax Rate Specified

When the incremental tax rate is known, the calculation of taxes can proceed without reference to the type of corporation or the province/territory in which it operates. The only problem relates to determining the capital gains tax rate which, as we have seen, is not directly related to the corporate tax rate. *For our purposes, we will assume that the capital gains tax rate, t_{CG}, is merely three quarters of the specified incremental corporate tax rate.*[8]

**EXAMPLE
8.6**

Capital gains tax on land transactions

Calculate the capital gains tax on the Senstech land transactions in Example 8.1. Senstech is a private Canadian controlled manufacturing corporation with annual taxable income of several million dollars.

Solution

Given: Net capital gain from Example 8.1 of $7,000
Find: Capital gains tax

Senstech's incremental combined tax rate is the manufacturing federal rate of 22.12% plus the Manitoba rate of 17% or 39.12%. The capital gains tax rate is taken as 3/4 of this value, or 29.34%. The capital gains tax associated with the land transaction is

$$\text{capital gains tax} = 29.34\% \times \$7,000$$
$$= \$2,054.$$

Comments: The capital gains tax rate of 29.34% has been calculated as ¾ of the incremental combined tax rate for a manufacturing business. If we had chosen to be more correct and recognize the capital gain as nonoperating income, the capital gains tax rate would be ¾ of a 35.8% federal rate and a 17% Manitoba rate, or 39.6%.

8.4.2   Claiming Maximum Deductions

The level of taxable income is dependent upon the gross income or revenues and the deductions that that can be claimed against revenues. These deductions include all the expenses associated with delivering a good or service as well as debt interest and CCA. Revenue Canada does not permit the total deductions to exceed the revenues. In other words, the taxable income must be greater than or equal to zero.

The constraint of nonnegative taxable income can limit the deductions actually used. For example, it may not be possible to claim all of the CCA in a particular year because there is insufficient income. In this case, the unused CCA is claimed in subsequent years.

---

[8] Effective February 28, 2000 the capital gains tax rate is two-thirds rather than three-quarters of the incremental corporate tax rate.

Such a limitation adds considerable complexity to the calculation of taxes. *However for our purposes, we will always assume that for any investment all available deductions can be claimed at their maximum level in any given year.* This means that negative taxable income will arise if the available deductions exceed the revenue. The income taxes for such an investment then become a positive cash flow. The underlying assumption is that there are other business operations which already pay taxes. The total taxes payable on these other business operations will be reduced by the "positive" income taxes for the new investment. These tax savings are a credit or cash inflow to the investment under consideration. This concept will be explained by way of example in Chapter 9.

8.4.3 Timing of Corporate Income Tax Payments

The taxation year for a corporation is its fiscal or business year. During its fiscal year, a corporation must make monthly payments of federal and provincial income tax based on an estimate of the taxes actually owed by the corporation. This estimate may be nothing more than one twelfth of taxes paid in the preceding fiscal year.

Although a corporation has up to 6 months after the end of its fiscal year to file federal and provincial tax returns, it must make a final tax payment within approximately 2 months of the end of its fiscal year. This payment represents the difference between total income tax payable and the amount paid in monthly installments. The actual deadline is dependent on various factors such as corporate size and type of industry, and these types of considerations can add considerable complexity to an engineering economic analysis. *For our purposes, corporate investments will be assumed to occur at the beginning of a fiscal year, and the associated income taxes will be paid as a lump sum at the end of each fiscal year unless indicated otherwise.*

8.5 Personal Income Tax

Individuals pay taxes on income from several sources: wages and salaries, investment income, profits from the sale of capital assets, and business income. Typical investment income may include dividends from stock and interest from bank deposits. Profits from the sale of capital assets such as stocks and bonds are also taxable. The business income from proprietorships and partnerships is taxed in the same manner as individual income.

8.5.1 Calculation of Taxable Income

In calculating taxable income, all taxpayers begin with **total income**, which represents their earnings as determined in accordance with the provisions of the federal income tax laws. To compute income, we add

    + Wages and salary

    + Pension income

    + Interest income

    + Dividend income

    + Rental/royalty income

    + Net capital gains (3/4 of net value)[9]

    + Other income[10]

    = **Total income**

From total income, the Canada Customs and Revenue Agency allows you to subtract the following amounts:

    — Retirement plan contributions (registered pension plans or registered retirement savings plans)

    — Annual union, professional, or like dues

    — Child care expenses

    — Attendant care expenses

    — Business investment losses

    — Moving expenses

    — Alimony or maintenance paid

    — Carrying charges and interest expense on money borrowed for investment purposes

    — Other deductions

    = **Net income before adjustments**

Adjustments subtracted from the net income before adjustments include

    — Employee home relocation loan

    — Stock option and shares deduction

    — Losses from previous years (includes capital losses)

    — Northern resident's deduction

    — Additional deductions

    = **Taxable income**

---

[9] Effective February 28, 2000 this become two-thirds of the net value.

[10] Other income includes items such as consulting fees.

8.5.2 Income Taxes on Ordinary Income

Canada has three rates for individual taxation. Rates are progressive—that is, the higher the income, the larger the percentage paid in taxes. (The federal income tax is computed using the taxable income and the rates in Table 8.3.) These rates are likely to change over the years, so you must refer to the most recent tax guides to find the presently applicable rates.

Table 8.3	Taxable income	Tax rates
Individual Federal Income Tax Structure for 1999	Less than $29,590	17%
	$29,590 to $59,180	$5,030 on first $29,590 plus 26% of taxable income over $29,590
	More than $59,180	$12,724 on first $59,180 plus 29% of taxable income over $59,180

Your federal income tax is reduced by **nonrefundable tax credits**. The total amount **qualified** for nonrefundable tax credits includes the following eligible items:

- Basic personal amount
- Age amount
- Spousal amount
- Canada or Quebec pension plan contributions
- Employment insurance premiums
- Pension income amount
- Disability amount
- Disability amount transferred from dependent
- Tuition fees
- Education amount
- Interest on student loans
- Tuition fees and education amount transferred from a child
- Amounts transferred from your spouse
- Medical expenses less adjustment
- Charitable donations[11]

The nonrefundable tax credit is 17% of the total qualified amount. Other tax credits such as the dividend tax credit may also be available.

The **basic federal tax** is calculated as federal income tax less the nonrefundable tax credits and other tax credits. You must then calculate your federal surtax and provincial/territorial taxes. The Canada Customs and Revenue Agency and the provincial governments frequently add individual surtaxes as a means of collecting additional revenue over the short term. However, some surtaxes last for several years and have become a longer term taxation method.

---

[11] Charitable donations in excess of $250 earn nonrefundable tax credits at 29% rather than 17%.

Table 8.4	Province	Tax Rates
Individual Provincial Income Tax Rates for 1999	British Columbia	49.5% of basic federal tax + 30% surtax on British Columbia income tax above $5300 + 26% surtax on British Columbia income tax above $8660
	Ontario	40.5% of basic federal tax + 20% surtax on Ontario income tax above $4058 + 33% surtax on Ontario income tax above $5218
	Newfoundland	69% of basic federal tax + 10% surtax on Newfoundland income tax above $7900

The federal individual surtax, calculated as 3% of the **basic federal tax** payable, was eliminated in 1999. However, the surtax of 5% on any amounts of federal tax payable above $12,500 continues.

In addition to the federal tax payable, provincial tax on personal income must be considered. Provincial/territorial income tax is usually based upon a percentage of the **basic federal tax** payable and provincial individual surtaxes are calculated from the provincial income tax. Quebec has its own personal income tax which is administered and collected directly by the province. For Quebec residents, the **basic federal tax** is reduced by 16.5% in recognition of a different type of financial arrangement between the federal and Quebec governments. Table 8.4 gives examples of provincial 1999 income tax rates for three provinces. Despite the complexity in the federal and provincial rate structures for personal income tax, the

EXAMPLE 8.7 $ Average tax rate for personal income

A person living in Ontario in 1999 had a taxable income of $50,000 deriving from wages and investment interest. Estimate his or her personal income tax amount and average tax rate. The total amount qualified for nonrefundable tax credits is $8000.

Solution

Given: Taxable income of $50,000
Find: Personal income tax amount and the average tax rate

Federal tax = $5030 + 26% × ($50,000 – $29,590)	$10,337
Minus non-refundable tax credits (17% of $8000)	$ 1,360
Basic federal tax payable	$ 8,977
and	
Provincial tax in Ontario (40.5% × $8977)	$ 3,636
Total personal income taxes payable	$12,613
Therefore the average tax rate is ($12,613/$50,000)	= 25.2%

tax system is progressive: the more you earn, the higher the tax rate and the relative amount of taxes. Personal income taxes basically depend upon three factors—where you live, total taxable income, and the type of income.

In a simplistic approach, we can identify four types of income which are most common and receive different treatment from a tax perspective. Salaries and wages have an accompanying nonrefundable tax credit which derives from the basic personal exemption, Canada and Quebec pension plan contributions, and employment insurance premiums. Interest income has no such refundable tax credits attributed to it. Capital gains are similar to interest in this regard but are included in income at 3/4 of their actual value.[12]

The fourth type of income is dividends from taxable Canadian corporations. Such dividends are included in income at 125% of their actual value but earn a dividend tax credit equal to 13.33% of the dividend amount included as income. This tax credit reduces the basic federal tax as well as the provincial/territorial tax, which is calculated as a percentage of the basic federal tax.

We are not particularly interested in being able to calculate the total combined income taxes for an individual. However, we do want to be able to evaluate the tax consequences of planned personal investments by that individual. This requires the incremental combined tax rate for the investment.

By making some standard assumptions with respect to eligibility for personal deductions, KPMG, a national accounting firm, developed the data shown in Table 8.5. Combined marginal tax rates for the four types of income are shown as a function of total taxable income by province/territory, including Quebec. The rates shown for capital gains include the 3/4 factor.[13] As long as the income increment from a planned investment leaves the total taxable income within the same range, the marginal rate in the table is the incremental tax rate for the investment. When an investment moves the total taxable income into the next higher range, the incremental rate is a prorated value based on the marginal rates from each range as shown in Example 8.8.

While the actual incremental tax rates for a particular individual may be slightly different than those in Table 8.5 because of differences in personal exemptions, these rates are more than adequate for analyzing personal investment opportunities on an after-tax basis.

EXAMPLE 8.8
$

Incremental tax rate on new investment income

Linda Stephens is a software engineer for a Calgary-based firm. Her current annual taxable income is $55,000. A recent inheritance of $65,000 could be invested in corporate bonds which pay an annual interest rate of 8.5%. Determine the incremental combined tax rate and the income tax on this interest in 1999.

[12, 13] Effective February 28, 2000 capital gains are taxed at two-thirds rather than three-quarters of their value.

Table 8.5

Marginal Combined Personal Income Tax Rates by Province/Territory and Income Type for 1999[1]

Type of Income Increment: Salary/Wages

Existing Level of Taxable Income	Marginal Tax Rate on Income Increment[2]						
	$7300 to $29,590	$29,591 to $37,400	$37,401 to $39,000	$39,001 to $48,160	$48,161 to $59,180	$59,181 to $63,670	$63,671 and over
British Columbia	23.9%	37.3%	38.2%	38.9%	40.0/43.9%	49.0%	49.5/52.3%
Alberta	23.5%	36.5%	37.3%	37.9/38.9%	40.0%	44.6%	45.2%
Saskatchewan	26.6%	40.4%	41.3%	41.9/44.1%	45.3%	50.2%	50.8%
Manitoba	25.7%	39.1/41.1%	42.0%	42.6%	43.8%	48.4%	48.9%
Ontario	22.3%	34.8%	35.7%	36.3%	37.4/39.5%	44.1/48.2%	48.7%
Quebec	32.0/35.0%	42.5%	43.8%	44.8%	45.9/49.0%	51.6%	52.2%
New Brunswick	25.5%	39.9%	40.9%	41.6%	42.8%	47.7%	48.3/49.7%
Nova Scotia	25.2%	39.3%	40.3%	40.9%	42.1%	47.0%	47.6/49.2%
Prince Edward Island	25.3%	39.6%	40.5%	41.2%	42.4%/43.9%	49.0%	49.5%
Newfoundland	27.0%	42.2%	43.2%	43.9%	45.1%	50.3/52.3%	52.9%
Yukon Territory	24.0%	37.5%	38.3%	39.0%	40.2%	44.8/45.5%	46.1%
Northwest Territories/Nunavut	23.2%	36.2%	37.1%	37.7%	38.9%	43.4%	43.9%

Type of Income Increment: Interest

Existing Level of Taxable Income	Marginal Tax Rate on Income Increment[2]				
	$7,000 to $29,590	$29,591 to $46,740	$46,741 to $59,180	$59,181 to $62,400	$62,401 and over
British Columbia	25.4%	38.9%	40.0/43.9%	49.0%	49.5/52.3%
Alberta	25.0%	37.9/38.9%	40.0%	44.6%	45.2%
Saskatchewan	28.2%	41.9/44.1%	45.3%	50.2%	50.8%
Manitoba	27.2%	40.6/42.6%	43.8%	48.4%	48.9%
Ontario	23.7%	36.3%	37.4/39.5%	44.1/48.2%	48.7%
Quebec	34.3/37.3%	44.8%	45.9/49.0%	51.6%	52.2%
New Brunswick	27.2%	41.6%	42.8%	47.7%	48.3/49.7%
Nova Scotia	26.8%	41.0%	42.1%	47.0%	47.6/49.2%
Prince Edward Island	26.9%	41.2%	42.4/43.9%	49.0%	49.5%
Newfoundland	28.7%	43.9%	45.1/46.9%	52.3%	52.9%
Yukon Territory	25.5%	39.0%	40.2%	44.8/45.5%	46.1%
Northwest Territories/Nunavut	24.6%	37.7%	38.9%	43.4%	43.9%

Table 8.5 continued

Type of Income Increment: Dividends (Taxable Canadian Corporations)

Existing Level of Taxable Income	Marginal Tax Rate on Income Increment[2]				
	$7000 to $30,150	$30,151 to $59,180	$59,181 to $88,895	$88,896 to $115,500	$115,501 and over
British Columbia	6.8%	23.7%	29.3%	30.2/33.1%	33.5/35.3%
Alberta	7.2%	23.4%	28.8/29.5%	30.4%	30.8%
Saskatchewan	9.3%	25.9/26.9%	32.7/34.5%	35.3%	35.7%
Manitoba	9.3%	28.5%	34.1%	35.0%	35.3%
Ontario	6.4%	22.1%	27.3%	28.2/29.7/33.5%	32.9%
Quebec	16.5/20.3%	29.7/33.5%	36.6%	37.5%	37.9%
New Brunswick	7.3%	25.3%	31.3%	32.2%	32.6/33.5%
Nova Scotia	7.2%	24.9%	30.8%	31.7%	32.1/33.2%
Prince Edward Island	7.3%	25.1%	31.0%	31.9/33.1%	33.5%
Newfoundland	7.7%	26.8%	33.1%	34.0/35.3%	35.7%
Yukon Territory	6.9%	23.7%	29.4%	30.3/30.7%	31.1%
Northwest Territories/Nunavut	6.7%	23.0%	28.4%	29.3%	29.7%

Type of Income Increment: Capital Gains

Existing Level of Taxable Income	Marginal Tax Rate on Income Increment[2]				
	$7000 to $29,590	$29,591 to $46,740	$46,741 to $59,180	$59,181 to $62,400	$62,401 and over
British Columbia	19.1%	29.2%	30.0/32.9%	36.7%	37.2/39.2%
Alberta	18.7%	28.5/29.1%	30.0%	33.4%	33.9%
Saskatchewan	21.1%	31.5/33.1%	33.9%	37.7%	38.1%
Manitoba	20.4%	30.5/32.0%	32.8%	36.3%	36.7%
Ontario	17.8%	27.2%	28.1/29.6%	33.1/36.1%	36.6%
Quebec	25.7/27.9%	33.6%	34.5/36.7%	38.7%	39.1%
New Brunswick	20.4%	31.2%	32.1%	35.8%	36.2/37.3%
Nova Scotia	20.1%	30.7%	31.6%	35.2%	35.7/36.9%
Prince Edward Island	20.2%	30.9%	31.8%/32.9%	36.7%	37.2%
Newfoundland	21.5%	33.0%	33.8/35.2%	39.2%	39.7%
Yukon Territory	19.1%	29.3%	30.1%	33.6/34.1%	34.6%
Northwest Territories/Nunavut	18.5%	28.3%	29.1%	32.5%	32.9%

[1] From "Tax Facts 1999–2000" by KPMG, reprinted with permission

[2] Multiple values arise when provincial surtaxes increase within the indicated range of taxable income. Refer to KPMG reference for specific details.

Solution

Given: Annual bond interest payments of $5525 (8.5% of $65,000)
Find: 1999 combined income tax on this interest and the incremental tax rate

From Table 8.5, we obtain the following marginal tax rates on interest for Alberta residents.

Total Taxable Income	Marginal Tax Rate
$46,741 – $59,180	40.0%
$59,181 – $62,400	44.6%

Therefore, the income tax rate on the first $4180 of interest (the 40.0% rate limit of $59,180 – $55,000 of current taxable income) is 40.0% while the remaining $1345 of interest ($5525 – $4180) is taxed at 44.6%.

$$\text{Income tax on bond interest} = 40.0\% \times \$4180 + 44.6\% \times 1345.$$
$$= \$2272$$

The incremental combined tax rate on the bond interest is 41.1% ($2272/$5525).

EXAMPLE 8.9 $ Tax savings resulting from capital loss

In Example 8.2, Ken Smith, an Ontario resident, sold Imperial Oil and Air Canada stock and incurred a capital loss of $1100. If Mr. Smith's annual taxable income is about $48,000, what tax savings are realized as a result of this capital loss?

Solution

Given: Capital loss of $1100
Find: Tax savings

From Table 8.5, the Ontario marginal tax rate for capital gains at this level of taxable income is 28.1%. Therefore, the potential tax saving in 1999 arising from the capital loss is

$$\text{Capital loss tax saving} = 28.1\% \times \$1100$$
$$= \$309.$$

Comment: In order to realize this tax saving in 1999, Mr. Smith must have at least $1100 of capital gains on other investments.

Table 8.5 shows the marginal tax rates for capital gains in Ontario within a taxable income range, $46,741 to $59,180, as being multivalued (28.1%/29.6%). This is a result of provincial surtax changes within the specified taxable income range. For our purposes in such situations, we will always use the lower value. The error introduced with this approach is generally quite small. However, if it is of concern you can perform the calculations using both values and thereby measure the actual impact of the difference in rates.

8.5.3 Timing of Personal Income Tax Payments

In order to include tax effects when evaluating personal investments, a person needs to know when taxes are payable. Depending on personal circumstances, tax payment occurs in various ways and at different times. An employer will automatically withhold a portion of an employee's salary or wage from every paycheque and remit that portion to CCRA. However, if a person receives other income that is not taxed at the source like a salary or wage, he or she may be obliged to make monthly or quarterly payments of the estimated taxes on such income. Additionally, by April 30th of each year, an income tax return must be filed for the preceding calendar year. This return represents a formal calculation of the total taxes owed. Any difference between the taxes owed and tax payments already made is either returned as a refund or forwarded as the final payment.

The timing of the actual tax payments can be a significant complication in an economic analysis. *For our purposes, we will assume that annual personal income taxes are paid on the last day of each calendar year unless indicated otherwise.*

8.6 Depreciable Assets: Disposal Tax Effects

As in the disposal of capital assets, there are generally gains or losses on the sale (or exchange) of depreciable assets. To calculate a gain or loss, we first need to determine undepreciated capital cost (UCC) of the depreciable asset at the time of disposal. When a depreciable asset used in business is sold for an amount different from its UCC, this gain or loss has an effect on income taxes. The gain or loss is found as follows:

$$\text{Gains (losses)} = \text{salvage value} - \text{UCC}.$$

where the salvage value represents the proceeds from the sale minus any selling expense or removal cost.

These gains, known as **recaptured depreciation** or **recaptured CCA**, are taxable. In the unlikely event that an asset is sold for an amount greater than its capital cost, the gains (salvage value – UCC) are divided into two parts for tax purposes:

$$
\begin{aligned}
\text{Gains} \; &= \; \text{salvage value} - \text{UCC} \\
&= \; (\text{salvage value} - \text{cost base}) \\
&+ \; (\text{cost base} - \text{UCC}) \\
&= \; \text{capital gains} + \text{recaptured CCA}.
\end{aligned}
$$

Recall from Section 7.2.2 that cost base means the purchase cost of an asset plus any incidental costs, such as freight and installation. As illustrated in Figure 8.3, in this case

$$
\begin{aligned}
\text{Capital gains} \; &= \; \text{salvage value} - \text{cost base} \\
\text{Recaptured CCA} \; &= \; \text{cost base} - \text{UCC}.
\end{aligned}
$$

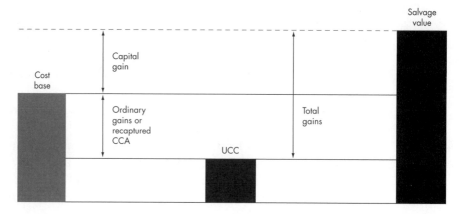

Figure 8.3 Capital gains and recaptured CCA

This distinction is necessary because capital gains are taxed at a capital gain tax rate and recaptured CCA is taxed at the ordinary income tax rate. As described previously, current tax law allows a lower rate of taxation for capital gains. When disposal of a depreciable asset results in a loss, the loss reduces taxes payable.

Section 8.6.1 describes an approximate representation of disposal tax effects for the declining balance and the straight-line depreciation methods. Appendix 8A provides a more rigorous analysis of disposal tax effects for the declining balance and straight-line depreciation methods. Unless otherwise stated, all examples and problems throughout this book will follow the approach described in Section 8.6.1.

8.6.1   Calculation of Disposal Tax Effects

The manner in which gains or losses from an asset disposal affects taxes depends upon the depreciation method used, the assumed timing of the disposal, and whether or not the asset disposal depletes the asset class. As was discussed in Chapter 7, Section 7.4.5, the disposal of an asset does not normally deplete the asset class. The situation is further complicated if there are new acquisitions added to the asset class in the year in which disposal of an asset occurs. The loss or recapture of CCA from the disposal is built into the undepreciated capital cost base for the entire asset class as a result of the asset pool accounting procedure. The tax effect attributable to the disposal will then be spread over a number of future years, which makes the calculation of the total disposal tax effect very tedious. Because of this difficulty and because disposal tax effects are normally not an important factor in determining the acceptability of a project, we assume that the tax implications of the disposal are completely realized in the year of disposal and there are no new acquisitions to the asset class in this year[14]. *Specifically, we assume that the disposal occurs just prior to the end of the last year of service and that the CCA in the year of asset disposal is calculated in the usual manner and without reference to the disposal.* The disposal tax effect, G, then represents the tax implications over and above the tax savings realized in the year of disposal from CCA. The effects of these assumptions and variations to them are analyzed in Appendix 8A.

Because of the above assumptions, any gains on disposal are added to income, and any losses incurred are treated as expenses for assets depreciated by any depreciation method. This greatly simplifies the calculation of the disposal tax effect because G is now merely the extra tax payment or the tax savings realized in the year of disposal, i.e.,

$$G = t(U_{\text{Disposal}} - S),$$

where $U_{\text{Disposal}} = UCC_N = UCC_{N-1} - CCA_N$. UCC_{N-1} is the undepreciated capital cost of the asset at the end of the year prior to disposal, and CCA_N is the capital cost allowance claimable in the year of disposal with the actual disposal ignored. When capital gains are involved, the total disposal tax effect G becomes

$$G = t(U_{\text{Disposal}} - P) - t_{CG}(S - P),$$

where P is the initial cost base (time zero installed cost), t_{CG} is the capital gains tax rate, and U_{Disposal} is the asset's UCC at the end of the year in which the disposal occurs. The choice of $(U_{\text{Disposal}} - S)$ rather than $(S - U_{\text{Disposal}})$ is arbitrary, but more convenient because it provides the correct sign for the after-tax cash flow correction in the last year of the asset's service life.

The net salvage value, NS, is the sum of the salvage value and the disposal tax effect:

$$NS = S + G.$$

EXAMPLE 8.10

Disposal tax effect on depreciable assets

A company purchased a drill press costing $250,000. The drill press is classified as a CCA Class 43 property with a declining balance rate of 30%. If it is sold at the end of 3 years, compute the gains (losses) for the following four salvage values: (a) $150,000, (b) $104,125, (c) $90,000, and (d) $270,000. Assume that the company's combined federal and provincial tax rate is 40% and that capital gains are taxed at 3/4 of their value, i.e., effectively taxed at 30% in this example.

Solution

Given: a CCA Class 43 asset, cost base = $250,000, sold 3 years after purchase

---

[14] This assumption may appear overly restrictive in situations which require asset replacement over the life of a project. However, it can be argued that these replacement assets are not available for use and therefore not added to the asset class until the beginning of the year following the disposal of the original asset.

Find: Disposal tax effects, and net salvage value from the sale if sold for $150,000, $104,125, $90,000, or $270,000

In this example, we first compute the UCC of the machine at the end of year 3.

Year	Capital Cost Allowance (30% CCA rate)	UCC
0		$250,000
1	$37,500	212,500
2	63,750	148,750
3	44,625	104,125

The CCA amount in year 1 is reduced by 50% because of the 50% rule. The UCC at the end of year 3 is $104,125.

(a) Case 1: UCC < salvage value < cost base

$$
\begin{aligned}
\text{Disposal tax effects} &= G = t(U_{Disposal} - S) \\
&= 0.4 \times (\$104,125 - \$150,000) \\
&= -\$18,350.
\end{aligned}
$$

$$
\begin{aligned}
\text{Net salvage value} &= \text{salvage value} + \text{disposal tax effects} \\
&= \$150,000 - \$18,350 \\
&= \$131,650.
\end{aligned}
$$

This situation (salvage value exceeds UCC) is denoted as Case 1 in Figure 8.4.

(b) Case 2: Salvage value = UCC

In Case 2, the UCC is again $104,125. Thus, if the drill press's salvage value is $104,125, equal to the UCC, no taxes are calculated on that salvage value. Therefore, the net proceeds equal the salvage value.

(c) Case 3: Salvage value < UCC

Case 3 illustrates a loss when the salvage value (say, $90,000) is less than the UCC. We compute the net salvage value after tax as follows:

$$
\begin{aligned}
\text{Disposal tax effects} = G &= t(U_{Disposal} - S) \\
&= 0.4 \times (\$104,125 - \$90,000) \\
&= \$5650.
\end{aligned}
$$

$$
\text{Net salvage value} = \$90,000 + \$5,650 = \$95,650.
$$

(d) Case 4: Salvage value > cost base

This situation is not likely for most depreciable assets (except for real property). Nevertheless, the tax treatment on this gain is as follows:

$$
\begin{aligned}
\text{Capital gains} &= \text{salvage value } (S) - \text{cost base } (P) \\
&= \$270,000 - \$250,000 \\
&= \$20,000.
\end{aligned}
$$

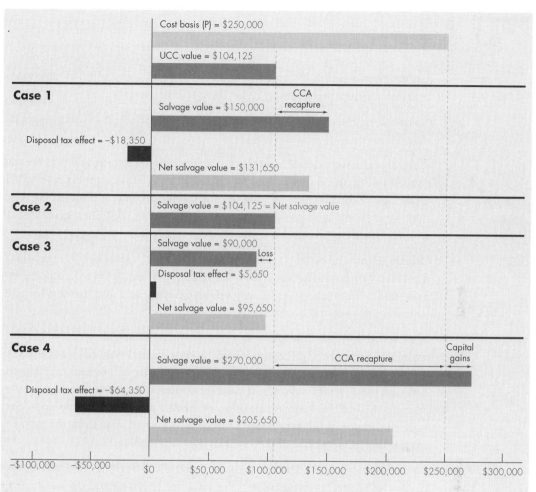

Figure 8.4 Calculations of gains or losses (Example 8.10)

$$\text{Capital gains tax} = \$20,000 \times 3/4^{14} \times 0.4$$
$$= \$6000$$

$$\text{Disposal tax effects} = -\text{capital gains tax} + t(U_{Disposal} - P)$$
$$= -\$6000 + 0.4 \times (\$104,125 - \$250,000)$$
$$= -\$6000 - \$58,350$$
$$= -\$64,350$$

$$\text{Net salvage value} = \$270,000 - \$64,350$$
$$= \$205,650$$

Comments: Note that in (c) the disposal tax effects are positive, which represents a reduction in tax due to the loss and increases the net proceeds. The corporation would be still paying tax, but less than if the asset had not been sold at a loss.

[14] Effective February 28, 2000 this value is two-thirds.

8.7 Summary

- Explicit consideration of taxes is a necessary aspect of any complete economic study of an investment project.

- Income taxes equal a taxable income amount multiplied by an appropriate tax rate.

- Three tax rates were defined in this chapter: **marginal tax rate**, which is the rate applied to the next dollar of income earned; **effective (average) tax rate**, which is the ratio of total income tax paid to total taxable income; and **incremental tax rate**, which is the average rate applied to the incremental income generated by a new investment project.

- **Capital gains** are taxed at three-quarters of their value (1999 regulations, decreases to two-thirds for 2000). **Capital losses** are deducted from capital gains and net remaining losses may be carried backward and forward for consideration in other tax years.

- Since we are interested primarily in the measurable financial aspects of depreciation, we consider the effects of depreciation or **capital cost allowance (CCA)** on two important measures of an organization's financial position, **net income** and **cash flow**. Once we understand that depreciation has a significant influence on the income and cash position of a firm, we will be able to appreciate fully the importance of utilizing CCA as a means to maximize the value of engineering projects and of the organization as a whole.

- For corporations, the Canadian tax system has the following characteristics:
  1. Tax rates are not progressive. The rates are constant and do not increase with earnings.
  2. Tax rates depend upon the type and characteristics of the corporation.
  3. The combined federal/provincial (territorial) tax rate is the sum of the federal and provincial/territorial rates which are set independently.

- An **investment tax credit** is a direct reduction of income taxes payable, arising from the acquisition of depreciable assets. Government uses the investment tax credit to stimulate investments in specific assets, in specific industries, or in specific areas of the country.

- For individuals, the Canadian tax system has the following characteristics:
  1. Tax rates are progressive. The more you earn, the higher your tax rate.
  2. Incremental tax rates on income from a new investment are dependent on where you live, your total taxable income, and the type of income from the investment.

3. Provincial/territorial taxes are calculated as a percentage of the basic federal tax and are therefore not independent of the federal tax structure.

- Disposal of depreciable assets—like disposal of capital assets—may result in gains (**recaptured depreciation**) or losses which must be considered in calculating taxes.

- Net salvage value of an asset is the actual salvage value adjusted for the **disposal tax effect.**

Problems

Level 1

8.1* Which of the following statements is correct?

(a) Over a project's life, a typical business will generate a greater amount of total project cash flows (undiscounted) if CCA is claimed more quickly.

(b) No matter how quickly CCA is claimed, total tax obligations over a project's life remain unchanged.

(c) CCA recapture equals the cost base minus an asset's UCC at the time of disposal.

(d) Cash flows normally include CCA since it represents a cost of doing business.

8.2* You purchased a computer system which cost $50,000 5 years ago. At that time, the system was estimated to have a service life of 5 years with salvage value of $5000. These estimates are still good. The property has been depreciated at a declining balance CCA rate of 30%. Now (at the end of year 5 from purchase) you are considering selling the computer at $10,000. What UCC should you use in determining the disposal tax effect?

8.3* Omar Shipping Company bought a tugboat for $75,000 (year 0) and expected to use it for 5 years, after which it will be sold for $12,000. Suppose the company estimates the following revenues and expenses for the first operating year.

Operating revenue	$200,000
Operating expenses	$84,000
CCA	$4,000

If the company pays taxes at the rate of 30% on its taxable income, what is the net income during the first year?

8.4* In Problem 8.3, assume for the moment that (1) all sales are for cash, and (2) all costs, except depreciation, were paid during year 1. How much cash would have been generated from operations?

8.5* Gilbert Corporation had a gross income of $500,000 in tax year 1, $150,000 in salaries, $30,000 in wages, $20,000 in interest charges, and $60,000 in CCA for an asset purchased 3 years ago. Ajax Corporation has a gross income of $500,000 in tax year 1, and $150,000 in salaries, $90,000 in wages, and $20,000 in interest expenses. Gilbert and Ajax are both Quebec based manufacturing companies. Apply the tax rates in Tables 8.1 and 8.2 and determine which of the following statements is correct.

(a) Both corporations will pay the same amount of income taxes in year 1.
(b) Both corporations will have the same amount of net cash flows in year 1.
(c) Ajax Corporation will have a larger net cash flow than Gilbert in year 1.
(d) Gilbert Corporation will have a larger taxable income than Ajax Corporation in year 1.

8.6* Tiger Construction Company had a gross income of $20,000,000 in tax year 1, $3,000,000 in salaries, $4,000,000 in wages, $800,000 in CCA, a loan principal payment of $200,000, and a loan interest payment of $210,000. Determine the net income of the company in tax year 1 if Tiger's tax rate is 40%.

8.7 A consumer electronics company was formed in Saskatoon, Saskatchewan to manufacture and sell a portable handset system that allows people with cellular car phones to receive calls up to 1000 feet from their vehicles. The company purchased a warehouse for $500,000 and converted it into a manufacturing plant. It completed installation of assembly equipment worth $1,500,000 on January 31. The plant began operation on February 1. The company had a gross income of $2,500,000 for the calendar year. Manufacturing costs and all operating expenses, excluding the capital expenditures, were $1,280,000.

The CCA for capital expenditures amounted to $128,000.

(a) Compute the taxable income of this company.
(b) How much will the company pay in income taxes for the year?

8.8 Consider a CCA Class 10 asset ($d = 30\%$), which was purchased at $60,000.

The applicable salvage values would be $20,000 in year 3, $10,000 in year 5, and $5000 in year 6, respectively. Compute the gain or loss amounts when the asset is disposed of

(a) in year 3,   (b) in year 5, and
(c) in year 6.

8.9* An electrical appliance company purchased an industrial robot costing $300,000 in year 0. The industrial robot to be used for welding operations is depreciated at a 30% declining balance rate for tax purposes. If the robot is to be sold after 5 years, compute the amounts of gains (losses) and the disposal tax effect for the following three salvage values with a corporate tax rate of 34%.

(a) $10,000   (b) $125,460
(c) $200,000

8.10 LaserMaster, Inc., an Edmonton, Alberta based laser-printing service company, had a sales revenue of $1,250,000 during the current tax year. The following represents the other financial information relating to the tax year.

Labor expenses	$550,000
Material costs	$185,000
CCA	$32,500
Interest income	$6,250
Interest expenses	$12,200
Rental expenses	$45,000
Proceeds from the sale of old printers	$23,000

(The printers had a combined UCC of $20,000 at the time of sale.) Assume interest income is taxed at the same rate as income deriving from sales.

(a) Determine the taxable income for the tax year.
(b) Determine the taxable gains for the tax year.
(c) Determine the amount of income taxes.
(d) Determine the disposal tax effect.

8.11 CanWest Inc., bought a machine for $50,000 on January 2, 1999. Management expects to use the machine for 10 years, at which time it will have a $1000 salvage value. Consider the following questions independently. Parts (a) through (d) relate to book depreciation.

(a) If CanWest uses straight-line depreciation, what is the book value of the machine on December 31, 2001?
(b) If CanWest uses double declining balance depreciation, what is the depreciation expense for 2001?
(c) If CanWest uses double declining balance depreciation switching to straight-line depreciation, when is the optimal time to switch?
(d) If CanWest uses sum-of-years'-digits depreciation, what is the total depreciation on December 31, 2004?
(e) If CanWest sold the machine on April 1, 2002, at a price of $30,000, what was the disposal tax effect? The corporate tax rate is 35% and the machine is depreciated at a 30% declining balance rate for tax purposes.

8.12* Buffalo Ecology Corporation is an Ontario based company which expects to generate a taxable income of $250,000 from its regular business in the current tax year. The company is considering a new venture: cleaning up oil spills made by fishing boats in lakes. This new venture is expected to generate additional taxable income of $150,000.

(a) Determine the firm's marginal tax rates before and after the venture.
(b) Determine the firm's average tax rates before and after the venture.

8.13 Repeat Problem 8.12 assuming Buffalo Ecology is not eligible for the small business deduction.

8.14 Precision Machining, a New Brunswick company, expects to have annual taxable incomes of $270,000 from its regular business over the next 6 years. The company is considering the proposed acquisition of a new milling machine. The machine's installed price is $200,000. The machine falls into a 30% declining balance CCA class, and it will have an estimated salvage value of $30,000 at the end of 6 years. The machine is expected to generate an additional before-tax revenue of $80,000 per year.

(a) What is the total amount of economic depreciation for the milling machine, if the asset is sold at $30,000 at the end of 6 years?
(b) Determine the company's marginal tax rates over the next 6 years without the machine.
(c) Determine the company's average tax rates over the next 6 years with the machine.

8.15 Major Electrical Company is located in Red Deer, Alberta. It expects to have an annual taxable income of $150,000 from its residential accounts over the next 2 years. The company is bidding on a 2-year wiring service for a large apartment complex. This commercial service requires the purchase of a new truck equipped with wire-pulling tools at a cost of $50,000. The equipment falls into CCA class 10 and will be retained for future use (instead of selling) after 2 years, indicating no gain or loss on this property. The project will bring in an additional annual revenue of $200,000, but it is expected to incur additional annual operating costs of $100,000. Compute the incremental tax rate applicable to the project's operating profits for the next 2 years.

8.16 Okanagan Juice Corporation is a public, manufacturing type company in British Columbia. It estimates its taxable income for next year at $2,000,000. The company is considering expanding its product line by introducing apple-peach juice for the next year. The market responses could be (1) good, (2) fair, or (3) poor. Depending on the market response, the expected additional taxable incomes are (1) $2,000,000 for a good response, (2) $500,000 for fair response, and (3) a loss of $100,000 for a poor response.

(a) Determine the incremental tax rate applicable to each situation.
(b) Determine the average tax rate that results from each situation.
(c) What is the marginal tax rate prior to the expansion and after the expansion with the three market scenarios?

8.17 A small manufacturing company in Aylmer, Quebec has an estimated annual taxable income of $95,000. Owing to an increase in business, the company is considering purchasing a new machine that will generate an additional (before-tax) net annual revenue of $50,000 over the next 5 years. The new machine requires an investment of $100,000. It will be depreciated at a declining balance CCA rate of 30%.

(a) What is the increment in income tax due to the purchase of the new machine in tax year 1?
(b) What is the incremental tax rate due to the purchase of the new equipment in year 1?

8.18 Simon Machine Tools Company is a Canadian controlled manufacturing company located in London, Ontario. It is considering the purchase of a new grinding machine to process special orders. The following financial information is available.

- Without the project: The company expects to have taxable income of $300,000 each year from its regular business over the next 3 years.

- With the project: This 3-year project requires the purchase of a new grinder at a cost of $50,000. The equipment falls into CCA class 43. The grinder will be sold at the end of project life for $10,000. The project will be bringing in an additional annual revenue of $80,000, but it is expected to incur additional annual operating costs of $20,000.

(a) What are the additional taxable incomes (due to undertaking the project) during years 1 through 3, respectively?
(b) What are the additional income taxes (due to undertaking the new orders) during years 1 through 3, respectively?
(c) Compute the disposal tax effect when the asset is disposed of at the end of year 3.

8.19 A company purchased a new forging machine to manufacture disks for airplane turbine engines. The new press cost $3,500,000, and is considered a CCA class 43 asset. The company has to pay annual city property taxes for ownership of this forging machine at a rate of 1.2% on the beginning of year undepreciated capital cost.

(a) Determine the UCC of the asset at the beginning of each tax year.
(b) Determine the total amount of property taxes over the machine's depreciable life.

$8.20 Julie Magnolia has $50,000 cash to invest for 3 years. Two types of investment opportunities are available for consideration. She can buy a corporate bond that pays interest of 9.5% per year. Alternately, Julie can invest the amount in a tract of land that could be sold at $75,000 (after paying the real estate commission) at the end of year 3.

Julie's marginal tax rate is 24% for interest income and 18% for capital gains. Calculate the income taxes payable on each investment.

$8.21* Anne Johnson completed her engineering degree in December of 1998. She moved to Hinton, Alberta and started work in January 1999 at an annual salary of $38,000. She does not have any other income sources so her taxable income for 1999 will be $38,000. Calculate Anne's expected income taxes. What are her average and marginal taxation rates?

$8.22 Robert Smith works in Nova Scotia and has a taxable income (from salary) of $50,000 per year. He has been offered a new job in Ontario with a salary increase of $15,000 and he may pay less tax.

(a) Calculate Roberts' average and marginal tax rate in Nova Scotia.
(b) If Robert accepts the job in Ontario, calculate his income taxes and average and marginal tax rate.
(c) If Robert were to find a job in Nova Scotia which paid $65,000 per year, calculate his income taxes and average and marginal tax rate.

$8.23* Harry McCarthy is a consulting engineer in Saskatchewan and his current taxable income is $70,000 per year derived exclusively from salary. If he moves to Vancouver he may be able to increase this taxable income by $5,000 as a result of a higher salary. He is concerned about the cost of housing and the tax changes between provinces.

(a) Based on his annual $70,000 of taxable income in Saskatchewan, calculate Harry's income taxes and average and marginal tax rates.
(b) With $75,000 of taxable income in British Columbia, what would Harry's income taxes and average and marginal tax rates be?

(c) What taxable income from salary would Harry need to earn in Saskatchewan in order to realize the same after tax income as in part b)?

$8.24 Frank Prout has job offers from two different companies. Company A has offered him a job at $40,000/year and he can choose to live in either Newfoundland or New Brunswick. Company B has offered a similar position at $38,000 and he may choose to live in either Ontario or Manitoba. Which company should he work for and where should he live from a tax perspective? (All of Frank's income comes from his salary.)

$8.25* Compare the taxes payable in Ontario and Newfoundland with the following taxable incomes:

(a) $25,000
(b) $45,000
(c) $65,000
(d) $100,000

(Assume all income derives from salary or wages.)

$8.26 What are the corresponding marginal tax rates on interest, dividends, and capital gains in Question 8.25?

$8.27* Mary Anderson is the president of a New Brunswick based electronics firm. Her current taxable income is $195,000 per year.

Assume that Mary has purchased 500 shares of each of the following stocks. The brokerage fees for the purchase and sale of these stocks total $2200.

Stock	Purchase Price	Current Market Value
Imperial Oil	$22.25	$31.95
Nortel	$32.50	$70.85
GEAC	$21.90	$34.65
Bombardier	$15.00	$21.80

Estimate the tax payable if Mary sells her stock.

$8.28 Nelson Xu is considering two personal investments. The first is Canada Savings Bonds (CSBs), which pay annual interest of 5.5%. The second is a stock priced at $25 per share with annual dividends per share of $1.50 per year.

Mr. Xu's annual taxable income before the additional investment is $55,000.

(a) Estimate the income taxes payable on bond interest, if Mr. Xu lives in Quebec and buys $10,000 worth of CSBs.

(b) Estimate the income taxes payable on dividends, if Mr. Xu lives in Quebec and buys $10,000 worth of stock.

(c) Repeat (a) and (b), if Mr. Xu lives in Saskatchewan.

Level 2

8.29 Quick Printing Company of Halifax, Nova Scotia had sales revenue of $1,250,000 from operations during tax year 1. Here are some operating data on the company:

Labor expenses	$550,000
Materials costs	185,000
CCA	32,500
Interest income on time deposit	6,250
Bond interest income on Apple Computer	4,500
Stock dividend income from Sears	3,900
Interest expense	12,200
Rental expenses	45,000
Dividend payment to Quick's shareholders	40,000
Proceeds from sale of old equipment that had a UCC of $20,000	23,000

(a) What isQuick's taxable income?

(b) What are Quick's taxable gains?

(c) What are Quick's marginal and effective (average) tax rates?

(d) What is Quick's net cash flow after tax?

(Note: Interest income received by a corporation is taxed at 52%. Dividends received by one corporation from another are taxed at 32%.)

8.30 Elway Aerospace Company of Winnipeg, Manitoba is a Canadian controlled manufacturing company. Last year, it had gross revenues of $1,200,000 from operations. The following financial transactions were posted during the year:

Manufacturing expenses (including CCA)	$450,000
Operating expenses (excluding interest expenses)	120,000
A new short-term loan from a bank	50,000
Interest expenses on borrowed funds (old and new)	40,000
Dividends paid to common stockholders	80,000
Old equipment sold	60,000

The old equipment had a UCC of $75,000 at the time of sale.

(a) Calculate Elway's income taxes.

(b) What is Elway's operating income?

8.31 Valdez Corporation commenced operations on January 1, 1999. The company's financial performance during its first year of operation was as follows:

• Sales revenues, $1,500,000.

• Labor, material, and overhead costs, $600,000.

• The company purchased a warehouse worth $500,000 in February. To finance this warehouse, the company issued $500,000 of long-term bonds on January 1, which carry an interest rate of 10%. The first interest payment occurred on December 31.

• For depreciation purposes, the purchase cost of the warehouse is divided into $100,000 in land and $400,000 in building. The building is a CCA

Class 1 asset and is depreciated accordingly.

- On January 5, the company purchased $200,000 of equipment, which falls into CCA Class 43.

- The corporate tax rate is 40%.

(a) Determine the total CCA allowed in 1999.

(b) Determine Valdez's income taxes for 1999.

8.32 A machine now in use that was purchased 3 years ago at a cost of $4000 has an undepreciated capital cost of $1800. It can be sold for $2500, but could be used for 3 more years, at the end of which time it would have no salvage value. The annual O&M costs amount to $10,000 for the old machine. A new machine can be purchased at an invoice price of $14,000 to replace the present equipment. Freight-in will amount to $800, and the installation cost will be $200. The new machine has an expected service life of 5 years and will have no salvage value at the end of that time. With the new machine, the expected direct cash savings amount to $8000 in the first year and $7000 for each of the next 2 years. Corporate income taxes are at an annual rate of 40%. The present machine has been depreciated at a declining balance CCA rate of 30% and the proposed machine would be depreciated at the same rate. (Note: Each question should be considered independently.)

(a) If the old asset is to be sold now, what would be the amount of its equivalent economic depreciation?

(b) For depreciation purposes, what would be the first cost of the new machine (cost base)?

(c) If the old machine is to be sold now, what would be the disposal tax effect?

(d) If the old machine can be sold for $5000 now instead of $2500, what would be the disposal tax effect?

The following questions relate to book depreciation:

(e) If the old machine had been depreciated using 175% DB and then switching to SL depreciation, what would be the current book value?

(f) If the machine were not replaced by the new one, when would be the time to switch from DB to SL?

Level 3

8.33 Van-Line Company, a small electronics repair firm located in Sydney, Nova Scotia, expects an annual taxable income of $70,000 from its regular business. The company is considering expanding its repair business to include personal computers. This expansion would bring in an additional annual income of $30,000, but will require an additional expense of $10,000 each year over the next 3 years. Using applicable current tax rates, answer the following:

(a) What is the marginal tax rate in tax year 1?

(b) What is the average tax rate in tax year 1?

(c) Suppose that the new business expansion requires a capital investment of $20,000 (CCA calculated at a 20% declining balance rate). What is the PE of the total income taxes to be paid over the project life, at $i = 10\%$?

$8.34 Chuck Robbins owns and operates a small electrical service business in Winnipeg, Manitoba. Chuck is married with two children, so he claims four dependents on his tax return. As business grows steadily, tax considerations are important to him. Therefore, Chuck is considering whether to incorporate the business. Under either form, the family will initially own 100% of the firm.

Chuck plans to finance the firm's expected growth by drawing a salary just sufficient for his family's living expenses and by retaining all other income in the business. He estimates the expected taxable income over the next 3 years to be as follows:

	Year 1	Year 2	Year 3
Taxable income for the business	$25,000	$35,000	$45,000
Personal taxable income	$30,000	$32,000	$35,000

The personal taxable income derives exclusively from salary and this salary amount is included as an expense in determining the taxable income for the business. Which form of business (corporation or sole proprietorship) will allow Chuck to pay the lowest taxes (and retain the most income) during the three year period? Note: as a sole proprietor **all** income is taxed at the personal tax rate.

8.35 Electronic Measurement and Control Company (EMCC) has developed a laser speed detector that emits infrared light invisible to humans and radar detectors alike. For full-scale commercial marketing, EMCC needs to invest $5 million in new manufacturing facilities. The system is priced at $3000 per unit. The company expects to sell 5000 units annually over the next 5 years. The new manufacturing facilities will be depreciated at a declining balance CCA rate of 30%. The expected salvage value of the manufacturing facilities at the end of 5 years is $1.6 million. The manufacturing cost for the detector is $1200 per unit, excluding depreciation expenses. The operating and maintenance costs are expected to run to $1.2 million per year. EMCC has a combined income tax rate of 35%, and undertaking this project will not change the current marginal tax rate.

(a) Determine the incremental taxable income, income taxes, and net income due to undertaking this new product for the next 5 years.

(b) Determine the gains or losses associated with the disposal of the manufacturing facilities at the end of 5 years.

8.36 Diamonid is a start-up diamond coating company planning to manufacture a microwave plasma reactor that synthesizes diamonds. Diamonid anticipates that the industry demand for diamonds will skyrocket over the next decade for use in industrial drills, high-performance microchips, and artificial human joints, among other things. Diamonid has decided to raise $50 million through issuing common stocks for investment in plant ($10 million) and equipment ($40 million) to establish a facility at Edmonton, Alberta. Each reactor can be sold at a price of $100,000 per unit. Diamonid can expect to sell 300 units per year during the next 8 years. The unit manufacturing cost is estimated at $30,000, excluding CCA. The operating and maintenance cost for the plant is estimated at $12 million per year. Diamonid expects to phase out the operation at the end of 8 years and revamp the plant and equipment to adopt a new diamond manufacturing technology. At that time, Diamonid estimates that the salvage values for the plant and equipment will be about 60% and 10% of the original investments, respectively. The plant and equipment will be depreciated at declining balance CCA rates of 20% and 30% respectively.

(a) If the 1999 corporate tax rates apply over the project life, determine the combined income tax rate each year.

(b) Determine the gains or losses at the time of plant revamping.

(c) Determine the net income each year over the plant life.

Disposal Tax Effects: Advanced Topics

The disposal tax effect calculations described in Section 8.6.1 incorporated, simplifying assumptions. Removal of these restrictions is relatively straighforward but complicates the calculations. The analysis presented in this appendix leads to more rigorous relationships for the determination of disposal tax effects.

8A.1 Additional Considerations in Disposal Tax Effect Calculations

Frequently, the disposal of an asset does not deplete the asset class. In this case, the treatment of disposal tax effects in Section 8.6.1 is only an approximation. Any difference between the salvage value and the UCC of an asset at the time of disposal affects future capital cost allowance claims for assets remaining in the class because of the asset pool accounting procedure. The actual tax effects over the future years after disposal are dependent on which depreciation method is used.

Removal of an asset from an asset class at any value other than its UCC will change the undepreciated capital cost base. The amount of the change will be equal to the difference between the UCC at disposal and the salvage value (or the cost base when capital gains are involved). We will define the change or offset in the undepreciated capital cost base as follows:

$$\Delta = (U_{\text{Disposal}} - S),$$

where Δ represents the change in undepreciated capital cost,

U_{Disposal} represents UCC at disposal, and
S represents the salvage value.

(This choice rather than $(S - U_{\text{Disposal}})$ is convenient because it provides the correct sign for the tax correction.)

The quantity Δ will be depreciated over time within the capital cost base for the entire asset class, but actually belongs to an asset that no longer exists. The depreciation of Δ will produce a series of annual cash flows due to capital cost allowance which may be positive ($U_{\text{Disposal}} > S$) or negative ($U_{\text{Disposal}} < S$). When U_{Disposal} equals S, there is no residual effect of the asset within the capital cost base for the asset class.

You will recall that:

Net income = taxable income − income taxes

Income taxes = tax rate × taxable income, and

Taxable income = gross income (revenues)
− cost of goods sold
− operating expenses
− CCA.

Therefore

Net income = (1 − tax rate) × (revenues − expenses)
+ tax rate × CCA
= $(1 - t)(R - E) + t \times$ CCA

and the tax rate times CCA in each year ($t \times CCA_n$) represents a positive cash flow contribution to net income.

When an asset disposal results in a nonzero value of Δ, the CCA derived from Δ by the depreciation process will result in annual cash flows beyond the point of disposal equal to $t \times CCA_n$. These cash flows will be positive or negative depending upon the sign of Δ itself. These are the disposal tax effects which must be taken into account. The disposal tax effect which we need to quantify is simply the present value measured at the time of disposal, of all the future $t \times CCA_n$ cash flows.

In Section 8.6.1, we assumed that the disposal occurs just prior to the end of the asset's last year of service. An equally valid assumption has the disposal occurring right after the end of the asset's last year of service, i.e., at the beginning of the following year. In this section, we will analyze the disposal tax effects that occur with different disposal timing assumptions, asset addition assumptions, and depreciation methods.

8A.2 Straight-Line Depreciation Method

Calculating disposal tax effects for assets depreciated on a straight-line basis is complicated by the slightly different rules applicable to the various straight line classes. Since these types of assets arise infrequently in problems of engineering interest, we will not attempt to deal with all possible situations. Our discussion is restricted to classes that are exempt from the 50% rule and for which the depreciation process takes the asset to a zero book value over period of N_T years. N_T is also called the tax life of the asset. We will use N to indicate the service life of the asset in years. The annual CCA claimed for N years, or for N_T years, whichever is smaller, is

$$\text{CCA} = \frac{P}{N_T}, \tag{8A.1}$$

where P represents the capital cost of the asset. When a disposal occurs at a salvage value which differs from the UCC at that point, the CCA amounts claimed in subsequent years to correct for this difference cannot exceed the annual amount claimed prior to disposal as given in Eq. (8A.1). Therefore, the period of time, N^*, over which the correction is required becomes

$$N^* = \frac{|\,U_{\text{Disposal}} - S\,|}{P/N_T} = N_T \frac{|\,U_{\text{Disposal}} - S\,|}{P} \tag{8A.2}$$

This is likely to be a noninteger quantity, which can be represented as

$$N^* = N_I^* + N_{NI}^*, \; 0 \le N_{NI}^* < 1 \tag{8A.3}$$

where the subscripts I and NI refer to the integer and noninteger nature of these quantities. The maximum value of the adjustment in any year after disposal for disposal tax effects is $t \times CCA$.

If the disposal occurs immediately after the last year of service (i.e, at the beginning of year $N + 1$), the tax effects will occur over the $N_I^* + 1$ years as shown in Figure 8A.1. Remember that the CCA for year N (the last year of the asset's service life) has been claimed. In this case, U_{Disposal} is the UCC at the end of year N, U_N. The present equivalent of the disposal tax effects, expressed at the time of disposal or the end of year N, is

$$G = \frac{t(U_N - S)}{\mid U_N - S \mid}\left[\sum_{n=1}^{N_I^*} \frac{P}{N_T}(P/F,\ i,\ n) + N_{NI}\frac{P}{N_T}(P/F,\ i,\ N_I^* + 1)\right] \qquad (8A.4)$$

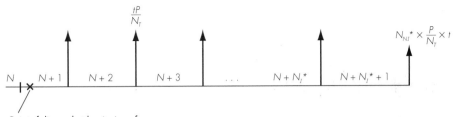

Point of disposal at beginning of year

Figure 8A.1 Straight-line disposal tax effects when disposal occurs at beginning of year $N + 1$ and $U_{\text{Disposal}} > S$

If disposal occurs just prior to the end of the asset's last year of service (i.e., in year N), the CCA for the asset's last year of service has not been claimed. The UCC at disposal, U_{Disposal}, is actually the UCC of the asset at the end of year $N - 1$, U_{N-1}. The disposal tax effects will be reflected in year N and the N_I^* years after the asset's service life. The series of tax effects due to the disposal in this case is shown in Figure 8A.2. The present equivalent of the disposal tax effects, expressed at time of disposal, is

$$G = \frac{t(U_{N-1} - S)}{\mid U_{N-1} - S \mid}\left[\sum_{n=0}^{N_I^*-1} \frac{P}{N_T}(P/F,\ i,\ n) + N_{NI}^*\frac{P}{N_T}(P/F,\ i,\ N_I^*)\right] \qquad (8A.5)$$

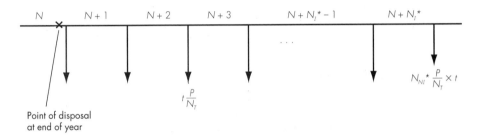

Point of disposal
at end of year

Figure 8A.2 Straight-line disposal tax effects when disposal occurs at end of year N and $U_{\text{Disposal}} < S$

Comparison of disposal tax assumptions for the straight-line depreciation method

(a) Calculate the disposal tax effects for straight-line depreciation with zero salvage value using the approximate method described in Section 8.6.1 for the following data:

$$P = \$20,000$$
$$N_T = 8$$
$$N = 5$$
$$S = \$2000$$
$$t = 40\%$$
$$i = 10\%.$$

(b) Compare these results to those obtained using the more rigorous methods developed in Section 8A.2.

Solution

The annual CCA claim with straight-line depreciation is

$$CCA = \frac{\$20,000}{8} = \$2500.$$

The UCC at $N = 5$ is calculated as

$$U_5 = U_{Disposal} = P - 5 \times CCA$$
$$= \$7500.$$

(a) The approximate method for handling disposal tax effects uses the relationship

$$G = t(U_{Disposal} - S).$$

Then

$$G = 0.4(\$7500 - \$2000) = \$2200.$$

(b) When using the more rigorous approach, the number of years, N^*, over which the disposal tax effects occur, needs to be determined:

$$N^* = \frac{|\,U_{Disposal} - S\,|}{CCA} = \frac{|\,\$7500 - \$2000\,|}{\$2500} = 2.2 \text{ years}.$$

Therefore

$$N_I^* = 2 \text{ and } N_{NI}^* = 0.2.$$

If disposal occurs at the beginning of year 6, then according to Eq. (8A.4):

$$G = 0.4 \times (+\ 1)[\$2500(P/F, 10\%, 1)$$
$$+\ \$2500(P/F, 10\%, 2) + 0.2 \times \$2500(P/F, 10\%, 3)]$$
$$=\ \$1886.$$

If the disposal occurs at the end of year 5, then $U_{Disposal} = U_4 = \$10,000$ and

$$N^* = \frac{|\,\$10,000 - \$2500\,|}{\$2500} = 3.2$$

or

$$N_I^* = 3 \text{ and } N_{NI}^* = 0.2.$$

Using Eq. (8.5):

$$G = 0.4 \times (+\ 1)[\$2500 + \$2500(P/F, 10\%, 1)$$
$$+\ \$2500(P/F, 10\%, 2) + 0.2 \times \$2500(P/F, 10\%, 3)]$$
$$=\ \$2886.$$

It should be noted that the G values cannot be compared directly because they include different effects. The approximate method used in (a) and the method based on the beginning of the year disposal assumption give G values that do not include another tax effect at year 5, namely, the $t \times CCA_5$ tax saving. However, the G value calculated with the end of year assumption includes all tax effects at year 5.

The $t \times CCA_5$ tax saving is $0.4 \times \$2500$ or $\$1000$, which needs to be added to the first two G values before comparing them with the third G value.

	Total Tax Effects at Year 5
Part (a):	$2200 + $1000 = $3200
Part (b) Beginning of year assumption:	$1886 + $1000 = $2886
Part (b) End of year assumption:	$2886

Comment: The $314 difference between the rigorous and approximate methods is small. When discounted back to time zero at 10%, the difference is only $195 in a present value context. This demonstrates the adequacy of the simpler approach described in Section 8.6.1.

8A.3 Declining Balance Depreciation Method

As stated previously, the disposal of an asset frequently does not deplete the asset class. The loss or gain due to disposal of the asset affects the capital cost allowances of the assets remaining in the class to be claimed in future years—a result of the asset pool accounting procedure. We will still use N to indicate the service life of the asset under consideration.

When the declining balance method is used for CCA computations, we need to consider the following questions:

1. Are new assets being added to the asset class in the same year as asset disposal occurs from the class?

2. Does the disposal of the asset under consideration occur at the end of year N or the beginning of year $N + 1$?

If new assets are purchased and added to the asset class, and there is a disposal from this class in the same year, the 50% rule applies to the disposal, as well as to the new purchases. It is the difference between the total new purchases and the total disposals in a year to which the 50% rule applies (refer to Section 7.4.4). If the disposal occurs at the end of year N (the last year of the asset's service life), then in theory, the CCA for this year will not be claimed, and the first disposal tax effect will occur at the end of year N. If the disposal occurs at the beginning of year $N + 1$ (the year after the asset's service life of N years), the CCA for year N has been claimed, and the first disposal tax effect will occur at the end of year $N + 1$. Based on the above sets of assumptions, we will discuss four cases of disposal tax effects calculation.

Case I

1. The disposal of an asset occurs at the beginning of year $N + 1$.
2. No new purchases of assets belonging to the same asset class occur in the year of disposal.

The undepreciated capital cost to use in determining whether there is a gain or loss is the UCC at the end of year N, since the CCA for year N has been claimed already. **Case I** is based on the same assumptions as those used by Edge and Irvine[4] in their derivation of capital tax factors.

When the original asset belongs to a declining balance asset class with a declining balance rate of d, the CCA values in future years resulting from the disposal are given by the usual depreciation relations:

$$CCA_n = d(1 - d)^{n-1}(U_{\text{Disposal}} - S)$$ (8A.6)

where $n(\geq 1)$ represents the number of years from the year of disposal. The disposal tax effect (G) at the time of disposal is calculated as the present equivalent of the $t \times CCA_n$ tax effects (see Figure 8A.5):

$$G = td(U_{\text{Disposal}} - S)\sum_{n=1}^{\infty}(1 - d)^{n-1}(P/F, i, n).$$ (8A.7)

---

[4] C. Geoffrey Edge and V. Bruce Irvine, *A Practical Approach to the Appraisal of Capital Expenditures*, Second Ed., The Society of Management Accountants of Canada, 1981.

It can be shown using the summation relationship for a geometric series that the summation term in Eq. (8A.7) is equal to

$$\frac{1}{(i + d)}$$

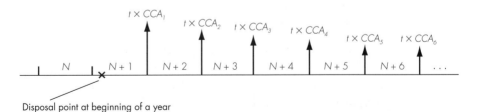

Figure 8A.3 Timing of disposal tax effects for **Cases I** and **II**.

Therefore,

$$G = \frac{td}{i + d}(U_{\text{Disposal}} - S)$$

$$= \frac{td}{i + d}(U_N - S), \tag{8A.8}$$

for assets depreciated on a declining balance basis under **Case I**.

Case II

1. The disposal of an asset occurs at the beginning of year $N + 1$.
2. New purchases of assets belonging to the same asset class occur in the year of disposal.

In this case, we also have $U_{\text{Disposal}} = U_N$. As discussed in Section 7.4.4, whenever there are disposals and new purchases in the same year, the 50% rule is applied to the difference between the total acquisition and the total salvage values from disposal. That is, the 50% rule has to be applied not only to new purchases but also to disposals. The CCA values arising from the disposal are as follows and reflect the 50% rule effect on the salvage value:

$$CCA_1 = dU_{\text{Disposal}} - \frac{d}{2}S,$$

$$CCA_n = (1 - d)^{n-1}dU_{\text{Disposal}} - (1 - \frac{d}{2})(1 - d)^{n-2}dS \text{ for } n \geq 2. \tag{8A.9}$$

where n indicates the number of years from the year of disposal. The tax effects due to these CCAs are again represented in Figure 8A.3. The present equivalent of these tax effects, expressed at time of disposal (end of year N), is

$$G = \frac{td}{i+d}\left(U_N - S \times \frac{1 + i/2}{1 + i}\right),$$ (8A.10)

since $U_{\text{Disposal}} = U_N$.

For $i < 25\%$, $\frac{1 + i/2}{1 + i}$ is greater than 0.9 and can be considered approximately equal to 1. Under this condition, **Case I** and **Case II** give essentially the same result. Any differences are a direct result of the 50% rule on the salvage value.

Case III

1. The disposal of an asset occurs just prior to the end of year N.
2. No new purchases of assets belonging to the same asset class occur in the year of disposal.

In this case the undepreciated capital cost at disposal is equal to the UCC at the end of year $N - 1$, i.e., $U_{\text{Disposal}} = U_{N-1}$. The CCA for year N will not be claimed specifically in this case as it forms part of the disposal tax effects realized in the year of disposal and in years that follow. The CCA values resulting from this disposal can be calculated in the same way as those for **Case I**. However, U_{Disposal} is equal to U_{N-1} rather than U_N, and these tax effects start one year earlier. These tax effects are shown in Figure 8A.6. The present equivalent of the tax effects expressed at the end of year N is

$$G = \frac{td(1 + i)}{i + d}(U_{N-1} - S),$$ (8A.11)

since $U_{\text{Disposal}} = U_{N-1}$.

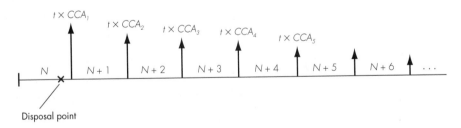

Figure 8A.4 Timing of disposal tax effects for **Cases III** and **IV**.

Case IV

1. The disposal of an asset occurs just prior to the end of year N.
2. New purchases of assets belonging to the same asset class occur in the year of disposal.

In this case, the undepreciated capital cost at disposal is equal to the UCC at the end of year $N - 1$ (i.e., $U_{\text{Disposal}} = U_{N-1}$). The CCA for year N will not be claimed in this case. The tax effects will be realized in the year of disposal and years that follow. The CCA values resulting from this disposal can be calculated in the same way as those for **Case II**. However, U_{Disposal} is equal to U_{N-1}, rather than U_N, and these tax effects start one year earlier as represented in Figure 8A.4. The present equivalent of the tax effects expressed at the end of year N is

$$G = \frac{td(1 + i)}{i + d} \left(U_{N-1} - S \times \frac{1 + i/2}{1 + i} \right),\qquad (8A.12)$$

since $U_{\text{Disposal}} = U_{N-1}$.

(a) UCC < salvage value < cost base

Case I: $U_{Disposal} = U_3$

$$G = \frac{td}{i + d}(U_3 - S) = \frac{0.4 \times 0.3}{0.15 + 0.3} \times (\$104{,}125 - \$150{,}000)$$

$$= -\$12{,}233.$$

Case II: $U_{Disposal} = U_3$

$$G = \frac{td}{i + d}\left(U_3 - S \times \frac{1 + i/2}{1 + i}\right)$$

$$= \frac{0.4 \times 0.3}{0.15 + 0.3}\left(\$104{,}125 - \$150{,}000 \times \frac{1.075}{1.15}\right)$$

$$= -\$9625.$$

Case III: $U_{Disposal} = U_2$

$$G = \frac{td(1 + i)}{i + d}(U_2 - S)$$

$$= \frac{0.4 \times 0.3 \times 1.15}{0.15 + 0.3} \times (\$148{,}750 - \$150{,}000)$$

$$= -\$383.$$

Case IV: $U_{Disposal} = U_2$

$$G = \frac{td(1 + i)}{i + d}\left(U_2 - S \times \frac{1 + i/2}{1 + i}\right)$$

$$= \frac{0.4 \times 0.3 \times 1.15}{0.15 + 0.3} \times \left(\$148{,}750 - \$150{,}000 \times \frac{1.075}{1.15}\right)$$

$$= \$2617.$$

Comparing the disposal tax effects on the basis of total tax effects at year 3:

	G	+	$t \times CCA_3$	Total Tax Effects at Year 3
Example 8.10	−$18,350		$17,850	−$500
Case I	−$12,233		$17,850	$5617
Case II	−$9,625		$17,850	$8225
Case III	−$383		0	−$383
Case IV	$2,617		0	$2,617

(b) Salvage value = undepreciated capital cost at year 3

Following the same procedure as in (a), but with $S = \$104,125$, we find the following total tax effects at year 3.

	G	+	$t \times CCA_3$	Total Tax Effects at Year 3
Example 8.10	0		$17,850	$17,850
Case I	0		$17,850	$17,850
Case II	$1,810		$17,850	$19,660
Case III	$13,685		0	$13,685
Case IV	$15,768		0	$15,768

(c) Salvage value < UCC

Following the same procedure as in part (a) but with $S = \$90,000$, we find the following total tax effects at year 3.

	G	+	$t \times CCA_3$	Total Tax Effects at = Year 3
Example 8.10	$5,650		$17,850	$23,500
Case I	$3,767		$17,850	$21,617
Case II	$5,331		$17,850	$23,181
Case III	$18,016		0	$18,016
Case IV	$19,816		0	$19,816

(d) Salvage value > cost base

For most depreciable assets, this situation is not likely (except for real property). Nevertheless, the tax treatment on this gain is as follows:

$$\text{Capital gains} = \text{salvage value } (S) - \text{cost base } (P)$$
$$= \$270,000 - \$250,000$$
$$= \$20,000$$

In Example 8.10 and **Cases III and IV**, this capital gain is realized in year 3. For **Cases I** and **II**, the disposal at the beginning of year 4 will not result in a capital gains tax until the end of year 4. Therefore the capital gains tax measured at the end of year 3 is as follows:

Example 8.10, **Cases III and IV**

$$\text{Capital gains tax} = \$20,000 \times 3/4 \times 0.4$$
$$= \$6000.$$

Cases I and **II** measured at the end of year 3

$$\text{Capital gains tax} = \$20,000 \times 3/4 \times 0.4(P/F, 15\%, 1)$$
$$= \$5217.$$

Substituting the cost base P for S and proceding as in part (a), we find the following total tax effects at year 3.

	Capital Gains Tax	+	G	+	t x CCA$_3$	=	Total Tax Effects at Year 3
Example 8.10	–$6000		–$50,350		$17,850		–$38,500
Case I	–$5217		–$38,900		$17,850		–$26,267
Case II	–$5217		–$34,552		$17,850		–$21,919
Case III	–$6000		–$31,050		0		–$37,050
Case IV	–$6000		–$26,050		0		–$32,050

Comment: The total tax effects at the point of disposal are dependent on the specific assumptions. However, given the uncertainty in predicting a salvage value several years into the future, the fact that tax effects tend to be small, and the discounting process used to move them to time zero, the additional complications demonstrated here are unwarranted in most instances. The simpler approach used in Example 8.10 is adequate for most purposes.

Problems

8A.1 With a corporate tax rate of 40% and MARR = 15%, calculate the disposal tax effects for the asset in Problem 8.8 after 3, 5, and 6 years, using:

a) Simplified approach of Section 8.6.1.

b) Case I assumptions

c) Case II assumptions

d) Case III assumptions

e) Case IV assumptions

8A.2 If MARR = 15%, what is the present equivalent at time zero of the year 6 disposal tax effects in Problem 8.A.1?

8A.3 If MARR = 10% calculate the disposal tax effects for the robot in Problem 8.9 for the three different salvage values using:

a) Simplified approach of Section 8.6.1.

b) Case I assumptions

c) Case II assumptions

d) Case III assumptions

e) Case IV assumptions

CHAPTER 9

Developing Project Cash Flows

R eynolds Metals' McCook plant produces aluminum coils, sheets, and plates. Its annual production runs at 400 million kilograms. In an effort to improve McCook's current production system, an engineering team, led by the divisional vice president, went on a fact-finding tour of Japanese aluminum and steel companies to observe their production systems and methods. The large fans, which the Japanese companies used to reduce the time that coils need to cool down after various processing operations, were cited among the observations. Cooling the hot process coils with the fans was estimated to significantly reduce the queue or work-in-process (WIP) inventory buildup allowed for cooling. The process also made possible reductions in production lead time and an improvement in delivery performance. The possibility of reducing production time and, as a consequence, the WIP inventory, excited the team members. After the trip, Neal Donaldson, the plant engineer, was asked to investigate the economic feasibility of installing cooling fans at the McCook plant. Neal's job was to justify the purchase of cooling fans for his plant. He was given one week to prove the idea was a good one. Essentially, all he knew were these brief background details: The number of fans, their locations, and the project cost. Everything else was left to Neal's devices. He alone was to determine any effects the purchase might have; what products would be involved; what would be the number of days by which the cooling queue would be reduced; and how much money would be saved. Can Neal logically explain how it all works?

Projecting cash flows is the most important—and the most difficult—step in the analysis of a capital project. Typically, a capital project requires investment outlays initially and only later produces annual net cash inflows. A great many variables are involved in forecasting cash flows, and many individuals, ranging from engineers to cost accountants and marketing people, participate in the process. This chapter provides the general principles on which determining a project's cash flows are based.

To help us imagine the range of activities that are typically initiated by project proposals, we begin this chapter with an overview of how firms classify projects. A whole variety of types of projects exists, each having its own characteristic set of economic concerns. We will next provide an overview of the typical cash flow elements of engineering projects in Section 9.2. Once we have defined these elements, we will examine how to develop cash flow statements that are used to analyze the economic value of projects. In Section 9.3, we will use several examples to demonstrate the development of after-tax project cash flow statements. The generalized cash flow approach is covered in Section 9.4. Then, in Section 9.5, we will present some techniques for automating the process of developing a cash flow statement with an electronic spreadsheet, Excel. By the time you have finished this chapter, you should be prepared not only to understand the format and significance of after-tax cash flow statements, but also how to develop them yourself.

9.1 Project Proposals and Classifications

In a company with capable, imaginative executives and employees, and an effective incentive system, many ideas for capital investment will be advanced. Since some ideas will be good and others not, procedures are usually established for screening projects. Firms generally classify projects as either **profit-adding** or **profit-maintaining** (Figure 9.1). Profit-adding projects include

- Expansion projects

- Product-improvement projects

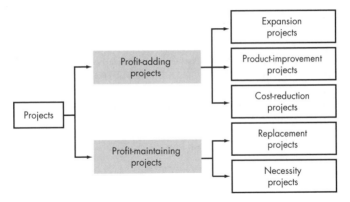

Figure 9.1 Classification of investment projects

- Cost-reduction projects.

Profit-maintaining projects include

- Replacement projects

- Necessity projects.

In practice, projects may contain elements of more than one of these categories. Such projects are often classified according to their primary purpose. As we briefly mentioned in Section 1.5, this classification scheme allows management to address key questions. Can the existing plant be used to achieve the new production level? Does the company have the knowledge and skill to undertake the investment? The answers to these questions help firms screen out proposals that are not feasible given a company's resources.

9.1.1 Profit-Adding Projects

All the capital investment problems in this category have this general form: It is proposed that a certain amount be invested now with the expectation that a return on the investment will be received in the future. The issue is whether the amount of anticipated future cash flows is large enough to justify investing the proposed funds in the project.

Expansion Projects

This project class includes expenditures intended to increase sales and profits by

1. **Introducing new products:** New products differ from existing products with respect to use, function, or size and are intended to increase sales by reaching new markets or customers or to meet previously unmet user requirements. Sales

of these products typically occur in addition to existing sales. Decisions regarding new projects are ultimately based on whether the expected cash inflows from the sale of the new product are large enough to warrant investment in new equipment and working capital and the other costs required to make and introduce the product.

2. **Providing facilities to meet current or forecasted sales opportunities for existing products:** The issue in this case is whether to build or purchase a new facility. The expected future cash inflows are the additional incomes generated from the goods and services produced by the new facility.

Product-Improvement Projects

This class of projects includes expenditures intended to improve the saleability of existing products and to provide products that will displace a competitor's products. The purpose of these expenditures is to maintain or improve the competitive position of existing products. The newly introduced products differ from existing products only with respect to design, quality, color, or style and are not intended to reach new markets or customers or to meet end-use requirements previously unmet.

Cost-Reduction Projects

This class includes projects designed to

1. Reduce costs and expenses of existing operations at current yearly production volume.

2. Avoid forecasted future cost increases that would be incurred at the current yearly production volume.

3. Avoid forecasted future cost increases that would be incurred if yearly production volume were to be increased.

For example, should we buy equipment to automate an operation now done manually? The expected future cash inflows on this investment would be the savings resulting from lower operating costs.

9.1.2 Profit-Maintaining Projects

Profit-maintaining projects are those whose primary purpose is not to reduce costs or increase sales, but simply to maintain ongoing operations. Projects in this category are often proposed by a statement that details both the reasons for the expenditure and the consequences of deferring it.

Replacement Projects

Projects in this class are those required to replace existing, but obsolete or worn-out, assets; failure to implement these projects results in a slowdown or shutdown of operations. For replacement projects, the replacement, rather than the repair of existing equipment, must be justified. Any incremental revenue resulting from replacement projects is considered to be an added benefit in the evaluation of the project. The future expected cash inflows from replacement projects are the cost savings that result from lower operating costs or the revenues that result from the additional volume produced by the new equipment, or both.

Necessity Projects

Some investments are made based on necessity, rather than on an analysis of their profitability. Such projects primarily yield intangible benefits because their economic advantages are not easily determined or are, perhaps, nonexistent. Typical examples include employee recreational facilities, child-care facilities, pollution-control equipment, and the installation of safety devices. The latter two examples are projects for which capital expenditures must be made to comply with environmental control, safety, or other statutory requirements, possibly to avoid penalties. These investments use capital, but do not provide accountable cash inflows.

9.2 Incremental Cash Flows

When a company purchases a fixed asset such as equipment, it makes an investment. The company commits funds today in the expectation of earning a return on those funds in the future. Such investments are similar to those made by a bank when it lends money. In the case of a bank loan, the future cash flow consists of interest plus repayment of the principal. For a fixed asset, the future return is in the form of cash flows generated by the profitable use of the asset. In evaluating a capital investment, we are concerned only with those cash flows that result directly from the investment. These cash flows, called **differential** or **incremental cash flows**, represent the change in the firm's total cash flow that occurs as a direct result of the investment. In this section, we will look into some of the cash flow elements common to most investments.

9.2.1 Elements of Cash Flows

Many variables are involved in the estimation of cash flows, and many individuals and departments participate in the process. For example, the capital outlays associated with a new product are generally obtained from the engineering staff, while operating costs are estimated by accountants and production engineers. On the cash inflow side, we may need to forecast unit sales and sales prices. These forecasts are usually made by the marketing department, which gives consideration to pricing, the effects of advertising, the state of the economy, what strategies competitors are using, and trends in the market and in consumer tastes.

We cannot overstate the importance and complexity of cash flow estimates. However, certain principles, if observed, will help to minimize errors. In this section, we will examine some of the important cash flow elements that must be considered in project evaluation.

New Investments and Existing Assets

A typical project usually involves a cash outflow in the form of an initial investment in equipment or other assets. The relevant investment costs are incremental costs, such as the cost of the asset, shipping and installation costs, and the cost of training employees to use the new asset.

If the purchase of a new asset results in the sale of an existing asset, the net proceeds from this sale reduce the amount of the incremental investment. In other words, the incremental investment represents the total amount of additional funds that must be committed to the investment project. When existing equipment is sold, the transaction results in either an accounting gain or loss. The gain or loss is dependent on whether the amount realized from the sale is greater or less than the equipment's undepreciated capital cost. In any event, when existing assets are sold, the relevant amount by which the new investment is reduced consists of the proceeds of the sale, adjusted for tax effects.

Salvage Value (or Net Selling Price)

In many cases, the estimated salvage value of a proposed asset is so small, and occurs so far in the future, that it may have no significant effect on the decision. Furthermore, any salvage value that is realized may be offset by removal and dismantling costs. In situations where the estimated salvage value is significant, the net salvage value is viewed as a cash inflow at the time of disposal. The **net salvage value** of the existing asset is its selling price minus any costs incurred in selling, dismantling, and removing it, and plus disposal tax effect.

Investments in Working Capital

Some projects require investment in nondepreciable assets. If a project increases a firm's revenues, for example, more funds will be needed to support the higher level of operations. Investment in nondepreciable assets is often called **investment in working capital**. In accounting, working capital means the amount carried in cash, accounts receivable, and inventory that is available to meet day-to-day operating needs. For example, additional working capital may be needed to meet the greater volume of business that will be generated by a project. Part of this increase in current assets may be supplied from increased accounts payable, but the remainder must come from permanent capital. This additional working capital is as much a part of the initial investment as the equipment itself. (We will explain the amount of working capital required for a typical investment project in Section 9.3.2.)

Working Capital Release

As a project approaches termination, inventories are sold off and receivables are collected, that is, at the end of the project, these items can be liquidated at their cost. As this occurs, the company experiences an end-of-project cash flow that is approximately equal to the net working capital investment that was made when the project began. This recovery of working capital is not taxable income, since it merely represents a return of investment funds to the company.

Cash Revenues/Savings

A project generally either increases revenues or reduces costs. Either way, the amount involved should be treated as a cash inflow for capital budgeting purposes. (A reduction in costs is equivalent to an increase in revenues, even though the actual sales revenues may remain unchanged.)

Cost of Operations

The cost associated with manufacturing a new product or delivering a new service needs to be determined. Investment in fixed assets results in ongoing expenditures or costs to operate, maintain and repair these assets. In this book, we use one of the following three approaches to list the total cost of operating a project:

1. It is expressed as specific cost types such as labor, materials, overhead, etc.

2. It is represented within the two broad categories of cost of goods sold and operating expenses.

3. It is broadly referred to as operating and maintenance (O&M) costs or simply operating costs (OC).

Leasing Expenses

When a piece of equipment or a building is leased (instead of purchased) for business use, leasing expenses become cash outflows. Many firms lease computers, automobiles, and industrial equipment subject to technological obsolescence. (We will discuss the leasing issue in Chapter 11.)

Interest and Repayment of Borrowed Amounts

When we borrow money to finance a project, we need to make interest payments as well as principal payments. Proceeds from both short-term borrowing (bank loans) and long-term borrowing (bonds) are treated as cash inflows, but repayments of debts (both principal and interest) are classified as cash outflows.

Income Taxes and Tax Credits

Any income tax payments following profitable operations should be treated as cash outflows for capital budgeting purposes. As we learned in Chapter 8, when an investment is made in depreciable assets, capital cost allowance (CCA) offsets part of what would otherwise be additional taxable income. This is called a **tax shield**, or **tax savings**, and we must take it into account when calculating income taxes. If any investment **tax credit** is allowed, this tax credit will directly reduce income taxes, resulting in cash inflows.

In summary, the following types of cash flows, depicted in Figure 9.2, are common in engineering investment projects.

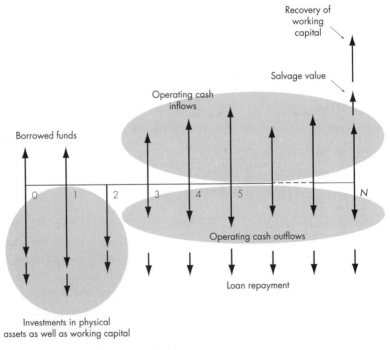

Figure 9.2   Types of cash flow elements used in project analysis

- Cash outflows
    - Initial investment (including installation and freight costs)
    - Investment in working capital
    - Cost of operations
    - Loan interest and principal repayments
    - Income taxes.
- Cash inflows
    - Incremental revenues
    - Reduction in costs (cost savings)
    - Allowed tax credits
    - Salvage value
    - Release of working capital investment
    - Loan amounts from both short-term and long-term borrowing.

Then, the net annual cash flow from the project is

$$\text{Net annual cash flow} = \text{annual cash inflow} - \text{annual cash outflow}.$$

9.2.2   Classification of Cash Flow Elements

Once the cash flow elements are determined (both inflows or outflows), we may group them into three areas: (1) cash flow elements associated with operations, (2) cash flow elements associated with investment activities (capital expenditures), and (3) cash flow elements associated with project financing (such as borrowing). The main purpose of grouping cash flows this way is to provide information about the operating, investing, and financing activities of a project.

Operating Activities

In general, cash flows from operations include current sales revenues, cost of goods sold, operating expenses, and income taxes. Cash flows from operations should generally reflect the cash effects of transactions entering into the determination of net income. The interest portion of a loan repayment is a deductible expense allowed when determining net income, and it is included in the operating activities. Since we usually look only at yearly flows, it is logical to express all cash flows on a yearly basis.

As we discussed in Section 8.2.4, we can determine the net cash flow from operations either (1) based on net income or (2) based on cash flow by computing income taxes directly. When we use net income as the starting point for cash flow determination, we should add any noncash expenses (mainly, capital cost allowance) to net income to estimate the net cash flow. Thus, on an annual basis,

$$\text{Net operating cash flow} = \text{net income} + \text{capital cost allowance}.$$

| **Approach 1** | **Approach 2** |
Income Statement Approach	**Direct Cash Flow Approach**
Cash revenues (savings)	Cash revenues (savings)
– Cost of goods sold	– Cost of goods sold
– Capital cost allowance (CCA)	
– Operating expense	– Operating expense
– Interest expense	– Interest expense
Taxable income	
– Income taxes	– Income taxes
Net income	Operating cash flow
+ Capital cost allowance (CCA)	

In business practice, accountants usually prepare cash flow statements based on net income, namely using Approach 1, whereas Approach 2 is commonly used in many traditional engineering economic texts. If you learn only Approach 2, it is more than likely that you need to be retrained to learn Approach 1 to communicate with the financing and accounting professionals within your organization. *Therefore, we will use the income statement approach (Approach 1) whenever possible throughout the text.*

Investing Activities

In general, three types of investment flows are associated with buying a piece of equipment: The original investment, salvage value at the end of its useful life, and the working capital investment or recovery. We will assume that our outflow for both capital investment and working capital investment take place in year 0. It is possible, however, that both investments will not occur instantaneously but, rather, over a few months as the project gets into gear; we could then use year 1 as an investment year. (Capital expenditures may occur over several years before a large investment project becomes fully operational. In this case, we should enter all expenditures as they occur.) For a small project, either method of timing these flows is satisfactory, because the numerical differences are likely to be insignificant.

Financing Activities

Cash flows classified as financing activities include (1) the amount of borrowing and (2) the repayment of principal. Since interest payments are tax deductible expenses, they are usually classified as operating, not financing, activities.

Net cash flow for a given year is simply the sum of the net cash flows from operating, investing, and financing activities. Table 9.1 can be used as a checklist when you set up a cash flow statement, because it groups each type of cash flow element into operating, investing, or financing activities.

Cash Flow Element	Direction
Operating activities	
Sales revenue	Inflow
Cost savings	Inflow
Cost of goods sold	Outflow
Operating expense	Outflow
Interest payments	Outflow
Lease expenses	Outflow
Income taxes	Outflow
Investing activities	
Capital investment	Outflow
Salvage value	Inflow
Working capital	Outflow
Working capital recovery	Inflow
Disposal tax effect	Inflow
Financing activities	
Borrowed amounts	Inflow
Principal repayments	Outflow

Table 9.1
Classifying Cash Flow Elements into Types of Activities

9.3 Developing Cash Flow Statements

In this section, we will illustrate through a series of numerical examples how we actually prepare a project's cash flow statement; a generic version is shown in Figure 9.3 where we first determine the net income from operations and then adjust the net income by adding any noncash expenses, mainly capital cost allowance. We will also consider a case in which a project generates a negative taxable income for an operating year.

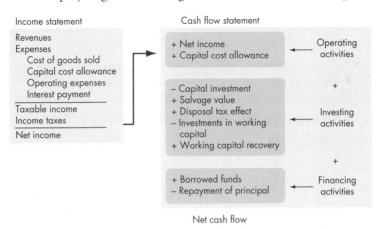

Figure 9.3   A popular format used for presenting a cash flow statement

9.3.1 When Projects Require Only Operating and Investing Activities

We will start with the simplest case of generating after-tax cash flows for an investment project with only operating and investment activities. In the sections ahead, we will add complexities to this problem by including working capital investments (Section 9.3.2) and borrowing activities (Section 9.3.3).

EXAMPLE 9.1

Cash flow statement—operating and investing activities for an expansion project

A computerized machining center has been proposed for a small tool manufacturing company. If the new system, which costs $125,000, is installed, it will generate annual revenues of $100,000 and will require $20,000 in annual labor, $12,000 in annual material expenses, and another $8000 in annual overhead (power and utility) expenses. The automation facility would be classified as a Class 43 property. The company expects to phase out the facility at the end of 5 years, at which time it will be sold for $50,000. Find the year-by-year after-tax net cash flow for the project at a 40% marginal tax rate based on the net income (Approach 1), and determine the after-tax present equivalent of the project at the company's MARR of 15%.

Discussion: We can approach the problem in two steps by using the format shown in Fig. 9.3 to generate an income statement and then a cash flow statement. We will follow this form in our listing of givens and unknowns below. In year 0 (that is, at present) we have an investment cost of $125,000 for the equipment.[1] Capital cost allowances will be calculated based on this cost for years 1 to 5. The revenues and costs are uniform annual flows in years 1 to 5. We can see below that once we find the capital cost allowance for each year, we can easily compute the net cash flows for years 1 to 4, which have fixed revenue and expense entries along with the variable capital cost allowances. In year 5, we will need to incorporate the salvage value and any disposal tax effect from the asset's disposal.

We will use the business convention that no signs (positive or negative) are used in preparing the income statement, except in the situation where we have a negative taxable income or tax savings. In this situation we will use ( ) to denote a negative entry. However, in preparing the cash flow statement, we will observe explicitly the sign convention: A positive sign indicates a cash inflow; a negative sign indicates a cash outflow.

---

[1] We will assume that the asset is purchased and placed in service at the beginning of year 1 (i.e., time 0), and the first year's capital cost allowance will be claimed at the end of year 1.

Income Statement (Year n)

Revenues	$100,000
Expenses	
Labor	$20,000
Material	$12,000
Overhead	$8000
Capital cost allowance	CCA_n
Taxable income	TI_n
Income tax (40%)	T_n
Net income	NI_n

In years 1–5

Cash Flow Statement (Year n)

Operating activities		
Net income	NI_n	
Capital cost allowance	$+CCA_n$	
Investing activities		
Machining center	$-P_0$	Only in year 0
Salvage value	$+S_5$	Only in year 5
Disposal tax effect	$+G_5$	Only in year 5
Net cash flow	NCF_n	

Solution

Given: Cash flow information stated above
Find: After-tax cash flow

Before presenting the cash flow table, we need to do some preliminary calculations. The following notes explain the essential items in Table 9.2.

- Calculation of capital cost allowance
  From Table 7.1, we find that a class 43 property has a declining balance rate of 30%. We should also apply the 50%-rule for the first year. Using the following equations

$$CCA_1 = Pd/2, \text{ for } n = 1$$

$$CCA_n = Pd\left(1 - \frac{d}{2}\right)(1 - d)^{n-2}, \text{ for } n \geq 2$$

we find that the capital cost allowances for years 1 through 5 are $18,750, $31,875, $22,313, $15,619, and $10,933, respectively. These calculated CCA values are used in Table 9.2 to complete the income statement.

- Salvage value and disposal tax effect
  At the end of year 5, the asset is disposed of. A salvage value in the amount of $50,000 is received. However, we need to calculate the disposal tax effect as we may either have to pay some additional income tax or realize a tax saving.

Year	0	1	2	3	4	5
Income Statement						
Revenues		$100,000	$100,000	$100,000	$100,000	$100,000
Expenses						
Labor		20,000	20,000	20,000	20,000	20,000
Material		12,000	12,000	12,000	12,000	12,000
Overhead		8,000	8,000	8,000	8,000	8,000
CCA (30%)		18,750	31,875	22,313	15,619	10,933
Taxable income		$ 41,250	$ 28,125	$ 37,688	$ 44,381	$ 49,067
Income taxes (40%)		16,500	11,250	15,075	17,753	19,627
Net income		$ 24,750	$ 16,875	$ 22,613	$ 26,629	$ 29,440
Cash Flow Statement*						
Operating activities						
Net income		$ 24,750	$ 16,875	$ 22,613	$ 26,629	$ 29,440
CCA		18,750	31,875	22,313	15,619	10,933
Investment activities						
Investment	−$125,000					
Salvage						50,000
Disposal tax effect						−9,796
Net cash flow	−$125,000	$ 43,500	$ 48,750	$ 44,925	$ 42,248	$ 80,578

Table 9.2

Cash Flow Statement for the Automated Machining Center Project Using Approach 1 (Example 9.1)

* Cash flows associated with asset acquisition and disposal are not considered within the definition of operating income. Therefore, in preparing the income statement, the capital expenditures or related items such as disposal tax effect and salvage value are not included. Nevertheless, these items represent actual cash flows in the year they occur and must be shown in the cash flow statement.

1. The total capital cost allowance that has been claimed over the five year period is equal to

$$\sum_{n=1}^{5} CCA_n = \$18,750 + \$31,875 + \$22,313 + 15,619 + \$10,933$$
$$= \$99,489.$$

2. The undepreciated capital cost at the end of year 5 is

$$U_5 = P - \sum_{n=1}^{5} CCA_n = \$125,000 - \$99,489$$
$$= \$25,511.$$

3. The disposal tax effect G is

$$G = t(U_5 - S) = 40\%(\$25,511 - \$50,000)$$
$$= -\$9,796.$$

Since the disposal tax effect is negative, it represents a negative cash flow. This means that additional tax in the amount of $9796 has to be paid due to the disposal.

- Investment analysis

  Once we obtain the project's after-tax net cash flows, we can determine their present equivalent at the firm's interest rate. The after-tax cash flow series is shown in Figure 9.4. Since this series does not contain any patterns to simplify our calculations, we must find the present equivalent of each payment. Using $i = 15\%$, we have

$$
\begin{aligned}
PE(15\%) \ = \ & -\$125,000 + \$43,500(P/F, 15\%, 1) \\
& + \$48,750(P/F, 15\%, 2) + \$44,925(P/F, 15\%, 3) \\
& + \$42,248(P/F, 15\%, 4) + \$80,578(P/F, 15\%, 5) \\
= \ & \$43,443.
\end{aligned}
$$

This means that investing $125,000 in this automated facility would bring in enough revenue to recover the initial investment and the cost of funds at 15% per annum, with a surplus of $43,443.

Comment: As a variation to Table 9.2 which uses Approach 1, a tabular format widely used in traditional engineering economics texts is shown in Table 9.3. The approach taken in developing Table 9.3 is based on Approach 2, which computes income taxes directly. Without seeing the footnotes in Table 9.3, however, it is not intuitively clear how the last column (net cash flow) is obtained. Therefore, we will use the cash flow statement based on the income (Approach 1) whenever possible throughout this text.

Figure 9.4   Cash flow diagram (Example 9.1)

Table 9.3

Net Cash Flow Table Generated by Traditional Method Using Approach 2 (Example 9.1)

	A	B	C	D	E	F	G	H	I	J
	Year End	Investment and Salvage Value	Revenue	Labor	Expenses Materials	Overhead	CCA	Taxable Income	Income Taxes	Net Cash Flow
	0	$(125,000)								$(125,000)
	1		$100,000	$(20,000)	$(12,000)	$(8,000)	$(18,750)	$41,250	$(16,500)	43,500
	2		100,000	(20,000)	(12,000)	(8,000)	(31,875)	28,125	(11,250)	48,750
	3		100,000	(20,000)	(12,000)	(8,000)	(22,313)	37,688	(15,075)	44,925
	4		100,000	(20,000)	(12,000)	(8,000)	(15,619)	44,381	(17,753)	42,248
	5		100,000	(20,000)	(12,000)	(8,000)	(10,933)	49,067	(19,627)	40,373
		50,000*						24,489	(9,796)	40,204

* Salvage value. Note that col. H = col. C + col. D + col. E + col. F + col. G except for salvage value; col. I = 0.4 × col. H; and col. J = col. B + col. C + col. D + col. E + col. F + col. I.

Information required to calculate income taxes

9.3.2   When Projects Require Working Capital Investments

In many cases, changing a production process by replacing old equipment or by adding a new product line will have an impact on cash balances, accounts receivable, inventory, and accounts payable. For example, if a company is going to market a new product, inventories of the product and larger inventories of raw materials will be needed. Accounts receivable from sales will increase, and management might also decide to carry more cash because of the higher volume of activities. These investments in working capital are investments just as are those in depreciable assets (except that they cannot be depreciated).

Consider the case of a firm that is planning a new product line. The new product will require a 2-month's supply of raw materials at a cost of $40,000. The firm could provide $40,000 in cash on hand to pay them. Alternatively, the firm could finance these raw materials via a $30,000 increase in accounts payable (60-day purchases) by buying on credit. The balance of $10,000 represents the amount of net working capital that must be invested.

Working capital requirements differ according to the nature of the investment project. For example, larger projects may require greater average investments in inventories and accounts receivable than would smaller ones. Projects involving the acquisition of improved equipment entail different considerations. If the production rate for the new equipment is higher than that of the old equipment, the firm may be able to decrease its average inventory holdings because new orders can be filled faster as a result of using the new equipment. (One of the main advantages cited in installing advanced manufacturing systems, such as flexible manufacturing systems, is the reduction in inventory made possible by the ability to respond to market demand more quickly.) Therefore, it is also possible for working capital needs to decrease because of an investment. If inventory levels were to decrease at the start of a project, the decrease would be considered a cash inflow, since the cash freed up from inventory could be put to use in other places. (See Example 9.5.)

Two examples will be provided to illustrate the effects of working capital on a project's cash flows. Example 9.2 will show how the net working capital requirement is computed, and Example 9.3 will examine the effects of working capital on the automated machining center project discussed in Example 9.1.

EXAMPLE 9.2   Working capital requirements

Consider Example 9.1. Suppose that the tool manufacturing company's annual revenue projection of $100,000 is based on an annual volume of 10,000 units (or 833 units per month). Assume the following accounting information:

Price (revenue) per unit	$10
Unit variable manufacturing costs	$4
Labor	$2
Material	$1.20
Overhead	$0.80
Monthly volume	833 units
Finished goods inventory	2-month supply
Raw materials inventory	1-month supply
Accounts payable	30 days
Accounts receivable	60 days

The accounts receivable period of 60 days means that revenues from the current month's sales will be collected 2 months later. Similarly, accounts payable of 30 days indicates that payment for materials will be made approximately 1 month after the materials are received. Determine the working capital requirement for this operation.

Solution

Given: Information stated above
Find: Working capital requirement

A policy of having 1 month's raw material and 2 months' finished goods inventory on hand means that $7665 will be tied up in inventory and not available for other use:

Raw materials inventory: ($1.20/unit × 1 month × 833 units/month) = $1000
Finished goods inventory: ($4.00/unit × 2 months × 833 units/month) = $6665

Conceptually, the company needs to build up these inventories even before filling the first order.

Figure 9.5 illustrates the working capital requirements for the first 12-month period. During the first month, the company produces and sells 833 units. Since customers will pay on a net 60-day basis, this will create an accounts receivable in the amount of $8333 that will be collected 2 months later. Since the variable unit cost is $4, the total manufacturing cost would be $3332. Since the supplier requires material payment within 30 days, the company can defer the payment until the next month. This

		Month									
	1	2	3	4	5	6	7	8	9	10	11
Current assets											
Inventory to maintain											
Raw material (1 month)	$1,000	$1,000	$1,000	$1,000	$1,000	$1,000	$1,000	$1,000	$1,000	$1,000	$1,000
Finished goods (2 months)	6,665	6,665	6,665	6,665	6,665	6,665	6,665	6,665	6,665	6,665	6,665
Accounts receivable											
Beginning balance	–	8,333	16,666								16,666
+ Sales	8,333	8,333	8,333								8,333
– Collected	–	–	(8,333)								(8,333)
Ending balance	8,333	16,666	16,666								16,666
Current liabilities											
Accounts payable											
Beginning balance	–	1,000	1,000								1,000
+ Purchases	3,332	3,332	3,332								3,332
– Paid	(2,332)	(3,332)	(3,332)								(3,332)
Ending balance	1,000	1,000	1,000								1,000
Working capital requirements											
Current asset											
+ Funds tied in inventory											
Raw materials	1,000	1,000	1,000	1,000	1,000	1,000	1,000	1,000	1,000	1,000	1,000
Finished goods	6,665	6,665	6,665	6,665	6,665	6,665	6,665	6,665	6,665	6,665	6,665
+ Increase in accounts receivable	8,333	16,666	16,666	16,666	16,666	16,666	16,666	16,666	16,666	16,666	16,666
Current liabilities											
– Increase in accounts payable	(1,000)	(1,000)	(1,000)	(1,000)	(1,000)	(1,000)	(1,000)	(1,000)	(1,000)	(1,000)	(1,000)
Working capital required	$ 14,998	$ 23,331	$ 23,331	$ 23,331	$ 23,331	$ 23,331	$ 23,331	$ 23,331	$ 23,331	$ 23,331	$ 23,331

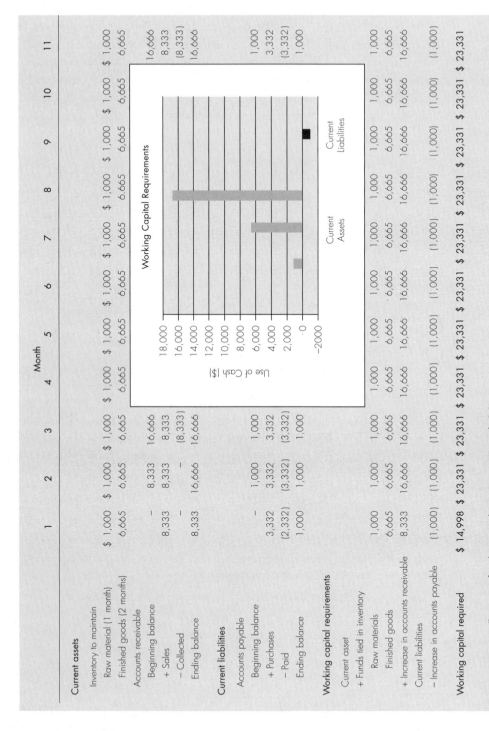

Figure 9.5   Illustration of working capital requirements (Example 9.2)

will create an accounts payable in the amount of $1000 ($1.2/unit × 833 units/month). Therefore, the net payment (labor and overhead) during the first month would be $2332. During the second month, the company must pay off the accounts payable created during the first month. Immediately after the payment, the company can purchase another $1000 worth raw materials on credit to produce and sell the second batch of 833 units. This will result in no change in accounts payable, but it will increase accounts receivable in the amount of $8333 (or the total credit sales of $16,666). The net cost will be $3332 during the second month ($2332 for labor and overhead and $1000 for the raw materials purchased during the first month). During the third month, with receipt of the first payment of $8333, the balance of the accounts receivable will remain unchanged. This payment cycle will repeat over the project.

Now we can look at the working capital requirements on an annual basis. Accounts receivable of $16,666 (2 months' sales) means that in year 1 the company will have cash inflows of $83,333, which is less than the projected sales of $100,000 ($8333 × 12). In years 2 to 5, collections will be $100,000, equal to sales, because beginning and ending accounts receivable will be $16,666, with sales of $100,000. Collections of the final accounts receivable of $16,666 would occur in the first 2 months of year 6, but can be added to the year 5 revenue to simplify the calculations. The important point is that cash inflow lags sales by $16,666 in the first year.

Assuming the company wishes to build up 2 months' inventory during the first year, it must produce 833 × 2 = 1666 more units than are sold the first year. The extra cost of these goods in the first year is 1666 units ($4 variable cost per unit), or $6665. The finished goods inventory of $6665 represents the variable cost incurred to produce 1666 more units than are sold in the first year. In years 2 to 4, the company will produce and sell 10,000 units per year, while maintaining its 1666 units supply of finished goods. In the final year of operations, the company will produce only 8334 units (for 10 months) and will use up the finished goods inventory. As 1666 units of the finished goods inventory get liquidated during the last year, a working capital release in the amount of $6665 will occur. Along with the collections of the final accounts receivable of $16,666, a total working capital release of $23,331 will remain when the project terminates. Now we can calculate the working capital requirements as follows:

Accounts receivables	
(833 units/month × 2 months × $10)	$16,666
Finished good inventory	
(833 units/month × 2 months × $4)	6,665
Raw materials inventory	
(833 units/month × 1 month × $1.20)	1,000
Accounts payable (raw material purchase)	
(833 units/month × 1 month × $1.20)	(1,000)
Net working capital required	$23,331

Comments: In our example, during the first year, the company produces 11,666 units to maintain 2 months' finished good inventory, but it sells only 10,000 units. On what basis should the company calculate the net income during the first year (use 10,000 or 11,666 units)? Any increases in inventory expenses will reduce the taxable income; therefore, this calculation is based on 10,000 units. The reason is that accounting measure of net income is based on the **matching concept**. If we report revenue when it is *earned* (whether it is actually received or not), and we report expenses when they are *incurred* (whether they are paid or not), we are using the *accrual method* of accounting. By tax law, this accrual method must be used for purchases and sales whenever business transactions involve an inventory. Therefore, most manufacturing and merchandising businesses use the accrual method in recording revenues and expenses. Any cash inventory expenses not accounted for in the net income calculation will be reflected in the working capital requirement.

EXAMPLE 9.3 Cash flow statement—including working capital

Reconsider Example 9.1. Update the after-tax cash flows for the automated machining center project by including a working capital requirement of $23,331 in year 0 and full recovery of the working capital at the end of year 5.

Solution

Given: Flows as in Example 9.1, with the addition of a working capital requirement = $23,331
Find: Net after-tax cash flows with working capital and present equivalent

Using the procedure outlined above, the net after-tax cash flows for this machining center project are grouped as shown in Table 9.4. As the table indicates, investments in working capital are cash outflows when they are expected to occur, and recoveries are treated as cash inflows at the times they are expected to materialize. In this example, we assume that the investment in working capital made at period 0 will be recovered at the end of the project's life.[2] Moreover, we also assume a full recovery of the initial working capital. However, many situations occur in which the investment in working capital may not be fully recovered (e.g., inventories may deteriorate in value or become obsolete). The equivalent net present equivalent of the after-tax cash flows including the effects of working capital is calculated as

$$
\begin{aligned}
PE(15\%) &= -\$148{,}331 + \$43{,}500(P/F, 15\%, 1) \\
&\quad + \$48{,}750(P/F, 15\%, 2) + \$44{,}925(P/F, 15\%, 3) \\
&\quad + \$42{,}248(P/F, 15\%, 4) + \$103{,}909(P/F, 15\%, 5) \\
&= \$31{,}712.
\end{aligned}
$$

---

[2] In fact, we could assume that the investment made in working capital would be recovered at the end of the first operating cycle (say, year 1). However, the same amount of investment in working capital has to be made again at the beginning of year 2 for the second operating cycle, and the process repeats until the project terminates. Therefore, the net cash flow transaction looks as though the initial working capital will be recovered at the end of project life (Figure 9.6.)

Period	0	1	2	3	4	5
Investment	−$23,331	−$23,331	−$23,331	−$23,331	−$23,331	0
Recovery	0	23,331	23,331	23,331	23,331	23,331
Net flow	−$23,331	0	0	0	0	$23,331

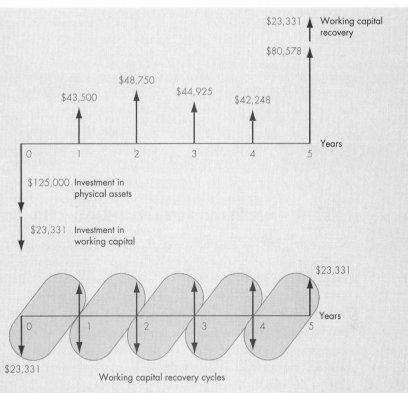

Figure 9.6 Cash flow diagram (Example 9.3)

	Year	0	1	2	3	4	5
Table 9.4 Cash Flow Statement for Automated Machining Center Project with Working Capital Requirement (Example 9.3)	**Income Statement**						
	Revenues		$100,000	$100,000	$100,000	$100,000	$100,000
	Expenses						
	Labor		20,000	20,000	20,000	20,000	20,000
	Materials		12,000	12,000	12,000	12,000	12,000
	Overhead		8,000	8,000	8,000	8,000	8,000
	CCA (30%)		18,750	31,875	22,313	15,619	10,933
	Taxable income		$ 41,250	$ 28,125	$ 37,688	$ 44,381	$ 49,067
	Income taxes (40%)		16,500	11,250	15,075	17,753	19,627
	Net income		$ 24,750	$ 16,875	$ 22,613	$ 26,629	$ 29,440
	Cash Flow Statement						
	Operating activities						
	Net income		$ 24,750	$ 16,875	$ 22,613	$ 26,629	$ 29,440
	CCA		18,750	31,875	22,313	15,619	10,933
	Investment activities						
	Investment	−$125,000					
	Salvage						50,000
	Disposal tax effects						−9,796
	Working capital	−23,331					23,331
	Net cash flow	−$148,331	$ 43,500	$ 48,750	$ 44,925	$ 42,248	$103,909

This present equivalent is $11,731 less than in the situation with no working capital requirement (Example 9.1). This example demonstrates that working capital requirements must be considered when properly assessing a project's worth.

Comment: The $11,731 reduction in present equivalent is just the present equivalent of an annual series of 15% interest payments on the working capital which is borrowed by the project at time 0 and repaid at the end of year 5:

$$\$23,331(15\%)(P/A, 15\%, 5) = \$11,731.$$

The investment tied up in working capital results in lost earnings.

9.3.3   When Projects Are Financed with Borrowed Funds

Many companies use a mixture of debt and equity to finance their physical plant and equipment. The ratio of total debt to total investment, generally called the **debt ratio**, represents the percentage of the total initial investment provided by borrowed funds. For example, a debt ratio of 0.3 indicates that 30% of the initial investment is borrowed, and the rest is provided from the company's earnings (also known as **equity**). Since interest is a tax-deductible expense, companies in high tax brackets may incur lower after-tax financing costs by financing through debt. (Along with the effect on taxes, the method of loan repayment can also have a significant impact. We will discuss the issue of project financing in Chapter 10.)

EXAMPLE 9.4

Cash flow statement—with financing (borrowing)

Rework Example 9.3, assuming that $62,500 of the $125,000 paid for the investment is obtained through debt financing (debt ratio = 0.5). The loan is to be repaid in equal annual installments at 10% interest over 5 years. The remaining $62,500 will be provided by equity (e.g., from retained earnings).

Solution

Given: Same as in Example 9.3, but $62,500 is borrowed, repaid in equal installments over 5 years at 10%
Find: Net after-tax cash flows in each year

We first need to compute the size of the annual loan repayment installments:

$$\$62,500(A/P, 10\%, 5) = \$16,487.$$

Next, we determine the repayment schedule of the loan by itemizing both the interest and principal represented in each annual repayment. Using Eqs. (3.16) and (3.17), we obtain

Year	Beginning Balance	Interest Payment	Principal Payment	Ending Balance
1	$62,500	$6,250	$10,237	$52,263
2	52,263	5,226	11,261	41,002
3	41,002	4,100	12,387	28,615
4	28,615	2,861	13,626	14,989
5	14,989	1,499	14,988	0

The resulting after-tax cash flow is detailed in Table 9.5. The present equivalent of the after-tax cash flow series is

$$
\begin{aligned}
PE(15\%) =\ & -\$85,831 + \$29,513(P/F, 15\%, 1) \\
& + \$34,353(P/F, 15\%, 2) + \$30,078(P/F, 15\%, 3) \\
& + \$26,905(P/F, 15\%, 4) + \$88,021(P/F, 15\%, 5) \\
=\ & \$44,729.
\end{aligned}
$$

Table 9.5

Cash Flow Statement for Automated Machining Center Project with Debt Financing (Example 9.4)

Year	0	1	2	3	4	5
Income Statement						
Revenues		$100,000	$100,000	$100,000	$100,000	$100,000
Expenses						
Labor		20,000	20,000	20,000	20,000	20,000
Materials		12,000	12,000	12,000	12,000	12,000
Overhead		8,000	8,000	8,000	8,000	8,000
Debt interest		6,250	5,226	4,100	2,861	1,499
CCA (30%)		18,750	31,875	22,313	15,619	10,933
Taxable income		$ 35,000	$ 22,899	$ 33,587	$ 41,520	$ 47,568
Income tax (40%)		14,000	9,159	13,435	16,608	19,027
Net income		$ 21,000	$ 13,739	$ 20,152	$ 24,912	$ 28,541
Cash Flow Statement						
Operating activities						
Net income		$ 21,000	$ 13,739	$ 20,152	$ 24,912	$ 28,541
CCA		18,750	31,875	22,313	15,619	10,933
Investment activities						
Investment	−$125,000					
Salvage						50,000
Disposal tax effect						−9,796
Working capital	−23,331					23,331
Financing activities						
Principal repayment	$ 62,500	−10,237	−11,261	−12,387	−13,626	−14,988
Net cash flow	−$ 85,831	$ 29,513	$ 34,353	$ 30,078	$ 26,905	$ 88,021

When this amount is compared with that found in the case that involved no borrowing ($31,712), we see that debt financing actually increases present equivalent by $13,017. This surprising result is largely due to the fact that the firm is able to borrow the funds at an after-tax rate of 6% (i.e., 10%(1 − t)) which is less than its MARR (opportunity cost rate) of 15%. To some extent firms can usually borrow money at lower rates than their MARR. If so, why don't they borrow all the funds they need for capital investment? We will address this question in Chapter 10.

9.3.4 When Projects Result in Negative Taxable Income

In a typical project year, revenues may not be large enough to offset expenses, thereby resulting in a negative taxable income. A negative taxable income from a single project does *not* mean that a firm does not need to pay income tax: Rather, the negative figure can be used to reduce the taxable incomes generated by other business operations.[3] Therefore, a negative taxable income usually results in **tax savings**. When we evaluate an investment project using an incremental tax rate, we also assume that the firm has sufficient taxable income from other activities so that changes due to the project under consideration will not change the incremental tax rate.

When we compare **cost-only** mutually exclusive projects (service projects), we have no revenues to consider in their cash flow analysis. In this situation, we typically assume no revenue (zero), but proceed as before when constructing the after-tax cash flow statement. With no revenue to match expenses, we have a negative taxable income, resulting in tax savings as before. Example 9.5 illustrates how we may develop an after-tax cash flow statement for this type of project.

EXAMPLE 9.5 **After-tax cash flow analysis for a cost-only project**

Reconsider the McCook plant's cooling fan project that was introduced in the opening of this chapter. Suppose that Mr. Donaldson compiled the following financial data:

- The project will require an investment of $563,000 in cooling fans now.

- The cooling fans would provide 16 years of service with negligible salvage values considering the removal costs.

- It is expected that the amount of time required between hot rolling and the next operation would be reduced from 5 days to 2 days. Cold rolling queue time would be reduced from 2 days to 1 day for each cold roll pass. The net effect of these changes would be a reduction of WIP inventory at a value of $2,121,000. Because of the lead time involved in installing the fans, as well as the consumption of stock-piled WIP inventory, this working capital release will be realized 1 year after the fans are installed.

- The cooling fans have a declining balance capital cost allowance rate of 20%.

- Annual electricity costs are estimated to be increased by $86,000.

- Reynolds' required after-tax rate of return is known to be 20% for this type of cost reduction project.

---

[3] Even if the firm does not have any other taxable income to offset in the current tax year, the operating loss can be carried back to each of the preceding 3 years and forward for the following 7 years to offset taxable income in those years.

- Reynolds' incremental tax rate is 40%.

- Develop the project cash flows over the service period, and determine if the investment is a wise one.

Solution

Given: Required investment = $563,000; service period = 16 years; salvage value = $0; CCA rate for cooling fans = 20%; marginal tax rate = 40%; working capital release = $2,121,000 one year later; annual operating cost (electricity) = $86,000 Find: (a) annual after-tax cash flows and (b) accept or reject the investment

(a) Because we can assume that the annual revenues would stay the same as before and after the installation of the cooling fans, we can treat these unknown revenue figures as zero. Table 9.6 summarizes the cash flow statement for the cooling fan project. With no revenue to offset the expenses, the taxable income will be negative, resulting in tax savings. Note that the working capital recovery (as opposed to working capital investment for a typical investment project) is shown in year 1. Note that the disposal tax effect is not zero even though the salvage value of the cooling fans is zero.

(b) At MARR = 20%, the PE of this investment would be

$$
\begin{aligned}
PE(20\%) \quad = \quad & - \$563,000 + \$2,091,920(P/F, 20\%, 1) \\
& - \$11,064(P/F, 20\%, 2) - \ldots \\
& - \$42,686(P/F, 20\%, 16) \\
= \quad & \$1,063,864.
\end{aligned}
$$

Even with only one-time savings in WIP, this cooling fan project is economically justifiable.[4]

Comments: As the cooling fans reach the end of their respective service lives, it is tempting to add the working capital investment ($2,120,000) at the end of year 16 working with the assumption that the plant will return to the system without the cooling fans and thus will require the additional investment in working capital. How sensible is this assumption? In fact, not at all. If the system has proven to be effective, and the plant will remain in definite service, we need to make another investment to purchase a new set of cooling fans at the end of year 16. However, this investment should bring benefits to the second cycle of the operation, so that it should be charged against the cash flows for the second cycle, not the first cycle. Therefore, in developing the cash flows for the first life cycle, we should avoid adding this working capital investment figure at the end of the service life.

---

[4] This is not a simple investment. If you desire to find the true IRR for this project, you need to follow the procedures outlined in Appendix 4A. At a MARR of 20% (or external interest rate), the IRR (or RIC) is 267.56%, which is significantly larger than the MARR.

Table 9.6
Cash Flow Statement for the Cooling Fan Project without Revenue (Example 9.5)

Year	0	1	2	3	4	5	6	7	8	9	10	11	12	13	14	15	16
Income Statement																	
Revenues																	
Expenses																	
CCA		$ 56,300	$ 101,340	$ 81,072	$ 64,858	$ 51,886	$ 41,509	$ 33,207	$ 26,566	$ 21,253	$ 17,002	$ 13,602	$ 10,881	$ 8,705	$ 6,964	$ 5,571	$ 4,457
Electricity cost		86,000	86,000	86,000	86,000	86,000	86,000	86,000	86,000	86,000	86,000	86,000	86,000	86,000	86,000	86,000	86,000
Taxable income		(142,300)	(187,340)	(167,072)	(150,858)	(137,886)	(127,509)	(119,207)	(112,566)	(107,253)	(103,002)	(99,602)	(96,881)	(94,705)	(92,964)	(91,571)	(90,457)
Income taxes (40%)		(56,920)	(74,936)	(66,829)	(60,343)	(55,154)	(51,004)	(47,683)	(45,026)	(42,901)	(41,201)	(39,841)	(38,753)	(37,882)	(37,186)	(36,628)	(36,183)
Net income		$ (85,380)	$(112,404)	$(100,243)	$ (90,515)	$ (82,732)	$ (76,505)	$ (71,524)	$ (67,539)	$ (64,352)	$ (61,801)	$(59,761)	$(58,129)	$(56,823)	$(55,778)	$(54,943)	$(54,274)
Cash Flow Statement																	
Operating activities																	
Net income		$ (85,380)	$(112,404)	$(100,243)	$(100,243)	$ (90,515)	$ (82,732)	$ (76,505)	$ (71,524)	$ (67,539)	$ (64,352)	$ (61,801)	$(59,761)	$(58,129)	$(56,823)	$(55,778)	$(54,274)
CCA		56,300	101,340	81,072	64,858	51,886	41,509	33,207	26,566	21,253	17,002	13,602	10,881	8,705	6,964	5,571	4,457
Investment activities																	
Cooling fans	$(563,000)																
Salvage value																	
Disposal tax effects																	0
Working capital		2,121,000															7131
Net cash flow	$(563,000)	$2,091,920	$ (11,064)	$ (19,171)	$ (25,657)	$ (30,846)	$ (34,996)	$ (38,317)	$ (40,974)	$ (43,099)	$ (44,799)	$ (46,159)	$(47,247)	$(48,118)	$(48,814)	$(49,372)	$(42,686)

Note: The working capital release attributable to reduction in work-in-process inventories will be materialized at the end of year 1.

9.3.5   When Projects Require Multiple Assets

Up to this point, our examples have been limited to situations where only one asset was employed in a project. In many situations, however, a project may require purchase of multiple assets with different property classes. For example, a typical engineering project may involve more than just the purchase of equipment—it may need a building in which to carry out manufacturing operations. Even the various assets may become available-for-use at different points in time. What we have to do is to itemize the timing of the investment requirement and the capital cost allowances according to the asset introduction. Example 9.6 illustrates the development of project cash flows that require multiple assets.

EXAMPLE
9.6

A project requiring multiple assets

Langley Manufacturing Company (LMC), a manufacturer of fabricated metal products, is considering the purchase of a new computer-controlled milling machine for $90,000 to produce a custom-ordered metal product. The costs for installation of the machine, site preparation, and wiring are expected to be $10,000. The machine also needs special jigs and dies, which will cost $12,000. The milling machine is expected to last 10 years, and the jigs and dies to last 5 years. The machine will have a $10,000 salvage value at the end of its life. The special jigs and dies are worth only $1000 as scrap metal at any time in their lives. From Table 7.1, page 439, we find that the milling machine has a CCA rate of 30% while the jigs and dies have a CCA rate of 100%. LMC needs to either purchase or build an 800 m² warehouse in which to store the product before shipping to the customer. LMC has decided to purchase a building near the plant at a cost of $160,000. For CCA purposes, the warehouse cost of $160,000 is divided into $120,000 for the building (its CCA rate is 4%) and $40,000 for land. At the end of 10 years, the building will have a salvage value of $80,000, but the value of the land will have appreciated to $110,000. The revenue from increased production is expected to be $150,000 per year. The additional annual production costs are estimated as follows: materials, $22,000, labor, $32,000, energy $3500, and other miscellaneous costs, $2500. For the analysis, a 10-year life will be used. LMC has an incremental and marginal tax rate of 40% and a MARR of 18%, respectively. No money is borrowed to finance the project. Capital gains will be taxed at 30%.[5]

Discussion: Three types of assets are to be considered in this problem: The milling machine, the jigs and dies, and the warehouse. The cost basis for each asset has to be determined separately. For the milling machine, we need to add the site-preparation expense to the cost basis, whereas we need to subtract the land cost from the warehouse cost to establish the correct cost basis for the building.

---

[5] Capital gains tax rate is 75% of the company's marginal tax rate.

- The milling machine: $90,000 + $10,000 = $100,000.

- Jigs and dies: $12,000.

- Warehouse (building): $120,000.

- Warehouse (land): $40,000.

Since the jigs and dies last only 5 years, we need to make a specific assumption regarding the replacement cost at the end of 5 years. In this problem, we will assume that the replacement cost would be approximately equal to the cost of the initial purchase.

Solution

Given: Cash flow elements provided above, $t = 40\%$, MARR = 18%
Find: Net after-tax cash flow, PE

Table 9.7 and Figure 9.7 summarize the net after-tax cash flows associated with the multiple-asset investment. We assume that the building will be disposed of on December 31 of the tenth year. The undepreciated capital cost and the disposal tax effect for the building are calculated as follows:

$$U_{10} = \$120,000(1 - 2\%)(1 - 4\%)^9 = \$81,442$$

$$G_{10} = t(U_{10} - S_{10}) = 0.4(\$81,442 - \$80,000)$$
$$= \$577$$

We show the disposal tax effects associated with disposal of each asset as follows:

Property (Asset)	Cost Base	Salvage Value (S)	Undepreciated Capital Cost (U)	Losses (Gains)(U–S)	Disposal Tax Effects
Land	$ 40,000	$110,000	$40,000	$(70,000)	$(21,000)
Building	120,000	80,000	81,442	1,442	577
Milling machine	100,000	10,000	3,430	(6,570)	(2,628)
Jigs and dies	12,000	1,000	0	(1,000)	(400)

The PE of the project is

$$
\begin{aligned}
PE(18\%) &= -\$272,000 + \$63,360(P/F, 18\%, 1) + \$68,482(P/F, 18\%, 2) \\
&\quad + \ldots + \$233,494(P/F, 18\%, 10) \\
&= \$34,880 > 0,
\end{aligned}
$$

and the IRR for this investment is about 21%, which exceeds the MARR. Therefore, this project is acceptable.

Comment: Note that the losses (gains) posted in the table above can be classified into two types: CCA losses (gains) and capital gains. Only $70,000 for land represents capital gains, whereas others represent CCA losses (gains) upon disposal. The tax rate for capital gains is 75% of the tax rate for CCA gains. Note also that the disposal tax effect is positive if there is a loss and negative if there is a gain. It is equal to the losses (gains) column multiplied by the corresponding tax rate.

Table 9.7

Cash Flow Statement for the LMC's Machining Center Project with Multiple Assets (Example 9.6)

Year	0	1	2	3	4	5	6	7	8	9	10
Income Statement											
Revenues		$150,000	$150,000	$150,000	$150,000	$150,000	$150,000	$150,000	$150,000	$150,000	$150,000
Expenses											
Materials		22,000	22,000	22,000	22,000	22,000	22,000	22,000	22,000	22,000	22,000
Labor		32,000	32,000	32,000	32,000	32,000	32,000	32,000	32,000	32,000	32,000
Energy		3,500	3,500	3,500	3,500	3,500	3,500	3,500	3,500	3,500	3,500
Others		2,500	2,500	2,500	2,500	2,500	2,500	2,500	2,500	2,500	2,500
CCA											
Building (4%)		2,400	4,704	4,516	4,335	4,162	3,995	3,836	3,682	3,535	3,393
Machines (30%)		15,000	25,500	17,850	12,495	8,747	6,123	4,286	3,000	2,100	1,470
Tools (100%)		6,000	6,000	—	—	—	6,000	6,000	—	—	—
Taxable income		66,600	53,796	67,634	73,170	77,092	73,882	75,879	83,318	84,365	85,137
Income taxes (40%)		26,640	21,518	27,054	29,268	30,837	29,553	30,351	33,327	33,746	34,055
Net income		$ 39,960	$ 32,278	$ 40,580	$ 43,902	$ 46,255	$ 44,329	$ 45,527	$ 49,991	$ 50,619	$ 51,082
Cash Flow Statement											
Operating activities											
Net income		$ 39,960	$ 32,278	$ 40,580	$ 43,902	$ 46,255	$ 44,329	$ 45,527	$ 49,991	$ 50,619	$ 51,082
CCA		23,400	36,204	22,366	16,830	12,908	16,118	14,121	6,682	5,635	4,863
Investment activities											
Land	$ (40,000)										110,000
Building	(120,000)										80,000
Machines	(100,000)										10,000
Tools (1st cycle)	(12,000)					1,000					
Tools (2nd cycle)						(12,000)					1,000
Disposal tax effects											
Land											(21,000)
Building											577
Machines											(2,628)
Tools						(400)					(400)
Net cash flow	$(272,000)	$ 63,360	$ 68,482	$ 62,946	$ 60,732	$ 47,763	$ 60,447	$ 59,649	$ 56,673	$ 56,254	$233,494

Note: Investment in tools (jigs and dies) will be repeated at the end of year 5 with the same initial purchase costs.

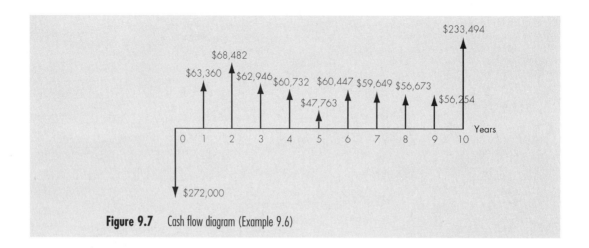

Figure 9.7 Cash flow diagram (Example 9.6)

9.3.6 When Investment Tax Credits Are Allowed

As discussed in Section 8.3.5, the investment tax credit (ITC) has been used as a vehicle to encourage investments of certain types or in certain areas of Canada. A corporation may earn investment tax credit by acquiring certain property or making certain expenditures in certain geographical areas.

How do we adjust our after-tax cash flow analysis to consider the effects of tax credit? The current tax regulation indicates the following.[6] The investment tax credit reduces the company's income tax payment by that amount at the end of the first year the asset is in service. At the same time, the cost base for calculation of capital cost allowances has to be reduced by the amount of ITC. Thus, this will affect the capital cost allowances to be claimed for this asset and the disposal tax effect of this asset.

EXAMPLE **Considering an investment tax credit**

9.7

Suppose, in Example 9.6, that a 35% ITC was allowed on the purchase of the milling machine and the jigs and dies only. How does the ITC affect the profitability of the investment?

Solution

Given: Net cash flows in Table 9.6, ITC = 35%.
Find: PE

---

[6] Check with Canada Customs and Revenue Agency for details on ITC. The website of Canada Customs and Revenue Agency is at *www.ccra-adrc.gc.ca/menu.html*.

The ITC amount for each asset is

Milling machine: $0.35 \times \$100,000 = \$35,000$
Jigs and dies: $\quad 0.35 \times \$12,000 = \$4200.$

The ITC for the milling machine occurs in year 1 while the ITC for jigs and dies occurs in both year 1 and year 6. The new cost base for each asset is:

Milling machine: $\$100,000 - \$35,000 = \$65,000$
Jigs and dies: $\quad \$12,000 - \$4,200 = \$7800.$

The capital cost allowances for these two assets will be changed and the disposal tax effects of the two assets will be changed too. The updated net cash flows are given in Table 9.8. We can then find the new PE with ITC considered as follows:

$$
\begin{aligned}
\text{New PE} &= -\$272,000 + \$99,620(P/F, 18\%, 1) \\
&\quad + \$64,072(P/F, 18\%, 2) + \dots + \$232,808(P/F, 18\%, 10) \\
&= \$59,649 \\
\text{New IRR} &= 24\%.
\end{aligned}
$$

Note that the ITC increases the profitability of the project. (For a certain investment situation, the ITC could increase the profitability of the project from somewhat marginal to moderate.) This increase will certainly encourage the firm to undertake the investment.

9.4 Generalized Cash Flow Approach

If we are analyzing project cash flows for a corporation that consistently operates in the highest tax bracket, we can assume that the firm's marginal tax rate will remain the same, whether the project is accepted or rejected. In this situation, we may apply the top marginal tax rate to each taxable item in the cash profile, thus obtaining the after-tax cash flows. By aggregating individual items, we obtain the project's net cash flows. This approach is referred to as the **generalized cash flow approach.** As we shall see in later chapters, this approach affords several analytical advantages. In particular, when we compare service projects, the generalized cash flow method is computationally efficient. (Examples are provided in Appendix 9A and Chapter 11.)

9.4.1 Setting Up Net Cash Flow Equations

To produce the generalized cash flow table, we first examine each cash flow element. We can do this as follows, using the scheme for classifying cash flows that we have just developed:

Table 9.8

Cash Flow Statement for the LMC's Machining Center Project with Multiple Assets Considering Investment Tax Credit (Example 9.7)

Year	0	1	2	3	4	5	6	7	8	9	10
Income Statement											
Revenues		$150,000	$150,000	$150,000	$150,000	$150,000	$150,000	$150,000	$150,000	$150,000	$150,000
Expenses											
Materials		22,000	22,000	22,000	22,000	22,000	22,000	22,000	22,000	22,000	22,000
Labor		32,000	32,000	32,000	32,000	32,000	32,000	32,000	32,000	32,000	32,000
Energy		3,500	3,500	3,500	3,500	3,500	3,500	3,500	3,500	3,500	3,500
Others		2,500	2,500	2,500	2,500	2,500	2,500	2,500	2,500	2,500	2,500
CCA											
Building (4%)		2,400	4,704	4,516	4,335	4,162	3,995	3,836	3,682	3,535	3,393
Machines (30%)		9,750	16,575	11,603	8,122	5,685	3,980	2,786	1,950	1,365	956
Tools (100%)		3,900	3,900				3,900	3,900			
Taxable income		73,950	64,821	73,882	77,543	80,153	78,125	79,479	84,368	85,100	85,651
Income taxes (40%)		29,580	25,928	29,553	31,017	32,061	31,250	31,791	33,747	34,040	34,260
Net income		$ 44,370	$ 38,893	$ 44,329	$ 46,526	$ 48,092	$ 46,875	$ 47,687	$ 50,621	$ 51,060	$ 51,391
Cash Flow Statement											
Operating activities											
Net income		$ 44,370	$ 38,893	$ 44,329	$ 46,526	$ 48,092	$ 46,875	$ 47,687	$ 50,621	$ 51,060	$ 51,391
CCA		16,050	25,179	16,118	12,457	9,847	11,875	10,521	5,632	4,900	4,349
Investment activities											
Land	(40,000)										110,000
Building	(120,000)										80,000
Machines	(100,000)										10,000
Tools (1st cycle)	(12,000)					1,000					
Tools (2nd cycle)						(12,000)					1,000
ITC											
Machines		35,000									
Tools		4,200					4,200				
Disposal tax effects											
Land											(21,000)
Building											577
Machines											(3,108)
Tools						(400)					(400)
Net cash flow	($272,000)	$ 99,620	$ 64,072	$ 60,447	$ 58,983	$ 46,539	$ 62,950	$ 58,209	$ 56,253	$ 55,960	$232,808

Note: Investment in tools (jigs and dies) will be repeated at the end of year 5 with the same initial purchase costs.

A_n = + Revenues at time n (R_n)

    − Expenses excluding CCA and debt interest at time n (E_n)

    − Debt interest at time n (I_n)

    − Income taxes at time n (T_n)

          Operating activities

    − Investment at time n (P_n)

    + Net proceeds from sale at time n ($S_n + G_n$)

    − Working capital investment at time n (W_n)

    + Working capital recovery at time n (W'_n)

          Investing activities

    + Proceeds from loan (L_n)

    − Repayment of principal (PP_n)

          Borrowing activities

where A_n is the net after-tax cash flow at the end of period n.

Capital cost allowance is *not* a cash flow and is therefore excluded from E_n (although it must be considered when calculating income taxes). Note also that $(S_n + G_n)$ represents the net salvage value after adjustments for disposal tax effect (G_n). Not all terms are relevant in calculating a cash flow in every year. For example, the term $(S_n + G_n)$ appears only when the asset is disposed of.

In terms of symbols, we can express A_n as

$$
\begin{aligned}
A_n \quad = \quad & R_n - E_n - I_n - T_n \quad \longleftarrow \text{Operating activities} \\
& - P_n + (S_n + G_n) - W_n + W'_n \quad \longleftarrow \text{Investing activities} \\
& + L_n - PP_n \quad \longleftarrow \text{Financing activities} \quad (9.1)
\end{aligned}
$$

If we designate T_n as the total income taxes paid at time n and t as the marginal tax rate, income taxes on this project are

$$
\begin{aligned}
T_n \quad = \quad & (\text{Taxable income})(\text{marginal tax rate}) \\
= \quad & (R_n - E_n - I_n - CCA_n)t \\
= \quad & (R_n - E_n)t - (I_n + CCA_n)t. \quad (9.2)
\end{aligned}
$$

The term $(I_n + CCA_n)t$ is known as the tax shield (or tax savings) from financing and capital cost allowance. Now, substituting the result of Eq. (9.2) into Eq. (9.1) we obtain

$$
\begin{aligned}
A_n \quad = \quad & (R_n - E_n - I_n)(1 - t) + t \times CCA_n \\
& - P_n + (S_n + G_n) - W_n + W'_n \\
& + L_n - PP_n. \quad (9.3)
\end{aligned}
$$

9.4.2   Presenting Cash Flows in Compact Tabular Formats

After-tax cash flow components over the project life can be grouped by type of activities in a compact tabular format as follows:

Cash Flow Elements	End of Period 0 1 2 ... N
Operating activities	
$+ (1 - t)(R_n)$	
$- (1 - t)(E_n)$	
$- (1 - t)(I_n)$	
$+ t \times CCA_n$	
Investment activities	
$- P_n$	
$+ (S_n + G_n)$	
$- W_n$	
$- W'_n$	
Financing activities	
$+ L_n$	
$- PP_n$	
Net cash flow	
A_n	

However, in preparing their after-tax cash flow, most business firms adopt the income statement approach presented in previous sections because they want to know the accounting income along with the cash flow statement.

EXAMPLE 9.8

Generalized cash flow approach

Reconsider Example 9.4. Use the generalized cash flow approach to obtain the after-tax cash flows:

Solution

Given:

Investment cost (P_n) = $125,000	
Investment in working capital (W_n) = $23,331	
Annual revenues (R_n) = $100,000	
Annual expenses other than capital cost allowance and debt interest (E_n) = $40,000	
Debt interest (I_n), years 1 to 5 = $6250, $5226, $4100, $2861, $1499	
Principal repayment (PP_n), years 1 to 5 = $10,237; $11,261; $12,387; $13,626; $14,988	
Capital cost allowance (CCA_n), years 1 to 5 = $18,750; $31,875; $22,313; $15,619; $10,933	
Marginal tax rate (t) = 40%	
Salvage value (S_n) = $50,000	

Find: Annual after-tax cash flows (A_n)

Step 1: Find the cash flow at year 0.

1. Investment in depreciable asset (P_0) = \$125,000
2. Investment in working capital (W_0) = \$23,331
3. Borrowed funds (L_0) = \$62,500
4. Net cash flow $(A_0) = -P_0 - W_0 + L_0 = -\$125,000 - \$23,331 + \$62,500$
   $= -\$85,831.$

Step 2: Find the cash flow at years 1 to 4.

$$A_n = (R_n - E_n - I_n)(1 - t) + t \times CCA_n - PP_n$$

n	Net Cash Flow (\$)
1	$(100,000 - 40,000 - 6250)(0.60) + 18,750(0.40) - 10,237 = \$29,513$
2	$(100,000 - 40,000 - 5226)(0.60) + 31,875(0.40) - 11,261 = \$34,353$
3	$(100,000 - 40,000 - 4100)(0.60) + 22,313(0.40) - 12,387 = \$30,078$
4	$(100,000 - 40,000 - 2861)(0.60) + 15,619(0.40) - 13,626 = \$26,905$

Step 3: Find the cash flow for year 5.

1. Operating cash flow:
   $(100,000 - 40,000 - 1499)(0.60) + 10,933(0.40) = \$39,474$
2. Net salvage value (Refer to Example 9.1) = $S + G$ = \$50,000 − \$9796
   = \$40,204.
3. Recovery of working capital, W'_5 = \$23,331
4. Principal repayment, PP_5 = \$14,988
5. Net cash flow in year 5, A_5 = \$39,474 + \$40,204 + \$23,331 − \$14,988
   = \$88,021.

Our results and overall calculations are summarized in Table 9.9. Checking this versus the results we obtained in Table 9.5 confirms our result.

Table 9.9 Cash Flow Statement for Example 9.4 Using the Generalized Cash Flow Approach (Example 9.8)	0	1	2	3	4	5
(1 − 0.40) (Revenue)		\$ 60,000	\$ 60,000	\$ 60,000	\$ 60,000	\$ 60,000
−(1 − 0.40) (Expenses)		(24,000)	(24,000)	(24,000)	(24,000)	(24,000)
−(1 − 0.40) (Debt interest)		(3,750)	(3,136)	(2,460)	(1,717)	(899)
+(0.40) (CCA)		7,500	12,750	8,925	6,248	4,373
Investment	\$(125,000)					
Net proceeds from sale						\$40,204
Investment in working capital	(23,331)					
Recovery of working capital						23,331
Borrowed funds	62,500					
Principal repayment		(10,237)	(11,261)	(12,387)	(13,626)	(14,988)
Net cash flow	\$ (85,831)	\$ 29,513	\$ 34,353	\$ 30,078	\$ 26,905	\$ 88,021

9.5 Computer Notes

As you have learned in the previous sections, creating an after-tax cash flow table can be tedious and time-consuming. An electronic spreadsheet is a perfect choice to automate much of the computation. Alternatively, **EzCash** can automate the entire analysis with its built-in computational tools for calculating depreciation and loan interest. To illustrate the use of electronic spreadsheet software, we will use Example 9.4.

The most popular spreadsheet applications used in engineering economic analysis are those that help in after-tax cash flow analysis.[7] Figure 9.8 shows how to prepare a cash flow statement for Example 9.4 with Excel. First you create cell blocks for input as well as output data. In our example, we treated the income tax rate and the required rate of return (MARR) as the varying inputs, and the PE and IRR as varying outputs that need to be measured. Then, in column B, you create the list of elements required to calculate the net income as well as the net cash flow. You enter the cell values for **Revenue, Labor, Material, Overhead, CCA,** and **Debt Interest.** Then you let Excel calculate the cell values for **Taxable Income, Income Taxes,** and **Net Income.** Here the income tax will be a function of the tax rate in cell D3. On the cash flow statement side, you enter the cell values for **Investment, Salvage, Working Capital, Borrowed Funds,** and **Principal Repayment.** Then Excel will calculate the cell values for **Disposal Tax Effect** and **Net Cash Flow.** The amount of the Disposal Tax Effect in cell H38 may be calculated based on the following cell formula

Cell H38 = (tax rate)(undepreciated capital cost – salvage value)
= (tax rate)(cost base – total CCA – salvage value)
= D3 * (–C32 – SUM(D18:H18) – H35)

Only the cell formula is entered into the box. Of course, you can automate the loan interest and principal calculations using the worksheet developed in Chapter 3. Similarly, you can also automate the CCA calculations using the worksheet developed in Chapter 7.

Once we obtain the net cash flow of the investment, the next step is to measure the investment worth. Using the NPV and IRR functions, we can calculate the PE as well as the rate of return of the investment. We have to note the IRR function in Excel gives us the interest rate at which the present equivalent is equal to zero. It will give us the internal rate of return only if the project is a simple investment. These calculations are also shown as outputs in Fig. 9.8. This financial calculation is only the beginning of our spreadsheet utility. If we are unsatisfied with the projections for any of the income statement items, we can change the cell values. The changes will

---

[7] **EzCash** has an Excel-like spreadsheet utility with many built-in financial functions to facilitate after-tax cash flow analysis. For example, **EzCash** can automate the entire analysis with built-in computational tools for calculating depreciation and loan interest.

carry over to adjust net income, net cash flow, which will yield the revised PE and IRR figures. Obviously, the potential for "what-if" analysis is almost unlimited. For this purpose, we normally group the input parameters at the beginning of the worksheet and express the cell formulas as functions of these input parameters. As you change one or more of these parameters, any changes in the net cash flows are immediately posted in the worksheet. For example, what will happen to the profitability of the project if the firm's MARR changes from 18% to 20%? To answer this question, simply change the MARR value in the input parameter. Excel will recalculate the PE value at this new interest rate.

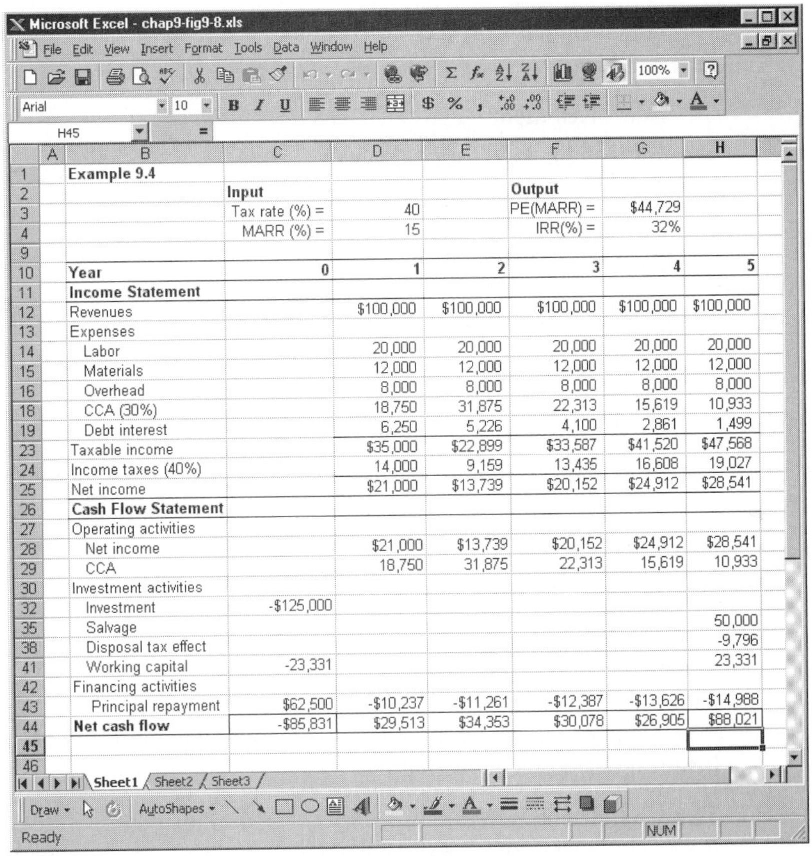

	A	B	C	D	E	F	G	H
1		**Example 9.4**						
2			**Input**			**Output**		
3			Tax rate (%) =	40		PE(MARR) =	$44,729	
4			MARR (%) =	15		IRR(%) =	32%	
9								
10		**Year**	0	1	2	3	4	5
11		**Income Statement**						
12		Revenues		$100,000	$100,000	$100,000	$100,000	$100,000
13		Expenses						
14		Labor		20,000	20,000	20,000	20,000	20,000
15		Materials		12,000	12,000	12,000	12,000	12,000
16		Overhead		8,000	8,000	8,000	8,000	8,000
18		CCA (30%)		18,750	31,875	22,313	15,619	10,933
19		Debt interest		6,250	5,226	4,100	2,861	1,499
23		Taxable income		$35,000	$22,899	$33,587	$41,520	$47,568
24		Income taxes (40%)		14,000	9,159	13,435	16,608	19,027
25		Net income		$21,000	$13,739	$20,152	$24,912	$28,541
26		**Cash Flow Statement**						
27		Operating activities						
28		Net income		$21,000	$13,739	$20,152	$24,912	$28,541
29		CCA		18,750	31,875	22,313	15,619	10,933
30		Investment activities						
32		Investment	-$125,000					
35		Salvage						50,000
38		Disposal tax effect						-9,796
41		Working capital	-23,331					23,331
42		Financing activities						
43		Principal repayment	$62,500	-$10,237	-$11,261	-$12,387	-$13,626	-$14,988
44		**Net cash flow**	-$85,831	$29,513	$34,353	$30,078	$26,905	$88,021

Figure 9.8 Automating the process of developing a cash flow statement with Excel for data in Example 9.4

9.6 Summary

- Identifying and estimating relevant project cash flows is perhaps the most challenging aspect of engineering economic analysis. All cash flows can be organized into one of the following three categories:

  1. Operating activities
  2. Investing activities
  3. Financing a¡ctivities.

- The following cash flow types account for the most common flows a project may generate:

  1. New investment
  2. Salvage value and disposal of existing assets (or net selling price)
  3. Working capital
  4. Working capital release
  5. Cash revenues/savings
  6. Cost of operations
  7. Interest and loan payments
  8. Taxes and tax credits.

  In addition, although not cash flows, capital cost allowances may exist in a project analysis and must be accounted for.

  Table 9.1 summarizes these elements and organizes them into operating, investing, and financing activities.

- The **income statement approach** is typically used in organizing project cash flows. This approach groups cash flows according to whether they are from operating, investing, or financing activities.

- The **generalized cash flow approach** (illustrated in Table 9.8) to organizing cash flows can also be used. The cash flows can be generated more quickly and the formatting of the results is less elaborate than with the income statement approach. There are also analytical advantages, which we will discover in later chapters. However, the generalized approach is less intuitive and not commonly understood by business people.

Problems

Note: *1. Unless indicated otherwise, the quoted marginal tax rate does not change as a result of the proposed investments. 2. The discount rate stated in each problem is the after-tax MARR.*

Level 1

9.1* Consider the following financial data for an investment project:

- Required capital investment at $n = 0$: $100,000
- Project service life: 10 years
- Salvage value at $N = 10$: $15,000
- Annual revenue: $150,000
- Annual O&M costs (not including CCA): $50,000
- Class 8 asset for tax purpose ($d = 20\%$)
- Income tax rate: 40%.

Determine the project cash flow at the end of year 10.

9.2* Suppose in Problem 9.1 the firm borrowed the entire capital investment at 10% interest over 10 years. If the required principal and interest payments in year 10 are

- Principal payment in year 10: $14,795, and
- Interest payment in year 10: $1480,

what would be the net cash flow at the end of year 10?

9.3* You are considering purchasing industrial equipment to expand one of your production lines. The equipment costs $100,000 and has an estimated service life of 6 years. Assuming that the equipment will be financed entirely from your business-retained earnings (equity funds), a fellow engineer has calculated the expected after-tax cash flows, including the net salvage value, at the end of its project life as follows:

End of Year	Net Cash Flow
0	– $100,000
1 – 5	500,000
6	600,000

Now you are pondering the possibility of financing the entire amount by borrowing from a local bank at 12% interest. You can make an arrangement to pay only the interest each year over the project period by deferring the principal payment until the end of 6 years. Your firm's interest rate is also 12%. The expected marginal income tax rate over the project period is known to be 40%. What is the amount of economic gain (or loss) in present equivalent by using debt financing over equity financing?

9.4* A corporation is considering purchasing a machine that will save $130,000 per year before taxes. The cost of operating the machine, including maintenance, is $20,000 per year. The machine will be needed for 4 years, after which it will have a zero salvage value. The machine is a class 43 property with a CCA rate of 30%. If the firm wants a 12% rate of return after taxes, how much can it afford to pay for this machine? The firm's income tax rate is 40%.

9.5* Your company needs an air compressor and has narrowed the choice to two alternatives, A and B. The following financial data have been collected:

	Model A	Model B
First cost	$25,000	$35,000
Annual O&M cost	5,600	3,500
Salvage value	0	4,000
Service life	10 years	10 years
CCA rate	$d = 30\%$	$d = 30\%$

The marginal tax rate is 40%. Which of the following statements are incorrect? Assume the MARR = 20%. Select all incorrect ones.

(a) Select Model A because you can save $506 annually.
(b) Select Model A because the incremental rate of return (Model A – Model B) exceeds 20%.
(c) Select Model A because you can save $2121 in present equivalent.
(d) Select Model A because the incremental rate of return (Model B – Model A) is 13%, which is less than 20%.

Level 2

9.6 Kelowna Construction Company builds residential solar homes. Because of an anticipated increase in business volume, the company is considering acquisition of a loader at a cost of $54,000. This acquisition cost includes delivery charges and applicable taxes. The firm has estimated that if the loader is acquired, the following additional revenues and operating costs (excluding CCA) should be expected:

End of Year	Additional Operating Revenue	Additional Operating Costs Excluding CCA	Allowed CCA
1	$66,000	$29,000	$10,800
2	70,000	28,400	17,280
3	74,000	32,000	10,368
4	80,000	38,800	6,221
5	64,000	31,000	5,320
6	50,000	25,000	3,110

The projected revenue is assumed to be in cash in the year indicated, and all the additional operating costs are expected to be paid in the year in which they are incurred. The estimated salvage value for the loader at the end of the sixth year is $8000. The firm's incremental (marginal) tax rate is 35%. What is the after-tax cash flow if the loader is acquired? Using this data, (a) develop the cash flow statement based on net income (Approach 1 as shown in Table 9.2) and (b) develop the cash flow statement based on the conventional tabular method (Approach 2 as shown in Table 9.3).

9.7 An automobile manufacturing company is considering the purchase of an industrial robot to do spot welding, which is currently done by skilled labor. The initial cost of the robot is $235,000, and the annual labor savings are projected to be $122,000. If purchased, the robot's CCA rate is 30%. This robot will be used for 7 years, at the end of which time the firm expects to sell the robot for $50,000. The company's marginal tax rate is 38% over the project period. Determine the net after-tax cash flows for each period over the project life.

9.8 A Calgary company is planning to market an answering device for people working alone, who want the prestige that comes with having a secretary, but who cannot afford one. The device, called Tele-Receptionist, is similar to a voice-mail system. It uses digital recording technology to create the illusion that a person is operating the switchboard at a busy office. The company purchased a 4000 m² building and converted it to an assembly plant for $600,000 ($100,000 of land and $500,000 worth of building). Installation of the assembly equipment worth $500,000 was completed on December 31. The plant will begin operation on January 1. The company expects to have a gross annual income of $2,500,000 over the next 5 years. Annual operating costs (excluding CCA) are projected to be $1,280,000. For CCA purposes, the assembly plant building will be classified as CCA class 1 property ($d = 4\%$) and the assembly equipment as CCA class 43 property ($d =$

30%). The property value of the land and the building at the end of year 5 would appreciate as much as 15% over the initial purchase cost. The residual value of the assembly equipment is estimated to be about $50,000 at the end of year 5. The firm's marginal tax rate is expected to be about 40% over the project period. Determine the project's after-tax cash flows over the period of 5 years.

9.9 A highway contractor is considering buying a new trench excavator that costs $200,000 and can dig a 1-meter wide trench at the rate of 5 meters per hour. With adequate maintenance, the production rate will remain constant for the first 3 years of operation, then decrease by 0.5 meter per hour for each additional year thereafter. The excavator has to dig a trench of 2000 metres long each year. The operating and maintenance costs will be $15 per hour. The equipment is a class 38 property with $d = 30\%$. At the end of 5 years, the excavator will be sold for $40,000. Assuming that the contractor's marginal tax rate is 34% per year, determine the annual after-tax cash flow.

9.10 A small children's clothing manufacturer is considering an investment to computerize its management information system for material requirement planning, piece-goods coupon printing, and invoice and payroll. An outside consultant has been asked to estimate the initial hardware requirement and installation costs. He suggests the following:

PC systems (10 PCs, 4 printers)	$65,000
Local area networking system	15,000
System installation and testing	4,000

The expected life of these computer systems is 6 years, with an estimated salvage value of $1000. The proposed system is classified as class 10 property with $d = 30\%$. A group of computer consultants needs to be hired to develop various customized software packages to run on these systems. Software development costs will be $20,000 and can be expensed during the first tax year. The new system will eliminate two clerks, whose combined annual payroll costs would be $52,000. Additional annual costs to run this computerized system are expected to be $12,000. Borrowing is not considered an option for this investment. No tax credit is available for this system. The firm's expected marginal tax rate over the next 6 years will be 40%. The firm's interest rate is 18%. Compute the after-tax cash flows over the investment's life by using the income statement approach.

9.11 The Manufacturing Division of Windsor Vending Machine Company is considering its Toronto Plant's request for an automatic screw-cutting machine to be included in the division's 2000 capital budget:

- Name of project: Mazda Automatic Screw Machine
- Project cost: $48,018
- Purpose of project: To reduce the cost of some of the parts that are now being subcontracted by this plant, to cut down on inventory due to a shorter lead time, and to better control the quality of the parts. The proposed equipment includes the following cost basis:

Machine cost	$35,470
Accessory cost	6,340
Tooling	2,356
Freight	980
Installation	1,200
Sales tax	1,672
Total cost	$48,018

- Anticipated savings: As shown in the table below
- CCA class 43 with $d = 30\%$
- Marginal tax rate: 40%
- MARR: 15%.

(a) Determine the net after-tax cash flows over the project life of 6 years. Assume a salvage value of $3500.

(b) Is this project acceptable based on the PE criterion?

(c) Determine the IRR for this investment.

Item	Hours		Present Method	Proposed Method
	Present M/C Labour	Proposed M/C Labour		
Setup		350		$7,700
Run ·	2,410	800		17,600
Overhead				
Indirect labour				3,500
Fringe benefits				8,855
Maintenance				1,350
Tooling				6,320
Repair				890
Supplies				4,840
Power				795
Taxes and insurance				763
Other relevant costs				
Floor space				3,210
Subcontracting			$122,468	
Material				27,655
Other				210
Total			$122,468	$83,688
Operating advantage				$38,780

9.12 A firm has been paying a print shop $18,000 annually to print the company's monthly newsletter. The agreement with this print shop has now expired, but it could be renewed for a further 5 years. The new subcontracting charges are expected to be 12% higher than they were in the previous contract. The company is also considering the purchase of a desktop publishing system with a high-quality laser printer driven by a microcomputer. With appropriate word/graphics software, the newsletter can be composed and printed in near typeset quality. A special device is also required to print photos in the newsletter. The following estimates have been quoted by a computer vendor:

Microcomputer	$5,500
Laser printer	8,500
Photo device/scanner	10,000
Software	2,000
Total cost basis	$26,000
Annual O&M costs	10,000

The salvage value of each piece of equipment at the end of 5 years is expected to be only 10% of the original cost. The company's marginal tax rate is 40%, and the whole desktop publishing system is classified as CCA class 10 property with $d = 30\%$.

(a) Determine the projected net after-tax cash flows for the investment.

(b) Compute the IRR for this project.

(c) Is this project justifiable at MARR = 12%?

9.13* An asset in class 8 ($d = 20\%$) costs $100,000 and has a zero estimated salvage value after 6 years of use. The asset will generate annual revenues of $300,000 and will require $100,000 in annual labor and $50,000 in annual material expenses. There are no other revenues and expenses. Assume a tax rate of 40%.

(a) Compute the after-tax cash flows over the project life.

(b) Compute the PE at MARR = 12%. Is this an acceptable investment?

9.14 Alberta Aluminum Company is considering making a major investment of $150 million ($5 million for land, $45 million for buildings, and $100 million for manufacturing equipment and facilities) to develop a stronger, lighter material, called aluminum lithium, which will make aircraft sturdier and more fuel efficient. Aluminum lithium, which has been sold commercially for only a few years as an alternative to composite materials, will likely be the material of choice for the next generation of commercial and military aircraft because it is so much lighter than conventional aluminum alloys, which use a combination of copper, nickel, and magnesium to harden aluminum. Another advantage of aluminum lithium is that it is cheaper than composites. The firm predicts that aluminum lithium will account for about 5% of the structural weight of the average commercial aircraft within 5 years and 10% within 10 years. The proposed plant, which has an estimated service life of 12 years, would have a capacity of about 10 million kilograms of aluminum lithium, although the actual production of the material is expected to be only 3 million kilograms during the first 4 years, 5 million for the next 3 years, and 8 million for the remaining plant life. Aluminum lithium costs $12 per kg to produce, and the firm would expect to sell it at $17 per kg. The building will be depreciated with $d = 4\%$ with the building placed in service in July 1 during the first year. All manufacturing equipment and facilities will be classified class 43 property with $d = 30\%$. At the end of project life, the land will be worth $8 million, the building $30 million, and the equipment $10 million. Assuming that the firm's marginal tax rate is 40%, and its capital gains tax rate is 30%,

(a) Determine the net after-tax cash flows.
(b) Determine the IRR for this investment.
(c) Determine whether the project is acceptable if the firm's MARR is 15%.

9.15 An automaker is considering installing a three-dimensional (3-D) computerized car-styling system at a cost of $200,000 (including hardware and software). With the 3-D computer modeling system, designers will have the ability to view their design from many angles and to fully account for the space required for the engine and passengers. The digital information used to create the computer model can be revised in consultation with engineers, and the data can be used to run milling machines that make physical models quickly and precisely. The automaker expects to decrease turnaround time by 22% for designing a new automobile model (from configuration to final design). The expected savings is $250,000 per year. The training and operating and maintenance cost for the new system is expected to be $50,000 per year. The system has a 5-year useful life and can be depreciated as CCA class 43 property with $d = 30\%$. The system will have an estimated salvage value of $5000. The automaker's marginal tax rate is 40%. Determine the annual cash flows for this investment. What is the return on investment for this project?

9.16 A facilities engineer is considering a $50,000 investment in an energy management system (EMS). The system is expected to save $10,000 annually in utility bills for N years. After N years, the EMS will have a zero salvage value. In an after-tax analysis, how many years would N need to be to earn 10% return on the investment? Assume a class 10 property with $d = 30\%$ and a 35% tax rate.

9.17 A corporation is considering purchasing a machine that will save $130,000 per year before taxes. The cost of operating the machine, including maintenance, is $20,000 per year. The machine will be needed for 5 years, after which it will have a zero salvage value. Assume that the machine's CCA rate is 30%. The marginal income tax rate is 40%. If the firm wants 12% IRR after taxes, how much can it afford to pay for this machine?

9.18 Ampex Corporation produces a wide variety of tape cassettes for commercial and government markets. Due to increased competition in VHS cassette production, Ampex is concerned about pricing its product competitively. Currently, Ampex has 18 loaders that load cassette tapes in half-inch VHS cassette shells. Each loader is manned by one operator per shift. Ampex currently produces 25,000 half-inch tapes per week and operates 15 shifts per week, 50 weeks per year.

As a means of reducing the unit cost, Ampex can purchase cassette shells for $0.15 less (each) than it can currently produce them. A supplier has guaranteed a price of $0.77 per cassette for the next three years. However, Ampex's current loaders will not be able to load the proposed shells properly. To accommodate the vendor's shells, Ampex would have to purchase eight KING-2500 VHS loaders at a cost of $40,000 each. For these new machines to operate properly, Ampex must also purchase $20,827 worth of conveyor equipment, the cost of which will be included in the overall capital cost of $340,827. The new machines are much faster and will handle more than the current demand of 25,000 cassettes per week. The new loaders will require two people per machine per shift, three shifts per day, five days a week. The new machines will fall into class 43 with $d = 30\%$ and will have an approximate life of eight years. At the end of the project life, Ampex expects the market value for each loader to be $3000.

The average pay of the needed new employees is $8.27 per hour, and adding 23% for benefits will make it $10.17 per hour. The new loaders are simple to operate; therefore, the training impact of the alternative is minimal. The operating cost, including maintenance, is expected to stay the same for the new loaders. This cost will not be considered in the analysis. The cash inflows from the project will be a material savings per cassette of $0.15, and the labor savings of two employees per shift. This gives an annual savings in materials and labor costs of $187,500 and $122,065, respectively. If the new loaders are purchased, the old machines will be shipped to other plants for standby use. That is, no disposal of asset occurs. Ampex's marginal tax rate is running at 40%.

(a) Determine the after-tax cash flows over the project life.
(b) Determine the IRR for this investment.
(c) Is this investment profitable at MARR = 15%?

9.19* The Motch Machinery Company is planning to expand its current spindle product line. The required machinery would cost $500,000. The building to house the new production facility would cost $1.5 million. The land would cost $250,000, and $150,000 working capital would be required. The product is expected to result in additional sales of $675,000 per year for 10 years, at which time the land can be sold for $500,000, the building for $700,000, and the equipment for $50,000. All of the working capital will be recovered. The annual disbursements for labor, materials, and all other expenses are estimated to be $425,000. The firm's income tax rate is 40%, and any capital gains will be taxed at 30%. The building will be depreciated with $d = 4\%$. The manufacturing facility will be classified as class 43 property with $d = 30\%$. The firm's MARR is also known to be 15% after taxes.

(a) Determine the projected net after-tax cash flows from this investment. Is the expansion justified?
(b) Repeat (a) above using the traditional tabular method (Approach 2).
(c) Compare the IRR of this project with the situation with no working capital.

9.20 An industrial engineer at Winnipeg Textile Mill proposed the purchase of scanning equipment for the company's warehouse and weave rooms. The engineer felt that the purchase would ensure a better system of locating cartons in the warehouse by recording the location of the cartons and storing the data in the computer. The estimated investment, annual operating and maintenance costs, and expected annual savings are as follows:

- Cost of equipment and installation: $44,500
- Project life: 6 years
- Expected salvage value: $0
- Investment in working capital (fully recoverable at the end of project life): $10,000
- Expected annual labor and materials savings: $62,800
- Expected annual expenses: $8120
- Depreciation method: CCA class 8 with $d = 20\%$.

The firm's marginal tax rate is 35%.

(a) Determine the net after-tax cash flows over the project life.

(b) Compute the IRR for this investment.

(c) At MARR = 18%, is this project acceptable?

9.21 Sarnia Chemical Corporation is considering investing in a new composite material. R&D engineers are investigating exotic metal-ceramic and ceramic-ceramic composites to develop materials that will withstand high temperatures, such as those to be encountered in the next generation of jet fighter engines. The company expects a 3-year R&D period before these new materials can be applied to commercial products. The following financial information is presented for management review:

- R&D cost: $5 million over a 3-year period. R&D expenditure of $0.5 million at the end of year 1, $2.5 million at the end of year 2, and $2 million at the end of year 3. These R&D expenditures will be expensed rather than amortized for tax purposes.
- Capital investment: $5 million at the beginning of year 4. This investment consists of $2 million in a building, and $3 million in plant equipment. The company already owns a piece of land as the building site.
- Depreciation method: The building

(class 1 property) and plant equipment (class 43 property).

- Project life: 10 years after a 3-year R&D period.
- Salvage value: 10% of the initial capital investment for the equipment and 50% for the building (at the end of their 10 year life).
- Total sales: $50 million (at the end of year 4) with an annual sales growth rate of 10% per year (compound growth) during the first 6 years (year 4 through year 9) and −10% (negative compound growth) per year for the remaining project life.
- Out-of-pocket expenditures: 80% of annual sales.
- Working capital: 10% of annual sales. This amount is dependent on the sales in the year. It is required at the beginning of the year and recovered at the end of the same year.
- Marginal tax rate: 40%.

(a) Determine the net after-tax cash flows over the project life.

(b) Determine the IRR for this investment.

(c) Determine the annual equivalent over 13 years for this investment at MARR = 20%.

9.22 Refer to the data for the Kelowna Construction Company in Problem 9.6. If the firm expects to borrow the initial investment ($54,000) at 10% over 2 years (equal annual payments of $31,114), determine the project's net cash flows.

9.23 In Problem 9.7, to finance the industrial robot, the company will borrow the entire amount from a local bank, and the loan will be paid off at the rate of $50,000 per year, plus 12% on the unpaid balance until the loan is paid off. Determine the net after-tax cash flows over the project life.

9.24 Refer to the data for the children's clothing company in Problem 9.10. Suppose that the initial investment of $84,000 will be borrowed from a local bank at an interest rate of 11% over 5 years (five equal annual payments). Recompute the after-tax cash flow.

9.25 Montreal Die Casting Company is considering the installation of a new process machine for their manufacturing facility. The machine costs $250,000 installed, will generate additional revenues of $80,000 per year, and will save $50,000 per year in labor and material costs. The machine will be financed by a $150,000 bank loan repayable in three equal, annual principal installments, plus 9% interest on the outstanding balance. The machine is a class 43 property with $d = 30\%$. The useful life of this process machine is 10 years, at which time it will be sold for $20,000. The combined marginal tax rate is 40%.

(a) Find the year-by-year after-tax cash flow for the project using the income statement approach.
(b) Compute the IRR for this investment.
(c) At MARR = 18%, is this project economically justifiable?

9.26 Consider the following financial information about a retooling project at a computer manufacturing company:

- The project costs $2 million and has a 5-year service life.
- The retooling project can be classified as a class 43 property with $d = 30\%$.
- At the end of fifth year, any assets held for the project will be sold. The expected salvage value will be about 10% of the initial project cost.
- The firm will finance 40% of the investment from an outside financial institution at an interest rate of 10%. The firm is required to repay the loan with five equal annual payments.
- The firm's incremental (marginal) tax rate on this investment is 35%.
- The firm's MARR is 18%.

- With the financial information above,

(a) Determine the after-tax cash flows.
(b) Compute the annual equivalent worth for this project.

9.27* An auto-parts manufacturing plant wants to add two delivery trucks to its current fleet system. The trucks would cost $50,000 each, and the firm expects to use these trucks for 5 years. The salvage value of the trucks at the end of year 5 would be about $5000 each. The system qualifies for the class 10 property with $d = 30\%$. Financing for 40% of the initial cost (i.e. $40,000) will be obtained from a local bank at an annual interest rate of 12% over 3 years. The borrowed money will be repaid in three equal annual payments. These trucks will speed up auto-part delivery to customers, and the firm expects additional revenue over the next 5 years. The firm already computed the projected net income and recorded it as shown in the cash flow statement below.

Cash Flow Elements	End of Period					
	0	1	2	3	4	5
Cash from operations						
Net income		30,000	40,000	42,500	40,000	35,800
CCA		15,000	—	17,850	—	—
Investment/ salvage	−100,000					10,000
Disposal tax effect						—
Loan/ principal repayment	+40,000	—	—	—		
Net cash flow	—	—	—	—	—	—

(a) Complete the cash flow statement based on the information above, assuming a tax rate of 35%.
(b) Compute the PE for this project at MARR = 18%.
(c) Compute the IRR for this project.

9.28 A fully automatic chucker and bar machine is to be purchased for $35,000. This amount is to be borrowed with the stipulation that it be repaid by six equal end-of-year payments at 12%, compounded annually. The machine is expected to provide an annual revenue of $10,000 for 6 years and is to be depreciated with $d = 30\%$. The salvage value at the end of 6 years is expected to be $3000. Assuming a marginal tax rate of 36% and a MARR of 15%.

(a) Determine the after-tax cash flow for this asset through 6 years.

(b) Determine if this project is acceptable based on the PE criterion.

9.29 A manufacturing company is considering the acquisition of a new injection molding machine at a cost of $100,000. Because of a rapid change in product mix, the need for this particular machine is expected to last only 8 years, after which time the machine is expected to have a salvage value of $10,000. The annual operating cost is estimated to be $5000. The addition of this machine to the current production facility is expected to generate an annual revenue of $40,000. The firm has only $60,000 available from its equity funds, so it must borrow the additional $40,000 required at an interest rate of 10% per year with repayment of principal and interest in eight equal annual amounts. The applicable marginal income tax rate for the firm is 40%. Assume that the asset is class 43 property with $d = 30\%$.

(a) Determine the after-tax cash flows.

(b) Determine the PE of this project at MARR = 14%.

9.30* Suppose an asset has a first cost of $6000, a life of 5 years, a salvage value of $2000 at the end of 5 years, and a net annual before-tax revenue of $1500. The firm's marginal tax rate is 35%. The asset will be classified as class 8 property with $d = 20\%$.

(a) Using the generalized cash flow approach, determine the cash flow after taxes.

(b) Rework part (a) assuming that the entire

investment would be financed by a bank loan at an interest rate of 9%.

(c) Given a choice between the financing methods of parts (a) and (b), show calculations to justify your choice of which is the better one at a discount rate of 9%.

9.31 A construction company is considering the proposed acquisition of a new earthmover. The purchase price is $100,000, and an additional $25,000 is required to modify the equipment for special use by the company. The equipment falls into the class 38 with $d = 30\%$, and it will be sold after 5 years (project life) for $50,000. Purchase of the earthmover will have no effect on revenues, but it is expected to save the firm $60,000 per year in before-tax operating costs, mainly labor. The firm's marginal tax rate is 40%. Assume that the initial investment is to be financed by a bank loan at an interest rate of 10%, payable annually. Determine the after-tax cash flows by using the generalized cash flow approach and the present equivalent of investment for this project if the firm's MARR is known to be 12%.

9.32 Air North, a leading regional airline that is now carrying 54% of all the passengers that pass through central Canada, is considering the possibility of adding a new long-range aircraft to its fleet. The aircraft being considered for purchase is the McDonnell Douglas DC-9-532 "Funjet," which is quoted at $60 million per unit. McDonnell Douglas requires a 10% down payment at the time of delivery, and the balance is to be paid over a 10-year period at an interest rate of 12% compounded annually. The actual payment schedule calls for only interest payments over the 10-year period, with the original principal amount to be paid off at the end of the tenth year. Air North expects to generate $35 million per year by adding this aircraft to its current fleet but also estimates an operating and maintenance cost of $20 million per year. The aircraft is expected to have a 15-year service life with a salvage value of 15% of the original purchase price. If the aircraft is bought, it will be depreciated as a class 9 property with $d = 25\%$. The firm's combined federal and provincial marginal tax rate is 38%, and its required minimum attractive rate of return is 18%.

(a) Determine the cash flow associated with the debt financing using the generalized cash flow approach.
(b) Is this project acceptable?

Level 3

9.33 Saskatoon Machinery Inc. manufactures drill bits. One of the production processes for a drill bit is called tipping, where carbide tips are inserted into the bit to make it stronger and more durable. This tipping process usually requires four or five operators depending on the weekly work load. The same operators were assigned to the stamping operation, in which the size of the drill bit and the company's logo are imprinted into the bit. Saskatoon is considering acquiring three automatic tipping machines to replace the manual tipping and stamping operations. If the tipping process is automated, Saskatoon engineers will have to redesign the shapes of the carbide tips to be used in the machine. The new design requires less carbide, resulting in material savings. The following financial data has been compiled:

- Project life: 6 years.
- Expected annual savings: reduced labor, $56,000; reduced material, $75,000; other benefits (reduced carpal tunnel syndrome and related problems), $28,000; reduced overhead, $15,000.
- Expected annual O&M costs: $22,000.
- Tipping machines and site preparation:

equipment (three machines) costs including delivery, $180,000; site preparation, $20,000.
- Salvage value: $30,000 (three machines) at the end of 6 years.
- Depreciation method: class 43 with $d = 30\%$.
- Investment in working capital: $25,000 at the beginning of the project; that same amount will be fully recovered at the end of project.
- Other accounting data: Marginal tax rate of 39%, MARR of 18%.

To raise $200,000, Saskatoon is considering the following financing options:

- Option 1: Finance the tipping machines using their retained earnings.
- Option 2: Secure a 12% term loan over 6 years (six equal annual installments).

(a) Determine the net after-tax cash flows for each financing option.
(b) What is Saskatoon's present equivalent of owning the equipment by borrowing?
(c) Recommend the better course of action for Saskatoon.

9.34 An international manufacturer of prepared food items needs 50,000,000 kWh of electrical energy a year with maximum demand of 10,000 kW. The local utility presently charges $0.085 per kWh, a rate considered high throughout the industry. Because the firm's power consumption is so large, its engineers are considering installing a 10,000 kW steam-turbine plant. Three types of plant have been proposed ($ units in thousand):

	Plant A	Plant B	Plant C
Total investment (boiler/turbine/ electrical/ structures)	$8530	$9498	$10,546
Annual operating cost			
Fuel	1128	930	828
labor	616	616	616
Maintenance	150	126	114
Supplies	60	60	60
Insurance and property taxes	10	12	14

The service life of each plant is expected to be 20 years. The plant investment will be subject to a 10% CCA rate. The expected salvage value of the plant at the end of its useful life is about 10% of its original investment. The firm's MARR is known to be 12%. The firm's marginal income tax rate is 39%.

(a) Determine the unit power cost ($/kWh) for each plant.
(b) Which plant would provide the most economical power?

9.35* Morgantown Mining Company is considering a new mining method at its Blacksville mine. The method, called longwall mining, is carried out by a robot. Coal is removed by the robot, not by tunneling like a worm through an apple, which leaves more of the target coal than is removed, but rather by methodically shuttling back and forth across the width of the deposit and devouring nearly everything. The method can extract about 75% of the available coal, compared with 50% for conventional mining, which is largely done with machines that dig tunnels. Moreover, the coal can be recovered far more inexpensively. Currently, at Blacksville alone, the company mines 5 million tonnes a year with 2200 workers. By installing two longwall robot machines, the company can mine 5 million tonnes with only 860 work-

ers. (A robot miner can dig more than 6 tonnes of coal every minute.) Despite the loss of employment, the United Mine Workers union generally favors longwall mines for two reasons: The union officials are quoted as saying, (1) "It would be far better to have highly productive operations that were able to pay our folks good wages and benefits than to have 2200 shovelers living in poverty," and (2) "Longwall mines are inherently safer in their design." The company projects the following financial data upon installation of the longwall mining:

Robot installation (2 units)	$9.3 million
Total amount of usable coal deposit	50 million tonnes
Annual mining capacity	5 million tonnes
Project life	10 years
Estimated salvage value	$0.5 million
Working capital requirement	$2.5 million
Expected additional revenues:	
Labor savings	$6.5 million
Accident prevention	$0.5 million
Productivity gain	$2.5 million
Expected additional costs:	
O&M costs	$ 2.4 million

(a) Estimate the firm's net after-tax cash flows over the project life, if the firm is allowed to use the units-of-production method for CCA calculations. The firm's marginal tax rate is 40%. Find the PE value with MARR = 15%.
(b) Estimate the firm's net after-tax cash flows, if the CCA rate for the robots is 30%. Find the PE value with MARR = 15%.

9.36 The National Parts Inc., an auto-parts manufacturer, is considering purchasing a rapid prototyping system to reduce prototyping time for form, fit, and function applications in automobile parts manufacturing. An outside consultant has been called in to estimate the initial hardware requirement and installation costs. He suggests the following:

- Prototyping equipment: $187,000.
- Postcuring apparatus : $25,000.
- Maintenance: $36,000 per year by the equipment manufacturer.
- Resin: Annual liquid polymer consumption: 1600 litres at $87.50 per litre.
- Site preparation: Some facility changes are required when installing the rapid prototyping system. (Certain liquid resins contain a toxic substance, so the work area must be well vented.)

The expected life of the system is 6 years with an estimated salvage value of $30,000. The proposed system is classified as a class 43 property. A group of computer consultants must be hired to develop customized software to run on these systems. These software development costs will be $20,000 and can be expensed during the first tax year. The new system will reduce prototype development time by 75% and the material waste (resin) by 25%. This reduction in development time and material waste will save the firm $114,000 and $35,000 annually, respectively. The firm's expected marginal tax rate over the next 6 years will be 40%. The firm's MARR is 20%.

(a) Assuming that the entire initial investment will be financed from the firm's retained earnings (equity financing), determine the after-tax cash flows over the investment life. Compute the PE of this investment.

(b) Assuming the entire initial investment will be financed through a local bank at an interest rate of 13% compounded annually, determine the net after-tax cash flows for the project. Compute the PE of this investment.

(c) Select the better financing option based on the rate of return on incremental investment.

The After-Tax Cash Flow Diagram Approach and Capital Tax Factors

T he income statement and generalized cash flow approaches in Sections 9.3 and 9.4 require a year-by-year estimate of each cash flow element. These estimates are then combined in a manner that yields a net after-tax cash flow for each year. The resulting net cash flows are the basis for the subsequent discounted cash flow analysis.

The advantage of these approaches is that they detail the magnitude of each annual net cash flow and indicate whether its value is positive or negative. Prior to the availability of spreadsheets, problems which had a large number of cash flow elements and a long service life were very time-consuming. Therefore, an alternative approach was often employed in which the problem was represented on an after-tax cash flow diagram.

With this approach, interest factors are used to develop a relationship for the desired present equivalent, annual equivalent, or future equivalent. This alternative provides us with the required measure of investment attractiveness without making it necessary to calculate the detailed net cash flows on an annual basis.

Although most problems will be analyzed using **EzCash** or a computer spreadsheet, the after-tax cash flow diagram approach is a very useful technique which can save a considerable amount of time when a computer is not available.

9A.1 The After-Tax Cash Flow Diagram

Net cash flows on an after-tax basis can be represented using the generalized cash flow approach [see Eq. (9.3)]. For any specific problem, the relevant cash flow elements that comprise the net cash flow can be portrayed on a cash flow diagram.

To illustrate, we will assume a typical problem, which has investment at time zero and disposal at year N. In order to simplify the presentation we will also assume that there are no debt or working capital considerations, although these can be included if required.

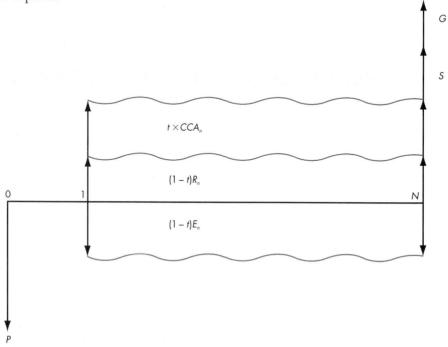

Figure 9A.1 A general after-tax cash flow diagram

The problem can now be represented on a cash flow diagram as shown in Figure 9A.1. Here the wavy lines indicate an arbitrary pattern of cash flows and N indicates the life of the project.

The present equivalent of cash flows in the diagram is

$$
\begin{aligned}
PE \quad = \quad & -P + (1-t)\sum_{n=1}^{N} R_n(P/F,\, i,\, n) \\
& -(1-t)\sum_{n=1}^{N} E_n(P/F,\, i,\, n) \\
& + t\sum_{n=1}^{N} CCA_n(P/F,\, i,\, n) \\
& + (S+G)(P/F,\, i,\, N).
\end{aligned} \tag{9A.1}
$$

If R_n and E_n follow a standard pattern (uniform series, linear gradient series, or geometric series) or a combination thereof, as they often do, we will be able to replace the summation terms for R_n and E_n with interest factors. However, the $t \times CCA_n$ terms do not exhibit a nice cash flow pattern because of the 50%-rule for the first year. If the 50%-rule did not apply, we would have a geometric gradient series with $g = -d$ in these terms. Even with the 50%-rule, these $t \times CCA_n$ terms for $n > 1$ still exhibit such a geometric gradient pattern with $g = -d$. By adding and then subtracting a cash flow in the amount of

$$
\frac{tPd}{2(1-d)}
$$

to the first year's cash flow, we can calculate the present equivalent of all $t \times CCA_n$ terms for $n \geq 1$ using the following equation:

$$
PE_{t \times CCA_n} = tPd\,\frac{1-\frac{d}{2}}{1-d}\,(P/A_1,\, -d,\, i,\, N) - \frac{tPd}{2(1-d)}(P/F,\, i,\, 1), \tag{9A.2}
$$

where d is the declining balance rate.

For example, if $R_n = R$ and $E_n = E$ are both constant, equation (9A.1) can be written as:

$$
\begin{aligned}
PE \quad = \quad & -P + (1-t)R(P/A,\, i,\, N) - (1-t)E(P/A,\, i,\, N) \\
& + (S+G)(P/F,\, i,\, N) \\
& + tPd\,\frac{1-\frac{d}{2}}{1-d}\,(P/A_1,\, -d,\, i,\, N) \\
& - \frac{tPd}{2(1-d)}(P/F,\, i,\, 1).
\end{aligned} \tag{9A.3}
$$

If an amortized loan at an interest rate of i_d is included in project financing, the project cash flow diagram is as shown in Figure 9A.2. In Figure 9A.2, P_d is the amount of debt financing, $A = P_d \times (A/P,\, i_d,\, N)$ represents the constant annual debt payment including both principal and interest payments, and I_n is the interest portion in the nth annual payment. I_n can be calculated with Eq. (3.16) in Chapter 3. With the after-tax cash flow diagram given in Figure 9A.2, one can easily find the present equivalent of the project with this kind of debt financing.

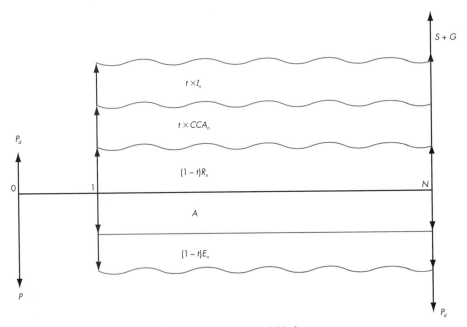

Figure 9A.2   General after-tax cash flow diagram with amortized debt financing

If the amount of debt, P_d, can be paid off with a single lump sum at the end of the project and only interest charges are paid from year to year during the project life, there is a constant interest payment I in each year:

$$I = P_d \times i_d$$

The after-tax cash flow diagram of the project with this kind of debt financing is shown in Figure 9A.3. One can then easily find the present equivalent of the project with this kind of debt financing using Figure 9A.3.

For a constant level of working capital requirement, we simply add the working capital investment cash flow at time 0 and its recovery at the end of year N. The present equivalent evaluation can be conducted based on the corresponding after-tax cash flow diagram.

Since the concept of asset class accounting and the declining balance method are used in calculation of capital cost allowances in Canada, the concept of capital tax factors may be used in project evaluations as shown in the following sections.

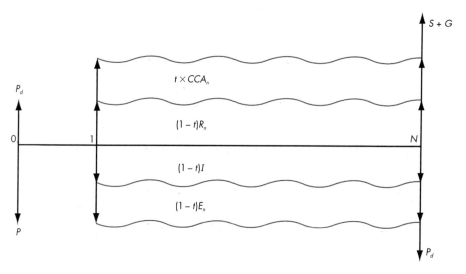

Figure 9A.3   General after-tax cash flow diagram with constant debt-interest payment

9A.2 Capital Tax Factors

In situations where CCA is calculated on the basis of straight-line depreciation, the annual tax savings, $t \times CCA_n$, represents a uniform series of cash flows. This fact makes calculating a present equivalent value of these cash flow elements relatively simple. However, when CCA is calculated on a declining balance basis, $t \times CCA_n$ changes each year. Including these tax savings explicitly on a year-by-year basis and calculating the disposal tax effect at the end of project make the present equivalent determination quite tedious. Fortunately we can employ **capital tax factors**, which handle the tax savings implicitly, to avoid such problems. Another reason is that when an asset is disposed of, the salvage value is deducted from the undepreciated capital cost of the whole class to which this asset belongs. In other words, the tax effect on CCA gain or loss is not completely realized in the year of disposal. Instead, this effect is spread over the following years because of the asset class accounting procedure. The capital tax factors handle these tax effects more accurately than the methods that are used elsewhere in this book.

Let us consider an asset of value P, which qualifies for CCA on a declining balance basis. If we ignore the 50%-rule for the moment, the tax savings generated by this capital expenditure in year n is given by

$$t \times CCA_n = td(1-d)^{n-1}P. \qquad (9A.4)$$

We will allow the tax savings due to this capital expenditure to last for infinite years, assuming that a correction will be exercised at the time of disposal. This situation can be represented as shown in Figure 9A.4.

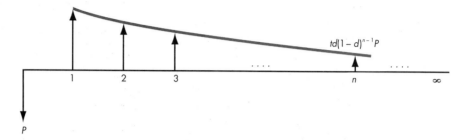

Figure 9A.4 Capital investment and the corresponding tax savings for the declining balance method

The present value of all the cash flows shown to time infinity is thus

$$-P + \frac{tdP}{1 + i} + \frac{td(1 - d)P}{(1 + i)^2} + \frac{td(1 - d)^2P}{(1 + i)^3} + \ldots + \frac{td(1 - d)^{n-1}P}{(1 + i)^n} + \ldots$$

$$= -P + \frac{tdP}{1 + i} \times \left[1 + \frac{(1 - d)}{1 + i} + \frac{(1 - d)^2}{(1 + i)^2} + \ldots + \frac{(1 - d)^n}{(1 + i)^n} + \ldots \right]$$

$$= -P + \frac{tdP}{1 + i} \times \frac{1 + i}{i + d}$$

$$= -P \times \left(1 - \frac{td}{i + d} \right)$$

$$= -P \times CTF \tag{9A.5}$$

where

$$CTF = 1 - \frac{td}{i + d}$$

is the capital tax factor (*CTF*). Note that the terms in the square brackets form an infinite geometric series with a summed value of

$$\frac{1 + i}{i + d}.$$

A similar type of analysis can be made when the 50%-rule is used. In this case the tax savings in each year are

$$t \times CCA_1 = t\frac{d}{2}P \tag{9A.6}$$

and

$$t \times CCA_n = td\left[1 - \frac{d}{2} \right](1 - d)^{n-2}P, \text{ for } n > 1. \tag{9A.7}$$

The present equivalent of all future tax savings from year 1 to time infinity can now be shown to be

$$\frac{td}{i+d}\left(\frac{1+\frac{i}{2}}{1+i}\right)P$$

and the present equivalent of the capital expenditure adjusted for all future tax savings is

$$-P + \frac{td}{i+d} \times \frac{1+\frac{i}{2}}{1+i} \times P$$

$$= -P \times \left(1 - \frac{td}{i+d} \times \frac{1+\frac{i}{2}}{1+i}\right)$$

$$= -P \times CTF_{50\%}$$

where

$$CTF_{50\%} = 1 - \frac{td}{i+d} \times \frac{1+\frac{i}{2}}{1+i} \tag{9A.8}$$

is the 50%-rule capital tax factor ($CTF_{50\%}$).

It is worth noting that CTF and $CTF_{50\%}$ are independent of any time frame considerations, as they are developed on the basis of time going to infinity. (CTF and $CTF_{50\%}$ tables are included in Appendix E for various values of t, d and i.)

9A.3 Application of Capital Tax Factors

Consider the investment represented in the general after-tax cash flow diagram of Figure 9A.1. The $t \times CCA_n$ cash flows and the disposal tax effects, G, have been included explicitly and appear in Eq. 9A.1. The disposal tax effect, G, depends on the specific assumptions used. As discussed in Section 8A.3, there are four possible cases where an asset disposal does not deplete the asset class.

Alternatively, the problem can be formulated with capital tax factors. Using the 50%-rule capital tax factor, we can calculate a capital cost adjusted for all future tax savings to time infinity as $P \times CTF_{50\%}$. However, the investment generates tax savings only over N years, so the tax savings included for the period from N (Cases III and IV as discussed in Section 8A.3) or $N + 1$ (Cases I and II) to time infinity must be eliminated in some manner. If no new assets are added to the same class in the year of disposal, the asset class accounting procedure merely subtracts the salvage value from the undepreciated capital cost base of the asset class. The reduction of the undepreciated capital cost by S in this case will reduce the future tax savings that could be realized from this asset class. The adjustments to the future tax savings due to S are calculated with the same declining balance scheme. However, when these adjustments are made depends on the disposal timing assumption. The situation is further complicated when new assets are added to the class in the year in which a disposal occurs. This development leads to the same four cases considered in Appendix 8A.3.

Under Case I, the following assumptions are made:

1. The disposal of an asset occurs at the beginning of year $N + 1$.
2. There are no new purchases of assets belonging to the same asset class in the year of disposal.

With these assumptions, the first extra tax payment will occur in year $N + 1$ in the amount of tdS. For the nth year after N years of service of the asset, the extra tax payment will be

$$td(1 - d)^{n-1}S.$$

We don't have to consider U_{Disposal} at disposal time because its effects in the subsequent years have been incorporated by applying the capital tax factor $CTF_{50\%}$ to the initial asset value P (Section 9A.2). This is how capital tax factors allow us to simplify our numerical calculations.

The salvage value and its tax effects can thus be represented as shown in Figure 9A.5. The present equivalent of all the cash flows in Figure 9A.5, expressed at the end of year N, is

$$S\sum_{n=1}^{\infty} \frac{td(1 - d)^{n-1}S}{(1 + i)^n}$$

$$= S\left(1 - \frac{td}{i + d}\right)$$

$$= S \times CTF_{\text{I}}$$

where

$$CTF_{\text{I}} = 1 - \frac{td}{i + d} \tag{9A.9}$$

is the capital tax factor to be applied to S under Case I. It is equal to the CTF developed in Section 9A.2 and tabulated in Appendix E.

The assumptions under Case I, although not stated explicitly therein, are the basis for the capital tax factor approach recommended by a number of other authors, including Edge and Irvine[1] and Sprague and Whittaker[2].

---

[1] C. Geoffrey Edge and V. Bruce Irvine, *A Practical Approach to the Appraisal of Capital Expenditures*, Second Edition, The Society of Management Accountants of Canada, 1981.

[2] J.C. Sprague and J.D. Whittaker, *Economic Analysis for Engineers and Managers: The Canadian Context*, Prentice Hall of Canada Ltd., 1986.

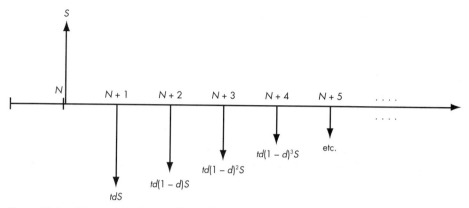

Figure 9A.5 Salvage value and its tax effects under Case I

The salvage value and corresponding tax effects under Case II (Refer to Section 8A.3 for the model assumptions) are shown in Figure 9A.6.

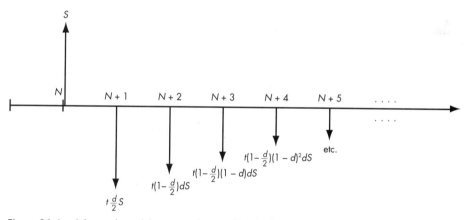

Figure 9A.6 Salvage value and the corresponding tax effects for Case II

The present equivalent of the cash flows in Figure 9A.6, expressed at the end of year N, is

$$S - \frac{t\frac{d}{2}S}{1+i} - \sum_{n=2}^{\infty} \frac{t(1-\frac{d}{2})(1-d)^{n-2}dS}{(1+i)^n}$$

$$= S \times \left(1 - \frac{td}{i+d} \times \frac{1+\frac{i}{2}}{1+i}\right)$$

$$= S \times CTF_{II}$$

where

$$CTF_{\text{II}} = 1 - \frac{td}{i + d} \times \frac{1 + \frac{i}{2}}{1 + i} \tag{9A.10}$$

is the capital tax factor to be applied to S under Case II. This factor is equal to the 50%-rule capital tax factor derived in Section 9A.2 and tabulated in Appendix E.

We can also easily verify that the capital tax factors for Cases III and IV, as defined in Section 8A.3, are as follows:

$$CTF_{\text{III}} = 1 - \frac{td}{i + d} (1 + i) \tag{9A.11}$$

$$CTF_{\text{IV}} = 1 - \frac{td}{i + d} \left(1 + \frac{i}{2}\right) \tag{9A.12}$$

Using these and the other capital tax factors derived previously, the after-tax cash flow diagram for a capital expenditure, depreciated on a declining balance basis, can be portrayed as shown in Figure 9A.7. In this diagram, the capital tax factor CTF_s for the salvage value may be replaced by CTF_{I}, CTF_{II}, CTF_{III}, or CTF_{IV}, depending on which set of assumptions apply to the disposal. The tax savings arising from CCA are no longer represented explicitly in Figure 9A.7, but are implicit through their representations in the capital tax factors.

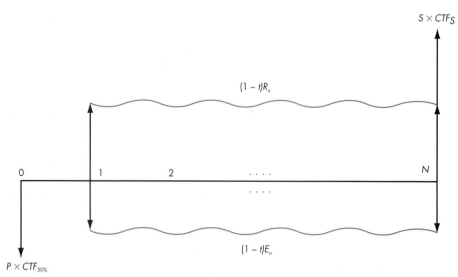

Figure 9A.7 The after-tax cash flow diagram for an investment using capital tax factors

When capital tax factors are used, the after-tax cash flow diagram should be developed as shown in Figure 9A.7. No $t \times CCA_n$ cash flows should appear in the diagram.

If you prefer to include the $t \times CCA_n$ cash flows explicitly, then the disposal tax effects *must* be calculated as discussed in Section 8A.3. Including the $t \times CCA_n$ cash flows in the diagram precludes the use of capital tax factors.

If a capital gain occurs when the asset is disposed, P should replace S in Figure 9A.7. In addition, the capital gains amount subtracted by the capital gains tax payment, i.e. $(S - P)(1 - t_c)$, where t_c is the capital gains tax rate, should be added to the after-tax cash flow diagram either at the end of year N (Cases III and IV) or at the end of year $N + 1$ (Cases I or II), depending on the timing of the disposal.

Coverage of capital tax factors in other texts is generally limited to Case I, with little explanation of the specific disposal timing assumptions or means of treating additions to the asset class. The broader consideration given to these issues in this text yields the four cases in which the capital tax factor applied to the salvage value takes different forms.

EXAMPLE 9A.1 Using capital tax factors

SKS Inc. is considering the construction of a new plant to manufacture mining equipment for Alberta's oil sands industries. The following estimates apply to this investment:

Investment Category	Investment (P)	Salvage Value in 10 years (S)
Building (B)	$1,000,000	$700,000
Land (L)	400,000	600,000
Machinery (M)	1,200,000	400,000

The company plans to operate the plant for 10 years. Annual operating costs are estimated to be $800,000 and a major overhaul needed in year 6 is estimated to cost $100,000. The declining balance method with the 50%-rule is used for CCA purposes. The CCA rates for the building and equipment are 4% and 30%, respectively. The company's tax rate is 40% and the minimum acceptable rate of return (MARR) is 15%. What is the annual equivalent operating revenue that the plant must generate to justify the investment? Assume that disposal occurs at the beginning of year 11, and that no other purchases of building or equipment class assets take place in year 11.

Solution:

Given: Operating cost $(E) = \$800,000$ per year; Additional operating cost, year 6 $(E_6) = \$100,000$; $d_B = 4\%$, $d_M = 30\%$, $t = 40\%$, MARR = 15%; Case I applies.
Find: The annual equivalent revenue (AER) needed to generate a rate of return of 15%.

The after-tax cash flow diagram of the investment is

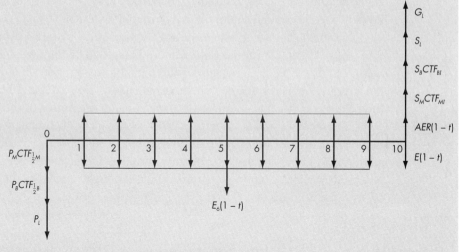

Figure 9A.8 After-tax cash flow diagram (Example 9A.1)

where

$$CTF_{50\%M} = 1 - \frac{td_M}{i + d_M} \times \frac{1 + \frac{i}{2}}{1 + i} = 1 - \left[\frac{(0.4)(0.3)}{0.15 + 0.3}\right]\left[\frac{(1 + 0.075)}{1 + 0.15}\right] = 0.7507$$

$$CTF_{IM} = 1 - \frac{td_M}{i + d_M} = 0.7333$$

$$CTF_{50\%B} = 1 - \frac{td_B}{i + d_B} \times \frac{1 + \frac{i}{2}}{1 + i} = 1 - \left[\frac{(0.4)(0.04)}{0.15 + 0.04}\right]\left[\frac{1 + 0.075}{1 + 0.15}\right] = 0.9213$$

$$CTF_{IB} = 1 - \frac{td_B}{i + d_B} = 0.9158.$$

Since the disposal of land occurs at the beginning of year 11, the capital gains tax will be paid at the end of year 11. As a result, G_L expressed at the end of year 10 is calculated by moving the capital gains tax payment backward by one year with a factor of $(1 + i)^{-1}$.

Thus

$$G_L = -\frac{3}{4}t\left(\frac{S_L - P_L}{1 + i}\right) \text{ (capital gains tax)}$$

$$= -\frac{3}{4} \times 0.4\left(\frac{\$600{,}000 - \$400{,}000}{1.15}\right)$$

$$= -\$52{,}174.$$

Then

$$AE = AER(1 - t) - E(1 - t)$$

$$- (P_M CTF_{50\%M} + P_B CTF_{50\%B} + P_L)(A/P,\ 15\%,\ 10)$$

$$- E_6(1 - t)(P/F,\ 15\%,\ 6)(A/P,\ 15\%,\ 10)$$

$$+ (S_B CTF_{IB} + S_M CTF_{IM} + S_L + G_L)(A/F,\ 15\%,\ 10)$$

$$= 0.6\ AER - 800,000 \times 0.6$$

$$- (1,200,000 \times 0.7507 + 1,000,000 \times 0.9213 + 400,000) \times 0.1993$$

$$- 100,000 \times 0.6 \times 0.4323 \times 0.1993$$

$$+ (700,000 \times 0.9158 + 400,000 \times 0.7333 + 600,000 - 52,174) \times 0.0493$$

$$= 0.6\ AER - 854,969 = 0.$$

Therefore

$$AER = \frac{\$854,969}{0.6} = \$1,424,949.$$

The required annual equivalent revenue is approximately 1.425 million dollars to justify the investment.

Comment: Although this solution provides no detailed information on the net after-tax cash flows, we are able to determine minimum annual equivalent revenue. This detail is implicit in the solution formulation.

Problems (for Appendix 9A)

9A.1 Reconsider Problem 9.4. Assume that the disposal will occur at the end of year 4 and there are no acquisitions of other assets in class 43 in year 4. Use capital tax factors to solve the problem. Is this result significantly different from the one obtained in Problem 9.4? Explain.

9A.2 Reconsider Problem 9.12. Assume MARR = 12%. Use capital tax factors to compute the present equivalent of the project under each of the four cases of assumptions on the timing of the disposal and whether new assets in the same class are acquired in the year of the disposal. Are these PE values significantly different from one another? Is this project acceptable?

9A.3 Reconsider Problem 9.19. Calculate the PE value for each of the four cases of asset disposal discussed in the text. Is it easy to find the IRR of the project using capital tax factors? What is the IRR of the project?

CHAPTER 10

Capital Budgeting Decisions

T he Southeastern Coca-Cola Bottling Company stores the gasoline used by its fleet of trucks in underground storage tanks.[1] Current U.S. Environmental Protection Agency (EPA) regulations require all such tanks to be upgraded to meet year 2002 standards or to be replaced by that same year. Tanks installed prior to 1981 cannot be upgraded and must be replaced by 2002. Tanks installed between 1981 and 1987 may be upgraded by 2002, and then these must be replaced within a 10-year period. Tanks installed during 1987 and thereafter may be completely upgraded to 2002 standards. These tanks would then have an expected life of 30 years.

[1]This story was written by Professor V. E. Unger of Auburn University with Mr. William Garvin of Hazclean Environmental Consultants, Inc., and Mr. Elbert Mullis of the Coca-Cola Bottling Company. (Reprinted with thanks.)

Several tanks may occupy one underground pit. All tanks in a pit are either to be upgraded or replaced simultaneously. Prior to the upgrade or replacement, yearly inspections will be required of each tank at a cost of approximately $900 per tank. Some pit locations may have more than one pit. The pits on a given site may be upgraded or replaced individually, or as a group. Certain economies of scale are associated with upgrading or replacing all tanks on a site at the same time.

The company has over 17 pits in 11 different locations throughout the southeast. Given the upgrade/replacement option for each pit, as well as the option of working on pits individually, or in combination with others at the same site, over 340 alternatives must be considered. The company wants to meet all EPA requirements over a 25-year time horizon.

Without budget limitations, the replacement problem would be easy: Simply select for each pit, or combination of pits, the least-cost alternative. However, to replace or upgrade all tanks over the 25-year time horizon would cost in excess of $1,500,000, and the company anticipates having only an annual tank-replacement budget of $200,000 or less. Because the project is to be implemented over an extended period of time, the firm's cost of capital will tend to fluctuate during the project period. In this circumstance, the choice of an appropriate interest rate (MARR) for use in the project evaluation becomes a critical issue. Given these budget and other restrictions, the company would like to determine the least-cost replacement/upgrade strategy for the 25-year period. The company is also interested in minimizing the maximum budget allocations required in any one year.

In this chapter, we will present the basic framework of **capital budgeting**, which involves investment decisions related to fixed assets. Here, the term **capital budget** includes planned expenditures on fixed assets; **capital budgeting** encompasses the entire process of analyzing projects and deciding whether or not they should be included in the capital budget. In previous chapters, we focused on how to evaluate and compare investment projects—the analysis aspect of capital budgeting. In this chapter, we will focus on the budgeting aspect. Proper capital budgeting decisions require choice of the method of project financing, the schedule of investment opportunities, and an estimate of the minimum attractive rate of return (MARR).

10.1 Methods of Financing

In previous chapters, we have focused on problems relating to investment decisions. In reality, investment decisions are not always independent of the source of finance. For convenience, however, in economic analysis, investment decisions are usually separated from finance decisions—first the investment project is selected, and then the choice of financing sources is considered. After the source is chosen, appropriate modifications to the investment decision are made.

We have also assumed that the assets employed in an investment project are obtained with the firm's own capital (retained earnings) or from short-term borrowings. In practice, this arrangement is not always attractive, or even possible. If the investment calls for a significant infusion of capital, the firm may raise the needed capital by issuing stock. Alternatively, the firm may borrow the funds by issuing bonds to finance such purchases. In this section, we will first discuss how a typical firm raises new capital from external sources. Then we will discuss how external financing affects after-tax cash flows and how the decision to borrow affects the investment decision.

The two broad choices a firm has for financing an investment project are **equity financing** and **debt financing**.[2] We will look briefly at these two options for obtaining external investment funds, and also examine their effects on after-tax cash flows. First we present basic terminology and definitions relating to stocks, to which we will refer in the following sections and in Chapter 11.

10.1.1   Stock Terminology

Equity: Equity is the value of an enterprise beyond the total debt it may have. It is the ownership of a company held by its common and preferred shareholders.

Preferred Stock: Preferred shares are entitled to dividends at a predetermined rate which must be paid before any dividends can be paid to the common shareholders. This preference as to dividend payment is usually cumulative, in other words, if the company is unable to pay preferred dividends when due, they accumulate and must be paid sometime in the future before any profits can be distributed to the common shareholders. In the event that the company decides to or has to terminate its business, the preferred shares are usually entitled to a stipulated portion of the assets in priority to the common shares. However, the preferred shareholders do not have a voice in management unless the company fails to pay dividends on the preferred shares. A sample preferred share certificate is given in Figure 10.1.

Common stock or common shares: Common shares represent ownership in a company and give the holders voting power in the selection of management. They have a proportionate but unspecified claim on profits. Common shareholders may receive dividends when earned and declared by the board of directors. The dividends may be in the form of additional stock. The most attractive features of common shares to individual investors are the potential of growth in share price, liquidity in the stock market, and favorable treatment in Canada of dividend income and capital gains (see Chapter 8). A sample common share certificate is shown in Figure 10.2.

Dividends: A company's net income after the preferred dividends are paid may be distributed as dividends to the common shareholders or reinvested in the business. The amount of dividends to be distributed to the common shareholders is determined by the board of directors. If any part of the net earnings is retained in the company, the value of the common shares should increase.

---

[2] A hybrid financing method, known as *lease financing*, is discussed in Chapter 11.

(a)

(b)

Figure 10.1 A sample preferred share certificate: (a) front, (b) back. Source: The Canadian Securities Course, The Canadian Securities Institute, Suite 360, 33 Yonge Street, Toronto, Ontario, M5E 1G4. 1990. Pages 176–177 (Courtesy of Ontario Banknote Ltd.)

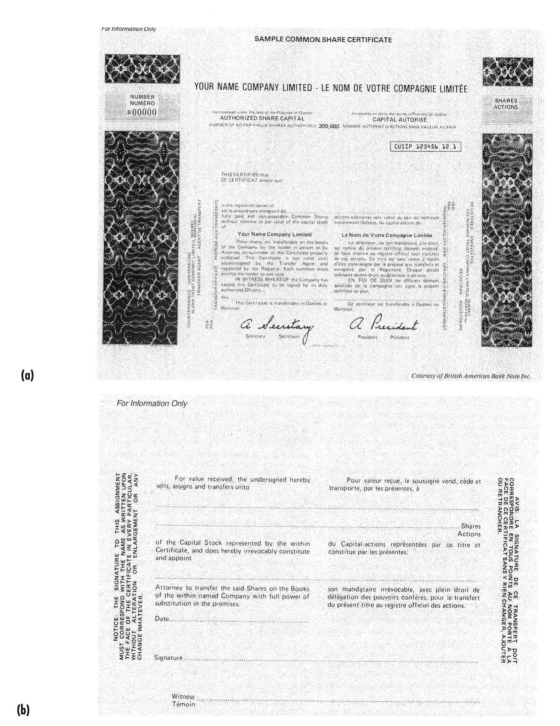

Figure 10.2 A sample common share certificate: (a) front, (b) back. Source: The Canadian Securities Course, The Canadian Securities Institute, Suite 360, 33 Yonge Street, Toronto, Ontario, M5E 1G4. 1990. Pages 176–177 (Courtesy of Ontario Banknote Ltd.)

10.1.2 Equity Financing

Equity financing can take two forms: (1) Use of retained earnings otherwise paid to stockholders or (2) issuance of stock. Both forms of equity financing use funds invested by the current or new owners of the company.

Until now, many of our economic analyses presumed that companies had cash on hand to make capital investments — implicitly, we were dealing with cases of financing by retained earnings. A simplified view of a company's cash flows in Figure 10.3 illustrates the source of **retained earnings**. Most of these stages are familiar to you from previous chapters. Several deductions are applied to the revenue to arrive at the taxable income, including interest payments owed on borrowed money, I, and the capital cost allowance, CCA. After the income tax T has been calculated, the CCA is added back in, leaving the firm with available cash. Once the debts for the period are repaid, the firm may allocate a portion of the remaining money to the company's owners — the shareholders — in the form of a dividend, and use the remaining retained earnings to reinvest in the company now, or keep some on hand for future needs.

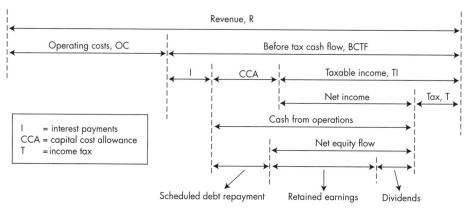

Figure 10.3   Retained earnings

If a company does not have sufficient cash on hand to make an investment and does not wish to borrow in order to fund it, financing can be arranged by selling common stock to raise the required funds. (Many small biotechnology and computer firms raise capital by going public and selling common stock.) To do this, the company has to decide how much money to raise, the type of securities to issue (common stock or preferred stock), and the basis for pricing the issue.

Once the company has decided to issue common stock, the firm must estimate **flotation costs**—the expenses it will incur in connection with the issue, such as investment bankers' fees, lawyers' fees, accountants' costs, and printing and engraving. Usually an investment banker will buy the issue from the company at a discount, below the price at which the stock is to be offered to the public. (The discount usually represents the *flotation costs*.) If the company is already publicly owned, the offering price will commonly be based on the existing market price of the stock. If the company is going public for the first time, no established price will exist, so investment bankers have to estimate the expected market price at which the stock will sell after the stock issue. Example 10.1 illustrates how the flotation cost affects the cost of issuing common stock.

EXAMPLE 10.1 **Issuing common stock**

Scientific Sports, Inc. (SSI), a golf club manufacturer, has developed a new metal club (Driver). The club is made out of titanium alloy—an extremely light and durable metal available with good vibration-damping characteristics (Figure 10.4). The company expects to acquire considerable market penetration with this new product. To produce it, the company needs a new manufacturing facility, which will cost $10 million. The company decided to raise this $10 million by selling common stock. The firm's current stock price is $30 per share. Investment bankers have informed management that the new public issue must be priced at $28 per share because of decreasing demand, which will occur as more shares become available on the market. The flotation costs will be 6% of the issue price, so SSI will net $26.32 per share. How many shares must SSI sell to net $10 million after flotation expenses?

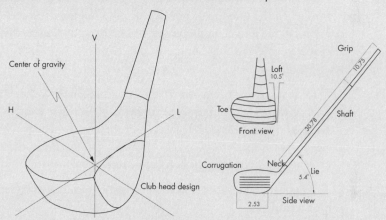

Figure 10.4 SSI's new golf club (Driver) design developed using advanced engineering materials (Example 10.1)

Solution

Let X be the number of shares to be sold. The total flotation cost will be

$$(0.06)(\$28)(X) = 1.68X.$$

To net $10 million, we establish the following relationship:

Sales proceeds – flotation cost	=	net proceeds
$28X - 1.68X$	=	$10,000,000
$26.32X$	=	$10,000,000
X	=	379,940 shares.

Now we can figure out the flotation costs for issuing the common stock as follows:

$$1.68(379,940) = \$638,300.$$

10.1.3 Debt Financing

In addition to equity financing, the second major type of financing a company can select is **debt financing**. Debt financing includes both short-term borrowing from financial institutions and the sale of long-term bonds, where money is borrowed from investors for a fixed period. With debt financing, the interest paid on the loans or bonds is treated as an expense for income-tax purposes. Since interest is a tax-deductible expense, companies in high tax brackets may incur lower after-tax financing costs with a debt. In addition to influencing the borrowing interest rate and tax bracket, a loan-repayment method can also affect financing costs.

When the debt-financing option is used, we need to separate the interest payments from the repayment of the loan for our analysis. The interest-payment schedule depends on the total repayment schedule established at the time of borrowing. The two common debt-financing methods are as follows:

1. **Bond Financing**
   This type of debt financing does not involve partial payment of principal; only interest is paid each year (or semiannually). The principal is paid in a lump sum when the bond matures. (See Section 3.8 for bond terminologies and valuation.) Bond financing is similar to equity financing in that flotation costs are involved when issuing bonds.

2. **Term Loans**
   Term loans involve an equal repayment arrangement, where the sum of the interest payment and the principal payment is uniform; interest payments decrease while principal payments increase over the life of the loan. Term loans are usually negotiated directly between the borrowing company and a financial institution —generally a commercial bank, an insurance company, or a pension fund.

Example 10.2 illustrates how these different methods can affect the cost of issuing bond or term loans.

EXAMPLE Debt financing

10.2 Refer to Example 10.1. Suppose SSI has instead decided to raise the $10 million by debt financing. SSI could issue a mortgage bond or secure a term loan. Conditions for each option are as follows:

- Bond financing: The flotation cost is 1.8% of the $10 million issue. The company's investment bankers have indicated that a 5-year bond issue with a face value of $1000 can be sold for $985. The bond would require annual interest payments of 12%.

- Term loan: A $10 million bank loan can be secured at an annual interest rate of 11% for 5 years; it would require five equal annual installments.

(a) How many $1000 par value bonds would SSI have to sell to raise the $10 million?

(b) What are the annual payments (interest and principal) for the bond?

(c) What are the annual payments (interest and principal) for the term loan?

Solution

(a) To net $10 million, SSI would have to sell

$$\$10,000,000/(1 - 0.018) = \$10,183,300$$

worth of bonds and pay $183,300 in flotation costs. Since the $1000 bond will be sold at a 1.5% discount, the total number of bonds to be sold would be

$$\$10,183,300/\$985 = 10,338.38.$$

(b) For the bond financing, the annual interest is equal to

$$\$10,338,380 \, (0.12) = \$1,240,606.$$

Only the interest is paid each year, and thus the principal amount owed remains unchanged.

(c) For the term loan, the annual payments are

$$\$10,000,000(A/P, 11\%, 5) = \$2,705,703.$$

The principal and interest components of each annual payment are summarized in Table 10.1. Note that, although SSI would be responsible for a $10,338,380 debt with the bond issue, they only net $10,000,000 new capital because of the flotation cost and the discount sale price.

	0	1	2	3	4	5
Table 10.1						
Two Common Methods of Debt Financing (Example 10.2)	1. Bond financing: No principal repayments until end of life					
Beginning balance	$10,338,380	$10,338,380	$10,338,380	$10,338,380	$10,338,380	$ 10,338,380
Interest owed		1,240,606	1,240,606	1,240,606	1,240,606	1,240,606
Repayment						
Interest payment		1,240,606	1,240,606	1,240,606	1,240,606	1,240,606
Principal payment						(10,338,380)
Ending balance	$10,338,380	$10,338,380	$10,338,380	$10,338,380	$10,338,380	0
	2. Term loan: Equal annual repayment [$10,000,000 (A/P, 11%, 5) = $2,705,703]					
Beginning balance	$10,000,000	$10,000,000	$ 8,394,297	$ 6,611,967	$ 4,633,580	$ 2,437,571
Interest owed		1,100,000	923,373	727,316	509,694	268,133
Repayment						
Interest payment		(1,100,000)	(923,373)	(727,316)	(509,694)	(268,133)
Principal payment		(1,605,703)	(1,782,330)	(1,978,387)	(2,196,009)	(2,437,570)
Ending balance	$10,000,000	$ 8,394,297	$ 6,611,967	$ 4,633,580	$ 2,437,571	0

10.1.4 Capital Structure

The ratio of total debt to total capital, generally called the **debt ratio**, or **capital structure**, represents the percentage of the total capital provided by borrowed funds. For example, a debt ratio of 0.4 indicates that 40% of the capital is borrowed, and the remaining funds are provided from the company's equity (retained earnings or stock offerings). This type of financing is called **mixed financing**.

Borrowing affects a firm's capital structure, and firms must determine the effects of a change in the debt ratio on their market value before making an ultimate financing decision. Even if debt financing is attractive, you should understand that companies do not simply borrow funds to finance projects. A firm usually establishes a **target capital structure**, or **target debt ratio**, after considering the effects of various financing methods. This target may change over time as business conditions vary, but a firm's management always strives to achieve this target whenever individual financing decisions are considered. If the actual debt ratio is below the target level, any new capital will probably be raised through debt financing. On the other hand, if the debt ratio is currently above the target, expansion capital would be raised by issuing stock.

How does a typical firm set the target capital structure? This is a rather difficult question to answer, but we can list several factors that affect the capital structure policy. First, capital structure policy involves a trade-off between risk and return. As you take on more debt for business expansion, the inherent business risk[3] also increases, but investors view business expansion as a healthy indicator for a corporation with higher expected earnings. When investors perceive higher business risk, the firm's stock price tends to be depressed. On the other hand, when investors perceive higher expected earnings, the firm's stock price tends to increase. The optimal capital structure is thus the one that strikes a balance between business risk and expected future earnings. The greater the firm's business risk, the lower its optimal debt ratio.

Second, a major reason for using debt is that interest is a deductible expense for business operations, which lowers the effective cost of borrowing. On the other hand, dividends paid to common stockholders are not deductible. If a company uses debt, it must pay interest on this debt, whereas if it uses equity, it pays dividends to its equity investors (shareholders). A company needs $1 in before-tax (B/T) income to pay $1 of interest, but if the company's tax rate is 34%, it needs $1/(1 - 0.34) = $1.52 of B/T income to pay $1 of dividend.

Third, financial flexibility, the ability to raise capital on reasonable terms from the financial market, is an important consideration. Firms need a steady supply of capital for stable operations. When money is tight in the economy, investors prefer

---

[3] Unlike equity financing, where dividends are optional, debt interest and principal (face value) must be repaid on time. Also uncertainty is involved in making projections of future operating income as well as expenses. In bad times, debt can be devastating, but in good times, the tax deductibility of interest payments increases profits to owners.

to advance funds to companies with a healthy capital structure (lower debt ratio). These three elements (business risk, taxes, and financial flexibility) are major factors that determine the firm's optimal capital structure. Example 10.3 illustrates how a typical firm finances a large-scale engineering project by maintaining the predetermined capital structure.

EXAMPLE 10.3

Project financing based on an optimal capital structure

Reconsider SSI's $10 million venture project in Example 10.1. Suppose that SSI's optimal capital structure calls for a debt ratio of 0.5. After reviewing SSI's capital structure, the investment banker convinced management that it would be better off, in view of current market conditions, to limit the stock issue to $5 million and to raise the other $5 million as debt by issuing bonds. Because the amount of capital to be raised in each category is reduced by half, the flotation cost would also change. The flotation cost for common stock would be 8.1%, whereas the flotation cost for bonds would be 3.2%. As in Example 10.2, the 5-year, 12% bond will have a par value of $1000 and will be sold for $985.

Assuming that the $10 million capital would be raised from the financial market, the engineering department has detailed the following financial information.

- The new venture will have a 5-year project life.

- The $10 million capital will be used to purchase land for $1 million, a building for $3 million, and equipment for $6 million. The plant site and building are already available, and production can begin during the first year. The building and the equipment fall into declining balance Class 1 ($d = 4\%$) and Class 43 ($d = 30\%$), respectively. The 50% rule is applicable. At the end of year 5, the salvage value for each asset is as follows: the land $1.5 million, the building $2 million, and the equipment $2.5 million.

- For common stockholders, an annual cash dividend in the amount of $2 per share is planned over the project life. This steady cash dividend payment is deemed necessary to maintain the market value of the stock.

- The unit production cost is $50.31 (material, $22.70; labor and overhead (excluding depreciation), $10.57; and tooling, $17.04).

- The unit price is $250, and SSI expects an annual demand of 20,000 units.

- The operating and maintenance cost, including advertising expense, would be $600,000 per year.

- An investment of $500,000 in working capital is required at the beginning of the project, and the amount will be fully recovered when the project terminates.

- The firm's marginal tax rate is 40%, and this rate will remain constant throughout the project period.

- The firm's MARR is 20%.

(a) Determine the after-tax cash flows for this investment with external financing.

(b) What is the rate of return on this investment?

Discussion: As the amount of financing and flotation costs change, we need to recalculate the number of shares (or bonds) to be sold in each category. For a $5 million common stock issue, the flotation cost increases to 8.1%.[4] The number of shares to be sold to net $5 million is 5,000,000/(0.919)(28) = 194,311 shares (or $5,440,708). For a $5 million bond issue, the flotation cost is 3.2%. Therefore, to net $5 million, SSI has to sell 5,000,000/(0.968)(985) = 5243.95 units of $1000 par value. This implies that SSI is effectively borrowing $5,243,948, upon which figure the annual bond interest will be calculated. The annual bond interest payment is $5,243,948(0.12) = $629,274.

Solution

(a) **After-tax cash flows**

Table 10.2 summarizes the after-tax cash flows for the new venture. The following calculations and assumptions were used in developing the cash flow table.

- Revenue: $250 × 20,000 = $5,000,000 per year

- Costs of goods: $50.31 × 20,000 = $1,006,200 per year

- Bond interest: $5,243,948 × 0.12 = $629,274 per year

- Capital cost allowance: Assuming that the building is placed in service in January, the first year's CCA percentage is 2.0%. Therefore, the allowed CCA amount is $3,000,000 × 0.02 = $60,000. The CCA in each subsequent year would be 4.0% of the undepreciated capital cost at the beginning of that year. For the second year, the CCA would be

$$($3,000,000 - $60,000) × 0.04 = $117,600.$$

CCA on equipment is calculated in a similar manner using a 30% rate (Year 1 = 15%).

---

[4] Flotation costs are higher for small issues than for large ones due to existence of fixed costs — certain costs must be incurred regardless of the size of the issue, so the percentage of flotation costs increases as the size of issue gets smaller.

Table 10.2
Effects of
Project
Financing on
After-Tax Cash
Flows
(Example
10.3)

Year	0	1	2	3	4	5
Income Statement						
Revenue		$5,000,000	$5,000,000	$5,000,000	$5,000,000	$ 5,000,000
Expenses						
Cost of goods		1,006,200	1,006,200	1,006,200	1,006,200	1,006,200
O&M		600,000	600,000	600,000	600,000	600,000
Bond interest		629,274	629,274	629,274	629,274	629,274
CCA						
Building		60,000	117,600	112,896	108,380	104,045
Equipment		900,000	1,530,000	1,071,000	749,700	524,790
Taxable income		1,804,526	1,116,926	1,580,630	1,906,446	2,135,691
Income taxes		721,810	446,770	632,252	762,578	854,726
Net income		1,082,716	670,156	948,378	1,143,868	1,281,415
Cash Flow Statement						
Operating						
Net income		1,082,716	670,156	948,378	1,143,868	1,281,415
CCA		960,000	1,647,600	1,183,896	858,080	628,835
Investment						
Land	($1,000,000)					1,500,000
Building	(3,000,000)					2,000,000
Equipment	(6,000,000)					2,500,000
Working capital	(500,000)					500,000
Disposal tax effects						(461,365)
Financing						
Common stock	5,000,000					(5,440,708)
Bond	5,000,000					(5,243,948)
Cash dividend		(388,622)	(388,622)	(388,622)	(388,622)	(388,622)
Net cash flow	($500,000)	$1,654,094	$1,929,134	$1,743,652	$1,613,326	($3,124,393)

- Disposal tax effect:

   Capital gain on land = $1,500,000 − $1,000,000 = $500,000
   Tax on capital gain = −0.4 × 3/4 × $500,000 = −$150,000

For the building and equipment, the tax effect at the time of disposal equals (UCC − salvage value) × t

Property	UCC	Salvage Value	Gains (Losses)	Disposal Tax Effect, G
Building	$2,497,079	$2,000,000	$ 497,079	$ 198,831
Equipment	1,224,510	2,500,000	(1,275,490)	(510,196)

The total disposal tax effect, G, is then

$$G = -\$150,000 + \$198,831 - \$510,196 = -\$461,365$$

- Cash dividend: 194,311 shares × $2 = $388,622 per year.

- Common stock: When the project terminates and the bonds are retired, the debt ratio is no longer 0.5. If SSI wants to maintain the constant capital structure (0.5), SSI would repurchase the common stock in the amount of $5,440,708 at the prevailing market price. In developing Table 10.2, we assumed this repurchase of common stock had taken place at the end of project years. In practice, a firm may or may not repurchase the common stock. As an alternative means of maintaining the desired capital structure, the firm may use this extra debt capacity released to borrow for other projects.

- Bond: When the bonds mature at the end year 5, the total face value in the amount of $5,243,948 must be paid to the bondholders.

(b) **Measure of project worth**

The PE for this project is then

$$PE(20\%) = -\$500,000 + \$1,654,094(P/F, 20\%, 1) + \ldots$$
$$-\$3,124,393(P/F, 20\%, 5)$$
$$= \$2,749,554.$$

Since this is not a pure investment, we use the ERR method (with the external rate equal to MARR) to determine that the IRR is 330%. Even though the project requires a significant amount of cash expenditure at the end of project life, it appears to be a profitable one.

In Example 10.3, we neither determined the overall cost of capital for this project financing, nor did we explain the relationship between the cost of capital and the MARR. In the remaining sections, these issues will be discussed.

10.2 Cost of Capital

In most of the capital budgeting examples in the earlier chapters, we assumed that the projects under consideration were financed entirely with equity funds. In those cases, the cost of capital may have represented the firm's required return on equity. However, most firms finance a substantial portion of their capital budget with long-term debt (bonds), and many also use preferred stock as a source of capital. In these cases, a firm's cost of capital must reflect the average cost of the various sources of long-term funds that the firm uses, not only the cost of equity. In this section, we will discuss the ways in which the cost of each individual type of financing (retained earnings, common stock, preferred stock, and debt) can be estimated,[5] given a firm's target capital structure.

---

[5] Estimating or calculating the cost of capital in any precise fashion is a very difficult task.

10.2.1 Cost of Equity

Whereas debt and preferred stocks are contractual obligations that have easily determined costs, it is not easy to measure the cost of equity. In principle, the cost of equity capital involves an **opportunity cost**. In fact, the firm's after-tax cash flows belong to the stockholders. Management may either pay out these earnings in the form of dividends, or retain earnings and reinvest them in the business. If management decides to retain earnings, an opportunity cost is involved—stockholders could have received the earnings as dividends and invested this money in other financial assets. Therefore, the firm should earn on its retained earnings at least as much as the stockholders themselves could earn in alternative, but comparable, investments.

What rate of return can stockholders expect to earn on retained earnings? This question is difficult to answer, but the value sought is often regarded as the rate of return stockholders require on a firm's common stock. If a firm cannot invest retained earnings so as to earn at least the rate of return on equity, it should pay these funds to these stockholders and let them invest directly in other assets that do provide this return.

When investors are contemplating buying a firm's stock, they have two things in mind: (1) cash dividends and (2) gains (share appreciation) at the time of sale. From a conceptual standpoint, investors determine market values of stocks by discounting expected future dividends at a rate that takes into account any future growth. Since investors seek growth companies, a desired growth factor for future dividends is usually included in the calculation.

To illustrate, let's take a simple numerical example. Suppose investors in the common stock of ABC Corporation expect to receive a dividend of $5 by the end of the first year. The future annual dividends will grow at an annual rate of 10%. Investors will hold the stock for 2 more years and will expect the market price of the stock to rise to $120 by the end of the third year. Given these hypothetical expectations, ABC assumes that investors would be willing to pay $100 for this stock in today's market. What is the required rate of return on ABC's common stock (k_r)? We may answer this question by solving the following equation for k_r:

$$\$100 = \frac{\$5}{(1 + k_r)} + \frac{\$5(1 + 0.1)}{(1 + k_r)^2} + \frac{\$5(1 + 0.1)^2 + \$120}{(1 + k_r)^3}.$$

In this case, $k_r = 11.44\%$. This implies that if ABC finances a project by retaining its earnings or by issuing additional common stock at the going market price of $100 per share, it must realize at least 11.44% on new investment just to provide the minimum rate of return required by the investors. Therefore, 11.44% is the specific cost of equity that should be used when calculating the weighted average cost of capital. As flotation costs are involved in issuing new stock, the cost of equity will increase. If investors view ABC's stock as risky and, therefore, are willing to buy the stock at a lower price than $100 (but with the same expectations), the cost of equity will also increase. Now we can generalize the result as follows:

Cost of Retained Earnings

Let's assume the same hypothetical situation for ABC. Recall that ABC's retained earnings belong to holders of its common stock. If ABC's current stock is traded for a market price of P_0, with a first-year dividend[6] of D_1, but growing at the annual rate of g thereafter, the specific cost of retained earnings for an infinite period of holding (stocks will change hands over the years, but it does not matter who holds the stock) can be calculated as

$$P_0 = \frac{D_1}{(1 + k_r)} + \frac{D_1(1 + g)}{(1 + k_r)^2} + \frac{D_1(1 + g)^2}{(1 + k_r)^3} + \cdots$$

$$= \frac{D_1}{1 + k_r} \sum_{n=0}^{\infty} \left[\frac{(1 + g)}{(1 + k_r)} \right]^n,$$

$$= \frac{D_1}{1 + k_r} \left[\frac{1}{1 - \frac{1+g}{1+k_r}} \right], \text{ where } g < k_r.$$

Solving for k_r, we obtain

$$k_r = \frac{D_1}{P_0} + g. \tag{10.1}$$

If we use k as the discount rate for evaluating the new project, it will have a positive PE only if the project's IRR exceeds k_r. Therefore, any project with a positive PE, calculated at k_r, induces a rise in the market price of the stock. Hence, by definition, k_r is the rate of return required by shareholders and should be used as the cost of the equity component when calculating the weighted average cost of capital.

Issuing New Common Stock

As flotation costs are involved in issuing new stock, we can modify the cost of retained earnings (k_r) by

$$k_e = \frac{D_1}{P_0(1 - f_c)} + g, \tag{10.2}$$

where k_e = the cost of common equity, and f_c = the flotation cost as a percentage of stock price.

---

[6] When we check the stock listings in the newspaper, we do not find the expected first-year dividend, D_1. Instead, we find the dividend paid out most recently, D_0. So if we expect growth at a rate g, the dividend at the end of 1 year from now, D_1, may be estimated as

$$D_1 = D_0 (1 + g).$$

Either calculation is deceptively simple because in fact, several ways are available to determine the cost of equity. In reality, the market price fluctuates constantly, as do a firm's future earnings. Thus, future dividends may not grow at a constant rate as the model indicates. For a stable corporation with moderate growth, however, the cost of equity as calculated by evaluating either Eq. (10.1) or Eq. (10.2) serves as a good approximation.

Cost of Preferred Stock

A preferred stock is a hybrid security in the sense that it has some of the properties of bonds and other properties that are similar to common stock. Like bondholders, holders of preferred stock receive a fixed annual dividend. In fact, many firms view the payment of the preferred dividend as an obligation just like interest payments to bondholders. It is therefore relatively easy to determine the cost of preferred stock. For the purposes of calculating the weighted average cost of capital, the specific cost of a preferred stock will be defined as

$$k_p = \frac{D^*}{P^*(1 - f_c)}, \tag{10.3}$$

where D^* = the fixed annual dividend, P^* = the issuing price, and f_c as defined above.

Cost of Equity

Once we have determined the specific cost of each equity component, we can determine the weighted average cost of equity (i_e) for a new project:

$$i_e = ak_r + bk_e + ck_p, \tag{10.4}$$

where a = fraction of total equity financed from retained earnings, b = fraction of total equity financed from issuing new stock, c = fraction of equity financed from issuing preferred stock, and $a + b + c = 1$. Example 10.4 illustrates how we may determine the cost of equity.

**EXAMPLE
10.4**

Determining the cost of equity

Alpha Corporation needs to raise $10 million for plant modernization. Alpha's target capital structure calls for a debt ratio of 0.4, indicating that $6 million has to be financed from equity.

- Alpha is planning to raise $6 million from the following equity sources:

Source	Amount	Fraction of Total Equity
Retained earnings	$1 million	0.167
New common stock	4 million	0.666
Preferred stock	1 million	0.167

- Alpha's current common stock price is $40—the market price that reflects the firm's future plant modernization. Alpha is planning to pay an annual cash dividend of $5 at the end of the first year, and the annual cash dividend will grow at an annual rate of 8% thereafter.

- Additional common stock can be sold at the same price of $40, but there will be 12.4% flotation costs.

- Alpha can issue $100 par preferred stock with a 9% dividend. (This means that Alpha will calculate the dividend based on the par value, which is $9 per share.) The stock can be sold on the market for $95, and Alpha must pay flotation costs of 6% of the market price.

Determine the cost of equity to finance the plant modernization.

Solution

We will itemize the cost of each equity component.

- Cost of retained earnings: With $D_1 = \$5$, $g = 8\%$, and $P_0 = \$40$,

$$k_r = \frac{5}{40} + 0.08 = 20.5\%.$$

- Cost of new common stock: With $D_1 = \$5$, $g = 8\%$, $f_c = 12.4\%$,

$$k_e = \frac{5}{40(1 - 0.124)} + 0.08 = 22.27\%.$$

- Cost of preferred stock: With $D^* = \$9$, $P^* = \$95$ $f_c = 0.06$,

$$k_p = \frac{9}{95(1 - 0.06)} = 10.08\%.$$

- Cost of equity: With $a = 0.167$, $b = 0.666$, and $c = 0.167$,

$$i_e = (0.167)(0.205) + (0.666)(0.2227) + (0.167)(0.1008)$$
$$= 19.96\%.$$

10.2.2   Cost of Debt

Now let us consider the calculation of the specific cost that is to be assigned to the debt component of the weighted average cost of capital. The calculation is relatively straightforward and simple. As we discussed in Section 10.1.3, the two types of debt financing are term loans and bonds. Because the interest payments on both are tax deductible, the effective cost of debt will be reduced.

To determine the after-tax cost of debt (i_d), we can evaluate the following expression:

$$i_d = Sk_s(1 - t) + (1 - S)k_b(1 - t), \tag{10.5}$$

where S = the fraction of the term loan over the total debt, k_s = the before-tax interest rate on the term loan, t = the firm's marginal tax rate, and k_b = the before-tax interest rate on the bond.

As for bonds, a new issue of long-term bonds incurs flotation costs. These costs reduce the proceeds to the firm, thereby increasing the specific cost of the capital raised. For example, when a firm issues a $1000 par bond, but nets only $940, the flotation cost will be 6%. Therefore, the effective after-tax cost of the bond component will be higher than the nominal interest rate specified on the bond. We will examine this problem with an example.

EXAMPLE 10.5 Determining the cost of debt

Refer to Example 10.4, and suppose that Alpha has decided to finance the remaining $4 million by securing a term loan and issuing 20-year $1000 par bonds for the following condition.

Source	Amount	Fraction	Interest Rate	Flotation Cost
Term loan	$1 million	0.25	12% per year	
Bonds	3 million	0.75	10% per year	6%

If the bond can be sold to net $940 (after deducting the 6% flotation cost), determine the cost of debt to raise $4 million for the plant modernization. Alpha's marginal tax rate is 38%, and it is expected to remain constant in the future.

Solution

First, we need to find the effective after-tax cost of issuing the bond with a flotation cost of 6%. The before-tax specific cost is found by solving the following equivalence formula (see Section 3.8.3).

$$\$940 = \frac{\$100}{(1 + k_b)} + \frac{\$100}{(1 + k_b)^2} + \cdots + \frac{\$100 + \$1000}{(1 + k_b)^{20}}$$

$$= \$100 \, (P/A, k_b, 20) + \$1000 \, (P/F, k_b, 20).$$

Solving for k_b, we obtain k_b = 10.74%. Note that the cost of the bond component increases from 10% to 10.74% after considering the 6% flotation cost.

The after-tax cost of debt is the interest rate on debt, multiplied by $(1 - t)$. In effect, the government pays part of the cost of debt because interest is tax deductible. Now we are ready to compute the after-tax cost of debt as follows:

$$i_d = (0.25)(0.12)(1 - 0.38) + (0.75)(0.1074)(1 - 0.38)$$

$$= 6.85\%.$$

10.2.3 Calculating the Cost of Capital

With the specific cost of each financing component determined, now we are ready to calculate the tax-adjusted weighted average cost of capital based on total capital. Then, we will define the marginal cost of capital that should be used in project evaluation.

Weighted Average Cost of Capital

Assuming that a firm raises capital based on the target capital structure and that the target capital structure remains unchanged in the future, we can determine a **tax-adjusted weighted-average cost of capital** (or, simply stated, the **cost of capital**). This cost of capital represents a composite index reflecting the cost of raising funds from different sources. The cost of capital is defined as

$$k = \frac{i_d C_d}{V} + \frac{i_e C_e}{V}. \tag{10.6}$$

where C_d = Total debt capital (such as bonds) in dollars,

C_e = Total equity capital in dollars,

V = $C_d + C_e$,

i_e = Average equity interest rate per period considering all equity sources,

i_d = After-tax average borrowing interest rate per period considering all debt sources, and

k = Tax-adjusted weighted-average cost of capital.

Note that the cost of equity is already expressed in terms of after-tax cost, because any return to holders of either common stock or preferred stock is made after payment of income taxes.

Marginal Cost of Capital

Now we know how to calculate the cost of capital. Could a typical firm raise unlimited new capital at the same cost? The answer is no. As a practical matter, as a firm tries to attract more new dollars, the cost of raising each additional dollar will at some point rise. As this occurs, the weighted average cost of raising each additional new dollar also rises. Thus, the **marginal cost of capital** is defined as the cost of obtaining another dollar of new capital, and the marginal cost rises as more and more capital is raised during a given period. In evaluating an investment project, we are using the concept of marginal cost of capital. The formula to find the marginal cost of capital is exactly the same as Eq. (10.6). However, the costs of debt and equity in Eq. (10.6) are the interest rates on new debt and equity, not outstanding (or combined) debt or equity. Our primary concern with the cost of capital is to use it in evaluating a new investment project. The rate at which the firm has borrowed in the past is less important for this purpose. Example 10.6 works through the computations for finding the cost of capital (k).

EXAMPLE 10.6 **Calculating the marginal cost of capital**

Reconsider Examples 10.4 and 10.5. The marginal income tax rate (t) for Alpha is expected to remain at 38% in the future. Assuming that Alpha's capital structure (debt ratio) also remains unchanged in the future, determine the cost of capital (k) of raising $10 million in addition to existing capital.

Solution

With C_d = $4 million, C_e = $6 million, V = $10 million, i_d = 6.85%, i_e = 19.96%, and Eq. (10.6), we calculate

$$k = \frac{(0.0685)(4)}{10} + \frac{(0.1996)(6)}{10}$$

$$= 14.71\%.$$

This 14.71% would be the marginal cost of capital that a company with this financial structure would expect to pay to raise $10 million.

10.3 Choice of Minimum Attractive Rate of Return

Thus far, we have said little about what interest rate, or minimum attractive rate of return (MARR), is suitable for use in a particular investment situation. Choosing the MARR is a difficult problem; no single rate is always appropriate. In this section, we will discuss briefly how to select a MARR for project evaluation. Then, we will examine the relationship between capital budgeting and the cost of capital.

10.3.1 Choice of MARR when Project Financing Is Known

In Chapter 9, we focused on calculating after-tax cash flows, including situations involving debt financing. When cash flow computations reflect interest, taxes, and debt repayment, what is left is called **net equity flow**. If the goal of a firm is to maximize the wealth of its stockholders, why not focus only on the after-tax cash flow to equity, instead of on the flow to all suppliers of capital? Focusing on only the equity flows will permit us to use the cost of equity as the appropriate discount rate. In fact, we have implicitly assumed that all after-tax cash flow problems in earlier chapters, where the financing flows were explicitly stated, represent net equity flows, so the MARR used represents the **cost of equity** (i_e). Example 10.7 illustrates project evaluation by the net equity flow method.

EXAMPLE 10.7

Project evaluation by net equity flow

Suppose the Alpha Corporation, which has the capital structure described in Example 10.6, wishes to install a new set of machine tools, which is expected to increase revenues over the next 5 years. The tools require an investment of $150,000, to be financed with 60% equity and 40% debt. The equity interest rate (i_e), which combines both sources of common and preferred stocks, is 19.96%. Alpha will use a 12% short-term loan to finance the debt portion of the capital ($60,000), with the loan to be repaid in equal annual installments over 5 years. The Capital Cost Allowance rate is 30%, subject to the 50% rule, and zero salvage value is expected. Additional revenues and operating costs are expected to be

n	Revenues ($)	Operating cost
1	$68,000	$20,500
2	73,000	20,000
3	79,000	20,500
4	84,000	20,000
5	90,000	20,500

The marginal tax rate (combined federal and provincial rate) is 38%. Evaluate this venture by using net equity flows at $i_e = 19.96\%$.

Solution

The calculations are shown in Table 10.3. The PE and IRR calculations are as follows:

End of Year	0	1	2	3	4	5
Income Statement						
Revenue		$68,000	$73,000	$79,000	$84,000	$90,000
Expenses						
Operating cost		20,500	20,000	20,500	20,000	20,500
Interest payment		7,200	6,067	4,797	3,376	1,783
CCA		22,500	38,250	26,775	18,743	13,120
Taxable income		17,800	8,683	26,928	41,881	54,597
Income taxes (38%)		6,764	3,300	10,233	15,915	20,747
Net income		$11,036	$ 5,383	$16,695	$25,966	$33,850
Cash Flow Statement						
Cash from operation						
Net income		$11,036	$ 5,383	$ 16,695	$ 25,966	$33,850
CCA		22,500	38,250	26,775	18,743	13,120
Investment and salvage	$(150,000)					0
Disposal tax effects						11,633
Loan principal repayment	+60,000	(9,445)	(10,578)	(11,847)	(13,269)	(14,861)
Net cash flow	$ (90,000)	$24,091	$ 33,055	$ 31,623	$ 31,440	$ 43,742

$UCC_5 = \$150,000(0.85)(0.7)^4 = \$30,613; G = (\$30,613 - 0)(0.38) = \$11,633$

$$PE(19.96\%) = -\$90,000 + \$24,091(P/F, 19.96\%, 1)$$
$$+ \$33,055(P/F, 19.96\%, 2) + \$31,623(P/F, 19.96\%, 3)$$
$$+ \$31,440(P/F, 19.96\%, 4) + \$43,742(P/F, 19.96\%, 5)$$
$$= \$4163$$

$$IRR = 21.88\% > 19.96\%$$

The internal rate of return for this cash flow is 21.88%, which exceeds $i_e = 19.96\%$. Thus, the project would be profitable.

Comment: In this problem we assumed that the Alpha Corporation would be able to raise the additional equity funds at the same rate of 19.96%, so that this 19.96% can be viewed as the marginal cost of capital.

10.3.2 Choice of MARR when Project Financing is Unknown

You might well ask why, if we use the cost of equity (i_e) exclusively, what use is the k? The answer to this question is that by using the value of k, we may evaluate investments without explicitly treating the debt flows (both interest and principal). In this case, we make a tax adjustment to the discount rate by employing the effective after-tax cost of debt. This approach recognizes that the net interest cost is effectively transferred from the tax collector to the creditor in the sense that there is a dollar-for-dollar reduction in taxes up to this amount of interest payments. Therefore, debt financing is treated implicitly. This method would be appropriate when debt financing is not identified with individual investments, but rather enables the company to engage in a set of investments. (Except where financing flows are explicitly stated, all previous examples in this book have implicitly assumed the more realistic and appropriate situation where debt financing is not identified with individual investment. Therefore, the MARRs represent the weighted cost of capital [k].) Example 10.8 provides an illustration of this concept.

EXAMPLE 10.8

Project evaluation by marginal cost of capital

In Example 10.7, suppose that Alpha Corporation has not decided how the $150,000 will be financed. However, Alpha believes that the project should be financed according to its target capital structure, debt ratio of 40%. Evaluate Example 10.7 using k.

Solution

By not accounting for the cash flows related to debt financing, we calculate the after-tax cash flows as shown in Table 10.4. Notice that, when we use this procedure, interest and the resulting tax shield are ignored when deriving the net incremental after-tax cash flow. In other words, no cash flow is related to financing activity. Thus, taxable income is overstated; income taxes are also overstated. To compensate for these overstatements, the discount rate is reduced accordingly. The implicit assumption is that the tax overpayment is exactly equal to the reduction in interest implied by i_d.

The time 0 flow is simply the total investment—$150,000 in this example. Recall that Alpha's k was calculated to be 14.71%. The internal rate of return for the after-tax flow in the last line of Table 10.4 is calculated as follows:

$$PE(14.71\%) = -\$150,000 + \$38,000(P/F, 14.71\%, 1)$$
$$+ \$47,395(P/F, 14.71\%, 2) + \$46,445(P/F, 14.71\%, 3)$$

End of Year	0	1	2	3	4	5
Income Statement						
Revenue		$68,000	$73,000	$79,000	$84,000	$90,000
Expenses						
Operating cost		20,500	20,000	20,500	20,000	20,500
CCA		22,500	38,250	26,775	18,743	13,120
Taxable income		$25,000	$14,750	$31,725	$45,257	$56,380
Income tax (38%)		9,500	5,605	12,055	17,198	21,424
Net income		$15,500	$ 9,145	$19,670	$28,059	$34,956
Cash Flow Statement						
Cash from operation						
Net income		$15,500	$9,145	$19,670	$28,059	$34,956
CCA		22,500	38,250	26,775	18,743	13,120
Investment and salvage	$(150,000)					0
Disposal tax effects						11,663
Net cash flow	$(150,000)	$38,000	$47,395	$46,445	$46,802	$59,709

Table 10.4 After-Tax Cash Flow Analysis when Project Financing Is Known: Cost of Capital Approach (Example 10.8)

$$+ \ \$46,802(P/F,\ 14.71\%,\ 4) + \$59,709(P/F,\ 14.71\%,\ 5)$$
$$= \ \$7,010$$
$$IRR \ = \ 16.54\% > 14.71\%.$$

Since the IRR exceeds the value of k, the investment would be profitable. Here, we evaluated the after-tax flow by using the value of k, and we reached the same conclusion about the desirability of the investment.

Comments: Although the PE and IRR values calculated for Alpha's investment in Examples 10.7 and 10.8 differ in value, the net equity flow and the cost of capital methods lead to the same accept/reject decision. This is usually the case for independent projects (assuming the same amortization schedule for debt repayment, such as term loans), and usually ranks projects identically for mutually exclusive alternatives. Some differences may be observed, as special financing arrangements may increase (or even decrease) the attractiveness of a project by manipulating tax shields and the timing of financing inflows and payments.

In summary, in cases where the exact debt-financing and repayment schedules are known, we recommend the use of the net equity flow method. The appropriate MARR would be the cost of equity, i_e. If no specific assumption is made about the exact instruments that will be used to finance a particular project (but we do assume that the given capital structure proportions will be maintained), we may determine the after-tax cash flows without incorporating any debt cash flows. Then, we use the marginal cost of capital (k) as the appropriate MARR.

10.3.3   Choice of MARR under Capital Rationing

It is important to distinguish between the cost of capital (k), as calculated in Section 10.3.1, and the MARR (i) used in project evaluation under **capital rationing**. **Capital rationing** refers to situations where the funds available for capital investment are not sufficient to cover potentially acceptable projects. When investment opportunities exceed the available money supply, we must decide which opportunities are preferable. Obviously, we want to ensure that all the selected projects are more profitable than the best rejected project. The best rejected project (or the worst accepted project) is the best opportunity forgone, and its value is called the **opportunity cost**. When a limit is placed on capital, the MARR is assumed to be equal to this opportunity cost, which is usually greater than the marginal cost of capital. In other words, the value of i represents the corporation's time-value trade-offs and reflects partially the available investment opportunities. Thus, there is nothing illogical about borrowing money at k and evaluating cash flows using the different rate, i. Presumably, the money will be invested to earn a rate i, or greater. In the following example, we will provide guidelines for selecting a MARR for project evaluation under capital rationing.

A company may borrow funds to invest in profitable projects, or it may return (invest) to its **investment pool** any unused funds until they are needed for other investment activities. Here, we may view the borrowing rate as a marginal cost of capital (k), as defined in (Eq. 10.6). Suppose that all available funds can be placed in investments yielding a return equal to r, the **lending rate**. We view these funds as an investment pool. The firm may withdraw funds from this investment pool for other investment purposes, but if left in the pool, the funds will earn at the rate of r (which is thus the opportunity cost). The MARR is thus related to either the borrowing interest rate or the lending interest rate. To illustrate the relationship among the borrowing interest rate, lending interest rate, and MARR, let us define the following:

$$
\begin{aligned}
k &= \text{Borrowing rate (or the cost of capital)} \\
r &= \text{Lending rate (or opportunity cost)} \\
i &= \text{MARR.}
\end{aligned}
$$

Generally (but not always), we might expect k to be greater than, or equal to, r. We must pay more for the use of someone else's funds than we can receive for "renting out" our own funds (unless we are running a lending institution). Then, we will find that the appropriate MARR would be found between r and k. The concept for developing a discount rate (MARR) under capital rationing will be understood best by a numerical example.

EXAMPLE 10.9 Determining an appropriate MARR as a function of budget

Sand Hill Corporation has identified six investment opportunities that will last 1 year. The firm draws up a list of all potentially acceptable projects. The list shows each project's required investment, projected annual net cash flows, and IRR. Then it ranks the projects according to their IRR, listing the highest IRR first.

| Project | Cash Flow | | IRR |
	A0	A1	
1	–$10,000	$12,000	20%
2	–10,000	11,500	15
3	–10,000	11,000	10
4	–10,000	10,800	8
5	–10,000	10,700	7
6	–10,000	10,400	4

Suppose $k = 10\%$, which remains constant for the budget amount up to $60,000 and $r = 6\%$. Assuming that the firm has available (a) $40,000, (b) $60,000, and (c) $0 for investments, what is the reasonable choice for the MARR in each case?

Solution

We will consider the following steps to determine the appropriate discount rate (MARR) under a capital rationing environment:

- Step 1: Develop the firm's cost of capital schedule as a function of the capital budget. For example, the cost of capital can increase as the amount of financing increases. Also determine the firm's lending rate if any unspent money is lent out or remains invested in its investment pool.

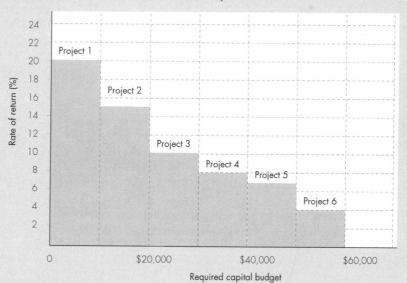

Figure 10.5 An investment opportunity schedule (Example 10.9)

- Step 2: Plot this investment opportunity schedule by showing how much money the company could invest at different rates of return as shown in Figure 10.5.

(a) If the firm has $40,000 available for investing, it should invest in projects 1, 2, 3, and 4. Clearly, it should not borrow at 10% to invest in either project 5 or 6. In these cases, the best rejected project is project 5. The worst accepted project is project 4. The opportunity cost for rejecting the best project is then 7%. Therefore, the MARR = 7% or $r < MARR < k$. (If you view the opportunity cost as the cost associated with accepting the worst project, the MARR could be 8%.)

(b) If the firm has $60,000 available, it should invest in projects 1, 2, 3, 4, and 5. It could lend the remaining $10,000 rather than invest these funds in project 6. For this new situation, we have $MARR = r = 6\%$.

(c) If the firm has no funds available, it probably would borrow to invest in projects 1 and 2. The firm might also borrow to invest in project 3, but it would be indifferent to this alternative, unless some other consideration was involved. In this case, $MARR = k = 10\%$; therefore, we can say that $r \leq MARR \leq k$ when we have a complete certainty about future investment opportunities. Figure 10.6 illustrates the concept of selecting a MARR under capital rationing.

Comments: In this example, for simplicity, we assumed that the timing of each investment is the same for all competing proposals, say period 0. If each alternative requires investments over several periods, the analysis will be significantly

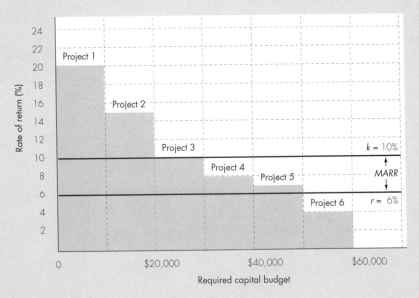

Figure 10.6 A choice of MARR under capital rationing (Example 10.9)

Now we can generalize what we have learned. If a firm finances investments through borrowed funds, it should use MARR = k; if the firm is a lender, it should use MARR = r. A firm may be a lender in one period and a borrower in another; consequently, the appropriate rate to use may vary from period to period. In fact, whether the firm is a borrower or a lender may well depend on its investment decisions.

In practice, most firms establish a MARR for all investment projects. Note the assumption that we made in Example 10.9: **Complete certainty** about investment opportunities was assumed. Under uncertain economic environments, generally, the MARR would be much greater than k, the firm's cost of capital. For example, if k = 10%, a MARR of 15% would not be considered excessive. Few firms are willing to invest in projects earning only slightly more than their cost of capital due to elements of *risk* in the project.

If the firm has a large number of current and future opportunities that will yield the desired return, we can then view the MARR as the minimum rate at which the firm is willing to invest, and we can also assume that proceeds from current investments can be reinvested to earn at the MARR. Furthermore, *if we choose the "do-nothing" alternative, all available funds are invested at the MARR.* In engineering economics, we also normally separate the risk issue from the concept of MARR. As we will show in Chapter 13, we treat the effects of risk explicitly when we must. Therefore, any reference to the MARR in this book refers strictly to the risk-free interest rate.

10.4 Capital Budgeting

In this section, we will examine the process of deciding whether projects should be included in the capital budget. In particular, we will consider decision procedures that should be applied when we have to evaluate a set of multiple investment alternatives for which we have a limited capital budget.

10.4.1 Evaluation of Multiple Investment Alternatives

In Chapter 5, we learned how to compare two or more mutually exclusive projects. Now, we shall extend the comparison techniques to a set of multiple decision alternatives which are not necessarily mutually exclusive. Here, we distinguish a **project** from an **investment alternative**, which is a decision option. For a single project, we have two investment alternatives: To accept or reject the project. For two independent projects, we can have four investment alternatives: (1) To accept both projects, (2) to reject both projects, (3) to accept only the first project, and (4) to accept only the second project. As we add interrelated projects, the number of investment alternatives to consider grows exponentially.

To perform a proper capital budgeting analysis, a firm must group all projects under consideration into decision alternatives. This grouping requires the firm to distinguish between projects that are independent of one another and those that are dependent on one another to formulate alternatives correctly.

Independent Projects

An **independent project** is one that may be accepted or rejected without influencing the accept-reject decision of another independent project. For example, the purchase of a milling machine, office furniture, and a forklift truck constitutes three independent projects. Only projects that are economically independent of one another can be evaluated separately. (Budget constraints may prevent us from selecting one or more of several independent projects; this external constraint does not alter the fact that the projects are independent.)

Dependent Projects

In many decision problems, several investment projects are related to one another such that the acceptance or rejection of one project influences the acceptance of others. The two such types of dependencies are as follows: Mutually exclusive projects and contingent projects. We say that two or more projects are **contingent** if the acceptance of one requires the acceptance of another. For example, the purchase of a computer printer is dependent upon the purchase of a computer, but the computer may be purchased without considering the purchase of the printer.

10.4.2 Formulation of Mutually Exclusive Alternatives

We can view the selection of investment projects as a problem of selecting a single decision alternative from a set of mutually exclusive alternatives. Note that each independent project is an investment alternative, but that a single investment alternative may entail a whole group of investment projects. The common method of handling various project relationships is to arrange the investment projects so that the selection decision involves only mutually exclusive alternatives. To obtain this set of mutually exclusive alternatives, we need to enumerate all of the feasible combinations of the projects under consideration.

Independent Projects

With a given number of independent investment projects, we can easily enumerate mutually exclusive alternatives. For example, in considering two projects, A and B, we have four decision alternatives, including a "do nothing" alternative.

Alternative	Description	X_A	X_B
1	Reject A, reject B	0	0
2	Accept A, reject B	1	0
3	Reject A, accept B	0	1
4	Accept A, accept B	1	1

In our notation, X_j is a decision variable associated with investment project j. If X_j = 1, project j is accepted; if X_j = 0, project j is rejected. Since the acceptance of one of these alternatives will exclude any other, the alternatives are considered to be mutually exclusive.

Mutually Exclusive Projects

Suppose we are considering two independent sets of projects (A and B). Within each independent set there are two mutually exclusive projects (A1, A2), and (B1, B2). The selection of either A1 or A2, however, is also independent of the selection of any project from the set (B1, B2). For this set of investment projects, the mutually exclusive alternatives are

Alternative	(X_{A1}, X_{A2})	(X_{B1}, X_{B2})
1	(0 , 0)	(0 , 0)
2	(1 , 0)	(0 , 0)
3	(0 , 1)	(0 , 0)
4	(0 , 0)	(1 , 0)
5	(1 , 0)	(1 , 0)
6	(0 , 1)	(1 , 0)
7	(0 , 0)	(0 , 1)
8	(1 , 0)	(0 , 1)
9	(0 , 1)	(0 , 1)

Note that, with two independent sets of two mutually exclusive projects, we can have nine different decision alternatives.

Contingent Projects

Suppose C is contingent on the acceptance of both A and B, and acceptance of B is contingent on acceptance of A. The possible number of decision alternatives can be formulated as follows:

Alternative	X_A	X_B	X_C
1	0	0	0
2	1	0	0
3	1	1	0
4	1	1	1

Thus, we can easily formulate a set of mutually exclusive investment alternatives with a limited number of projects that are independent, mutually exclusive, or contingent merely by arranging the projects in a logical sequence.

One difficulty with the enumeration approach is that, as the number of projects increases, the number of mutually exclusive alternatives increases exponentially. For example, for ten independent projects, the number of mutually exclusive alternatives is 2^{10}, or 1024. For 20 independent projects, 2^{20}, or 1,048,576, alternatives exist. As the number of decision alternatives increases, we may have to resort to mathematical programming to find the solution. Fortunately, in real-world business, the number of engineering projects to consider at any one time is usually manageable, so the enumeration approach is a practical one.

10.4.3 Capital Budgeting Decisions with Limited Budgets

Recall that capital rationing refers to situations where the funds available for capital investment are not sufficient to cover all the projects. In this situation, we enumerate all investment alternatives as before, but eliminate from consideration any mutually exclusive alternatives exceeding the budget. The most efficient way to proceed in a capital rationing situation is to select the group of projects that maximizes the total PE of future cash flows over required investment outlays. Example 10.10 illustrates the concept of an optimal capital budget under a rationing situation.

EXAMPLE 10.10

Four energy saving projects[7] under budget constraints

The facilities department at an electronic instrument firm had under consideration four energy-efficiency projects.

---

[7] The project descriptions (but not the analysis) are adapted from Khan AM, Fiorino D. "Case Study: The Capital Asset Pricing Model in Project Selection." *The Engineering Economist*, 1991. (Reprinted with permission.)

Project 1 (electrical): This project requires replacing the existing standard efficiency motors in the air conditioners and exhaust blowers of a particular building with high-efficiency motors.

Project 2 (building envelope): This project involves coating the inside surface of existing fenestration in a building with low-emissivity solar film.

Project 3 (air conditioning): This project requires the installation of heat exchangers between a building's existing ventilation and relief air ducts.

Project 4 (lighting): This project requires the installation of specular reflectors and delamping of a building's existing ceiling grid-lighting troffers.

These projects require capital outlays in the $50,000 to $140,000 range and have useful lives of about 8 years. The facilities department's first task was to estimate the annual savings that can be realized by these energy-efficiency projects. Currently, the company pays 7.80 cents per kilowatt-hour (kWh) for electricity and $4.85 per thousand cubic metres of air treated. Assuming that the current energy prices would continue for the next 8 years, the company has estimated the cash flows and the IRR for each project as follows.

Project	Investment	Annual O&M Cost	Annual Savings (Energy)	Annual Savings (Dollars)	IRR
1	$46,800	$1200	151,000 kWh	$11,778	15.43%
2	104,850	1050	513,077 kWh	40,020	33.48%
3	135,480	1350	6,700,000 m³	32,495	15.95%
4	94,230	942	470,740 kWh	36,718	34.40%

As each project could be adopted in isolation, at least as many alternatives as projects are possible. For simplicity assume that all projects are independent as opposed to being mutually exclusive, that they are equally risky, and that their risks are all equal to those of the firm's average existing assets.

(a) Determine the optimal capital budget for the energy savings projects.

(b) With $250,000 approved for energy improvement funds during the current fiscal year, the department did not have sufficient capital on hand to undertake all four projects without any additional allocation from headquarters. Enumerate the total number of decision alternatives and select the best alternative.

Discussion: The PE calculation cannot be shown yet, as we do not know the marginal cost of capital. Therefore, our first task is to develop the **marginal cost of capital (MCC)** schedule, a graph that shows how the cost of capital changes as more and more new capital is raised during a given year. The graph in Figure 10.7 is the company's marginal cost of capital schedule. The first $100,000 would be raised at 14%, the next $100,000 at 14.5%, the next $100,000 at 15%, and for

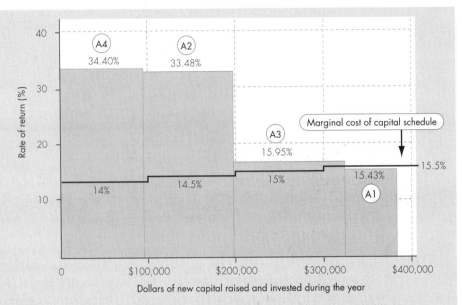

Figure 10.7 Combining the marginal cost of capital schedule and investment opportunity schedule curves to determine a firm's optimal capital budget (Example 10.10)

any amount over $300,000 at 15.5%. We then plot the IRR data for each project as the **investment opportunity schedule (IOS)** shown in the graph. The IOS schedule shows how much money the firm could invest at different rates of return.

	j	Alternative	Required Budget	Combined Annual Net Savings
Table 10.5 Mutually Exclusive Decision Alternatives (Example 10.10)	1	0	0	0
	2	A1	$ (46,800)	$ 10,578
	3	A2	(104,850)	38,970
	4	A3	(135,480)	31,145
	5	A4	(94,230)	35,776
	6	A4,A1	(141,030)	46,354
	7	A2,A1	(151,650)	49,548
	8	A3,A1	(182,280)	41,723
	9	A4,A2	(199,080)	74,746
	10	A4,A3	(229,710)	66,921
	11	A2,A3	(240,330)	70,115
Best alternative	⑫	A4,A2,A1	(245,880)	85,324
	13	A4,A3,A1	(276,510)	77,499
	14	A2,A3,A1	(287,130)	80,693
	15	A4,A2,A3	(334,560)	105,891
	16	A4,A2,A3,A1	(381,360)	116,469

Infeasible alternatives

Solution

(a) Optimal capital budget if projects can be accepted in part:

Consider project A4: Its IRR is 34.40%, and it can be financed with capital that costs only 14%. Consequently, it should be accepted. Projects A2 and A3 can be analyzed similarly; all are acceptable because the IRR exceeds the marginal cost of capital. Project A1, on the other hand, should be rejected because its IRR is less than the marginal cost of capital. Therefore, the firm should accept the three projects (A4, A2, and A3) which have rates of return in excess of the cost of capital that would be used to finance them if we end up with a capital budget of $334,560. This should be the amount of the *optimal capital budget*.

In Fig. 10.7, even though two rate changes occur in the marginal cost of capital in funding project A3 (first change from 14.5% to 15%, second change from 15% to 15.5%), the accept/reject decision for A3 remains unchanged, as its rate of return exceeds the marginal cost of capital. What would happen if the MCC cut through project A3? For example, suppose that the marginal cost of capital for any project raised above $300,000 would cost 16% instead of 15.5%, thereby causing the MCC schedule to cut through Project A3. Should we then accept A3? If we can take A3 in part, we would take on only part of it up to 74.49%.

(b) Optimal capital budget if projects cannot be accepted in part:

If projects can be accepted in part, the budget limit does not cause any difficulty in selecting the best alternatives. For example, in Fig. 10.7, we accept A4 and A2 in full and A3 in part as much as the $250,000 budget allows, or 37.58% of A3.

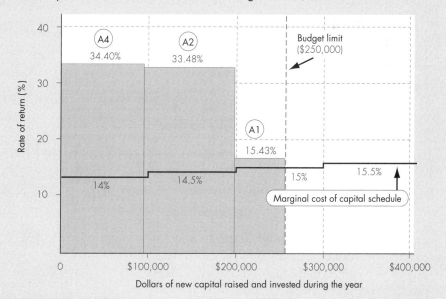

Figure 10.8 The appropriate cost of capital to be used in the capital budgeting process for decision alternative 12, with a $250,000 budget limit (Example 10.10) is 15%.

If projects cannot be funded partially, we first need to enumerate the number of feasible investment decision alternatives within the budget limit. As shown in Table 10.5, the total number of mutually exclusive decision alternatives that can be obtained from four independent projects is 16, including the "do-nothing" alternative. However, decision alternatives 13, 14, 15, and 16 are not feasible because of a $250,000 budget limit. So, we only need to consider alternatives 1 through 12.

Now, how do we compare these alternatives as the marginal cost of capital changes for each decision alternative? Consider again Fig. 10.7. If we take A1 first, it would be acceptable, because its 15.43% return would exceed the 14% cost of capital used to finance it. Why couldn't we do this? The answer is that we are seeking to maximize the excess of returns over costs, or the area above the MCC, but below the IOS. We accomplish this by accepting the most profitable projects first. This logic leads us to conclude that, as long as the budget permits, A4 should be selected first and A2 second. This will consume $199,080, which leaves us $50,920 in unspent funds. The question is what we are going to do with this left-over money. Certainly, it is not enough to take A3 in full, but we can take A1 in full. Full funding for A1 will fit the budget and the project's rate of return still exceeds the marginal cost of capital (15.43% > 15%). Unless the left-over funds earn more than 15.43% interest, alternative 12 becomes the best (Figure 10.8).

Comments: In Example 10.9, the MARR was found by applying a capital limit to the investment opportunity schedule. The firm was then allowed to borrow or lend the money as the investment situation dictates. In this example, no such borrowing is explicitly assumed.

10.5 Summary

- Methods of financing fall into two broad categories:

    1. **Equity financing** uses retained earnings or funds raised from an issuance of stock to finance a capital investment.

    2. **Debt financing** uses money raised through loans or by an issuance of bonds to finance a capital investment.

- Companies do not simply borrow funds to finance projects. Well-managed firms usually establish a **target capital structure** and strive to maintain the **debt ratio** when individual projects are financed.

- The **cost of capital** formula is a composite index reflecting the cost of funds raised from different sources. The formula is:

$$k = \frac{i_d C_d}{V} + \frac{i_e C_e}{V}, \qquad V = C_d + C_e$$

where i_e is the **cost of equity** and i_d is the after-tax **cost of debt**.

- The **marginal cost of capital** is defined as the cost of obtaining another dollar of new capital. The marginal cost rises as more and more capital is raised during a given period.

- Without a capital limit, the choice of MARR is dictated by the availability of financing information:

  1. In cases where the exact debt-financing and repayment schedules are known, we recommend the use of the net equity flow method. The proper MARR would be the cost of equity, i_e.

  2. If no specific assumption is made about the exact instruments that will be used to finance a particular project (but we do assume that the given capital structure proportions will be maintained), we may determine the after-tax cash flows without incorporating any debt cash flows. Then, we use the marginal cost of capital (k) as the proper MARR.

- Under conditions of capital rationing, the selection of MARR is more difficult, but generally the following possibilities exist:

Conditions	MARR
A firm borrows some capital from lending institutions at the borrowing rate, k, and some from its investment pool at the lending rate, r	$r < \text{MARR} < k$
A firm borrows all capital from lending institutions at the borrowing rate, k.	$\text{MARR} = k$
A firm borrows all capital from its investment pool at the lending rate, r.	$\text{MARR} = r$

- The cost of capital used in the capital budgeting process is determined at the intersection of the investment opportunity schedule (**IOS**) and the marginal cost of capital schedule (**MCC**). If the cost of capital at the intersection is used, then the firm will make correct accept/reject decisions, and its level of financing and investment will be optimal. This view assumes that the firm can invest and borrow at the rate where the two curves intersect.

- If a strict budget is placed in a capital budgeting problem and no projects can be taken in part, all feasible investment decision scenarios need to be enumerated. Depending upon each investment scenario, the cost of capital will also likely change. Our task is to find the best investment scenario in light of a changing cost of capital environment. As the number of projects to consider increases, we may eventually resort to a more advanced technique, such as a mathematical programming procedure.

- When dealing with multiple projects with various interdependencies, we need to organize them into a set of mutually exclusive investment alternatives that cover all feasible investment combinations.

- **Independent projects** are those for which acceptance or rejection of one project does not affect acceptance or rejection of the other(s). If only independent projects are being considered, the maximum number of investment alternatives is 2^x, where x = the number of independent projects.

- **Dependent projects** are those for which acceptance or rejection of one project affects acceptance or rejection of the other(s). Dependent projects can be:

  1. **Mutually exclusive**—accepting one project requires rejecting the other(s); or

  2. **Contingent**—accepting one project is contingent upon having first accepted another.

Problems

Note: *Assume all bond coupon payments are annual.*

Level 1

10.1* In order to finance a new project, a $1000 par value bond will be issued. The bond will be sold at discount with a market price of $970 and a coupon rate of 10%. The flotation costs for a new issue would be approximately 5%. The bond matures in 10 years and the corporate tax rate is 40%. Compute the after-tax cost of this debt financing (in percent).

10.2* The Northern Ontario Mining Company has the following capital structure.

Source	Amount
Debt	$665,000
Preferred stock	345,000
Common stock	1,200,000

The company wants to maintain the current capital structure in future project financing. If the company plans to use $800,000 debt for a new venture, what would be the scale of the total investment?

10.3* The capital structure for Okanagan Citrus Corporation is given as follows:

Sources	Amount
Long term bonds	$3,000,000
Preferred stock	2,000,000
Common stock	5,000,000

Assuming that the firm will maintain its capital structure in the future, determine the firm's weighted average cost of capital (k) if the firm has a 7.5% cost of debt (after-tax), a 12.8% cost of preferred stock, and a 20% cost of common stock (cost of equity).

10.4* Optical World Corporation, a manufacturer of peripheral vision storage systems, needs $10 million to market its new robotics-based vision systems. The firm is considering both financing options: common stock and bonds. If the firm decides to raise the capital through issuing common stock, the flotation costs will be 6% and the share price will be $25. If the firm decides to use debt financing, it can sell a 10-year, 12% bond with a par value of $1000. The bond flotation costs will be 1.9%.

(a) For equity financing, determine the flotation costs and the number of shares to be sold to net $10 million.

(b) For debt financing, determine the flotation costs and the number of $1000 par value bonds to be sold to net $10 million. What is the required annual interest payment?

10.5 Consider a project whose initial investment is $300,000, which must be financed at an interest rate of 12% per year. Assuming that the required repayment period is 6 years, determine the repayment schedule by identifying the principal as well as the interest payments for each of the following methods:

(a) Equal repayment of the principal
(b) Equal repayment of the interest
(c) Equal annual installments.

10.6* Calculate the after-tax cost of debt under each of the following conditions:

(a) Interest rate, 12%; tax rate, 25%
(b) Interest rate, 14%; tax rate, 34%
(c) Interest rate, 15%; tax rate, 40%.

10.7* Consider the 5 investment projects with project interdependencies as outlined below:

- (A1 and A2) are mutually exclusive.
- (B1 and B2) are mutually exclusive and

either project is contingent on the acceptance of A2.

- C is contingent on the acceptance of B1.

Assuming that "do-nothing" is not an option, which of the following statements is (are) correct?

(a) There is a total of 5 feasible decision alternatives.
(b) (A1, B2) is feasible.
(c) (A2, B1) and (A2, B2) are both feasible.
(d) (B1, C) is feasible.
(e) (B1, B2, C) is feasible.
(f) (A1, B1, C) is feasible.
(g) All of the above.
(h) None of the above.

10.8 Four investment proposals have been submitted by the Corporate Planning Committee. The cash flow profiles for the 4 investment proposals are summarized as follows: Each project has a 5-year service life. The firm's MARR is 10%. (All cash flows are given in after-tax figures.)

Project Profiles

	A	B	C	D
Initial investment	$60,000	$40,000	$80,000	$100,000
Annual net operating cash flow	20,000	10,000	15,000	25,000
Salvage value	0	20,000	20,000	30,000

(a) Suppose that the 4 projects are independent and there is no limit on the budget. Formulate all possible decision alternatives along with the composite cash flows.
(b) Suppose that projects A and B are mutually exclusive. Formulate all possible decision alternatives.

Level 2

10.9* The Beaver Corporation is considering three independent investments. The required investments and expected IRRs of these projects are as follows:

Investment Opportunity Schedule		
Project	Investment	IRR
A	$180,000	32%
B	250,000	18
C	120,000	15

The company intends to finance the projects by 40% debt and 60% equity. The after-tax cost of debt is 8% for the first $100,000, after which the cost will be 10%. Retained earnings (internally generated) in the amount of $150,000 are available for investment. The common stockholders' required rate of return is 20%. If new stock is issued, the cost, k_e, will be 23%. What would be the marginal cost of capital to raise the first $250,000 project cost?

10.10* Using the marginal cost of capital curve defined in Problem 10.9, decide which project(s) would be included.

(a) Project A only
(b) Projects A and B
(c) Projects A, B, and C
(d) None of the projects.

10.11* Based on the marginal cost of capital schedule and the investment opportunity schedule defined in Problem 10.9, what would be the reasonable MARR (or discount rate) to use in evaluating the investments?

(a) MARR < 17.8%
(b) 17.8% < MARR ≤ 18%
(c) 18% < MARR < 32%
(d) MARR = 15.2%.

10.12 Edison Power Company currently owns and operates a coal-fired combustion turbine plant, which was installed 20 years ago.

Making Turbines Efficient and Versatile

With improved technology, gas turbines can run on cheap, lower-quality fuels like coal or wood, or on wastes, and do so more efficiently and cleanly than existing gas turbines or coal-burning plants.

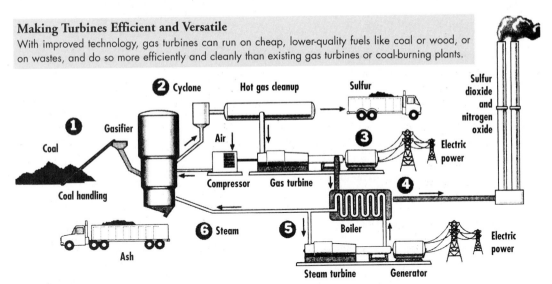

The Process

1. Coal or other hydrocarbon material is treated with steam under pressure, releasing combustible gases.
2. The gas is cleaned in a cyclone and with other processes, removing the sulfur.
3. Carbon monoxide and hydrogen are burned in a gas turbine, which spins a generator to make electricity and also powers the compressor.
4. Hot exhaust from the turbine goes to a boiler, which makes steam, and then goes up the stack.
5. The steam drives another turbine, which also turns a generator to make electricity.
6. The remaining steam and air from the compressor are sent to the gasifier to process more fuel into combusible gases.

Figure 10.9 A new coal-processing treatment that makes turbines efficient and versatile (Problem 10.12)

Because of degradation of the system, 65 forced outages occurred during the last year alone and two boiler explosions during the last 7 years. Edison is planning to scrap the current plant and install a new, improved gas turbine, which produces more energy per unit of fuel than typical coal-fired boilers (see Figure 10.9).

The 50-MW gas-turbine plant, which runs on gasified coal, wood, or agricultural wastes, will cost Edison $65 million. Edison wants to raise the capital from three financing sources: 45% common stock, 10% preferred stock (which carries a 6% cash dividend when declared), and 45% borrowed funds. Edison's investment banks quote Edison the following flotation costs:

Financing Source	Flotation Costs	Selling Price	Par Value
Common stock	4.6%	$32/share	$10
Preferred stock	8.1%	$55/share	$15
Bond	1.4%	$980	$1000

(a) What are the total flotation costs to raise $65 million?
(b) How many shares (both common and preferred) or bonds must be sold to raise $65 million?
(c) If Edison pays annual cash dividends of $2 per common share, and annual bond interest payments are at the rate of 12%, how much cash should Edison have available to meet both the equity

and debt obligation? (Note that whenever a firm declares cash dividends to its common stockholders, the preferred stockholders are entitled to receive dividends of 6% of par value.)

10.13 Sweeney Paper Company is planning to sell $10 million worth of long-term bonds with a 11% interest rate. The company believes that it can sell the $1000 par value bonds at a price that will provide a yield to maturity of 13%. The flotation costs will be 1.9%. If Sweeney's marginal tax rate is 35%, what is its after-tax cost of debt?

10.14 Mobil Appliance Company's earnings, dividends, and stock price are expected to grow at an annual rate of 12%. Mobil's common stock is currently traded at $18 per share. Mobil's last cash dividend was $1.00, and its expected cash dividend for the end of this year is $1.09. Determine the cost of retained earnings (k_r).

10.15 Refer to Problem 10.14. Mobil wants to raise capital to finance a new project by issuing new common stock. With the new project, the cash dividend is expected to be $1.10 at the end of the current year, and its growth rate is 10%. The stock now sells for $18, but new common stock can be sold to net Mobil $15 per share.

  (a) What is Mobil's flotation cost as a percentage?
  (b) What is Mobil's cost of new common stock (k_e)?

10.16 The Callaway Company's cost of equity is 22%. Its before-tax cost of debt is 13%, and its marginal tax rate is 40%. The firm's capital structure calls for a debt-to-equity ratio of 45%. Calculate Callaway's cost of capital.

10.17 Delta Chemical Corporation is expected to have the following capital structure for the foreseeable future.

Source of Financing	Percent of Total Funds	Before-Tax Cost	After-Tax Cost
Debt	30%		
Short-term	10	14%	
Long-term	20	12	
Equity	70%		
Common stock	55		30%
Preferred stock	15		12

The flotation costs are already included in each cost component. The marginal income tax rate (t) for Delta is expected to remain at 40% in the future. Determine the cost of capital (k).

10.18 DNA Corporation, a biotech engineering firm, has identified seven R&D projects for funding. Each project is expected to be in the R&D stage for 3 to 5 years, and the IRR figure represents the royalty income from selling the R&D results to pharmaceutical companies.

Project	Investment Type	Required IRR
1. Vaccines	$15 M	22%
2. Carbohydrate chemistry	25 M	40
3. Antisense	35 M	80
4. Chemical synthesis	5 M	15
5. Antibodies	60 M	90
6. Peptide chemistry	23 M	30
7. Cell transplant/ gene therapy	19 M	32

DNA Corporation can only raise $100 million. DNA's borrowing rate is 18%, and its lending rate is 12%. Which R&D projects should be included in the budget?

10.19 Gene Fowler owns a house that contains 20.2 square metres of windows and 4.0 square metres of doors. Electricity usage totals 46,502 kWh: 7960 kWh for lighting and appliances, 5500 kWh for water heating, 30,181 kWh for space heating to 20°C, and 2861 kWh for space cooling to 25°C. The 14 energy-savings alternatives have been suggested by the local power company for

Fowler's 162 square-metre home. Fowler can borrow money at 12% and lends money at 8%.

No.	Structural Improvement	Annual Savings	Estimated Costs	Payback Period
1	Add storm windows	$128–156	$455 –556	3.5 years
2	Insulate ceilings to R-30	149–182	408–499	2.7
3	Insulate floors to R-11	158–193	327–399	2.1
4	Caulk windows and doors	25–31	100–122	4.0
5	Weather-strip windows and doors	31–38	224–274	7.2
6	Insulate ducts	184–225	1677–2049	9.1
7	Insulate space heating water pipes	41–61	152–228	3.7
8	Install heat retardants on E, SE, SW, W windows	37–56	304–456	8.2
9	Install heat reflecting film on E, SE, SW, W windows	21–31	204–306	9.9
10	Install heat absorbing film on E, SE, SW, W windows	5–8	204–306	39.5
11	Upgrade 6.5 EER A/C to 9.5 EER unit	21–32	772–1158	36.6
12	Install heat pump water heating system	115–172	680–1020	5.9
13	Install water heater jacket	26–39	32–48	1.2
14	Install clock thermostat to reduce heat from 20°C to 16°C for 8 hours each night	96–144	88–132	1.1

Note: EER (Energy Efficiency Ratio). R-value indicates the degree of resistance to heat. The higher the number, the greater the insulating quality.

(a) If Fowler lives in the house for next 10 years, which alternatives would be selected with no budget constraint? Assume that his interest rate is 8%. Assume also that all installations would last 10 years. Fowler will be conservative in calculating the net present equivalent of each alternative (using the minimum annual savings at the maximum cost).

(b) If he wants to limit his energy savings investments to $1800, which alternatives should he include in his budget?

10.20 Consider the following set of investment projects, each having a service life of 10 years.

Projects	First Cost	Net Annual Revenue	Net Salvage Value
A1	$10,000	$2,000	$ 1,000
A2	12,000	2,100	2,000
B1	20,000	3,100	5,000
B2	30,000	5,000	8,000
C	35,000	4,500	10,000

- A1 and A2 are mutually exclusive.
- B1 and B2 are mutually exclusive.

With a budget limit of $50,000, which investment projects should be selected? Assume that the firm's MARR is 8%.

10.21* Consider the following investment projects:

n	Project Cash Flows			
	A	B	C	D
0	–$2000	–$3000	–$1000	
1	1000	4000	1400	–$1000
2	1000		–100	1090
3	1000			
i^*	23.38%	33.33%	32.45%	9%

Suppose that you have only $3500 available at period 0. Neither additional budgets nor borrowing are allowed in any future budget period. However, you can lend out any remaining funds (or available funds) at 10% interest per period.

(a) If you want to maximize the future worth at period 3, which projects would you select? What is the future worth (the total amount available for lending at the end of period 3)? No partial projects are allowed.

(b) Suppose in (a) that, at period 0, you are allowed to borrow $500 at an interest rate of 13%. The loan has to be repaid at the end of year 1. Which project would you select to maximize your future worth at period 3?

(c) Considering a lending rate of 10% and a borrowing rate of 13%, what would be the reasonable MARR for project evaluation?

10.22 Consider the following set of investment projects.

Project	Initial Investment	PE(10%)
A	$100	55
B	200	63
C	400	95
D	300	70
E	150	60
F	250	75

- Projects A and E are mutually exclusive.
- Project B is contingent upon project A.
- Project D is contingent upon project E.
- Project C is contingent upon project A.
- Project F is an independent project.
- Projects B and D are mutually exclusive.

(a) Formulate all mutually exclusive alternatives.

(b) With no restriction on budget, which alternative is the best?

10.23* Consider the following set of investment projects.

Project	Initial Investment	PE(10%)
A	$400	65
B	550	70
C	620	95
D	580	75
E	380	60
F	600	50

- Projects A and B are mutually exclusive.
- Project C is contingent upon project A.
- Project D is contingent upon project C.
- Project F is contingent upon project B.
- Project E is contingent upon project F.

With no capital limit, list all combinations of decision alternatives and then find the best decision alternative based on the present worth criterion.

10.24 Consider the following set of investment projects.

Project	Initial Investment	PE(10%)
A	$500	570
B	700	630
C	900	980
D	650	770
E	600	650
F	750	820

- Projects A and E are mutually exclusive.
- Project B is contingent upon project F.
- Project D is contingent upon project E.
- Project C is contingent upon project B.
- Projects B and D are mutually exclusive.

(a) With no capital limit, list all combinations of decision alternatives.

(b) Find the best decision alternative based on the present equivalent criterion.

10.25 Eight investment proposals have been identified for cost reduction in a certain metal fabricating plant. All have lives of ten years, zero salvage values, and no disposal tax effects. The required investment and the estimated after-tax reduction in annual disbursements are given for each proposal. Gross rates of return are also shown for each proposal: Proposals A1, A2, A3, and A4 are mutually exclusive because they are alternative ways of changing operation A. Similarly B1, B2, and B3 are mutually exclusive.

Proposal	Required Investment	Annual Savings	Rate of Return
A1	$20,000	$ 5,600	25%
A2	30,000	6,900	19
A3	40,000	8,180	16
A4	50,000	10,070	15
B1	10,000	1,360	6
B2	20,000	3,840	14
B3	30,000	5,200	12
C	20,000	5,440	24

(a) Which proposals would you select if it is stated that the MARR is 10% after taxes and if there is no limitation on available funds, using the rate of return criterion?

(b) Which proposals would you select at a MARR of 10% using the present equivalent on total investment?

Level 3

10.26 A chemical plant is considering the purchase of a computerized control system. The initial cost is $200,000 and will produce net savings of $100,000 per year. If purchased, the system will be written off at a declining balance CCA rate of 30%. This system will be used for 4 years, at the end of which time the firm expects to sell the system for $30,000. The firm's marginal tax rate on this investment is 35%. The firm will finance the purchase either through its retained earnings or by borrowing from a local bank. Two commercial banks are willing to lend the $200,000 at an interest rate of 10%, but they have different repayment plans. Bank A requires 4 equal annual principal payments with interest calculated based on the unpaid balance.

Bank A's Repayment Plan

End of Year	Principal	Interest
1	$50,000	$20,000
2	50,000	15,000
3	50,000	10,000
4	50,000	5,000

Bank B offers a payment plan that extends over 5 years with 5 equal annual payments.

Bank B's Repayment Plan

End of Year	Principal	Interest	Total
1	$32,759	$20,000	$52,759
2	36,035	16,724	52,759
3	39,638	13,121	52,759
4	43,602	9,157	52,759
5	47,962	4,796	52,759

(a) Determine the cash flows if the computer-control system is bought through the company's retained earnings (equity financing).

(b) Determine the cash flows if the asset is financed through either bank A or bank B.

(c) Recommend the best course of financing action. (Assume that the firm's MARR is also known to be 10%.)

10.27 Maritime Textile Company is considering the acquisition of a new knitting machine at a cost of $200,000. Because of a rapid change in fashion styles, the need for this particular machine is expected to last only 5 years, after which the machine is expected to have a salvage value of $50,000. The annual operating cost is estimated at $10,000. The addition of this machine to the current production facility is expected to generate an additional revenue of $90,000 annually. The declining balance CCA rate for this asset is 30%. The income tax rate applicable for Maritime is 36%. The initial investment will be financed with 60% equity and 40% debt. The before-tax debt interest rate, which combines both short-term and long-term financing, is 12%, with the loan to be repaid in equal annual installments. The equity interest rate (i_e), which combines both sources of common and preferred stocks, is 18%.

(a) Evaluate this investment project by using net equity flows.

(b) Evaluate this investment project by using k.

10.28 The Huron Development Company is considering buying an overhead pulley system. The new system has a purchase price of $100,000, an estimated useful life of 5 years, and an estimated salvage value of $30,000. It is expected to allow the company to economize on electric power usage, labor, and repair costs, as well as to reduce the number of defective products. A total annual savings of $45,000 will be realized if the new pulley system is installed. The company is in the 30% marginal tax bracket, and the system is subject to a 30% CCA rate. The initial investment will be financed with 40% equity and 60% debt. The before-tax debt interest rate, which combines both short-term and long-term financing, is 15%, with the loan to be repaid in equal annual installments over the project life. The equity interest rate (i_e), which combines both sources of common and preferred stocks, is 20%.

(a) Evaluate this investment project by using net equity flows.
(b) Evaluate this investment project by using k.

10.29 Consider the following mutually exclusive machines.

	Machine A	Machine B
Initial investment	$40,000	$60,000
Service life	6 years	6 years
Salvage value	$4,000	$8,000
Annual O&M cost	$8,000	$10,000
Annual revenues	$20,000	$28,000
CCA rate	30%	30%

The initial investment will be financed with 70% equity and 30% debt. The before-tax debt interest rate, which combines both short-term and long-term financing, is 10%, with the loan to be repaid in equal annual installments over the project life. The equity

interest rate (i_e), which combines both sources of common and preferred stock, is 15%. The firm's marginal income tax rate is 35%.

(a) Compare the alternatives using $i_e = 15\%$. Which alternative should be selected?
(b) Compare the alternatives using k. Which alternative should be selected?
(c) Compare the results obtained in (a) and (b).

10.30 Anglo Chemical Corporation (ACC) is a multinational manufacturer of industrial chemical products. ACC has made great progress in energy-cost reduction and has implemented several cogeneration projects in the United States and Canada, including the completion of a 35 megawatt (MW) unit in Chicago and a 29 MW unit at Baton Rouge. The division of ACC being considered for one of its more recent cogeneration projects is a chemical plant located in Sarnia. The plant has a power usage of 80 million kilowatt hours (kWh) annually. However, on the average, it uses 85% of its 10 MW capacity, which would bring the average power usage to 68 million kWh annually. Ontario Hydro presently charges $0.09 per kWh of electric consumption for the ACC plant, a rate that is considered high throughout the industry. Because ACC's power consumption is so large, the purchase of a cogeneration unit is considered to be desirable. Installation of the cogeneration unit would allow ACC to generate their own power and to avoid the annual $6,120,000 expense to Ontario Hydro. The total initial investment cost would be $10,500,000: $10,000,000 for the purchase of the power unit itself, a gas fired 10 MW Allison 571, and engineering, design and site preparation. The remaining $500,000 includes the purchase of interconnection equipment, such as poles and distribution lines that will be used to interface the cogenerator with the existing utility facilities. ACC is considering two financing options:

- ACC could finance $2,000,000 through the manufacturer at 10% for 10 years and will finance the remaining $8,500,000 through issuing common stock. The flotation cost for a common stock offering is 8.1%, and the stock will be priced at $45 per share.
- Investment bankers have indicated that 10-year 9% bonds could be sold at a price of $900 for each $1000 bond. The flotation costs would be 1.9% to raise $10.5 million.

(a) Determine the debt-repayment schedule for the term loan from the equipment manufacturer.

(b) Determine the flotation costs and the number of common shares to be sold to raise the $8,500,000.

(c) Determine the flotation costs and the number of $1000 par value bonds to be sold to raise $10.5 million.

10.31 (Continuation of Problem 10.30) As ACC management has decided to raise the $10.5 million by selling bonds, the company's engineers have estimated the operating costs of the cogeneration project.

The annual cash flow comprises many factors: maintenance, standby power, overhaul costs, and other miscellaneous expenses. Maintenance costs are projected to be approximately $500,000 per year. The unit must be overhauled every 3 years at a cost of $1.5 million. Miscellaneous expenses, such as additional personnel and insurance, are expected to total $1 million. Another annual expense is that for standby power, which is the service provided by the utility in the event of a cogeneration unit trip or scheduled maintenance outage. Unscheduled outages are expected to occur four times annually, each outage averaging 2 hours in duration at an annual expense of

$6400. Overhauling the unit takes approximately 100 hours and occurs every 3 years, requiring another triennial power cost of $100,000. Fuel (spot gas) will be consumed at a rate of 8.44 megajoules per kWh, including the heat recovery cycle. At $1.896 per gigajoule, the annual fuel cost will reach $1,280,000. Due to obsolescence, the expected life of the cogeneration project will be 12 years, after which Allison will pay ACC $1 million for salvage of all equipment.

Revenue will be realized from the sale of excess electricity to the utility company at a negotiated rate. Since the chemical plant will consume on average 85% of the unit's 10 MW output, 15% of the output will be sold at $0.04 per kWh, bringing in an annual revenue of $480,000. ACC's marginal tax rate (combined federal and provincial) is 36%, and the minimum required rate of return for any cogeneration project is 27%. The anticipated costs and revenues are summarized as follows:

Initial investment	
Cogeneration unit, engineering, design, and site preparation (CCA rate = 10%)	$10,000,000
Interconnection equipment (CCA rate = 30%)	500,000
Salvage after 12-year's use	
Cogeneration unit	975,000
Interconnection equipment	25,000
Annual expenses	
Maintenance	500,000
Misc. (additional personnel and insurance)	1,000,000
Standby power	6,400
Fuel	1,280,000
Other operating expenses	
Overhaul every 3 years	1,500,000
Standby power during overhaul	100,000
Revenues	
Sale of excess power to Ontario Hydro	480,000

If the cogeneration unit and other connecting equipment could be financed by issuing corporate bonds at an interest rate of 9%, compounded annually, with the flotation expenses as indicated in Problem 10.30, determine the net cash flow from the cogeneration project.

10.32 Ten potential investments in energy conservation have been identified for possible funding. All have lives of 10 years and are assumed to have 100% resale value after 10 years. All investments are made in nondepreciable assets. The required investment, the estimated after-tax savings in annual expense, and the rates of return on total investment are also shown for each proposal. The proposals related to a particular activity are identified by the same letter and they are mutually exclusive.

Proposal	Required Investment	Annual Savings	Rate of Return
A1	$20,000	$5,000	25%
A2	30,000	5,700	19
A3	40,000	8,400	16
A4	50,000	9,300	18.6
B1	10,000	600	6
B2	20,000	2,800	14
B3	30,000	4,800	16
C1	20,000	4,800	24
C2	60,000	10,200	17
D	5,000	700	14

(a) List all possible decision alternatives.
(b) At a MARR of 12%, which alternative would you select?

10.33 National Food Processing Company is considering investments in plant modernization and plant expansion. These proposed projects would be completed in 2 years, with varying requirements of money and plant engineering. Although some uncertainty exists in the data, management is willing to use the following data in selecting the best set of proposals. The resource limitations are as follows:

- First-year expenditures: $450,000
- Second-year expenditures: $420,000
- Engineering hours: 11,000 hours

No.	Project	Investment Year 1	Year 2	IRR	Engineering Hours
1	Modernize production line	$300,000	0	30%	4000
2	Build new production line	100,000	$300,000	43%	7000
3	Provide numerical control for new production line	0	200,000	18%	2000
4	Modernize maintenance shops	50,000	100,000	25%	6000
5	Build raw material processing plant	50,000	300,000	35%	3000
6	Buy present subcontractor's facilities for raw material processing	200,000	0	20%	600
7	Buy a new fleet of delivery trucks	70,000	10,000	16%	0

The situation requires that a new or modernized production line be provided (projects 1 or 2). The optional numerical control (Project 3) is applicable only to the new line. The company obviously does not want to both buy (Project 6) and build (Project 5) raw material processing facilities; it can, if desirable, rely on the present supplier as an independent firm. Neither the maintenance shop project (Project 4) nor the delivery-truck purchase (Project 7) is mandatory.

(a) Enumerate all possible mutually exclusive alternatives without considering the budget and engineering-hour constraints.
(b) Identify all feasible mutually exclusive alternatives given the budget and engineering–hour constraints.
(c) Suppose that the firm's marginal cost of capital will be 14% for raising the required capital up to $1 million. Which projects would be included in the firm's budget?

CHAPTER 11

Income Tax Effects on Some Specific Types of Investment Decisions

Jim and Patti Johnston are recent engineering graduates who have good jobs with a large telecommunications company. They have disposable income that they wish to invest wisely to provide some short-term returns. This extra income can help finance their plan for an around-the-world trip in three years' time. In addition, they are aware that even though retirement is 30 to 40 years away, they should start putting money away for it now. Obviously, Jim and Patti could meet both these goals by placing money in savings accounts. However, they also believe that they should invest a portion of their savings in stocks and bonds that can provide better returns. How do income taxes affect the returns on their investments? Can Jim and Patti avoid paying income tax on money that they are saving for retirement? In this chapter, we consider personal investments including income tax effects with respect to bonds, stocks, and registered retirement savings plans (RRSPs). Registered education savings plans (RESPs) will also be discussed.

To this point of the book, the acquisition of an asset has always equated to ownership. In some instances money has to be borrowed to finance all or part of the asset because of the magnitude of the initial cost of the asset. The purchaser does have the freedom of retaining or disposing of the asset as dictated by business considerations because it is owned. Can assets be acquired for use without owning them? The answer is yes, many kinds of fixed assets can be leased. Rapid obsolescence of the asset was once a primary reason for entering into a lease. More and more companies are now requesting leases that offer economic flexibility, whether in the form of early cancellation or of upgrades, which places the burden of ownership on the lessor. Another reason for the propensity for leasing shown by many companies is that to them leasing is an inexpensive way of acquiring capital assets for use without the initial capital outlay. In other words, leasing can be considered another source of financing in addition to equity and debt as we discussed earlier. In this chapter we will include income tax considerations in making lease-versus-buy decisions.

Replacement analysis was discussed in Chapter 6 on a before-tax basis. In most real-world situations, income taxes cannot be eliminated from such analysis even though they add considerable complexity. The replacement analysis section in this chapter incorporates tax effects into the kinds of replacement analysis topics covered in Chapter 6.

11.1 Personal Investments

With today's Internet technology, it is easy for individual investors to access information related to investment decision making. Purchasing and selling stocks, bonds, and mutual funds have never been easier. Information on public companies is available on line. With one mouse click, you can complete a transaction of purchasing any number of shares of stocks, bonds, or mutual funds.

As discussed in Chapter 8, individual income tax rates are progressive and certain types of income receive preferential tax treatment. Such treatment is offered as an incentive for individuals to invest, not only for personal gain, but also as a way of fostering overall economic growth. For example, capital gains are now taxed at ¾ of their actual value. Prior to 1994, each individual had a lifetime capital gains exemption of $100,000, that is, the first $100,000 of capital gains was tax-free. However, the $100,000 lifetime capital gains exemption is no longer available for capital property sold after February 22, 1994. Dividends from Canadian corporations also end up being taxed at a lower rate because of a special tax credit.

The tax payable on money from an investment in a given calendar year is based on the money or benefit realized in that year and is independent of the point in the year when it was received. For example, $100 of interest received on January 1st has the same tax implication as $100 of interest received on December 31st of the same year. By receiving $100 of interest on January 1st, you may choose to spend it or make another investment with it. In the latter case, this second investment may also earn income that results in taxes. These taxes are attributable to the second investment. *In evaluating personal investments in this book, we will assume that the tax owed as a result of an investment is paid at the end of the calendar year unless specified otherwise.*

11.1.1 Investment in Bonds

Bond terminology and bond evaluations are discussed on a before-tax basis in Section 3.8. There are two ways for an investor to earn income from a bond—interest and a capital gain on disposal. Most government bonds (Canada Savings Bonds, Alberta Capital Bonds, etc.) can only be purchased and redeemed at their face value. Therefore, there is no possibility of a capital gain (or loss) for these bonds. On the other hand, corporate bonds and certain government bonds are publicly traded and the market price of such a bond is determined by its coupon rate relative to the prevailing money market interest rates and the perceived credibility and financial stability of the issuing company. If the bond coupon rate is greater than the prevailing money market interest rate, investors *may* pay a premium over the face value of the bond to obtain the higher coupon rate. When the bond coupon rate is less than the prevailing money market rate, investors *may* pay less than the face value. Evaluation of bond investments, including tax effects, is based on the usual after-tax cash flow analysis.

EXAMPLE 11.1
$

Evaluation of bond investment

A corporate bond with a face value of $1000 was purchased for this amount at its January 1st time of issue. It is a simple interest bond that pays 6% interest on June 30th of each year. Determine the required selling price on December 31st after three years of ownership in order to realize an annual 10% after-tax rate of return. Taxes are paid at the end of each year. The marginal tax rates on interest income and capital gains are 40% and 30% respectively.

Solution

Given: $1,000 6% bond, purchased at face value
Find: The selling price, S, after 3 years to provide a 10% after-tax rate of return

$$\text{Annual interest payment} = \$1000 \times 0.06 = \$60/\text{year on June 30.}$$
$$\text{Income tax on interest} = 0.4 \times \$60 = \$24/\text{year at the end of year.}$$
$$\text{Capital gain on sale} = S - \$1000$$
$$\text{Capital gains tax} = 0.3\,(S - \$1000) \text{ at the end of year.}$$

The after-tax cash flow diagram for this investment is given in Figure 11.1. The required rate of return = 10% on 12 month basis, which is $[(1.10)^{1/2} - 1]\ 100\% = 4.88\%$ on a 6 month basis.

$$PE = -\$1000 + \$60(P/A, 10\%, 3)(F/P, 4.88\%, 1) - \$24(P/A, 10\%, 3)$$
$$+ [S - 0.3(S - \$1000)](P/F, 10\%, 3)$$
$$= 0$$

Solving the above equation for S yields

$$S = \$1288.80$$

Comments: This means that selling the bond at $1288.80 will yield a 10% after-tax rate of return. If the selling price is higher than $1288.80, the after-tax rate of return will be higher than 10%.

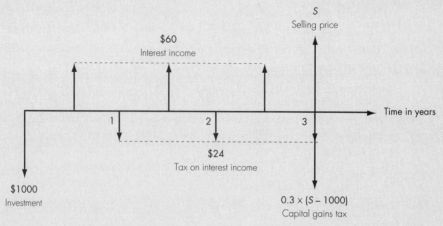

Figure 11.1   Cash flow diagram (Example 11.1)

EXAMPLE 11.2 $ How much to pay for a bond

A publicly traded bond series has a coupon rate of 9%, a face value of $1000, and a maturity date of December 31, 2005. The interest is paid semiannually. Assume that today's prevailing coupon rate for bonds with similar quality is 7% with interest paid semiannually. How much would an investor be willing to pay to buy this bond on January 1, 2000? Assume that the investor's marginal tax rate is 45%.

Solution

Given: $1,000 9% bond, semiannual interest payment
Find: The purchase price, P, which provides a return equal to that of comparable quality bonds
The bond under consideration pays a higher interest rate than newly issued bonds of similar quality. Thus, investors would be willing to pay more than its face value for

this bond for the higher interest rate. The value of the bond on January 1, 2000 is determined by its earning potential if it is bought and kept to maturity. If the bond is bought for a premium and later redeemed for its face value, there will be a capital loss. We assume that there are other capital gains to offset the capital loss when the bond is redeemed at maturity in year 2005.

$$\text{Interest cash flow} = \$1000 \times 4.5\%$$
$$= \$45 \text{ every 6 months.}$$

$$\text{Annual tax payment on interest income} = \$45 \times 2 \times 45\%$$
$$= \$40.50 \text{ every year.}$$

$$\text{Tax saving on capital loss at maturity} = 0.75 \times 45\% \times (P - \$1000)$$
$$= 0.3375\,P - \$337.5.$$

The prevailing bond interest rate is 7% per annum compounded semiannually. The corresponding after-tax rate of return is 7%(1 − 0.45) = 3.85% per annum. The semi-annual rate of return is then 3.85%/2 = 1.925% and the annual effective interest rate is $i_a = (1+1.925\%)^2 - 1 = 3.887\%$.

The after-tax cash flow diagram shown in Figure 11.2 is used to find the present value of the bond under consideration. By setting the present equivalent of all cash flows in the after-tax cash flow diagram equal to zero, we obtain the following equation:

$$\begin{aligned} PE = {} & -P - \$40.50(P/A, 3.887\%, 6) + \$45(P/A, 1.925\%, 12) \\ & + (\$1000 + 0.3375\,P - \$337.5)(P/F, 1.925\%, 12) \\ = {} & 0. \end{aligned}$$

Solving for P, we find P = $1082.73.

Comments: If an investor pays $1082.73 to buy the bond, he or she would earn the same rate of return as a newly issued bond series on January 1, 2000. If such a bond can be acquired at a price lower than the calculated $1082.73, the rate of return would be higher than the prevailing bond investments.

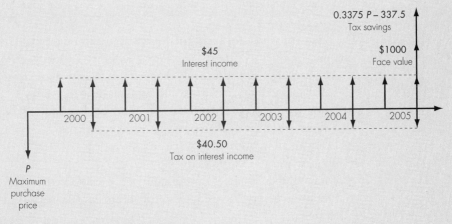

Figure 11.2 Cash flow diagram (Example 11.2)

11.1.2 Investment in Stocks

Stock investment provides an investor with equity ownership in a company. In the case of public companies, stock shares can be purchased and sold through a broker. With the Internet technology, the information on stock prices, dividends, and companies is available online. Individual investors can also purchase and sell stocks with a single mouse click on the Internet. Unlike bonds, which have a guaranteed redemption value, i.e., the face value, at a specified redemption date, the price of stock is determined by the market and reflects general market conditions and investors' perception of the company's present and future performance. Most of the terms and concepts used in this section are defined in Appendix A.

Information on Stock Prices

The prices of publicly traded stocks are listed in all major newspapers. They are also available at many Websites on the Internet. For example, the Website of the Toronto Stock Exchange (www.tse.com) provides 15 minute delayed quotes on all stocks traded at the exchange. Each stock has a stock symbol. You can search the name of a company to find its stock symbol. Prices and dividends for the common shares of a few well-known Canadian companies are listed below.

Company (Stock Symbol)	Price per Share on Dec. 3, 1999	Annual Dividend per Share (Payment Frequency)
Air Canada (AC)	$10.15	0
Alberta Energy (AEC)	$42.85	$0.40 (annual)
Bell Canada International (BI)	$25.75	0
CIBC (CM)	$33.00	$1.20 (quarterly)
Imperial Oil (IMO)	$34.80	$0.78 (quarterly)
Nortel Networks (NT)	$120.30	$0.22 (quarterly)
Power Corp (POW)	$27.80	$0.50 (quarterly)
Westcoast Energy (W)	$23.50	$1.28 (quarterly)

The quotes provided for each stock at www.tse.com often include the following information:

- Last traded price and the net change from the previous day's closing price
- Last bid price and size (price and number of shares to buy)
- Last ask price and size (price and number of shares to sell)
- Volume (number of shares that have changed hands)
- Daily opening price, high price and low price
- Highest and lowest prices during the past 52 weeks
- Total number of shares outstanding, i.e., owned by the public
- Quoted market value (product of the number of shares outstanding and the closing price)

- Dividend yield (annual dividends divided by the closing price expressed as a percentage)
- Dividend timing indicator (Q for quarterly payment, A for annual payment, etc.)
- P/E ratio (stock price divided by the annual earnings per share)
- Earnings per share.

Major Financial Statements

Suppose that you have found all information on the stock price of a company. To assess the growth potential of the stock, you need to read its financial reports. All public companies must provide the following financial statements at least once a year:

1. A balance sheet describing the financial position of the company at a specific date

2. An income statement showing revenues and costs over a financial period

3. A cash flow statement showing the sources and the uses of funds over a financial period.

These financial statements are prepared or verified by an independent auditor. Any reservations the auditor may have about these statements are also stated in the report.

These financial statements, together with the complete annual report, are the most important documents that potential investors should look into to determine the attractiveness of the company for investment purposes. Appendix A describes in detail these financial statements and provides the balance sheet, consolidated income statement and consolidated statement of cash flows of the Gillette Company. In the following, we will use these sample statements to calculate some indicators of the company's operation.

The **working capital ratio** is defined as the ratio between the company's current assets and its current liabilities. This ratio shows the company's ability to meet its short-term debt obligations. The excess of current assets over current liabilities is called the **net working capital**. Using the data given in Exhibit A.1 we find that in 1998, Gillette had a working capital ratio of $5440M/$3478M = 1.56. The adequacy of this ratio depends on the company's type of business, composition of its current assets, its inventory turn-over rate, and credit terms. Generally speaking, a working capital ratio of 2 to 1 is considered adequate[1]. If it is too high, it may show that the management lacks ability to put cash resources to profitable use. If it is too low, the company may not be able to pay its short term debts. From Exhibit A.1, we also see that $508M (=$5529 − $5021) of the total earnings from 1998 was retained and invested in the business. The total stockholders' equity had decreased, while the long-term debt had increased from 1997 to 1998.

A consolidated income statement is given in Exhibit A.2. From this state-

---

[1] Canadian Securities Institute, Investment Terms and Definitions: A guide to Canadian stocks, bonds and other securities —plus an investment glossary. 1985.

ment, we can see that the Gillette Company's **net profit margin** in 1998 was equal to (net income)/(net sales) = $1081M/$10,056M = 11%. This ratio shows the efficiency of the company's management if it is compared to other companies in similar industry. The **income per common share** (basic) increased from $0.85 in 1996 to $0.96 in 1998. This shows how well the investors' money was utilized to generate income by the management.

The cash flow statement in Exhibit A.3 shows that the net cash position remained relatively constant from 1997–1998 ($105M in 1997 and $102M in 1998). This indicates that the net cash provided from operations are mostly spent in investment activities and financing activities. The investment in property, plant and equipment and the dividend payments had been steady and growing over the three years.

Judging the Investment Quality of Stocks

To evaluate the quality of stocks, one needs to look at the company's track record and/or the market forecast for the industry the company is in. For example, some high technology stocks do not have a steady track record. However, they may be in high demand simply because of the potential of these companies. The following factors should be considered when judging the investment quality of stocks:

1. The business that the company is in. Is there a continuing or growing demand for the products and/or services that the company is providing? Are there changes in consumer preferences? Are there tighter government controls or incentives? Are the industries using this company's products or services growing?

2. The company's historical growth: Examine the dividend payments, net profit margins, returns on shareholder's equity, and sales growth.

3. The company's current financial situation: Examine debt ratio, working capital ratio, etc.

4. The company's current management team. Has the management team been changed lately? What are the philosophies of the management?

5. The investment projects that the company is developing: New product development, product or service diversification, cost reduction projects, brand-name product development, and so on.

Another important factor is the demand for the stock of a certain company. This depends on the perception or the expectation of the public on the price and/or earning potential of the stock of a company. This can be reflected in the **price/earnings ratio**. If this ratio is higher than usual, it indicates on one hand that the stock price is too high and on the other hand there is a high demand on this stock. Sometimes demand for and investor interest in a certain stock drives its price to a ridiculous level. That is why investment in stocks has risks and there may be a speculative nature in it. Some of the Internet related stocks in the late 1990s were chased by the public purely based on potential, without any track record.

Cash Flow Analysis of Stock Investment

Stock investments with cash flow predictions or estimates may be analyzed in a way similar to bond investment. This is also based on after-tax cash flow analysis. We use an example to illustrate.

EXAMPLE 11.3 $

After-tax rate of return calculation of stock investment

You purchased a newly issued stock on January 1, 1993 for $20 per share. The shares have paid quarterly dividends of 25¢/share since their acquisition. You sold this stock on January 1, 2000 for $29/share. Determine the annual after-tax rate of return on this investment if dividends are taxed at 28% and capital gains are taxed at 30%. Taxes are payable on December 31 each year.

Solution

Given: Stock transactions as described above
Find: The annual after-tax rate of return on the stock

$$\text{Annual taxes on dividends} = 0.28 \times (4 \text{ quarters} \times \$0.25/\text{quarter})$$
$$= \$0.28.$$

$$\text{Capital gains} = \$29 - \$20 = \$9 \text{ on January 1, 2000.}$$

$$\text{Tax on capital gains} = 9 \times 0.3 = \$2.7 \text{ on December 31, 2000.}$$

The after-tax cash flow diagram for this investment is shown in Figure 11.3.

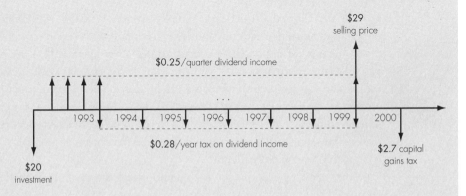

Figure 11.3 Cash flow diagram (Example 11.3)

Let i represent the effective quarterly after-tax rate of return. To find i, we need to solve the following equation:

$$PE = -\$20 + \$0.25(P/A, i, 28) - \$0.28(A/F, i, 4)(P/A, i, 28)$$
$$+ \$29(P/F, i, 28) - \$2.7(P/F, i, 32)$$
$$= 0.$$

Using the interest factor tables in the Appendix and linear interpolation we can find the interest rate i.

If $i = 1.75\%$, then $PE = + \$0.29$
If $i = 2\%$, then $PE = - \$0.90$.

By linear interpolation, we find that $PE = 0$ at

$$i = 1.75\% + \frac{2\% - 1.75\%}{0.29 + 0.90} \times 0.29$$
$$= 1.81\% \text{ per quarter.}$$

The effective annual after-tax rate of return is

$$i_a = (1 + i_{quarter})^4 - 1 = (1 + 1.81\%)^4 - 1 = 7.44\%.$$

Comments: As described in Section 8.5.2, taxes on dividends are actually calculated by adding 125% of the value of the dividends in income and then subtracting 13.3% of the value of the dividends from the total tax owed. However, using a single dividend tax rate as shown in Table 8.5 is often sufficient for evaluating investment alternatives.

11.1.3   Registered Retirement Savings Plans (RRSPs)

One of the allowed deductions for tax purposes is RRSP contributions. This tax deductibility means that contributions are made with before-tax dollars. Furthermore, income earned on these contributions is not taxed within the plan. Taxes are payable on any withdrawals from the plan at the time of withdrawal. In general, individuals fall into a lower income tax bracket on retirement so withdrawals after retirement tend to be taxed at a lower rate. In addition, deferring tax payment makes sense even though the individual will be in the same tax bracket on retirement because of the time value of the tax savings today. The maximum annual RRSP contribution is regulated by the government.

The following example serves to demonstrate the advantage of RRSP investment compared to a savings account.

EXAMPLE 11.4 $

Comparison of an RRSP account with a savings account

An engineer wishes to save for retirement by making annual contributions over a period of 30 years. These savings would be drawn down over a 10-year period after retirement.

| Marginal tax rate | 45% | years 1 to 30 |
| Average tax rate | 35% | years 31 to 40 |

The savings will earn 10% per year and the investor can afford contributions of $2000 per year provided there is no tax on this money. Compare the annual after-tax cash from these savings after retirement if the savings were in an RRSP versus a savings account.

Solution

Given: Annual contributions to 1) RRSP and 2) savings account for 30 years and then withdrawals over 10 years

Find: The annual after-tax cash available after retirement from each of the two options.

1. RRSP contributions are $2000 annually because they can be made with before-tax dollars. The cash flow diagram for this option is given in Figure 11.4.

$$F \text{ at year } 30 = \$2000(F/A, 10\%, 30)$$
$$= \$329K.$$

$$\text{Annual withdrawals, } A, \text{ between years 31 and 40} = \$329K(A/P, 10\%, 10)$$
$$= \$53.6K.$$

RRSP withdrawals are taxable at 35%

$$\text{After-tax cash} = (1 - 0.35) \times \$53.6K$$
$$= \$34.8K/\text{year}.$$

2. Savings account contributions involve after-tax dollars. Annual contributions available are $2000 \times (1 - 0.45) = \$1100$.

The cash flow diagram looks the same as the RRSP option except that the annual deposit over the first 30 years is only $1100 per year.

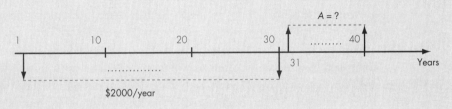

Figure 11.4 Cash flow diagram for the RRSP account (Example 11.4)

The interest on a savings account in any year is taxable. Assume year-end interest and tax payments.

$$i_{bt} = \text{before-tax interest rate} = 10\%$$
$$i_{at} = \text{after-tax interest rate.}$$

Over one year, the future equivalent, F, of a present sum, P, after-tax is

$$
\begin{aligned}
F &= (1 + i_{bt})P - t(i_{bt})P \\
&= P[1 + (1 - t)i_{bt}] \\
&= P(1 + i_{at}).
\end{aligned}
$$

The after-tax interest rate is:

$$i_{at} = (1 - t)i_{bt}. \tag{11.1}$$

$$
\begin{aligned}
i_{at} &= (1 - 0.45) \times 10\% \\
&= 5.5\% \text{ for 30 years}
\end{aligned}
$$

$$
\begin{aligned}
F \text{ at year } 30 &= \$1100(F/A, 5.5\%, 30) \\
&= \$79.7K.
\end{aligned}
$$

Between years 31 and 40

$$i_{at} = (1 - 0.35) \times 10\% = 6.5\%$$

$$
\begin{aligned}
A &= \$79.7K(A/P, 6.5\%, 10) \\
&= \$11.1K
\end{aligned}
$$

which is the annual after-tax cash available, as the taxes have already been paid!

Note that the RRSP provides a significantly larger annual income after retirement.

11.1.4   Registered Education Savings Plan (RESP)

As we discussed in the previous section, an RRSP is an effective savings vehicle for one's own retirement. On the other hand, an RESP is a savings vehicle for one's children's or grandchildren's post-secondary education. Thus, one's children or grandchildren are the beneficiaries of the RESP. Unlike the RRSP, the contributions made into an RESP are not tax deductible and as a result, when the principal (the amount of the original contributions) is withdrawn from the plan, it is not taxable. The growth of the contributions in the RESP is tax-free until money is withdrawn from the plan. When money is withdrawn from the plan by the beneficiary for post-secondary education, it is taxable in the hands of the beneficiary. As we know, the tax to be paid by a student is often zero or very small. The contribution limits for RESP are $4000 per year for 21 years with a lifetime limit of $42,000 per beneficiary. The money in an RESP can be invested in stocks, bonds, and/or mutual funds.

Another very attractive feature of RESP is that the contributions in it may qualify for a Canadian Education Saving Grant (CESG). The CESG was introduced in

1998. For the contributions made into each beneficiary's RESP, the federal government will pay a grant equal to 20% of annual contributions to a maximum of $400 per year. The total lifetime grant for each beneficiary is $7,200. This grant is deposited into the RESP directly, is not a taxable income, and is not included in the calculation of RESP contribution limits. Of course, the growth of this grant in the plan is also tax-free until money is withdrawn from the plan. This grant and its growth are strictly for the beneficiary's post-secondary education. If the money is withdrawn from the plan for purposes other than the beneficiary's post-secondary education, the grant and its growth will have to be paid back to the federal government.

Interested students may find more information on RESP and CESG on the Internet. The evaluation of the RESP is similar to that of the RRSP.

11.2 Lease versus Buy Decisions

In this section, we will examine the alternative way of obtaining facilities and equipment through leasing, and demonstrate the generalized cash flow approach to comparing cost-only mutually exclusive projects.

11.2.1 General Elements of Leases

Before explaining leasing methods, we must define the terms lessee and lessor. The **lessee** is the party leasing the property, and the **lessor** is the owner of property that is leased. Leasing takes several different forms, the most important of which are **operating (or service) leases** and **financial (or capital) leases**.

Operating Leases

Automobiles and trucks are frequently acquired under operating leases, and the operating-lease contract for office equipment, such as computers and copying machines, is becoming more common. These operating leases have three unique characteristics.

1. The lessor maintains and services the leased equipment, and the cost of this maintenance is figured into the lease payments.

2. Operating leases are not fully amortized; in other words, the payments required under the lease contract are not sufficient to recover the full cost of the equipment.

3. Operating leases frequently contain a cancellation clause, which gives the lessee the right to cancel the lease before the expiration of the basic agreement.

These operating-lease contracts are written for a period considerably shorter than the expected economic service life of the leased equipment, and the lessor expects to recover all investment costs through subsequent renewal payments, through subsequent leases to other lessees, or by sale of the leased equipment. The cancellation clause is an important consideration to the lessee, because the equipment can be returned if it is rendered obsolete by technological developments or is no longer needed because of a decline in business.

Financial Leases

In contrast to operating leases, financial leases are usually drawn up as **net leases**; that is, the lessee assumes responsibility for paying most of the operating and maintenance costs of the equipment, and payment to the lessor is primarily for principal and interest. The lessee's obligations under a net financial lease are, therefore, similar to obligations incurred under a debt financial instrument. However, the difference is that leasing does not increase the debt ratio of the company and the lessee is not the owner of the asset. Companies utilizing financial leases treat them as a cheap way of containing fixed assets for long-term use.

11.2.2   Lease-or-Buy Decision by the Lessee

A lease-or-buy decision begins only after a company has decided that the acquisition of a piece of equipment is necessary to carry out an investment project. Having made this decision, the company may be faced with several alternative methods of financing the acquisition: cash purchase, debt purchase, or acquisition via a lease.

In a situation of debt purchase, the present equivalent expression is similar to that of a purchase for cash, except that it has additional items—loan repayments and a tax shield on interest payments. The only way that the lessee can evaluate the cost of a lease is to compare it against the best available estimate of what the cost would be if the lessee owned the equipment.

To lay the groundwork for a more general analysis, we shall first consider how to analyze the lease-or-buy decision for a project with a single fixed asset for which the company expects a service life of N periods. Since the net after-tax revenue is the same for both alternatives, we need only consider the incremental cost of owning the asset and the incremental cost of leasing. Using the generalized cash flow approach presented in Section 9.4, the incremental cost of owning the asset by 100% debt financing may be expressed as

$$
\begin{aligned}
PEC(i)_{\text{Buy}} = \ & + \text{PE of loan repayment} \\
& + \text{PE of after-tax O\&M costs} \\
& - \text{PE of tax credit on capital cost allowance and interest} \\
& - \text{PE of net proceeds from sale.} \quad\quad\quad (11.2)
\end{aligned}
$$

Note that the asset acquisition (investment) cost is offset by the same amount of borrowing at time 0, so that we only need to consider the loan repayment series.

Assume that the firm can lease the asset at a constant amount per period. The project's incremental cost of leasing, $PEC(i)_{\text{Lease}}$, then becomes

$$PEC(i)_{\text{Lease}} = \text{PE of after-tax lease expenses.}$$

If the lease does not provide for the maintenance of the equipment leased, then the lessee must assume this responsibility. In this situation, the maintenance term in Eq. (11.2) can be dropped when calculating the incremental cost of owning the asset.

The criterion for the decision to lease as opposed to purchase thus reduces to a comparison between $PEC(i)_{\text{Buy}}$ and $PEC(i)_{\text{Lease}}$. In our terms, purchase is preferred if the combined present equivalent of loan repayment series and after-tax O&M cost, reduced by the present equivalent of the savings due to CCA and interest payment and the net proceeds from disposal of the asset, is less than the present equivalent of the net lease costs.

EXAMPLE
11.5

Lease-or-buy decision

The Montreal Electronics Company is considering replacing an old, 1-tonne-capacity industrial forklift truck. The truck has been used primarily to move goods from production machines into storage. The company is working nearly at capacity and is operating on a two-shift basis, 6 days per week. Montreal management is considering the possibility of either owning or leasing the new truck. The plant engineer has compiled the following data for the management:

- The capital cost of a gas-powered truck is $20,000. The new truck would use about 18 litres of gasoline (per 8-hour shift) at a cost of 49¢ per litre. If the truck is operated 16 hours per day, its expected life will be 4 years, and an engine overhaul at a cost of $1500 will be required at the end of 2 years.

- The Windsor Industrial Truck Company was servicing the old forklift truck, and Montreal would buy the new truck through Windsor. Windsor offers a service agreement to users of its trucks that provides for a monthly visit by an experienced service representative to lubricate and tune the trucks and costs $120 per month. Insurance and property taxes for the truck are $650 per year.

- The truck is classified as Class 10 property with a declining balance rate of 30%. Montreal has a 40% tax rate. The estimated resale value of the truck at the end of 4 years will be 15% of the original cost.

- Windsor also has offered to lease a truck to Montreal. Windsor will maintain the equipment and guarantee to keep the truck in serviceable condition at all times. In the event of a major breakdown, Windsor will provide a replacement truck, at its expense. The cost of the operating lease plan is $10,200 per year. The contract term is 3 years' minimum, with the option to cancel on 30 days' notice thereafter.

- The company can secure short-term debt at 10% interest rate.

- Based on recent experience, the company expects that funds committed to new investments should earn at least a 12% rate of return after taxes.

Compare the cost of owning versus leasing the truck.

Discussion: We may calculate the fuel costs for two-shift operations as

$$(18 \text{ litres/shift})(2 \text{ shifts/day})(49¢/\text{litre}) = \$17.64 \text{ per day.}$$

The truck will operate 300 days per year, so the annual fuel cost will be $5292. However, both alternatives require the company to supply its own fuel, so the fuel cost is not relevant for our decision-making. Therefore, we may drop this common cost item from our calculation.

Solution

Given: Cost information as given, MARR = 12%, marginal tax rate = 40%
Find: Incremental cost of owning versus leasing the truck

(a) Owning the truck

To compare the incremental cost of ownership with the incremental cost of leasing, we make the following additional estimates and assumptions.

- Step 1: The preventive-maintenance contract, which costs $120 per month (or $1440 per year) will be adopted. With annual insurance and taxes of $650, the present equivalent of the after-tax O&M cost is

$$P_1 = -(\$1440 + \$650)(1 - 0.40)(P/A, 12\%, 4) = -\$3809.$$

- Step 2: The engine overhaul is not expected to increase either the salvage value or the service life. Therefore, the overhaul cost ($1500) will be expensed all at once rather than capitalized. The present equivalent of this after-tax overhaul expense is

$$P_2 = -\$1500(1 - 0.40)(P/F, 12\%, 2) = -\$717.$$

- Step 3: If Montreal decided to purchase the truck through debt financing, the first step in determining financing costs would be to compute the annual installments of the debt-repayment schedule. Assuming that the entire investment of $20,000 is financed at a 10% interest rate, the annual payment would be

$$A = \$20,000(A/P, 10\%, 4) = \$6309.$$

The present equivalent of this loan-payment series is

$$P_3 = -\$6309(P/A, 12\%, 4) = -\$19,163.$$

- Step 4: The interest payment each year (10% of the beginning balance) is calculated as follows:

Year	Beginning Balance	Interest Charged	Annual Payment	Ending Balance
1	$20,000	$2,000	-$6,309	$15,691
2	15,691	1,569	-6,309	10,951
3	10,951	1,095	-6,309	5,737
4	5,737	573	-6,309	0

With a 30% CCA rate, the combined tax savings due to capital cost allowances and interest payments can be calculated as follows:

n	CCA_n	I_n	Combined Tax Savings		
1	$3,000	$2,000	$5,000 × 0.40	=	$2,000
2	5,100	1,569	6,669 × 0.40	=	2,668
3	3,570	1,095	4,665 × 0.40	=	1,866
4	2,499	573	3,072 × 0.40	=	1,229

Therefore, the present equivalent of the combined tax credit is

$$P_4 = \$2000(P/F, 12\%, 1) + \$2668(P/F, 12\%, 2)$$
$$+ \$1866(P/F, 12\%, 3) + \$1229(P/F, 12\%, 4)$$
$$= \$6022.$$

- Step 5: With the estimated salvage value of 15% of the initial investment ($3000) and the undepreciated capital cost, we compute the net proceeds from the sale of the truck at the end of 4 years as follows:

$$\text{Undepreciated capital cost} = \$20,000(1 - 15\%)(1 - 30\%)^3 = \$5831$$
$$\text{Losses} = U - S = \$5831 - \$3000 = \$2831$$
$$\text{Tax savings} = 0.40 \times \$2831 = \$1132$$
$$\text{Net proceeds from sale} = \$3000 + \$1132 = \$4132.$$

The present equivalent amount of the net salvage value is

$$P_5 = \$4132(P/F, 12\%, 4) = \$2626.$$

- Step 6: Therefore, the net present equivalent of owning the truck through 100% debt financing is

$$PE(12\%)_{Buy} = P_1 + P_2 + P_3 + P_4 + P_5$$
$$= -\$14,847.$$

(b) Leasing the truck

How does the cost of acquiring a forklift truck under the lease compare with the cost of owning the truck?

- Step 1: The lease payment is also a tax-deductible expense. The net cost of leasing has to be computed explicitly on an after-tax basis. The calculation of the annual incremental leasing costs is as follows.

$$
\begin{aligned}
\text{Annual lease payments (12 months)} &= \$10{,}200 \\
\text{Less 40\% taxes} &= \phantom{\$10{,}}4{,}080 \\
\text{Annual net costs after taxes} &= \$6{,}120.
\end{aligned}
$$

- Step 2: The total present equivalent of leasing is

$$
\begin{aligned}
PE(12\%)_{\text{Lease}} &= -\$6120(P/A,\ 12\%,\ 4) \\
&= -\$18{,}589.
\end{aligned}
$$

Purchasing the truck with debt financing would save Montreal $3742 in PE.

Comments: In our example, leasing the truck appears to be more expensive than purchasing the truck with debt financing. You should not conclude, however, that leasing is always more expensive than owning. In many cases, analysis favors a lease option. The interest rate, salvage value, lease-payment schedule, and debt financing all have an important effect on decision-making.

In Example 11.5, we have assumed that the lease payments occur at the end of year. However, many leasing contracts require the payments to be made at the beginning of every year. As we have stated before, tax payments or tax savings are assumed to be realized only at the end of year. In this case, the after-tax cash flow diagram shown in Figure 11.5 should be used for analysis of the lease option.

Using the after-tax cash flow diagram in Figure 11.5, we can find the present equivalent cost of lease financing when lease payment occurs at beginning of year as follows:

$$
PEC = LC + LC(P/A,\ i,\ N-1) - LC \times t\ (P/A,\ i,\ N). \tag{11.3}
$$

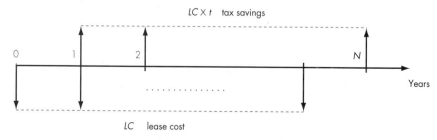

Figure 11.5 After tax cash flow diagram for lease option when lease payment occurs at beginning of year

11.2.3 Lessor's Point of View

Why are operating leases more expensive in the preceding example? To answer this question, we need to look at the transaction from the lessor's point of view. The lessor is assured of receiving revenue only over a finite period. At the expiration of the lease, the lessee may return the equipment, and the lessor will receive no further revenue from it unless the lessor is able to rent it to someone else or sell it as used equipment. The risk to the lessor is that a particular lessee will not have a continued need for the truck and that the lessor will not be able to dispose of the returned equipment profitably. We describe this function as absorbing the risk of obsolescence.

Another reason an operating lease option may be expensive is that the lessor may accept responsibility for maintaining the equipment in good operating condition. The lessor must charge a price sufficient to provide a cushion to absorb any unusual maintenance costs that might be incurred. If Montreal intends to use the truck for 4 years, it would be wiser to seek a financial lease arrangement instead of the operating lease.

11.3 Replacement Analysis

In Chapter 6, we covered various concepts and techniques that are useful in replacement analysis. In this section, we will illustrate how to use those concepts and techniques to conduct replacement analysis on an after-tax basis.

To apply the concepts and methods covered in Section 6.6 through 6.8 in an after-tax comparison of defender and challenger, we have to incorporate the disposal tax effects whenever an asset is disposed of. Whether the defender is kept or the challenger is purchased, we also need to incorporate the tax effects of capital cost allowances in our analysis.

Replacement studies require knowledge of the capital cost allowance schedule and of taxable gains or losses at disposal. Note that the capital cost allowance schedule is determined at the time of asset acquisition, whereas the disposal tax effects are determined by the relevant tax law at the time of disposal. In this section, we will use examples to illustrate how to do the following analyses on an after-tax basis:

1. Calculate the net market value of defender
2. Use the cash flow approach and the opportunity cost approach in comparison of defender and challenger
3. Calculate the economic life of defender or challenger
4. Conduct replacement analysis under the infinite planning horizon.

EXAMPLE 11.6

Net market value from disposal of defender

Reconsider the asset analyzed in Example 6.9. In addition, assume that Macintosh has been claiming capital cost allowances on the machine using a CCA rate of 20% for the past two years, the firm's marginal tax rate is 40%, and its MARR is 12%. Determine the net market value of the asset if it is disposed of today.

Solution

We compute the following:

$$\text{Undepreciated capital cost, } U = \$20,000(1 - 10\%)(1 - 20\%) = \$14,400$$

$$\text{Current market value, } S = \$10,000$$

$$\text{Losses} = U - S = \$14,400 - \$10,000 = \$4400$$

$$\text{Disposal tax effect} = (\text{Losses}) \times t$$
$$= \$4400 \times 40\%$$
$$= \$1760$$

$$\text{Net salvage value} = \$10,000 + \$1760$$
$$= \$11,760$$

This calculation is illustrated in Figure 11.6.

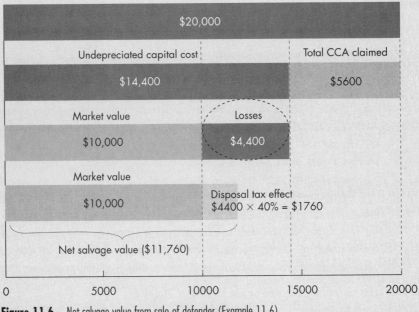

Figure 11.6 Net salvage value from sale of defender (Example 11.6)

EXAMPLE 11.7 Replacement analysis using the cash flow approach

Compare the defender analyzed in Examples 6.9 and 11.6 and the challenger given in Example 6.10. Assume that the challenger would fall into the same CCA class as the defender (d = 20%). The firm's marginal tax rate is 40% and its MARR is 12%. Use the cash flow approach to determine whether the replacement is justified now.

Solution

• *Option 11: Keep the defender*

Table 11.1 shows the worksheet format the company uses to analyze the defender. Each line is numbered, and a line-by-line description of the table follows.

Table 11.1

Replacement Analysis Worksheet Option 1 — Keep the Defender (Example 11.7)

n	-2	-1	0	1	2	3
Financial data (cost information)						
(1) CCA		$ 2,000	$ 3,600	$ 2,880	$2,304	$1,843
(2) Undepreciated capital cost	$20,000	$18,000	$14,400	$11,520	$9,216	$7,373
(3) Salvage value						$2,000
(4) Loss from sale						$5,373
(5) Repair cost			$ 5,000			
(6) O&M costs				$ 8,000	$8,000	$8,000
Income statement						
(7) Revenue						
(8) Costs						
(9) CCA				$ 2,880	$ 2,304	$1,843
(10) O&M costs				$ 8,000	$ 8,000	$8,000
(11) Taxable income				($10,880)	($10,304)	($9,843)
(12) Income taxes				($ 4,352)	($ 4,122)	($3,937)
(13) Net income				($ 6,528)	($ 6,182)	($5,906)
Cash flow statement						
(14) Operating activities						
Net income				-$6,528	-$6,182	-$5,906
CCA				$2,880	$2,304	$1,843
(15) Investment activities						
Investment						$2,000
(16) Salvage value						
(17) Disposal tax effect						$2,149
(18) Net cash flow				-$3,648	-$3,878	$86

- *Lines 1–4:* The company paid $20,000 to acquire the defender two years ago. The CCA claimed in each of the past two years was calculated with d = 20% and shown in the table. Note that the 50%-rule was used in calculating the CCA for the first year of its use. If the defender is kept, the CCA schedule for the next three years would be $2880, $2304, and $1843, respectively. This would result in a total CCA of $12,267 over five years and an undepreciated capital cost of $20,000 − $12,267 = $7373 three years from today.

- *Lines 5–6:* The repair cost in the amount of $5000 was already incurred before the replacement decision. This is a sunk cost and should not be considered in our replacement analysis. If a repair in the amount of $5000 were required to keep the defender in serviceable condition, this would show as an expense at time 0. If the old machine is retained for the next 3 years, the before-tax annual O&M costs are as shown in Line 6.

- *Line 7:* This is a service project, and revenue will remain unchanged regardless of the replacement decisions.

- *Lines 8–10:* Two items are listed under the costs category: CCA and O&M costs.

- *Lines 11–13*: Without revenues, the net income figures are negative.

- *Line 15*: With the cash flow approach, no new investment is required to retain the defender.

- *Line 17*: Since the defender would be sold at less than the undepreciated capital cost, the disposal would create a loss, which would reduce the firm's taxable income and, hence, its tax payment. The disposal tax effect is equal to (undepreciated capital cost − salvage value) (0.40) = ($5373)(40%) = $2149. This loss is an operating loss because it reflects the fact that inadequate capital cost allowance was taken on the defender. An operating loss results in tax savings. Therefore, the net proceeds from sale of the old printer at the end of three more years is $2000 + $2149 = $4149.

- *Line 18*: Since a new investment is not required to keep the old machine, the net cash flows would consist of the operating cash flows and the net proceeds from the sale of the old machine at the end of year 3.

• Option 2: Replace the defender

Table 11.2 shows the worksheet format the company uses to analyze the challenger. Each line is numbered, and a line-by-line description of the table follows.

- *Line 1*: The net proceeds from the sale of the old printer at $n = 0$ is shown. As shown in Example 11.6, the net proceeds from the sale are the sum of the salvage value and the disposal tax effect.

- *Line 2*: The purchase price of the new machine, including installation and freight charges, is shown.

- *Lines 3–4*: The CCA schedule along with the undepreciated capital costs for the new machine (d = 20%) is shown. The 50%-rule is used for the first year.

- *Lines 5–6*: With the salvage value estimated at $5500, we expect a loss of $3140 (= $8640 − $5500) on the sale of the new machine at the end of year 3.

- *Line 7*: The O&M costs for the new machine are listed.

- *Lines 8–14*: The net income figures are tabulated here to determine the net cash flows.

- *Line 15*: The net operating cash flows over the project's 3-year life are shown. These flows are found by adding the noncash expense (CCA) to the net income.

- *Lines 16–17*: The total net cash outflow at the time the replacement is made is shown. The company writes a check for $15,000 to pay for the new machine. However, this outlay is partially offset by the proceeds from the sale of the old equipment in the amount of $11,760.

- *Lines 18–19*: These lines show the cash flows associated with the disposal of the new machine. To begin, Line 18 shows the estimated salvage value of the new

Table 11.2
Replacement Analysis Worksheet (Challenger) Option 2— Replace the Defender (Example 11.7)

n:	0	1	2	3
Financial data (cost information)				
(1) Net proceeds from sale of old printer	$11,760			
(2) Cost of new printer	$15,000			
(3) CCA		$ 1,500	$ 2,700	$2,160
(4) Undepreciated capital cost		$13,500	$10,800	$8,640
(5) Salvage value				$5,500
(6) Loss from sale of new printer				$3,140
(7) O&M costs		$ 6,000	$ 6,000	$6,000
Income statement				
(8) Revenue				
(9) Costs				
(10) CCA		$1,500	$2,700	$2,160
(11) O&M costs		$6,000	$6,000	$6,000
(12) Taxable income		($7,500)	($8,700)	($8,160)
(13) Income taxes		($3,000)	($3,480)	($3,264)
(14) Net income		($4,500)	($5,220)	($4,896)
Cash flow statement				
(15) Operating activities				
Net income		–$4,500	–$5,220	–$4,896
CCA		$1,500	$2,700	$2,160
(16) Investment activities				
Investment	–$15,000			
(17) Net proceeds from sale of old printer	$11,760			
(18) Salvage value				$5,500
(19) Disposal tax effect				$1,256
(20) Net cash flow	–$3,240	–$3,000	–$2,520	$4,020

Note: The computational procedukre to determine the net proceeds from sale of the old printer is shown in Example 11.6.

machine at the end of its 3-year life, $5500. Since the undepreciated capital cost of the new machine at the end of year 3 is $8640, the company will have a disposal tax effect of ($8640 – 5500) (0.40) = $1256.

- *Line 20*: This line represents the net cash flows associated with replacement of the old machine with the new one.

The actual cash flow diagrams are shown in Figure 11.7. Since both the defender and challenger have the same service life, we can use either PE or AE analysis.

$$PE(12\%)_{old} = -\$3648(P/F, 12\%, 1) - \$3878(P/F, 12\%, 2) + \$86(P/F, 12\%, 3)$$
$$= -\$6288.$$

$$AE(12\%)_{old} = -\$6288(A/P, 12\%, 3)$$
$$= -\$2618.$$

$$PE(12\%)_{new} = -\$3240 - \$3000(P/F, 12\%, 1) - \$2520(P/F, 12\%, 2)$$
$$+ \$4020(P/F, 12\%, 3)$$
$$= -\$5066.$$

$$AE(12\%)_{new} = -\$5066(A/P, 12\%, 3)$$
$$= -\$2109.$$

Because of the annual difference of $509 in favour of the challenger, the replacement should be made now.

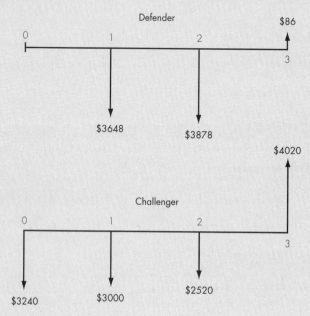

Figure 11.7 Comparison of defender and challenger based on cash flow approach (Example 11.7)

EXAMPLE 11.8 Replacement analysis using an opportunity cost approach

Rework Example 11.7 using the opportunity cost approach.

Solution

Recall that the cash flow approach in Example 11.7 credited the net proceeds in the amount of $11,760 from the sale of the defender toward the $15,000 purchase price of the challenger, and no initial outlay would have been required had the decision been to keep the defender. If the decision to keep the defender had been made the opportunity cost approach treats the $11,760 current net salvage value of the defender as a cost incurred. Figure 11.8 illustrates the cash flows related to these decision options.

Since the lifetimes are the same, we can use either PE or AE analysis.

$$PE(12\%)_{old} = -\$11,760 - \$3648(P/F, 12\%, 1) - \$3878(P/F, 12\%, 2)$$
$$+\$86(P/F, 12\%, 3)$$
$$= -\$18,048.$$

$$AE(12\%)_{old} = -\$18,048(A/P, 12\%, 3) = -\$7514.$$

$$PE(12\%)_{new} = -\$15,000 - \$3000(P/F, 12\%, 1)$$
$$-\$2520(P/F, 12\%, 2) + \$4020(P/F, 12\%, 3)$$
$$= -\$16,826.$$

$$AE(12\%)_{new} = -\$16,826(A/P, 12\%, 3) = -\$7006.$$

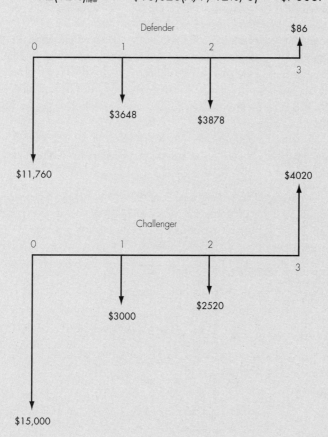

Figure 11.8   Comparison of defender and challenger based on opportunity cost approach (Example 11.8)

The decision is the same as in Example 11.7. Since both the challenger and defender cash flows were adjusted by the same amount, $11,760, at time 0, this should not come as a surprise.

EXAMPLE
11.9

Economic service life for a lift truck

Consider Example 6.12 with the following additional data. The asset belongs to Class 43 with a CCA rate of 30%. The firm's marginal tax rate is 40%. The after-tax MARR is 15%.

Solution

Two approaches may be used to find the economic life of an asset: (a) The generalized cash flow approach and (b) the tabular approach. We will now demonstrate both approaches.

(a) Generalized Cash Flow Approach: First we could determine the relevant after-tax cash flows if the lift truck were to be retained for just 1 year by means of the income statement/cash flow statement approach as shown in Table 11.3(a). Since we have only a few cash flow elements (O&M, CCA, and salvage value), a more efficient way to obtain the after-tax cash flow is to use the generalized cash flow approach discussed in Section 9.4. Recall that the capital cost allowances result in a tax reduction equal to the CCA amount multiplied by the tax rate. The operating expenses are multiplied by the factor of (1 – the tax rate) to obtain the after-tax O&M. For a situation in which the asset is retained for 1 year, Table 11.3(b) summarizes the cash flows obtained by using the generalized cash flow approach.

If we use the expected operating costs and the net salvage values given in Table 11.4, we can continue to generate yearly after-tax entries for the asset's remaining physical life. For the first 2 options, we compute the equivalent annual costs of owning and operating as follows.

- $n = 1$: One–year replacement cycle:

$$AE(15\%) = \{-\$18,000 + [-(0.6)(\$4000) \\ + (0.40)(\$2700) \\ + \$12,120](P/F, 15\%, 1)\}(A/P, 15\%, 1) \\ = -\$9900.$$

- $n = 2$: Two-year replacement cycle:

$$AE(15\%) = \{-\$18,000 + [- 0.6(\$4000) \\ + 0.4(\$2700)](P/F, 15\%, 1) + [-0.6(\$5600) \\ + 0.4(\$4590) + \$8784](P/F, 15\%, 2)\}(A/P, 15\%, 2) \\ = -\$8401.$$

Similarly, the annual equivalents for other replacement cycles or holding periods can be computed as shown in Table 11.5 (column 12). If the truck were to be sold after 3 years, it would have a minimum annual cost of $8040 per year, and this is the life most favorable for comparison purposes. By replacing the asset

Table 11.3	Year	0	1
After-Tax Cash Flow Calculation for Owning and Operating the Asset for 1 Year (Example 11.9)	**(a) Income Statement Approach**		
	Income Statement		
	Revenue		
	Operating expense		
	O&M		$ 4,000
	Capital cost allowance		2,700
	Taxable income		$ (6,700)
	Income taxes (40%)		(2,680)
	Net income		$ (4,020)
	Cash flow statement		
	Operating activities		
	Net income		$ (4,020)
	Capital cost allowance		2,700
	Investment activities		
	Investment	$(18,000)	
	Salvage value		10,000
	Disposal tax effect		2,120
	Net cash flow	$(18,000)	$10,800
	(b) Generalized Cash Flow Approach		
	Investment	$(18,000)	
	+ (0.4) (capital cost allowance)		$ 1,080
	− (0.6) (O&M)		(2,400)
	Net proceeds from sale		12,120
	Net cash flow	$(18,000)	$10,800

perpetually according to an economic life of 3 years, we obtain the minimum infinite AE cost stream. Figure 11.9 illustrates this concept. Of course, we should envision a long period of required service for the asset—this life no doubt being heavily influenced by market values, O&M costs, and capital cost allowances.

(b) Tabular Approach: The tabular approach separates the annual cost elements into two parts: One associated with the capital recovery of the asset and the other associated with operating the asset. In computing the capital recovery cost, we need to determine the after-tax salvage values at the end of each holding period as calculated previously in Table 11.4. Then we compute the total annual equivalent costs of the asset for any given year's operation:

Table 11.4

Forecasted Operating Costs and Net Proceeds from Sale as a Function of Holding Period (Example 11.9)

Holding Period	O&M	Permitted Annual Capital Cost Allowances Over the Holding Period							Total CCA	Undepreciated Capital Cost	Expected Market Value	Disposal Tax Effect	Net A/T Salvage Value
		1	2	3	4	5	6	7					
1	$4,000	$2,700							$ 2,700	$15,300	$10,000	$2,120	$12,120
2	$5,600	2,700	$4,590						7,290	10,710	7,500	1,284	8,784
3	$7,840	2,700	4,590	$3,213					10,503	7,497	5,625	749	6,374
4	$10,976	2,700	4,590	3,213	$2,249				12,752	5,248	4,219	412	4,630
5	$20,366	2,700	4,590	3,213	2,249	$1,574			14,326	3,674	3,164	204	3,368
6	$21,513	2,700	4,590	3,213	2,249	1,574	$1,102		15,429	2,571	2,373	79	2,452
7	$30,118	2,700	4,590	3,213	2,249	1,574	1,102	$771	16,200	1,800	1,780	8	1,788

Note: The asset costs $18,000 and has a CCA rate of 30%. In year 5, O&M is equal to normal operating expense $15,366 + overhaul $5,000.

Table 11.5

(1) Holding Period	(2)	(3) Before-Tax Operating Expenses	(4)	(5)	(6) After-Tax Flow if the Asset Is Kept for *N* More Years	(7)	(8)	(9)	(10) Annual Equivalent Cost	(11)	(12)
N	*n*	O&M	CCA	A/T O&M	Net CCA Credit	Investment Operating Cost	Net A/T and Net Salvage	Cash Flow	Capital Cost	Operating Cost	Total Cost
0							$(18,000)	$(18,000)			
1	1	$4,000	$2,700	$ (2,400)	$1,080	$ (1,320)	12,120	10,800	$(8,580)	(1,320)	$(9,900)
	0						(18,000)	(18,000)			
	1	4,000	2,700	(2,400)	1,080	(1,320)	–	(1,320)			
2	2	5,600	4,590	(3,360)	1,836	(1,524)	8,784	7,260	(6,987)	(1,415)	(8,401)
	0						(18,000)	(18,000)			
	1	4,000	2,700	(2,400)	1,080	(1,320)	–	(1,320)			
	2	5,600	4,590	(3,360)	1,836	(1,524)	–	(1,524)			
3 (Economic Life)	3	7,840	3,213	(4,704)	1,285	(3,419)	6,374	2,955	(6,048)	(1,992)	(8,040) ← Minimum cost
	0						(18,000)	(18,000)			
	1	4,000	2,700	(2,400)	1,080	(1,320)	–	(1,320)			
	2	5,600	4,590	(3,360)	1,836	(1,524)	–	(1,524)			
	3	7,840	3,213	(4,704)	1,285	(3,419)	–	(3,419)			
4	4	10,976	2,249	(6,586)	900	(5,686)	4,630	(1,056)	(5,378)	(2,732)	(8,109)
	0						(18,000)	(18,000)			
	1	4,000	2,700	(2,400)	1,080	(1,320)	–	(1,320)			
	2	5,600	4,590	(3,360)	1,836	(1,524)	–	(1,524)			
	3	7,840	3,213	(4,704)	1,285	(3,419)	–	(3,419)			
	4	10,976	2,249	(6,586)	900	(5,686)	–	(5,686)			
5	5	20,366	1,574	(12,220)	630	(11,590)	3,368	(8,222)	(4,870)	(4,046)	(8,916)
	0						(18,000)	(18,000)			
	1	4,000	2,700	(2,400)	1,080	(1,320)	–	(1,320)			
	2	5,600	4,590	(3,360)	1,836	(1,524)	–	(1,524)			
	3	7,840	3,213	(4,704)	1,285	(3,419)	–	(3,419)			
	4	10,976	2,249	(6,586)	900	(5,686)	–	(5,686)			
	5	20,366	1,574	(12,220)	630	(11,590)	–	(11,590)			
6	6	21,513	1,102	(12,908)	441	(12,467)	2,452	(10,015)	(4,476)	(5,008)	(9,484)
	0						(18,000)	(18,000)			
	1	4,000	2,700	(2,400)	1,080	(1,320)	–	(1,320)			
	2	5,600	4,590	(3,360)	1,836	(1,524)	–	(1,524)			
	3	7,840	3,213	(4,704)	1,285	(3,419)	–	(3,419)			
	4	10,976	2,249	(6,586)	900	(5,686)	–	(5,686)			
	5	20,366	1,574	(12,220)	630	(11,590)	–	(11,590)			
	6	21,513	1,102	(12,908)	441	(12,467)	–	(12,467)			
7	7	30,118	771	(18,071)	309	(17,762)	1,788	(15,974)	(4,165)	(6,160)	(10,325)

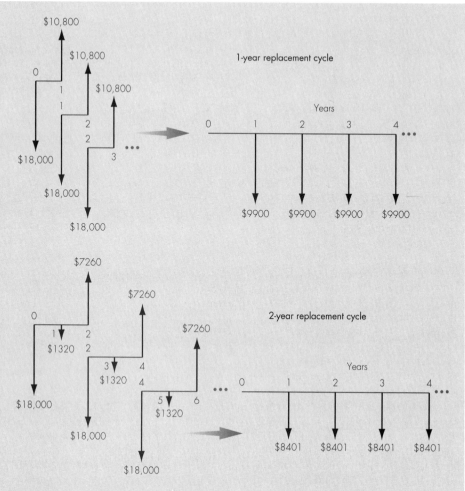

Figure 11.9 Conversion of an infinite number of replacement cycles to infinite AE cost streams (Example 11.9)

$$\text{Total equivalent annual costs} = \begin{array}{l}\text{Capital recovery cost} \\ + \text{ equivalent annual} \\ \text{operating costs.}\end{array}$$

If we examine the equivalent annual costs itemized in Table 11.5 (columns 10 and 11), we see that, as the asset ages, equivalent annual operating costs increase. At the same time, capital recovery cost decreases with prolonged use of the asset. The combination of decreasing capital recovery costs and increasing annual O&M costs results in the total annual equivalent cost taking on a form similar to that depicted in Figure 11.10.

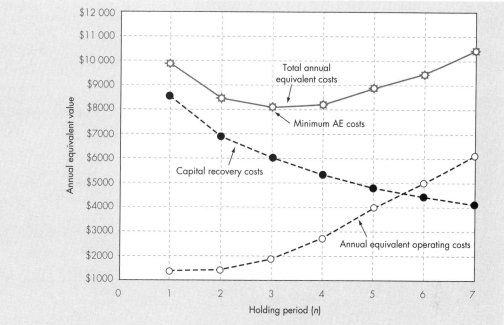

Figure 11.10 Economic service life obtained by finding the minimum AE cost (Example 11.9)

EXAMPLE
11.10

Replacement analysis under the infinite planning horizon

Consider Example 6.13, page 387, with the following additional data. The old machine was bought 4 years ago for $9000. Both defender and challenger are in asset class 43 with a CAA rate of 30%. The company's MARR and tax rate are 15% and 40%, respectively.

Solution

1. Economic Service Life

*   Defender

The defender has been used for 4 years. We can use the declining balance method with the 50% rule to calculate its current undepreciated cost. The following shows the net salvage values of the old machine if it is sold in a number of years.

Holding Period (n)	Undepreciated Capital Cost (U)	Salvage Value (S)	Disposal Tax Effect	Net Salvage Value (NS)
0	$2624	$5000	$(950)	$4050
1	1837	4000	(865)	3135
2	1286	3000	(686)	2314
3	900	2000	(440)	1560
4	630	1000	(148)	852
5	441	–	176	176
6	309	–	123	123

If the company retains the inspection machine, it is in effect deciding to overhaul the machine and, in addition, invest the machine's current market value (after taxes) in that alternative. Although no physical investment cash flow transaction will occur, the firm is withholding from the investment the net market value of the inspection machine (opportunity cost). Similarly, the after-tax O&M costs are as follows:

n	Overhaul	Forecasted O&M Cost	After-Tax O&M Cost
0	$1200		$1200(1 – 0.40) = $720
1	0	$2000	2000(1 – 0.40) = 1200
2	0	3500	3500(1 – 0.40) = 2100
3	0	5000	5000(1 – 0.40) = 3000
4	0	6500	6500(1 – 0.40) = 3900
5	0	8000	8000(1 – 0.40) = 4800

Using the current year's net market value as the investment required to retain the defender, we obtain the data in Table 11.6, which indicates that the remaining economic life of the defender is 2 years, *in the absence of future challengers*. The overhaul (repair) cost of $1200 in year 0 can be treated as a deductible operating expense for tax purposes, as long as it does not add value to the property. (Any repair or improvement expenses that increase the value of the property must be capitalized by depreciating them over the estimated service life.)

* Challenger

For the challenger, we must determine the undepreciated capital cost of the asset at the end of each holding period to compute the net salvage value. This is shown in Table 11.7(a). With the net salvage values computed in Table 11.7(a), we are now ready to find the economic service life of the challenger by generating AE value entries. These calculations are summarized in Table 11.7(b). The economic life of the challenger is 4 years with an AE(15%) value of $3946.

2. Optimal Time to Replace the Defender

Since the annual equivalent cost for the defender's remaining useful life (2 years) is $3205, which is less than the annual equivalent cost of the challenger at its economic life ($3946), the decision will be to keep the defender for now. The defender's remaining useful life of 2 years does not necessarily imply that the defender should be kept for 2 years before switching to the challenger. The reason for this is that the defender's remaining useful life of 2 years was calculated without considering what type of challenger would be available in the future. When a challenger's financial data is available, we need to enumerate all replacement timing possibilities. Since the defender can be used for another 5 years, six replacement strategies exist:

- Replace now with the challenger.

- Replace in year 1 with the challenger.

- Replace in year 2 with the challenger.

- Replace in year 3 with the challenger.

- Replace in year 4 with the challenger.

- Replace in year 5 with the challenger.

If the costs and efficiency of the current challenger remain unchanged in the future years, we know right away that the first two strategies will be rejected. In other words, we will keep the defender for at least two years. Whether we should replace the defender with the challenger at the end of year 2, year 3, year 4, or year 5 can be answered with the marginal analysis method used in Example 6.13. To determine if we should do the replacement at the end of year 2, we need to find the incremental cost of keeping the defender for one more year, i.e., the 3rd year. If this one-year incremental cost is smaller than the AEC* of the challenger ($3946) at its own economic life), the defender should not be replaced at the end of year 2. Whether it should be replaced at the end of year 3, we need to conduct a similar marginal analysis for keeping the defender for the 4th year.

The annual cost of keeping the defender for the 3rd year is

$$NSV_2(1+i) - NSV_3 - t \times CCA_3 + OC_3(1-t)$$
$$= \$2314 \times 1.15 - \$1560 - 0.4 \times \$386 + \$5000 \times 0.6$$
$$= \$3947.$$

This cost is about the same as the cost ($3946) of the challenger. Thus, it does not make any economic difference whether the defender is replaced at the end of year 2 or kept for one more year and replaced at the end of year 3. It should not be kept for the 4th year since its cost in that year will be higher than AEC* for the challenger.

Table 11.6
Economics of Retaining the Defender for N More Years (Example 11.10)

(1)	(2)	(3)	(4)	(5)	(6)	(7)	(8)	(9)	(10)	(11)	(12)
Holding Period		Before-Tax Operating Expenses			After-Tax Flow if the Asset Is Kept for N More Years				Equivalent Annual Cost		
N	n	O&M	CCA	A/T O&M	CCA Credit	Net Operating Cost	Investment and Net Salvage	Net A/T Cash Flow	Capital Cost	Operating Cost	Total Cost
1	0	$1200	$787	$ (720)	$315	$ (720)	$(4050)	$(4770)	$(1523)	$(1713)	$(3236)
	1	2000		(1200)		(885)	3135	2250			
②	0	1200	787	(720)	315	(720)	(4050)	(4770)	(1415)	(1791)	(3205) Minimum AE cost
	1	2000	551	(1200)	220	(885)	–	(885)			
	2	3500		(2100)		(1880)	2314	434			
Economic Life 3	0	1200	787	(720)	315	(720)	(4050)	(4770)	(1325)	(2094)	(3419)
	1	2000	551	(1200)	220	(885)	–	(885)			
	2	3500	386	(2100)	154	(1880)	–	(1880)			
	3	5000		(3000)		(2846)	1560	(1286)			
4	0	1200	787	(720)	315	(720)	(4050)	(4770)	(1248)	(2434)	(3682)
	1	2000	551	(1200)	220	(885)	–	(885)			
	2	3500	386	(2100)	154	(1880)	–	(1880)			
	3	5000	270	(3000)	108	(2846)	–	(2846)			
	4	6500		(3900)		(3792)	852	(2940)			
5	0	1200	787	(720)	315	(720)	(4050)	(4770)	(1182)	(2774)	(3956)
	1	2000	551	(1200)	220	(885)	–	(885)			
	2	3500	386	(2100)	154	(1880)	–	(1880)			
	3	5000	270	(3000)	108	(2846)	–	(2846)			
	4	6500	189	(3900)	76	(3792)	–	(3792)			
	5	8000		(4800)		(4724)	176	(4548)			

Table 11.7(a)
Forecasted Operating Costs and Net Proceeds from Sale as a Function of Holding Period — Challenger (Example 11.10)

| Holding Period | O&M | Permitted Annual Capital Cost Allowances Over the Holding Period | | | | | | | Total CCA | Undepreciated Capital Cost | Expected Market Value | Disposal Tax Effect | Net A/T Salvage Value |
		1	2	3	4	5	6	7					
1	$2000	$1500							$1500	$8500	$6000	$1000	$7000
2	3000	1500	$2550						4050	5950	5100	340	5440
3	4000	1500	2550	$1785					5835	4165	4335	-68	4267
4	5000	1500	2550	1785	$1250				7085	2916	3685	-308	3377
5	6000	1500	2550	1785	1250	$875			7959	2041	3132	-436	2696
6	7000	1500	2550	1785	1250	875	$612		8571	1429	2662	-493	2169
7	8000	1500	2550	1785	1250	875	612	$429	9000	1000	2263	-505	1758

Note: The asset costs $10,000 and has a CCA rate of 30%. The 50% rule is used for the first year.

Table 11.7(b)

Economics of Owning and Operating the Challenger for N More Years (Example 11.10)

(1) Holding Period N	(2) n	(3) O&M	(4) CCA	(5) A/T O&M	(6) CCA Credit	(7) Net Operating Cost	(8) Investment and Net Salvage	(9) Net A/T Cash Flow	(10) Capital Cost	(11) Operating Cost	(12) Total Cost
1	0						$(10,000)	$(10,000)	(4500)	(600)	(5100)
	1	$2000	$1500	$(1200)	$600	$(600)	7000	6400			
2	0						(10,000)	(10,000)	(3621)	(628)	(4249)
	1	2000	1500	(1200)	600	(600)	–	(600)			
	2	2800	2550	(1680)	1020	(660)	5440	4780			
3	0						(10,000)	(10,000)	(3151)	(863)	(4014)
	1	2000	1500	(1200)	600	(600)	–	(600)			
	2	2800	2550	(1680)	1020	(660)	–	(660)			
	3	3600	1785	(2160)	714	(1446)	4267	2821			
④	0						(10,000)	(10,000)	(2826)	(1119)	(3946)
	1	2000	1500	(1200)	600	(600)	–	(600)			
	2	2800	2550	(1680)	1020	(660)	–	(660)			
	3	3600	1785	(2160)	714	(1446)	–	(1446)			
	4	4400	1250	(2640)	500	(2140)	3377	1237			
5	0						(10,000)	(10,000)	(2583)	(1364)	(3947)
	1	2000	1500	(1200)	600	(600)	–	(600)			
	2	2800	2550	(1680)	1020	(660)	–	(660)			
	3	3600	1785	(2160)	714	(1446)	–	(1446)			
	4	4400	1250	(2640)	500	(2140)	–	(2140)			
	5	5200	875	(3120)	350	(2770)	2696	74			
6	0						(10,000)	(10,000)	(2395)	(1591)	(3986)
	1	2000	1500	(1200)	600	(600)	–	(600)			
	2	2800	2550	(1680)	1020	(660)	–	(660)			
	3	3600	1785	(2160)	714	(1446)	–	(1446)			
	4	4400	1250	(2640)	500	(2140)	–	(2140)			
	5	5200	875	(3120)	350	(2770)	–	(2770)			
	6	6000	612	(3600)	245	(3355)	2169	(1186)			
7	0						(10,000)	(10,000)	(2245)	(1801)	(4046)
	1	2000	1500	(1200)	600	(600)	–	(600)			
	2	2800	2550	(1680)	1020	(660)	–	(660)			
	3	3600	1785	(2160)	714	(1446)	–	(1446)			
	4	4400	1250	(2640)	500	(2140)	–	(2140)			
	5	5200	875	(3120)	350	(2770)	–	(2770)			
	6	6000	612	(3600)	245	(3355)	–	(3355)			
	7	6800	429	(4080)	171	(3909)	1758	(2151)			

Column group headers: (3)–(4) Before-Tax Operating Expenses; (5)–(9) After-Tax Flow if the Asset Is Kept for N More Years; (10)–(12) Equivalent Annual Cost.

Economic Life (arrow to circled ④)

Minimum AE cost → (3946)

11.4 Computer Notes

Perhaps the most tedious aspect of replacement analysis is the calculation of economic service lives.[2] By now you should have a good idea of how to tailor an electronic spreadsheet method to a specific problem, so in this section, we will focus on an Excel screen format that can be used to determine economic service lives. (Here we use the data from Example 11.10.)

Even programming the procedure on the spreadsheet can take a considerable amount of time. Figure 11.11 shows a typical Excel application that can be used to determine the economic service life for the challenger in Example 11.10. On the top portion of the spreadsheet, you enter the financial data, such as investment, tax rate, current UCC, and MARR. In the middle of the screen, you enter other financial data, such as O&M costs, permitted capital cost allowance amounts over the holding period, and market values. From these, you calculate the total capital cost allowance, the undepreciated capital cost, disposal tax effects, and the net after-tax salvage value (column N) as a function of the holding period.

In the lower portion of the spreadsheet, you will find the PE figures on operating costs, CCA tax credits, and annualized capital cost and operating costs. (Note the sign convention: Cost is expressed as a positive number.) Specifically, we list

- Holding period (column A): List the holding period of the asset.

- After-tax market value (column B): Copy the after-tax salvage values that were calculated in column N.

- Present equivalent of market value (column C): List the present equivalent values of the after-tax market values at various holding periods.

- After-tax O&M cost (column D): The before-tax O&M costs in cells B11–B17 are converted into the equivalent after-tax O&M costs.

- Present equivalent of O&M cost (column E): Compute the equivalent present values of the after-tax O&M costs over each holding period.

- Cumulative present value of O&M cost (column F): Here the present equivalent figure for the total operating cost up to the holding period is listed.

- Cumulative present value of CCA credit (column G): Here the total present equivalent of the CCA tax credits (savings) up to the holding period is computed. For example, to find the cell entry G28, we first identify the permitted

---

[2] **EzCash** has a built-in function for automating this calculation. First go to the opening menu and select **Special** followed by **Economic Service Life. EzCash** will display the blank worksheet with data columns of (1) Period, (2) Salvage, (3) O&M costs, (4) Depreciation, and (5) Revenue, along with input parameters (such as investment, current book value, tax rate, and physical life). Just as with a spreadsheet, you can use the cell pointer to indicate the cell and enter the desired data.

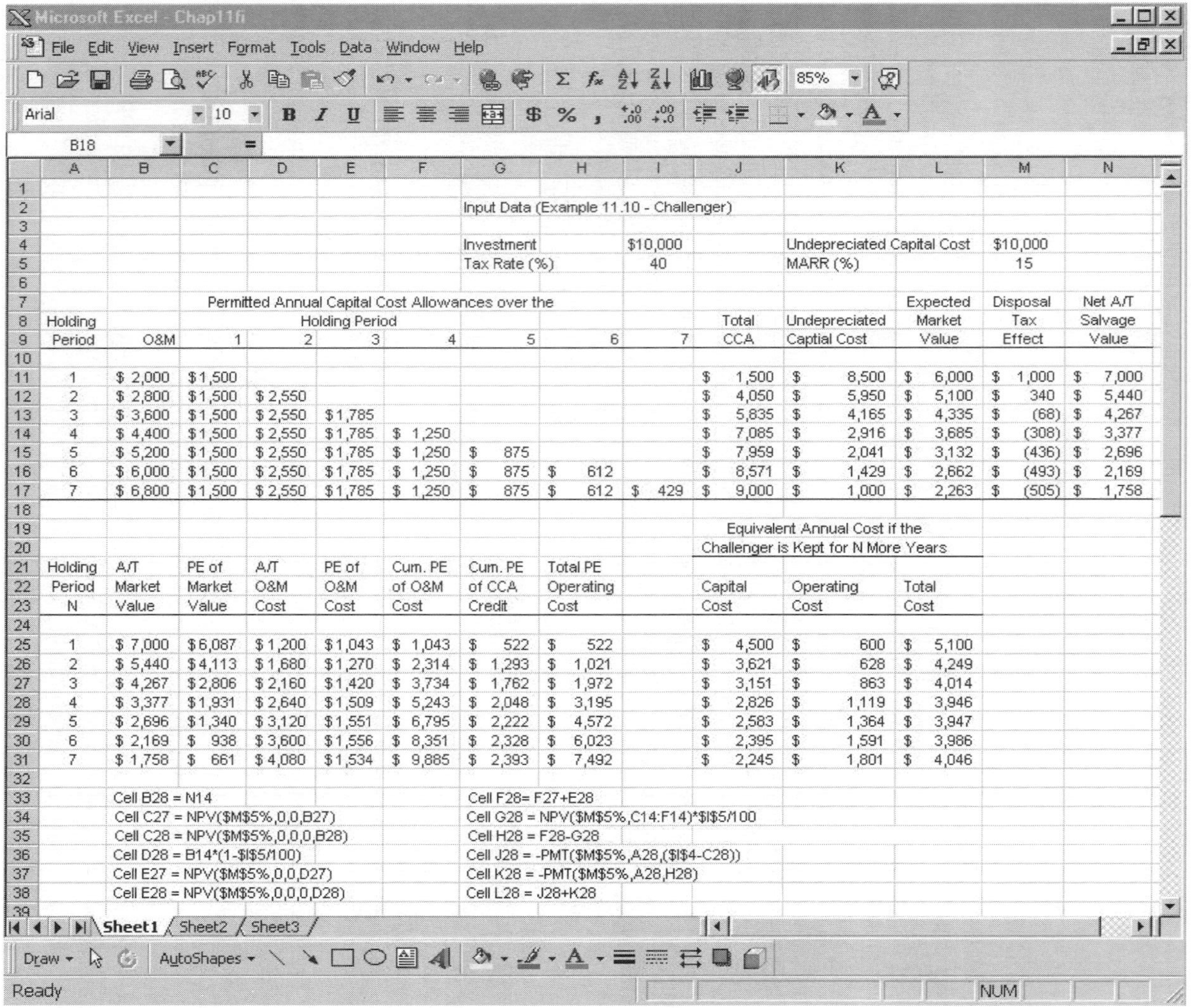

Input Data (Example 11.10 - Challenger)

				Undepreciated Capital Cost	$10,000
Investment	$10,000			MARR (%)	15
Tax Rate (%)	40				

Permitted Annual Capital Cost Allowances over the Holding Period

Holding Period	O&M	1	2	3	4	5	6	7	Total CCA	Undepreciated Capital Cost	Expected Market Value	Disposal Tax Effect	Net A/T Salvage Value
1	$2,000	$1,500							$1,500	$8,500	$6,000	$1,000	$7,000
2	$2,800	$1,500	$2,550						$4,050	$5,950	$5,100	$340	$5,440
3	$3,600	$1,500	$2,550	$1,785					$5,835	$4,165	$4,335	$(68)	$4,267
4	$4,400	$1,500	$2,550	$1,785	$1,250				$7,085	$2,916	$3,685	$(308)	$3,377
5	$5,200	$1,500	$2,550	$1,785	$1,250	$875			$7,959	$2,041	$3,132	$(436)	$2,696
6	$6,000	$1,500	$2,550	$1,785	$1,250	$875	$612		$8,571	$1,429	$2,662	$(493)	$2,169
7	$6,800	$1,500	$2,550	$1,785	$1,250	$875	$612	$429	$9,000	$1,000	$2,263	$(505)	$1,758

Equivalent Annual Cost if the Challenger is Kept for N More Years

Holding Period N	A/T Market Value	PE of Market Value	A/T O&M Cost	PE of O&M Cost	Cum. PE of O&M Cost	Cum. PE of CCA Credit	Total PE Operating Cost	Capital Cost	Operating Cost	Total Cost
1	$7,000	$6,087	$1,200	$1,043	$1,043	$522	$522	$4,500	$600	$5,100
2	$5,440	$4,113	$1,680	$1,270	$2,314	$1,293	$1,021	$3,621	$628	$4,249
3	$4,267	$2,806	$2,160	$1,420	$3,734	$1,762	$1,972	$3,151	$863	$4,014
4	$3,377	$1,931	$2,640	$1,509	$5,243	$2,048	$3,195	$2,826	$1,119	$3,946
5	$2,696	$1,340	$3,120	$1,551	$6,795	$2,222	$4,572	$2,583	$1,364	$3,947
6	$2,169	$938	$3,600	$1,556	$8,351	$2,328	$6,023	$2,395	$1,591	$3,986
7	$1,758	$661	$4,080	$1,534	$9,885	$2,393	$7,492	$2,245	$1,801	$4,046

Cell B28 = N14
Cell C27 = NPV($M$5%,0,0,B27)
Cell C28 = NPV($M$5%,0,0,0,B28)
Cell D28 = B14*(1-$I$5/100)
Cell E27 = NPV($M$5%,0,0,D27)
Cell E28 = NPV($M$5%,0,0,0,D28)

Cell F28 = F27+E28
Cell G28 = NPV($M$5%,C14:F14)*$I$5/100
Cell H28 = F28-G28
Cell J28 = -PMT($M$5%,A28,($I$4-C28))
Cell K28 = -PMT($M$5%,A28,H28)
Cell L28 = J28+K28

Figure 11.11 Economic service life calculated with Excel for Example 11.10 — Annotated cell formulas

CCA amounts from cells C14 to F14. Then these amounts are multiplied by the tax rate (say, 40%) to determine the CCA tax credits. Compute the present equivalent for these tax credits and sum these amounts to determine the cumulative present equivalent of the CCA tax credit up to the holding period.

- Total operating cost (column H): This column represents the cumulative present equivalent figures on the total after-tax operating costs obtained by subtracting the CCA tax credits (column G) from the cumulative present value of the O&M costs (column F).

- Capital recovery cost (column J): To determine the annualized capital recovery cost, first find the net capital cost by subtracting the present value of net salvage value (column C) from the initial investment (cell I4) at the end of each holding period. Then find the annual equivalent cost by multiplying the net capital cost by the capital recovery factor up to each holding period.

- Equivalent annual operating cost (column K): Find the annualized operating cost—Multiply the cell values in column H by the capital recovery factor up to each holding period.

- Total equivalent annual cost (column L): Sum the cell values of columns J and K.

From the total cost column in L, the minimum cost occurs at the end of year 4, which is the economic service life. The specific cell formulas used to determine the entries at the economic service life (year 4) are shown at the bottom of the spreadsheet.

11.5 Summary

- **Personal investments** can provide returns in the form of **interest** (for example, savings accounts, guaranteed investment certificates, and bonds), **dividends** (stocks), and **capital gains** (bonds, stocks, and properties).
- Returns earned on personal investments are taxable at rates appropriate to the **types of income**. Analyses of personal investments must include income tax payments at the appropriate times.
- Major **financial reports** are an excellent source of information useful in assessment of the stock quality of public companies.
- **Registered retirement savings plans**, RRSPs, represent a highly attractive investment for retirement. Contributions to an RRSP account and the income earned by the contributions are not taxable until money is withdrawn from the account.

- **Registered education savings plans**, RESPs, represent a savings vehicle for the post-secondary education of one's children or grandchildren. The contributions made in an RESP account are not tax deductible. However, the growth of the money in the account is tax-free until money is withdrawn for post-secondary education purposes. In addition, the federal government offers a grant for contributions in RESPs.

- **Leasing** represents an alternative means of acquiring assets to meet a particular business need. Lease details with respect to duration, cancellations, maintenance expenses, payment timing, etc., are dependent on the individual lease; however, ownership of the asset always remains with the lessor.

- In after-tax **replacement analyses**, the **net market value**, which includes the disposal tax effect, represents the opportunity cost of keeping the defender. Analyses of the defender and challenger must include the tax effects on operating costs, capital cost allowances, and the disposals of assets at the end of the holding period. The decision criteria and rules in after-tax replacement analyses are identical to those for the before-tax situations discussed in Chapter 6.

Problems

Note: *The following assumptions apply to the problems in this chapter unless otherwise specified.*

1. After-tax analyses are to be conducted.
2. The MARR value given is after-tax MARR.
3. Capital gains tax rate is ¾ of the marginal tax rate stated.
4. The 50%-rule is used for CCA purposes.
5. For personal financing questions, cash flows may be more frequent than annual cas flows. If this is specified, don't use the end-of-year cash flow convention. However, tax payment is made once a year.
6. For problems 11.27 through 11.50, assume the MARR values given in the corresponding problems in Chapter 6 are after-tax MARR values.

Level 1

$11.1* Mr. Jackson bought a bond for $1800 on January 1, 1994. The bond had a face value of $2000 and a coupon rate of 9% with interest paid semiannually. He sold the bond for $2000 on December 31, 1999. Mr. Jackson's marginal tax rate was 47%. What was the annual after-tax rate of return that Mr. Jackson made from this investment?

$11.2* Cindy Smith bought XYZ Company's stocks at $45.50 per share 10 years ago. Over the years, the dividend payments had been at $0.10 per share paid at the end of each year. Recently, she sold the stocks at $250 per share. Her marginal tax rate was 23% for dividend income and 30% for capital gains income. What was her annual after-tax rate of return from this investment?

11.3* Consider the following lease of a trailer truck. The lease term is for 5 years with a lease payment of $12,000 at the end of each year. The lessor is responsible for regular maintenance of the truck. The lessee's marginal tax rate is 40%. The MARR is equal to 12%. What is the annual equivalent cost of leasing the truck? What if the lease payment has to be made at the beginning of each year?

11.4* Consider a defender that is 4 years old. It was bought for $100,000. It is a class 9 asset with a CCA rate of 25%. Its current market value is $60,000. The company's marginal tax rate is 40%. Its MARR is 12%. What is its net market value if it is sold today?

11.5* Consider the asset discussed in Problem 11.4. If the defender is kept, its salvage value will decline by 20% per year. Its operating cost next year is estimated to be $10,000 and will increase by $3000 per year thereafter. What is the remaining economic life of the defender? What is the annual equivalent cost corresponding to the economic life of this asset?

11.6* Consider the following challenger: purchase price = $25,000, asset class = 43, CCA rate = 30%, salvage value = $20,000 in year 1 and then decreases by 20% each year, and operating cost = $7,000 in year 1 and then increases by 25% each year. MARR = 20%. Marginal tax rate = 43%. What is the economic life of the asset considered in Problem 11.8? What is the annual equivalent cost of this asset at its economic life?

Level 2

$11.7* Reconsider Problem 3.82. Assume that the investor's marginal tax rate for interest income is 46%. Note that the tax rate for capital gains is 75% of the marginal tax rate. Answer all three questions based on after-tax cash flow analysis.

$11.8 Reconsider Problem 3.83. Assume that the investor's marginal tax rate is 40%. What is the required selling price to yield a 7% after-tax rate of return?

$11.9 Reconsider Problem 3.84. What must be the minimum selling price in order to realize a 6% after-tax rate of return? Assume that the investor's marginal tax rate is 42%. Why is the investor now only requiring a 6% rate of return, which is smaller than the quoted 9% rate of return in Problem 3.84?

$11.10 Reconsider Problem 3.85. Assume that the investor's marginal tax rate is 42%. For the calculated selling price in Problem 3.85, what is the investor's after-tax rate of return? Can you explain why it is much smaller than the required 9% rate of return quoted in Problem 3.85?

$11.11 Reconsider Problem 3.86. Assume that your marginal tax rate is 42%.

$11.12 Reconsider Problem 3.87. Assume that your marginal tax rate is 46%.

$11.13 Reconsider Problem 3.88. Assume that an investor is interested in these bonds and her marginal tax rate is 42%. Is your answer different from the one you get when tax effects are ignored?

$11.14 Reconsider Problem 3.89. Assume that your marginal tax rate is 42%.

$11.15 Reconsider Problem 3.90. Assume that the investor's marginal tax rate is 46%.

$11.16*An investor bought a stock 5 years ago for $20.20 per share. Recently he sold it for $42.40 per share. In the past 5 years, he has received dividends of $0.50 per share at the end of each year. Suppose that his marginal tax rate is 29% for dividend income and 34% for capital gains. What is the

after-tax rate of return he has earned from this investment?

$11.17 An investor with a marginal tax rate of 29% for dividend income and 34% for capital gains is comparing the stocks of the following two companies.

Company A: Current price = $10.00/share, dividends = $0.20/share/year, expected price in 5 years = $45.00

Company B: Current price = $5.00/share, dividends = $0.10/share/year, expected price in 5 years = $15.00

All dividends are paid once per year. Which stock would provide a higher rate of return?

$11.18 Suppose that you have $10,000 available for investment. You are comparing the following three options:

(a) A bond series with $1000 face value, 10% coupon rate (semiannual interest payment), selling price of $980, and maturity in 10 years.
(b) A stock with a price of $10/share, dividends of $0.10/share/year, and estimated price of $30/share in 10 years.
(c) A stock with price of $20/share, no dividends, and estimated price of $100/share in 10 years.

All dividends are paid once per year. Assume your marginal tax rate is 23% for dividend income, 38% for interest income, and 29% for capital gains. Which one would be a better choice? Why?

11.19* The Jacob Company needs to acquire a new lift truck for transporting its final product to the warehouse. One alternative is to purchase the lift truck for $40,000, which will be financed by the bank at an interest rate of 12%. The loan must be repaid in four equal installments, payable at the end of each year. Under this borrow-to-purchase arrangement, Jacob would have to maintain the truck at a cost of $1200 payable at year-end. Alternatively, Jacob could lease the truck on a 4-year contract for a lease payment of $11,000 per year. Each annual lease payment must be made at the beginning of each year. The truck would be maintained by the lessor. The truck is a class 10 asset with a CCA rate of 30%; its expected market value after 4 years is $10,000. At that time Jacob plans to replace the truck irrespective of whether it leases or buys. Jacob has a marginal tax rate of 40% and a MARR of 15%.

(a) What is Jacob's present equivalent cost of leasing?

(b) What is Jacob's present equivalent cost of owning?

(c) Should the truck be leased or purchased?

11.20 Janet Wigandt, an electrical engineer for Instrument Control Inc. (ICI), has been asked to perform a lease-buy analysis on a new pin-inserting machine for its PC-board manufacturing.

- Buy Option: The equipment costs $120,000. To purchase it, ICI could obtain a term loan for the full amount at 10% of interest with four equal annual installments (end-of-year payment). The machine has a CCA rate of 30%. Annual revenues of $200,000 and operating costs of $40,000 are anticipated. The machine requires annual maintenance at a cost of $10,000. Because technology is changing rapidly in pin-inserting machinery, the salvage value of the machine is expected to be only $20,000.

- Lease Option: Business Leasing Inc. (BLI) is willing to write a 4-year operating lease on the equipment for payments of $44,000, at the beginning of each year. Under this operating-lease arrangement, BLI will maintain the asset, so that the maintenance cost of $10,000 will be saved annually.

ICI's marginal tax rate is 40%, and its MARR is 15% during the analysis period.

(a) What is ICI's present equivalent cost of owning the equipment?

(b) What is ICI's present equivalent cost of leasing the equipment?

(c) Should ICI buy or lease the equipment?

11.21 Consider the following lease versus borrow-and-purchase problem:

- Borrow-and-Purchase Option:

    1. Jensen Manufacturing Company plans to acquire sets of special industrial tools with a 4-year life and a cost of $200,000, delivered and installed. The tools have a CCA rate of 30%.

    2. Jensen can borrow the required $200,000 at a rate of 10% over 4 years. Four equal annual payments (end-of-year) would be made in the amount of $63,094 = $200,000(A/P,10%,4). The annual interest and principal payment schedule is as follows:

End of Year	Interest	Principal
1	$20,000	$43,094
2	15,691	47,403
3	10,950	52,144
4	5,736	57,358

    3. The estimated salvage value for the tool sets at the end of 4 years is $20,000.

    4. If Jensen borrows and buys, it will have to bear the cost of maintenance, which will be performed by the tool manufacturer at a fixed contract rate of $10,000 per year.

- Lease Option:

    1. Jensen can lease the tools for 4 years at an annual rental charge of $70,000, payable at the end of each year.

    2. The lease contract specifies that the lessor will maintain the tools at no additional charge to Jensen.

Jensen's tax rate is 40%.

(a) What is Jensen's PE of after-tax cash flow of leasing at MARR = 15%?

(b) What is Jensen's PE of after-tax cash flow of owning at MARR = 15%?

$11.22 Tom Hagstrom has decided to acquire a new car for his small business. One alternative is to purchase the car outright for $16,170, financing with a bank loan for the net purchase price. The bank loan calls for 36 equal monthly payments of $541.72 at an interest rate of 12.6% compounded monthly. Payments must be made at the end of each month. A CCA rate of 30% may be applied to the vehicle. It has an expected salvage value of $5800 after 3 years.

If Tom takes the lease option, he is required to pay $500 for a security deposit refundable at the end of the lease, and $425 a month at the beginning of each month for 36 months. Use monthly cash flows in this analysis.

Tom plans to replace the car after 3 years irrespective of whether he leases or buys. Tom's marginal tax rate is 35%. His MARR is known to be 13% per year.

(a) Determine the net cash flows for each option.

(b) Which option is the better choice?

11.23 The Boggs Machine Tool Company has decided to acquire a pressing machine. One alternative is to lease the machine on a 3-year contract for a lease payment of $15,000 per year, with payments to be made at the beginning of each year. The lease would include maintenance. The second alternative is to purchase the machine outright for $100,000, financing with a bank loan for the net purchase price and amortizing the loan over a 3-year period at an interest rate of 12% per year (annual payment = $41,635).

Under the borrow-to-purchase arrangement, the company would have to maintain

the machine at an annual cost of $5000, payable at year-end. The machine falls in asset class 43; and it has an expected salvage value of $50,000, at the end of year 3. At that time the company plans to replace the machine irrespective of whether it leases or buys. Boggs has a tax rate of 40% and a MARR of 15%.

(a) What is Boggs' PE cost of leasing?

(b) What is Boggs' PE cost of owning?

(c) From the financing analysis in (a) and (b) above, what are the advantages and disadvantages of leasing and owning?

11.24 An asset is to be purchased for $25,000. The asset is expected to provide revenue of $10,000 a year, and have operating costs of $2500 a year. The asset is considered to be class 43 property with a 30% CCA rate. The company is planning to sell the asset at the end of year 5 for $5000. Given that the company's marginal tax rate is 30% and that it has a MARR of 10% for any project undertaken, answer the following questions:

(a) What is the net cash flow for each year given the asset is purchased with borrowed funds at an interest rate of 12% with repayment in five equal end-of-year payments?

(b) What is the net cash flow for each year, given the asset is leased at a rate of $3500 at the end of a year (financial lease)?

(c) Which method (if either) should be used to obtain the new asset?

11.25 The headquarters building owned by a rapidly growing company is not large enough for current needs. A search for enlarged quarters revealed two new alternatives that would provide sufficient room, enough parking, and the desired appearance and location.

- Option 1: Lease for $144,000 per year.
- Option 2: Purchase for $800,000, including a $150,000 cost for land.

- Option 3: Remodel the current head-quarters building.

It is believed that land values will not change over the ownership period, but the value of all structures will decline to 10% of the purchase price in 30 years. Annual property tax payments (income tax deductible) are expected to be 5% of the purchase price. The present headquarters building is already paid for and is now valued at $300,000. Assume that the unde-preciated capital cost for the present build-ing is $150,000. The land it is on is appraised at $60,000. Assume that the company paid $60,000 a few years ago to purchase this piece of land. The structure can be remodelled at a cost of $300,000 to make it comparable to other alterna-tives. However, the remodelling will occu-py part of the existing parking lot. An adja-cent, privately owned parking lot can be leased for 30 years under an agreement that the first year's rental of $9000 will increase by $500 each year. The annual property taxes on the remodelled property will again be 5% of the present valuation plus the cost to remodel. The study period for the comparison is 30 years, and the desired rate of return on investments is 12%. Assume that the firm's marginal tax rate is 40% and the new building and remodelled structure will be depreciated under class 1 property with $d = 4\%$. If the annual upkeep costs are the same for all three alternatives, which one is preferable?

11.26 Enterprise Capital Leasing Company is in the business of leasing tractors for construc-tion companies. The firm wants to set a 3-year lease payment schedule for a tractor purchased at $53,000 from the equipment manufacturer. The asset is classified as a class 38 property with a 30% CCA rate. The tractor is expected to have a salvage value of $22,000 at the end of 3 years' rental. Enter-prise will require a lessee to make a security deposit in the amount of $1500 that is refundable at the end of the lease term. Enterprise's marginal tax rate is 35%. If Enterprise wants an after-tax return of 10%, what lease-payment schedule should be set?

11.27* Problem 6.32 with the following additional information: The asset is a class 10 proper-ty with a CCA rate of 30%, the firm's marginal tax rate is 40%, and its after-tax MARR is 15%. In addition, assume that the tax savings on the immediate overhaul cost can be realized at time 0.

11.28 Problem 6.33 with the following additional information: The machines are class 43 property with a CCA rate of 30% and the firm's marginal tax rate is 40%.

11.29 Problem 6.34 with the following additional information: The old machine was pur-chased 10 years ago for $100,000, both the old and the new machines have a CCA rate of 30%, and the firm's marginal tax rate is 40%.

11.30 Problem 6.35 with the following additional information: Both machines are class 10 property with a CCA rate of 30% and the firm's marginal tax rate is 35%.

11.31 Problem 6.36 with the following additional information: The old machine was bought 8 years ago at a price of $40,000, both machines have a CCA rate of 30%, and the firm's marginal tax rate is 40%.

11.32 Problem 6.37 with the following additional information: Both machines are class 43 property with a CCA rate of 30% and the firm's marginal tax rate is 35%.

11.33 Problem 6.38 with the following additional information: The machine's CCA rate is 20% and the firm's marginal tax rate is 40%.

11.34 Problem 6.39 with the following additional information:

Undepreciated capital cost at the end of year n: $U_n = 20,000 - 2500\ n$.

Marginal tax rate: $t = 40\%$.

11.35 Problem 6.41 with the following additional information: The asset's CCA rate is 30% and the firm's marginal tax rate is 40%.

11.36* Problem 6.42 with the following additional information: The old machine was bought 7 years ago with a purchase price of $120,000, both machines are class 43 property with a CCA rate of 30%, and the firm's marginal tax rate is 40%.

11.37 Problem 6.43 with additional information given in Problem 11.36 and below: Model B is also a class 43 property with a CCA rate of 30%.

11.38 Problem 6.44 with the following additional information: Both machines are class 43 property with a CCA rate of 30% and the firm's marginal tax rate is 30%.

11.39 Problem 6.45 with the following additional information: The old machine's undepreciated capital cost is $3000, both machines are class 43 property with a CCA rate of 30%, and the firm's marginal tax rate is 40%.

11.40 Problem 6.46 with the following additional information: The old asset has an undepreciated capital cost of $4000, both machines are class 43 property with a CCA rate of 30%, and the firm's marginal tax rate is 40%.

11.41 Problem 6.47 with the following additional information: Both machines are class 43 property with a CCA rate of 30% and the firm's marginal tax rate is 40%.

11.42 Problem 6.49 with the following additional information: Both machines are class 8 property with a CCA rate of 20% and the firm's marginal tax rate is 35%.

11.43 Problem 6.53 with the following additional information: Both assets are class 43 property with a CCA rate of 30% and the firm's marginal tax rate is 40%.

11.44 Problem 6.54 with the following additional information: The old machine's undepreciated capital cost is $2000, both machines are class 43 property with a CCA rate of 30%, and the firm's marginal tax rate is 40%.

11.45 Problem 6.55 with the following additional information: Both machines are class 43 property with a CCA rate of 30% and the firm's marginal tax rate is 30%.

11.46 Problem 6.58 with the following additional information: Both machines are class 43 property with a CCA rate of 30% and the firm's marginal tax rate is 30%.

Level 3

11.47 Problem 6.60 with the following additional information: The storage facility will be depreciated by the straight-line method with N = 10 and S = 0 according to Kazakhstan's tax law and the firm's marginal tax rate is 30%. Assume that the 50% rule does not apply in Kazakhstan. Assume that the existing storage facility cost $250,000 to build 5 years ago and now has a market value of zero.

11.48 Problem 6.61 with the following additional information: The current asset has an undepreciated capital cost of $100,000, both systems are class 43 property with a CCA rate of 30%, and the firm's marginal tax rate is 40%.

11.49 Problem 6.62 with the following additional information: The old edger's undepreciated capital cost is $20,000, the buildings have a 4% CCA rate, the equipment has a 30% CCA rate, and the firm's marginal tax rate is 40%.

11.50 Problem 6.63 with the following additional information: The current undepreciated capital cost of the old machine is $14,286, both machines are class 43 property with a CCA rate of 30%, and the firm's marginal tax rate is 40%.

11.51 National Office Automation Inc. (NOAI) is a leading developer of imaging systems, controllers, and related accessories. The company's product line consists of systems for desktop publishing, automatic identification, advanced imaging, and office information markets. The firm's manufacturing plant in Windsor, Ontario, consists of eight different functions: cable assembly, board assembly, mechanical assembly, controller integration, printer integration, production repair, customer repair, and shipping. The process to be considered is the transportation of pallets loaded with eight packaged desktop printers from printer integration to the shipping department. Several alternatives have been examined to minimize operating and maintenance costs. The two most feasible alternatives are:

- Option 1: Use of gas-powered lift trucks to transport pallets of packaged printers from printer integration to shipping. The truck also is used to return printers that must be reworked. The trucks can be leased at a cost of $5465 per year. With a maintenance contract costing $6317 per year, the dealer will maintain the trucks. A fuel cost of $1660 per year is also expected. The truck requires a driver for each of the three shifts, at a total cost of $58,653 per year for labor. It is also estimated that transportation by truck would cause damages to material and equipment totalling $10,000 per year.

- Option 2: Installing an automatic guided vehicle system (AGVS) to transport pallets of packaged printers from printer integration to shipping and to return products that require rework. The AGVS, using an electrical powered cart and embedded wire-guidance system, would do the same job that the truck currently does, but without drivers. The total investment costs, including installation, are itemized as follows:

Vehicle and system installation	$97,255
Staging conveyor	24,000
Power supply lines	5,000
Transformers	2,500
Floor surface repair	6,000
Batteries and charger	10,775
Shipping	6,500
Sales tax	6,970
Total AGVS system cost	$159,000

NOAI could obtain a term loan for the full investment amount ($159,000) at a 10% interest rate. The loan would be amortized over 5 years, with payments made at the end of each year. The AGVS are considered class 43 property, and it has an estimated service life of 10 years and no salvage value. If the AGVS is installed, a maintenance contract would be obtained at a cost of $20,000, payable at the beginning of each year. The firm's marginal tax rate is 35%, and its MARR is 15%.

(a) Determine the net cash flows for each alternative over 10 years.
(b) Compute the incremental cash flows (Option 2 – Option 1) and determine the rate of return on this incremental investment.
(c) Determine the best course of action based on the rate of return criterion.

CHAPTER 12

Inflation and Economic Analysis

You may have heard your grandparents fondly remembering the "good old days" of penny candy or 35 cent-per-gallon (i.e., 7.7¢/litre) gasoline. But even a college student in his or her early twenties can relate to the phenomenon of escalating costs. Do you remember when a postage stamp cost less than half its current price? When the admission for a movie was under $5?

Item	1979 Price	1998 Price	% Increase
Consumer price index (CPI) [1992=100]	47.6	108.6	128.2
Annual housing expense	$3429	$7629	122.5
Annual automobile expense	1067	2948	176.3
Store brand bread (675g)	0.60	1.30	116.7
Large eggs (1 doz, grade A)	1.06	1.78	67.9
Milk (1L)	0.61	1.42	132.8
Soft drink (per litre)	0.68	0.60	−11.8
Round steak (kg)	6.42	9.50	48.0
Postage	0.17	0.45	164.7
College tuition	895	4427	394.7

The accompanying table demonstrates price differences between 1979 and 1998 for some commonly bought items. For example, a loaf of bread cost 60 cents in 1979, whereas it cost $1.30 in 1998. In 1998, the same 60 cents bought only a fraction of the bread it would have bought in 1979—specifically, about 46.15% of a loaf. From 1979 to 1998, the 60 cent sum had lost 53.85% of its purchasing power. Of course, wages have also increased during this period. In the following table, statistics about pay rates for several engineering disciplines are shown[1]. The average salary of an engineer employed full-time in Canada was $29,831 in 1980, according to the 1981 Census[2]. By 1995, the average engineer's employment income had risen to $55,400, which is equivalent to an annual increase of 4.21% over 15 years. During the same period, the annual inflation rate remained at 4.69%, indicating that engineers' pay did not quite keep up with the rise in prices in Canada.

Professional Area	1980 Average Pay	1995 Average Pay	% Increase
Chemical	$32,113	$59,029	83.8
Civil	31,181	53,606	71.9
Electrical	29,211	57,054	95.3
Industrial	26,505	53,128	100.4
Mechanical	28,791	54,081	87.8
Metallurgical	31,229	60,251	92.9
Mining	33,669	62,032	84.2
Petroleum	36,519	72,543	98.6
Total	$29,831	$55,400	85.7

Up to this point, we have demonstrated how to develop cash flows in a variety of ways and how to compare them under constant conditions in the general economy. In other words, we have assumed that prices remain relatively unchanged over long periods. As you know from personal experience, this is not a realistic assumption. In this chapter, we will define and quantify **inflation** and then go on to apply it in several economic analyses. We will demonstrate inflation's effect on capital cost allowance, borrowed funds, rate of return of a project, and working capital, within the bigger picture of developing project cash flows.

---

[1] You may be tempted to rate the financial merits of different engineering disciplines based solely on these values. This could be misleading for many reasons. For example, some specialties may have fewer senior engineers in the workforce, either because the demands have changed over time, or people get promoted into nonengineering managerial positions. Most provincial engineering associations can provide detailed current salary data broken down by experience, levels of responsibility, and other factors.

[2] Statistics Canada, 1981 Census, Cat. No. 92-930 and Cat. No. 92-919.

12.1 Meaning and Measure of Inflation

Historically, the general economy has usually fluctuated in such a way as to experience **inflation**, a loss in the purchasing power of money over time. Inflation means that the cost of an item tends to increase over time, or, to put it another way, the same dollar amount buys less of an item over time. **Deflation** is the opposite of inflation in that prices usually decrease over time, and hence, a specified dollar amount gains in purchasing power. Inflation is far more common than deflation in the real world, so our consideration in this chapter will be restricted to accounting for inflation in economic analyses.

12.1.1 Measuring Inflation

Before we can introduce inflation into an engineering economic problem, we need a means of isolating and measuring its effect. Consumers usually have a relative, if not a precise, sense of how their purchasing power is declining. This sense is based on their experience of shopping for food, clothing, transportation, and housing over the years. Economists have developed a measure called the **consumer price index** (CPI), which is based on a typical **market basket** of goods and services required by the average consumer. This market basket consists of hundreds of items from eight major groups: (1) food, (2) shelter, (3) household operations and furnishings, (4) clothing and footwear, (5) transportation, (6) health and personal care, (7) recreation, education and reading, and (8) alcoholic beverages and tobacco products.

The CPI compares the cost of the typical market basket of goods and services in a current month with its cost 1 month before, 1 year before, or 10 years before. The point in the past to which current prices are compared is called the **base period**. The index value for this base period is set at 100. The CPI for any other year can then be calculated in relation to the base year:

$$\frac{\text{CPI}_{\text{Any.year}}}{\text{CPI}_{\text{Base.year}}} = \frac{\text{Cost}_{\text{Any.year}}}{\text{Cost}_{\text{Base.year}}} \quad \text{or} \quad \text{CPI}_{\text{Any.year}} = \frac{\text{Cost}_{\text{Any.year}}}{\text{Cost}_{\text{Base.year}}} \times 100$$

At the time of writing, the current base period in Canada is 1992. That is, the total cost of the prescribed market basket in 1992 is given a price index of 100. The *Consumer Price Index* (Catalogue No 62-001), a monthly publication prepared by Statistics Canada, lists CPIs for the country, provinces and select cities. From Table 12.1, the CPI for the year 1998 is 108.6, indicating an increase of $(108.6 - 100)/100 = 8.6\%$ over 1992 aggregate prices. Given the CPI, the consumer product inflation rate can be calculated just as easily between *any* two years, such as 1988–1998:

$$\frac{\text{CPI}_{1998} - \text{CPI}_{1988}}{\text{CPI}_{1988}} = \frac{108.6 - 84.8}{84.8} = 0.2807 = 28.07\%$$

showing a modest increase over that decade.

Year	CPI	IPPI	RMPI	Capital Equipment	Lumber and Timber	Primary Steel Products
1986	78.1	91.7	94.6	89.4	89.1	102.9
1987	81.5	94.2	101.5	90.4	88.2	104.5
1988	84.8	98.3	98.2	90.8	88.2	110.6
1989	89.0	100.3	101.4	93.0	88.9	113.3
1990	93.3	100.6	105.6	95.0	87.4	111.0
1991	98.5	99.5	99.0	96.7	86.3	107.8
1992	100.0	100.0	100.0	100.0	100.0	100.0
1993	101.8	103.6	105.3	105.0	140.5	103.8
1994	102.0	109.9	114.7	109.4	160.2	116.1
1995	104.2	118.1	124.7	113.2	135.8	128.8
1996	105.9	118.6	129.0	115.6	159.5	124.8
1997	107.6	119.5	126.9	117.9	167.1	126.7
1998	108.6	119.4	108.4	123.6	148.9	127.8

Table 12.1 Selected Price Indexes (Base Year 1992 = 100)

This method of assessing inflation does not imply, however, that consumers actually purchase the same goods and services year after year. Consumers tend to adjust their shopping practices to changes in relative prices and to substitute other items for those whose prices have greatly increased in relative terms. We must understand that the CPI does not take into account this sort of consumer behaviour, because it is predicated on the purchase of a fixed market basket of the same goods and services, in the same proportion, month after month. For this reason, the CPI is called a **price index** rather than a **cost-of-living index**, although the general public often refers to it as a cost-of-living index. Furthermore, the CPI is not a measure of cost-of-living in the traditional sense of "necessities" for survival.

On the other hand, adjustments are made to the basket of goods every four years or so to update expenditure weights as consumers change their spending patterns, and to incorporate occasional technical improvements. A significant change in 1995 was to extend the price sampling from cities of 30,000 population or more to *all* areas of Canada, rural and urban. Despite tremendous changes in the consumer world, the CPI has served its function well since its inception in the early 1900s.

Industry Price Indexes

The consumer price index is a good measure of the general price increase of consumer products. However, it is not a good measure of industrial price increases. When performing engineering economic analysis, the appropriate price indexes must be selected to estimate the price increases of raw materials, finished products, and operating costs. *Industrial Price Indexes* (Catalogue 62-011), a monthly publication prepared by Statistics Canada, provides the industrial product price index (IPPI) and the raw material price index (RMPI) for various industrial goods[3]. Table 12.1 lists the IPPI and RMPI together with select other price indexes over a number of years.

From Table 12.1, we can easily calculate the inflation rate of lumber and timber from 1997 to 1998 as follows:

$$\frac{148.9 - 167.1}{167.1} = -0.1089 = -10.89\%.$$

Since the inflation rate calculated is negative (deflation), the price of lumber and timber decreased at an annual rate of 10.89% over the year 1998. Although the long-term trend is upwards, the prices of some commodities can be quite volatile in the short term as shown in the table.

Average Inflation Rate (*f*)

To account for the effect of varying yearly inflation rates over a period of several years, we can compute a single rate that represents an **average inflation rate**. Since each individual year's inflation rate is based on the previous year's rate, these rates have a compounding effect. As an example, suppose we want to calculate the average inflation rate for a 2-year period: The first year's inflation rate is 4%, and the second year's rate is 8%, using a base price of $100.

- Step 1: To find the price at the end of the second year we use the process of compounding:

First year

$$\$100(1 + 0.04)(1 + 0.08) = \$112.32.$$

Second year

- Step 2: To find the average inflation rate *f*, we establish the following equivalence equation:

$$\$100(1 + f)^2 = \$112.32 \text{ or } \$100(F/P, f, 2) = \$112.32.$$

Solving for *f* yields

$$f = 5.98\%.$$

We can say that the price increases in the 2 years are equivalent to an average annual percentage rate of 5.98% per year. Note that the average is a geometric, not an arithmetic, average over a several-year period. Our computations are simplified by using a single average rate such as this, rather than a different rate for each year's cash flows.

---

[3] Price indexes are now available for a nominal fee via the Internet through Statistics Canada Website: http://www.statscan.ca. Some university libraries may provide access to this data as a service.

EXAMPLE
12.1

Average inflation rate

Reconsider the price increases for the nine items listed in the table at the opening of this chapter. Determine the average inflation rate for each item over the 19-year period.

Solution

Let's take the first item, the annual housing expense, for a sample calculation. Since we know the prices during both 1979 and 1998, we can use the appropriate equivalence formula (single-payment compound amount factor or growth formula).

Given: $P = \$3429$, $F = \$7629$, and $N = 1998 - 1979 = 19$
Find: f

Equation: $F = P(1 + f)^N$

$$\$7629 = \$3429(1 + f)^{19}$$
$$f = \sqrt[19]{2.2248} - 1$$
$$= 0.0430 = 4.30\%.$$

In a similar fashion, we can obtain the average inflation rates for the remaining items as follows:

Item	1979 Price	1998 Price	Average Inflation Rate
Annual housing expense	$3429.00	$7629.00	4.30%
Annual automobile expense	1067.00	2948.00	5.49
Store brand bread (675g)	0.60	1.30	4.15
Large eggs (1doz, grade A)	1.06	1.78	2.77
Milk (1L)	0.61	1.42	4.55
Soft drink (per litre)	0.68	0.60	−0.66
Round steak (kg)	6.42	9.50	2.08
Postage	0.17	0.45	5.26
College tuition	895.00	4427.00	8.78

The cost of college tuition increased most among the items listed in the table. Family-size soft drinks actually dropped in price!

General Inflation Rate ($\bar{f}$) versus Specific Inflation Rate (f_j)

When we use the CPI as a base to determine the average inflation rate, we obtain the **general inflation rate**. We need to distinguish carefully between the general inflation rate and the average inflation rate for specific goods:

- **General inflation rate ($\bar{f}$) :** This average inflation rate is calculated based on the CPI for all items in the market basket. The market interest rate is expected to respond to this general inflation rate.

- **Specific inflation rate (f_j):** This rate is based on an index (or the CPI) specific to segment j of the economy. For example, we must often estimate the future cost for an item, e.g., labor, material, housing, or gasoline. (When we refer to the average inflation rate for just one item, we will drop the subscript j for simplicity.)

In terms of CPI, we define the general inflation rate as

$$CPI_{n2} = CPI_{n1} (1 + \bar{f})^{n2 - n1},$$ (12.1)

or

$$\bar{f} = \left[\frac{CPI_{n2}}{CPI_{n1}} \right]^{\frac{1}{n2 - n1}} - 1,$$ (12.2)

where
$\bar{f}$ = The general inflation rate,
CPI_{n1} = The consumer price index for the initial period $n1$,
CPI_{n2} = The consumer price index for the end period $n2$.

Knowing the CPI values for 2 consecutive years, we can calculate the annual general inflation rate:

$$\bar{f}_n = \frac{CPI_n - CPI_{n-1}}{CPI_{n-1}},$$ (12.3)

where $\bar{f}_n$ = the general inflation rate for period n.

As an example, let us calculate the general inflation rate for the year 1998, where $CPI_{1997} = 107.6$ and $CPI_{1998} = 108.6$:

$$\frac{108.6 - 107.6}{107.6} = 0.0093 = 0.93\%,$$

which was an unusually good year for the Canadian economy, when compared with the average general inflation rate of 4.44%[4] over the last 19 years.

---

[4] Using 1992 as the base year, the $CPI_{1979} = 47.6$, and from Table 12.1 $CPI_{1998} = 108.6$. To calculate the average general inflation rate from 1979 to 1998, we evaluate the following:

$$\bar{f} = \left[\frac{108.6}{47.6} \right]^{1/19} - 1 = 4.44\%.$$

EXAMPLE
12.2

Yearly and average inflation rates

The following table shows a utility company's cost to supply a fixed amount of power to a new housing development; the indices are specific to the utilities industry. Assume that year 0 is the base period.

Year	Cost
0	$504,000
1	538,400
2	577,000
3	629,500

Determine the inflation rate for each period, and calculate the average inflation rate over the 3 years.

Solution

Inflation rate during year 1 (f_1):

$$(\$538,400 - \$504,000)/\$504,000 = 6.83\%.$$

Inflation rate during year 2 (f_2):

$$(\$577,000 - \$538,400)/\$538,400 = 7.17\%.$$

Inflation rate during year 3 (f_3):

$$(\$629,500 - \$577,000)/\$577,000 = 9.10\%.$$

The average inflation rate over 3 years is

$$f = \left(\frac{\$629,500}{\$504,000}\right)^{1/3} - 1 = 0.0769 = 7.69\%$$

$$\text{or } f = [(1 + 0.0683)(1 + 0.0717)(1 + 0.0910)]^{1/3} = 7.69\%$$

Note that, although the average inflation rate[5] is 7.69% for the period taken as a whole, none of the years within the period had this rate.

12.1.2 Actual versus Constant Dollars

To introduce the effect of inflation into our economic analysis, we need to define several inflation-related terms.[6]

- **Actual (current) dollars** (A_n): Actual dollars are estimates of future cash flows for year n that take into account any anticipated changes in amount due to inflationary or deflationary effects. Usually these amounts are determined by applying an inflation rate to base year dollar estimates.

---

[5] Since we obtained this average rate based on costs that are specific to the utility industry, this rate is not the general inflation rate. This is a specific inflation rate for this utility.

[6] Based on the ANSI Z94 Standards Committee on Industrial Engineering Terminology, *The Engineering Economist.* 1988; 33(2): 145–171.

- **Constant (real) dollars (A'_n):** Constant dollars represent constant purchasing power independent of the passage of time. In situations where inflationary effects were assumed when cash flows were estimated, these estimates can be converted to constant dollars (base year dollars) by adjustment using some readily accepted **general inflation rate**. We will assume that the base year is always time 0 unless we specify otherwise.

Conversion from Constant to Actual Dollars

Since constant dollars represent dollar amounts expressed in terms of the purchasing power of the base year, we may find the equivalent dollars in year n using the general inflation rate ($\bar{f}$)

$$A_n = A'_n(1 + \bar{f})^n = A'_n(F/P, \bar{f}, n), \qquad (12.4)$$

where A'_n = Constant-dollar expression for the cash flow occurring at the end of year n, and

A_n = Actual-dollar expression for the cash flow at the end of year n.

If the future price of a specific cash flow element (j) is not expected to follow the general inflation rate, we will need to use the appropriate average inflation rate applicable to this cash flow element, f_j, instead of $\bar{f}$.

EXAMPLE 12.3 **Conversion from constant to actual dollars**

Transco Company is considering making and supplying computer-controlled traffic-signal switching boxes to be used throughout Western Canada. Transco has estimated the market for its boxes by examining data on new road construction and on deterioration and replacement of existing units. The current price per unit is $550; the before-tax manufacturing cost is $450. The start-up investment cost is $250,000. The projected sales and net before-tax cash flows in constant dollars are as follows:

Period	Unit Sales	Net Cash Flow in Constant $
0		−$250,000
1	1000	100,000
2	1100	110,000
3	1200	120,000
4	1300	130,000
5	1200	120,000

Assume that the price per unit as well as the manufacturing cost keeps up with the general inflation rate, which is projected to be 5% annually. Convert the project's before-tax cash flows into the equivalent actual dollars.

Solution

We first convert the constant dollars into actual dollars. Using Eq. (12.4), we obtain the following (note that the cash flow in year 0 is not affected by inflation):

Period	Net Cash Flow in Constant $	Conversion Factor	Cash Flow in Actual $
0	−$250,000	$(1 + 0.05)^0$	−$250,000
1	100,000	$(1 + 0.05)^1$	105,000
2	110,000	$(1 + 0.05)^2$	121,275
3	120,000	$(1 + 0.05)^3$	138,915
4	130,000	$(1 + 0.05)^4$	158,016
5	120,000	$(1 + 0.05)^5$	153,154

Figure 12.1 illustrates this conversion process graphically.

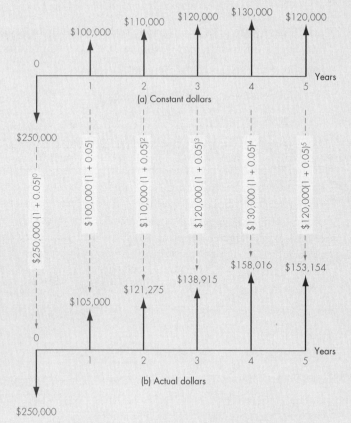

Figure 12.1 Converting constant dollars (a) to equivalent actual dollars (b) (Example 12.3)

Conversion from Actual to Constant Dollars

This is the reverse process of converting from constant to actual dollars. Instead of using the compounding formula, we use a discounting formula (single-payment present worth factor):

$$A'_n = \frac{A_n}{(1 + \bar{f})^n} = A_n(P/F, \bar{f}, n).$$

(12.5)

Once again, we may substitute f_j for $\bar{f}$ if future prices are not expected to follow the general inflation rate.

EXAMPLE 12.4

Conversion from actual to constant dollars

Jagura Creek Fish Company, an aquacultural production firm, has negotiated a 5-year lease on 20 acres of land, which will be used for fish ponds. The annual cost stated in the lease is $20,000, to be paid at the beginning of each of the 5 years. The general inflation rate $\bar{f}$ = 5%. Find the equivalent cost in constant dollars in each period.

Discussion: Although the $20,000 annual payments are *uniform*, they are not expressed in constant dollars. Unless an inflation clause is built into a contract, any stated amounts refer to *actual dollars*.

Solution

Using Eq. (12.5), the equivalent lease payments in constant dollars can be determined as follows:

End of Period	Cash Flow in Actual $	Conversion at f	Cash Flow in Constant $	Loss in Purchasing Power
0	−$20,000	$(1 + 0.05)^0$	−$20,000	0%
1	−20,000	$(1 + 0.05)^{-1}$	−19,048	4.76
2	−20,000	$(1 + 0.05)^{-2}$	−18,141	9.30
3	−20,000	$(1 + 0.05)^{-3}$	−17,277	13.62
4	−20,000	$(1 + 0.05)^{-4}$	−16,454	17.73

Note that, under the inflationary environment, the lender's receipt of the lease payment in year 5 is worth only 82.27% of the first lease payment.

12.2 Equivalence Calculations under Inflation

In previous chapters, our equivalence analyses took into consideration changes in the **earning power** of money, i.e., interest effects. To factor in changes in **purchasing power** as well, i.e., inflation, we may use either (1) constant dollar analysis or (2) actual dollar analysis. Either method produces the same solution; however, each method requires use of a different interest rate and procedure. Before presenting the two procedures for integrating interest and inflation, we will give a precise definition of the two interest rates.

12.2.1 Market and Inflation-Free Interest Rates

Two types of interest rates are used in equivalence calculations: (1) the market interest rate and (2) the inflation-free interest rate. The rate to apply depends on the assumptions used in estimating the cash flow.

- **Market interest rate (i):** This rate takes into account the combined effects of the earning value of capital (earning power) and any anticipated inflation or deflation (purchasing power). Virtually all interest rates stated by financial institutions for loans and savings accounts are market interest rates. Most firms use a market interest rate (also known as **inflation-adjusted MARR**) in evaluating their investment projects.

- **Inflation-free interest rate (i'):** This rate is an estimate of the true earning power of money when the inflation effects have been removed. This rate is commonly known as **real interest rate**, and it can be computed if the market interest rate and inflation rate are known. As you will see later in this chapter, in the absence of inflation, the market interest rate is the same as the inflation-free interest rate. Therefore, all the interest rates mentioned in previous chapters are inflation-free interest rates.

In calculating any cash flow equivalence, we need to identify the nature of project cash flows. The three common cases follow:

Case 1: All cash flow elements are estimated in constant dollars.

Case 2: All cash flow elements are estimated in actual dollars.

Case 3: Some of the cash flow elements are estimated in constant dollars, and others are estimated in actual dollars.

For case 3, we simply convert all cash flow elements into one type—either constant or actual dollars. Then we proceed with either constant-dollar analysis as for case 1 or actual-dollar analysis as for case 2.

12.2.2 Constant Dollar Analysis

Suppose that all cash flow elements are already given in constant dollars, and that you want to compute the present equivalent of the constant dollars (A_n') occurring in year n. In the absence of an inflationary effect, we should use i' to account for only the earning power of the money. To find the present equivalent of this constant-dollar amount at i', we use

$$P_n = \frac{A_n'}{(1 + i')^n} \tag{12.6}$$

Since income taxes are levied based on taxable incomes in actual dollars, constant dollar analysis can only be used under certain conditions. One common application is the evaluation of many long-term public projects, because governments do not pay income taxes.

EXAMPLE 12.5

Equivalence calculation when flows are stated in constant dollars

Consider the constant dollar flows originally given in Example 12.3. If Transco managers want the company to earn a 12% inflation-free rate of return (i') before tax on any investment, what would be the present equivalent of this project?

Solution

Since all values are in constant dollars, we can use the inflation-free interest rate. We simply discount the cash flows at 12% to obtain the following:

$$
\begin{aligned}
PE\,(12\%) \;=\; & -\$250{,}000 + \$100{,}000(P/A,12\%,\,5)\\
& + \$10{,}000(P/G,12\%,\,4) + \$20{,}000(P/F,12\%,\,5)\\
=\; & \$163{,}099
\end{aligned}
$$

Since the equivalent net receipts exceed the investment, the project can be justified before considering any tax effect.

12.2.3 Actual Dollar Analysis

Now let us assume that all cash flow elements are estimated in actual dollars. To find the equivalent present worth of this actual dollar amount (A_n) in year n, we may use either the **deflation method** or the **adjusted-discount method**.

Deflation Method

The deflation method requires two steps to convert actual dollars into equivalent present worth dollars. First we convert actual dollars into equivalent constant dollars by discounting by the general inflation rate, a step that removes the inflationary effect. Then we find the present equivalent using i'.

EXAMPLE
12.6

Equivalence calculation when cash flows are in actual dollars: Deflation method

Applied Instrumentation, a small manufacturer of custom electronics, is contemplating an investment to produce sensors and control systems that have been requested by a fruit drying company. The work would be done under a proprietary contract, which would terminate in 5 years. The project is expected to generate the following cash flows in actual dollars.

n	Net Cash Flow in Actual Dollars
0	–$75,000
1	32,000
2	35,700
3	32,800
4	29,000
5	58,000

(a) What are the equivalent year 0 dollars (constant dollars) of these cash flows, if the general inflation rate $(\bar{f})$ is 5% per year?

(b) Compute the present equivalent of these cash flows in constant dollars at $i' = 10\%$.

Solution

The net cash flows in actual dollars can be converted to constant dollars by deflating them, again assuming a 5% yearly deflation factor. The deflated or constant dollar cash flows can then be used to determine the PE at i'.

(a) We convert the actual dollars into constant dollars as follows:

n	Cash Flows in Actual Dollars	Multiplied by Deflation Factor	Cash Flows in Constant Dollars
0	–$75,000	1	–$75,000
1	32,000	$(1 + 0.05)^{-1}$	30,476
2	35,700	$(1 + 0.05)^{-2}$	32,381
3	32,800	$(1 + 0.05)^{-3}$	28,334
4	29,000	$(1 + 0.05)^{-4}$	23,858
5	58,000	$(1 + 0.05)^{-5}$	45,445

(b) We compute the equivalent present worth of constant dollars using $i' = 10\%$:

n	Cash Flows in Constant Dollars	Multiplied by Discounting Factor	Equivalent Present Worth
0	−$75,000	1	−$75,000
1	30,476	$(1 + 0.10)^{-1}$	27,706
2	32,381	$(1 + 0.10)^{-2}$	26,761
3	28,334	$(1 + 0.10)^{-3}$	21,288
4	23,858	$(1 + 0.10)^{-4}$	16,295
5	45,445	$(1 + 0.10)^{-5}$	28,218
			$45,268

Figure 12.2 shows the conversion from the cash flows stated in year 0 dollars to equivalent present worth.

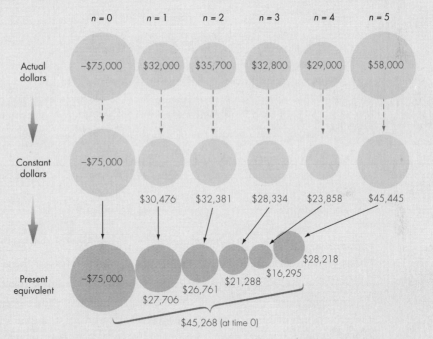

Figure 12.2 Illustration of deflation method: Converting actual dollars to constant dollars and then to equivalent present worth (Example 12.6)

Adjusted-Discount Method

The two-step process shown in Example 12.6 can be greatly streamlined by the efficiency of the **adjusted-discount method**, which performs deflation and discounting in one step. Mathematically we can combine this two-step procedure into one by:

$$P_n = \cfrac{\cfrac{A_n}{(1+\bar{f})^n}}{(1+i')^n}$$

$$= \cfrac{A_n}{(1+\bar{f})^n(1+i')^n}$$

$$= \cfrac{A_n}{[(1+\bar{f})(1+i')]^n}. \qquad (12.7)$$

Since the market interest rate (i) reflects both the earning power and the purchasing power, we have the following relationship:

$$P_n = \frac{A_n}{(1+i)^n}. \qquad (12.8)$$

The equivalent present worth values in Eqs. (12.7) and (12.8) must be equal at year 0. Therefore,

$$\frac{A_n}{(1+i)^n} = \frac{A_n}{[(1+\bar{f})(1+i')]^n}.$$

This leads to the following relationship among $\bar{f}$, i', and i:

$$(1+i) = (1+\bar{f})(1+i')$$
$$= 1 + i' + \bar{f} + i'\bar{f}.$$

Simplifying the terms yields:

$$i = i' + \bar{f} + i'\bar{f}.$$

or equivalently, $\qquad i = (1+i')(1+\bar{f}) - 1. \qquad (12.9)$

This implies that the market interest rate is a function of two terms, i' and $\bar{f}$. Note that without an inflationary effect, the two interest rates are the same ($\bar{f} = 0 \rightarrow i = i'$). As either i' or $\bar{f}$ increases, i also increases. For example, we can easily observe that, when prices increase due to inflation, bond rates climb, because lenders (i.e., anyone who invests in a money-market fund, a bond, or a guaranteed investment certificate of deposit) demand higher rates to protect themselves against erosion in the value of their dollars. If inflation were to remain at 3%, you might be satisfied with an interest rate of 7% on a bond because your return would more than beat inflation. If inflation were running at 10%, however, you would not buy a 7% bond; you might insist instead on a return of at least 14%. On the other hand, when prices are coming down, or at least are stable, lenders do not fear the loss of purchasing power with the loans they make, so they are satisfied to lend at lower interest rates.

EXAMPLE 12.7 Equivalence calculation when flows are in actual dollars: Adjusted-discounted method

Consider the cash flows in actual dollars in Example 12.6. Compute the equivalent present worth of these cash flows using the adjusted-discount method.

Solution

First, we need to determine the market interest rate i. With $\bar{f} = 5\%$ and $i' = 10\%$, we obtain

$$
\begin{aligned}
i &= i' + \bar{f} + i'\,\bar{f} \\
&= 0.10 + 0.05 + (0.10)(0.05) \\
&= 15.5\%.
\end{aligned}
$$

n	Cash Flows in Actual Dollars	Multiplied by	Equivalent Present Worth
0	−$75,000	1	−$75,000
1	32,000	$(1 + 0.155)^{-1}$	27,706
2	35,700	$(1 + 0.155)^{-2}$	26,761
3	32,800	$(1 + 0.155)^{-3}$	21,288
4	29,000	$(1 + 0.155)^{-4}$	16,295
5	58,000	$(1 + 0.155)^{-5}$	28,218
			$45,268

The conversion process is shown in Figure 12.3. Note that the equivalent present worth that we obtain using the adjusted-discount method ($i = 15.5\%$) is exactly the same as the result we obtained in Example 12.6.

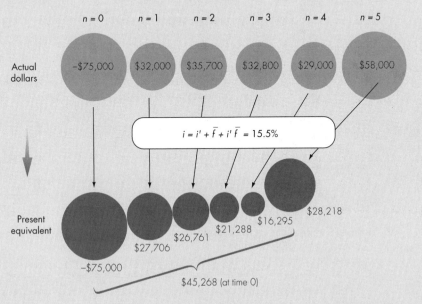

Figure 12.3 Illustration of adjusted-discount method: Converting actual dollars to present equivalent dollars by applying the market interest rate (Example 12.7)

12.2.4   Mixed Dollar Analysis

We will consider another situation in which some cash flow elements are expressed in constant (or today's) dollars and the other elements in actual dollars. In this situation, we can convert all cash flow elements into the same dollar units (either constant or actual). If the cash flow is converted into actual dollars, the market interest rate (i) should be used in calculating the equivalence value. If the cash flow is converted in terms of constant dollars, the inflation-free interest rate (i') should be used. Example 12.8 illustrates this situation.

EXAMPLE
12.8
$

Equivalence calculation with composite cash flow elements

A couple wishes to establish a college fund at a bank for their 5-year-old child. The college fund will earn 8% interest, compounded quarterly. Assuming that the child enters college at age 18, the couple estimates that an amount of $30,000 per year, in terms of today's dollars, will be required to support the child's college expenses for 4 years. The college expense is estimated to increase at the annual rate of 6%. Determine the equal quarterly deposits the couple must make until they send their child to college. Assume that the first deposit will be made at the end of the first quarter, and deposits will continue until the child reaches age 17. The child will enter college at age 18, and the annual college expense will be paid at the beginning of each college year. In other words, the first withdrawal will be made at age 18.

Discussion: In this problem, future college expenses are expressed in terms of today's dollars, whereas the quarterly deposits are in actual dollars. Since the interest rate quoted on the college fund is a market interest rate, we may convert the future college expenses into actual dollars.

Age	College expenses (in today's dollars)	College expenses (in actual dollars)		
18 (1st year)	$30,000	$30,000(F/P, 6%,13)	=	$63,988
19 (2nd year)	30,000	30,000(F/P, 6%,14)	=	67,827
20 (3rd year)	30,000	30,000(F/P, 6%,15)	=	71,897
21 (4th year)	30,000	30,000(F/P, 6%,16)	=	76,211

Solution

College expenses as well as the quarterly deposit series in actual dollars are shown in Figure 12.4.

We first select $n = 12$ years or age 17 as the base period for our equivalence calculation. Then, calculate the accumulated total amount at the base period at 2% interest per quarter. Since the deposit period is 12 years, we have a total of 48 quarterly deposits. If the first deposit is made at the end of first quarter, we have a 48-quarter deposit period. Therefore, the total deposit balance at age 17 would be

$$V_1 = C(F/A, 2\%, 48)$$
$$= \boxed{79.3535C}.$$

The equivalent lump-sum worth of the total college expenditure at the base period would be

$$V_2 = \$63{,}988(P/F, 2\%, 4) + \$67{,}827(P/F, 2\%, 8)$$
$$+ \$71{,}897(P/F, 2\%, 12) + \$76{,}211(P/F, 2\%, 16)$$
$$= \$229{,}211.$$

By setting $V_1 = V_2$ and solving for C, we obtain

$$79.3535C = \$229{,}211$$
$$C = \$2888.48 \text{ per quarter.}$$

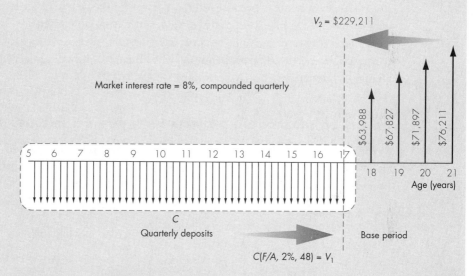

Figure 12.4   Establishing a college fund under an inflationary economy for a 5-year-old child by making 48 quarterly deposits (Example 12.8)

Comments: In this problem, we have ignored income taxes which may be owed on the interest earned in the fund. If the money is invested in a registered education savings plan (RESP), the interest is normally tax-exempt.

12.3 Effects of Inflation on Project Cash Flows

We will now introduce inflation into some investment projects. We are especially interested in two elements of project cash flows—capital cost allowances and interest expenses. These two elements are essentially immune to the effects of inflation, as they are always given in actual dollars. We will also consider the complication of how to proceed when multiple price indexes have been used to generate various project cash flows.

12.3.1   Capital Cost Allowances under Inflation

Because capital cost allowances are calculated on some base-year purchase amount, they do not increase over time to keep pace with inflation. Thus, they lose some of their value to defer taxes because inflation drives up the general price level and hence taxable income. Similarly, the salvage values of depreciable assets can increase with the general inflation rate and, because any gains on salvage values are taxable, they can result in increased taxes. Example 12.9 illustrates how a project's profitability changes under an inflationary economy.

EXAMPLE 12.9

Effects of inflation on projects with depreciable assets

Reconsider the automated machining center investment project described in Example 9.1, page 540. The summary of the financial facts in the absence of inflation is as follows:

Item	Description or Data
Project	Automated machining center
Required investment	$125,000
Project life	5 years
Salvage value	$50,000
CCA rate	20% declining balance
Annual revenues	$100,000 per year
Annual expenses	
Labor	$20,000 per year
Material	$12,000 per year
Overhead	$8000 per year
Marginal tax rate	40%
Inflation-free interest rate (i')	15%

The after-tax cash flow for the automated machining center project was given in Table 9.2, page 542, and the net present equivalent of the project in the absence of inflation was calculated to be $41,231.

What will happen to this investment project if the general inflation rate during the next 5 years is expected to be 5% annually? Sales and operating costs are assumed to increase accordingly. Capital cost allowance will remain unchanged, but taxes, profits, and thus cash flow, will be higher. The firm's inflation-free interest rate (i') is known to be 15%.

(a) Determine the PE of the project using the deflation method.

(b) Compare the PE with that in the inflation-free situation.

Discussion: All cash flow elements, except capital cost allowances, are assumed to be in constant dollars. Since income taxes are levied on actual taxable income, we will use the actual-dollar analysis, which requires that all cash flow elements be expressed in actual dollars.

- For the purposes of this illustration, all inflationary calculations are made as of year end.

- Cash flow elements such as sales, labor, material, overhead, and selling price of the asset will be inflated at the same rate as the general inflation rate.[7] For example, whereas annual sales had been estimated at $100,000, under conditions of inflation they become 5% greater in year 1, or $105,000; 10.25% greater in year 2, and so forth.

Period	Sales in Constant $	Conversion using f	Sales in Actual $
1	$100,000	$(1 + 0.05)^1$	$105,000
2	100,000	$(1 + 0.05)^2$	110,250
3	100,000	$(1 + 0.05)^3$	115,763
4	100,000	$(1 + 0.05)^4$	121,551
5	100,000	$(1 + 0.05)^5$	127,628

Future cash flows in actual dollars for other elements can be obtained in a similar way.

- No change occurs in the investment in year 0 or in capital cost allowances since these items are unaffected by expected future inflation.

- The selling price of the asset is expected to increase at the general inflation rate. Therefore, the salvage in actual dollars will be

$$\$50,000(1 + 0.05)^5 = \$63,814.$$

This increase in salvage value will also worsen the tax impact at the time of sale as the undepreciated capital cost remains unchanged. The calculations for both the capital cost allowance and disposal tax effect are shown in Table 12.2.

---

[7] This is a simplistic assumption. In practice, these elements may have price indexes other than the CPI. Differential price indexes will be treated in Example 12.10.

Table 12.2	Inflation Rate	0	1	2	3	4	5
Income Statement							
Revenues	5%		$105,000	$110,250	$115,763	$121,551	$127,628
Expenses							
Labor	5%		21,000	22,050	23,153	24,310	25,526
Material	5%		12,600	13,230	13,892	14,586	15,315
Overhead	5%		8,400	8,820	9,261	9,724	10,210
CCA			12,500	22,500	18,000	14,400	11,520
Taxable income			$ 50,500	$ 43,650	$ 51,457	$ 58,531	$ 65,057
Income taxes (40%)			20,200	17,460	20,583	23,412	26,023
Net income			$ 30,300	$ 26,190	$ 30,874	$ 35,119	$ 39,034
Cash Flow Statement							
Operating activities							
Net income			30,300	26,190	30,874	35,119	39,034
CCA			12,500	22,500	18,000	14,400	11,520
Investment activities							
Investment		(125,000)					
Salvage	5%						63,814
Disposal tax effects							(7,094)
Net cash flow (in actual dollars)		$(125,000)	$ 42,800	$ 48,690	$ 48,874	$ 49,519	$107,274

Cash Flow Statement for the Automated Machining Center Project under Inflation (Example 12.9)

Solution

Table 12.2 shows after-tax cash flows in actual dollars. Using the deflation method, we convert the cash flows to constant dollars with the same purchasing power as those used to make the initial investment (year 0), assuming a general inflation rate of 5%. Then, we discount these constant-dollar cash flows at i' to determine the PE.

Year	Net Cash Flow in Actual $	Conversion using f	Net Cash Flow in Constant $	PE at 15%
0	-$125,000	$(1 + 0.05)^0$	-$125,000	-$125,000
1	42,800	$(1 + 0.05)^{-1}$	40,762	35,445
2	48,690	$(1 + 0.05)^{-2}$	44,163	33,394
3	48,874	$(1 + 0.05)^{-3}$	42,219	27,760
4	49,519	$(1 + 0.05)^{-4}$	40,739	23,293
5	107,274	$(1 + 0.05)^{-5}$	84,052	41,789
				$ 36,681

Since PE = $36,681 > 0, the investment is still economically attractive.

Comments: Note that the PE in the absence of inflation was $41,231 in Example 9.1. The $4550 decline (known as inflation loss) in the PE under inflation, illustrated above, is due entirely to income tax considerations. The capital cost allowance is a charge against taxable income, which reduces the amount of taxes paid, and as a result, increases the cash flow attributable to an investment by the amount of taxes saved. But the capital cost allowance under existing tax laws is based on historic cost. As time goes by, the capital cost allowance is charged to taxable income in dollars of declining purchasing power; as a result, the "real" cost of the asset is not totally reflected in the capital cost allowance. Depreciation costs are thereby understated, and the taxable income is overstated, resulting in higher taxes. In "real" terms, the amount of this additional income tax is $4550, which is also known as the **inflation tax**.[8] In general, any investment that, for tax purposes, is depreciated and expensed over time, rather than expensed immediately, is subject to the inflation tax.

12.3.2 Multiple Inflation Rates

As we noted previously, the inflation rate (f_j) represents a rate applicable to a specific segment j of the economy. For example, if we were estimating the future cost of a piece of machinery, we should use the inflation rate appropriate for that item. Furthermore, we may need to use several rates to accommodate the different costs and revenues in our analysis. The following example introduces the complexity of multiple inflation rates.

EXAMPLE 12.10 **Applying specific inflation rates**

We will rework Example 12.9 using different annual indexes (differential inflation rates) in the prices of cash flow components. Suppose that we expect the general rate of inflation $(\bar{f})$ to average 6% during the next 5 years. We also expect that the selling price of the equipment will increase 3% per year, that wages (labor) and overhead will increase 5% per year; and that the cost of material will increase 4% per year. We expect sales revenue to climb at the general inflation rate. Table 12.3 shows the relevant calculations based on the income statement format. For simplicity, all cash flows and inflation effects are assumed to occur at year's end. Determine the net present equivalent of this investment, using the adjusted-discount method.

---

[8] This term and the explanation given here were introduced by Professor Brandt Allen in his article, "Evaluating Capital Expenditures under Inflation: A Primer." *Business Horizon*, 1976.

	Inflation Rate	0	1	2	3	4	5
Income Statement							
Revenues	6%		$106,000	$112,360	$119,102	$126,248	$133,823
Expenses							
Labor	5%		21,000	22,050	23,153	24,310	25,526
Material	4%		12,480	12,979	13,498	14,038	14,600
Overhead	5%		8,400	8,820	9,261	9,724	10,210
CCA			12,500	22,500	18,000	14,400	11,520
Taxable income			$ 51,620	$ 46,011	$ 55,190	$ 63,775	$ 71,967
Income taxes (40%)			20,648	18,404	22,076	25,510	28,787
Net income			$ 30,972	$ 27,606	$ 33,114	$ 38,265	$ 43,180
Cash Flow Statement							
Operating activities							
Net income			30,972	27,606	33,114	38,265	43,180
CCA			12,500	22,500	18,000	14,400	11,520
Investment activities							
Investment		(125,000)					
Salvage	3%						57,964
Disposal tax effects							(4,753)
Net cash flow (in actual dollars)		$(125,000)	$ 43,472	$ 50,106	$ 51,114	$ 52,665	$107,910

Table 12.3
Cash Flow Statement for the Automated Machining Center Project under Inflation, with Multiple Price Indexes (Example 12.10)

Solution

The after-tax cash flows in actual dollars are shown in the last line of Table 12.3. To evaluate the present equivalent using actual dollars, we must adjust the original discount rate of 15%, which is an inflation-free interest rate, i'. The appropriate interest rate to use is the market interest rate:[9]

$$\begin{aligned} i &= i' + \bar{f} + i'\,\bar{f} \\ &= 0.15 + 0.06 + (0.15)(0.06) \\ &= 21.90\%. \end{aligned}$$

The present equivalent is obtained as follows:

$$\begin{aligned} PE\,(21.90\%) &= -\$125,000 + \$43,472(P/F,\ 21.90\%,\ 1) \\ &\quad + \$50,106(P/F,\ 21.90\%,\ 2) + \ldots \\ &\quad + \$107,910(P/F,\ 21.90\%,\ 5) \\ &= \$36,542. \end{aligned}$$

---

[9] In practice, the market interest rate is usually given and the inflation-free interest rate can be calculated when the general inflation rate is known for years in the past or is estimated for time in the future. In our example, we are considering the opposite situation.

12.3.3 Effects of Borrowed Funds under Inflation

Loan repayment is based on the historical contract amount; the payment size does not change with inflation. Yet inflation greatly affects the value of these future payments, which are computed in year 0 dollars. First, we shall look at how the values of loan payments change under inflation. Interest expenses are usually already stated in the loan contract in actual dollars and need not be adjusted. Under the effect of inflation, the constant-dollar costs of both interest and debt principal repayments are reduced. Example 12.11 illustrates the effects of inflation on payments with project financing.

EXAMPLE 12.11

Effects of inflation on payments with financing (borrowing)

Let us rework Example 12.9 with a debt-to-equity ratio of 0.50, where the debt portion of the initial investment is borrowed at 10% annual interest and repaid in five equal payments. Assume, for simplicity, that the general annual inflation rate $(\bar{f})$ of 5% during the project period will affect all revenues, expenses (except depreciation and loan payments), and salvage value. Determine the PE of this investment.

Solution

For equal future payments, the actual dollar cash flows for the financing activity are represented by the circles in Figure 12.5. If inflation were to occur, the cash flow, measured in year 0 dollars, would be represented by the shaded circles in Fig. 12.5. Table 12.4 summarizes the after-tax cash flows under this situation. For simplicity, assume that all cash flows and inflation effects occur at year's end. To evaluate the present equivalent using actual dollars, we must adjust the original discount rate (MARR) of 15%, which is an inflation-free interest rate i'. The appropriate interest rate to use is thus the market interest rate:

$$
\begin{aligned}
i &= i' + \bar{f} + i'\bar{f} \\
&= 0.15 + 0.05 + (0.15)(0.05) \\
&= 20.75\%.
\end{aligned}
$$

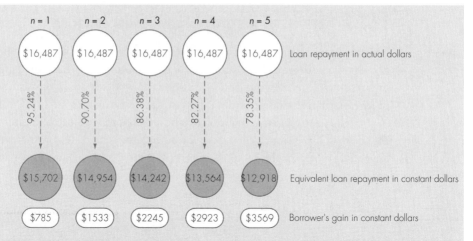

Figure 12.5 Equivalent loan repayment cash flows measured in year 0 dollars and borrower's gain over the loan life (Example 12.11)

Then, from Table 12.4, we compute the equivalent present worth of the after-tax cash flow as follows:

Table 12.4
Cash Flow Statement for the Automated Machining Center Project under Inflation, with Borrowed Funds (Example 12.11)

	Inflation Rate	0	1	2	3	4	5
Income Statement							
Revenues	5%		$105,000	$110,250	$115,763	$121,551	$127,628
Expenses							
Labor	5%		21,000	22,050	23,153	24,310	25,526
Material	5%		12,600	13,230	13,892	14,586	15,315
Overhead	5%		8,400	8,820	9,261	9,724	10,210
CCA			12,500	22,500	18,000	14,400	11,520
Debt interest			6,250	5,226	4,100	2,861	1,499
Taxable income			$ 44,250	$ 38,424	$ 47,357	$ 55,669	$ 63,558
Income taxes (40%)			17,700	15,369	18,943	22,268	25,423
Net income			$ 26,550	$ 23,054	$ 28,414	$ 33,401	$ 38,135
Cash Fow Statement							
Operating activities							
Net income			26,550	23,054	28,414	33,401	38,135
CCA			12,500	22,500	18,000	14,400	11,520
Investment activities							
Investment		$(125,000)					
Salvage	5%						63,814
Disposal tax effects							(7,094)
Financing activities							
Borrowed funds		62,500					
Principal repayment			(10,237)	(11,261)	(12,387)	(13,626)	(14,988)
Net cash flow (in actual dollars)		$ (62,500)	$ 28,813	$ 34,293	$ 34,027	$ 34,175	$ 91,387

The present equivalent is

$$PE (20.75\%) = -\$62{,}500 + \$28{,}813(P/F, 20.75\%, 1) + \dots$$
$$+ 91{,}387(P/F, 20.75\%, 5)$$
$$= \$55{,}884.$$

Comments: In the absence of inflation, the project would have a net present worth of $54,246 at an interest rate of 15%. (This is not shown here but can be calculated easily.) When compared with the result of $55,884 (with inflation), the present worth gain due to inflation is $55,884 − $54,246 = $1638. This increase in PE is due to the debt-financing. An inflationary trend decreases the purchasing power of future dollars, which helps long-term borrowers because they can repay a loan with dollars of reduced buying power. That is, the debt-financing cost is reduced in an inflationary environment. In this case, the benefits of financing under inflation have more than offset the *inflation tax* effect on depreciation and salvage value. The gain may not be totally realistic: The interest rate for borrowing is also generally higher during periods of inflation because it is a market driven rate.

12.4 Rate of Return Analysis under Inflation

In addition to affecting individual aspects of a project's income statement, inflation can have a profound effect on its overall return, i.e., the very acceptability of an investment project. In this section, we will explore the effects of inflation on return on investments and show several examples.

12.4.1 Effects of Inflation on Return on Investment

The effect of inflation on the rate of return for an investment depends on how future revenues respond to inflation. Under inflation, a company is usually able to compensate for increasing material and labor prices by raising its selling prices. However, even if future revenues increase to match the inflation rate, the allowable CCA, as we have seen, does not increase. The result is increased taxable income and higher income-tax payments. This increase reduces the available constant dollar after-tax benefits and, therefore, the inflation-free after-tax rate of return (IRR′). The next example will help us to understand this situation.

EXAMPLE 12.12 IRR analysis with inflation

Hartsfield Company, a manufacturer of auto parts, is considering the purchase of a set of machine tools at a cost of $30,000. The purchase is expected to generate increased sales of $24,500 per year and increased operating costs of $10,000 per year in each of the next 4 years. Additional profits will be taxed at a rate of 40%. The asset falls into CCA class 43 (rate = 30%) for tax purposes. The project has a 4-year life with zero salvage value. (All dollar figures represent constant dollars.)

(a) Assuming zero inflation, what is the expected internal rate of return?

(b) What is the expected IRR' if the general inflation rate is 10% during each of the next 4 years? (Here also assume that $f_j = \bar{f} = 10\%$.)

(c) If this is an independent alternative and the company has an inflation-free MARR (i.e., MARR') of 18%, should the company invest in the equipment?

Solution

(a) Rate of Return Analysis without Inflation
We find the expected rate of return by first computing the after-tax cash flow by the income statement approach, as shown in Table 12.5. The first part of the table shows the revenue from additional sales, operating costs, capital cost allowance, and taxes. The asset will be depreciated fully over 4 years, with no expected salvage value.

Table 12.5 Rate of Return Calculation without Inflation (Example 12.12)

	0	1	2	3	4
Income Statement					
Revenues		$24,500	$24,500	$24,500	$24,500
Expenses					
O & M		10,000	10,000	10,000	10,000
CCA		4,500	7,650	5,355	3,749
Taxable income		$10,000	$ 6,850	$ 9,145	$10,752
Income taxes (40%)		4,000	2,740	3,658	4,301
Net income		$ 6,000	$ 4,110	$ 5,487	$ 6,451
Cash Flow Statement					
Operating activities					
Net income		6,000	4,110	5,487	6,451
CCA		4,500	7,650	5,355	3,749
Investment activities					
Machine center	$(30,000)				
Salvage					0
Disposal tax effects					3,499
Net cash flow (in actual dollars)	$(30,000)	$10,500	$11,760	$10,842	$13,698

Thus, if the investment is made, we expect to receive additional annual cash flows of $10,500, $11,760, $10,842, and $13,698. This is a simple investment so we can calculate the IRR for the project as follows:

$$
\begin{aligned}
PE\,(i') &= -\$30,000 + \$10,500(P/F, i', 1) + \$11,760(P/F, i', 2) \\
&\quad + \$10,842(P/F, i', 3) + \$13,698(P/F, i', 4) \\
&= 0.
\end{aligned}
$$

Solving for i' yields

$$
IRR' = i'^{*} = 19.66\%.
$$

The project has an inflation-free rate of return of 19.66%, i.e., the company will recover its original investment ($30,000) plus interest at 19.66% each year for each dollar still invested in the project. Since the IRR' > MARR' of 18%, the company should buy the equipment.

(b) Rate of Return Analysis under Inflation

With inflation, we assume that sales, operating costs, and future selling price of the asset will increase. Capital cost allowance will be unchanged, but taxes, profits, and cash flow will be higher. We might think that higher cash flows will mean an increased rate of return. Unfortunately, this is not the case. We must recognize that cash flows for each year are stated in dollars of declining purchasing power. When the net after-tax cash flows are converted to dollars with the same purchasing power as those used to make the original investment, the resulting rate of return decreases. These calculations, assuming an inflation rate of 10% in sales and operating expenses, that is, 10% annual decline in the purchasing power of the dollar, are shown in Table 12.6. For example, whereas additional sales had been $24,500 yearly, under conditions of inflation they would be 10% greater in year 1, or $26,950; 21% greater in year 2; and so forth. No change in investment or capital cost allowance will occur, since these items are unaffected by expected future inflation. We have restated the after-tax cash flows (actual dollars) in dollars of a common purchasing power (constant dollars) by deflating them, again assuming an annual deflation factor of 10%. The constant-dollar cash flows are then used to determine the real rate of return.

$$
\begin{aligned}
PE\,(i') &= -\$30,000 + \$10,336(P/F, i', 1) + \$11,229(P/F, i', 2) \\
&\quad + \$10,309(P/F, i', 3) + \$12,114(P/F, i', 4) \\
&= 0.
\end{aligned}
$$

Solving for i' yields

$$
i' = 16.90\%.
$$

The rate of return for the project's cash flows in constant dollars (year 0 dollars) is 16.90%, which is less than the 19.66% return in the inflation-free case. Since IRR' < MARR', the investment is no longer acceptable.

Table 12.6
Rate of Return Calculation under Inflation (Example 12.12)

	Inflation rate	0	1	2	3	4
Income Statement						
Revenues	10%		$26,950	$29,645	$32,610	$35,870
Expenses						
O & M	10%		11,000	12,100	13,310	14,641
CCA			4,500	7,650	5,355	3,749
Taxable income			$11,450	$9,895	$13,945	$17,481
Income taxes (40%)			4,580	3,958	5,578	6,992
Net Income			$6,870	$5,937	$8,367	$10,489
Cash Flow Statement						
Operating activities						
Net Income			6,870	5,937	8,367	10,489
CCA			4,500	7,650	5,355	3,749
Investment activities						
Machine center		$(30,000)				
Salvage	10%					0
Disposal tax effects						3,499
Net cash flow (in actual dollars)		$(30,000)	$11,370	$13,587	$13,722	$17,736
Net cash flow (in constant dollars)		$(30,000)	$10,336	$11,229	$10,309	$12,114

PE (18%) = $ (653); IRR (actual dollars) = 28.59%; IRR′ (constant dollars) = 16.90%

Comments: We could also calculate the rate of return by setting the PE of the actual dollar cash flows to 0. This would give a value of IRR = 28.59%, but this is an inflation-adjusted IRR. We could then convert it to an IRR′ by deducting the amount due to inflation:

$$i' = \frac{(1 + i)}{(1 + \bar{f})} - 1$$

$$= \frac{(1 + 0.2859)}{(1 + 0.10)} - 1$$

$$= 16.90\%,$$

which gives the same final result of IRR′ = 16.90%.

12.4.2 Effects of Inflation on Working Capital

The loss of tax savings from capital cost allowance is not the only way that inflation may distort an investment's rate of return. Another source of decrease in a project's rate of return is working-capital drain. Capital projects requiring increased levels of working capital suffer from inflation because additional cash must be invested to maintain new price levels. For example, if the cost of inventory increases, additional outflows of cash are required to maintain appropriate inventory levels over time. A similar phenomenon occurs with funds committed to accounts receivable. These additional working-capital requirements can significantly reduce a project's rate of return. The next example will illustrate the effects of working-capital drain on a project's rate of return.

Effect of inflation on profits with working capital

Consider Example 12.12. Suppose that a $1000 investment in working capital is expected and that all the working capital will be recovered at the end of the project's 4-year life. Determine the rate of return on this investment.

Solution

Using the data in the upper part of Table 12.7, we can calculate the IRR′ = IRR of 18.83% in the absence of inflation. The PE (18%) = $524. The lower part of Table 12.7 includes the effect of inflation on the proposed investment. As illustrated in Figure 12.6, working-capital levels can be maintained only by additional infusions

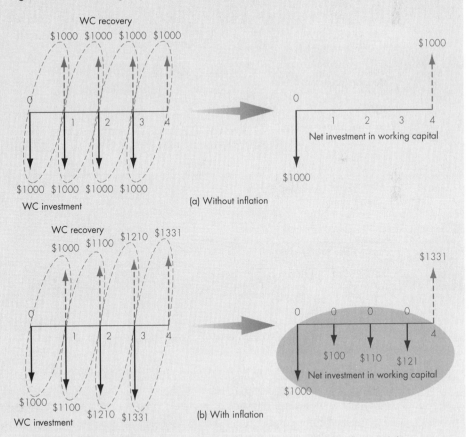

Figure 12.6 Working capital requirements under inflation: (a) Requirements without inflation and (b) requirements with inflation, assuming a 1-year recovery cycle (Example 12.13)

Table 12.7

Effects of Inflation on Working Capital and After-Tax Rate of Return (Example 12.13)

Case 1: Without Inflation

Income Statement	0	1	2	3	4
Revenues		$24,500	$24,500	$24,500	$24,500
Expenses					
O & M		10,000	10,000	10,000	10,000
CCA		4,500	7,650	5,355	3,749
Taxable income		$10,000	$ 6,850	$ 9,145	$10,752
Income taxes (40%)		4,000	2,740	3,658	4,301
Net income		$ 6,000	$ 4,110	$ 5,487	$ 6,451

Cash Flow Statement	0	1	2	3	4
Operating activities					
Net income		6,000	4,110	5,487	6,451
CCA		4,500	7,650	5,355	3,749
Investment activities					
Machine center	$(30,000)				
Working capital	(1,000)				1,000
Salvage					0
Disposal tax effects					3,499
Net cash flow (in actual dollars)	$(31,000)	$10,500	$11,760	$10,842	$14,698

PE (18%) = $524; IRR' = 18.83%

Case 2: With Inflation

Income Statement	Inflation Rate	0	1	2	3	4
Revenues	10%		$26,950	$29,645	$32,610	$35,870
Expenses						
O & M	10%		11,000	12,100	13,310	14,641
CCA			4,500	7,650	5,355	3,749
Taxable income			$11,450	$ 9,895	$13,945	$17,481
Income taxes (40%)			4,580	3,958	5,578	6,992
Net income			$ 6,870	$ 5,937	$ 8,367	$10,489

Cash Flow Statement	Inflation Rate	0	1	2	3	4
Operating activities						
Net income			6,870	5,937	8,367	10,489
CCA			4,500	7,650	5,355	3,749
Investment activities						
Machine center		$(30,000)				
Working capital	10%	(1,000)	(100)	(110)	(121)	1,331
Salvage	10%					0
Disposal tax effects						3,499
Net cash flow (in actual dollars)		$(31,000)	$11,270	$13,477	$13,601	$19,067
Net cash flow (in constant dollars)		$(31,000)	$10,245	$11,138	$10,218	$13,023

PE (18%) = $1382; IRR (actual dollars) = 27.34%; IRR' (constant dollars) = 15.76%

of cash—the working-capital drain also appears in the lower part of Table 12.7. For example, the $1000 investment in working capital made in year 0 will be recovered at the end of the first year assuming a 1-year recovery cycle. However, due to 10% inflation, the required working capital for the second year increases to $1100. In addition to reinvesting the $1000 recovered working capital, an additional investment of $100 must be made. This $1100 will be recovered at the end of the second year. However, the project will need a 10% increase, or $1210 for the third year, and so forth.

As Table 12.7 illustrates, the effect of the working-capital drain is significant. Given an inflationary economy and investment in working capital, the project's IRR' drops to 15.76%, or PE (18%) = −$1382. By using either IRR analysis or PE analysis we end up with the same result (as we must): Alternatives that are attractive when inflation does not exist may not be acceptable when inflation does exist.

12.5 Computer Notes

Many of the examples in this chapter were presented in tabular format—for example, Table 12.2. This same tabular format lends itself very easily to electronic spreadsheet analysis. The effects of inflation can be easily taken into account by using an electronic spreadsheet. We will seek an Excel solution to the following example.

EXAMPLE 12.14   **Excel spreadsheet analysis including differential inflation**

A construction firm is offered a fixed-price contract for a 5-year period. The firm will be paid $23,500 per year over the contract period. In order to accept the contract, the firm must purchase equipment costing $15,000 and requiring $13,000 (constant dollars) per year to operate. The equipment has a 30% capital cost allowance rate and is expected to have a salvage value of $1000 (constant dollars) at the end of 5 years. Use a tax rate of 40% and an inflation-free interest rate of 20%. If the general inflation rate $(\bar{f})$ is expected to average 5% over the next 5 years, salvage value inflates at an annual rate of 5%, and operating expenses at 8% per year for the project duration, should the contractor accept the contract?

Solution

(a) Constructing an after-tax cash table with Excel

The analysis of this situation should explicitly consider inflation because the contractor is being offered a fixed-price contract and cannot increase the fee to compensate for increased costs. In this problem, as is common practice, we assume that all estimated costs will be expressed in today's dollars, and actual costs will be increased due to inflation.

The Excel spreadsheet analysis is shown in Figure 12.7, where the first four rows are designated as input fields. Column C in both the income statement and the cash flow statement is reserved for entering the specific inflation rate for the item listed in that row. In row 12, operating expenses are increased at 8% due to inflation. We can automate the O&M cell entries by using the following cell formulas:

$$\text{Cell E12} = \$D\$2*(1 + \$C\$12)$$
$$\text{Cell F12} = E12*(1 + \$C\$12)$$
$$\vdots \qquad \qquad \vdots$$
$$\text{Cell I12} = H12*(1 + \$C\$12)$$

Note that the salvage value in cell I27 is also increased at 5%. (When no inflation rate is given for a specific cost category, it is assumed that the general inflation rate holds.) Disposal tax effects are based on the difference between undepreciated capital cost and salvage value, in this case, $3061 – $1276 = ($1785), or a loss of $1785. This results in a tax savings of ($1785)(0.40) = $714, as shown in cell I28.

As explained in this chapter, actual-dollar cash flows are converted to constant-dollar flows by "deflating" at the general inflation rate. The IRR' is calculated at 21.04%, based on constant-dollar cash flows. This amount is compared to a MARR' of 20%, the inflation-free interest rate, indicating the contract is acceptable as long as the 5% and 8% inflation rates are correct.

Figure 12.7 Excel example of an after-tax cash flow analysis including differential inflation (Example 12.14)

(b) Break-even analysis with Excel's Goal Seek command

Since it is impossible to predict inflation rates in any precise manner, it is always wise to investigate the effects of changes in these rates. This is very easy to do with a spreadsheet. For example, we can easily determine the value of the general inflation rate at which this project exactly earns 20%. The value of $\bar{f}$ could be adjusted manually until the IRR′ was exactly 20% or the PE becomes zero.

A more efficient way to solve such a break-even problem is to use the **Goal Seek** command on the Tools menu. (To find the optimum value for a particular cell by adjusting the values of one or more cells or to apply specific limitations to one or more values involved in the calculation, you can use Microsoft Excel Solver.)

To illustrate the steps to be taken to use the Goal Seek command, consider the Goal Seek menu screen in Figure 12.8.

Step 1: Select the target cell containing the formula. In our example, cell D35 contains the PE formula, so enter $D$35 in the **Set cell** box.

Step 2: Enter 0 in the **To value** box, because we are seeking the PE to be zero.

Step 3: Enter $H$2 in the **By changing cell** box, because the value of the general inflation rate is stored in cell H2. Note that H2 is the reference or name of the cell containing the variable that you want adjusted until the goal is reached.

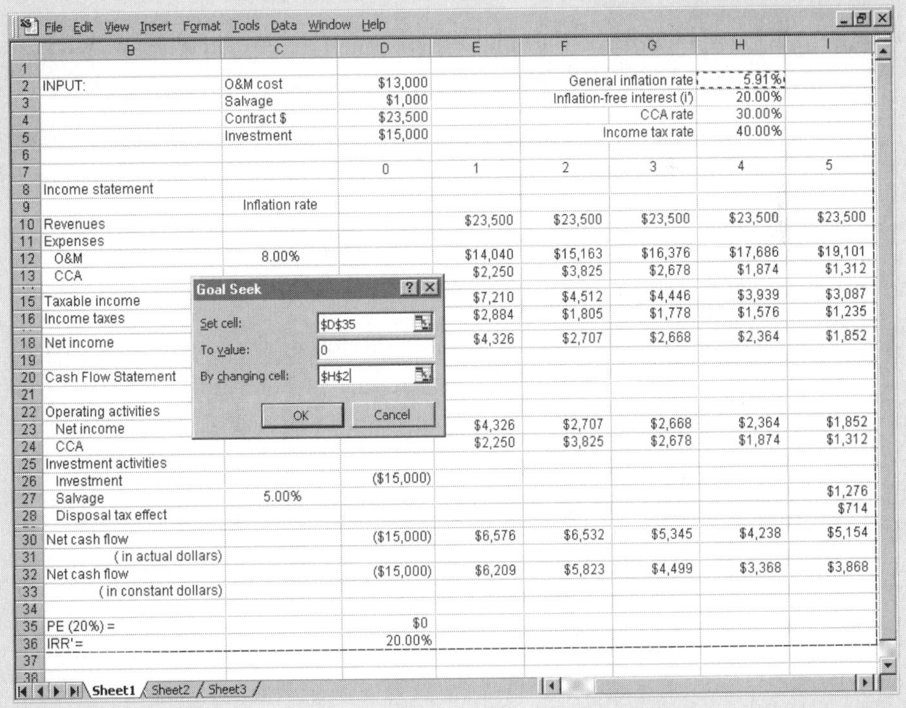

	B	C	D	E	F	G	H	I
1								
2	INPUT:	O&M cost	$13,000			General inflation rate	5.91%	
3		Salvage	$1,000			Inflation-free interest (i')	20.00%	
4		Contract $	$23,500			CCA rate	30.00%	
5		Investment	$15,000			Income tax rate	40.00%	
6								
7			0	1	2	3	4	5
8	Income statement							
9		Inflation rate						
10	Revenues			$23,500	$23,500	$23,500	$23,500	$23,500
11	Expenses							
12	O&M	8.00%		$14,040	$15,163	$16,376	$17,686	$19,101
13	CCA			$2,250	$3,825	$2,678	$1,874	$1,312
15	Taxable income			$7,210	$4,512	$4,446	$3,939	$3,087
16	Income taxes			$2,884	$1,805	$1,778	$1,576	$1,235
18	Net income			$4,326	$2,707	$2,668	$2,364	$1,852
19								
20	Cash Flow Statement							
21								
22	Operating activities							
23	Net income			$4,326	$2,707	$2,668	$2,364	$1,852
24	CCA			$2,250	$3,825	$2,678	$1,874	$1,312
25	Investment activities							
26	Investment		($15,000)					
27	Salvage	5.00%						$1,276
28	Disposal tax effect							$714
30	Net cash flow		($15,000)	$6,576	$6,532	$5,345	$4,238	$5,154
31	(in actual dollars)							
32	Net cash flow		($15,000)	$6,209	$5,823	$4,499	$3,368	$3,868
33	(in constant dollars)							
34								
35	PE (20%) =		$0					
36	IRR'=		20.00%					
37								
38								

Figure 12.8 Excel application to "what if" questions—at what general inflation rate does the project break even? (Example 12.14)

Step 4: Press the OK button to seek the solution. From Fig. 12.8, we see that this would not be a good project if $\bar{f}$ were greater than 5.91% (cell H2). (Note that at the exact value of $\bar{f}$ (5.9126%), the PE in cell D35 will be zero, or its rate of return will be 20%.) Similarly, you can vary the O&M cost and salvage value to see how the project's profitability changes.

12.6 Summary

- The **Consumer Price Index (CPI)** is a statistical measure of change, over time, of the prices of goods and services in major expenditure groups—such as food, housing, clothing, transportation, and health care—typically purchased by Canadian consumers. Essentially, the CPI compares the cost of a sample "market basket" of goods and services in a specific period relative to the cost of the same "market basket" in an earlier reference period. This reference period is designated as the **base period**.

- **Inflation** is the term used to describe a **decline in purchasing power** evidenced in an economic environment of rising prices.

- **Deflation** is the opposite of inflation: An increase in purchasing power evidenced by falling prices.

- The **general inflation rate** $(\bar{f})$ is an average inflation rate based on the CPI. An annual general inflation rate $(\bar{f})$ can be calculated using the following equation:

$$\bar{f}_n = \frac{CPI_n - CPI_{n-1}}{CPI_{n-1}}.$$

- Specific, individual commodities do not always reflect the general inflation rate in their price changes. We can calculate an **average inflation rate** $(\bar{f}_j)$ for a specific commodity (j) if we have an index (that is, a record of historical costs) for that commodity, for example primary steel products.

- Project cash flows may be stated in one of two forms:

    Actual dollars (A_n): Dollars that reflect the inflation or deflation rate; the dollar amounts that are actually exchanged

    Constant dollars (A'_n): Year 0 (i.e., base year) dollars; dollars with the same purchasing power

- Interest rates for project evaluation may be stated in one of two forms:

    Market interest rate (i): A rate which combines the effects of interest and inflation; used with **actual dollar** analysis

Inflation-free interest rate (i'): A rate from which the effects of inflation have been removed; this rate is used with **constant dollar** analysis

- To calculate the present equivalent of actual dollars, we can use a two-step or a one-step process:

  Deflation method—two steps:
  1. Convert actual dollars by deflating with the general inflation rate of $\bar{f}$.
  2. Calculate the PE of constant dollars by discounting at i'.

  Adjusted-discount method—one step (use the market interest rate):

  $$P_n = \frac{A_n}{[(1 + \bar{f})(1 + i')]^n}$$

  $$= \frac{A_n}{(1 + i)^n}.$$

- A number of individual elements of project evaluations can be distorted by inflation. These are summarized in Table 12.8.

Table 12.8

Effects of Inflation on Project Cash Flows and Return

Item	Effects of Inflation
Capital cost allowance	Capital cost allowance is charged to taxable income in dollars of declining value; taxable income is overstated, resulting in higher taxes.
Salvage values	Inflated salvage values combined with undepreciated capital cost values based on historical costs result in a less beneficial disposal tax effect (i.e., smaller loss or larger gain), hence higher taxes.
Loan repayments	Borrowers repay historical loan amounts with dollars of decreased purchasing power, reducing the debt-financing cost.
Working capital requirement	Known as a *working capital drain*, the cost of working capital increases in an inflationary economy.
Rate of return and PE	Unless revenues are sufficiently increased to keep pace with inflation, tax effects and/or a working capital drain result in lower rate of return or lower PE.

Problems

Level 1

In problem statements, the term "market interest rate" represents the "inflation-adjusted MARR" for project evaluation or the "interest rate" quoted by a financial institution for commercial loans.

Year	CPI
1994	102.0
1995	104.2
1996	105.9
1997	107.6
1998	108.6

12.1* A series of five constant-dollar (or real dollar) payments, beginning with $6000 at the end of the first year, are increasing at the rate of 5% per year. Assume that the average general inflation rate is 4%, and the market (inflation-adjusted) interest rate is 11% during this inflationary period. What is the equivalent present worth of the series?

12.2* "At a market interest rate of 7% per year and an inflation rate of 5% per year, a series of three equal annual receipts of $100 in constant dollars is equivalent to a series of three annual receipts of $105 in actual dollars." Which of the following statements is correct? Justify your answer.

(a) The amount of actual dollars is overstated.

(b) The amount of actual dollars is understated.

(c) The amount of actual dollars is about right.

(d) Sufficient information is not available to make a comparison.

12.3* The following figures represent the CPI indexes (base period 1992 = 100) for urban consumers in Canadian cities. Determine the average general inflation rate between 1994 and 1998.

12.4* How many years will it take for the dollar's purchasing power to be one-half what it is now, if the average inflation rate is expected to continue at the rate of 9% for an indefinite period? (Hint: You may apply the Rule of 72 which states that the number of years for an investment to double is approximately equal to $72/i\%$; see Example 2.8.)

12.5* Which of the following statements is incorrect?

(a) A negative inflation rate implies that you are experiencing a deflationary economy.

(b) Under an inflationary economy, debt financing is always a preferred option because you are paying back with cheaper dollars.

(c) If a project requires some investment in working capital under an inflationary economy, its rate of return will decrease when compared with the same project without inflation.

(d) Under an inflationary economy, in general, your tax burden will not increase, as long as inflationary adjustments are made in tax brackets.

$12.6* You are considering purchasing a $1000 bond with a coupon rate of 9.5%, interest payable annually. If the current inflation rate is 4% per year, which will continue in the foreseeable future, what would be the real rate of return if you sold the bond at $1080 after 2 years?

12.7* Vermont Casting has received an order to supply 250 units of casted fireplace inserts each year for Construction Sherbrooke over a 2-year period. The current sales price is $1500 per unit, and the current manufacturing cost per unit is $1000. Vermont is taxed at a rate of 40%. Both prices and costs are expected to rise at a rate of 6% per year. Vermont will produce these units on existing machines that have negligible undepreciated capital cost. Since the orders will be filled at the end of each year, the unit sale price and unit cost during the first year would be $1590 and $1060, respectively. Vermont's market interest rate is 15%. Which of the following present equivalent calculations for a single unit is incorrect?

(a) PE = $500(1 − 0.4)(*P/F*, 8.49%, 1) + $561.8(1 − 0.4)(*P/F*, 15%, 2)
(b) PE = $530(1 − 0.4)[(*P/F*, 15%, 1) + (1.06)(*P/F*, 15%, 2)]
(c) PE = $500(1 − 0.4)(*P/A*, 9%, 2)
(d) PE = $500(1 − 0.4)(*P/A*, 8.49%, 2).

$12.8 The following data indicate the median prices of houses in a major city price during the last 6 years:

Period	Price ($)
1992	$78,200
1998	98,500

Assuming that the base period (price index = 100) is period 1992, compute the average price index for the median house price for the year 1998.

12.9 The following data indicate the price indexes of a lumber product (base period 1992 = 100) during the last 5 years:

Period	Index
1994	155.2
1995	137.7
1996	153.1
1997	165.5
1998	159.8
1999	?

(a) Assuming that the base period (price index = 100) is reset to the year 1994 period, compute the new lumber price index for each year from 1994 to 1998.
(b) Calculate the average annual inflation rate over the 1994–1998 period.
(c) If the past trend is expected to continue, how would you estimate the lumber product at time period 1999? How confident are you in this projection?

12.10 For prices that are increasing at the annual rate of 5% the first year and 8% the second year, determine the average inflation rate ($\bar{f}$) over these 2 years.

$12.11 (a) In example 12.1, we showed that the inflation rate for tuition has averaged 8.78% per year in Canada over the last couple of decades. If Jim and Elizabeth just had their first child, and current tuition is $5000 per year, what will the cost be for the first year of university in 18 years if the inflation rate remains constant?

b) If Jim and Elizabeth have $5000 available now, at what annual interest rate would they have to invest it to have sufficient funds to cover the tuition for the first year of university in 18 years? Assume that the interest earned is tax-free.

12.12* An annuity provides for 10 consecutive end-of-year payments of $4500. The average general inflation rate is estimated to be 5% annually, and the market interest rate is 12% annually. What is the annuity worth in terms of a single equivalent amount of today's dollars?

12.13 Given the following cash flows in actual dollars, convert to equivalent cash flows in constant dollars if the base year is time 0. Keep cash flows at same point in time, i.e., year 0, 4, 5, and 7. Assume that the market interest rate is 16% and that the general inflation rate $(\bar{f})$ is estimated at 4% per year.

n	Cash Flow (in actual $)
0	$1500
4	2500
5	3500
7	4500

$12.14* The purchase of a car requires a $25,000 loan to be repaid in monthly installments for 4 years at 12% interest, compounded monthly. If the general inflation rate is 6%, compounded monthly, find the actual and constant dollar value of the 20th payment.

$12.15 A 10-year $1000 bond pays a nominal rate of 9%, compounded semiannually. If the market interest rate is 12%, compounded annually, and the general inflation rate is 6% per year, find the actual and constant dollar amount (time 0 dollars) of the 16th interest payment on the bond.

$12.16* Suppose that you borrow $20,000 at 12%, compounded monthly, over 5 years. Knowing that the 12% represents the market interest rate, the monthly payment in actual dollars will be $444.90. If the average monthly general inflation rate is expected to be 0.5%, determine the equivalent equal monthly payment series in constant dollars.

$12.17 Suppose that you just purchased a used car worth $6000 in today's dollars. Assume also that you borrowed $5000 from a local bank at 9%, compounded monthly, over 2 years. The bank calcu-

lated your monthly payment at $228. Assuming that average general inflation will run at 0.5% per month over the next 2 years,

(a) Determine the annual inflation-free interest rate (i') for the bank.
(b) What equal monthly payments, in terms of constant dollars over the next 2 years, are equivalent to the series of actual payments to be made over the life of the loan?

Level 2

12.18 A company is considering buying workstation computers to support its engineering staff. In today's dollars, it is estimated that the maintenance costs for the computers (paid at the end of each year) will be $25,000, $30,000, $32,000, $35,000 and $40,000 for years 1 to 5, respectively. The general inflation rate $(\bar{f})$ is estimated to be 8% per year, and the company will receive 15% per year on its invested funds during the inflationary period. The company wants to pay for maintenance expenses in equivalent equal payments (in actual dollars) at the end of each of the 5 years. Find the amount of the company's payment.

12.19 A series of four annual constant-dollar payments beginning with $7000 at the end of the first year is growing at the rate of 8% per year [assume that the base year is the current year $(n = 0)$]. If the market interest rate is 13% per year, and the general inflation rate $(\bar{f})$ is 7% per year, find the present equivalent of this series of payments based on

(a) constant-dollar analysis
(b) actual-dollar analysis.

12.20 Consider the cash flow diagrams, where the equal-payment cash flow in constant dollars (a) is converted to the equal-payment cash flow in actual dollars (b), at an annual general inflation rate of $\bar{f} = 3.8\%$ and $i = 9\%$. What is the amount A in actual dollars equivalent to $A' = \$1000$ in constant dollars?

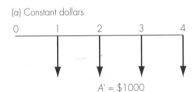

(a) Constant dollars

$A' = \$1000$

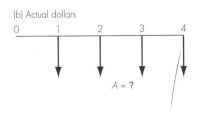

(b) Actual dollars

$A = ?$

12.21 The annual fuel costs to operate a small solid-waste treatment plant are projected to be $1.5 million, without considering any future inflation. The best estimates indicate that the annual inflation-free interest rate (i') will be 6% and the general inflation rate ($\bar{f}$) = 5%. If the plant has a remaining useful life of 5 years, what is the present equivalent of its fuel costs using actual dollar analysis?

$12.22 A father wants to save for his 8-year-old child's college expenses. The child will enter college 10 years from now. An annual amount of $30,000 in constant dollars will be required to support the child's college expenses for 4 years. Assume that these college payments will be made at the beginning of the school year. The future general inflation rate is estimated to be 6% per year, and the market interest rate on the savings account will average 8%, compounded annually. Given this information,

(a) What is the amount of the child's first-year expense in terms of actual dollars?

(b) What is the equivalent single-sum amount at the present time for these college expenses?

(c) What is the equal amount, in actual dollars, the father must save each year until his child goes to college?

12.23 Consider the following project's after-tax cash flow and the expected annual general inflation rate during the project period:

End of Year	Cash Flow (in actual $)	Expected General Inflation Rate
0	−$45,000	
1	26,000	6.5%
2	26,000	7.7
3	26,000	8.1

(a) Determine the average annual general inflation rate over the project period.

(b) Convert the cash flows in actual dollars into equivalent constant dollars with the base year 0.

(c) If the annual inflation-free interest rate is 5%, what is the present equivalent of the cash flow? Is this project acceptable?

12.24 Gentry Machines Inc. has just received a special job order from one of its clients. The following financial data has been collected:

- This 2-year project requires purchase of a special purpose equipment of $50,000. The declining balance CCA rate for the equipment is 30%.
- The machine will be sold at the end of 2 years for $30,000 (today's dollars).
- The project will bring in an additional annual revenue of $100,000 (actual dollars), but it is expected to incur an additional annual operating cost of $40,000 (today's dollars).
- The project requires an investment in working capital in the amount of $10,000 at $n = 0$. In each following year, additional working capital needs to be provided at the same rate as general inflation. Any investment in working capital will be recovered at the end of project termination.

- To purchase the equipment, the firm expects to borrow $50,000 at 10% over a 2-year period (equal annual payments of $28,810 [actual dollars]).
- The firm expects a general inflation rate of 5% per year during the project period. The firm's marginal tax rate is 40%, and its market interest rate is 18%.

(a) Compute the after-tax cash flows in actual dollars.

(b) What is the equivalent present value at time 0?

12.25* The J. F. Manning Metal Co. is considering the purchase of a new milling machine during year 0. The machine's base price is $180,000, and it will cost another $20,000 to modify it for special use by the firm. This results in a $200,000 cost base for capital cost allowance. The machine falls into CCA class 43 ($d = 30\%$). The machine will be sold after 3 years for $80,000 (actual dollars). Use of the machine will require an increase in net working capital (inventory) of $10,000 at the beginning of the project year. The machine will have no effect on revenues, but it is expected to save the firm $80,000 (today's dollars) per year in before-tax operating costs, mainly labor. The firm's marginal tax rate is 40%, and this rate is expected to remain unchanged over the project's duration. However, the company expects labor costs to increase at an annual rate of 5%, and the working capital requirement to grow at an annual rate of 8% due to inflation. The salvage value of the milling machine is not affected by inflation. The general inflation rate is estimated to be 6% per year over the project period. The firm's market interest rate is 20%.

(a) Determine the project cash flows in actual dollars.

(b) Determine the project cash flows in constant (time 0) dollars.

(c) Is this project acceptable?

$12.26 Sonja Jensen is considering the purchase of a fast-food franchise. Sonja will be operating on property that is to be converted into a parking lot in 5 years, but that may be rented in the interim for $800 per month. The franchise and necessary equipment will have a total first cost of $55,000 and a salvage value of $10,000 (in today's dollars) after 5 years. Sonja is told that the future annual general inflation rate will be 5%. The projected operating revenues and expenses in actual dollars, other than rent and depreciation for the business, are as follows:

End of Year	Revenue	Expenses
1	$30,000	$15,000
2	35,000	21,000
3	55,000	25,000
4	70,000	30,000
5	70,000	30,000

Assume that the initial investment has a CCA rate of 20%, and Sonja's tax rate will be 30%. Sonja can invest her money at a rate of at least 10% in other investment activities.

(a) Determine the cash flows associated with the investment over its investment life.

(b) Compute the projected after-tax rate of return (real) for this investment opportunity.

$12.27 You have $10,000 cash, which you want to invest. Normally, you would deposit the money in a savings account that pays an annual interest rate of 6%. However, you are now considering the possibility of investing in a bond. Your alternatives are either a provincial bond paying 9% or a corporate bond paying 8.4%. Your marginal tax rate is 40% for interest income and 30% for capital gains. You expect the general inflation to be 3% during the investment period. Both bonds pay interest once at the end of each year. The provincial bond matures at the end of year 5. The corporate bond is expected to have a selling price of

$11,000 at the end of year 5.

(a) Determine the real (inflation-free) rate of return for each bond.
(b) Without knowing your MARR, can you make a choice between these two bonds?

12.28 Air Canajet is considering two types of engines for use in its planes. Each has the same life, same maintenance, and repair record.

- Engine A costs $150,000 and uses 140,000 litres per 1000 hours of operation at the average service load encountered in passenger service.
- Engine B costs $225,000 and uses 100,000 litres per 1000 hours of operation at the same service load.

Both engines are estimated to have 10,000 service hours before any major overhaul of the engines is required. If fuel costs $0.45 per litre currently, and its price is expected to increase at the rate of 8% per year due to inflation, which engine should the firm install for an expected 2000 hours of operation per year? The firm's marginal income tax rate is 40%, and the engine falls into Class 9 (CCA rate = 25%) for which the 50% rule applies. Assume the firm's market interest rate is 20%. It is estimated that both engines will retain a market value of 35% of their initial cost (actual dollars) if they are sold on market after 10,000 hours of operation.

(a) Using the present equivalent criterion, which project would you select?
(b) Using the annual equivalent criterion, which project would you select?
(c) Using the future worth criterion, which project would you select?

12.29* Johnson Chemical Company has just received a special subcontracting job from one of its clients. This 2-year project requires purchase of a special-purpose painting sprayer for $60,000. This equipment has a 30% declining balance rate for CCA cal-

culations. After the subcontracting work is completed, the painting sprayer will be sold at the end of 2 years for $40,000 (actual dollars). The painting system will require an increase in net working capital (spare parts inventory such as spray nozzles) of $5000. This investment in working capital will be fully recovered at the end of project termination. The project will bring in additional annual revenue of $120,000 (today's dollars), but it is expected to incur additional annual operating costs of $60,000 (today's dollars). It is projected that, due to inflation, there will be sales price increases at an annual rate of 5%. (This implies that revenues will increase at an annual rate of 5%.) An annual increase of 4% for expenses and working capital requirement is expected. The company has a marginal tax rate of 30% and it uses a market interest rate of 15% for project evaluation during the inflationary period. If the firm expects a general inflation of 8% during the project period:

(a) Compute the after-tax cash flows in actual dollars.
(b) What is the rate of return on this investment (real earnings)?
(c) Is this special project profitable?

Level 3

$12.30 A man is planning to retire in 20 years. Money can be deposited at 6%, compounded monthly, and it is also estimated that the future general inflation ($\bar{f}$) rate will be at 5%, compounded annually. What monthly deposit must be made each month until the man retires so that he can make annual withdrawals of $40,000 in terms of today's dollars over the 15 years following his retirement? (Assume that his first withdrawal occurs at the end of the first 6 months after his retirement.)

$12.31 On her 23rd birthday a young female engineer decides to start saving toward building up a retirement fund that pays 8% interest, compounded quarterly (market interest rate). She feels that $600,000 worth of purchasing power in today's dollars will be adequate to see her through her sunset years after her 63rd birthday. Assume a general inflation rate of 6% per year.

(a) If she plans to save by making 160 equal quarterly deposits, what should be the amount of her quarterly deposit in actual dollars?

(b) If she plans to save by making end-of-year deposits, increasing by $1000 over each subsequent year, how much would her first deposit be in actual dollars?

12.32 Hugh Health Product Corporation is considering the purchase of a computer to control plant packaging for a spectrum of health products. The following data have been collected:

- First cost = $100,000 to be borrowed at 10% interest where only interest is paid each year, and the principal is due in a lump sum at end of year 2
- Economic service life (project life) = 6 years
- Estimated salvage value at the end of project life, in year 0 dollars = $15,000
- CCA rate (DB) = 30%
- Marginal income tax rate = 40%
- Annual revenue = $150,000 (today's dollars)
- Annual expense (not including CCA and interest) = $80,000 (today's dollars)
- Market interest rate = 20%

(a) With an average general inflation rate of 5% expected during the project period, which will affect all revenues, expenses, and the salvage value, determine the cash flows in actual dollars.

(b) Compute the net present equivalent of the project under inflation.

(c) Compute the net present equivalent loss (gain) due to inflation.

(d) In (c), how much of the present equivalent loss (or gain) is due to borrowing?

12.33 Longueil Textile Company is considering automating their piece-goods screen printing system at a cost of $20,000. The firm expects to phase out this automated printing system at the end of 5 years due to changes in style. At that time, the firm could scrap the system for $2000 in today's dollars. The expected net savings due to the automation are in today's dollars (constant dollars) as follows:

End of Year	Cash Flow (in constant $)
1	$15,000
2	17,000
3–5	14,000

The system qualifies for a 30% declining balance rate for tax purposes. The expected average general inflation rate over the next 5 years is approximately 5% per year. The firm will finance the entire project by borrowing at 10%. The scheduled repayment of the loan will be as follows:

End of Year	Principal Payment	Interest Payment
1	$6042	$2000
2	6647	3396
3	7311	731

The firm's market interest rate for project evaluation is 20%. Assume that the net savings and the selling price will be responsive to this average inflation rate. The firm's marginal tax rate is known to be 40%.

(a) Determine the after-tax cash flows of this project in actual dollars.

(b) Determine the net present equivalent reduction (or gains) in profitability due to inflation.

12.34 Fuller Ford Company is considering purchasing a vertical drill machine. The machine will cost $50,000 and will have an 8-year service life. The salvage value of the machine at the end of 8 years is expected to be $5000 in today's dollars. The machine will generate annual revenues of $20,000 (today's dollars), but it expects to have annual operating costs (excluding CCA) of $8000 (today's dollars). The asset has a CCA rate of 30%. The project requires a working capital investment of $10,000 at year 0. The marginal income tax rate for the firm is 35%. The firm's market interest rate is 18%.

(a) Determine the internal rate of return of this investment.

(b) Assume that the firm expects a general inflation rate of 5%, but that it also expects an 8% annual increase in revenue and working capital and a 6% annual increase in operating costs due to inflation. Compute the real (inflation-free) internal rate of return. Is this project acceptable?

12.35 Land Development Corporation is considering the purchase of a bulldozer. The bulldozer will cost $100,000 and will have an estimated salvage value of $30,000 at the end of 6 years. The asset will generate annual before-tax revenues of $80,000 over the next 6 years. The asset is depreciated at a CCA rate of 30%. The marginal tax rate is 40%, and the firm's market interest rate is known to be 18%. All dollar figures represent constant dollars at time 0 and are responsive to the general inflation rate $\bar{f}$.

(a) With $\bar{f}$ = 6%, compute the after-tax cash flows in actual dollars.

(b) Determine the real rate of return of this project on after-tax basis.

(c) Suppose that the initial cost of the project will be financed through a local bank at an interest rate of 12%, with an annual payment of $24,323

over 6 years. With this additional condition, answer part (a) above.

(d) In part (a), determine the present equivalent loss due to inflation.

(e) In part (c), determine how much the project has to generate in additional before-tax annual revenue in actual dollars (equal amount) to make up the inflation loss.

12.36 Wilson Machine Tools Inc., a manufacturer of fabricated metal products, is considering the purchase of a high-tech computer-controlled milling machine at a cost of $95,000. The cost of installing the machine, preparing the site, wiring, and rearranging other equipment is expected to be $15,000. This installation cost will be added to the machine cost to determine the total capital cost basis for CCA. Special jigs and tool dies for the particular product will also be required at a cost of $10,000. The milling machine is expected to last 10 years, and the jigs and dies only for 5 years. Therefore, another set of jigs and dies has to be purchased at the end of 5 years. The milling machine will have a $10,000 salvage value at the end of its life, and the special jigs and dies are worth only $300 as scrap metal at any time in their lives. The machine has a 30% CCA rate, and the special jigs and dies have a 100% CCA rate. With the new milling machine, Wilson expects an additional annual revenue of $80,000 due to increased production. The additional annual production costs are estimated as follows: Materials, $9000; labor, $15,000; energy, $4500; and miscellaneous O&M costs, $3000. Wilson's marginal income tax rate is expected to remain at 35% over the project life of 10 years. All dollar figures represent today's dollars. The firm's market interest rate is 18%, and the expected general inflation rate during the project period is estimated at 6%.

(a) Determine the project cash flows in the absence of inflation.

(b) Determine the after-tax internal rate of return for the project in (a).

(c) Suppose that Wilson expects price increases during the project period: material at 4% per year, labor at 5% per year, and energy and other O&M costs at 3% per year. To compensate for these increases in prices, Wilson is planning to increase annual revenue at the rate of 7% per year by charging its customers a higher price. No changes in salvage value are expected for the machine or for the jigs and dies. Determine the project cash flows in actual dollars.

(d) In (c), determine the real (inflation-free) rate of return of the project.

(e) Determine the economic loss (or gain) in present equivalent due to inflation.

12.37 Recent biotechnological research has made possible the development of a sensing device that implants living cells on a silicon chip. The chip is capable of detecting physical and chemical changes in cell processes. Proposed uses include researching the mechanisms of disease on a cellular level, developing new therapeutic drugs, and replacing the use of animals in cosmetic and drug testing. Biotech Device Corporation (BDC) has just perfected a process for mass-producing the chip. The following information has been compiled for the board of directors.

- BDC's marketing department plans to target sales of the device to larger chemical and drug manufacturers. BDC estimates that annual sales would be 2000 units, if the device were priced at $95,000 per unit (dollars of the first operating year).

- To support this level of sales volume, BDC would need a new manufacturing plant. Once the "go" decision is made, this plant could be built and made ready for production within 1 year. BDC would need a 10-hectare tract of land that would cost $1.5 million. If the decision were to be made, the land could be purchased on December 31, 2000. The building would cost $5 million and has a 4% capital cost allowance rate. The first payment of $1 million would be due to the contractor on December 31, 2001, and the remaining $4 million on December 31, 2002.

- The required manufacturing equipment would be installed late in 2002 and would be paid for on December 31, 2002. BDC would have to purchase the equipment at an estimated cost of $8 million, which includes transportation and $500,000 for installation. The equipment has a 30% CCA rate.

- The building and equipment become available-for-use at the start of the year 2003.

- The project would require an initial investment of $1 million in working capital. This initial working capital investment would be made on December 31, 2002, and on December 31 of each following year, net working capital would be increased by an amount equal to 15% of any sales increase expected during the coming year. The investments in working capital would be fully recovered at the end of project year.

- The project's estimated economic life is 6 years (excluding the 2-year construction period). At that time, the land is expected to have a market value of $2 million, the building a value of $3 million, and the equipment a value of $1.5 million. The estimated variable manufacturing costs would total 60% of the dollar sales. Fixed costs, excluding capital cost allowance, would be $5 million for the first year of operations. Since the plant would begin operations on January 1, 2003, the first operating cash flows would occur on December 31, 2003. (If the project does proceed, this latter should be treated as a taxable opportunity cost.)

- Sales prices and fixed overhead costs, other than CCA, are projected to increase with general inflation, which is expected to average 5 percent per year over the 6-year life of the project.
- To date, BDC has spent $5.5 million on research and development (R&D) associated with the cell implanting research. The company has already expensed $4 million R&D costs. The remaining $1.5 million will be amortized over 6 years (i.e., the annual amortization expense would be $250,000; like CCA, this amortization is not a cash flow, but must be deducted before calculating taxes). If BDC decides not to proceed with the project, the $1.5 million R&D cost could be written off on December 31, 2000. (If the project does proceed, this latter should be treated as a taxable opportunity cost.)
- BDC's marginal tax rate is 40%, and its market interest rate is 20%. Any capital gains will be taxed at 30%.

(a) Determine the after-tax cash flows of the project in actual dollars.

(b) Determine the inflation-free (real) IRR of the investment.

(c) Would you recommend that the firm accept the project?

12.38 Tiger Construction Company purchased its current bulldozer (Caterpillar D8H) for $350,000 and placed it in service 6 years ago. Since the purchase of the Caterpillar, new technology has produced changes in machines, which resulted in an increase in productivity of approximately 20%. The Caterpillar worked in a system with a fixed (required) production level to maintain overall system productivity. As the Caterpillar aged and logged more downtime, more hours had to be scheduled to maintain the required production. Tiger is considering the purchase of a new bulldozer

(Komatsu K80A) to replace the Caterpillar. The following data have been collected by Tiger's civil engineer (ignoring inflation.)

	Defender (Caterpillar D8H)	Challenger (Komatsu K80A)
Useful life	Not known	Not known
Purchase price		$400,000
Salvage value, if kept for		
0 year	$75,000	$400,000
1 year	60,000	300,000
2 year	50,000	240,000
3 year	30,000	190,000
4 year	30,000	150,000
5 year	10,000	115,000
Fuel use (litres/hour)	30	40
Maintenance costs		
1	$46,800	$35,000
2	46,800	38,400
3	46,800	43,700
4	46,800	48,300
5	46,800	58,000
Operating hours (hours/year)		
1	1,800	2,500
2	1,800	2,400
3	1,700	2,300
4	1,700	2,100
5	1,600	2,000
Productivity index	1.00	1.20
Other relevant information		
Fuel cost ($/litre)		$0.45
Operator's wages ($/hour)		$23.40
Market interest rate (MARR)		15%
Marginal tax rate		40%
CCA rate		30%

(a) A civil engineer notices that both machines have different working hours and hourly production capacities. To compare the different units of capacity, the engineer needs to devise a combined index to reflect the new machine's productivity as well as actual operating hours relative to the existing machine. Develop such a combined productivity index for each period.

(b) Adjust the operating and maintenance costs by this index.

(c) Compare the two alternatives. Should the defender be replaced now?

(d) If the following price index were forecasted for the next 5 project years, should the defender be replaced now? Assume that salvage values are unchanged.

Forecasted Price Index

Year	General Inflation	Fuel	Wage	Maintenance
0	100	100	100	100
1	108	110	115	108
2	116	120	125	116
3	126	130	130	124
4	136	140	135	126
5	147	150	140	128

Project Risk and Uncertainty

T he prospective profits or losses of a mining investment depend upon the vagaries of such diverse matters as metal prices, ore quality, labor negotiations, and various choices made by the mine operator. Consider the choices faced by Westmin Resources Limited when it evaluated its H-W copper-zinc deposit prior to a decision to proceed with development. The potential profitability of this project based on metal prices alone is illustrated at three alternate production levels below.

H-W MINERAL DEPOSIT—IMPACT OF METAL PRICE VARIATIONS

Production rate (tonnes per day)	2700	1800	1350
Capital cost ($ millions)	150.20	128.40	98.60
Operating cost ($/tonne)	39.02	48.33	56.69
Ore reserve* (million tonnes)	13.31	9.72	7.42
Optimistic price forecast			
Total revenue§ ($ millions)	1,022.2	877.2	728.2
Total profits§ ($ millions)	503.0	407.7	307.7
Gross cash flow** ($ millions)	352.8	279.3	209.1
PE¶ ($ millions)	56.3	49.3	34.1
IRR (percent)	19.5	20.8	19.0
Payback (years)	3.2	2.8	3.3
Pessimistic price forecast			
Total revenue§ ($ millions)	742.6	637.3	529.0
Total profits§ ($ millions)	223.4	168.0	108.5
Gross cash flow** ($ millions)	73.2	39.6	9.9
PE¶ ($ millions)	(23.5)	(26.1)	(33.9)
IRR (percent)	5.6	5.0	1.6
Payback (years)	5.7	6.9	9.3

* Represents the total ore reserve that can be economically utilized at the specified production level.

§ Represents the simple sums of the revenues and profits, respectively, over a 15-year project life.

** The total after-tax cash flows over the project life.

¶ Assumes an interest rate of 10% per annum.

Acknowledgement: As related by Mr. Carl C. Hunter of Dalcor Consultants Ltd., West Vancouver, B.C. (Reprinted by permission.)

While exhibiting a wide range of potential profit outcomes, the possibilities listed above are by no means definitive. Other factors that affect the profitability of a mining investment, estimated with varying degrees of reliability, include the following:

- **Metal prices:** The price of metal is the most important and uncertain factor on the revenue side of metal mining investment analysis. Metal prices fluctuate from highs of five times the long term "economic price" to lows of half this value. These low values are roughly determined by the marginal production cost curve of the "average industry", while price peaks have no limit in the short term.

- **Ore grades:** It is impossible to know either the quality or quantity of ore in the ground exactly. The reliability of mining tonnage and grade estimates relates directly to the amount of drilling, sampling, and testing done on the site. The extent to which such preliminary work is performed is related to the investor's appetite for risk.

- **Capital and operating costs:** Capital cost estimates are based upon experience and the chosen level of engineering design. Typically, the investor chooses to impose a degree of accuracy on the estimate that in turn determines the level of engineering work.

    Operating costs depend on production rates and processes, the ratio of waste rock to ore, the haul distances involved, and the cost of process materials, energy, and labor. Estimating with accuracy is a function of experience, and is consequently performed by experts in various fields.

As this example illustrates, a project's economic assessment must be based on profitability calculations, which in turn are derived from estimates of variation in the significant factors that can affect profitability. Risk analysis takes into account the range of variation as well as the likelihood of the changes.

In previous chapters, cash flows from projects were assumed to be known with complete certainty; our analysis was concerned with measuring the economic worth of projects and selecting the best investment projects. Although these types of analyses can provide a reasonable decision basis in many investment situations, we should certainly consider the more usual situation where forecasts of cash flows are subject to some degree of uncertainty. In this type of situation, management rarely has precise expectations about the future cash flows to be derived from a particular project. In fact, the best that a firm can reasonably expect to do is to estimate the range of possible future costs and benefits and the relative chances of achieving a reasonable return on the investment. We use the term **risk** to describe an investment project whose cash flow is not known in advance with absolute certainty, but for which an array of alternative outcomes and their probabilities (odds) are known. We will also use the term **project risk** to refer to variability in a project's PE. A greater project risk usually means a greater variability in a project's PE, or alternatively that the *risk is the potential for loss*. This chapter begins by exploring the origins of project risk.

13.1 Origins of Project Risk

The decision to make a major capital investment such as introducing a new product requires cash flow information over the life of a project. The profitability estimate of an investment depends on cash flow estimations, which are generally uncertain. The factors to be estimated include the total market for the product; the market share that the firm can attain; the growth in the market; the cost of producing the product, including labor and materials; the selling price; the life of the product; the cost and life of the equipment needed; and the effective tax rates. Many of these factors are subject to substantial uncertainty. A common approach is to make single-number "best estimates" for each of the uncertain factors and then to calculate measures of profitability, such as PE or rate of return for the project. This approach has two drawbacks:

1. No guarantee can ever ensure that the "best estimates" will ever match actual values.

2. No provision is made to measure the risk associated with an investment or the project risk. In particular, managers have no way of determining either the probability that a project will lose money or the probability that it will generate large profits.

Because cash flows can be so difficult to estimate accurately, project managers frequently consider a range of possible values for cash flow elements. If a range of values for individual cash flows is possible, it follows that a range of values for the PE of a given project is also possible. Clearly, the analyst will want to try to gauge the probability and reliability of individual cash flows occurring and, consequently, the level of certainty about overall project worth.

Quantitative statements about risk are given as numerical probabilities or as values for likelihood (odds) of occurrence. Probabilities are given as decimal fractions in the interval 0.0 to 1.0. An event or outcome that is certain to occur has a probability of 1.0. As the probability of an event approaches 0, the event becomes increasingly less likely to occur. The assignment of probabilities to the various outcomes of an investment project is an aspect of **risk analysis**. Example 13.1 illustrates some important probability concepts that are easily demonstrated in daily life.

EXAMPLE 13.1   Improving the odds—all it takes is $7 million and a dream

In Virginia's six-number lottery, or lotto, players pick six numbers from 1 to 44. The winning combination is determined by a machine that looks like a popcorn machine, except that it is filled with numbered table-tennis balls. On February 15, 1992, the Virginia lottery drawing offered the following prizes,[1] assuming the first prize is not shared:

Number of Prizes	Prize Category	Total Amount
1	First prize	$27,007,364
228	Second prizes ($899 each)	204,972
10,552	Third prizes ($51 each)	538,152
168,073	Fourth prizes ($1 each)	168,073
	Total winnings	$27,918,561

Common among regular lottery players is this dream: Waiting until the jackpot reaches an astronomical sum and then buying every possible number, thereby guaranteeing a winner. Sure it would cost millions of dollars, but the payoff would be much greater. Is it worth trying? How do the odds of winning the first prize change as you increase the number of ticket purchases?

Discussion: In Virginia, one investment group came tantalizingly close to cornering the market on all possible combinations of six numbers from 1 to 44. State lottery officials say that the group bought 5 million of the possible 7 million tickets (precisely 7,059,052). Each ticket cost $1 each. The lottery had a more than $27 million jackpot.[2]

---

[1] Prizes are based on before-tax values and on the actual number of second- and third-prize winning tickets sold.

[2] Source: *The New York Times*, February 25, 1992.

- Economic Logic: If the jackpot is big enough, provided nobody else buys a winning ticket, it makes economic sense to buy one lottery ticket for every possible combination of numbers and be sure to win. A group in Australia apparently tried to do this in the February 15 (1992) Virginia lottery drawing.

- The Cost: Since 7,059,052 combinations of numbers are possible[3] and each ticket costs $1, it would cost $7,059,052 to cover every combination. The total cost remains the same regardless of the size of the jackpot.

- The Risk: The first prize jackpot is paid out in 20 equal yearly installments, so the actual payoff on all prizes is $2,261,565 the first year and $1,350,368 per year for the next 19 years. If more than one first prize-winning ticket is sold, the prize is shared so that the maximum payoff depends on an ordinary player not buying a winning ticket. Since this lottery began in January 1990, 120 of the 170 drawings have not yielded a first-prize winner.

Solution

In the Virginia game, 7,059,052 combinations of numbers are possible. The following table summarizes the winning odds for various prizes for a one ticket-only purchase.

Number of Prizes	Prize Category	Winning Odds
1	First prize	0.0000001416
228	Second prizes	0.0000323
10,552	Third prizes	0.00149
168,073	Fourth prizes	0.02381

So a person who buys one ticket has odds of 1 in slightly more than 7 million. Holding more tickets increases the odds of winning, so that 1000 tickets have odds of 1 in 7000 and 1 million tickets have odds of 1 in 7. Since each ticket costs $1, it would receive at least a share in the jackpot and many of the second, third, and fourth place prizes. Together these combined prizes were worth $911,197 on February 15. Suppose that the Australian group bought all the tickets (7,059,052). We may consider two separate cases.

- Case 1: If none of the prizes was shared, the rate of return on this lottery investment, with prizes paid at the end of each year would be

$$PE\ (i) = -\$7,059,052 + \$911,197(P/F, i, 1)$$
$$+ \$1,350,368(P/A, i, 20)$$
$$= 0.$$
$$i^* = 20.94\%.$$

---

[3] One ticket each for every possible combination of 6 numbers from 1 to 44: Let $C(n,k)$ = the number of combinations of n distinct numbers taken k at a time. Then

$$C(44,6) = \frac{44!}{6!(44-6)!} = 7,059,052.$$

The first-prize payoff over 20 years is equivalent to putting the same $7,059,052 in a more conventional investment that pays a guaranteed 20.94% return before taxes for 20 years, a rate available only from speculative investments with fairly high risk. (If the prizes are paid at the beginning of each year, the rate of return would be 27.88%.)

- Case 2: If the first prize is shared with one other ticket, the rate of return on this lottery investment would be 8.87%. (With the prizes paid at the beginning of each year, the rate of return would be 10.48%.) Certainly, if the first prize is shared by more than one, the rate of return would be far less than 8.87%.

Comments: Only lack of time prevented the group from buying the extra 2 million tickets. On February 15, the winning number, yielding a prize of $1,350,368 a year for 20 years, was pulled. Officials checked their records and found that one winning ticket, with the numbers 8, 11, 13, 15, 19, and 20, had been sold. Several clues pointed to an Australian investment group as being the winner. The investment group from Australia was able to reduce the uncertainty inherent to any lottery game by purchasing the bulk of the lottery tickets. Conceptually, we can entirely reduce the uncertainty (or risk) by purchasing the entire ticket pool. However, ordinarily in a project investment environment, it is not feasible to reduce project risk in the same way that this lottery example illustrated.

13.2 Methods of Describing Project Risk

We may begin analyzing project risk by first determining the uncertainty inherent in a project's cash flows. We can do this analysis in a number of ways, which range from making informal judgments to calculating complex economic and statistical analyses. In this section, we will introduce three methods of describing project risk: (1) sensitivity analysis, (2) break-even analysis, and (3) scenario analysis. Each method will be explained with reference to a single example (Windsor Metal Company).

13.2.1 Sensitivity Analysis

One way to glean a sense of the possible outcomes of an investment is to perform a sensitivity analysis. Sensitivity analysis determines the effect on the PE of variations in the input variables (such as revenues, operating cost, and salvage value) used to estimate after-tax cash flows. A **sensitivity analysis** reveals how much the PE will change in response to a given change in an input variable. In cash flow calculations, some items have a greater influence on the final result than others. In some problems, the most significant item may be easily identified. For example, the estimate of sales volume is often a major factor in a problem in which the quantity sold varies among the alternatives. In other problems, we may want to locate the items that have an important influence on the final results so that they can be subjected to special scrutiny.

Sensitivity analysis is sometimes called "what-if" analysis because it answers questions such as: What if incremental sales are only 1000 units, rather than 2000 units? Then what will the PE be? Sensitivity analysis begins with a base-case situation, which is developed using the most-likely values for each input. We then change the specific variable of interest by several specified percentages above and below the most-likely value, while holding other variables constant. Next, we calculate a new PE for each of these values. A convenient and useful way to present the results of a sensitivity analysis is to plot **sensitivity graphs**. The slopes of the lines show how sensitive the PE is to changes in each of the inputs: The steeper the slope, the more sensitive the PE is to a change in a particular variable. Sensitivity graphs identify the crucial variables that affect the final outcome most. We will use Example 13.2 to illustrate the concept of sensitivity analysis.

EXAMPLE 13.2 Sensitivity analysis

Windsor Metal Company (WMC), a small manufacturer of fabricated metal parts, must decide whether to enter the competition to become the supplier of transmission-housings for Gulf Electric. Gulf Electric produces transmission-housings in its own in-house manufacturing facility, but it has almost reached its maximum production capacity. Therefore, Gulf is looking for an outside supplier. To compete, WMC must design a new fixture for the production process and purchase a new forge. The new forge would cost $125,000. This total includes retooling costs for the transmission-housings. If WMC gets the order, it may be able to sell as many as 2000 units per year to Gulf Electric for $50 each, and variable production costs,[4] such as direct labor and direct material costs, will be $15 per unit. The increase in fixed costs,[5] other than capital cost allowance, will amount to $10,000 per year. The firm expects that the proposed transmission-housings project will have about a 5-year product life. The firm also estimates that the amount ordered by Gulf Electric for the first year will be ordered in each of the subsequent 4 years. (Due to the nature of contracted production, the annual demand and unit price would remain the same over the project after the contract is signed.) The capital investment is eligible for a declining balance CCA rate of 30%, and the marginal income tax rate is expected to remain at 40%. At the end of 5 years, the forge is expected to retain a market value of about 32% of the original investment. Based on this information, the engineering and marketing staffs of WMC have prepared the cash flow forecasts shown in Table 13.1. Since the PE = $40,460 is positive at the 15% opportunity cost of capital (MARR), the project appears to be worth undertaking.

---

[4] Expenses that change in direct proportion to the change in volume of sales or production, as defined in Section 1.7.

[5] Expenses that do not vary as the volume of sales or production changes. For example, property taxes, insurance, depreciation, and rent are usually fixed expenses.

Table 13.1		0	1	2	3	4	5
After-Tax Cash Flow for WMC's Transmission- Housings Project (Example 13.2)	**Income Statement**						
	Revenues						
	Unit price	$ 50	$ 50	$ 50	$ 50	$ 50	
	Demand (units)		2,000	2,000	2,000	2,000	2,000
	Sales revenue		$100,000	$100,000	$100,000	$100,000	$100,000
	Expenses						
	Unit variable cost		$ 15	$ 15	$ 15	$ 15	$ 15
	Variable cost		30,000	30,000	30,000	30,000	30,000
	Fixed cost		10,000	10,000	10,000	10,000	10,000
	CCA		18,750	31,875	22,313	15,619	10,933
	Taxable income		$ 41,250	$ 28,125	$ 37,688	$ 44,381	$ 49,067
	Income taxes (40%)		16,500	11,250	15,075	17,753	19,627
	Net income		$ 24,750	$ 16,875	$ 22,613	$ 26,629	$ 29,440
	Cash Flow Statement						
	Operating activities						
	Net income		$ 24,750	$ 16,875	$ 22,613	$ 26,629	$ 29,440
	CCA		18,750	31,875	22,313	15,619	10,933
	Investment activities						
	Investment	$(125,000)					
	Salvage						$ 40,000
	Disposal tax effects						(5,796)
	Net cash flow	$(125,000)	$ 43,500	$ 48,750	$ 44,925	$ 42,248	$ 74,578

PE(15%) = $40,460   AE(15%) = $12,070   IRR = 27.1%

However, WMC's managers are uneasy about this project because of too many uncertain elements that have not been considered in the analysis. If it decided to take on the project, WMC must make the investment in the forging machine to provide Gulf Electric with some samples as a part of the bidding process. If Gulf Electric does not like WMC's sample, WMC stands to lose its entire investment in the forging machine. Another issue is that, if Gulf likes WMC's sample, but it is overpriced, WMC would be under pressure to bring the price in line with competing firms. Even the possibility that WMC would get a smaller order must be considered, as Gulf may utilize their overtime capacity to produce some extra units. It is also not certain about the variable and fixed cost figures. Recognizing these uncertainties, the managers want to assess the various potential future outcomes before making a final decision. Put yourself in WMC's management position and describe how you may resolve the uncertainty associated with the project. In doing so, perform a sensitivity analysis for each variable and develop a sensitivity graph.

Discussion: Table 13.1 shows WMC's expected cash flows—but a guarantee that they will indeed materialize cannot be assumed. WMC is not particularly confident in its revenue forecasts. The managers think that, if competing firms enter the market, WMC will lose a substantial portion of the projected revenues by not being able to increase its bidding price. Before undertaking the project described, the company needs to identify the key variables that will determine whether the project will succeed or fail. The marketing department has estimated revenue as follows:

$$\begin{aligned} \text{Annual revenue} \quad &= \quad \text{(Product demand)(unit price)} \\ &= \quad (2000)(\$50) = \$100,000. \end{aligned}$$

The engineering department has estimated variable costs such as labor and material per unit at $15. Since the projected sales volume is 2000 units per year, the total variable cost is $30,000.

After first defining the unit sales, unit price, unit variable cost, fixed cost, and salvage value, we conduct a sensitivity analysis with respect to these key input variables. This is done by varying each of the estimates by a given percentage and determining what effect the variation in that item will have on the final results. If the effect is large, the result is sensitive to that item. Our objective is to locate the most sensitive item(s).

Solution

Sensitivity analysis: We begin the sensitivity analysis with a consideration of the "base-case" situation, which reflects the best estimate (expected value) for each input variable. In developing Table 13.2, we changed a given variable by 20% in 5% increments, above and below the base-case value and calculated new PEs, while other variables were held constant. The values for both sales and operating costs were the expected, or base-case, values, and the resulting $40,460 is the base-case PE. Now we ask a series of "what-if" questions: What if sales are 20% below the expected level? What if operating costs rise? What if the unit price drops from $50 to $45? Table 13.2 summarizes the results of varying the values of the key input variables.

Sensitivity graph: Figure 13.1 shows the transmission project's sensitivity graphs for five of the key input variables. The base-case PE value of $40,460 is plotted against 0% deviation on the abscissa. Next, the value of product demand is reduced to 0.95 (−5% deviation) of its base-case value, and the PE is recomputed with all other variables held at their base-case value. We repeat the process by either decreasing or increasing the relative deviation from the base case. The lines for the unit price, unit variable cost, fixed cost, and salvage value are obtained in the same manner. In Figure 13.1, we see that the project's PE is (1) very sensitive to changes in product demand and unit price, (2) fairly sensitive to changes in the variable costs, and (3) relatively insensitive to changes in the fixed cost and the salvage value.

Table 13.2
Sensitivity Analysis for Five Key Input Variables (Example 13.2)

Deviation	−20%	−15%	−10%	−5%	0%	5%	10%	15%	20%
Unit price	$ 234	$10,291	$20,347	$30,404	$40,460	$50,517	$60,573	$70,630	$80,686
Demand	12,302	19,342	26,381	33,421	40,460	47,500	54,539	61,579	68,618
Unit variable cost	52,528	49,511	46,494	43,477	40,460	37,443	34,426	31,410	28,393
Fixed cost	44,483	43,477	42,472	41,466	40,460	39,455	38,449	37,443	36,438
Salvage value	38,074	38,671	39,267	39,864	40,460	41,057	41,654	42,250	42,847

(Base marker above 0% column)

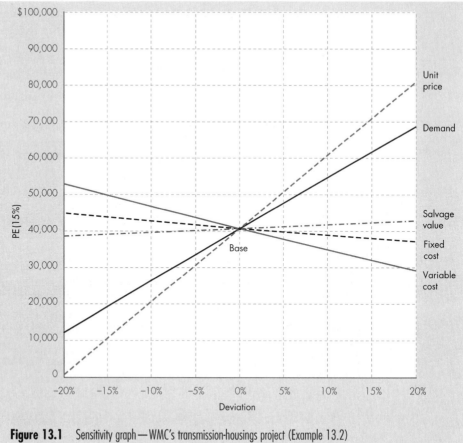

Figure 13.1 Sensitivity graph—WMC's transmission-housings project (Example 13.2)

Graphic displays such as the one in Figure 13.1 provide a useful means to communicate the relative sensitivities of the different variables on the corresponding PE value. However, sensitivity graphs do not explain any interactions among the variables or the likelihood of realizing any specific deviation from the base case. Certainly, it is conceivable that an answer might not be very sensitive to changes in either of two items, but very sensitive to combined changes in them.

13.2.2 Sensitivity Analysis for Mutually Exclusive Alternatives

In Figure 13.1, each variable is uniformly adjusted by ± 20% and all variables are plotted on the same chart. This uniform adjustment assumption can be too simplistic; in many situations, each variable can have a different range of uncertainty. Plotting all variables on the same chart could be confusing if there are too many variables to consider. When we perform sensitivity analysis for mutually exclusive alternatives, it may be more effective to plot the PEs (or any other measures such as AEs) of all alternatives over the range of each variable, basically one plot for each variable, with units of the variable on the horizontal axis. Example 13.3 illustrates this.

EXAMPLE
13.3

Sensitivity analysis for mutually exclusive alternatives

A local Canada Post Service Office is considering purchasing a 2000 kg capacity forklift truck, which will be used primarily for processing incoming as well as outgoing postal packages. Forklift trucks traditionally have been fueled by either gasoline, liquid propane gas (LPG), or diesel fuel. Battery-powered electric forklifts, however, are increasingly popular in many industrial sectors due to the economic and environmental benefits that accrue from their use. Therefore, the postal service is interested in comparing the four different types of fuel. Annual fuel and maintenance costs are measured in terms of number of shifts per year, where one shift is equivalent to 8 hours of operation.

The postal service is unsure of the number of shifts per year, but it expects it should be somewhere between 200 and 260 shifts. Canada Post uses 10% as an interest rate for any project evaluation of this nature. This analysis will be conducted on a before-tax basis. Develop a sensitivity graph that shows how the choice of alternatives changes as a function of number of shifts per year.

	Electrical Power	LPG	Gasoline	Diesel Fuel
Life expectancy	7 year	7 years	7 years	7 years
Initial cost	$29,739	$21,200	$20,107	$22,263
Salvage value	$3000	$2000	$2000	$2200
Maximum shifts per year	260	260	260	260
Fuel consumption/shift	31.25 kWh	26.1 L	29.6 L	18.5 L
Fuel cost/unit	$0.05/kWh	$0.43/L	$0.45/L	$0.44/L
Fuel cost per shift	$1.56	$11.22	$13.32	$8.14
Annual maintenance cost				
Fixed cost	$500	$1000	$1000	$1000
Variable cost/shift	$3.5	$7	$7	$7

Solution

Two annual cost components are pertinent to this problem: (1) Ownership cost (capital cost) and (2) operating cost (fuel and maintenance cost). Since the operating cost is already given on an annual basis, we only need to determine the equivalent annual ownership cost for each alternative.

(a) Ownership cost (capital cost): Using the capital recovery with return formula developed in Eq. (4.6), we compute

Electrical power: CR(10%) = ($29,739 – $3000)(A/P, 10%, 7) + (0.10)$3000
= $5792.

LPG: CR(10%) = ($21,200 – $2000)(A/P, 10%, 7) + (0.10)$2000
= $4144.

Gasoline: CR(10%) = ($20,107 – $2000)(A/P, 10%, 7) + (0.10)$2000
= $3919.

Diesel fuel: CR(10%) = ($22,263 – $2200)(A/P, 10%, 7) + (0.10)$2200
= $4341.

(b) Annual operating cost: We can express the annual operating cost as a function of number of shifts per year (M) by combining the variable and fixed cost portions of fuel and maintenance expenditures.

Electrical power: $500 + (1.56 + 3.5)M = $500 + 5.06M.

LPG: $1000 + (11.22 + 7)M = $1000 + 18.22M.

Gasoline: $1000 + (13.32 + 7)M = $1000 + 20.32M.

Diesel fuel: $1000 + (8.14 + 7)M = $1000 + 15.14M.

(c) Total equivalent annual cost: This is the sum of the ownership cost and operating cost.

Electrical power: $AE(10\%) = 6292 + 5.06M.$

LPG: $AE(10\%) = 5144 + 18.22M.$

Gasoline: $AE(10\%) = 4919 + 20.32M.$

Diesel fuel: $AE(10\%) = 5341 + 15.14M.$

In Figure 13.2, these four annual equivalent costs are plotted as a function of number of shifts, M. It appears that the economics of the electric forklift truck can be justified as long as the number of annual shifts exceeds approximately 95.

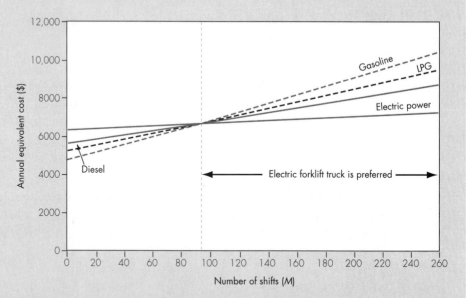

Figure 13.2 Sensitivity analysis for mutually exclusive alternatives (Example 13.3)

13.2.3 Break-Even Analysis

When we perform a sensitivity analysis of a project, we are asking how serious the effect of lower revenues or higher costs will be on the project's profitability. Managers sometimes prefer to ask how much sales can decrease below forecasts before the project begins to lose money. This type of analysis is known as **break-even analysis**. In other words, break-even analysis is a technique for studying the effect of variations in output on a firm's PE (or other measures). We will present an approach to break-even analysis based on the project's cash flows.

To illustrate the procedure of break-even analysis based on PE, we use the generalized cash flow approach we discussed in Section 9.4. We compute the PE of cash inflows as a function of an unknown variable (say X)—this variable could be annual sales. For example:

$$\text{PE of cash inflows} = f_1(x)$$

Next, we compute the PE of cash outflows as a function of X:

$$\text{PE of cash outflows} = f_2(x)$$

The net PE is of course the difference between these two numbers. Then, we look for the break-even value of x that makes

$$f_1(x) = f_2(x).$$

Note that this break-even value calculation is similar to that used to calculate the internal rate of return where we want to find the interest rate that makes the net PE equal zero and many other similar "cutoff values" where a choice changes.

EXAMPLE 13.4

Break-even analysis

Through the sensitivity analysis in Example 13.2, WMC's managers are convinced that the PE is most sensitive to changes in annual sales volumes. Determine the break-even PE value as a function of that variable.

Solution

The analysis is shown in Table 13.3, where the revenues and costs of the WMC transmission-housings project are set out in terms of an unknown quantity of annual sales, X.

Table 13.3		0	1	2	3	4	5
After-Tax Cash	Cash inflow						
Flow for	Net salvage						34,204
WMC's	Revenue						
Transmission-	$X(1 - 0.4)(\$50)$		30X	30X	30X	30X	30X
Housings	CCA credit						
Project	0.4 (CCA)		7,500	12,750	8,925	6,248	4,373
(Example	Cash outflow						
13.4)	Investment	–125,000					
	Variable cost						
	$-X(1 - 0.4)(\$15)$		–9X	–9X	–9X	–9X	–9X
	Fixed cost						
	$-(1 - 0.4)(\$10,000)$		–6,000	–6,000	–6,000	–6,000	–6,000
	Net cash flow	–125,000	21X + 1,500	21X + 6,750	21X + 2,925	21X + 248	21X + 32,577

We calculate both the PEs of cash inflow and outflows as

- PE of cash inflows

$$PE\ (15\%)_{\text{Inflow}} = \text{(PE of after-tax net revenue)}$$
$$+ \text{(PE of net salvage value)}$$
$$+ \text{(PE of tax savings from CCA)}.$$

$$= 30X\ (P/A,\ 15\%,\ 5) + \$34,204(P/F,\ 15\%,\ 5)$$
$$+ \$7500(P/F,\ 15\%,\ 1) + \$12,750(P/F,\ 15\%,\ 2)$$
$$+ \$8925(P/F,\ 15\%,\ 3) + \$6248(P/F,\ 15\%,\ 4)$$
$$+ \$4373(P/F,\ 15\%,\ 5)$$
$$= 30X(P/A,\ 15\%,\ 5) + \$44,782$$
$$= 100.5650X + \$44,782.$$

- PE of cash outflows:

$$PE\ (15\%)_{\text{Outflow}} = \text{(PE of capital expenditure)}$$
$$+ \text{(PE of after-tax expenses)}.$$

$$= \$125,000 + (9X + \$6000)(P/A,\ 15\%,\ 5)$$
$$= 30.1694X + \$145,113.$$

The net PE of cash flows for the WMC is thus

$$PE\ (15\%) = 100.5650X + \$44,782$$
$$- (30.1694X + \$145,113)$$
$$= 70.3956X - \$100,331.$$

In Table 13.4, we compute the PE of the inflows and the PE of the outflows as a function of demand (X).

Table 13.4	Demand (X)	PE of Inflow (100.5650X + $44,782)	PE of Outflow (30.1694X + $145,113)	Net PE (70.3956X − 100,331)
Determination	0	$ 44,782	$145,113	$(100,331)
of Break-Even	500	95,065	160,198	(65,133)
Volume	1000	145,347	175,282	(29,935)
Based on	1425	188,087	188,104	(17)
Project's PE	1426	188,188	188,135	53
(Example	1500	195,630	190,367	5,262
13.4)	2000	245,912	205,452	40,460
	2500	296,195	220,537	75,658

Break-even volume = 1426 units.

The PE will be just slightly positive if the company sells 1426 units. Precisely calculated, the zero-PE point (break-even volume) is 1425.29 units:

$$PE\,(15\%) = 70.39356X - \$100,623$$
$$= 0$$
$$X_b = 1426 \text{ units.}$$

In Figure 13.3, we have plotted the PEs of the inflows and outflows under various assumptions about annual sales. The two lines cross when sales are 1426 units, the point at which the project has a zero PE. Again we see that, as long as sales are greater or equal to 1426, the project has a positive PE.

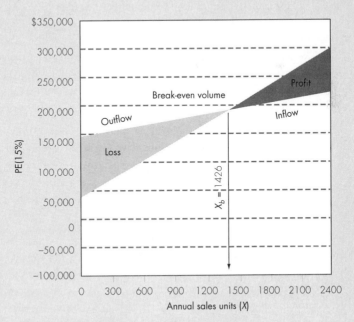

Figure 13.3 Break-even analysis based on net cash flow (Example 13.4)

13.2.4   Scenario Analysis

Although both sensitivity and break-even analyses are useful, they have limitations. Often it is difficult to specify precisely the relationship between a particular variable and the PE. The relationship is further complicated by interdependencies among the variables. Holding operating costs constant while varying unit sales may simplify the analysis, but in reality, operating costs do not behave in this manner. Yet, it may complicate the analysis too much to permit movement in more than one variable at a time.

A scenario analysis is a technique that does consider the sensitivity of PE to simultaneous changes in key variables over the range of likely variable values. For example, the decision-maker may consider two extreme cases, a "worst-case" scenario (low unit sales, low unit price, high variable cost per unit, high fixed cost, and so on) and a "best-case" scenario. The PEs under the worst and the best conditions are then calculated and compared to the expected, or base-case, PE. Example 13.5 will illustrate a plausible scenario analysis for WMC's transmission-housings project.

EXAMPLE 13.5   **Scenario analysis**

Consider again WMC's transmission-housings project in Example 13.2. Assume that the company's managers are fairly confident of their estimates of all the project's cash flow variables, except the estimates for sales demand. Further, assume that they regard a decline in unit sales to below 1600 or a rise above 2400 as extremely unlikely. Thus, decremental annual sales of 400 units defines the lower bound, or the worst-case scenario, whereas incremental annual sales of 400 units defines the upper bound, or the best-case scenario. (Remember that the most-likely value was 2000 in annual unit sales.) Discuss the worst- and best-case scenarios, assuming that the unit sales for all 5 years would be equal.

Discussion: To carry out the scenario analysis, we ask the marketing and engineering staffs to give optimistic (best-case) and pessimistic (worst case) estimates for the key variables. Then we use the worst-case variable values to obtain the worst-case PE and the best-case variable values to obtain the best-case PE.

Solution

The results of our analysis are summarized below. We see that the base-case produces a positive PE, the worst-case produces a negative PE, and the best-case produces a large positive PE.

Variable Considered	Worst-Case Scenario	Most-Likely-Case Scenario	Best-Case Scenario
Unit demand	1,600	2,000	2,400
Unit price ($)	48	50	53
Variable cost ($)	17	15	12
Fixed cost ($)	11,000	10,000	8,000
Salvage value ($)	30,000	40,000	50,000
PE(15%)	–$5,564	$40,460	$104,587

By just looking at the results in the table, it is not easy to interpret the scenario analysis or to make a decision based on it. For example, we could say that there is a chance of losing money on the project, but we do not yet have a specific probability for this possibility. Clearly, we need estimates of the probabilities of occurrence of the worst-case, the best-case, the base-case (most-likely), and all the other possibilities. The need to estimate probabilities leads us directly to our next step, developing a probability distribution (or the probability that the variable in question takes on a certain value). If we can predict the effects on the PE of variations in the parameters, why should we not assign a probability distribution to the possible outcomes of each parameter and combine these distributions in some way to produce a probability distribution for the possible outcomes of the PE? We shall consider this issue in the following sections.

13.3 Probability Concepts for Investment Decisions

In this section, we shall assume that the analyst has available the probabilities (likelihoods) of future events from either previous experience in a similar project or a market survey. The use of probability information can provide management with a range of possible outcomes and the likelihood of achieving different goals under each investment alternative.

13.3.1 Assessment of Probabilities

We will first define terms related to probability concepts, such as random variable, probability distribution, and cumulative probability distribution.

Random Variables

A **random variable** is a parameter or variable that can have more than one possible value. The value of a random variable at any one time is unknown until the event occurs, but the probability that the random variable will have a specific value is known in advance. In other words, associated with each possible value of the random variable is a likelihood, or probability, of occurrence. For example, when your school team plays a football game, only three events regarding the game outcome are possible: Win, lose, or tie (if allowed). The game outcome is a random variable, which is largely dictated by the strength of your opponent.

To indicate random variables, we will adopt the convention of a capital italic letter (for example, X). To denote the situation where the random variable takes a specific value, we will use a lower-case italic letter (for example, x). Random variables are classified as either discrete or continuous. Any random variables that take on only isolated (countable) values are **discrete random variables**. **Continuous random variables** may have any value in a certain interval. For example, the game outcome described above is a discrete random variable. But suppose you are interested in the amount of beverage sold on a given game day. The quantity (or volume) of beverage sold will depend on the weather conditions, the number of people attending the game, and other factors. In this case, the quantity is a random variable with a continuous range of values.

Probability Distributions

For a discrete random variable, the probability figure for each random event needs to be assessed. For a continuous random variable, the probability function needs to be assessed since the event takes place over a continuous domain. In either case, a range of possibilities over the feasible outcomes exists. These together make up a **probability distribution**.

Probability assessments may be based on past observations or historical data, if the same trends or characteristics of the past are expected to prevail in the future. Forecasting weather or predicting a game outcome in many professional sports is done based on the compiled statistical data. Any probability assessments based on objective data are called **objective probabilities**. We are not restricted to objective probabilities. In many real investment situations, no objective data are available to consider. In these situations, we assign **subjective probabilities** that we think are appropriate to the possible states of nature. As long as we act consistently with our beliefs about the possible events, we may reasonably account for the economic consequences of such uncertain events in our profitability analysis.

For a continuous random variable, we usually try to establish a range of values, i.e., we try to determine a **minimum value** (L) and a **maximum value** (H). Next, we determine whether any value within these limits might be *more likely* to occur than are the other values, i.e., does the distribution have a **mode** (M_o), or a most frequently occurring value?

If the distribution does have a mode, we can represent the variable by a **triangular distribution**, such as that shown in Figure 13.4. If we have no reason to assume that one value is any more likely to occur than any other, perhaps the best we can do is to describe the variable as a **uniform distribution**, as shown in Figure 13.5. These two distributions are frequently used to represent the variability of a random variable when the only information we have is its minimum, its maximum, and whether or not the distribution has a mode. For example, suppose the best judgment of the analyst was that the sales revenue could vary anywhere from $2000 to $5000 per day, and any value within the range would be equally likely. This judgment about the variability of sales revenue could be represented by a uniform distribution.

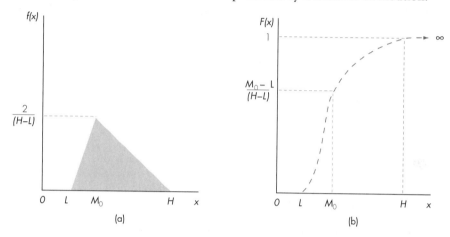

Figure 13.4 A triangular probability distribution: (a) Probability function and (b) cumulative probability distribution

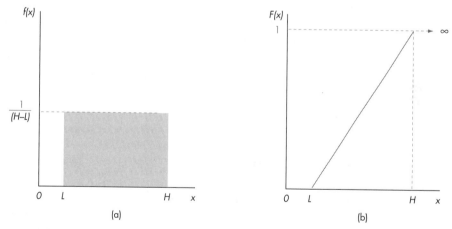

Figure 13.5 A uniform probability distribution: (a) Probability function and (b) cumulative probability distribution

For WMC's transmission-housings project, we can think of the discrete random variables (X and Y) as two quantities whose values cannot be predicted with certainty at the time of decision making. Let us assume the probability distributions in Table 13.5: We see that the product demand level with the highest probability is 2000 units, whereas the unit sales price with the highest probability is $50. These, therefore, are the most-likely values. We also see a substantial probability that a unit demand, other than 2000 units, will be realized. When we use only the most-likely values in an economic analysis, we are in fact ignoring these other outcomes.

Table 13.5
Probability Distributions for Unit Demand (X) and Unit Price (Y) for WMC's Project

Product Demand (X)		Unit Sale Price (Y)	
Units (x)	P(X = x)	Unit Price (y)	P(Y = y)
1600	0.20	$48	0.30
2000	0.60	50	0.50
2400	0.20	53	0.20

X and Y are assumed to be independent random variables. In competitive markets, price generally affects sales quantity. In this problem however, WMC assumes that within the assumed range price will not affect demand, although both are uncertain outcomes of the contract negotiations.

Cumulative Distribution

As we have observed in the previous section, the probability distribution provides information regarding the probability that a random variable will be some value, x. We can use this information, in turn, to define the cumulative distribution function. The **cumulative distribution** function shows the probability that the random variable will attain a value smaller than, or equal to, some value, x. A common notation for the cumulative distribution is

$$F(x) = P(X \le x) = \begin{cases} \sum_{j:\, x_j \le x} p_j \text{ (for a discrete random variable)} \\ \int_{L}^{x} f(x)dx \text{ (for a continuous random variable)} \end{cases}$$

where p_j is the probability of occurrence of the x_jth value of the discrete random variable, and $f(x)$ is a probability function for a continuous variable. With respect to a continuous random variable, the cumulative distribution rises continuously in a smooth (rather than stepwise) fashion.

In Example 13.6, we will explain the method by which probabilistic information can be incorporated into our analysis. Again, WMC's transmission-housing project will be used. In the next section we will show you how to compute some composite statistics using all the data.

EXAMPLE 13.6 **Cumulative probability distributions**

Suppose that the only parameters subject to risk are the number of unit sales (X) to Gulf Electric each year and the unit sales price (Y). From experience in the market, WMC assesses the probabilities of outcomes for each variable as shown in Table

13.5. Determine the cumulative probability distributions for these random variables.

Solution

Consider the demand probability distribution (X) given previously in Table 13.5 for WMC's transmission-housings project:

Unit Demand (X)	Probability, $P(X = x)$
1600	0.2
2000	0.6
2400	0.2

If we want to know the probability that demand will be less than, or equal to, any particular value, we can use the following cumulative probability function:

$$F(x) = P(X \le x) = \begin{cases} 0.2 & x \le 1600 \\ 0.8 & x \le 2000 \\ 1.0 & x \le 2400. \end{cases}$$

For example, if we want to know the probability that the demand will be less than or equal to 2000, we can examine the appropriate interval ($x \le 2000$), and we shall find that the probability is 80%.

We can find the cumulative distribution for Y in a similar fashion. Graphic represen-

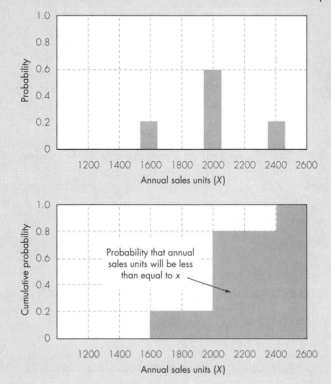

Figure 13.6 Probability and cumulative probability distributions for random variable X (annual sales) (Example 13.6)

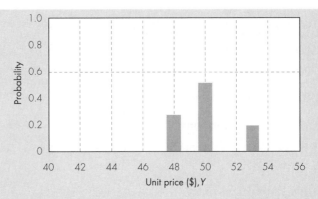

Figure 13.7a Probability distributions for random variable Y (unit price) from Table 13.5

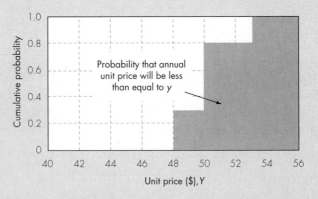

Figure 13.7b Cumulative probability distributions for random variable Y (unit price) from Table 13.5

tations of the probability distributions and the cumulative probability distributions for X and Y are given in Figures 13.6 and 13.7, respectively.

13.3.2 Summary of Probabilistic Information

Although knowledge of the probability distribution of a random variable allows us to make a specific probability statement, a single value that may characterize the random variable and its probability distribution is often desirable. Such a quantity is the **expected value** of the random variable. We also want to know something about how the values of the random variable are dispersed about the expected value (i.e., the **variance**). In investment analysis, this dispersion information is interpreted as the degree of project risk. Expected value represents the weighted average of the random variable, and the variance captures the variability of the random variable.

Measure of Expectation

The **expected value** (also called the **mean**) is a weighted average value of the random variable where the weighting factors are the probabilities of occurrence. All distributions (discrete and continuous) have an expected value. We will use $E[X]$ (or μ) to denote the expected value of random variable X. For a random variable X that has either discrete or continuous values, we compute an expected value with

$$E[X] = \mu = \begin{cases} \displaystyle\sum_{j=1}^{J} p_j x_j & \text{(discrete case)} \\[2em] \displaystyle\int_{L}^{H} xf(x)dx & \text{(continuous case),} \end{cases} \qquad (13.1)$$

where J is the number of discrete events and L and H are the lower and upper bounds of the continuous probability distribution.

The expected value of a distribution gives us important information about the "average" value of a random variable, such as the PE, but it does not tell us anything about the variability on either side of the expected value. Will the range of possible values of the random variable be very small, and will all the values be located at, or near, the expected value? For example, the following represents the temperatures recorded on February 28, 1999, for two cities in Canada.

Location	Low	High	Average
Calgary	−6°C	10°C	2°C
Toronto	0°C	4°C	2°C

Even though both cities had identical mean (average) temperatures on that particular day, they had different variations in extreme temperatures. We shall examine this variability issue in the following section.

Measure of Variation

Another measure needed when we are analyzing probabilistic situations is a measure of the risk due to the variability of the outcomes. There are several measures of the variation of a set of numbers that are used in statistical analysis—the **range** and the **variance** (or **standard deviation**), among others. The variance and the standard deviation are used most commonly in the analysis of risk situations. We will use $Var[X]$ or σ_x^2 to denote the variance and σ_x to denote the standard deviation of random variable X. (If there is only one random variable in an analysis, we normally omit the subscript.)

The variance tells us the degree of spread, or dispersion, of the distribution on either side of the mean value. As the variance increases, the spread of the distribution increases; the smaller the variance, the narrower the spread about the expected value.

To determine the variance, we first calculate the deviation of each possible outcome x_j from the expected value $(x_j - \mu)$. Then we square each result and multiply it by the probability of x_j occurring (that is, p_j). The summation of all these products serves as a measure of the distribution's variability. For a random variable that has only discrete values, the equation to compute variance[6] is

$$Var[X] = \sigma_x^2 = \sum_{j=1}^{J} (x_j - \mu)^2 p_j, \tag{13.2}$$

where p_j is the probability of occurrence of the jth value of the random variable (x_j), and μ is as defined by Eq. (13.1). To be most useful, any measure of risk should have a definite value (unit). One such measure is the **standard deviation**. To calculate the standard deviation, we take the positive square root of $Var[X]$, which is measured in the same units as X:

$$\sigma_x = \sqrt{Var[X]}. \tag{13.3}$$

The standard deviation is a probability-weighted deviation (more precisely, the square root of the weighted sum of squared deviations) from the expected value. Thus, it gives us an idea of how far above, or below, the expected value the actual value is likely to be. For most probability distributions, the value of the actual outcome will be observed within the $\pm 3\sigma$ range.

In practice, the calculation of the variance is somewhat easier if we use the following formula:

$$\begin{aligned} Var[X] &= \sum p_j x_j^2 - (\sum p_j x_j)^2 \\ &= E[X^2] - (E[X])^2 \end{aligned} \tag{13.4}$$

The term $E[X^2]$ in Eq. (13.4) is interpreted as the mean value of the squares of the random variable (i.e., the actual values squared). The second term is simply the mean value squared. Example 13.7 will illustrate how we compute measures of variation.

EXAMPLE 13.7 Mean and variance calculation

Consider WMC's transmission-housings project. Unit sales (X) and unit price (Y) are estimated as in Table 13.5. Now compute the means, variances, and standard deviations for the random variables X and Y.

Solution

For the product demand variable (X):

---

[6] For a continuous random variable, we compute the variance as follows:

$$Var[X] = \int_L^H (x - u)^2 f(x) dx.$$

x_j	p_j	$x_j p_j$	$(x_j - E[X])^2$	$(x_j - E[X])^2 p_j$
1600	0.20	320	$(-400)^2$	32,000
2000	0.60	1200	0	0
2400	0.20	480	$(400)^2$	32,000
		$E[X] = 2000$		$Var[X] = 64,000$
				$\sigma_x = 252.98$

For the variable unit price (Y):

y_j	p_j	$y_j p_j$	$(y_j - E[Y])^2$	$(y_j - E[Y])^2 p_j$
$48	0.30	$14.40	$(-2)^2$	1.20
50	0.50	25.00	$(0)^2$	0
53	0.20	10.60	$(3)^2$	1.80
		$E[Y] = 50.00$		$Var[Y] = 3.00$
				$\sigma_y = 1.73$

13.3.3   Joint and Conditional Probabilities

Thus far, we have not looked at how some variables can influence the outcomes of others. It is, however, entirely possible—indeed it is likely—that the values of some parameters will be dependent on the values of others. We commonly express these dependencies in terms of conditional probabilities. An example is product demand, which will generally be influenced by unit price.

We define a **joint probability** as

$$P(x, y) = P(X = x \mid Y = y)P(Y = y), \tag{13.5}$$

where $P(x, y)$ is the probability of observing those specific outcomes for X and Y, $P(X = x \mid Y = y)$ is the **conditional probability** of observing x, given $Y = y$, and $P(Y = y)$ is the **marginal probability** of observing $Y = y$. Certainly important cases exist where a knowledge of the occurrence of event X does not change the probability of an event Y. That is, if X and Y are **independent**, the joint probability is simply

$$P(x, y) = P(x)P(y). \tag{13.6}$$

The concepts of joint, marginal, and conditional distributions are best illustrated by numerical examples.

EXAMPLE

13.8

$

Probability distribution for 2 dependent variables

Janice and Paul are planning to invest $7000 now for their 1-year-old daughter's university education. The impact of their investment depends on the performance of the Education Savings Plan, and the cost of tuition in seventeen years when Mabel begins her studies. Their financial advisor, Suzanne, estimates probabilities of some likely scenarios, noting that a high rate of return in the fund would be driven in part by a high general inflation rate, which would also induce higher tuition fees. For example, Suzanne believes that the likelihood of the annual (actual) rate of return (RoR) being 10% is 0.20. If this occurs, she estimates that the conditional probability of the tuition being $14,000 per year[7] in 17 years is 0.60. Therefore, the probability of this joint event (RoR =10% and tuition=$14,000) is:

$$
\begin{aligned}
P(x, y) \quad &= \quad P(10\%, \$14,000) \\
&= \quad P(y = \$14,000 \mid x = 10\%)\,P(x = 10\%) \\
&= \quad (0.60)(0.20) \\
&= \quad 0.12
\end{aligned}
$$

The probabilities for other joint events are obtained in a similar manner, as presented in Table 13.6:

Table 13.6

Assessments of Conditional and Joint Probabilities (Example 13.8)

Fund annual RoR (X)	Probability	Annual Tuition (Y)	Conditional Probability	Joint Probability
6%	0.30	$10,000	0.30	0.09
		$12,000	0.50	0.15
		$14,000	0.20	0.06
8%	0.50	$10,000	0.20	0.10
		$12,000	0.40	0.20
		$14,000	0.40	0.20
10%	0.20	$10,000	0.10	0.02
		$12,000	0.30	0.06
		$14,000	0.60	0.12

From Table 13.6, we can see that the fund return (X) ranges from 6% to 10% per year, the tuition fees (Y) range from $10,000 to $14,000 per year, and nine joint events (X, Y) are possible. The sum of these joint probabilities must equal 1, as shown in Table 13.7.

---

[7] For simplicity, we will assume that the tuition remains constant over the 4 years of Mabel's studies.

Table 13.7	Joint Event (x, y)	P(x, y)	FE of Fund in 17 Years (@x%)	PE (@x%) at n=17 of 4 Years' Tuition	Proportion of Tuition Covered by Education Fund
Likelihoods and Expected Values for Joint Events (Example 13.8)	(6%, $10,000)	0.09	$18,849	$36,730	51.3%
	(6%, $12,000)	0.15	18,849	44,076	42.7%
	(6%, $14,000)	0.06	18,849	51,422	36.7%
	(8%, $10,000)	0.10	25,900	35,771	72.4%
	(8%, $12,000)	0.20	25,900	42,925	60.3%
	(8%, $14,000)	0.20	25,900	50,079	51.7%
	(10%, $10,000)	0.02	35,381	34,869	100.0%
	(10%, $12,000)	0.06	35,381	41,842	84.6%
	(10%, $14,000)	0.12	35,381	48,816	72.5%
	Sum = 1.00		E(X) = $25,681	E(Y) = $44,246	E(X/Y) = 58.6%

When X and Y are independent variables, the ratio of the means can be approximated as follows: $E(X/Y) \approx E(X)/E(Y)$. In general, the same approximation holds true for dependent variables. Although there is a 2% chance that Paul and Janice's investment will be sufficient to cover all of the expenses [i.e., event (10%, $10,000) from Table 13.7] , because of the risk in this decision it is possible in the "worst-case" scenario that only about a third of the fees can be paid through this investment.

The marginal distribution for the tuition fees, y, can be developed from the joint event probabilities by fixing y and summing over x:

y_i	$P(y_i) = \sum_x P(x, y)$
$10,000	P(6%, $10,000) + P(8%, $10,000) + P(10%, $10,000) = 0.21
$12,000	P(6%, $12,000) + P(8%, $12,000) + P(10%, $12,000) = 0.41
$14,000	P(6%, $14,000) + P(8%, $14,000) + P(10%, $14,000) = 0.38

This marginal distribution tells us that 41% of the time we can expect the tuition fees to be $12,000 in 17 years, and that 21% and 38% of the time, we can expect them to be $10,000 and $14,000 respectively.

13.4 Probability Distribution of Net PE

After we have identified the random variables in a project and assessed the probabilities of the possible events, the next step is to develop the project's PE distribution.

13.4.1 Procedure for Developing a PE Distribution

We will consider the situation where all the random variables used in calculating the PE are independent. To develop the PE distribution, we may follow these steps:

- Express the PE as functions of unknown random variables.
- Determine the probability distribution for each random variable.
- Determine the joint events and their probabilities.
- Evaluate the PE equation at these joint events.
- Order the PE values in increasing order of PE.

These steps can best be illustrated by Example 13.9.

EXAMPLE 13.9 **Procedure for developing a net PE distribution**

Consider WMC's transmission-housings project in Example 13.2. Use the unit demand (X) and price (Y) given in Table 13.5, and develop the PE distribution for the WMC project. Then, calculate the mean and variance of the PE distribution.

Solution

Table 13.8 summarizes the after-tax cash flow for the WMC's transmission-housings project as functions of random variables X and Y. From this table, we can compute the PE of cash inflows as follows:

$$PE\,(15\%) \quad = \quad 0.6XY\,(P/A,\,15\%,\,5) + 44{,}782$$
$$= \quad 2.0113XY + \$44{,}782.$$

The PE of cash outflows is

$$PE\,(15\%) \quad = \quad \$125{,}000 + (9X + \$6000)(P/A,\,15\%,\,5)$$
$$= \quad 30.1694X + \$145{,}113.$$

Thus, the net PE is

$$PE\,(15\%) \quad = \quad 2.0113X\,(Y - \$15) - \$100{,}331.$$

If the product demand X and the unit price Y are independent random variables, the PE (15%) will also be a random variable. To determine the PE distribution, we need to consider all the combinations of possible outcomes. The first possibility is the event

---

Item	0	1	2	3	4	5
Cash inflow						
Net salvage						34,204
Revenue						
$X(1-0.4)Y$		0.6XY	0.6XY	0.6XY	0.6XY	0.6XY
CCA credit						
0.4 (CCA)		7,500	12,750	8,925	6,248	4,373
Cash outflow						
Investment	−125,000					
Variable cost						
$-X(1-0.4)(\$15)$		−9X	−9X	−9X	−9X	−9X
Fixed cost						
$-(1-0.4)(\$10,000)$		−6,000	−6,000	−6,000	−6,000	−6,000
Net cash flow	−125,000	0.6X(Y − 15) + 1,500	0.6X(Y − 15) + 6,750	0.6X(Y − 15) + 2,925	0.6X(Y − 15) + 248	0.6X(Y − 15) + 32,577

Table 13.8 After-Tax Cash Flow as a Function of Unknown Unit Demand (X) and Unit Price (Y) (Example 13.9)

where $x = 1600$ and $y = \$48$. Since X and Y are considered to be independent random variables, the probability of this joint event is

$$P(x = 1600, y = \$48) = P(x = 1600)P(y = \$48)$$
$$= (0.20)(0.30)$$
$$= 0.06.$$

With these values as input, we compute the possible net PE outcome as follows:

$$PE\,(15\%) = 2.0113X\,(Y - \$15) - \$100,331$$
$$= 2.0113(1600)(\$48 - \$15) - \$100,331$$
$$= \$5866.$$

Eight other outcomes are possible: they are summarized with their joint probabilities in Table 13.9 and depicted in Figure 13.8.

The probability distribution in Table 13.9 indicates that the project's PE varies between $5866 and $83,100, but that no loss under any of the circumstances exam-

Table 13.9 The PE Probability Distribution with Independent Random Variables (Example 13.9)

Event No.	x	y	P(x,y)	Cumulative Joint Probability	PE
1	1600	$48.00	0.06	0.06	$ 5,866
2	1600	50.00	0.10	0.16	12,302
3	1600	53.00	0.04	0.20	21,956
4	2000	48.00	0.18	0.38	32,415
5	2000	50.00	0.30	0.68	40,460
6	2000	53.00	0.12	0.80	52,528
7	2400	48.00	0.06	0.86	58,964
8	2400	50.00	0.10	0.96	68,618
9	2400	53.00	0.04	1.00	83,100

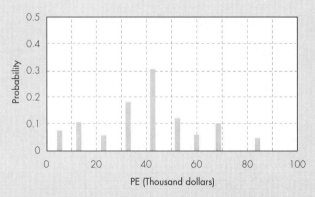

Figure 13.8 PE probability distributions when X and Y are independent (Example 13.9)

ined occurs. From the cumulative distribution, we further observe that there is a 0.38 probability that the project would realize a PE less than that forecast for the base-case situation ($40,460). On the other hand, there is a 0.32 probability that the PE will be greater than this value. Certainly, the probability distribution provides much more information on the likelihood of each possible event as compared with the scenario analysis presented in Section 13.2.4.

We have developed a probability distribution for the PE by considering random cash flows. As we observed, a probability distribution helps us to see what the data imply in terms of project risk. Now, we can see how to summarize the probabilistic information—the mean and the variance. For WMC's transmission-housings project, we compute the expected value of the PE distribution as shown in Table 13.10. Note that this expected value is the same as the most likely value of the PE distribution, as shown by the base case in Example 13.2. This equality was expected because both X and Y have a symmetrical probability distribution.

Table 13.10 Calculation of the Mean of PE Distribution (Example 13.9)	Event No.	x	y	P(x, y)	Cumulative Joint Probability	PE(15%)	Weighted PE
	1	1600	$48	0.06	0.06	$ 5,866	$ 352
	2	1600	50	0.10	0.16	12,302	1,230
	3	1600	53	0.04	0.20	21,956	878
	4	2000	48	0.18	0.38	32,415	5,835
	5	2000	50	0.30	0.68	40,460	12,138
	6	2000	53	0.12	0.80	52,528	6,303
	7	2400	48	0.06	0.86	58,964	3,538
	8	2400	50	0.10	0.96	68,618	6,862
	9	2400	53	0.04	1.00	83,100	3,324
						E[PE] =	$40,460

We obtain the variance of the PE distribution, assuming independence between X and Y and using Eq. (13.2), as shown in Table 13.11. We could obtain the same result more easily by using Eq. (13.4).

Table 13.11 Calculation of the Variance of PE Distribution (Example 13.9)	Event No.	x	y	P(x, y)	PE(15%)	(PE − E[PE])²	Weighted (PE − E[PE])²
	1	1600	$48	0.06	$ 5,866	1,196,761,483	$71,805,689
	2	1600	50	0.10	12,302	792,878,755	79,287,875
	3	1600	53	0.04	21,956	342,394,172	13,695,767
	4	2000	48	0.18	32,415	64,724,796	11,650,463
	5	2000	50	0.30	40,460	0	0
	6	2000	53	0.12	52,528	145,630,792	17,475,695
	7	2400	48	0.06	58,964	342,394,172	20,543,650
	8	2400	50	0.10	68,618	792,878,755	79,287,875
	9	2400	53	0.04	83,100	1,818,119,528	72,724,781

$$\text{Var[PE]} = \$366,471,797$$
$$\sigma = \$19,143$$

13.4.2 Decision Rules

Once the expected value has been located from the PE distribution, it can be used to make an accept-reject decision in much the same way that a single PE is used when a single possible outcome for an investment project is considered. The decision rule is called the **expected value criterion** and using it, we may accept a single project if its expected PE value is positive. In the case of mutually exclusive alternatives, we select the one with the highest expected PE. The use of expected PE has an advantage over the use of a point estimate, such as the most likely value, because it includes all possible cash flow events and their probabilities.

The justification for the use of the expected value criterion is based on the **law of large numbers**, which states that, if many repetitions of an experiment are performed, the average outcome will tend toward the expected value. This justification may seem to negate the usefulness of the expected value criterion, since most often in project evaluation we are concerned with a single, nonrepeatable "experiment"—i.e., an investment alternative. However, if a firm adopts the expected value criterion as a standard decision rule for *all* its investment alternatives, over the long term, the law of large numbers predicts that accepted projects tend to meet their expected values. Individual projects may succeed or fail, but the average project result tends to meet the firm's standard for economic success.

The expected-value criterion is simple and straightforward to use, but it fails to reflect the variability of investment outcome. Certainly, we can enrich our decision by incorporating the variability information along with the expected value. Since the variance represents the dispersion of the distribution, it is desirable to minimize it. In other words, the smaller the variance, the less the variability (hence potential for loss) associated with the PE. Therefore, when we compare mutually exclusive projects, we may select the alternative with the smaller variance if its expected value is the same as, or larger than, those of other alternatives. In cases where preferences are not clear cut, the ultimate choice depends on the decision-maker's tradeoffs—how far he or she is willing to accept a greater variability to achieve a higher expected value. In other words, the challenge is to decide what level of risk you are willing to accept and then, having decided on your risk tolerance, to understand the implications of that choice. Example 13.10 illustrates some of the critical issues that need to be considered in evaluating mutually exclusive risky projects.

EXAMPLE 13.10 Comparing risky mutually exclusive projects

With ever-growing concerns about air pollution, the greenhouse effect, and stricter emission standards, Green Engineering has developed a prototype conversion unit that allows a motorist to switch from gasoline to compressed natural gas (CNG) or vice versa. Driving a car equipped with Green's conversion kit is not much different from driving a conventional model. A small dial switch on the dashboard controls which fuel is to be used. Four different configurations are available according to types of vehicles: compact, mid-size, large-size, and trucks. In the past, Green has built a few prototype vehicles powered by alternative fuels other than gasoline, but has been reluctant to go into higher volume production without more evidence of public demand. Therefore, Green Engineering initially would like to target one market segment (one configuration model) in offering the conversion kit. Green Engineering's marketing group has compiled the potential PE distribution for each different configuration when marketed independently.

Profit PE (10%) (unit: thousands)	Probabilities			
	Model 1	Model 2	Model 3	Model 4
$1000	0.35	0.10	0.40	0.20
1500	0	0.45	0	0.40
2000	0.40	0	0.25	0
2500	0	0.35	0	0.30
3000	0.20	0	0.20	0
3500	0	0	0	0
4000	0.05	0	0.15	0
4500	0	0.10	0	0.10

Evaluate the expected return and risk for each model configuration and recommend which one, if any, should be selected.

Solution

For model 1, we calculate the mean and variance of the PE distribution as follows:

$$E[PE]_1 = \$1000(0.35) + \$2000(0.40)$$
$$+ \$3000(0.20) + \$4000(0.05)$$
$$= \$1950;$$

$$Var[PE]_1 = 1000^2(0.35) + 2000^2(0.40)$$
$$+ 3000^2(0.20) + 4000^2(0.05) - (1950)^2$$
$$= 747{,}500.$$

In a similar manner, we compute the other values, with the following results:

Configuration	E [PE]	Var [PE]
Model 1	$1,950	747,500
Model 2	2,100	915,000
Model 3	2,100	1,190,000
Model 4	2,000	1,000,000

Results are plotted in Figure 13.9. If we make a choice based solely on the expected value, we may select either model 2 or model 3, because they have the highest expected PE.

If we consider the variability along with the expected PE, however, the correct choice is not obvious. We will first eliminate the alternatives that are clearly inferior to other alternatives.

Model 2 versus Model 3: We see that models 2 and 3 have the same mean of $2100, but model 2 has a much lower variance. In other words, model 2 dominates model 3, so we eliminate model 3 from further consideration.

Model 2 and Model 4: Similarly, we see that model 2 is preferred over model 4, because model 2 has a higher expected value and a smaller variance—model 2 again

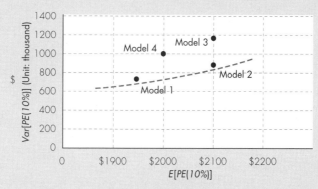

Figure 13.9 Mean-variance chart showing project dominance. Both model 3 and model 4 are dominated by model 2 (Example 13.10)

dominates model 4, so we eliminate model 4. In other words, model 2 dominates both models 3 and 4 as we consider the variability of PE.

Model 1 and Model 2: Even though the mean and variance rule has enabled us to narrow down our choices to only two models (models 1 and 2), it does not indicate what the ultimate choice should be. In other words, comparing models 1 and 2, we see that E[PE] increases from $1950 to $2100 at the price of a higher Var [PE], increasing from 747,500 to 915,000. That choice will depend on the decision-maker's tradeoffs between the incremental expected return ($150) and the incremental risk (167,500). We cannot choose between the two simply on the basis of mean and variance, so we must resort to other probabilistic information.[9]

Comments: When we need to compare mutually exclusive risky projects where no project dominance is observed, an incremental analysis can provide additional probabilistic information regarding the likelihood that the PE of one project would exceed that of the other project, say $P\{PE_1 \geq PE_2\}$. If this probability is 1, model 1 dominates model 2, so model 1 is preferred. If this probability is equal to zero, the reverse is true.

We first consider all the possible joint events that satisfy the statement $P\{PE_1 > PE_2\}$. (Note the strict inequality.) For example, when we realize the event of $1000 from model 1, no events from model 2 can satisfy the condition, because the smallest value we can observe from model 2 is $1000. But with the $2000 event observed for model 1, an event of either $1000 or $1500 from model 2 will meet the condition. Assuming statistical independence between the two models, we can easily compute the joint probability for such a joint event: For the joint event ($2000, $1000), the joint probability is (0.4)(0.10) = 0.04; for the event ($2000, $1500), it is (0.40)(0.45) = 0.18. Similarly, we can enumerate other possible joint events, as follows:

Model 1	Model 2	Joint Probability
$1000	No event	(0.35)(0.00) = 0.000
2000	1000	(0.40)(0.10) = 0.040
	1500	(0.40)(0.45) = 0.180
3000	1000	(0.20)(0.10) = 0.020
	1500	(0.20)(0.45) = 0.090
	2500	(0.20)(0.35) = 0.070
4000	1000	(0.05)(0.10) = 0.005
	1500	(0.05)(0.45) = 0.023
	2500	(0.05)(0.35) = 0.018
		0.445

[9] As we seek further refinement in our decision under risk, we may consider the expected utility theory, or stochastic dominance rules, which are beyond the scope of our text. See Park CS and Sharp-Bette GP. *Advanced Engineering Economics*. New York: John Wiley, 1990 (Chapters 10 and 11).

The probability that the PE of model 1 will exceed that of model 2 is calculated to be only 44.5%. This result implies that there is a 52% probability that the reverse situation may hold true, indicating a possible preference for model 2. (A 3.5% probability exists that both models would result in the identical PE.) The ultimate decision is again up to the investor, but this type of additional information helps the decision-maker to discern the best alternative in a complex decision environment.

13.5 Computer Notes

The electronic spreadsheet is an ideal tool for performing sensitivity analysis and break-even analysis. By changing any number of input values, the impact on one or more calculated quantities can be easily observed. However, if we wish to examine the effects of changes over ranges of input variables, the task of generating and organizing the results becomes much more onerous. Fortunately, software producers have responded to this need by incorporating simple commands to automate the preparation of sensitivity analyses. Three of these features are presented below.

13.5.1   Scenario Manager

This tool permits the user to easily define and compare a set of scenarios. Each scenario must specify values for a set of input variables. For the Base-Case, we will use Example 13.5 from the Windsor Metal Company. The income and cash flow statements are shown in Figure 13.10. Note that only the six shaded cells (C5, C6, C10, C12, B25, G26) contain input data. The remaining numerical cells hold formulas, which is important for sensitivity analysis so that all values get updated appropriately. For example, if the unit price in cell C5 is changed to $48, then cells D5 to G5 also drop to $48 and all tabulated values are adjusted automatically.

The scenarios are defined by selecting *Tools/Scenarios* which creates the Scenario Manager window, as shown in Figure 13.10. To *add* a scenario, you name it (WMC Worst-Case), and give a new value to each input cell that you want to change. A second scenario (WMC Best-Case) is also defined, using the same deviations from the Base-Case as in Example 13.5. A report is created by selecting *Summary*, at which time you provide a list of output cells that you wish to see in the summary table. The results for three scenarios are reproduced in Figure 13.11, showing the impact on net cash flows each year, and on the present equivalent.

13.5.2   One-Variable Data Tables

A data table permits the analyst to evaluate the sensitivity to changes in a single input quantity. We will select "unit price" as the independent variable in this

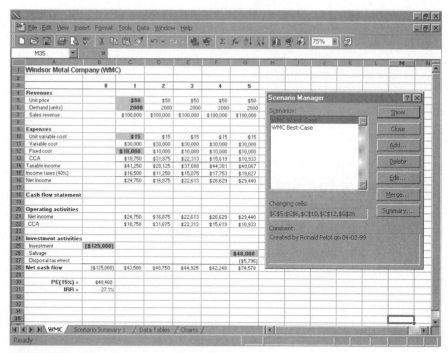

Figure 13.10 Applying the Scenario Manager to the WMC problem

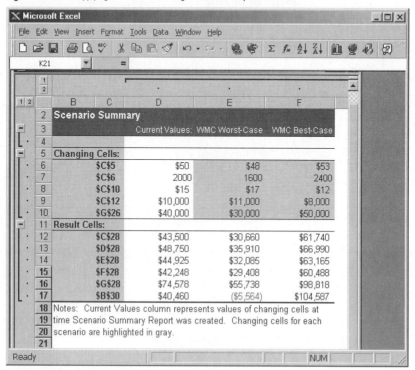

Figure 13.11 Scenario analysis summary report for the WMC problem

illustration. In Figure 13.12, a range of unit prices is entered into a column (I5:I15), and formulas are typed across the top of the table for each output quantity that we wish to consider. The formulas must refer to cells that are affected by changes in the chosen input variable. In this case, the output values of interest, and the corresponding formulas are:

- Present equivalent—cell J4: =B30

- Internal rate of return—cell K4: =B31

Highlight the block containing the column of input values and row of output formulas (I4:K15), select the command *Data/Table*, and in the *Column input cell* field, type C5. This indicates that the column of input values (I5:I15) will be substituted in sequence into the cell C5 in the income statement, generating new output values each time. These output values are used to automatically fill the data table, as shown in cells J5 to K15 of Figure 13.12. The results of a one-variable sensitivity analysis are easily plotted using an XY graph (Figure 13.13), where in this example two vertical axes are used to accommodate the different scales for PE and IRR.

This procedure could also be used to create Table 13.2 of Example 13.2, in which the independent variables are entered across the top of the table. Other features of this data management tool, such as adding a new formula to an existing table, are explained in the software's Help file.

13.5.3   Two-Variable Data Tables

A two-variable data table displays the changes in a single output variable over a range of two independent input variables. In Figure 13.12, a range of values for the first input variable (unit price) is entered in the column I21:I31, and the set of values for the second input variable (demand) are inserted into row J20:N20. The formula for the output variable of interest, which must depend on both input variables, is typed into the upper left corner cell of the table. In this case, the cell I20 contains the reference =B30, reflecting the PE value from the cash flow statement.

Highlight the block containing the column and row of input values (I20:N31), and select the command *Data/Table*. To indicate which values in the cash flow statement are replaced by the range of input data values during the analysis, type C6 in the *Row input cell* field, and C5 in the *Column input cell* field, representing the demand and unit price quantities, respectively. The formula in cell I20 is recalculated for every combination of input variables, filling the data table as shown in cells J21 to N31 of Figure 13.12. The results of a two-variable sensitivity analysis can be graphed using a surface chart (Figure 13.14). Multicolors and different perspectives facilitate the interpretation of such charts. They are very useful for determining the optimal output value over the range of input variables, particularly when the function is highly nonlinear.

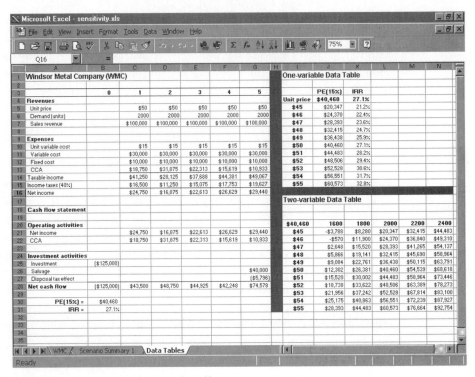

Figure 13.12 Sensitivity analysis using data tables

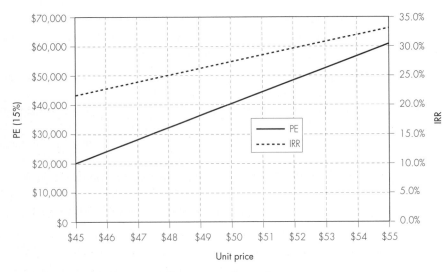

Figure 13.13 XY graph of the one-variable sensitivity analysis results

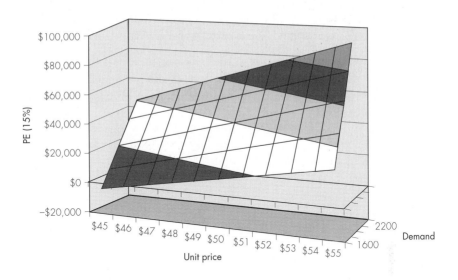

Figure 13.14 Surface chart of the two-variable sensitivity analysis results

13.6 Summary

- Often, cash flow amounts and other aspects of investment project analysis are uncertain. Whenever such uncertainty exists, we are faced with the difficulty of **project risk**—the possibility that an investment project will not meet our minimum requirements for acceptability and success.

- Three of the most basic tools for assessing project risk are as follows:
    1. **Sensitivity analysis**—a means of identifying the project variables which, when varied, have the greatest effect on project acceptability.
    2. **Break-even analysis**—a means of identifying the value of a particular project variable that causes the project to exactly break even.
    3. **Scenario analysis**—a means of comparing a base-case or expected project measurement (such as PE) to one or more additional scenarios, such as best and worst case, to identify the extreme and most-likely project outcomes.

- Sensitivity, break-even, and scenario analyses are reasonably simple to apply, but also somewhat simplistic and imprecise in cases where we must deal with multifaceted project uncertainty. **Probability concepts** allow us to further refine the analysis of project risk by assigning numerical values to the likelihood that project variables will have certain values.

- The end goal of a probabilistic analysis of project variables is to produce a PE distribution. From the distribution, we can extract such useful information as the **expected PE value**, the extent to which other PE values vary from, or are clustered around, the expected value (**variance**), and the best- and worst-case PEs.

- Our real task is not to find risk-free projects—they don't exist in real life. The challenge is to decide what level of risk we are willing to assume and then, having decided on one's risk tolerance, to understand the implications of that choice.

Problems

Level 1

13.1* For a certain investment project, the net present equivalent can be expressed as functions of sales price (X) and variable production cost Y: PE = $10,450(2X - Y) - 7890$. The base values for X and Y are $20 and $10, respectively. If the sales price is increased 10% over the base price, how much change in PE can be expected?

$13.2* An investor bought 100 shares of stock at a cost of $10 per share. She held the stock for 15 years and wants to sell it now. For the first 3 years, she received no dividends. For each of the next 7 years she received total dividends of $100 per year. For each of the remaining 5 years, no dividends were paid. In the last 15 years, the investor's marginal tax rate and capital gain tax rate was averaging about 30% and 20% respectively. What would be the break-even selling price to earn a 15% return on investment after tax?

13.3* Project A has the following probability distribution of expected future returns: ·

Probability	Net Future Equivalent
0.1	−$12,000
0.2	4,000
0.4	12,000
0.2	20,000
0.1	30,000

What is the expected FE for Project A?

13.4* In Problem 13.3, what is the standard deviation of expected future equivalent for Project A?

$13.5* In a resort location, you find a slot machine that costs you $1.00 per play. Odds for potential payoffs are as follows:

Payoff	Probability
$25	0.01
$1	0.50
$0	0.49

With 100 plays, what is the expected value of net payoff?

13.6* Alberta Metal Forming Company has just invested $500,000 of fixed capital in a manufacturing process, which is estimated to generate an after-tax annual cash flow of $200,000 in each of the next 5 years. At the end of year 5, no further market for the product and no salvage value for the manufacturing process is expected. If a manufacturing problem delays plant start-up for 1 year (leaving only 4 years of process life), what additional annual after-tax cash flow will be needed to maintain the same internal rate of return as would be experienced if no delay occurred?

13.7 Mike Lazenby, an industrial engineer at the Energy Conservation Service, has found that the anticipated profitability of a newly developed water-heater temperature control device can be measured by present worth as follows:

$$PE = 4.028V(2X - \$11) - \$77,860,$$

where V is the number of units produced and sold, and X is the sales price per unit. Mike also found that the V parameter value could occur anywhere over the range of 1000 to 6000 units and the X parameter value anywhere between $20 to $45 per unit. Develop a sensitivity graph as a function of number of units produced and sales price per unit.

13.8* A plant engineer wishes to know which of two types of light bulbs should be used to light a warehouse. The bulbs currently used cost $45.90 per bulb and last 13,870 hours before burning out. The new bulb ($60 per bulb) provides the same amount of light at the same power, but lasts twice as long. The labor cost to change a bulb is $16.00. The lights are on 19 hours a day, 365 days a year. If the firm's MARR is 15%, what is the maximum price (per bulb) the engineer should be willing to pay to switch to the new bulb? (Assume the firm's marginal tax rate is 40%.)

13.9 Rocky Mountain Publishing Company is considering introduction of a new morning newspaper in Edmonton. Its direct competitor charges $0.50 at retail, with $0.05 going to the retailer. For the level of news coverage the company desires, it determines the fixed cost of editors, reporters, rent, press room expenses, and wire service charges to be $300,000 per month. The variable cost of ink and paper is $0.10 per copy, but advertising revenues of $0.05 per paper will be generated. To print the morning paper, the publisher has to purchase a new printing press, which will cost $600,000. The press machine, with a 30% CCA rate, will be used for 10 years, at which time its salvage value would be about $100,000. Assume 25 weekdays in a month, a 40% tax rate, and a 13% after-tax MARR. How many copies per day must be sold to break even at a selling price of $0.50 per paper at retail?

13.10 A corporation is trying to decide whether or not to buy the patent for a product designed by another company. The decision to buy will mean an investment of $8 million, and the demand for the product is not known. If demand is light, the company expects a return of $1.3 million each year for 3 years. If the demand is moderate, the return will be $2.5 million each

year for 4 years, and a high demand means a return of $4 million each year for 4 years. It is estimated the probability of a high demand is 0.4 and the probability of a light demand is 0.2. The firm's interest rate (risk-free) is 12%. Calculate the expected present equivalent. On this basis should the company make the investment? (All figures represent after-tax values.)

13.11* A manufacturing firm is considering two mutually exclusive projects. Both projects have an economic service life of 1 year with no salvage value. The first cost and the present equivalent of the revenue for each project are given as follows:

First Cost	Project 1 ($1000)		Project 2 ($800)	
	Probability	Revenue	Probability	Revenue
PE revenue	0.2	$2000	0.3	$1000
	0.6	3000	0.4	2500
	0.2	3500	0.3	4500

We assume that both projects are statistically independent from each other.

(a) If you are an expected value maximizer, which project would you select?
(b) If you also consider the variance of the project, which project would you select?

Level 2

13.12* A company is currently paying a sales representative $0.25 per km to drive her car for company business. The company is considering supplying the representative with a car, which would involve the following: A car costs $20,000, has a service life of 5 years, and a market value of $5000 at the end of that time. Monthly storage costs for the car are $80, and the cost of fuel, tires, and maintenance is 14 cents per km. The car falls into Class 10 (CCA rate = 30%). The firm's marginal tax rate is 40%. What annual distance must the salesperson travel by car for the cost of the two methods of providing transportation to be equal if the interest rate is 15%?

13.13* Ford Construction Company is considering the proposed acquisition of a new earthmover. The mover's basic price is $70,000, and it will cost another $15,000 to modify it for special use by the company. This earthmover falls into CCA class 38 ($d = 30\%$). It will be sold after 4 years for $30,000. The earthmover purchase will have no effect on revenues, but it is expected to save the firm $32,000 per year in before-tax operating costs, mainly labor. The firm's marginal tax rate (federal plus provincial) is 40%, and its MARR is 15%.

(a) Is this project acceptable based on the most-likely estimates given in the problem?

(b) Suppose that the project will require an increase in net working capital (spare parts inventory) of $2000, which will be recovered at the end of year 5. Is the project acceptable under these conditions?

(c) If the firm's MARR is increased to 20%, what would be the required annual savings in labor so that the project remains profitable?

13.14 A real estate developer seeks to determine the most economical height for a new office building. The building will be sold after 5 years. The relevant net annual revenues and salvage values are as follows:

	Height			
	2 Floors	3 Floors	4 Floors	5 Floors
First cost (net after-tax)	$500,000	$750,000	$1,250,000	$2,000,000
Lease revenue	199,100	169,200	149,200	378,150
Net resale value (after-tax)	600,000	900,000	2,000,000	3,000,000

(a) The developer is uncertain about the interest rate i to use, but is certain that it is in the range 5% to 30%. For each building height, find the range of values of i for which that building height is the most economical.

(b) Suppose that the developer's interest rate is known to be 15%. What would be the cost, in terms of net present equivalent, of an error in overestimation in resale value (i.e., the true value resulted in a value 10% lower than that of the original estimate)?

13.15 A special purpose milling machine was purchased 4 years ago for $40,000. It was estimated at that time that this machine would have a life of 10 years and a salvage value of $1000 with a cost of removal of $1500. These estimates are still good. This machine has annual operating costs of $5000. A new machine, which is more efficient, will reduce operating costs to $1000, but it will require an investment of $25,000. The life of the new machine is estimated to be 6 years with a salvage value of $2000. The CCA for both machines is based on a declining balance rate of 30%. An offer of $6000 for the old machine has been made, and the purchaser would pay for removal of the machine. The firm's marginal tax rate is 40%, and its required minimum rate of return is 10%.

(a) What incremental cash flows will occur at the end of years 0 through 6 as a result of replacing the old machine? Should the old machine be replaced now?

(b) Suppose that the annual operating costs for the old milling machine would increase at an annual rate of 9% over the remaining service life. With this change in future operating costs for the old machine, would the answer in (a) change?

(c) Assuming constant operating costs again, what is the minimum trade-in value for the old machine so that both alternatives are economically equivalent?

13.16 A local telephone company is considering the installation of a new phone line for a new row of apartment complexes. Two types of cables are being considered: conventional copper wire and fiber optics. Transmission by copper wire cables, although cumbersome, involves much less complicated and expensive support hardware than fiber optics. The local company may use five different types of copper wire cables: 100 pairs, 200 pairs, 300 pairs, 600 pairs and 900 pairs per cable. In calculating the first cost of a cable the following equation is used:

- Cost per length = [Cost per metre + cost per pair (number of pairs)](length)
- 22 gauge copper wire = $5.55 per metre
- Cost per pair = $0.043 per pair

The annual after-tax operating cost of the cable as a percent of the first cost is 18.4%. The life of the system is 30 years.

In fiber optics, a cable is referred to as a ribbon. One ribbon contains 12 fibers. The fibers are grouped in fours; therefore, one ribbon contains three groups of four fibers. Each group of four fibers can produce 672 lines (equivalent to 672 pairs of wires) and, since each ribbon contains three groups, the total capacity of the ribbon is 2016 lines. To transmit signals using fiber optics, many modulators, wave guides, and terminators are needed to convert the signals from electric currents to modulated light waves. Fiber optic ribbon costs $9321 per kilometre. At each end of the ribbon three terminators are needed, one for each group of four fibers at a cost of $40,000 per terminator. Twenty-one modulating systems are needed at each end of the ribbon at a cost of $24,000 for a unit in the central office and $44,000 for a unit in the field. Every 7000 metres, a repeater is required to keep the modulated light waves in the ribbon at an intelligible intensity for detection. The unit cost of this repeater is $15,000. The annual cost including income taxes for the 21 modulating systems is 12.5% of the first cost for the units. The annual operating cost of the ribbon itself is 17.8% for its first cost. The life of the whole system is 30 years. (All figures represent after-tax costs.)

(a) Suppose that the apartments are located 8 kilometres from the phone company's central switching system and about 2000 telephones will be required. This would require either 2000 pairs of copper wire or one fiber optics ribbon and related hardware. If the telephone company's interest rate is 15%, which option is more economical?

(b) In (a), suppose that the apartments are located 16 kilometres or 40 kilometres from the phone company's central switching system. Which option is more economically attractive under each scenario?

13.17 A small manufacturing firm is considering the purchase of a new boring machine to modernize one of its production lines. Two types of boring machine are available on the market. The lives of machine A and machine B are 8 years and 10 years, respectively. The machines have the following receipts and disbursements. Use a MARR (after-tax) of 10% and a marginal tax rate of 30%:

Item	Machine A	Machine B
First cost	$6000	$8500
Service life	8 years	10 years
Salvage value	500	1000
Annual O&M costs	700	520
CCA	DB(30%)	DB(30%)

(a) Which machine would be most economical to purchase under the infinite planning horizon? Explain any assumption that you need to make about future alternatives.

(b) Determine the break-even annual O&M costs for machine A so that the present equivalent of machines A and B is the same.

(c) Suppose that the required service life of the machine is only 5 years. The estimated salvages at the end of the required service period are estimated to be $3000 for machine A and $3500 for machine B, respectively. Which machine is more economical?

13.18 Susan Campbell is thinking about going into the motel business down in Niagara Falls. The cost to build the motel is $2,200,000. The lot costs $600,000. Furniture and furnishings cost $400,000 and have a 20% CCA rate, while the motel building has a 4% CCA rate. The land will appreciate at an annual rate of 5% over the project period, but the building and furnishings will have a zero salvage value after 25 years. When the motel is full (100% capacity), it takes in (receipts) $4000 per day for 365 days per year. The motel has fixed operating expenses, exclusive of CCA, of $230,000 per year. The variable operating expenses are $170,000 at 100% capacity, and these vary directly with percent capacity down to zero at 0% capacity. If the interest rate is 10%, compounded annually, at what percentage capacity must this motel operate to break even? (Assume Susan's tax rate is 31%.)

13.19 Robert Cooper is considering building a rental property containing stores and offices at a cost of $250,000 on land that he already owns. Cooper estimates that annual receipts from rentals will be $35,000 and that annual disbursements, other than income taxes, will be about $12,000. The

property is expected to appreciate at an annual rate of 5%. Cooper expects to retain the property for 20 years once it is acquired. The CCA rate for the building is 4%. Cooper's marginal tax rate is 30%, and his MARR is 15%. What would be the minimum annual total of rental receipts that would make the investment break even?

13.20 Two different methods of solving a production problem are under consideration. Both methods are expected to be obsolete in 6 years. Method A would cost $80,000 initially and have annual operating costs of $22,000 a year. Method B would cost $52,000 initially and costs $17,000 a year to operate. The salvage value realized with method A would be $20,000 and with method B would be $15,000. Method A would generate $16,000 revenue income a year more than method B. Investments in both methods have a 30% CCA rate. The firm's marginal income tax rate is 40%. The firm's MARR is 20%. What would be the required additional annual revenue for method A such that you would be indifferent to one method or the other?

$13.21 Juan Carlos is considering two investment projects whose present equivalents are described as follows:

- Project 1: $PE(10\%) = 2X(X-Y)$, where X and Y are statistically independent discrete random variables with the following distributions:

X		Y	
Event	Probability	Event	Probability
$20	0.6	$10	0.4
40	0.4	20	0.6

- Project 2:

PE(10%)	Probability
$ 0	0.24
400	0.20
1600	0.36
2400	0.20

Note: Cash flows between the two projects are also assumed to be statistically independent.

(a) Develop the PE distribution for project 1.
(b) Compute the mean and variance of the PE for project 1.
(c) Compute the mean and variance of the PE for project 2.
(d) Suppose that projects 1 and 2 are mutually exclusive. Which project would you select?

13.22 A business executive is trying to decide whether to undertake one of two contracts or neither one. She has simplified the situation somewhat and feels that it is sufficient to imagine that the contracts provide alternatives as follows:

Contract A		Contract B	
PE	Probability	PE	Probability
$100,000	0.2	$40,000	0.3
50,000	0.4	10,000	0.4
0	0.4	−10,000	0.3

(a) Should the executive undertake either one of the contracts? If so, which one? What would she do if she made decisions by maximizing her expected PE?
(b) What would be the probability that Contract A would result in a larger profit than that of Contract B?

13.23 Two alternative machines are being considered for a cost reduction project.

- Machine A has a first cost of $60,000 and a salvage value (after-tax) of $22,000 at the end of 6-years' service life. Probabilities of annual after-tax operating costs of this machine are estimated as follows:

Annual O&M Costs	Probability
$5,000	0.20
8,000	0.30
10,000	0.30
12,000	0.20

- Machine B has a first cost of $35,000, and its estimated salvage value (after-tax) at the end of 4 years service is to be negligible. The annual after-tax operating costs are estimated to be as follows:

Annual O&M Costs	Probability
$8,000	0.10
10,000	0.30
12,000	0.40
14,000	0.20

The MARR on this project is 10%. The required service period of these machines is estimated to be 12 years, and no technological advance in either machine is expected.

(a) Assuming independence, calculate the mean and variance for the equivalent annual cost of operating each machine.
(b) From the results of part (a), calculate the probability that the annual cost of operating machine A will exceed the cost of operating machine B.

13.24 Two mutually exclusive investment projects are under consideration. It is assumed that the cash flows are statistically independent random variables with means and variances estimated as follows:

End of Year	Project A Mean	Project A Variance	Project B Mean	Project B Variance
0	−$5,000	1,000²	−$10,000	2,000²
1	4,000	1,000²	6,000	1,500²
2	4,000	1,500²	8,000	2,000²

(a) For each project, determine the mean and standard deviation for PE using an interest rate of 15%.

(b) Based on the results of part (a), which project would you recommend?

Level 3

13.25 The management of Langdale Mill is considering replacing a number of old looms in the mill's weave room. The looms to be replaced are two 220-cm President looms, sixteen 135-cm President looms, and twenty-two 185-cm Draper X-P2 looms. The company may either replace the old looms with new ones of the same kind or buy 21 new shutterless Pignone looms. The first alternative requires the purchase of 40 new President and Draper looms and the scrapping of the old looms. The second alternative involves scrapping the 40 old looms, relocating 12 Picanol looms, and constructing a concrete floor, plus purchasing the 21 Pignone looms and various related equipment.

Description	Alternative 1	Alternative 2
Machinery/related equipment	$2,119,170	$1,071,240
Removal cost of old looms/site preparation	26,866	49,002
Salvage value of old looms	62,000	62,000
Annual sales increase with new looms	7,915,748	7,455,084

Annual labor	261,040	422,080
Annual O&M	1,092,000	1,560,000
CCA	DB(30%)	DB(30%)
Project life	8 years	8 years
Salvage value	169,000	54,000

The undepreciated capital cost of all old looms are negligible. The corporate executives feel that various investment opportunities available for the mill will guarantee a rate of return on investment of at least 18%. The mill's marginal tax rate is 40%.

(a) Perform a sensitivity analysis on the project's data, varying the net operating revenue, labor cost, annual O&M cost, and the MARR. Assume that each of these variables can deviate from its base case expected value by ±10%, by ±20%, and by ±30%.

(b) From the results of part (a), prepare sensitivity diagrams and interpret the results.

13.26 The City of Guelph was having a problem locating land for a new sanitary landfill site when the alternative of burning the solid waste to generate steam was proposed. At the same time, Uniroyal Tire Company seemed to be having a similar problem, disposing of solid waste in the form of rubber tires. It was determined that there would be about 200 tonnes per day of waste to be burned; this included municipal and industrial waste. The city is considering building a waste-fired steam plant, which would cost $6,688,800. To finance the construction cost, the city will issue resource recovery revenue bonds in the amount of $7,000,000 at an interest rate of 11.5%. Bond interest is payable annually. The differential amount between the actual construction costs and the amount of bond financing ($7,000,000 − $6,688,800 = $311,200) will be used to settle the bond discount and expenses associated with the bond financing. The expected life of the steam plant is 20 years. The expected salvage value is estimated to be about $300,000. The expected labor costs would be $335,000 per year. The annual operating and maintenance costs (including fuel, electricity, maintenance, and water) are expected to be $175,000. The city expects 20% downtime per year for the waste-fired steam plant. This downtime would result in 4245 kg of waste, which along with 3265 kg of waste after incineration, would have to be disposed of as land fill. At the present rate of $42.88 per kilogram, this will cost the city a total of $322,000 per year. The steam plant will generate revenues from two sources: (1) steam sales, and (2) disposal tipping fees. With an input of 200 tonnes per day and 6.637 kg of steam per kg refuse, a maximum of 1,327,453 kg of steam can be produced per day. However, with 20% down time, the actual output would be 1,061,962 kg of steam per day. The initial steam charge will be approximately $4.00 per tonne. This would bring in $1,550,520 in steam revenue the first year. The tipping fee is used in conjunction with the sale of steam to offset the total plant cost. It is the goal of the Guelph steam plant to phase out the tipping fee as soon as possible. The tipping fee will be $20.85 per tonne in the first year of plant operation and will be phased out in the eighth year. The scheduled tipping fee assessment would be as follows:

Year	Tipping Fee
1	$976,114
2	895,723
3	800,275
4	687,153
5	553,301
6	395,161
7	208,585

(a) At an interest rate of 10%, would the steam plant generate sufficient revenue to recover the initial investment?

(b) At an interest rate of 10%, what would be the minimum charge (per tonne) for steam sales to make the project break even?

(c) Perform a sensitivity analysis to determine the input variable of the plant's down time.

$13.27 A financial investor has an investment portfolio worth $350,000. A bond in his investment portfolio will mature next month and provide him with $25,000 to reinvest. The choices have been narrowed down to the following two options.

- Option 1: Reinvest in a foreign bond that will mature in one year. This will entail a brokerage fee of $150. For simplicity, assume that the bond will provide interest over the 1-year period of $2450, $2000, or $1675 and that the probabilities of these occurrences are assessed to be 0.25, 0.45, and 0.30, respectively.

- Option 2: Reinvest in a $25,000 guaranteed investment certificate with a bank. Assume this GIC has an effective annual rate of 7.5%.

(a) Which form of reinvestment should the investor choose in order to maximize his expected financial gain?

(b) If the investor can obtain professional investment advice from Solomon and Brothers Inc., what would be the maximum amount the investor should pay for this service?

13.28 Kellog Company is considering the following investment project and has estimated all cost and revenues in constant dollars. The project requires a purchase of a $9000 asset, which will be used for only 2 years (project life).

- The salvage value of this asset at the end of 2 years is expected to be $4000.

- The project requires an investment of $2000 in working capital, and this amount will be fully recovered at the end of the year.

- The annual revenue as well as general inflation are discrete random variables, described by the following probability distributions. The random variables are statistically independent.

Annual Revenue (X)	Probability	General Inflation Rate (Y)	Probability
10,000	0.30	3%	0.25
20,000	0.40	5%	0.50
30,000	0.30	7%	0.25

- The investment has a declining balance rate of 30% for CCA purposes.

- It is assumed that the revenues, salvage value, and working capital are responsive to this general inflation rate.

- The revenue and inflation rate dictated during the first year will prevail over the remaining project period.

- The marginal income tax rate for the firm is 40%. The firm's inflation-free interest rate (i') is 10%.

(a) Determine the PE as a function of X and Y.

(b) In (a), compute the expected PE of this investment.

(c) In (a), compute the variance of the PE of this investment.

13.29 Mount Manufacturing Company produces industrial and public safety shirts. As is done in most apparel manufacturing, the cloth must be cut into shirt parts by marking sheets of paper in the way that the particular cloth is to be cut. At present, these sheet markings are done manually and the annual labor cost is running around $103,718. Mount has the option of purchasing one of two automated marking systems. The two systems are the Lectra System 305 and the Tex Corporation Marking System. The comparative characteristics of the two systems are as follows:

| | Most Likely Estimates | |
	Lectra System	Tex System
Annual labor cost	$51,609	$51,609
Annual material savings	$230,000	$274,000
Investment cost	$136,150	$195,500
Estimated life	6 years	6 years
Salvage value	$20,000	$15,000
CCA rate (DB)	30%	30%

The firm's marginal tax rate is 40%, and the interest rate used for project evaluation is 12% after taxes.

(a) Based on the most likely estimates, which alternative is the best?

(b) Suppose that the company estimates the material savings during the first year for each system on the basis of following probability distribution:

Lectra System	
Material Savings	Probability
$150,000	0.25
230,000	0.40
270,000	0.35

Tex System	
Material Savings	Probability
$200,000	0.30
274,000	0.50
312,000	0.20

Further assume that the annual material savings for both Lectra and Tex are statistically independent. Compute the mean and variance for the annual equivalent of operating each system.

(c) In part (b), calculate the probability that the annual benefit of operating Lectra will exceed the benefit of operating Tex.

13.30 Burlington Motor Carriers, a trucking company, is considering the installation of a two-way mobile satellite messaging service on their 2000 trucks. Based on tests done last year on 120 trucks, the company found that satellite messaging could cut 60% from its $5 million bill for long-distance communications with truck drivers. More important, the drivers reduced the number of "deadhead" kilometres—those driven without paying loads—by 0.5%. Applying that improvement to all 230 million kilometres covered by the Burlington fleet each year would produce an extra $1.25 million savings.

Equipping all 2000 trucks with the satellite hook-up will require an investment of $8 million and the construction of a message-relaying system costing $2 million. The equipment and on-board devices will have a service life of 8 years and negligible salvage value; they will be depreciated with a 30% CCA rate. Burlington's marginal tax rate is about 38%, and its required minimum attractive rate of return is 18%.

(a) Determine the annual net cash flows from the project.
(b) Perform a sensitivity analysis on the project's data, varying savings in telephone bill and savings in deadhead kilometres. Assume that each of these variables can deviate from its base case value by ±10%, ±20%, and ±30%.
(c) Prepare sensitivity diagrams and interpret the results.

13.31 The following is a comparison of the cost structure of a conventional manufacturing technology (CMT) with a flexible manufacturing system (FMS) at one Canadian firm.

| | Most-Likely Estimates | |
	CMT	FMS
Number of part types	3,000	3,000
Number of pieces produced/year	544,000	544,000
Variable labor cost/part	$2.15	$1.30
Variable material cost/part	$1.53	$1.10
Total variable cost/part	$3.68	$2.40
Annual overhead	$3.15M	$1.95M
Annual tooling costs	$470,000	$300,000
Annual inventory costs	$141,000	$ 31,500
Total annual fixed operating costs	$3.76M	$2.28M
Investment	$3.5M	$10M
Salvage value	$0.5M	$1M
Service life	10 years	10 years
CCA rate	30%	30%

(a) The firm's marginal tax rate and MARR are 40% and 15%, respectively. Determine the incremental cash flow (FMS – CMT) based on the most likely estimates.
(b) Management feels confident about all input estimates in CMT. However, the firm does not have any previous experience in operating an FMS. Therefore, many input estimates, except the investment and salvage value, are subject to

variations. Perform a sensitivity analysis on the project's data, varying the elements of operating costs. Assume that each of these variables can deviate from its base case value by ±10%, ±20%, and ±30%.

(c) Prepare sensitivity diagrams and interpret the results.

(d) Suppose that probabilities of the variable material cost and the annual inventory cost for the FMS are estimated as follows:

Material Cost

Cost per Part	Probability
$1.00	0.25
1.10	0.30
1.20	0.20
1.30	0.20
1.40	0.05

Inventory Cost

Annual Inventory Cost	Probability
$25,000	0.10
31,000	0.30
50,000	0.20
80,000	0.20
100,000	0.20

What are the best and the worst cases of incremental PE?

(e) In part (d), assuming that the random variables of the cost per part and the annual inventory cost are statistically independent, find the mean and variance of the PE for the incremental cash flows.

(f) In parts of (d) and (e), what is the probability that the FMS would be a more expensive investment option?

13.32 Redo problem 13.16 using a spreadsheet. Using a one-variable data table, generate the present equivalents of the two alternatives for distances ranging from 5 to 40 kilometres (in 5 kilometre increments). Plot the output on an XY graph, and discuss the results from the point of view of sensitivity analysis.

13.33 Reconsider Problem 13.12. Since the maintenance cost per km and the number of kilometres driven per year are quite uncertain, the company wishes to conduct a sensitivity analysis. Using a two-variable data table, calculate the difference in PEs for the two alternatives as a function of these two variables. Let the maintenance costs range from 12 to 16 cents per kilometre in 1¢ increments, and allow the annual distance to vary from 60,000 to 80,000 kilometres a year in 5000 km increments. Plot the output on an 3D surface chart, and discuss the results from the point of view of sensitivity analysis.

APPENDIX 13A

Risk Simulation

In Chapter 13, we examined analytical methods of determining the PE distributions and computing their means and variances. As we saw in Section 13.4.1, the PE distribution offers numerous options for graphically presenting probabilistic information to the decision-maker, such as the range and likelihoods of occurrence of possible levels of PE. Where we can adequately evaluate the risky investment problem by analytical methods, it is generally preferable to do so. However, many investment situations cannot be solved easily by analytical methods. In these situations, we may develop the PE distribution through computer simulation.

13A.1 Computer Simulation

Before we examine the details of risk simulation, let us consider a situation where we wish to train a new astronaut for a future space mission. Several approaches exist for training this astronaut. One (remote) possibility is to place the trainee in an actual space shuttle and to launch her into space. This approach certainly would be expensive; it also would be extremely risky because any human error made by the trainee would have tragic consequences. As an alternative, we can place the trainee in a flight simulator designed to mimic the behavior of the actual space shuttle in space. The advantage of this approach is that the astronaut trainee learns all the essential functions of space operation in a simulated space environment. The flight simulator generates test conditions approximating operational conditions, and any human errors made during training cause no harm to the astronaut or to the equipment being used.

The use of computer simulation is not restricted to simulating a physical phenomenon such as the flight simulator. In recent years, techniques for testing the results of some investment decisions before they are actually executed have been developed. As a result, many phases of business investment decisions have been simulated with considerable success. Now we can make an analogy between a space shuttle flight simulator and an investment simulator—a model for testing the results of some business investment decisions before they are actually executed. In fact, we can analyze WMC's transmission-housings project by building a simulation model. The general approach is to assign a subjective (or objective) probability distribution to each unknown factor and to combine these into a probability distribution for the project profitability as a whole. The essential idea is that, if we can simulate the actual state of nature for unknown investment variables on a computer, we may be able to obtain the resulting PE distribution.

The unit demand (X) in our WMC's transmission-housing project was one of the random variables in the problem. We can know the exact value for this random variable only after the project is implemented. Is there any way to predict the actual value before we make any decision on the project?

The following logical steps are often suggested for a computer program that simulates investment scenarios:

Step 1: Identify all the variables that affect the measure of investment worth (e.g., PE after taxes).

Step 2: Identify the relationships among all the variables. The relationships of interest here are expressed by the equations or the series of numerical computations by which we compute the PE of an investment project. These equations make up the model we are trying to analyze.

Step 3: Classify the variables into two groups: The parameters whose values are known with certainty and the random variables for which exact values cannot be specified at the time of decision making.

Step 4: Define distributions for all the random variables.

Step 5: Perform Monte Carlo sampling and describe the resulting PE distribution.

Step 6: Compute the distribution parameters and prepare graphic displays of simulation results.

Figure 13A.1 illustrates the logical steps involved in simulating a risky investment project. The risk simulation process we have described has two important advantages when compared with the analytical approach discussed in Chapter 13:

1. The number of variables that can be considered is practically unlimited, and the distributions used to define the possible values for each random variable can be of any type and any shape. The distributions can be based on statistical data if they are available, or, more commonly, on subjective judgment.

2. The method lends itself to sensitivity analyses. By defining some factors that have the most significant effect on the resulting PE values and using different distributions (in terms of either shape or range) for each variable, we can observe the extent to which the PE distribution is changed.

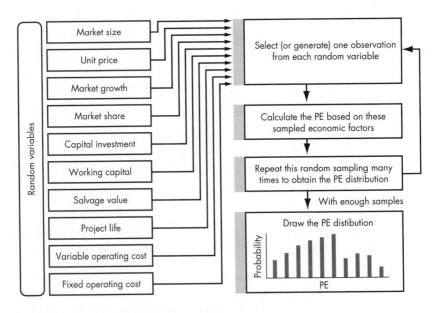

Figure 13A.1   Logical steps involved in simulating a risky investment

13A.2 Model Building

In this section, we shall present some of the procedural details related to the first three steps (model building) outlined in Section 13A.1. To illustrate the typical procedure involved, we shall work with an investment setting for WMC's transmission-housings project, described in Example 13.9.

The initial step is to define the measure of investment worth and the factors that affect that measure. For our presentation, we choose the measure of investment worth as an after-tax PE computed at a given interest rate i. In fact, we are free to choose any measure of worth, such as annual equivalent or future equivalent. In the second step, we must divide into two groups all the variables that we listed in Step 1 as affecting PE. One group consists of all the parameters for which values are known. The second groups includes all remaining parameters for which we do not know exact values at the time of analysis. The third step is to define the relationships that tie together all the variables. These relationships may take the form of a single equation or several equations.

EXAMPLE 13A.1

Developing a simulation model

Reconsider WMC's transmission-housings project in Example 13.9. Identify the input factors related to the project and develop the simulation model for the PE distribution.

Discussion: For the WMC project, the variables that affect the PE value are investment required, unit price, demand, variable production cost, fixed production cost, tax rate, CCA, and the firm's interest rate. Some of the parameters that might be included in the known group are investment cost and interest rate (MARR). If we have already purchased the equipment or have received a price quote, then we also know the CCA amount. Assuming that we are operating in a stable economy, we would probably know the tax rates for computing income taxes due.

The group of parameters with unknown values would usually include all the variables relating to costs and future operating expense and future demand and sales prices. These are the random variables for which we must assess the probability distributions.

For simplicity, we classify the input parameters or variables for WMC's 1 project as follows:

Assumed to Be Known Parameters	Assumed to Be Unknown Parameters
MARR	Unit price
Tax rate	Demand
CCA amount	Salvage value
Investment amount	
Project life	
Fixed production cost	
Variable production cost	

Note that, unlike the situation in Example 13.9, here we treat the salvage value as a random variable. With these assumptions, we are now ready to build the PE equation for the WMC project.

Solution

Recall that the basic investment parameters assumed for WMC's 5-year project in Example 13.9 were as follows.

- Investment = –$125,000,

- Marginal tax rate = 0.40,

- Annual fixed cost = $10,000,

- Variable unit production cost = $15/unit,

- MARR (i) = 15%,

- Annual CCA amounts:

n	CCA_n
1	$18,750
2	31,875
3	22,313
4	15,619
5	10,933

The after-tax annual revenue is expressed as a function of product demand (X) and unit price (Y):

$$R_n = XY(1 - t) = 0.6XY.$$

The after-tax annual expenses excluding CCA are also expressed as a function of product demand (X):

$$\begin{aligned} E_n &= (\text{Fixed cost} + \text{variable cost})(1 - t) \\ &= (\$10,000 + 15X)(0.60) \\ &= \$6000 + 9X. \end{aligned}$$

Then, the net after-tax cash revenue is

$$\begin{aligned} V_n &= R_n - E_n \\ &= 0.6XY - 9X - \$6000. \end{aligned}$$

The present equivalent of the net after-tax cash inflow from revenue is

$$\begin{aligned} \sum_{n=1}^{5} V_n(P/F, 15\%, n) &= [0.6X\,(Y - 15) - \$6000]\,(P/A, 15\%, 5) \\ &= 0.6X\,(Y - 15)(3.3522) - \$20,113. \end{aligned}$$

We will now compute the present equivalent of the total CCA credits:

$$t\sum_{n=1}^{5} CCA_n(P/F, i, n) = 0.40\,[\$18,750(P/F, 15\%, 1) + \$31,875(P/F, 15\%, 2)$$
$$+ \$22,313(P/F, 15\%, 3) + \$15,619(P/F, 15\%, 4)$$
$$+ \$10,933(P/F, 15\%, 5)]$$
$$= \$27,777.$$

Since the total CCA amount is \$99,490, the undepreciated capital cost at the end of year 5 is \$25,510 (\$125,000 – \$99,490). Any salvage value greater than this UCC is treated as a taxable gain, and this gain is taxed at t. In our example, the salvage value is considered to be a random variable. Thus, the amount of taxable gains (losses) also becomes a random variable. Therefore, the net salvage value after tax adjustment is

$$S - (S - \$25,510)t = S(1 - t) + 25,510t$$
$$= 0.6S + \$10,204.$$

Then, the present worth equivalent of this amount is

$$(0.6S + \$10,204)(P/F, 15\%, 5) = (0.6S + \$10,204)(0.4972)$$

Now, the PE equation can be summarized as

$$PE\,(15\%) = -\$125,000 + 0.6X(Y - 15)(3.3522) - \$20,113 + \$27,777$$
$$+ (0.6S + \$10,204)(0.4972)$$
$$= -\$112,263 + 2.0113X(Y - 15) + 0.2983S.$$

Note that the PE function is now expressed in terms of three random variables X, Y, and S.

13A.3 Monte Carlo Sampling

For some variables, we may base the probability distribution on objective evidence gleaned from the past if the decision-maker feels the same trend will continue to operate in the future. If not, we may use subjective probabilities as discussed in Section 13.3.1. Once we specify a distribution for a random variable, we need to determine ways to generate samples from this distribution. **Monte Carlo sampling** is a specific type of simulation method in which a random sample of outcomes is generated for a specified probability distribution. In this section, we shall discuss the Monte Carlo sampling procedure for an *independent* random variable.

13A.3.1 Random Numbers

The sampling process is the key part of the analysis. It must be done such that the sequence of values sampled will be distributed in the same way as the original distribution.

To accomplish this objective, we need a source of independent, identically distributed uniform random numbers between 0 and 1. We can use a table of random numbers but most digital computers have programs available to generate "equally likely (uniform)" random decimals between 0 and 1. We will use $U(0,1)$ to denote such a statistically reliable uniform random number generator, and we will use U_1, U_2, U_3, ... to represent uniform random numbers generated by this routine. (In Microsoft Excel, the RAND function can be used to generate such a random number sequence.)

13A.3.2 Sampling Procedure

For any given random numbers, the question is, how are they used to sample a distribution in a simulation analysis? The first task is to convert the distribution into its corresponding cumulative frequency distribution. Then, the random number generated is set equal to its numerically equivalent percentile and is used as the entry point on the $F(x)$ axis of the cumulative frequency graph. The sampled value of the random variable is the x value corresponding to this cumulative percentile entry point.

This method of generating random values works because choosing a random decimal between 0 and 1 is equivalent to choosing a random percentile of the distribution. Then, the random value distribution is used to convert the random percentile to a particular value. The method is general and can be used for any cumulative probability distribution, either continuous or discrete.

EXAMPLE **Monte Carlo Sampling**

13A.2

In Example 13A.1, we have developed a PE equation for WMC's transmission-housings project as a function of three random variables—demand (X), unit price (Y) and salvage value (S):

$$PE\,(15\%) = -\$112,263 + 2.0113X\,(Y - 15) + 0.2983S$$

- For random variable X, we will assume the same discrete distribution as defined in Table 13.5.

- For random variable Y, we will assume a triangular distribution with $L = \$48$, $H = \$53$, and $M_o = \$50$.

- For random variable S, we will assume a uniform distribution with $L = \$30,000$ and $H = \$50,000$.

With the random variables (X, Y, and S) distributed as above, and assuming that these random variables are *mutually independent* of each other, we need three uniform random numbers to sample one realization from each random variable. Determine the PE distribution based on 200 iterations.

Discussion: As outlined previously, a simulation analysis consists of a series of repetitive computations of PE. To perform the sequence of repeated simulation trials, we generate a sample observation for each random variable in the model and substitute these values into the PE equation. Each trial requires that we use a different random number in the sequence to sample each distribution. Thus, if three random variables affect the PE, we need three random numbers for each trial. After each trial, the computed PE is stored in the computer. As shown in Figure 13A.2, each value of PE computed in this manner represents one state of nature. The trials are continued until a sufficient number of PE values is available to define the PE distribution.

Solution

Suppose the following three uniform random numbers are generated for the first iteration: $U_1 = 0.12135$ for X, $U_2 = 0.82592$ for Y, and $U_3 = 0.86886$ for S.

- Demand (X): The cumulative distribution for X is already given in Example 13.6. To generate one sample (observation) from this discrete distribution, we first find the cumulative probability function, as depicted in Figure 13A.3(a). On a given trial, suppose the computer gives the random number 0.12135. We then enter the vertical axis at the 12.135 percentile (the percentile numerically equivalent to the random number), read across to the cumulative curve, then read down to

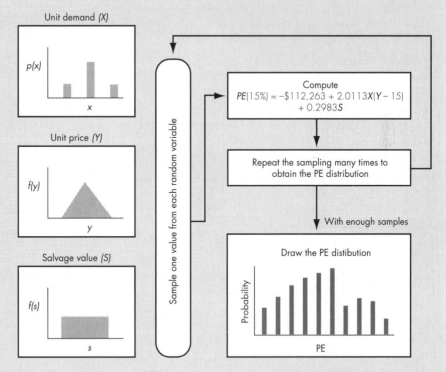

Figure 13A.2   A logical sequence of Monte Carlo simulation to obtain the PE distribution for WMC's transmission-housings project (Example 13A.2)

the x axis to find the corresponding value of the random variable X; this value 1600. This is the value of x that we use in the PE equation. On the next tr we sample another value of x by obtaining another random number, entering ordinate at the numerically equivalent percentile, and reading the correspond value of x from the x axis.

- Price (Y): Assuming that the unit price random variable can be estimated by three parameters, its probability distribution is shown in Fig.13A.3(b). Note Y takes a continuous value (unlike the discrete assumption in Table 13.5). cumulative distribution can be approximated numerically in a spreadsheet tal The sampling procedure is again similar to the discrete situation. Using the ranc number $U = 0.82592$, we can approximate $y = \$51.38$ by performing a lin interpolation between the two nearest values in the cumulative distribution tal

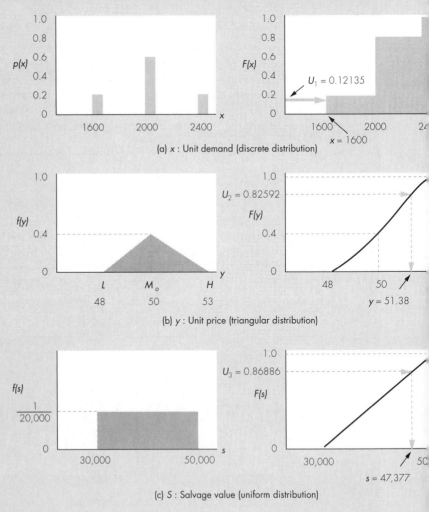

(a) x : Unit demand (discrete distribution)

(b) y : Unit price (triangular distribution)

(c) S : Salvage value (uniform distribution)

Figure 13A.3   Illustration of a sampling scheme for discrete and continuous random variables (Example 13A.

No.	U_1	U_2	U_3	x	y	s	PE
1	0.12135	0.82592	0.86886	1600	$51.38	$47,377	$18,957
2	0.72976	0.79885	0.41879	2000	$51.26	$38,376	$45,044
3	0.23145	0.58484	0.78720	2000	$50.50	$45,744	$44,202
4	0.67520	0.17786	0.71426	2000	$49.33	$44,285	$39,057
⋮	⋮	⋮	⋮	⋮	⋮	⋮	⋮
200	0.95953	0.70178	0.84848	2400	$50.88	$46,969	$74,969

Table 13A.1 Observed PE Values ($) for WMC's Simulation Project (Example 13A.2)

$18,957	$45,044	$44,202	$39,057	$71,736
36,369	17,840	67,478	40,753	42,239
42,459	35,708	39,624	12,535	9,968
35,024	65,755	42,994	68,907	44,485
60,623	15,053	69,225	36,067	61,402
11,589	32,746	68,015	42,129	47,656
41,946	67,002	34,236	77,236	43,142
39,484	42,678	34,848	47,353	71,315
60,681	71,514	38,540	44,435	8,890
18,071	44,182	65,181	46,359	12,030
73,988	47,975	9,329	41,763	20,554
36,129	42,403	11,597	65,551	40,034
39,728	40,378	8,086	76,318	67,673
51,819	63,063	52,028	14,677	14,616
10,843	37,815	39,093	47,737	34,661
77,240	40,536	43,540	41,768	43,886
41,481	12,312	40,272	43,634	60,644
43,863	39,637	9,035	43,568	68,213
68,937	18,021	12,831	63,691	37,972
41,042	16,140	45,007	43,560	46,653
62,297	50,912	40,781	49,947	42,839
13,007	34,089	39,758	65,079	41,965
69,448	12,223	43,275	17,208	34,167
12,493	42,280	76,754	30,754	75,072
42,833	78,019	72,864	32,782	45,630
12,217	36,548	75,412	37,354	46,015
39,434	45,845	8,127	44,344	38,522
72,550	45,618	39,300	41,816	44,292
45,735	70,083	43,591	66,513	11,656
40,944	42,755	39,983	35,732	14,628
32,043	10,031	46,821	47,448	36,499
43,683	73,704	41,446	40,931	11,638
38,667	65,243	36,958	69,238	38,658
41,642	42,189	38,494	72,450	73,042
48,075	43,216	34,298	14,778	34,031
36,012	67,325	74,614	12,795	17,090
45,000	68,793	39,600	67,268	42,771
18,049	41,051	36,344	46,269	42,956
37,718	71,594	10,198	51,866	34,321
11,207	39,140	36,741	16,672	74,969

Table 13A.2	Cell No.	Cell Interval	Observed Frequency	Relative Frequency	Cumulative Frequency
Simulated PE	1	$\$\ 8{,}086 \le PE \le \$11{,}583$	10	0.05	0.05
Frequency	2	$11{,}583 < PE \le \$15{,}079$	18	0.09	0,14
Distribution	3	$15{,}079 < PE \le \$18{,}576$	8	0.04	0.18
for WMC's	4	$18{,}576 < PE \le \$22{,}073$	2	0.01	0.19
Transmission-	5	$22{,}073 < PE \le \$25{,}569$	0	0.00	0.19
Housings	6	$25{,}569 < PE \le \$29{,}066$	0	0.00	0.19
Project	7	$29{,}066 < PE \le \$32{,}563$	2	0.01	0.20
(Example	8	$32{,}563 < PE \le \$36{,}060$	14	0.07	0.27
13A.2)	9	$36{,}060 < PE \le \$39{,}556$	23	0.12	0.39
	10	$39{,}556 < PE \le \$43{,}053$	37	0.19	0.57
	11	$43{,}053 < PE \le \$46{,}550$	27	0.14	0.71
	12	$46{,}550 < PE \le \$50{,}046$	9	0.05	0.75
	13	$50{,}046 < PE \le \$53{,}543$	4	0.02	0.77
	14	$53{,}543 < PE \le \$57{,}040$	0	0.00	0.77
	15	$57{,}040 < PE \le \$60{,}536$	0	0.00	0.77
	16	$60{,}536 < PE \le \$64{,}033$	7	0.04	0.81
	17	$64{,}033 < PE \le \$67{,}530$	10	0.05	0.86
	18	$67{,}530 < PE \le \$71{,}027$	10	0.05	0.91
	19	$71{,}027 < PE \le \$74{,}523$	10	0.05	0.96
	20	$74{,}523 < PE \le \$78{,}019$	9	0.05	1.00

Cell width = $3497; mean = $42,436; standard deviation = $18,705; minimum PE value = $8086; maximum PE value = $78,019

- Salvage (S): With the salvage value (S) distributed uniformly between $30,000 and $50,000, and a random number of $U = 0.86886$, the sample value is $s = \$47{,}377$, or $s = \$30{,}000 + (50{,}000 - 30{,}000)0.86886$. The sampling scheme is shown in Fig. 13A.3(c).

Now we can compute the PE equation with these sample values, yielding

$$
\begin{aligned}
PE\,(15\%) &= -\$112{,}263 + 2.0113(1600)(\$51.3841 - \$15) \\
&\quad + 0.2983(\$47{,}377) \\
&= \$18{,}957.
\end{aligned}
$$

This result completes the first iteration of PE_1 computation.

For the second iteration, we need to generate another set of three uniform random numbers (assume they are 0.72976, 0.79885, and 0.41879), to generate the respective sample from each distribution, and to compute $PE_2 = \$45{,}044$. If we repeat this process for 200 iterations, we obtain the PE values listed in Table 13A.1.

By ordering the observed data by increasing PE value and tabulating the ordered PE values, we obtain a frequency distribution shown in Table 13A.2. Such a tabulation results from dividing the entire range of computed PEs into a series of subranges (20 in this case), and then counting the number of computed values that fall in each of the 20 intervals. Note that the sum of all the frequencies of column 3 is the total number of trials that were made.

Column 4 simply expresses the frequencies of column 3 as a fraction of the total number of trials. At this point, all we have done is arrange the 200 numerical values of PE into a table of relative frequencies.

13A.4 Simulation Output Analysis

After a sufficient number of repetitive simulation trials has been run, the analysis is essentially completed. The only remaining tasks are to tabulate the computed PE values to determine the expected value and to make various graphic displays useful to management.

13A.4.1 Interpretation of Simulation Results

Once we obtain a PE frequency distribution (such as that shown in Table 13A.2), we need to make the assumption that the actual relative frequencies of column 4 in Table 13A.2 are representative of the probability of having a PE in each range. That is, we assume that the relative frequencies we observed in the sampling are representative of the proportions we would have obtained had we examined all the possible combinations.

This sampling is analogous to polling the opinions of voters about a candidate for public office. We could speak to every registered voter if we had the time and resources, but a simpler procedure would be to interview a smaller group of persons selected with an unbiased sampling procedure. If 60% of this scientifically selected sample supports the candidate, it probably would be safe to assume that 60% of all registered voters support the candidate. Conceptually, we do the same thing with simulation. As long as we ensure that a sufficient number of representative trials has been made, we can rely on the simulation results.

Once we have obtained the probability distribution of the PE, we face the crucial question: How do we use this distribution in decision-making? Recall that the probability distribution provides information regarding the probability that a random variable will attain some value x. We can use this information, in turn, to define the cumulative distribution, which expresses the probability that the random variable will attain a value smaller than or equal to some x, i.e., $F(x) = P(X \leq x)$. Thus, if the PE distribution is known, we can also compute the probability that the PE of a project will be negative. We use this probabilistic information in judging the profitability of the project.

With the assurance that 200 trials was a sufficient number for the WMC project, we may interpret the relative frequencies in column 4 of Table 13A.2 as probabilities. The PE values range between $8086 and $78,019, thereby indicating no loss for any situation. The PE distribution has an expected value of $42,436 and a standard deviation of $18,705.

13A.4.2 Creation of Graphic Displays

Using the output data in Table 13A.2, we can create the distribution in Figure 13A.4(a). A picture such as this can give the decision maker a feel for the ranges of possible PEs, for the relative likelihoods of loss versus gain, for the range of PEs that are most probable, and so on.

Another useful display is the conversion of the PE distribution to the equivalent cumulative frequency, as shown in Fig. 13A.4(b). Usually, a decision-maker is concerned with the likelihood of attaining at least a given level of PE. Therefore, we construct the cumulative distribution by accumulating the areas under the distribution as the PE increases. The decision maker can use Figure 13A.4(b) to answer many questions: For example, what is the likelihood of making at least a 15% return on investment, i.e., the likelihood that the PE will be at least 0? In our example, this probability is virtually 100%.

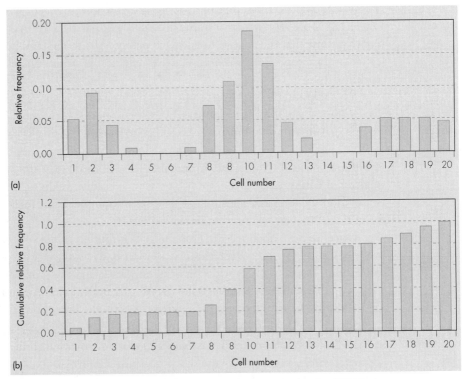

Figure 13A.4 Simulation result for WMC's transmission-housings project based on 200 iterations

13A.5 Dependent Random Variables

All our simulation examples have considered independent random variables. We must recognize that some of the random variables affecting the PE may be related to one another. If they are, we need to sample from distributions of the random variables in a manner that accounts for any dependency. For example, in the WMC project, both the demand and the unit price are not known with certainty. Both of these parameters would be on our list of variables for which we need to describe distributions, but they could be related inversely. When we describe distributions for these two parameters, we would have to account for the dependency. This issue can be critical, as the results obtained from a simulation analysis can be misleading if the analysis does not account for the dependent relationships. The sampling techniques for these dependent random variables are beyond the scope of this text, but can be found in many simulation textbooks.

13A.6 Summary

- **Risk simulation**, in general, is the process of modeling reality to observe and weigh the likelihood of the possible outcomes of a risky undertaking.

- **Monte Carlo sampling** is a specific type of randomized sampling method in which a random sample of outcomes is generated for a specified probability distribution. Such procedures can be built in a spreadsheet environment. Alternatively, specialized software packages are available which can simplify many of the complex tasks associated with large-scale simulations.

Problems

Level 1

13A.1 What is Monte Carlo simulation, and what are simulation's advantages and disadvantages as compared with the analytical approach discussed in Chapter 13?

13A.2 The following represents net cash flows and their respective probabilities for an investment project over its service life.

Time 0	Cash flow (A_0)	−$1000		
	Probability	1.0		
Time 1	Cash flow (A_1)	$300	$500	$1000
	Probability	0.3	0.4	0.3
Time 2	Cash flow (A_2)	$300	$1000	$2000
	Probability	0.2	0.7	0.1

The firm's MARR is known to be 10%.

(a) Express the $PE(i)$ as a function of A_0, A_1, and A_2.

(b) Using the following sequence of random numbers, obtain two samples of $PE(i)$.

Uniform random deviates = 0.024, 0.01, 0.13, 0.45, 0.56, 0.21, ...

Level 2

13A.3 In Problem 13A.2, use the random number generator on your computer to obtain 100 $PE(i)$ samples. Then construct a histogram based on these samples, and find the mean and variance of $PE(i)$ [Hint: You may consider the VLOOKUP function.]

13A.4 Consider the following investment project with three estimates for annual cash flows: Low, most-likely, and high estimates. Assume that these three estimates correspond to the parameters of a triangular distribution.

End of Year	Low	Most-Likely	High
0	−1500	−1000	−500
1	1000	2000	2500

You want to use risk simulation to estimate the mean and the standard deviation of the present equivalent of this project at an interest rate of 10%. Suppose that first two random numbers generated from a computer are 0.20 and 0.85. Compute the cash flows generated and the net present equivalent using these two random numbers.

13A.5 Consider the following investment project whose periodic cash flows are projected to follow a triangular distribution:

| | Project Cash Flow Estimates | | |
n	Low	Most-Likely	High
0	−$40	−$35	−$30
1	15	18	25
2	10	20	25
3	25	30	40

The firm's interest rate for project evaluation is 15%. Based on 100 Monte Carlo simulations, obtain a PE distribution of the project. Also estimate the mean and variance of the distribution and determine the probability that the PE is less than or equal to $30.

13A.6 Consider the following project, whose net cash flows over the next 5 years are given by three estimates. A quantity in parentheses represents a negative cash flow.

	Three Estimates		
End of		Most-	
Year	Low	Likely	High
0	($1,891,610)	($1,161,215)	($562,810)
1	(1,254,410)	(499,266)	106,170
2	(326,620)	150,595	707,430
3	227,350	607,372	918,290
4	753,750	803,229	918,680
5	657,300	720,800	980,400

These three estimates are equivalent to the pessimistic, most likely, and optimistic estimates in a triangular distribution. Using a spreadsheet package:

(a) Express the PE as a function of random cash flows at a given interest rate of i.

(b) At $i = 12\%$, obtain 100 PE samples on your computer based on the principle of Monte Carlo sampling and develop a PE probability distribution.

(c) From the results of part (b), find the expected value and the variance of the PE of this project.

(d) From the results of part (b), find the probability of the PE exceeding zero, and interpret the results.

Level 3

13A.7 A conventional muffler dampens the sound produced when an engine ignites fuel. It does this by using a series of baffles to redirect the sound of these explosions into an enclosed chamber where much of the sound dissipates. But this conventional muffler system tends to create a buildup of exhaust that causes back pressure on the engine, reducing power and efficiency. The electronic muffler would have the same purpose as its conventional counterpart—reducing noise—but would not create the back pressure. As a result, the electronic muffler would raise horsepower by as much as 7% and increase fuel efficiency.

A company is planning to build and market an electronic muffler on a trial basis for one or two car models. Based on initial responses from automakers, the firm has estimated the project's financial performance as follows:

Most-Likely Estimates	
Installed first cost	$9 million
Working capital	$1.5 million
Estimated life	5 years
Salvage value	$1 million
CCA rate	30%
Annual gross revenue	$8 million
Annual O&M costs	$3 million

The actual responses from the automakers when the product is out on the market are difficult to predict. The company has simplified the uncertain situation somewhat and feels that it is sufficient to imagine three possible outcomes: If product acceptance is poor, revenue will be only $3 million a year, but a strong response will produce gross profit of $16 million a year. The firm estimates that a 25% chance of poor acceptance, a 25% of excellent acceptance, and a 50% chance of average acceptance (the base case) is expected. The annual O&M costs as well as the working capital investment (in million dollars) will vary accordingly:

	Three States of Acceptance		
		Most-	
	Poor	Likely	Strong
Probability	0.25	0.50	0.25
Gross revenue	$3	$8	$16
Annual O&M cost	2	3	5
Working capital	1	1.5	3

The working capital investment will be recovered at the end of the project life. The firm's marginal tax rate is 35%, and the after-tax MARR is known to be 12%. (Assume that the first year market response will prevail for the remaining project years.)

(a) Develop a present equivalent expression for this investment.

(b) Assuming that the state of product acceptance is a discrete random variable with the probabilities assessed above, develop a PE probability distribution based on ten Monte Carlo samplings. Find the mean and variance of the PE distribution, and interpret the results.

(c) Repeat part (b) assuming instead that the state of product acceptance is a continuous triangular random variable with parameters assessed in the table above.

13A.8 MBQ Corporation is considering building a production line for very large scale integrated circuits (VLSIC). The small geometries involved demand that a "dry etch" technique be used rather than wet acid baths. The corporation has some experience with plasma etching, but the desired dimensions push the limits of conventional plasma etch technology, and newer reactive ion etch (RIE) equipment promises to provide higher process yields than the plasma etch equipment, but at a higher initial investment cost. Three dry etch processes are planned for the new semiconductor plant. While it is feasible to run multiple processes on one machine, separate machines will be purchased for each process to avoid possible cross-contamination during normal operations. A 3-year planning horizon is believed to be reasonable since it is anticipated that the next generation of integrated circuits will be introduced at that time. These future products may require new production technologies.

Two major product groups are planned for the new plant. One is a 64-bit microprocessor chip set (MCS); the second is a four-megabyte (4-MB) dynamic random access memory chip (DRAM). The 64-bit chip set includes a central processing unit, a floating point co-processor, a memory management unit, and other special purpose support chips. These chips will be sold as matched sets, which will operate at different clock speeds. The primary customers are automakers and appliance manufacturers who use these chips in their instrument controls. Those that can operate at more than 400 megahertz (MHz) will be sold as 400-MHz chip sets; those that cannot operate at 400 MHz, but can operate at more than 300 MHz, will be sold as 300-MHz chip sets; and those that can operate at more than 233 MHz but less than 300 MHz will be sold as 233-MHz chip sets. Those which cannot operate at 233 MHz or above are considered scrap and are not packaged.

Based upon market studies, anticipated process and packaging yields, the total production for the first year is planned to be 6000 wafer starts, with 12,000 the second year and 15,000 the third. Enough dynamic ram wafer starts will be made to fill out the planned production capacity. The annual revenue projections (in million dollars) are estimated to be *uniformly* distributed between the two bounds as follows:

| | Revenue Range: RIE Option | |
| | Microprocessor | 4 MB DRAM |
Year	Wafer Starts	Wafer Starts
1	$10M – $18M	$8M – $12M
2	35 – 45	20 – 26
3	30 – 40	25 – 32

| | Revenue Range: Plasma Option | |
| | Microprocessor | 4 MB DRAM |
Year	Wafer Starts	Wafer Starts
1	$9M – $12M	$7M – $10M
2	34 – 42	20 – 24
3	33 – 40	22 – 30

The investment required and other financial information for each option can be summarized as follows:

RIE Option: At $400,000 each for the RIE

systems, the RIE option has a total investment of $1,560,000 for three machines, including 30% of the system cost for installation, spares, and sales tax, subject to a 30% CCA rate. The estimated salvage value at the end of year 3 is $500,000 for the three systems. The estimated expenses for repair and maintenance are $20,000 for the first year and $40,000 for each of following 2 years. Operator labor is estimated to be $16,000 for each of the 3 years in the planning period. The RIE systems are expected to require $80,000 of process engineering support the first year, $60,000 the second year, and $50,000 the third year. Process gas consumption is expected to be $1000 the first year, $2000 the second year and $2500 the third year.

Plasma Etch Option: The plasma etch systems under consideration cost $250,000 each. The total initial cost of the plasma etch option is $858,000, including three machines plus 30% of the system cost for installation, spares, and sales tax. This investment also has a 30% CCA rate. At the end of year 3, it is believed that the three plasma etchers can be sold for $200,000. It is estimated that repair and maintenance expenses will total $20,000 the first year and $30,000 each year for the second and third year. The estimated operator labor is $20,000 per year. Process engineering support is expected to cost $80,000 the first year, $65,000 the second year and $60,000 the third year. Process gasses will cost $2000 the first year $4000 the second year, and $5000 the third year.

(a) Assuming that the firm's combined federal and provincial marginal tax rate is 40% and its MARR is 20%, find the PE expression for each option.
(b) Assume that all probabilities are independent. Using the Monte Carlo method for 200 trials, estimate the expected

PE, and the PE distribution for each option.
(c) With the results of part (b), which option would you recommend?

CHAPTER 14

Economic Analyses in the Public Sector

The Canadian government is investing hundreds of millions of dollars over several years to foster research and development in health information systems. The scope ranges from the creation of a National Health Surveillance Network, which would facilitate information transfer between health officers, to innovations in the area of telehealth. Telehealth builds on two Canadian strengths: information and communications technologies (ICT) and the universal health care system.

Telemedicine, the delivery of medical services at a distance, and telemonitoring, which enables remote follow-up to certain treatments, promise significant benefits to both patients and care givers. The most obvious advantage results from the time and cost savings of not having to commute. In urban centres, a stationary medical specialist may deliver diagnoses to many satellite clinics on the basis of electronically transmitted video, imaging data, or other information. For remote communities in northern Canada or elsewhere in the country, the value of such accessibility is magnified a thousandfold, as the cost or difficulty of travel can be insurmountable.

Through telehealth programs, the potential exists for improving health care delivery for tens of thousands of Canadians every year. For some applications, it could have a considerable impact on improving mortality and morbidity (the relative incidence of disease). Further advances in health and communications technology can reap even greater benefits. For example, the Sunnybrook Health Science Centre in Toronto is at the leading edge in the development of telemammography, permitting the transmission of digital imaging data for the detection of breast cancer. Benefits include improved speed and efficiency of the diagnosis, reduced need for repeat exams and travel for the patient, and improved access to mammography for women in sparsely populated areas.

Usually, public investment decisions involve a great deal of expenditure, and their benefits are expected to occur over an extended period of time. One of the important issues that the government must address is how it can determine whether its health care funding decisions, which affect the use of public funds, is in fact in the public interest. In identifying the benefits of projects of this nature, we need to consider both the primary benefits—the ones directly attributable to the project—and the secondary benefits—the ones indirectly attributable to the project. As an example, in addition to the travel costs savings realized with the introduction of telehealth services (primary benefit), improvements in imaging technology developed for telemedicine may be applicable in other areas of telecommunications (secondary benefits). Although not immediately evident, many other indirect benefits can be envisioned, such as less congestion, faster turnaround, and easier scheduling at some overcrowded major medical facilities.

Up to this point, we have focused attention on investment decisions in the private sector; the primary objective of these investments was to increase the wealth of businesses or individuals. In the public sector, federal, provincial, and local governments spend hundreds of billions of dollars annually on a wide variety of public activities, such as the telehealth initiatives described in this chapter opener. In addition, governments at all levels regulate the behavior of individuals and businesses by influencing the use of enormous quantities of productive resources. How can public decision makers determine whether their decisions, which affect the use of these productive resources, are, in fact, in the best public interest?

Benefit-cost analysis is a decision-making tool used to systematically develop useful information about the desirable and undesirable effects of public projects. In a sense, we may view benefit-cost analysis in the public sector as profitability analysis in the private sector. In other words, benefit-cost analysis attempts to determine whether the social benefits of a proposed public activity outweigh the social costs. Examples of benefit-cost analyses include studies of public transportation systems, environmental regulations on noise and pollution, public safety programs, education and training programs, public health programs, flood control, water resource development, and national defence programs.

The three types of benefit-cost analysis problems are as follows: (1) to maximize the benefits for any given set of costs (or budgets), (2) to maximize the net benefits when both benefits and costs vary, and (3) to minimize costs to achieve any given level of benefits (often called "cost-effectiveness" analysis). These three types of decision problems will be considered in this chapter.

14.1 Framework of Benefit-Cost Analysis

To evaluate public projects designed to accomplish widely differing tasks, we need to measure the benefits or costs in the same units in all projects so that we have a common perspective by which to judge different projects. In practice this generally means expressing both benefits and costs in monetary units, a process that often must be performed without accurate data. In performing benefit-cost analysis, we define "**users**" as the public and "**sponsors**" as the government.

The general framework for benefit-cost analysis can be summarized as follows:

1. Identify all users' benefits expected to arise from the project.

2. Quantify, as much as possible, these benefits in dollar terms so that different benefits may be compared against one another and against the costs of attaining them.

3. Identify sponsor's costs.

4. As much as possible, quantify these costs in dollar terms to allow comparisons.

5. Determine the equivalent benefits and costs at the base period; use an interest rate appropriate for the project.

6. Accept the project if the equivalent users' benefits exceed the equivalent sponsor's costs.

We can use benefit-cost analysis to choose among such alternatives as allocating funds for construction of a mass-transit system, a dam for irrigation, highways, or an air-traffic control system. If the projects are on the same scale with respect to cost, it is merely a question of choosing the project for which the benefits exceed the costs by the greatest amount.

14.2 Valuation of Benefits and Costs

The framework we just developed for benefit-cost analysis is no different from the one we have used throughout this text to evaluate private investment projects. The complications, as we shall discover in practice, arise in trying to identify and assign values to all the benefits and costs of a public project.

14.2.1 Users' Benefits

To begin a benefit-cost analysis, we identify all project **benefits** (favorable outcomes) and **disbenefits** (unfavorable outcomes) to the user. We should also consider the indirect consequences resulting from the project—the so-called **secondary effects**. For example, construction of a new highway will create new businesses such as gas stations, restaurants, and motels (benefits), but it will divert some traffic from the old road, and as a consequence, some businesses would be lost (disbenefits). Once the benefits and disbenefits are quantified, we define the users' benefits as follows:

Users' benefit (B) = Benefits – disbenefits

In identifying users' benefits, we should classify each one as a **primary benefit**—one directly attributable to the project—or a **secondary benefit**—one indirectly attributable to the project. As an example, at one time, the Canadian government was considering building a superconductor research laboratory in Ottawa. If it ever materializes, it could bring many scientists and engineers along with other supporting population to the region. Primary national benefits may include the long-term benefits that may accrue as a result of various applications of the research to Canadian businesses. Primary regional benefits may include economic benefits created by the research laboratory activities, which would generate many new supporting businesses. The secondary benefits might include the creation of new economic wealth as a consequence of a possible increase in international trade and any increase in the incomes of various regional producers attributable to a growing population.

The reason for making this distinction is that it may make our analysis more efficient. If primary benefits alone are sufficient to justify project costs, we can save time and effort by not quantifying the secondary benefits.

14.2.2 Sponsor's Costs

We determine the cost to the sponsor by identifying and classifying the expenditures required and any savings (or revenues) to be realized. The sponsor's costs should include both capital investment and annual operating costs. Any sales of products or services that take place on completion of the project will generate some revenues—

for example, toll revenues on highways. These revenues reduce the sponsor's costs. Therefore, we calculate the sponsor's costs by combining these cost elements:

Sponsor's cost = capital cost + operating and maintenance costs − revenues.

14.2.3   Social Discount Rate

As we learned in Chapter 10, the selection of an appropriate MARR for evaluating an investment project is a critical issue in the private sector. In public project analyses, we also need to select an interest rate, called the **social discount rate**, to determine equivalent benefits as well as the equivalent costs. Selection of the social discount rate in public project evaluation is as critical as selection of a MARR in the private sector.

When present equivalent calculations were initiated to evaluate public water resources and related land-use projects in the 1930s, relatively low discount rates were adopted compared to those existing in markets for private assets. The choice of a suitable **social discount rate** has undergone much debate over the intervening years. There are two prevailing views on the most suitable basis for establishing this rate.

One is referred to as the social rate of time preference (STP), which reflects the trade-off between current and future consumption. There are many reasons why the population may discount future benefits compared with earlier ones. It can be argued that people are nearsighted, preferring immediate satisfaction, or that they discount later benefits since they will not be around (eventually) to enjoy them. Conversely, if people are altruistic they would tend to weight benefits to future generations equally with gains in the near term.

A second perspective, referred to as the social opportunity cost of capital (SOCC), considers what other uses would be made with the money if the current public project investment(s) were not undertaken. One view is that the money would remain in the private sector, in which case the social discount rate is essentially based on expected rates of return averaged across the business sector. However, income taxes complicate this assessment. The government could select an average *before-tax* MARR from the private sector as the benchmark rate for project acceptance, or refer instead to the *after-tax* MARR expected by investors. Conversely, if public capital funds are considered rationed, then the relevant opportunity cost for a public project should be another public project rather than private sector activity. Furthermore, a baseline for public expenditures could be the cost of government debt, as reflected in long-term government bond rates.

In recent years, with the growing interest in performance budgeting and systems analysis in the 1960s, the tendency on the part of government agencies has been to examine the appropriateness of the discount rate in the public sector in relation to the efficient allocation of resources in the economic system as a whole.[1] Two views of the basis for determining the social discount rate prevail:

1. **Projects without private counterparts:** *The social discount rate should reflect only the prevailing government borrowing rate.* Projects such as dams designed purely for flood control, access roads for noncommercial uses, and reservoirs for community water supply may not have corresponding private counterparts. In those areas of government activity where benefit-cost analysis has been employed in evaluation, the rate of discount traditionally used has been the cost of government borrowing.

2. **Projects with private counterparts:** *The social discount rate should represent the rate that could have been earned had the funds not been removed from the private sector.* If all public projects were financed by borrowing at the expense of private investment, we may focus on the opportunity cost of capital in alternative investments in the private sector to determine the social discount rate. In the case of public capital projects, similar to some in the private sector that produce a commodity or a service (such as electric power) to be sold on the market, the rate of discount employed would be the average cost of capital as discussed in Chapter 10. The reasons for using the private rate of return as the opportunity cost of capital in projects similar to those in the private sector are (1) to prevent the public sector from transferring capital from higher-yielding to lower-yielding investments, and (2) to force public project evaluators to employ market standards in justifying projects.

The Treasury Board of Canada provides guidelines for the evaluation of public projects. Since the acceptability of some projects is very sensitive to the applied social discount rate, a base-case social discount rate is suggested, with a somewhat lower rate and commensurately higher rate suggested to conduct sensitivity evaluations (see Section 13.2.4).

14.2.4 Quantifying Benefits and Costs

Now that we have defined the general framework for benefit-cost analyses and discussed the appropriate discount rate, we will illustrate the process of quantifying the benefits and costs associated with a public project.[2]

Some provinces employ inspection systems for motor vehicles. Critics often charge that these programs lack efficacy and have a poor benefit-to-cost ratio in terms of reducing fatalities, injuries, accidents, and pollution.

---

[1] Mikesell RF. *The Rate of Discount for Evaluating Public Projects.* American Enterprise Institute for Public Policy Research, 1977.

[2] Based on Loeb PD and Gilad B. "The Efficacy and Cost Effectiveness of Vehicle Inspection," *Journal of Transport Economics and Policy,* May 1984: 145–164. The original cost data, which were given in 1981 dollars, were converted to the equivalent cost data in 1990 by using the prevailing consumer price indexes during the period.

Elements of Benefits and Costs

Primary and secondary benefits identified with a motor vehicle inspection program are as follows:

- **Users' Benefits**

   Primary benefits: Deaths and injuries related to motor-vehicle accidents impose specific financial costs on individuals and society. Preventing such costs through the inspection program has the following primary benefits.

   1. Retention of contributions to society that might be lost due to an individual's death.
   2. Retention of productivity that might be lost while an individual recuperates from an accident.
   3. Savings of medical, legal, and insurance services.
   4. Savings on property replacement or repair costs.

   Secondary benefits: Some secondary benefits are not measurable (for example, avoidance of pain and suffering); others can be quantified. Both types of benefits should be considered. A list of secondary benefits follows.

   1. Savings of income of families and friends of accidents victims who might otherwise be tending to accident victims.
   2. Avoidance of air and noise pollution and savings on fuel costs.
   3. Savings on enforcement and administrative costs related to the investigation of accidents.
   4. Pain and suffering.

- **Users' Disbenefits**

   1. Cost of spending time to have a vehicle inspected (including travel time), as opposed to devoting that time to an alternative endeavor (opportunity cost).
   2. Cost of inspection fees.
   3. Cost of repairs that would not have been made if the inspection had not been performed.
   4. Value of time expended in repairing the vehicle (including travel time).
   5. Cost in time and direct payment for reinspection.

- **Sponsor's Costs**

   1. Capital investments in inspection facilities.
   2. Operating and maintenance costs associated with inspection facilities. These include all direct and indirect labor, personnel, and administrative costs.

- **Sponsor's Revenues or Savings**

   1. Inspection fee.

Valuation of Benefits and Costs

The aim of benefit-cost analysis is to maximize the equivalent value of all benefits less that of all costs (expressed either in present values or annual values). This objective is in line with promoting the economic welfare of citizens. In general, the benefits of public projects are difficult to measure, whereas the costs are more easily determined. For simplicity, we will only attempt to quantify the primary users' benefits and sponsor's costs on an annual basis. Estimates for a population the size of Ontario are provided as an example.

(a) **Calculation of Primary Users' Benefits**

1. Benefits due to the reduction of deaths: The equivalent value of the average income stream lost by victims of fatal accidents[3] is estimated at $432,656 per victim. It is estimated that the inspection program would reduce the number of annual fatal accidents by 304 per year, resulting in a potential savings of

$$(304)(\$432,656) = \$131,527,424.$$

2. Benefits due to the reduction of damage to property: The average cost of damage to property per accident is estimated at $3398. This figure includes the cost of repairs for damages to the vehicle, the cost of insurance, the cost of legal and court administration, the cost of police accident investigation, and the cost of traffic delay due to accidents. Accidents are expected to be reduced by 37,910 per year, and about 63% of all accidents result in damage to property only. Therefore, the estimated annual value of benefits due to reduction of property damage would be estimated at

$$\$3398(37,910)(0.63) = \$81,155,453.$$

The overall annual primary benefits are estimated as the sum of

Value of reduction in fatalities	$131,527,424
Value of reduction in property damage	81,155,453
Total	$212,682,877

(b) **Calculation of Primary Users' Disbenefits**

1. Opportunity cost associated with time spent bringing vehicles for inspection: This cost is estimated as

$$C_1 = \text{(Number of cars inspected)}$$
$$\times \text{(average duration involved in travel)}$$
$$\times \text{(average wage rate).}$$

---

[3] These estimates were based on the total average income that these victims could have generated had they lived. This average value on human life was calculated by considering several factors, such as age, sex, and income group.

With an estimated average duration of 1.02 travel-time hours per car, an average wage rate of $11.25 per hour, and 5,136,224 inspected cars per year, we obtain

$$C_1 = 5,136,224 \, (1.02) \, (\$11.25)$$
$$= \$58,938,170.$$

2. Cost of inspection fee: This cost may be calculated as

$$C_2 = (\text{Inspection fee}) \times (\text{number of cars inspected}).$$

Assuming an inspection fee of $10 is to be paid for each car, the cost of the total annual inspection cost is estimated as

$$C_2 = (\$10)(5,136,224)$$
$$= \$51,362,240.$$

3. Opportunity cost associated with time spent waiting during the inspection process: This cost may be calculated by the formula

$$C_3 = (\text{Average waiting time in hour})$$
$$\times (\text{average wage rate per hour})$$
$$\times (\text{number of cars inspected}).$$

With an average waiting time of 9 minutes (or 0.15 hours),

$$C_3 = 0.15(\$11.25)(5,136,224) = \$8,667,378.$$

4. Vehicle usage costs for the inspection process: These costs are estimated as

$$C_4 = (\text{Number of inspected cars})$$
$$\times (\text{vehicle operating cost per kilometre})$$
$$\times (\text{average round trip distance to inspection station})$$

Assuming $0.30 operating cost per kilometre and 20 round-trip kilometres,

$$C_4 = 5,136,224 \, (\$0.30)(20) = \$30,817,344.$$

The overall primary annual disbenefits are estimated as

Item	Amount
C_1	$58,938,170
C_2	51,362,240
C_3	8,667,378
C_4	30,817,344
Total disbenefits, or $29.16 per vehicle	$149,785,132

(c) **Calculation of Primary Sponsor's Costs**

A provincially run program would incur an expenditure of $55,133,866 for inspection facilities (this value represents the annualized capital expenditure) and another annual operating expenditure of $18,080,000 for inspection, adding up to $73,213,866.

(d) **Calculation of Primary Sponsor's Revenues**

The sponsor's costs are offset to a large degree by annual inspection revenues; these must be subtracted to avoid double counting. Annual fee revenues are the same as the direct cost of inspection incurred by the users (C_2), which was calculated as $51,362,240.

Reaching a Final Decision

Finally, a discount rate of 6% was deemed appropriate because most provincial projects are financed by issuing 6% long-term bonds. The streams of costs and benefits are already discounted so as to obtain their present and annual equivalent values.

From the above estimates, the primary benefits of inspection are valued at $212,682,877, as compared to the primary disbenefits of inspection, which total $149,785,132. Therefore, the users' net benefits are

$$\text{User's net benefits} = \$212,682,877 - \$149,785,132$$
$$= \$62,897,745.$$

Now the sponsor's net costs are

$$\text{Sponsor's net costs} = \$73,213,866 - \$51,362,240$$
$$= \$21,851,626.$$

Since all benefits and costs are expressed in annual equivalents, we can use these values directly to compute the degree of benefits that exceeds the sponsor's costs:

$$\$62,897,745 - \$21,851,626 = \$41,046,119 \text{ per year.}$$

This positive AE amount indicates that the Ontario inspection system would be economically justifiable under the given assumptions. We can assume the AE amount would have been even greater had we also factored in secondary benefits. (For simplicity, we have not explicitly considered vehicle growth in the province of Ontario. For a complete analysis, this growth factor must be considered to account for all related benefits and costs in equivalence calculations.)

14.2.5   Difficulties Inherent in Public Project Analysis

As we observed in the motor-vehicle inspection program in the previous section, public benefits are very difficult to quantify in a convincing manner. For example, consider the valuation of a saved human life in any category. Conceptually, the total benefit associated with saving a human life may include the avoidance of the costs of insurance administration and legal and court costs. As well, the average potential income lost, considering the factors of age and sex, because of premature death must be included. Obviously, the difficulties associated with any attempt to put precise numbers on human life are insurmountable.

Consider this example: A few years ago, a 50-year-old business executive was killed in a plane accident. The investigation indicated that the plane was not properly maintained according to the federal guidelines. The executive's family sued the airline, and the court eventually ordered the airline to pay $5,250,000 to the victim's family. The judge calculated the value of the lost human life based on the assumption that, if the executive had lived and worked in the same capacity until his retirement, his remaining lifetime earnings would have been equivalent to $5,250,000 at the time of award. This is an example of how an individual human life was assigned a dollar value, but clearly any attempt to establish an average amount that represents the general population is potentially controversial. We might even take exception to this individual case: Does the executive's salary adequately represent his worth to his family? Should we also assign a dollar value to their emotional attachment to him, and if so, how much?

Consider a situation in which a local government is planning to widen a typical municipal highway to relieve chronic traffic congestion. Knowing that the project will be financed by local and provincial taxes, and that many out-of-province travelers are expected to benefit, should the project be justified solely on the benefits to local residents? Which point of view should we take in measuring the benefits—the municipal level, the provincial level, or both? It is important that any benefit measure be done from the appropriate *point of view*.

In addition to valuation and point-of-view issues, many possibilities for tampering with the results of benefit-cost analyses may exist. Unlike in the private sector, many public projects are undertaken based on political pressure rather than on their economic benefits alone. In particular, whenever the benefit-cost ratio becomes marginal, or less than one, a potential to inflate the benefit figures to make the project look good exists.

14.3 Benefit-Cost Ratios

An alternative way of expressing the worthiness of a public project is to compare the users' benefits (B) to the sponsor's cost (C) by taking the ratio B/C. In this section, we shall define the benefit-cost (B/C) ratio, and explain the relationship between the conventional PE criterion and the B/C ratio.

14.3.1 Definition of Benefit-Cost Ratio

For a given benefit-cost profile, let B and C be the present equivalents of benefits and costs defined by

$$B = \sum_{n=0}^{N} b_n (1 + i)^{-n} \tag{14.1}$$

$$C = \sum_{n=0}^{N} c_n (1 + i)^{-n}, \tag{14.2}$$

where b_n = Benefit at the end of period n, $b_n \geq 0$

c_n = Expense at the end of period n, $c_n \geq 0$

A_n = $b_n - c_n$

N = Project life

i = Sponsor's interest rate (discount rate).

The sponsor's costs (C) consist of the equivalent capital expenditure (P) and the equivalent annual operating costs (C') accrued in each successive period. (Note the sign convention we use in calculating a benefit-cost ratio. Since we are using a ratio, all benefits and cost flows are expressed in positive units. Recall that in previous equivalent worth calculations our sign convention was to explicitly assign "+" for cash inflows and "−" for cash outflows.) Let's assume that a series of initial investments is required during the first K periods, while annual operating and maintenance costs accrue in each following period. Then, the equivalent present value for each component is

$$P = \sum_{n=0}^{K} c_n (1 + i)^{-n} \tag{14.3}$$

$$C' = \sum_{n=K+1}^{N} c_n (1 + i)^{-n}, \tag{14.4}$$

and $C = P + C'$.

The *B/C* ratio[4] is defined as

$$BC(i) = \frac{B}{C} = \frac{B}{P + C'}, \qquad P + C' > 0. \tag{14.5}$$

If we are to accept a project, the $BC(i)$ must be greater than 1.

Note that we must express the values of B, C', and P in present equivalents. Alternatively, we can compute these values in terms of annual equivalents and use them in calculating the *B/C* ratio. The resulting *B/C* ratio is not affected.

EXAMPLE
14.1

Benefit-cost ratio

A public project being considered by a local government has the following estimated benefit-cost profile (Figure 14.1).

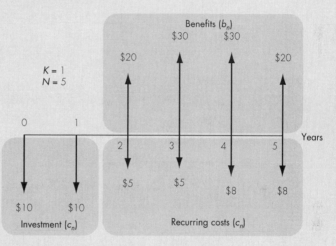

Figure 14.1   Classification of a project's cash flow elements (Example 14.1)

---

[4] An alternative measure, called the **net B/C ratio**, $B'C(i)$, considers only the initial capital expenditure as a cash outlay, and annual net benefits are used:

$$B'C(i) = \frac{B - C'}{P} = \frac{B'}{P}, \qquad P > 0.$$

The decision rule has not changed — the ratio must still be greater than one. It can be easily shown that a project with $BC(i) > 1$ will always have $B'C(i) > 1$, as long as both C and P are > 0, as they must be for the inequalities in the decision rules to maintain the stated senses. The magnitude of $BC(i)$ will generally be different than that for $B'C(i)$, but the magnitudes are irrelevant for making decisions. All that matters is whether the ratio exceeds the threshold value of one. However, some analysts prefer to use $B'C(i)$ because it indicates the net benefit (B') expected per dollar invested. But why do they care if the choice of ratio does not affect the decision? They may be trying to increase or decrease the magnitude of the reported ratio in order to influence audiences who do not understand the proper decision rule. People unfamiliar with benefit/cost analysis often assume that a project with a higher *B/C* ratio is better. This is not generally true, as is shown in 14.3.3. An incremental approach must be used to properly compare mutually exclusive alternatives.

n	b_n	c_n	A_n
0		$10	–$10
1		10	–10
2	$20	5	15
3	30	5	25
4	30	8	22
5	20	8	12

Assume that $i = 10\%$, $N = 5$, and $K = 1$. Compute B, C, P, C', and $BC(10\%)$.

Solution

$$
\begin{aligned}
B &= \$20(P/F, 10\%, 2) + \$30(P/F, 10\%, 3) \\
&\quad + \$30(P/F, 10\%, 4) + \$20(P/F, 10\%, 5) \\
&= \$71.98.
\end{aligned}
$$

$$
\begin{aligned}
C &= \$10 + \$10(P/F, 10\%, 1) + \$5(P/F, 10\%, 2) + \$5(P/F, 10\%, 3) \\
&\quad + \$8(P/F, 10\%, 4) + \$8(P/F, 10\%, 5) \\
&= \$37.41.
\end{aligned}
$$

$$
\begin{aligned}
P &= \$10 + \$10(P/F, 10\%, 1) \\
&= \$19.09.
\end{aligned}
$$

$$
\begin{aligned}
C' &= C - P \\
&= \$18.32.
\end{aligned}
$$

Using Eq. (14.5), we can compute the B/C ratio as

$$
\begin{aligned}
BC(10\%) &= \frac{\$71.98}{\$19.09 + \$18.32} \\
&= 1.92 > 1.
\end{aligned}
$$

The B/C ratio exceeds 1, so the users' benefits exceed the sponsor's costs.

14.3.2 Relationship Between B/C Ratio and PE

The B/C ratio yields the same decision for a project as does the PE criterion. Recall that the $BC(i)$ criterion for project acceptance can be stated as

$$
\frac{B}{P + C'} > 1. \tag{14.6}
$$

If we multiply the term $(P + C')$ on both sides of the equation and transpose the term $(P + C')$ to the left-hand side, we have

$$
\begin{aligned}
B &> (P + C') \\
B - (P + C') &> 0 \tag{14.7} \\
PE(i) = B - C &> 0, \tag{14.8}
\end{aligned}
$$

which is the same decision rule[5] as that which accepts a project by the PE criterion. This implies that we could use the benefit-cost ratio in evaluating private projects instead of using the PE criterion, or we could use the PE criterion in evaluating the public projects. Either approach will signal consistent project selection. Recall that in Example 14.1, $PE(10\%) = B - C = \$34.57 > 0$; the project would be acceptable under the PE criterion.

14.3.3 Comparing Mutually Exclusive Alternatives: Incremental Analysis

Let us now consider how we choose among mutually exclusive public projects. As we explained in Chapter 5, we must use the incremental investment approach in comparing alternatives based on any relative measure such as IRR or B/C.

Incremental Analysis Based on $BC(i)$

To apply incremental analysis, we compute the incremental differences between two alternatives for each term (B, P, and C') and take the B/C ratio based on these differences. To use the $BC(i)$ on incremental investment, we may proceed as follows:

1. If one or more alternatives have B/C ratios greater than 1, eliminate any alternatives with a B/C ratio less than 1.

2. Arrange the remaining alternatives in the increasing order of the denominator ($P + C'$). Thus, the alternative with the smallest denominator should be the first (j), the alternative with the second smallest (k), and so forth.

3. Compute the incremental differences for each term (B, P, and C') for the paired alternatives (j, k) in the list.

$$
\begin{aligned}
\Delta B &= B_k - B_j \\
\Delta P &= P_k - P_j \\
\Delta C' &= C'_k - C'_j.
\end{aligned}
$$

4. Compute the $BC(i)$ on incremental investment by evaluating

$$
BC(i)_{k-j} = \frac{\Delta B}{\Delta P + \Delta C'}.
$$

If $BC(i)_{k-j} > 1$, select the k alternative. Otherwise select the j alternative.

---

[5] We can easily verify a similar relationship between the net B/C ratio and the PE criterion.

5. Compare the selected alternative with the next one on the list by computing the incremental benefit-cost ratio.[6] Continue the process until you reach the bottom of the list. The alternative selected during the last pairing is the best one.

We may modify the decision procedures when we encounter the following situations:

- If $\Delta P + \Delta C' = 0$, we cannot use the benefit-cost ratio because this implies that both alternatives require the same initial investment and operating expenditure. When this happens, we simply select the alternative with the largest B value.

- In situations where public projects with unequal service lives are to be compared but they can be repeated, we may compute all component values (B, C', and P) on an annual basis and use them in incremental analysis.

EXAMPLE 14.2 Incremental benefit-cost ratios

Consider three investment projects, A1, A2, and A3. Each project has the same service life, and the present equivalent of each component value (B, P, and C') is computed at 10% as follows:

	A1	Projects A2	A3
B	$12,000	$35,000	$21,000
P	5,000	20,000	14,000
C'	4,000	8,000	1,000
$PE(i)$	$3,000	$7,000	$6,000

(a) If all three projects are independent, which projects would be selected based on $BC(i)$?

(b) If the three projects are mutually exclusive, which project would be the best alternative? Show the sequence of calculations that would be required to produce the correct results. Use the B/C ratio on incremental investment.

Solution

(a) Since $PE(i)_1$, $PE(i)_2$, and $PE(i)_3$ are positive, all projects would be acceptable if they were independent. Also, $BC(i)$ values for each project are greater than 1, so the use of the benefit-cost ratio criterion leads to the same accept-reject conclusion under the PE criterion.

	A1	A2	A3
$BC(i)$	1.33	1.25	1.40

[6] If we use the net B/C ratio as a basis, we need to order the alternatives in increasing order of P and compute the net B/C ratio on the incremental investment.

(b) If these projects are mutually exclusive, we must use the principle of incremental analysis. If we attempt to rank the projects according to the size of the B/C ratio, obviously we will observe a different project preference. For example, if we use the $BC(i)$ ratio on the total investment, we see that A3 appears to be the most desirable and A2 the least desirable, but selecting mutually exclusive projects on the basis of B/C ratios is incorrect. Certainly, with $PE(i)_2 > PE(i)_3 > PE(i)_1$, project A2 would be selected under the PE criterion. By computing the incremental B/C ratios, we will select a project that is consistent with the PE criterion.

We will first arrange the projects by increasing order of their denominator $(P + C')$ for the $BC(i)$ criterion:[7]

Ranking Base	A1	A3	A2
$P + C'$	$9,000	$15,000	$28,000

- A1 versus A3: With the do-nothing alternative, we first drop from consideration any project that has a B/C ratio smaller than 1. In our example, the B/C ratios of all three projects exceed 1, so the first incremental comparison is between A1 and A3:

$$BC(i)_{3-1} = \frac{\$21,000 - \$12,000}{(\$14,000 - \$5000) + (\$1000 - \$4000)}$$

$$= 1.5 > 1.$$

Since the ratio is greater than 1, we prefer A3 over A1. Therefore, A3 becomes the "current best" alternative.

- A3 versus A2: Next, we must determine whether the incremental benefits to be realized from A2 would justify the additional expenditure. Therefore, we need to compare A2 and A3 as follows:

$$BC(i)_{2-3} = \frac{\$35,000 - \$21,000}{(\$20,000 - \$14,000) + (\$8000 - \$1000)}$$

$$= 1.08 > 1.$$

The incremental B/C ratio again exceeds 1, and therefore we prefer A2 over A3. With no further projects to consider, A2 becomes the ultimate choice.[8]

---

[7] P is used as a ranking base for the $B'C(i)$ criterion. The order still remains unchanged — A1, A3, and A2.

[8] Using the net B/C ratio: If we had to use the net B/C ratio on this incremental investment decision, we would obtain the same conclusion. Since all $B'C(i)$ ratios exceed 1, all alternatives are viable. By comparing the first pair of projects on this list, we obtain:

$$B'C(i)_{3-1} = \frac{(\$21,000 - \$12,000) - (\$1000 - \$4000)}{(\$14,000 - \$5000)}$$

$$= 1.33 > 1.$$

Project A3 becomes the "current best." Next, a comparison of A2 and A3 yields

$$B'C(i)_{2-3} = \frac{(\$35,000 - \$21,000) - (\$8000 - \$1000)}{(\$20,000 - \$14,000)}$$

$$= 1.17 > 1.$$

Therefore, A2 becomes the best choice by the net B/C ratio criterion.

14.4 Analysis of Public Projects Based on Cost-Effectiveness

In evaluating public investment projects, we may encounter situations where competing alternatives have the same goals but the effectiveness with which those goals can be met may be, or may not be, measurable in dollars. In these situations, we compare decision alternatives directly based on their **cost-effectiveness**. Here we judge the effectiveness of an alternative in dollars or some nonmonetary measure by the extent to which that alternative, if implemented, will attain the desired objective. The preferred alternative is then either the one that produces the maximum effectiveness for a given level of cost, or the minimum cost for a fixed level of effectiveness.

14.4.1   General Procedure for Cost-Effectiveness Studies

A typical cost-effectiveness analysis procedure involves the following steps.

Step 1: Establish the goals to be achieved by the analysis.

Step 2: Identify the imposed restrictions on achieving the goals, such as budget.

Step 3: Identify all the feasible alternatives to achieve the goals.

Step 4: Identify the social discount rate to use in the analysis.

Step 5: Determine the equivalent life-cycle cost of each alternative, including research and development, testing, capital investment, annual operating and maintenance costs, and salvage value.

Step 6: Determine the basis for developing the cost-effectiveness index. Two approaches may be used: (1) the fixed-cost approach and (2) the fixed-effectiveness approach. If the fixed-cost approach is used, determine the amount of effectiveness obtained at a given cost. If the fixed-effectiveness approach is used, determine the cost to obtain the predetermined level of effectiveness.

Step 7: Compute the cost-effectiveness ratio for each alternative based on the selected criterion in Step 6.

Step 8: Select the alternative with the maximum cost-effective index.

14.4.2   A Cost-Effectiveness Case Example

To illustrate the procedures involved in cost-effectiveness analysis, we shall present an example in which the most cost-effective program for developing an adverse-weather precision-guided weapon system is selected.[9]

---

[9]   The case example is provided by Frederick A. Davis. All numbers used herein do not represent the actual values.

Problem Statement

In a recent international conflict, precision-guided weapons demonstrated remarkable success and accuracy against a wide array of fixed and mobile targets. Such weapons rely upon laser designation of the target by an aircraft. The aircraft illuminates the target by laser; the weapon, with its on-board laser sensor, locks onto the target and flies to it. During the war, aircraft were required to fly at moderate altitude (about 10,000 feet) to escape vulnerability to anti-aircraft artillery batteries. At these altitudes, the aircraft were above the cloud/smoke levels. Unfortunately, laser beams cannot penetrate cloud cover, smoke, or fog. As a result, on those days when clouds, smoke, and fog were present, the aircraft were unable to deliver the precision-guided weapons (Figure 14.2). This led to development of a weapon system that would correct these deficiencies.

Defining the Goals

Selection of the best system is based on cost/kill decision criteria. **Cost/kill** is defined as the unit cost of the weapon divided by the probability of the weapon achieving its target. Mission-effectiveness studies determine this probability.

The purposes of this study are to evaluate these alternatives from a cost viewpoint and to determine the best option, based on a cost/kill decision criterion. Also, anticipating governmental scrutiny of high weapon-system costs, any option costing more than the present value of a life-cycle cost of $120K per unit cannot be considered for selection. To respond rapidly to this critical military requirement, the

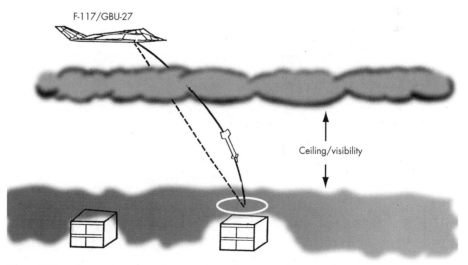

Figure 14.2 Conceptual use of an adverse weather guidance weapon by an aircraft

initial operational capability (IOC) date is assumed to be 7 years. IOC is defined as that point in the project life when the first block of weapons has been delivered to the field and is ready for operational use.

Description of Precision-Weapon Alternatives

Considering numerous weapon alternatives, preferred concepts were developed, and these are listed in Table 14.1. Also presented in the table are the qualitative characteristics of each alternative. These six weapon system alternatives are also considered to be mutually exclusive. Only one of the alternatives will be selected by the military to fill the mission capability void that currently exists. Each of the alternatives is at some level of technological maturity. Some are nearly off-the-shelf, while others are just emerging from laboratory development. Because of this, some of the alternatives will require considerably more up-front development funding prior to production than others. Table 14.1 also summarizes the results (probability of kill) by the mission studies for the six guidance systems before entering into laboratory research and development.

To consider the life-cycle costs associated with each option, the project begins with the Full Scale Development (FSD) phase, which accomplishes the up-front development prior to entering the production phase. To meet the 7-year IOC date,

Table 14.1
Weapon System Alternatives

Alternative A_i	Advantage	Disadvantage	Probability of Kill
A1: Inertial navigation system	Low cost, mature technology	Accuracy, target recognition	0.33
A2: Inertial navigation system: global positioning system	Moderate cost, mature technology	Target recognition	0.70
A3: Imaging infrared (I^2R)	Accurate, target recognition	High cost, bunkered target detection	0.90
A4: Synthetic aperture radar	Accurate, target recognition	High cost	0.99
A5: Laser detection/ranging	Accurate, target recognition	High cost, technical maturity	0.99
A6: Millimeter wave (MMW)	Moderate cost, accurate	Target recognition	0.80

the FSD phase must be completed in 4 years. A 10,000-unit buy over a 5-year production life is required to meet the military's needs. Because of the differing technological maturity of the six alternatives, widely varying FSD investments will be required.

Life-Cycle Cost for Each Alternative

The costs associated with FSD and production for each alternative will vary significantly. For systems incorporating the most current of emerging technologies, the FSD completion costs will be higher than those considered mature. Production costs will vary, depending upon the complexity of the system components. The objective of the FSD program is to rigorously test demonstrated system capability and correct any design flaws. The cost estimates used in both the FSD and the production phases were generated considering labor hours, material costs, equipment/tooling costs, subcontractor costs, travel, flight-test costs, documentation, and costs for program reviews. Table 14.2 summarizes the equivalent life-cycle cost for each program, estimated in constant (Year 0) dollars. The military periodically establishes an interest rate for equivalent life-cycle cost calculation, which is 10% at the time of this project evaluation.

Table 14.2
Life-Cycle Costs for Weapon Development Alternatives

| Phase | Year | Expenditures in Million Dollars | | | | | |
		A1*	A2	A3	A4	A5	A6
FSD	0	$15	$19	$50	$40	$75	$28
	1	18	23	65	45	75	32
	2	19	22	65	45	75	33
	3	15	17	50	40	75	27
IOC	4	90	140	200	200	300	150
	5	95	150	270	250	360	180
	6	95	160	280	275	370	200
	7	90	150	250	275	340	200
	8	80	140	200	200	330	170
PE(10%)		$315.92	$492.22	$884.27	$829.64	$1227.23	$613.70

*Sample calculation: Equivalent life-cycle cost for A1:

$$PE_1(10\%) = \$15 + \$18(P/F, 10\%, 1) + \ldots + \$80(P/F, 10\%, 8)$$
$$= \$315.92.$$

Cost-Effectiveness Index

The equivalent life-cycle costs of the system need to be divided by the 10,000 units to be purchased in order to arrive at a cost/unit figure. Then, the cost/kill for each alternative is computed. The following shows the resultant cost/kill and kill/cost figures:

Type	Cost/Unit	Probability of Kill	Cost/Kill	Kill/Cost
A1	$31,592	0.33	$95,733	0.0000104
A2	49,222	0.70	70,317	0.0000142
A3	88,427	0.90	98,252	0.0000102
A4	82,964	0.99	83,802	0.0000119
A5	122,723	0.99	123,963	0.0000081
A6	61,370	0.80	76,713	0.0000130

Figure 14.3 graphically presents the cost/kill values of the six guidance systems. Obviously, the A5 system is not a feasible solution, as its unit cost exceeds the constraint of $120K. Of the feasible alternatives, the lowest cost/kill value is that of the A2 weapon system. However, it is premature to conclude that the A2 option is the best. This case example is unique in the sense that neither the costs nor the benefits are fixed. The benefit is a kill, but the probability of kill is not the same across the alternatives. If we rely upon entirely on "cost per kill" criterion, a very cheap system with a low kill rate might come out the best.

To make a valid decision, we need to either (1) fix the cost of the system (then the objective would be to maximize the number of kills) or, equivalently, maximize the kill per cost, or (2) fix the number of kills (then the objective would be to minimize the cost of the system or, equivalently, minimize the cost per kill). The stated situation, fixing the 10,000 units to be purchased, is neither of the above. For example, if the military stipulates that the minimum probability of kill be 0.90 or above, in addition to the unit cost constraint of $120K and 10,000 units, A4 becomes the best. This is the system that will meet the military's requirement for an adverse-weather weapon system. (Note that, if the military had adopted the kill/cost criterion, the total budget commitment should have been spelled out.) Even though the cost estimates and the probability of kill were based on the best engineering judgment associated with the complexity of the system and their technological risk, project overrun costs are common, and therefore, uncertainty in the estimated costs for the six weapon alternatives can be expected.

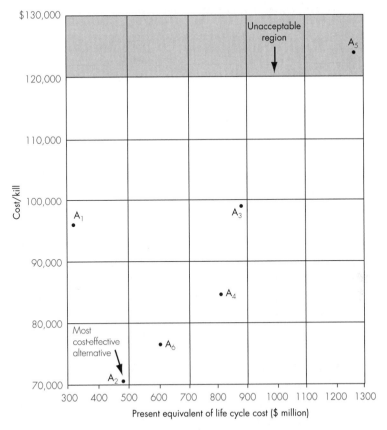

Figure 14.3 Cost effectiveness for six adverse weather guidance weapon systems

14.5 Summary

- **Benefit-cost analysis** is commonly used to evaluate public projects; several facets unique to public project analysis are neatly addressed by benefit-cost analysis:

  1. Benefits of a nonmonetary nature can be quantified and factored into the analysis.

  2. A broad range of project users distinct from the sponsor should be considered—benefits and disbenefits to *all* these users can (and should) be taken into account.

- Difficulties involved in public project analysis include the following:

  1. Identifying all the users of the project
  2. Identifying all the benefits and disbenefits of the project
  3. Quantifying all the benefits and disbenefits in dollars or some other unit of measure
  4. Selecting an appropriate **social discount rate** at which to discount benefits and costs to a present value.

- The *B/C* ratio is defined as

$$BC(i) = \frac{B}{C} = \frac{B}{P + C'}, \qquad P + C' > 0.$$

The decision rule is if $BC(i) \geq 1$, the project is acceptable.

- The net *B/C* ratio is defined as

$$B'C(i) = \frac{B - C'}{P} = \frac{B'}{P}, \qquad P > 0.$$

The net *B/C* ratio expresses the net benefit expected per dollar invested. The same decision rule applies as for the *B/C* ratio.

- When selecting among mutually exclusive alternatives based on B/C ratio, or net B/C ratio, incremental analysis must be used.

- The **cost-effectiveness method** allows us to compare projects on the basis of cost and nonmonetary effectiveness measures. We may either maximize effectiveness for a given cost criterion or minimize cost for a given effectiveness criterion.

Problems

Level 1

14.1* Which one of the following statements is incorrect?

(a) Both the PE criterion and the *B/C* ratio allows you to make a consistent accept/reject decision for a given investment as long as you use the same interest rate.

(b) As long as you use a *B/C* ratio as a base to compare mutually exclusive investment alternatives, it should be based on a *B/C* ratio on the incremental investment.

(c) When comparing mutually exclusive alternatives, using the B/C ratio may result in a different choice than if the *net* B/C ratio were used.

(d) Using the cost-effectiveness approach for comparing alternatives, maximizing effectiveness for a given cost, or minimizing cost for a given effectiveness, may result in different selections.

14.2* A city government is considering increasing the capacity of the current waste water treatment plant. The estimated financial data for the project are as follows:

Description	Data
Capital investment	$1,200,000
Project life	25 years
Incremental annual benefits	$250,000
Incremental annual costs	$100,000
Salvage value	$50,000
Discount rate	6%

Calculate the benefit-cost ratio and the net benefit-cost ratio for this expansion project.

14.3* Kingston Recreation and Parks Department is considering two mutually exclusive proposals for a new softball complex on a city-owned lot.

Alternative Design	Seating Capacity	Annual Benefits	Annual Costs	Required Investment
A1	3000	$194,000	$87,500	$800,000
A2	4000	224,000	105,000	1,000,000

The complex will be useful for 30 years and has no appreciable salvage value (regardless of seating capacity). Assuming an 8% discount rate, which of the following statements is incorrect?

(a) Select A1 because it has the largest *B/C* ratio.

(b) Select A1 because the costs are lower and it has the most benefits per seating capacity.

(c) Select A1 because it has the largest PE.

(d) Select A1 because the incremental benefits generated from A2 are not large enough to offset the additional investment ($200,000 over A1).

14.4* The province of Quebec is considering a bill that would ban the use of road salt on highways and bridges during icy conditions. Road salt is known to be toxic, costly, corrosive, and caustic. Carlon Chemical Company produces a calcium magnesium acetate de-icer (CMA) and sells it for $600 a ton as Ice-Away. Road salts, on the other hand, sold for an average of $14 a ton in 1999. Quebec needs about 600,000 tons of road salt each year. (Quebec spent $9.2 million on road salt in 1999.) Carlon estimates that each ton of salt on the road costs $650 in highway corrosion, $525 in rust on vehicles, $150 in corrosion to utility lines, and $100 in damages to water supplies, for a total of $1425. Unknown salt damage to vegetation and soil surrounding areas of highways has occurred. The province of Quebec would ban road salt (at least on expensive steel bridges or near sensitive lakes) if provincial studies support Carlon's cost claims.

(a) What would be the users' benefits and sponsor's costs if a complete ban on road salt were imposed in Quebec?

(b) How would you go about determining the salt damages (in dollars) to vegetation and soil?

Level 2

14.5 A public school board is considering the adoption of a 4-day school week as opposed to the current 5-day school week. The community is hesitant about the plan, but the superintendent of the school system envisions many benefits associated with the 4-day system, Wednesday being the "day off." The following pros and cons have been cited.

- Experiments with the 4-day system indicate that the "day off" in the middle of the week will cut down both on teacher and pupil absences.

- The longer hours on school days will require increased attention spans, which is not an appropriate expectation for younger children.

- The province bases its expenditures on its local school systems largely on the average number of pupils attending school in the system. Since the number of absences will decrease, provincial expenditures on the local school board should increase.

- Older students might want to work on Wednesdays. Unemployment is a problem in this region, however, and any influx of new job-seekers could aggravate an existing problem. Community

centers, libraries, and other public areas may also experience increased usage on Wednesdays.

- Parents who provide transportation for their children will see a savings in fuel costs. Primarily, only those parents whose children live fewer than 3 kilometres from the school would be involved. Children living more than 3 kilometres from school are eligible for free transportation provided by the school board.

- Decreases in both public and private transportation should result in fuel conservation, decreased pollution, and less wear on the roads. Traffic congestion should ease on Wednesdays in areas where congestion caused by school traffic is a problem.

- Working parents will be forced to make child-care arrangements (and possibly payments) for one weekday per week.

- Older students will benefit from wasting less time driving to and from school; Wednesdays will be available for study, thus taking the heavy demand off most nights. Bussed students will spend far less time per week waiting for buses.

- The local school board should see some ease in funding problems. The two areas most greatly impacted are the transportation system and school building operating costs.

(a) For this type of public study, what do you identify as the users' benefits and disbenefits?

(b) What items would be considered as the sponsor's costs?

(c) Discuss any other benefits or costs associated with the 4-day school week.

14.6 The Electric Department of the City of Prince George, British Columbia, operates generating and transmission facilities serving approximately 140,000 people in the city and surrounding county. The city has proposed construction of a $300 million 235-MW circulating fluidized bed combustor (CFBC) to power a turbine generator currently receiving steam from an existing boiler fueled by gas or oil. Among the advantages associated with the use of CFBC systems are the following:

- A variety of fuels can be burned, including inexpensive low-grade fuels with high ash and high sulfur content.

- The relatively low combustion temperatures inhibit the formation of nitrogen oxides. Acid-gas emissions associated with CFBC units would be expected to be significantly lower than emissions from conventional coal-fueled units.

- The sulfur-removal method, low combustion temperatures, and high-combustion efficiency characteristic of CFBC units result in solid wastes. These are physically and chemically more amenable to land disposal than the solid wastes resulting from conventional coal-burning boilers equipped with flue-gas desulfurization equipment.

Based on the Ministry of Energy and Mines projections of growth and expected market penetration, demonstration of a 235-MW unit could lead to as much as 41,000 MW of CFBC generation being constructed by the year 2010. The proposed project would reduce the city's dependency on oil and gas fuels by converting its largest generating unit to coal-fuel capability. Consequently, substantial reductions of local acid-gas emissions could be realized in comparison to the permitted emissions associated with oil fuel. The city has requested a $50 million cost share from the province. Cost sharing is considered attractive because the province's share would largely offset the risk to the utility of using such a new technology. To qualify

for the cost-sharing money, the city has to address the following the questions for the province.

(a) What is the significance of the project at local and provincial levels?
(b) What items would constitute the users' benefits and disbenefits associated with the project?
(c) What items would constitute the sponsor's costs?

By putting yourself in the city engineer's position, respond to these questions.

14.7 The following information appeared in *The New York Times*, April 12, 1990.

Predicting a doubling of traffic in the next three decades, federal highway officials are actively promoting a major program of computerization and automation that would fundamentally alter the designs of vehicles and highways. In the latest move to assist researchers, Transportation Secretary, Samuel K. Skinner, announced an $8 million project today to equip 100 cars in Orlando, Fla., with computerized displays that will receive instantaneous traffic updates and detour instructions from a traffic management center. The displays will suggest the best route to take to destinations selected by the drivers and will update the information quickly in response to accidents or traffic congestion. It is expected to go into operation in January 1992, and will be evaluated for a year. The project is being sponsored by the automobile association along with the General Motors Corporation, state and local governments, and the Department of Transportation. The federal share of the project's cost is $2.5 million.

Suppose that the experimental project proved to be successful and the city of Orlando would consider a full implementation of the computerized communication system.

(a) What would you identify as the primary elements of the users' benefits and disbenefits associated with the project?
(b) What would you identify as the secondary elements of the users' benefits and disbenefits?
(c) What would you identify as the elements of the sponsor's costs?

14.8 A city government is considering two types of town-dump sanitary systems. Design X requires an initial outlay of $400,000, with annual operating costs of $50,000 for the next 15 years; design Y calls for an investment of $300,000, with annual operating costs of $80,000 per year for the next 15 years. Fee collections from the residents would be $85,000 per year. The interest rate is 8%, and no salvage value is associated with either system.

(a) Using the net benefit-cost ratio $B'C(i)$, which system should be selected?
(b) If a new design (design Z), which requires an initial outlay of $350,000 and annual operating costs of $65,000, is proposed, would your answer in (a) change?

14.9* The Canadian Government is considering building apartments for government employees working in a foreign country and living in locally owned housing. A comparison of two possible buildings indicates the following:

	Building X	Building Y
Original investment by government agencies	$8,000,000	$12,000,000
Estimated annual maintenance costs	240,000	180,000
Savings in annual rent now being paid to house employees	1,960,000	1,320,000

Assume the salvage or sale value of the apartments to be 60% of the first investment. Use 10% discounting and a 20-year study period to compute the B/C ratio on incremental investment and make a recommendation.

14.10 Three public investment alternatives are available: A1, A2, and A3. Their respective total benefits, costs, and first costs are given in present equivalent. These alternatives have the same service life.

Present equivalent	Proposals		
	A1	A2	A3
P	100	300	200
B	400	700	500
C'	100	200	150

Assuming no do-nothing alternative, which project would you select based on the benefit-cost ratio $[BC(i)]$ on incremental investment?

14.11* A city that operates automobile parking facilities is evaluating a proposal that it erect and operate a structure for parking in a city's downtown area. Three designs for a facility to be built on available sites have been identified. (All dollar figures are in thousands.)

	Design E	Design F	Design G
Cost of site	$240	$180	$200
Cost of building	$2200	700	1400
Annual fee collection	$830	750	600
Annual maintenance cost	$410	360	310
Service life (years)	30	30	30

At the end of the estimated service life, whichever facility had been constructed would be torn down, and the land would be sold. It is estimated that the proceeds from the resale of the land will be equal to the cost of clearing the site. If the city's interest rate is known to be 10%, which design alternative would be selected based on the benefit-cost criterion?

14.12 Two different routes are under consideration for a new highway.

	Length of Highway	First Cost	Annual Upkeep
The "long" route	22 kilometres	$21 million	$140,000
Transmountain shortcut	10 kilometres	$45 million	$165,000

For either route, the volume of traffic will be 400,000 cars per year. These cars are assumed to operate at $0.25 per kilometre. Assume a 40-year life for each road and an interest rate of 10%. Determine which route should be selected.

14.13 The government is considering undertaking the four projects listed below. These projects are mutually exclusive, and the estimated present equivalent of their costs and benefits are shown in millions of dollars. All projects have the same duration.

Projects	PE of Benefits	PE of Costs
A1	40	85
A2	150	110
A3	70	25
A4	120	73

Assuming there is no do-nothing alternative, which alternative would you select? Justify your choice by using a benefit-cost [$BC(i)$] on incremental investment.

Level 3

14.14 The federal government is planning a hydroelectric project for a river basin. In addition to the production of electric power, this project will provide flood control, irrigation, and recreation benefits. The estimated benefits and costs expected to be derived from the three alternatives under consideration are listed below.

	Decision Alternatives		
	A	B	C
Initial cost	$8,000,000	$10,000,000	$15,000,000
Annual benefits or costs			
Power sales	$1,000,000	$1,200,000	$1,800,000
Flood control savings	250,000	350,000	500,000
Irrigation benefits	350,000	450,000	600,000
Recreation benefits	100,000	200,000	350,000
O&M costs	200,000	250,000	350,000

The interest rate is 10%, and the life of each of the projects is estimated to be 50 years.

(a) Find the benefit-cost ratio for each alternative.
(b) Select the best alternative based on the $BC(i)$ ratio.
(c) Select the best alternative based on the net $B'C$ ratio.

14.15 Fast growth in the city of Ottawa-Carleton and surrounding areas has resulted in increasing traffic congestion, for both vehicles and pedestrians. Aside from normal road maintenance and minor projects, the city has recently approved the allocation of $24 million to be used for major capital improvements to the road system in the capital. This extra funding allows the option of spreading that amount among many smaller projects or concentrating on fewer larger projects. The city engineers and planners were asked to prepare a priority list outlining which roads and facilities could be improved with the extra money. The engineers also computed the possible public benefits associated with each construction project; they accounted for possible reduction in travel time, a reduction in the accident rate, land appreciation, pedestrian safety, and savings in operating costs of vehicles.

District	No.	Project	Type of Improvement	Construction Cost	Annual O&M	Annual Benefits
I South	1	Hawthorne Road	4-lane	$ 980,000	$ 9,800	$ 313,600
	2	Walkley Road	New 4-lane extension	3,500,000	35,000	850,000
	3	Hunt Club Road East	4-lane	2,800,000	28,000	672,000
	4	Conroy Road	4-lane	1,400,000	14,000	490,000
II East	5	St. Joseph Boulevard	4-lane	2,380,000	47,600	523,600
	6	Place d'Orléans Station	New transitway	5,040,000	100,800	1,310,400
	7	Blackburn Hamlet	New 4-lane bypass	2,520,000	50,400	831,600
	8	Tenth Line Road	4-lane and interchange	4,900,000	98,000	1,021,000
III West	9	Baseline Road	4-lane	1,365,000	20,475	245,700
	10	Hunt Club Road West	New 4-lane extension	2,100,000	31,500	567,000
	11	Knoxdale Road	Realign	1,170,000	17,550	292,000
	12	March Road	4-lane	1,120,000	16,800	358,400
IV Central	13	Mackenzie King Bridge	Rehabilitate	2,800,000	56,000	980,000
	14	Elgin Street	Reconstruct	1,690,000	33,800	507,000
	15	Wellington Street	Reconstruct	975,000	15,900	273,000
	16	Plaza Bridge	Reconstruct	1,462,500	29,250	424,200

Assume a 20-year planning horizon, and that the municipality uses an interest rate of 10%. Which projects would be considered for funding in (a) and (b)?

(a) Due to political pressure, each district will have the same amount of funding, say $6 million.

(b) To compensate for previous imbalances in budget allocations, Districts I and II combined will get $15 million, while Districts III and IV combined will get $9 million. At least 2 properties from each district must be included.

14.16* A municipal Sanitation Department is responsible for the collection and disposal of all solid waste within the city limits. The city must collect and dispose of an average of 300 tonnes of garbage each day (7 days a week). The city is considering ways to improve the current solid-waste collection and disposal system.

• The present collection and disposal system uses Dempster Dumpmaster

Frontend Loaders for collection, and incineration or landfill for disposal. Each collecting vehicle has a load capacity of 10 tonnes, or 24 cubic metres, and dumping is automatic. The incinerator in use was manufactured in 1942. It was designed to incinerate 150 tonnes per 24 hours. A natural-gas afterburner has been added in an effort to reduce air pollution; however, the incinerator still does not meet provincial air-pollution requirements, and it is operating under a permit from the provincial Pollution Control Board. Prison-farm labor is used for the operation of the incinerator. Because the capacity of the incinerator is relatively low, some trash is not incinerated, but is taken to the city landfill. The trash landfill is located approximately 11 kilometres, and the incinerator approximately 5 kilometres, from the center of the city. The travel distance and costs in man-hours for delivery to the disposal sites is excessive; a high percentage of empty vehicle kilometres and man-hours are required because separate methods of disposal are used, and the destination sites are remote from the collection areas. The annual operating cost for the present system is $905,400. This includes $624,635 to operate the prison-farm incinerator, $222,928 to operate the existing landfill, and $57,837 to maintain the current incinerator.

- The proposed system locates a number of portable incinerators for the disposal of refuse waste within the city. Collection vehicles will also be staged at these incineration/disposal sites, with the necessary plant and support facilities for incineration operation, collection-vehicle fuelling and washing, support building for stores, and shower and locker rooms for collection and site crew personnel. The pick-up and collection procedure remains essentially the same as in the existing system. The disposal/staging sites, however, will be strategically located in the city based on the volume and location of wastes collected, thus eliminating long hauls and reducing the number of kilometres the collection vehicles must retravel from pick-up to disposal site.

Four versions of the proposed system are being considered, depending on how the required number of incinerator units are distributed among each of three available sites in the city. The type of incinerator is a modular prepackaged unit which can be installed at any one of the sites in the city. Such units meet all government standards on their exhaust emissions. The city needs 24 units, each with a rated capacity of 12.5 tonnes of garbage per 24 hours. The price per unit is $137,600, which means a capital investment of about $3,302,000. At each site, a single plant will be built, which can house up to 12 units. The capital cost of

plant facilities, annex buildings for supplies, and additional plant features such as landscaping, depend on the number of units installed at each site, because there are differences in geography and access.

The annual operating costs of the proposed system would also vary according to the distribution of incinerator units across sites. The O&M costs include the cost of electrical requirements, plant and building maintenance, and labour. Since there are fixed and variable costs associated with each of these elements, the total annual O&M costs across the system depend on which variation of the proposed system is being considered. As these centralized facilities will result in shorter transportation distances for the refuse than the present system, it is necessary to consider the savings accruing from this operating advantage. A labour savings is also realized because of the shorter routes, which permit more pick-ups during the day. The following table summarizes all costs and savings in thousands of dollars associated with the present and proposed systems.

Item	Present System	Costs for Proposed Systems			
		1	2	3	4
Capital costs					
Incinerators		$3302	$3302	$3302	$3302
Plant facilities		600	900	1260	1920
Annex buildings		91	102	112	132
Additional features		60	80	90	100
Total		$4053	$4384	$4764	$5454
Annual O&M costs	$905.4	$342	$480	$414	$408
Annual savings					
Pick-up transportation		$13.2	$14.7	$15.3	$17.1
Labor		87.6	99.3	103.5	119.4

A bond will be issued to provide the necessary capital investment at an interest rate of 8% with a maturity date 20 years in the future. The proposed systems are expected to last 20 years with negligible salvage values. If the current system is to be retained, the annual O&M costs would be expected to increase at an annual rate of 10%. The city will use the bond interest rate as the interest rate for any public project evaluation.

(a) Determine the operating cost of the current system in terms of dollars per tonne of solid waste.

(b) Determine the economics of each solid-waste disposal alternative in terms of dollars per tonne of solid waste.

14.17* Due to a rapid growth in population, a small town in British Columbia is considering several options to establish a waste water treatment facility that can handle a waste water flow of 8 million litres per day. The town has five treatment options available:

Option 1—No action: This option will lead to continued deterioration of the environment. If growth continues and pollution results, fines imposed (as high as $10,000 per day) would soon exceed construction costs.

Option 2—Land treatment facility: Provide a system for land treatment of wastewater to be generated over the next 120 years. This option will require the utilization of the most land for treatment of the wastewater. In addition to finding a suitable site, pumping of the wastewater for a considerable distance out of town will be required. The land cost in the area is $3000 per hectare. The system will use spray irrigation to distribute wastewater over the site. No more than one centimetre of wastewater can be applied in one week per hectare.

Option 3—Activated sludge-treatment facility: Provide an activated sludge-treatment facility at a site near the planning area. No pumping will be required for this alternative. Only 7 hectares of land will be needed for construction of the plant at a cost of $7000 per hectare.

Option 4—Trickling filter-treatment facility: Provide a trickling filter-treatment facility at the same site selected for the activated sludge plant of option 3. The land required will be the same as used for option 3. Both facilities will provide similar levels of treatment using different units.

Option 5—Lagoon treatment system: Utilize a three-cell lagoon system for treatment. The lagoon system requires substantially more land than options 3 and 4, but less than option 2. Due to the larger land requirement, this treatment system will have to be located some distance outside of the planning area and will require pumping of the wastewater to reach the site.

The following summarizes the capital expenditures and O&M costs associated with each option:

Option Number	Land Cost for Each Option		
	Land Required (hectares)	Land Cost ($)	
2	800	$2,400,000	
3	7	49,000	
4	7	49,000	
5	80	400,000	

The price of land is assumed to be appreciating at an annual rate of 3%.

Option Number	Capital Expenditures			
	Equipment	Structure	Pumping	Total
2	$500,000	$700,000	$100,000	$1,300,000
3	500,000	2,100,000	0	2,600,000
4	400,000	2,463,000	0	2,863,000
5	175,000	1,750,000	100,000	2,025,000

The equipment installed will require a replacement cycle of 15 years. Its replacement cost will increase at an annual rate of 5% (over the initial cost), and its salvage value at the end of the planning horizon will be 50% of the replacement cost. The structure requires replacement after 40 years and will have a salvage value of 60% of the original cost.

Option Number	Energy	Annual O&M Costs Labor	Repair	Total
2	$200,000	$95,000	$30,000	$325,000
3	125,000	65,000	20,000	210,000
4	100,000	53,000	15,000	168,000
5	50,000	37,000	5,000	92,000

The cost of energy and repair will increase at an annual rate of 5% and 2%, respectively. The labor cost will increase at an annual rate of 4%.

With the following sets of assumption, answer (a) and (b).

(a) If the interest rate (including inflation) is 10%, which option is the most cost-effective?

(b) Suppose a household discharges about 800 litres of waste water per day through the facility selected in (a). What should be the monthly assessed bill for this household?

- Assume analysis period of 120 years.

- Replacement costs for the equipment as well as pumping facilities will increase at an annual rate of 5%.

- Replacement cost for the structure will remain constant over the planning period. However, its salvage value will be 60% of the original cost. (Because it has a 40-year replacement cycle, any increase in the future replacement cost will have very little impact on the solution.)

- The equipment's salvage value at the end of its useful life will be 50% of the original replacement cost. For example, the equipment installed for option 2 will cost $500,000. It's salvage value at the end of 15 years will be $250,000.

- All O&M cost figures are given in today's dollars. For example, the annual energy cost of $200,000 for option 2 means that the actual energy cost during the first operating year will be $200,000(1.05) = $210,000.

- Option 1 is not considered a viable alternative, as its annual operating cost exceeds $36,500,00.

14.18 The U.S. Federal Highway Administration predicts that by the year 2005, Americans will be spending 8.1 billion hours per year in traffic jams. Most traffic experts believe that adding and enlarging highway systems will not alleviate the problem. As a result, current research on traffic management is focusing on three areas: (1) development of computerized dashboard navigational systems, (2) development of roadside sensors and signals that monitor and help manage the flow of traffic, and (3) development of automated steering and speed controls that might allow cars to drive themselves on certain stretches of highway.

In Los Angeles, perhaps the most traffic-congested city in the United States, a Texas Transportation Institute study found that traffic delays cost motorists $8 billion per year. But Los Angeles has already implemented a system of computerized traffic-signal controls that, by some estimates, has reduced travel time by 13.2%, fuel consumption by 12.5%, and pollution by 10%. And between Santa Monica and downtown Los Angeles, testing of an electronic traffic and navigational system—including highway sensors and cars with computerized dashboard maps—is being sponsored by federal, state, and local governments and General Motors Corporation. This test program costs $40 million; to install it throughout Los Angeles could cost $2 billion.

On a national scale, the estimates for implementing "smart" roads and vehicles is even more staggering: It would cost $18 billion to build the highways, $4 billion per year to maintain and operate them, $1 billion for research and development of driver-information aids, and $2.5 billion for vehicle-control devices. Advocates say the rewards far outweigh the costs.

(a) On a national scale, how would you identify the users' benefits and disbenefits for this type of public project?
(b) On a national scale, what would be the sponsors' cost?
(c) Suppose that the users' net benefits grow at 3% per year and the sponsor's costs grow at 4% per year. Assuming a social discount rate of 10%, what would be the B/C ratio over a 20-year study period?

APPENDIX A

Basic Financial Reports

Of the various reports corporations issue to their stockholders, the annual report is by far the most important. The annual report contains basic financial statements as well as management's opinion of the past year's operations and the firm's future prospects. There are three basic financial reports provided by any business firm: (1) the **balance sheet,** a picture of the financial status at a given point in time, (2) the **income statement,** which indicates whether the company is making or losing money during a stated period of time, and the **cash flow statement,** which details where the cash was obtained and where the cash was spent. The fiscal year can be any 12-month term. Note that the income and cash flow statements are for a period of time, often a quarter or a year. These statements measure the financial status of the company at that time point. Balance sheets are usually prepared at the end of a period for which the income and cash flow statements are prepared.

A.1 The Balance Sheet

The balance sheet contains a summary listing of the assets (items owned by the firm) and the liabilities (debts owed by the firm) on a particular date, which aligns to the end of a period for which an income statement has been prepared. One of the important features of any balance sheet statement is that the assets of the firm equal the liabilities of the firm plus the owner's equity in the business.

A.1.1   Assets

Asset items are listed in the order of current assets, fixed assets, and other assets. This ordering sequence represents the length of time it typically takes to convert them into cash. **Current assets** can be converted to cash or its equivalent in less than one year. These generally include three major accounts. The first is cash. A firm typically has a cash account at a bank to provide the funds needed to conduct day-to-day business. If a company's short-term cash position is negative, it is called short-term borrowing or credit line and is shown as a liability. The second account is accounts receivable, which is money that is owed to the firm but has not yet been paid. For example, when a company receives an order from a retail store, it sends an invoice along with the shipment to the retailer. The unpaid invoice falls into a category called "accounts receivable." As this bill is paid, it will be deducted from the accounts receivable account and placed into the cash account. A typical firm will have a 30- to 60-day accounts receivable, depending on the frequency of its bills and the payment terms for its customers. The third account is inventories, which constitute materials and supplies on hand.

Fixed assets are relatively permanent and would take time to convert into cash. The most common fixed assets include the physical investment in the business, such as land, buildings, factory machinery, office equipment, and automobiles. With the exception of land, most fixed assets have a finite useful life. For example, buildings and equipment are used up over a period of years. Each year, a portion of the usefulness of these assets expires, and a portion of their total cost should be recognized as a depreciation expense. The term **depreciation** means the accounting process for this gradual conversion of fixed assets into expenses.

Finally the balance sheet shows **other assets.** Typical assets in this category include investments made in other companies and intangible assets, such as goodwill, copyrights, and franchises. Goodwill appears on the balance sheet only when a going business is purchased in its entirety. This indicates any additional amount paid for the business above the book value of the business. (The fair market value is defined as the price that a buyer is willing to pay when the business is offered for sale. The fair market value is often larger than the book value of a business.)

A.1.2   Forms of Capital

Any business needs assets to operate. To acquire the assets, the firm must raise capital. When the firm finances its long-term needs externally, it may obtain funds from the capital markets. Capital comes in two basic forms, **debt** and **equity.** Debt capital refers to the borrowed capital from financial institutions. Equity capital refers to the capital obtained from the owners of the company.

The basic methods of **debt financing** include obtaining a bank loan and issuing bonds. Suppose, for example, that a firm needs $10,000 to purchase a computer. In this situation, the firm might borrow money from a bank and repay the loan with specified interest in a few years, a procedure known as **short-term debt financing.** Now, consider a firm that needs $100 million for a construction project. It would normally be very expensive (or require a substantial amount of mortgage) to borrow this sum directly from a bank. In this situation, the firm might go to the public to borrow money on a long-term basis. The document that records the nature of such an arrangement between a company and its investors is called a **bond.** Raising capital through issuing bonds is called a **long-term debt financing.**

Similarly, there are different types of **equity capital.** The equity of a proprietorship represents the money provided by the owner. For a corporation, equity capital comes in two forms: **preferred** and **common stock.** Investors provide capital to a corporation, and the company agrees to provide the investor with fractional ownership in the corporation. Preferred stock pays a stated **dividend,** much like the interest payment on bonds. Preferred stock has preference over common stocks with respect to the receipt of dividends and to the distribution of assets in the event of liquidation. When a company makes profits, it has to decide what to do with the profits. It can retain some of the profits for future investment and pay out the remaining profits to its common stockholders.

A.1.3 Liabilities

The liabilities of a company indicate where the funds used to acquire the assets and to operate the business are owed. Here, **liability** means an obligation to an outside party (or creditor). We usually divide liabilities into **current** and **other liabilities.** Current liabilities are those debts that must be paid in the near future (payable within one year). The major current liabilities include **accounts** and **notes payable** within a year and the "current portion" of any long-term debt, i.e., principal payments due within one year. They also include **accrued expenses** (wages, salaries, interest, rent, and taxes, owed but not yet due for payment) and advance payments and deposits from customers. Other liabilities include long-term liabilities, such as bonds, mortgages, and long-term notes, which are due and payable more than one year in the future.

A.1.4 Working Capital

Current assets and current liabilities are collectively known as **working capital.** The term **net working capital** refers to the difference between current assets and current liabilities. This figure indicates the extent to which current assets can be converted to cash to meet current obligations. Therefore, we view a firm's net working capital as a measure of its liquidity position. Here liquidity means that the capital can be converted rapidly to cash. Consequently, the more net working capital a firm has, the greater its ability to satisfy creditors' demands in the short-term.

A.1.5   Net Equity

Stockholders' equity (or **owner's equity** for a proprietorship) indicates the portion of the assets of a company that are provided by the investors (owners). Therefore, stockholders' equity is the liability of the company to its owners. This represents the book value that is available to the owners after all other debts have been paid. It generally consists of preferred and common stock, treasury stock, capital surplus, and retained earnings. **Paid-in capital** (**capital surplus**) is the amount of money received from the sale of stock in excess of the par (stated) value of the stock. **Outstanding stock** is the number of shares issued that actually are held by the public. The corporation can buy back part of its issued stock and hold it as **treasury stock.** The amount of **retained earnings** represents the cumulative net income of the firm since its beginning, less the total dividends that have been paid to stockholders. In other words, retained earnings indicate the amount of assets that have been financed by plowing profits back into the business. These retained earnings belong to the stockholders.

A.2   The Income Statement

The **income statement** is the second key financial report; it summarizes the firm's net income (revenue minus expenses) for a specified period of time. Every business prepares an annual income statement. Most businesses prepare quarterly and monthly income statements as well. The company's **accounting period** refers to the period of time covered by an income statement.

A.2.1   Reporting Format

At the top of the income statement are revenues from the business operation. **Revenue** is the money received from goods sold and services rendered during a given accounting period. **Net sales** represents the gross sales less any sales returns and allowances. Shown on the next several lines are the expenses and costs of doing business as deductions from the revenues. The largest expense for a typical manufacturing firm is its production costs, including such items as labor, materials, depreciation and overhead, known as **cost of goods sold.** Net sales less the cost of goods sold indicates the **net operating profit.** Operating expenses (such items as leases and administrative expenses) are subtracted from the net operating profit; the result is **taxable income** (or income before income taxes). Finally, we determine **net income** (or **net profit**) by subtracting **income taxes** from the taxable income. Net income is also known as **accounting income.**

Internal income statements (used by company management to monitor business performance) will separate costs and expenses into cost of goods sold, which are related to production costs and are to a large extent variable, and selling, general and administrative expenses that are, at least in the short term, usually fixed. Companies usually do not want competitors to know their fixed versus variable costs, since this information might help the competitor to design a pricing strategy. Hence, in annual reports, it is normal for all costs and expenses to be lumped into a common category, often called "cost of goods" or just "expenses."

A.2.2   Retained Earnings

As a part of the income statement, many corporations also report their retained earnings during the accounting period. When a corporation earns profits, it must decide what to do with them. The corporation may decide to pay out the profits as **dividends** to its stockholders. Or it may retain the profits in the business to finance expansion or support other business activities. When the company decides to do a combination of dividends and retained earnings, the category "available for common stockholders" reflects the net earnings of the corporation less the preferred stock dividends. When preferred and common stock dividends are subtracted from net income, the remainder is retained earnings (profits) for the year.

A.2.3   Earnings Per Share

Additional important financial information is provided in the income statement, namely, the **earnings per share** (EPS). In simple situations, we compute this by dividing the "available earnings to common stockholders" by the number of shares of common stock outstanding. Stockholders and potential investors want to know what their share of profits is, not just the total dollar amount. Presentation of profits on a per-share basis allows the stockholders to relate earnings to what they paid for a share of stock. Naturally, companies want to report a higher EPS to their investors as a means of summarizing how well they managed their businesses for the benefits of owners.

A.3 The Cash Flow Statement

The income statement explained in the previous section only indicates whether the company was making or losing money during the reporting period. Its emphasis is on determining the net income (profits) of the firm, for **operating activities.** However, the income statement ignores two other important business activities for the period—**financing** and **investing activities.** The third financial statement, the cash flow statement, details how the company generated the cash and how the cash was used during the reporting period.

A.3.1 Sources and Uses of Cash

The difference between sources (inflows) and uses (outflows) of cash represents the change in the cash position (or cash and cash equivalents) during the reporting period. The cash position at the end of the reporting period is equal to the sum of this change and the cash position at the end of the previous reporting period. This cash position at the end of a reporting period forms part of the working capital needed to cover the day-to-day operations for the next reporting period. For a company with a constant scale of operation, this amount of cash and cash equivalents is often maintained at a relatively constant level. It usually increases as the scale of operation increases. Thus, the change in cash and cash equivalents is often pretty small from year to year. This change in cash and cash equivalents sometimes is also called the net cash flow. As we have addressed throughout the book, net cash flows generated by a project should be used in assessment of the profitability of the project. However, net cash flows for the whole company are not as useful in judging the profitability of the company.

The more important information we can get from the cash flow statement is how cash flows generated from operations are used. For example, how much is paid to shareholders as dividends and how much is reinvested in the company in the form of retained earnings? How much in new investments did the company make during the reporting period? Where did the money for the new projects come from? Were they mainly through equity financing or debt financing? Has the company's debt structure changed? The answers to these questions can be found in the cash flow statements. These answers will help investors understand the management philosophy of the company and assess the health condition of the company. For example, if a company financed all new projects with debts and all or most income from operations were paid as dividends, a potential investor would doubt if this company will sustain a long-term growth.

A.3.2 Reporting Format

In preparing the cash flow statement, companies identify the sources and uses of cash according to types of business activities, specifically, operating activities, investing activities, and financing activities. We start with the net change in operating cash flows from the income statement. Here operating cash flows represent those cash flows related to production and sales of goods or services. All **noncash expenses** are simply added back to net income (or after-tax profits). For example, an expense such as depreciation is only an **accounting expense** (bookkeeping entry). While they are charged against current income as an expense, accounting expenses do not involve an actual cash outflow. The actual cash flow occurred when the asset was purchased.

Once we determine the operating cash flows, we consider any cash flow transactions related to the investment activities. Investment activities include such items as purchasing new fixed assets (cash outflow) or reselling old equipment (cash inflow). Finally, we detail any cash transactions related to financing any capital used in business. For example, the company could borrow or sell more stock, resulting in cash inflows. Paying off existing debt will result in cash outflows. By summarizing cash inflows and outflows from these activities for a given accounting period, we obtain the net changes in cash flow position of the company.

A.4 The Gillette Company's Financial Statements

Exhibits A.1, A.2, and A.3 show, respectively, the balance sheet, the income statement, and the cash flow statement obtained from The Gillette Company's 1998 Annual Report[1]. The explanations of these financial statements in the following sections derive from information in this annual report.

A.4.1 The Balance Sheet

The $11,902 million of assets shown in the balance sheet (Exhibit A.1) were necessary to support sales. Gillette obtained the bulk of the funds used to buy assets by: (1) buying on credit from their suppliers (accounts payable, $2170 million), (2) borrowing from financial institutions (long-term debts, $2256 million), (3) selling preferred and common stock to investors ($90 million and $1358 million, respectively), and (4) reinvestment into the business as reflected in the retained earnings account ($5529 million). As of the reporting date in 1998, the company had a book value of $3472 million in property, plant and equipment. The book value of these fixed assets at the end of 1997 was $3104 million. This means that after adjustment for depreciation and asset disposals, there was a net addition of $368 million in fixed assets in 1998.

From Exhibit A.1, we see that common shareholders had provided the company with total capital of $1979 million ($1358 million of common stock plus $621 million of additional paid-in capital). Since the company was incorporated, total accumulated earnings reinvested in the business amounted to $5529 million by the end of 1998. Since the total accumulated earnings reinvested in the business were $5021 million at the end of 1997, the retained earnings from the operations in 1998 were $508 million ($5529 million – $5021 million). These retained earnings also belonged to the common shareholders. Therefore, the common shareholders had a total investment of $7508 million ($1979 million plus $5529 million). The weighted average number of common shares outstanding (basic) in 1998 was 1117 million

---

[1] Source: The Gillette Company's 1998 Annual Report. Courtesy of Gillette Company.

shares (Exhibit A.2). Thus, the total investment value per common share was $6.72 (= $7508 million/1117 million shares). This figure, $6.72/share, is known as the **stock's book value**. In the year 1998, the company's stock traded in the general range of $35 to $65 per share. Note that this market price was different from the stock's book value. This difference represents the goodwill of the company if it is bought out. For example, if Gillette were bought out for $50 per share, the goodwill would be $43.28 per share ($50 − $6.72). Many factors affect the market price; the most important is how investors expect the company to do in the future. Certainly, the company's Sensor razor project and the actual performance of the project has had a major influence on the market value of its stock.

A.4.2   Income Statement

The company's income statement for 1998 is shown in Exhibit A.2. We can see that the net sales were $10,056 million in 1998, compared with $10,062 million in 1997, a slight decline of 0.06%. However, profits from operations dropped by 23% to $1789 million, compared to $2324 million in 1997. As explained in the annual report, this decline in profits was mainly because of the expenses of $535 million in company reorganization and realignment. The dividends paid to holders of preferred shares were $4 million in the year. Thus, the net income for common shareholders was $1077 million after the preferred stock dividends were paid. The net income per common share (basic) was $0.96 (=$1077/1117). The declared dividends for 1998 were $0.51/share. Thus, the total dividends paid out in 1998 were approximately $0.51/share × 1117 million shares = $570 million and the total retained earnings from this year's operation were approximately $507 million (= $1077 million − $570 million). This amount of retained earnings for 1998 was roughly equal to the amount we calculated from the balance sheet in Section A.4.1.

As we noted previously, an income statement provides information on how the firm's operations affect retained earnings; it does not say, however, how and where these cash flows were invested. In addition, it ignores the inflows and outflows from nonoperating activities. The cash flow statement is a tool for overcoming these deficiencies in the basic financial statements.

A4.3   Cash Flow Statement

Exhibit A.3 contains The Gillette Company's cash flow statement for 1998. As shown in the exhibit, cash flows from operations amounted to $1289 million. Note that this is significantly less than that found under the shorthand rule for calculating net cash flow [that is, adding back depreciation and amortization expenses ($459 million) to net income, or $1081 + $459 = $1540 million]. The main reason for the difference is that the changes in noncash working-capital items have been taken into account in the cash flow statement. To put this in another way, some of the cash generated from operations was reused by operations to carry higher levels of inventory and receivables less payables and accrued expenses. For example, the accounts receivable ($435 million) represented the amount of total sales on credit. Since this figure was included in the total sales in determining the net income, we need to subtract this figure to determine the true cash amount generated from operations. Similar adjustments were necessary for inventories, accounts payable, and other working capital items as summarized in Exhibit A.3.

For investing activities there was a net cash outflow of $798 million consisting of new investments in plant and equipment of $1000 million and $91 million in new businesses which were offset by cash inflows of $293 million associated with the disposal of property, plant and equipment, sale of businesses, and other cash recovery. On the other hand, financing activities produced a net outflow of $492 million. This net number is the result of expenditures of $1066 million in purchase of treasury stocks, $252 million in funding of German pension plans, and $552 million in cash dividends, which are partially offset by $708 million in cash inflows through increase of loans payable and additional $500 million of long-term debt. Finally, there was the effect of exchange-rate changes on cash at foreign subsidiaries. This amounted to a net decrease of $2 million. Together, the cash flows from the three types of business activities generated a negative cash flow of $3 million. The cash position at the end of 1997 was $105 million and at the end of 1998 it became $102 million. This net decrease of $3 million denotes the change in The Gillette Company's cash position as shown in the cash accounts in the balance sheet. Although this is already known from the balance sheet, one cannot explain this change in cash without the aid of a cash flow statement.

Exhibit A.1

The Gillette Company and Subsidiary Companies Consolidated Balance Sheet[2]
(Millions of dollars)

December 31, 1998 and 1997	1998	1997
Assets		
Current Assets		
Cash and cash equivalents	**$102**	$105
Receivables, less allowances: 1998—$79; 1997—$74	**2,943**	2,522
Inventories	**1,595**	1,500
Deferred income taxes	**517**	320
Other current assets	**283**	243
Total Current Assets	**5,440**	4,690
Property, Plant and Equipment, at cost less accumulated depreciation	**3,472**	3,104
Intangible Assets, less accumulated amortization	**2,448**	2,423
Other Assets	**542**	647
	$11,902	$10,864
Liabilities and Stockholders' Equity		
Current Liabilities		
Loans payable	**$981**	$552
Current portion of long-term debt	**9**	9
Accounts payable and accrued liabilities	**2,170**	1,794
Income taxes	**318**	286
Total Current Liabilities	**3,478**	2,641
Long-Term Debt	**2,256**	1,476
Deferred Income Taxes	**411**	359
Other Long-Term Liabilities	**898**	1,101
Minority Interest	**39**	39
Contingent Redemption Value of Common Stock Put Options	**277**	407
Stockholders' Equity		
8.0% Cumulative Series C ESOP Convertible Preferred, without par value,		
Issued: 1998—148,627 shares; 1997—154,156 shares	**90**	93
Unearned ESOP compensation	**(10)**	(17)
Common stock, par value $1 per share		
Authorized: 2,320,000,000 shares		
Issued: 1998—1,357,913,938 shares; 1997—1,352,581,842 shares	**1,358**	1,353
Additional paid-in capital	**621**	309
Earnings reinvested in the business	**5,529**	5,021
Accumulated other comprehensive income		
Foreign currency translation	**(826)**	(790)
Pension adjustment	**(47)**	(20)
Treasury stock, at cost:		
1998—252,507,187 shares; 1997—231,643,130 shares	**(2,172)**	(1,108)
Total Stockholders' Equity	**4,543**	4,841
	$11,902	$10,864

[2] Courtesy of The Gillette Company

Exhibit A.2

The Gillette Company and Subsidiary Companies
Consolidated Statement of Income[3]
(Millions of dollars, except per share amounts)

Years Ended December 31, 1998, 1997 and 1996	1998	1997	1996
Net Sales	**$10,056**	$10,062	$9,698
Cost of Sales	**3,853**	3,831	3,682
Gross Profit	**6,203**	6,231	6,016
Selling, General and Administrative Expenses	**3,879**	3,907	3,967
Reorganization and Realignment Expenses	**535**	—	—
Merger-Related Costs	**—**	—	413
Profit from Operations	**1,789**	2,324	1,636
Nonoperating Charges (Income)			
Interest income	**(8)**	(9)	(10)
Interest expense	**94**	78	77
Other charges — net	**34**	34	44
	120	103	111
Income before Income Taxes	**1,669**	2,221	1,525
Income Taxes	**588**	794	576
Net Income	**$ 1,081**	$1,427	$ 949
Net Income per Common Share, basic	**$.96**	$1.27	$ .85
Net Income per Common Share, assuming full dilution	**$.95**	$1.24	$ .83
Weighted average number of common shares outstanding (millions)			
Basic	**1,117**	1,118	1,107
Assuming full dilution	**1,144**	1,148	1,140

[3] Courtesy of The Gillette Company.

Exhibit A.3

The Gillette Company and Subsidiary Companies
Consolidated Statement of Cash Flows[4]
(Millions of dollars)

Years Ended December 31, 1998, 1997 and 1996	1998	1997	1996
Operating Activities			
Net income	$1,081	$1,427	$949
Adjustments to reconcile net income to net cash provided by operating activities:			
Provision for reorganization and realignment	535	—	—
Merger-related costs	—	—	413
Depreciation and amortization	459	422	381
Other	(46)	(23)	—
Changes in assets and liabilities, net of effects from acquisition of businesses:			
Accounts receivable	(435)	(340)	(459)
Inventories	(123)	(157)	(105)
Accounts payable and accrued liabilities	72	29	67
Other working capital items	(121)	80	(227)
Other noncurrent assets and liabilities	(133)	(158)	(11)
Net cash provided by operating activities	1,289	1,280	1,008
Investing Activities			
Additions to property, plant and equipment	(1,000)	(973)	(830)
Disposals of property, plant and equipment	88	59	41
Acquisition of businesses, less cash acquired	(91)	(3)	(299)
Sale of business	200	—	—
Other	5	12	(1)
Net cash used in investing activities	(798)	(905)	(1,089)
Financing Activities			
Purchase of treasury stock	(1,066)	(53)	(11)
Proceeds from sale of put options	56	27	—
Proceeds from exercise of stock option and purchase plans	126	210	150
Funding German pension plans	(252)	—	—
Proceeds from long-term debt	500	300	—
Decrease in long-term debt	(12)	(6)	(165)
Increase (decrease) in loans payable	708	(383)	578
Dividends paid	(552)	(466)	(451)
Net cash provided by (used in) financing activities	(492)	(371)	101
Effect of Exchange Rate Changes on Cash	(2)	(7)	(19)
Net Cash from Harmonization Period	—	24	—
Increase (Decrease) in Cash and Cash Equivalents	(3)	21	1
Cash and Cash Equivalents at Beginning of Year	105	84	83
Cash and Cash Equivalents at End of Year	$102	$105	$84
Supplemental disclosure of cash paid for:			
Interest	$120	$101	$94
Income taxes	$478	$451	$586
Noncash investing and financing activities:			
Acquisition of businesses			
Fair value of assets acquired	$100	$3	$361
Cash paid	91	3	300
Liabilities assumed	$9	$—	$61

[4] Courtesy of The Gillette Company.

Electronic Spreadsheets: Summary of Built-In Financial Functions

Most financial functions are available in electronic spreadsheets, namely, Excel, Lotus 123, and QuattroPro. As you will see, the keystrokes and choices are similar for all three. Each of these spreadsheet packages now supports dozens of financial functions. A selection of the most common ones is presented here.

B.1 Nominal Versus Effective Interest Rates

A conversion from a nominal interest rate to an effective interest rate (or vice versa) is easily obtained with spreadsheets.

Function Description	Excel	Lotus 123	QuattroPro
Effective interest rate	=EFFECT(**nominal_rate, npery**)		@EFFECT (**NomRate, Nper**)
Nominal interest rate	=NOMINAL (**effect_rate,npery**)		@NOMINAL(**EffectRate, Nper**)

Effect_rate is the effect interest rate; **npery** is the number of compounding periods per year; **Nominal_rate** is the nominal interest rate.

B.2 Single-Sum Compounding Functions

The single-sum compounding functions deal with either the single-payment compound amount factor or the present worth factor. With these functions, the future amount when the present single-sum is given, present amount when its future amount is specified, or unknown interest (growth) rate and the number of interest periods can be compounded.

Function Description	Excel	Lotus 123	QuattroPro
1. Calculating the future worth of a single payment	=FV(**i, N, O, P**)	@FVAMOUNT(**P, i, N**)	@FV(**Pmt, Rate, Nper**)
2. Calculating the present worth of a single payment	=PV(**i, N, O, F**)	@PVAMOUNT(**F, i, N**)	
3. Calculating an unknown number of payment periods (N)	=NPER(**i, O, P ,F**)	@CTERM(**i, F, P**)	@CTERM(**i, F, P**)
4. Calculating an unknown interest rate	+RATE(**N, O, P, F, O,** guess)	@RATE(**F, P, N**)	@RATE(**F, P, N**)

i is the interest rate, *N* is the number of the interest periods; *F* is the future worth specified at the end of period *N*, *P* is the present worth at period 0, **Guess** is your estimate of interest rate.

B.3 Annuity Functions

Annuity functions provide a full range capability to calculate the future value of an annuity, the present value of an annuity, and the interest rate used in an annuity payment. If a cash payment is made at the end of the period, it is an **ordinary annuity.** If a cash payment is made at the beginning of a period, it is known as an **annuity due**. When using annuity functions, you can specify the choice of annuity by setting type parameters; if type = 0, or is omitted, it is an ordinary annuity. If type = 1, it is an annuity due. In annuity functions, the parameters in bold face must be specified by the user.

Function Description	Excel	Lotus 123	QuattroPro
1. Calculating the number of payment periods (N) in an annuity	=NPER(i, A, O, F, type)	@TERM(A, i, F)	@TERM(A, i, F)
2. Calculating the number of payment periods (N) of an annuity when its present worth is specified	=NPER(i, A, P, F, type)	@NPER(A, i, F, type, P)	@NPER(i, A, P, F, type)
3. Calculating the future worth of an annuity (equal-payment series compound amount factor)	=FV(i, N, A)	@FV(A, i, N)	@FV(A, i, N)
4. Calculating the future worth of an annuity when its present worth is specified	=FV(i, N, A, P, type)	@FVAL(A, i, N, type, P)	@FVAL(i, N, A, P, type)
5. Calculating the periodic equal payments of an annuity (capital recovery factor)	=PMT(i, N, P)	@PMT(P, i, N)	@PMT(P, i, N)
6. Calculating the periodic equal payment of an annuity when its future worth is specified	=PMT(i, N, P, F, type)	@PAYMT(P, i, N, type, F)	@PAYMT(i, N, P, F, type)
7. Calculating the present worth of an annuity (equal-payment series present worth factor)	=PV(i, N, A)	@PV(A, i, N)	@PV(A, i, N)
8. Calculating the present worth of an annuity when its optional future worth is specified	=PV(i, N, A, F, type)	@PVAL(A, i, N, type, F)	@PVAL(i, N, A, F, type)
9. Calculating the interest rate used in an annuity	=RATE(N, A, P, F, type, guess)	@IRATE(N, A, P, type, F, guess)	@IRATE(N, A, P, F, type)

i is the interest rate per period; A is the payment made each period and cannot change over the life of the annuity; N is the total number of payment periods in an annuity; P is the present value, or the lump-sum (starting) amount that series of future payments is worth right now; F is the future value, or a cash balance you want to attain after the last payment is made. If F is omitted, it is assumed to be 0 (the future value of a loan, for example, is 0); *Type* is the number of 0 or 1 and indicates when payments are due. If type is omitted, it is assumed to be 0.

Set *type* equal to	If payments are due
0	At the end of the period
1	At the beginning of the period

Guess is your estimate of the interest rate.

B.4 Loan Analysis Functions

When you need to compute the monthly payments, interest and principal payments, several commands are available to facilitate a typical loan analysis.

Function Description	Excel	Lotus 123	QuattroPro
1. Calculating the periodic loan payment size (*A*)	=PMT(*i*, *N*, *P*, type)	@PMT(*P*, *i*, *N*) @PAYMT(*P*, *i*, *N*, type, *F*)	@PMT(*P*, *i*, *N*)
2. Calculating the portion of loan interest payment for a given period *n*	=IPMT(*i*, *n*, *N*, *P*, *F*, type)	@IPAYMT(*P*, *i*, *N*, *n*)	@IPAYMT(*i*, *n*, *N*, *P*, *F*, type)
3. Calculating the cumulative interest payment between two interval periods	=CUMIMPT(*i*, *N*, *P*, *start*, *end*, type)	@IPAYMT(*P*, *i*, *N*, *n*, *start*, *end*, type)	@AMAINT(*P*, *i*, *N*, *n*)
4. Calculating portion of loan principal payment for a given period *n*	=PPMT(*i*, *n*, *N*, *P*, *F*, type)	@PPAYMT(*P*, *i*, *N*, *n*)	@PPAYMT(*i*, *n*, *N*, *P*, *F*, type)
5. Calculating the cumulative principal payment between two interval periods	=CUMPRINC(*i*, *N*, *P*, *start*, *end*, type)	@PPAYMT(*P*, *i*, *N*, *n*, *start*, *end*, type)	@CUMPRINC(*i*, *N*, *P*, *start*, *end*, type)

i is the interest rate; *N* is the total number of payment periods; *P* is the present value; **start** is the first period in the calculation. Payment periods are numbered beginning with 1; **End** is the last period in the calculation; *Type* is the timing of the payment.

Type	Timing
0	Payment at the end of the period
1	Payment at the beginning of the period

B.5 Bond Functions

Several financial functions are available to evaluate investments in bonds. In particular, the yield calculation at bond maturity and bond pricing decisions can be easily made.

Function Description	Excel	Lotus 123	QuattroPro
1. Calculating accrued interest	=ACCRINT(**issue**, **first_interest**, **settlement**, **rate**, par, frequency, basis)	@ACCRUED(**settlement**, **maturity**, **coupon**, par, frequency, basis)	@ACCRINT(**settlement**, **maturity**, **coupon**, **issue**, **firstcpn**, par, frequency, calendar)
2. Calculating bond price	=PRICE (**settlement**, **maturity**, **rate**, **yield**, redemption, frequency, basis)	@ACCRUED(**settlement**, **maturity**, **coupon**, **yield**, redemption, frequency, basis)	@PRICE(**settlement**, **maturity**, **coupon**, **yield**, redemption, frequency, calendar)
3. Calculating maturity yield	=YIELD(**settlement**, **maturity**, **rate**, **price**, redemption, frequency, basis)	@YIELD(**settlement**, **maturity**, **coupon**, **price**, redemption, frequency, basis)	@YIELD(**settlement**, **maturity**, **coupon**, **price**, issue, calendar)

Settlement is the security's settlement date, expressed as a serial date number. Use *NOW* command to convert a date to a serial number; *Maturity* is the security's maturity date, expressed as a serial date number; *First_interest* is the security's first interest date, expressed as a serial number; *Coupon* is the security's annual coupon rate; *Issue* is a number representing the issue date; *Par* is the security's par value. If you omit par, ACCRINT uses $1000; *Price* is the security's price per $100 face value; *Redemption* is the security's redemption value per $100 face value; *Firstcpn* is a number representing the first coupon date. *Frequency* is the number of coupon payments per year. For annual payments, frequency = 1; for semi-annual payment, frequency = 2; *Yield* is the security's annual yield; *Basis* (or calendar) is the type of day count basis to use:

Basis (or Calendar)	Day count basis
0 or omitted	NASD* 30/360
1	Actual/actual
2	Actual/360
3	Actual/365
4	European 30/360

*National (U.S.) Association of Securities Dealers

B.6 Project Evaluation Tools

Several measures of investment worth are available to calculate the PE, IRR, and annual equivalent of a project's cash flow series.

Function Description	Excel	Lotus 123	QuattroPro
1. Net present equivalent calculation	=NPV(*i*, *range*)	@NPV(*i*, *range*)	@NPV(*i*, *range*, *type*)
2. Rate of return calculation	=IRR(*range*, *guess*)	@IRR(*guess*, *range*)	@IRR(*guess*, *range*)
3. Annual equivalent calculation	=PMT[*i*, *N*, *NPV* (*i*, *range*), type]	@PMT[*NPV* (*i*, *range*), *i*, *N*]	@PMT[*NPV*(*i*, *range*), *i*, *N*]

i is the minimum attractive rate of return (MARR); *Range* is the cell address where the cash flow streams are stored; *Guess* is the estimated interest rate in solving IRR.

B.7 Depreciation Functions

Function Description	Excel	Lotus 123	QuattroPro
1. Straight-line method	=SLN(*cost*, *salvage*, *life*)	@SLN(*cost*, *salvage*, *life*)	@SLN(*cost*, *salvage*, *life*)
2 Declining balance method	=DB(*cost*, *salvage*, *life*, *period*, *month*)		@DB(*cost*, *salvage*, *life*, *period*, *month*)
3 Double declining balance method: Calculates 200% declining balance depreciation	=DDB(*cost*, *salvage*, *life*, *period*, month)	@DDB(*cost*, *salvage*, *life*, *period*)	@DDB(*cost*, *salvage*, *life*, *period*)
4 Variable declining balance method: Calculates the depreciation by using the variable-rate declining balance method	=VDB(*cost*, *salvage*, *life*, *start*, *end*, factor, no_switch)	@VDB(*cost*, *salvage*, *life*, *start*, *end*, depreciation_percent, switch)	@VDB(*cost*, *salvage*, *life*, *start*, *end*, *factor*, *switch*)
5 Sum of the years' digits' method: Calculates sum of the years' digits depreciation	=SYD(*cost*, *salvage*, *life*, *period*)	@SYD(*cost*, *salvage*, *life*, *period*)	@SYD(*cost*, *salvage*, *life*, *period*)

cost is the initial cost (cost basis) of the asset; *Salvage* is the value at the end of the depreciable life; *Life* is the number of periods over which the asset is being depreciated (known as tax life or depreciable life); *Month* is the number of months in the first year; *Period* is the period for which you want to calculate the depreciation; *Factor* (or *depreciation percent*) is the rate at which the balance declines. If factor is omitted, it is assumed to be 2 (double-declining balance method). For 150% declining-balance method, enter factor = 1.5; *Start* is the starting period for which you want to calculate the depreciation; *End* is the ending period for which you want to calculate the depreciation; *No_switch* is a logical value specifying whether to switch to straight-line depreciation even when depreciation is greater than the declining balance method. If *no_switch* (or *switch* for Lotus) is TRUE, it does not switch to straight-line even when the depreciation is greater than the declining balance calculation; If *no_switch* is FALSE or omitted, it switches to straight-line depreciation when depreciation is greater than the declining balance calculation.

APPENDIX C

Interest Factors for Discrete Compounding

Table C.1 Interest Rate Factors (0.25%)

	Single Payment		Equal Payment Series				Gradient Series		
N	Compound Amount Factor (F/P,i,N)	Present Worth Factor (P/F,i,N)	Compound Amount Factor (F/A,i,N)	Sinking Fund Factor (A/F,i,N)	Present Worth Factor (P/A,i,N)	Capital Recovery Factor (A/P,i,N)	Gradient Uniform Series (A/G,i,N)	Gradient Present Worth (P/G.i.N)	N
1	1.0025	0.9975	1.0000	1.0000	0.9975	1.0025	0.0000	0.0000	1
2	1.0050	0.9950	2.0025	0.4994	1.9925	0.5019	0.4994	0.9950	2
3	1.0075	0.9925	3.0075	0.3325	2.9851	0.3350	0.9983	2.9801	3
4	1.0100	0.9901	4.0150	0.2491	3.9751	0.2516	1.4969	5.9503	4
5	1.0126	0.9876	5.0251	0.1990	4.9627	0.2015	1.9950	9.9007	5
6	1.0151	0.9851	6.0376	0.1656	5.9478	0.1681	2.4927	14.8263	6
7	1.0176	0.9827	7.0527	0.1418	6.9305	0.1443	2.9900	20.7223	7
8	1.0202	0.9802	8.0704	0.1239	7.9107	0.1264	3.4869	27.5839	8
9	1.0227	0.9778	9.0905	0.1100	8.8885	0.1125	3.9834	35.4061	9
10	1.0253	0.9753	10.1133	0.0989	9.8639	0.1014	4.4794	44.1842	10
11	1.0278	0.9729	11.1385	0.0898	10.8368	0.0923	4.9750	53.9133	11
12	1.0304	0.9705	12.1664	0.0822	11.8073	0.0847	5.4702	64.5886	12
13	1.0330	0.9681	13.1968	0.0758	12.7753	0.0783	5.9650	76.2053	13
14	1.0356	0.9656	14.2298	0.0703	13.7410	0.0728	6.4594	88.7587	14
15	1.0382	0.9632	15.2654	0.0655	14.7042	0.0680	6.9534	102.2441	15
16	1.0408	0.9608	16.3035	0.0613	15.6650	0.0638	7.4469	116.6567	16
17	1.0434	0.9584	17.3443	0.0577	16.6235	0.0602	7.9401	131.9917	17
18	1.0460	0.9561	18.3876	0.0544	17.5795	0.0569	8.4328	148.2446	18
19	1.0486	0.9537	19.4336	0.0515	18.5332	0.0540	8.9251	165.4106	19
20	1.0512	0.9513	20.4822	0.0488	19.4845	0.0513	9.4170	183.4851	20
21	1.0538	0.9489	21.5334	0.0464	20.4334	0.0489	9.9085	202.4634	21
22	1.0565	0.9466	22.5872	0.0443	21.3800	0.0468	10.3995	222.3410	22
23	1.0591	0.9442	23.6437	0.0423	22.3241	0.0448	10.8901	243.1131	23
24	1.0618	0.9418	24.7028	0.0405	23.2660	0.0430	11.3804	264.7753	24
25	1.0644	0.9395	25.7646	0.0388	24.2055	0.0413	11.8702	287.3230	25
26	1.0671	0.9371	26.8290	0.0373	25.1426	0.0398	12.3596	310.7516	26
27	1.0697	0.9348	27.8961	0.0358	26.0774	0.0383	12.8485	335.0566	27
28	1.0724	0.9325	28.9658	0.0345	27.0099	0.0370	13.3371	360.2334	28
29		0.9301	30.0382	0.0333	27.9400	0.0358	13.8252	386.2776	29
30		0.9278	31.1133	0.0321	28.8679	0.0346	14.3130	413.1847	30
31		0.9255	32.1911	0.0311	29.7934	0.0336	14.8003	440.9502	31
32		0.9232	33.2716	0.0301	30.7166	0.0326	15.2872	469.5696	32
33		0.9209	34.3547	0.0291	31.6375	0.0316	15.7736	499.0386	33
34		0.9186	35.4406	0.0282	32.5561	0.0307	16.2597	529.3528	34
35	1.09	0.9163	36.5292	0.0274	33.4724	0.0299	16.7454	560.5076	35
36	1.0941	0.9140	37.6206	0.0266	34.3865	0.0291	17.2306	592.4988	36
40	1.1050	0.9050	42.0132	0.0238	38.0199	0.0263	19.1673	728.7399	40
48	1.1273	0.8871	50.9312	0.0196	45.1787	0.0221	23.0209	1040.0552	48
50	1.1330	0.8826	53.1887	0.0188	46.9462	0.0213	23.9802	1125.7767	50
60	1.1616	0.8609	64.6467	0.0155	55.6524	0.0180	28.7514	1600.0845	60
72	1.1969	0.8355	78.7794	0.0127	65.8169	0.0152	34.4221	2265.5569	72
80	1.2211	0.8189	88.4392	0.0113	72.4260	0.0138	38.1694	2764.4568	80
84	1.2334	0.8108	93.3419	0.0107	75.6813	0.0132	40.0331	3029.7592	84
90	1.2520	0.7987	100.7885	0.0099	80.5038	0.0124	42.8162	3446.8700	90
96	1.2709	0.7869	108.3474	0.0092	85.2546	0.0117	45.5844	3886.2832	96
100	1.2836	0.7790	113.4500	0.0088	88.3825	0.0113	47.4216	4191.2417	100
108	1.3095	0.7636	123.8093	0.0081	94.5453	0.0106	51.0762	4829.0125	108
120	1.3494	0.7411	139.7414	0.0072	103.5618	0.0097	56.5084	5852.1116	120
240	1.8208	0.5492	328.3020	0.0030	180.3109	0.0055	107.5863	19398.9852	240
360	2.4568	0.4070	582.7369	0.0017	237.1894	0.0042	152.8902	36263.9299	360

Table C.2 Interest Rate Factors (0.50%)

	Single Payment		Equal Payment Series				Gradient Series		
N	Compound Amount Factor (F/P,i,N)	Present Worth Factor (P/F,i,N)	Compound Amount Factor (F/A,i,N)	Sinking Fund Factor (A/F,i,N)	Present Worth Factor (P/A,i,N)	Capital Recovery Factor (A/P,i,N)	Gradient Uniform Series (A/G,i,N)	Gradient Present Worth (P/G.i.N)	N
1	1.0050	0.9950	1.0000	1.0000	0.9950	1.0050	0.0000	0.0000	1
2	1.0100	0.9901	2.0050	0.4988	1.9851	0.5038	0.4988	0.9901	2
3	1.0151	0.9851	3.0150	0.3317	2.9702	0.3367	0.9967	2.9604	3
4	1.0202	0.9802	4.0301	0.2481	3.9505	0.2531	1.4938	5.9011	4
5	1.0253	0.9754	5.0503	0.1980	4.9259	0.2030	1.9900	9.8026	5
6	1.0304	0.9705	6.0755	0.1646	5.8964	0.1696	2.4855	14.6552	6
7	1.0355	0.9657	7.1059	0.1407	6.8621	0.1457	2.9801	20.4493	7
8	1.0407	0.9609	8.1414	0.1228	7.8230	0.1278	3.4738	27.1755	8
9	1.0459	0.9561	9.1821	0.1089	8.7791	0.1139	3.9668	34.8244	9
10	1.0511	0.9513	10.2280	0.0978	9.7304	0.1028	4.4589	43.3865	10
11	1.0564	0.9466	11.2792	0.0887	10.6770	0.0937	4.9501	52.8526	11
12	1.0617	0.9419	12.3356	0.0811	11.6189	0.0861	5.4406	63.2136	12
13	1.0670	0.9372	13.3972	0.0746	12.5562	0.0796	5.9302	74.4602	13
14	1.0723	0.9326	14.4642	0.0691	13.4887	0.0741	6.4190	86.5835	14
15	1.0777	0.9279	15.5365	0.0644	14.4166	0.0694	6.9069	99.5743	15
16	1.0831	0.9233	16.6142	0.0602	15.3399	0.0652	7.3940	113.4238	16
17	1.0885	0.9187	17.6973	0.0565	16.2586	0.0615	7.8803	128.1231	17
18	1.0939	0.9141	18.7858	0.0532	17.1728	0.0582	8.3658	143.6634	18
19	1.0994	0.9096	19.8797	0.0503	18.0824	0.0553	8.8504	160.0360	19
20	1.1049	0.9051	20.9791	0.0477	18.9874	0.0527	9.3342	177.2322	20
21	1.1104	0.9006	22.0840	0.0453	19.8880	0.0503	9.8172	195.2434	21
22	1.1160	0.8961	23.1944	0.0431	20.7841	0.0481	10.2993	214.0611	22
23	1.1216	0.8916	24.3104	0.0411	21.6757	0.0461	10.7806	233.6768	23
24	1.1272	0.8872	25.4320	0.0393	22.5629	0.0443	11.2611	254.0820	24
25	1.1328	0.8828	26.5591	0.0377	23.4456	0.0427	11.7407	275.2686	25
26	1.1385	0.8784	27.6919	0.0361	24.3240	0.0411	12.2195	297.2281	26
27	1.1442	0.8740	28.8304	0.0347	25.1980	0.0397	12.6975	319.9523	27
28	1.1499	0.8697	29.9745	0.0334	26.0677	0.0384	13.1747	343.4332	28
29	1.1556	0.8653	31.1244	0.0321	26.9330	0.0371	13.6510	367.6625	29
30	1.1614	0.8610	32.2800	0.0310	27.7941	0.0360	14.1265	392.6324	30
31	1.1672	0.8567	33.4414	0.0299	28.6508	0.0349	14.6012	418.3348	31
32	1.1730	0.8525	34.6086	0.0289	29.5033	0.0339	15.0750	444.7618	32
33	1.1789	0.8482	35.7817	0.0279	30.3515	0.0329	15.5480	471.9055	33
34	1.1848	0.8440	36.9606	0.0271	31.1955	0.0321	16.0202	499.7583	34
35	1.1907	0.8398	38.1454	0.0262	32.0354	0.0312	16.4915	528.3123	35
36	1.1967	0.8356	39.3361	0.0254	32.8710	0.0304	16.9621	557.5598	36
40	1.2208	0.8191	44.1588	0.0226	36.1722	0.0276	18.8359	681.3347	40
48	1.2705	0.7871	54.0978	0.0185	42.5803	0.0235	22.5437	959.9188	48
50	1.2832	0.7793	56.6452	0.0177	44.1428	0.0227	23.4624	1035.6966	50
60	1.3489	0.7414	69.7700	0.0143	51.7256	0.0193	28.0064	1448.6458	60
72	1.4320	0.6983	86.4089	0.0116	60.3395	0.0166	33.3504	2012.3478	72
80	1.4903	0.6710	98.0677	0.0102	65.8023	0.0152	36.8474	2424.6455	80
84	1.5204	0.6577	104.0739	0.0096	68.4530	0.0146	38.5763	2640.6641	84
90	1.5666	0.6383	113.3109	0.0088	72.3313	0.0138	41.1451	2976.0769	90
96	1.6141	0.6195	122.8285	0.0081	76.0952	0.0131	43.6845	3324.1846	96
100	1.6467	0.6073	129.3337	0.0077	78.5426	0.0127	45.3613	3562.7934	100
108	1.7137	0.5835	142.7399	0.0070	83.2934	0.0120	48.6758	4054.3747	108
120	1.8194	0.5496	163.8793	0.0061	90.0735	0.0111	53.5508	4823.5051	120
240	3.3102	0.3021	462.0409	0.0022	139.5808	0.0072	96.1131	13415.5395	240
360	6.0226	0.1660	1004.5150	0.0010	166.7916	0.0060	128.3236	21403.3041	360

Table C.3 Interest Rate Factors (0.75%)

	Single Payment		Equal Payment Series				Gradient Series		
N	Compound Amount Factor (F/P,i,N)	Present Worth Factor (P/F,i,N)	Compound Amount Factor (F/A,i,N)	Sinking Fund Factor (A/F,i,N)	Present Worth Factor (P/A,i,N)	Capital Recovery Factor (A/P,i,N)	Gradient Uniform Series (A/G,i,N)	Gradient Present Worth (P/G.i.N)	N
1	1.0075	0.9926	1.0000	1.0000	0.9926	1.0075	0.0000	0.0000	1
2	1.0151	0.9852	2.0075	0.4981	1.9777	0.5056	0.4981	0.9852	2
3	1.0227	0.9778	3.0226	0.3308	2.9556	0.3383	0.9950	2.9408	3
4	1.0303	0.9706	4.0452	0.2472	3.9261	0.2547	1.4907	5.8525	4
5	1.0381	0.9633	5.0756	0.1970	4.8894	0.2045	1.9851	9.7058	5
6	1.0459	0.9562	6.1136	0.1636	5.8456	0.1711	2.4782	14.4866	6
7	1.0537	0.9490	7.1595	0.1397	6.7946	0.1472	2.9701	20.1808	7
8	1.0616	0.9420	8.2132	0.1218	7.7366	0.1293	3.4608	26.7747	8
9	1.0696	0.9350	9.2748	0.1078	8.6716	0.1153	3.9502	34.2544	9
10	1.0776	0.9280	10.3443	0.0967	9.5996	0.1042	4.4384	42.6064	10
11	1.0857	0.9211	11.4219	0.0876	10.5207	0.0951	4.9253	51.8174	11
12	1.0938	0.9142	12.5076	0.0800	11.4349	0.0875	5.4110	61.8740	12
13	1.1020	0.9074	13.6014	0.0735	12.3423	0.0810	5.8954	72.7632	13
14	1.1103	0.9007	14.7034	0.0680	13.2430	0.0755	6.3786	84.4720	14
15	1.1186	0.8940	15.8137	0.0632	14.1370	0.0707	6.8606	96.9876	15
16	1.1270	0.8873	16.9323	0.0591	15.0243	0.0666	7.3413	110.2973	16
17	1.1354	0.8807	18.0593	0.0554	15.9050	0.0629	7.8207	124.3887	17
18	1.1440	0.8742	19.1947	0.0521	16.7792	0.0596	8.2989	139.2494	18
19	1.1525	0.8676	20.3387	0.0492	17.6468	0.0567	8.7759	154.8671	19
20	1.1612	0.8612	21.4912	0.0465	18.5080	0.0540	9.2516	171.2297	20
21	1.1699	0.8548	22.6524	0.0441	19.3628	0.0516	9.7261	188.3253	21
22	1.1787	0.8484	23.8223	0.0420	20.2112	0.0495	10.1994	206.1420	22
23	1.1875	0.8421	25.0010	0.0400	21.0533	0.0475	10.6714	224.6682	23
24	1.1964	0.8358	26.1885	0.0382	21.8891	0.0457	11.1422	243.8923	24
25	1.2054	0.8296	27.3849	0.0365	22.7188	0.0440	11.6117	263.8029	25
26	1.2144	0.8234	28.5903	0.0350	23.5422	0.0425	12.0800	284.3888	26
27	1.2235	0.8173	29.8047	0.0336	24.3595	0.0411	12.5470	305.6387	27
28	1.2327	0.8112	31.0282	0.0322	25.1707	0.0397	13.0128	327.5416	28
29	1.2420	0.8052	32.2609	0.0310	25.9759	0.0385	13.4774	350.0867	29
30	1.2513	0.7992	33.5029	0.0298	26.7751	0.0373	13.9407	373.2631	30
31	1.2607	0.7932	34.7542	0.0288	27.5683	0.0363	14.4028	397.0602	31
32	1.2701	0.7873	36.0148	0.0278	28.3557	0.0353	14.8636	421.4675	32
33	1.2796	0.7815	37.2849	0.0268	29.1371	0.0343	15.3232	446.4746	33
34	1.2892	0.7757	38.5646	0.0259	29.9128	0.0334	15.7816	472.0712	34
35	1.2989	0.7699	39.8538	0.0251	30.6827	0.0326	16.2387	498.2471	35
36	1.3086	0.7641	41.1527	0.0243	31.4468	0.0318	16.6946	524.9924	36
40	1.3483	0.7416	46.4465	0.0215	34.4469	0.0290	18.5058	637.4693	40
48	1.4314	0.6986	57.5207	0.0174	40.1848	0.0249	22.0691	886.8404	48
50	1.4530	0.6883	60.3943	0.0166	41.5664	0.0241	22.9476	953.8486	50
60	1.5657	0.6387	75.4241	0.0133	48.1734	0.0208	27.2665	1313.5189	60
72	1.7126	0.5839	95.0070	0.0105	55.4768	0.0180	32.2882	1791.2463	72
80	1.8180	0.5500	109.0725	0.0092	59.9944	0.0167	35.5391	2132.1472	80
84	1.8732	0.5338	116.4269	0.0086	62.1540	0.0161	37.1357	2308.1283	84
90	1.9591	0.5104	127.8790	0.0078	65.2746	0.0153	39.4946	2577.9961	90
96	2.0489	0.4881	139.8562	0.0072	68.2584	0.0147	41.8107	2853.9352	96
100	2.1111	0.4737	148.1445	0.0068	70.1746	0.0143	43.3311	3040.7453	100
108	2.2411	0.4462	165.4832	0.0060	73.8394	0.0135	46.3154	3419.9041	108
120	2.4514	0.4079	193.5143	0.0052	78.9417	0.0127	50.6521	3998.5621	120
240	6.0092	0.1664	667.8869	0.0015	111.1450	0.0090	85.4210	9494.1162	240
360	14.7306	0.0679	1830.7435	0.0005	124.2819	0.0080	107.1145	13312.3871	360

Table C.4 Interest Rate Factors (1.0%)

	Single Payment		Equal Payment Series				Gradient Series		
N	Compound Amount Factor (F/P,i,N)	Present Worth Factor (P/F,i,N)	Compound Amount Factor (F/A,i,N)	Sinking Fund Factor (A/F,i,N)	Present Worth Factor (P/A,i,N)	Capital Recovery Factor (A/P,i,N)	Gradient Uniform Series (A/G,i,N)	Gradient Present Worth (P/G.i.N)	N
1	1.0100	0.9901	1.0000	1.0000	0.9901	1.0100	0.0000	0.0000	1
2	1.0201	0.9803	2.0100	0.4975	1.9704	0.5075	0.4975	0.9803	2
3	1.0303	0.9706	3.0301	0.3300	2.9410	0.3400	0.9934	2.9215	3
4	1.0406	0.9610	4.0604	0.2463	3.9020	0.2563	1.4876	5.8044	4
5	1.0510	0.9515	5.1010	0.1960	4.8534	0.2060	1.9801	9.6103	5
6	1.0615	0.9420	6.1520	0.1625	5.7955	0.1725	2.4710	14.3205	6
7	1.0721	0.9327	7.2135	0.1386	6.7282	0.1486	2.9602	19.9168	7
8	1.0829	0.9235	8.2857	0.1207	7.6517	0.1307	3.4478	26.3812	8
9	1.0937	0.9143	9.3685	0.1067	8.5660	0.1167	3.9337	33.6959	9
10	1.1046	0.9053	10.4622	0.0956	9.4713	0.1056	4.4179	41.8435	10
11	1.1157	0.8963	11.5668	0.0865	10.3676	0.0965	4.9005	50.8067	11
12	1.1268	0.8874	12.6825	0.0788	11.2551	0.0888	5.3815	60.5687	12
13	1.1381	0.8787	13.8093	0.0724	12.1337	0.0824	5.8607	71.1126	13
14	1.1495	0.8700	14.9474	0.0669	13.0037	0.0769	6.3384	82.4221	14
15	1.1610	0.8613	16.0969	0.0621	13.8651	0.0721	6.8143	94.4810	15
16	1.1726	0.8528	17.2579	0.0579	14.7179	0.0679	7.2886	107.2734	16
17	1.1843	0.8444	18.4304	0.0543	15.5623	0.0643	7.7613	120.7834	17
18	1.1961	0.8360	19.6147	0.0510	16.3983	0.0610	8.2323	134.9957	18
19	1.2081	0.8277	20.8109	0.0481	17.2260	0.0581	8.7017	149.8950	19
20	1.2202	0.8195	22.0190	0.0454	18.0456	0.0554	9.1694	165.4664	20
21	1.2324	0.8114	23.2392	0.0430	18.8570	0.0530	9.6354	181.6950	21
22	1.2447	0.8034	24.4716	0.0409	19.6604	0.0509	10.0998	198.5663	22
23	1.2572	0.7954	25.7163	0.0389	20.4558	0.0489	10.5626	216.0660	23
24	1.2697	0.7876	26.9735	0.0371	21.2434	0.0471	11.0237	234.1800	24
25	1.2824	0.7798	28.2432	0.0354	22.0232	0.0454	11.4831	252.8945	25
26	1.2953	0.7720	29.5256	0.0339	22.7952	0.0439	11.9409	272.1957	26
27	1.3082	0.7644	30.8209	0.0324	23.5596	0.0424	12.3971	292.0702	27
28	1.3213	0.7568	32.1291	0.0311	24.3164	0.0411	12.8516	312.5047	28
29	1.3345	0.7493	33.4504	0.0299	25.0658	0.0399	13.3044	333.4863	29
30	1.3478	0.7419	34.7849	0.0287	25.8077	0.0387	13.7557	355.0021	30
31	1.3613	0.7346	36.1327	0.0277	26.5423	0.0377	14.2052	377.0394	31
32	1.3749	0.7273	37.4941	0.0267	27.2696	0.0367	14.6532	399.5858	32
33	1.3887	0.7201	38.8690	0.0257	27.9897	0.0357	15.0995	422.6291	33
34	1.4026	0.7130	40.2577	0.0248	28.7027	0.0348	15.5441	446.1572	34
35	1.4166	0.7059	41.6603	0.0240	29.4086	0.0340	15.9871	470.1583	35
36	1.4308	0.6989	43.0769	0.0232	30.1075	0.0332	16.4285	494.6207	36
40	1.4889	0.6717	48.8864	0.0205	32.8347	0.0305	18.1776	596.8561	40
48	1.6122	0.6203	61.2226	0.0163	37.9740	0.0263	21.5976	820.1460	48
50	1.6446	0.6080	64.4632	0.0155	39.1961	0.0255	22.4363	879.4176	50
60	1.8167	0.5504	81.6697	0.0122	44.9550	0.0222	26.5333	1192.8061	60
72	2.0471	0.4885	104.7099	0.0096	51.1504	0.0196	31.2386	1597.8673	72
80	2.2167	0.4511	121.6715	0.0082	54.8882	0.0182	34.2492	1879.8771	80
84	2.3067	0.4335	130.6723	0.0077	56.6485	0.0177	35.7170	2023.3153	84
90	2.4486	0.4084	144.8633	0.0069	59.1609	0.0169	37.8724	2240.5675	90
96	2.5993	0.3847	159.9273	0.0063	61.5277	0.0163	39.9727	2459.4298	96
100	2.7048	0.3697	170.4814	0.0059	63.0289	0.0159	41.3426	2605.7758	100
108	2.9289	0.3414	192.8926	0.0052	65.8578	0.0152	44.0103	2898.4203	108
120	3.3004	0.3030	230.0387	0.0043	69.7005	0.0143	47.8349	3334.1148	120
240	10.8926	0.0918	989.2554	0.0010	90.8194	0.0110	75.7393	6878.6016	240
360	35.9496	0.0278	3494.9641	0.0003	97.2183	0.0103	89.6995	8720.4323	360

Table C.5 Interest Rate Factors (1.25%)

	Single Payment		Equal Payment Series				Gradient Series		
N	Compound Amount Factor (F/P,i,N)	Present Worth Factor (P/F,i,N)	Compound Amount Factor (F/A,i,N)	Sinking Fund Factor (A/F,i,N)	Present Worth Factor (P/A,i,N)	Capital Recovery Factor (A/P,i,N)	Gradient Uniform Series (A/G,i,N)	Gradient Present Worth (P/G.i.N)	N
1	1.0125	0.9877	1.0000	1.0000	0.9877	1.0125	0.0000	0.0000	1
2	1.0252	0.9755	2.0125	0.4969	1.9631	0.5094	0.4969	0.9755	2
3	1.0380	0.9634	3.0377	0.3292	2.9265	0.3417	0.9917	2.9023	3
4	1.0509	0.9515	4.0756	0.2454	3.8781	0.2579	1.4845	5.7569	4
5	1.0641	0.9398	5.1266	0.1951	4.8178	0.2076	1.9752	9.5160	5
6	1.0774	0.9282	6.1907	0.1615	5.7460	0.1740	2.4638	14.1569	6
7	1.0909	0.9167	7.2680	0.1376	6.6627	0.1501	2.9503	19.6571	7
8	1.1045	0.9054	8.3589	0.1196	7.5681	0.1321	3.4348	25.9949	8
9	1.1183	0.8942	9.4634	0.1057	8.4623	0.1182	3.9172	33.1487	9
10	1.1323	0.8832	10.5817	0.0945	9.3455	0.1070	4.3975	41.0973	10
11	1.1464	0.8723	11.7139	0.0854	10.2178	0.0979	4.8758	49.8201	11
12	1.1608	0.8615	12.8604	0.0778	11.0793	0.0903	5.3520	59.2967	12
13	1.1753	0.8509	14.0211	0.0713	11.9302	0.0838	5.8262	69.5072	13
14	1.1900	0.8404	15.1964	0.0658	12.7706	0.0783	6.2982	80.4320	14
15	1.2048	0.8300	16.3863	0.0610	13.6005	0.0735	6.7682	92.0519	15
16	1.2199	0.8197	17.5912	0.0568	14.4203	0.0693	7.2362	104.3481	16
17	1.2351	0.8096	18.8111	0.0532	15.2299	0.0657	7.7021	117.3021	17
18	1.2506	0.7996	20.0462	0.0499	16.0295	0.0624	8.1659	130.8958	18
19	1.2662	0.7898	21.2968	0.0470	16.8193	0.0595	8.6277	145.1115	19
20	1.2820	0.7800	22.5630	0.0443	17.5993	0.0568	9.0874	159.9316	20
21	1.2981	0.7704	23.8450	0.0419	18.3697	0.0544	9.5450	175.3392	21
22	1.3143	0.7609	25.1431	0.0398	19.1306	0.0523	10.0006	191.3174	22
23	1.3307	0.7515	26.4574	0.0378	19.8820	0.0503	10.4542	207.8499	23
24	1.3474	0.7422	27.7881	0.0360	20.6242	0.0485	10.9056	224.9204	24
25	1.3642	0.7330	29.1354	0.0343	21.3573	0.0468	11.3551	242.5132	25
26	1.3812	0.7240	30.4996	0.0328	22.0813	0.0453	11.8024	260.6128	26
27	1.3985	0.7150	31.8809	0.0314	22.7963	0.0439	12.2478	279.2040	27
28	1.4160	0.7062	33.2794	0.0300	23.5025	0.0425	12.6911	298.2719	28
29	1.4337	0.6975	34.6954	0.0288	24.2000	0.0413	13.1323	317.8019	29
30	1.4516	0.6889	36.1291	0.0277	24.8889	0.0402	13.5715	337.7797	30
31	1.4698	0.6804	37.5807	0.0266	25.5693	0.0391	14.0086	358.1912	31
32	1.4881	0.6720	39.0504	0.0256	26.2413	0.0381	14.4438	379.0227	32
33	1.5067	0.6637	40.5386	0.0247	26.9050	0.0372	14.8768	400.2607	33
34	1.5256	0.6555	42.0453	0.0238	27.5605	0.0363	15.3079	421.8920	34
35	1.5446	0.6474	43.5709	0.0230	28.2079	0.0355	15.7369	443.9037	35
36	1.5639	0.6394	45.1155	0.0222	28.8473	0.0347	16.1639	466.2830	36
40	1.6436	0.6084	51.4896	0.0194	31.3269	0.0319	17.8515	559.2320	40
48	1.8154	0.5509	65.2284	0.0153	35.9315	0.0278	21.1299	759.2296	48
50	1.8610	0.5373	68.8818	0.0145	37.0129	0.0270	21.9295	811.6738	50
60	2.1072	0.4746	88.5745	0.0113	42.0346	0.0238	25.8083	1084.8429	60
72	2.4459	0.4088	115.6736	0.0086	47.2925	0.0211	30.2047	1428.4561	72
80	2.7015	0.3702	136.1188	0.0073	50.3867	0.0198	32.9822	1661.8651	80
84	2.8391	0.3522	147.1290	0.0068	51.8222	0.0193	34.3258	1778.8384	84
90	3.0588	0.3269	164.7050	0.0061	53.8461	0.0186	36.2855	1953.8303	90
96	3.2955	0.3034	183.6411	0.0054	55.7246	0.0179	38.1793	2127.5244	96
100	3.4634	0.2887	197.0723	0.0051	56.9013	0.0176	39.4058	2242.2411	100
108	3.8253	0.2614	226.0226	0.0044	59.0865	0.0169	41.7737	2468.2636	108
120	4.4402	0.2252	275.2171	0.0036	61.9828	0.0161	45.1184	2796.5694	120
240	19.7155	0.0507	1497.2395	0.0007	75.9423	0.0132	67.1764	5101.5288	240
360	87.5410	0.0114	6923.2796	0.0001	79.0861	0.0126	75.8401	5997.9027	360

Table C.6 Interest Rate Factors (1.50%)

	Single Payment		Equal Payment Series				Gradient Series		
N	Compound Amount Factor (F/P,i,N)	Present Worth Factor (P/F,i,N)	Compound Amount Factor (F/A,i,N)	Sinking Fund Factor (A/F,i,N)	Present Worth Factor (P/A,i,N)	Capital Recovery Factor (A/P,i,N)	Gradient Uniform Series (A/G,i,N)	Gradient Present Worth (P/G.i.N)	**N**
1	1.0150	0.9852	1.0000	1.0000	0.9852	1.0150	0.0000	0.0000	1
2	1.0302	0.9707	2.0150	0.4963	1.9559	0.5113	0.4963	0.9707	2
3	1.0457	0.9563	3.0452	0.3284	2.9122	0.3434	0.9901	2.8833	3
4	1.0614	0.9422	4.0909	0.2444	3.8544	0.2594	1.4814	5.7098	4
5	1.0773	0.9283	5.1523	0.1941	4.7826	0.2091	1.9702	9.4229	5
6	1.0934	0.9145	6.2296	0.1605	5.6972	0.1755	2.4566	13.9956	6
7	1.1098	0.9010	7.3230	0.1366	6.5982	0.1516	2.9405	19.4018	7
8	1.1265	0.8877	8.4328	0.1186	7.4859	0.1336	3.4219	25.6157	8
9	1.1434	0.8746	9.5593	0.1046	8.3605	0.1196	3.9008	32.6125	9
10	1.1605	0.8617	10.7027	0.0934	9.2222	0.1084	4.3772	40.3675	10
11	1.1779	0.8489	11.8633	0.0843	10.0711	0.0993	4.8512	48.8568	11
12	1.1956	0.8364	13.0412	0.0767	10.9075	0.0917	5.3227	58.0571	12
13	1.2136	0.8240	14.2368	0.0702	11.7315	0.0852	5.7917	67.9454	13
14	1.2318	0.8118	15.4504	0.0647	12.5434	0.0797	6.2582	78.4994	14
15	1.2502	0.7999	16.6821	0.0599	13.3432	0.0749	6.7223	89.6974	15
16	1.2690	0.7880	17.9324	0.0558	14.1313	0.0708	7.1839	101.5178	16
17	1.2880	0.7764	19.2014	0.0521	14.9076	0.0671	7.6431	113.9400	17
18	1.3073	0.7649	20.4894	0.0488	15.6726	0.0638	8.0997	126.9435	18
19	1.3270	0.7536	21.7967	0.0459	16.4262	0.0609	8.5539	140.5084	19
20	1.3469	0.7425	23.1237	0.0432	17.1686	0.0582	9.0057	154.6154	20
21	1.3671	0.7315	24.4705	0.0409	17.9001	0.0559	9.4550	169.2453	21
22	1.3876	0.7207	25.8376	0.0387	18.6208	0.0537	9.9018	184.3798	22
23	1.4084	0.7100	27.2251	0.0367	19.3309	0.0517	10.3462	200.0006	23
24	1.4295	0.6995	28.6335	0.0349	20.0304	0.0499	10.7881	216.0901	24
25	1.4509	0.6892	30.0630	0.0333	20.7196	0.0483	11.2276	232.6310	25
26	1.4727	0.6790	31.5140	0.0317	21.3986	0.0467	11.6646	249.6065	26
27	1.4948	0.6690	32.9867	0.0303	22.0676	0.0453	12.0992	267.0002	27
28	1.5172	0.6591	34.4815	0.0290	22.7267	0.0440	12.5313	284.7958	28
29	1.5400	0.6494	35.9987	0.0278	23.3761	0.0428	12.9610	302.9779	29
30	1.5631	0.6398	37.5387	0.0266	24.0158	0.0416	13.3883	321.5310	30
31	1.5865	0.6303	39.1018	0.0256	24.6461	0.0406	13.8131	340.4402	31
32	1.6103	0.6210	40.6883	0.0246	25.2671	0.0396	14.2355	359.6910	32
33	1.6345	0.6118	42.2986	0.0236	25.8790	0.0386	14.6555	379.2691	33
34	1.6590	0.6028	43.9331	0.0228	26.4817	0.0378	15.0731	399.1607	34
35	1.6839	0.5939	45.5921	0.0219	27.0756	0.0369	15.4882	419.3521	35
36	1.7091	0.5851	47.2760	0.0212	27.6607	0.0362	15.9009	439.8303	36
40	1.8140	0.5513	54.2679	0.0184	29.9158	0.0334	17.5277	524.3568	40
48	2.0435	0.4894	69.5652	0.0144	34.0426	0.0294	20.6667	703.5462	48
50	2.1052	0.4750	73.6828	0.0136	34.9997	0.0286	21.4277	749.9636	50
60	2.4432	0.4093	96.2147	0.0104	39.3803	0.0254	25.0930	988.1674	60
72	2.9212	0.3423	128.0772	0.0078	43.8447	0.0228	29.1893	1279.7938	72
80	3.2907	0.3039	152.7109	0.0065	46.4073	0.0215	31.7423	1473.0741	80
84	3.4926	0.2863	166.1726	0.0060	47.5786	0.0210	32.9668	1568.5140	84
90	3.8189	0.2619	187.9299	0.0053	49.2099	0.0203	34.7399	1709.5439	90
96	4.1758	0.2395	211.7202	0.0047	50.7017	0.0197	36.4381	1847.4725	96
100	4.4320	0.2256	228.8030	0.0044	51.6247	0.0194	37.5295	1937.4506	100
108	4.9927	0.2003	266.1778	0.0038	53.3137	0.0188	39.6171	2112.1348	108
120	5.9693	0.1675	331.2882	0.0030	55.4985	0.0180	42.5185	2359.7114	120
240	35.6328	0.0281	2308.8544	0.0004	64.7957	0.0154	59.7368	3870.6912	240
360	212.7038	0.0047	14113.5854	0.0001	66.3532	0.0151	64.9662	4310.7165	360

Table C.7 Interest Rate Factors (1.75%)

	Single Payment		Equal Payment Series				Gradient Series		
N	Compound Amount Factor (F/P,i,N)	Present Worth Factor (P/F,i,N)	Compound Amount Factor (F/A,i,N)	Sinking Fund Factor (A/F,i,N)	Present Worth Factor (P/A,i,N)	Capital Recovery Factor (A/P,i,N)	Gradient Uniform Series (A/G,i,N)	Gradient Present Worth (P/G.i.N)	N
1	1.0175	0.9828	1.0000	1.0000	0.9828	1.0175	0.0000	0.0000	1
2	1.0353	0.9659	2.0175	0.4957	1.9487	0.5132	0.4957	0.9659	2
3	1.0534	0.9493	3.0528	0.3276	2.8980	0.3451	0.9884	2.8645	3
4	1.0719	0.9330	4.1062	0.2435	3.8309	0.2610	1.4783	5.6633	4
5	1.0906	0.9169	5.1781	0.1931	4.7479	0.2106	1.9653	9.3310	5
6	1.1097	0.9011	6.2687	0.1595	5.6490	0.1770	2.4494	13.8367	6
7	1.1291	0.8856	7.3784	0.1355	6.5346	0.1530	2.9306	19.1506	7
8	1.1489	0.8704	8.5075	0.1175	7.4051	0.1350	3.4089	25.2435	8
9	1.1690	0.8554	9.6564	0.1036	8.2605	0.1211	3.8844	32.0870	9
10	1.1894	0.8407	10.8254	0.0924	9.1012	0.1099	4.3569	39.6535	10
11	1.2103	0.8263	12.0148	0.0832	9.9275	0.1007	4.8266	47.9162	11
12	1.2314	0.8121	13.2251	0.0756	10.7395	0.0931	5.2934	56.8489	12
13	1.2530	0.7981	14.4565	0.0692	11.5376	0.0867	5.7573	66.4260	13
14	1.2749	0.7844	15.7095	0.0637	12.3220	0.0812	6.2184	76.6227	14
15	1.2972	0.7709	16.9844	0.0589	13.0929	0.0764	6.6765	87.4149	15
16	1.3199	0.7576	18.2817	0.0547	13.8505	0.0722	7.1318	98.7792	16
17	1.3430	0.7446	19.6016	0.0510	14.5951	0.0685	7.5842	110.6926	17
18	1.3665	0.7318	20.9446	0.0477	15.3269	0.0652	8.0338	123.1328	18
19	1.3904	0.7192	22.3112	0.0448	16.0461	0.0623	8.4805	136.0783	19
20	1.4148	0.7068	23.7016	0.0422	16.7529	0.0597	8.9243	149.5080	20
21	1.4395	0.6947	25.1164	0.0398	17.4475	0.0573	9.3653	163.4013	21
22	1.4647	0.6827	26.5559	0.0377	18.1303	0.0552	9.8034	177.7385	22
23	1.4904	0.6710	28.0207	0.0357	18.8012	0.0532	10.2387	192.5000	23
24	1.5164	0.6594	29.5110	0.0339	19.4607	0.0514	10.6711	207.6671	24
25	1.5430	0.6481	31.0275	0.0322	20.1088	0.0497	11.1007	223.2214	25
26	1.5700	0.6369	32.5704	0.0307	20.7457	0.0482	11.5274	239.1451	26
27	1.5975	0.6260	34.1404	0.0293	21.3717	0.0468	11.9513	255.4210	27
28	1.6254	0.6152	35.7379	0.0280	21.9870	0.0455	12.3724	272.0321	28
29	1.6539	0.6046	37.3633	0.0268	22.5916	0.0443	12.7907	288.9623	29
30	1.6828	0.5942	39.0172	0.0256	23.1858	0.0431	13.2061	306.1954	30
31	1.7122	0.5840	40.7000	0.0246	23.7699	0.0421	13.6188	323.7163	31
32	1.7422	0.5740	42.4122	0.0236	24.3439	0.0411	14.0286	341.5097	32
33	1.7727	0.5641	44.1544	0.0226	24.9080	0.0401	14.4356	359.5613	33
34	1.8037	0.5544	45.9271	0.0218	25.4624	0.0393	14.8398	377.8567	34
35	1.8353	0.5449	47.7308	0.0210	26.0073	0.0385	15.2412	396.3824	35
36	1.8674	0.5355	49.5661	0.0202	26.5428	0.0377	15.6399	415.1250	36
40	2.0016	0.4996	57.2341	0.0175	28.5942	0.0350	17.2066	492.0109	40
48	2.2996	0.4349	74.2628	0.0135	32.2938	0.0310	20.2084	652.6054	48
50	2.3808	0.4200	78.9022	0.0127	33.1412	0.0302	20.9317	693.7010	50
60	2.8318	0.3531	104.6752	0.0096	36.9640	0.0271	24.3885	901.4954	60
72	3.4872	0.2868	142.1263	0.0070	40.7564	0.0245	28.1948	1149.1181	72
80	4.0064	0.2496	171.7938	0.0058	42.8799	0.0233	30.5329	1309.2482	80
84	4.2943	0.2329	188.2450	0.0053	43.8361	0.0228	31.6442	1387.1584	84
90	4.7654	0.2098	215.1646	0.0046	45.1516	0.0221	33.2409	1500.8798	90
96	5.2882	0.1891	245.0374	0.0041	46.3370	0.0216	34.7556	1610.4716	96
100	5.6682	0.1764	266.7518	0.0037	47.0615	0.0212	35.7211	1681.0886	100
108	6.5120	0.1536	314.9738	0.0032	48.3679	0.0207	37.5494	1816.1852	108
120	8.0192	0.1247	401.0962	0.0025	50.0171	0.0200	40.0469	2003.0269	120
240	64.3073	0.0156	3617.5602	0.0003	56.2543	0.0178	53.3518	3001.2678	240
360	515.6921	0.0019	29410.9747	0.0000	57.0320	0.0175	56.4434	3219.0833	360

Table C.8 Interest Rate Factors (2.0%)

	Single Payment		Equal Payment Series				Gradient Series		
N	Compound Amount Factor (F/P,i,N)	Present Worth Factor (P/F,i,N)	Compound Amount Factor (F/A,i,N)	Sinking Fund Factor (A/F,i,N)	Present Worth Factor (P/A,i,N)	Capital Recovery Factor (A/P,i,N)	Gradient Uniform Series (A/G,i,N)	Gradient Present Worth (P/G.i.N)	N
1	1.0200	0.9804	1.0000	1.0000	0.9804	1.0200	0.0000	0.0000	1
2	1.0404	0.9612	2.0200	0.4950	1.9416	0.5150	0.4950	0.9612	2
3	1.0612	0.9423	3.0604	0.3268	2.8839	0.3468	0.9868	2.8458	3
4	1.0824	0.9238	4.1216	0.2426	3.8077	0.2626	1.4752	5.6173	4
5	1.1041	0.9057	5.2040	0.1922	4.7135	0.2122	1.9604	9.2403	5
6	1.1262	0.8880	6.3081	0.1585	5.6014	0.1785	2.4423	13.6801	6
7	1.1487	0.8706	7.4343	0.1345	6.4720	0.1545	2.9208	18.9035	7
8	1.1717	0.8535	8.5830	0.1165	7.3255	0.1365	3.3961	24.8779	8
9	1.1951	0.8368	9.7546	0.1025	8.1622	0.1225	3.8681	31.5720	9
10	1.2190	0.8203	10.9497	0.0913	8.9826	0.1113	4.3367	38.9551	10
11	1.2434	0.8043	12.1687	0.0822	9.7868	0.1022	4.8021	46.9977	11
12	1.2682	0.7885	13.4121	0.0746	10.5753	0.0946	5.2642	55.6712	12
13	1.2936	0.7730	14.6803	0.0681	11.3484	0.0881	5.7231	64.9475	13
14	1.3195	0.7579	15.9739	0.0626	12.1062	0.0826	6.1786	74.7999	14
15	1.3459	0.7430	17.2934	0.0578	12.8493	0.0778	6.6309	85.2021	15
16	1.3728	0.7284	18.6393	0.0537	13.5777	0.0737	7.0799	96.1288	16
17	1.4002	0.7142	20.0121	0.0500	14.2919	0.0700	7.5256	107.5554	17
18	1.4282	0.7002	21.4123	0.0467	14.9920	0.0667	7.9681	119.4581	18
19	1.4568	0.6864	22.8406	0.0438	15.6785	0.0638	8.4073	131.8139	19
20	1.4859	0.6730	24.2974	0.0412	16.3514	0.0612	8.8433	144.6003	20
21	1.5157	0.6598	25.7833	0.0388	17.0112	0.0588	9.2760	157.7959	21
22	1.5460	0.6468	27.2990	0.0366	17.6580	0.0566	9.7055	171.3795	22
23	1.5769	0.6342	28.8450	0.0347	18.2922	0.0547	10.1317	185.3309	23
24	1.6084	0.6217	30.4219	0.0329	18.9139	0.0529	10.5547	199.6305	24
25	1.6406	0.6095	32.0303	0.0312	19.5235	0.0512	10.9745	214.2592	25
26	1.6734	0.5976	33.6709	0.0297	20.1210	0.0497	11.3910	229.1987	26
27	1.7069	0.5859	35.3443	0.0283	20.7069	0.0483	11.8043	244.4311	27
28	1.7410	0.5744	37.0512	0.0270	21.2813	0.0470	12.2145	259.9392	28
29	1.7758	0.5631	38.7922	0.0258	21.8444	0.0458	12.6214	275.7064	29
30	1.8114	0.5521	40.5681	0.0246	22.3965	0.0446	13.0251	291.7164	30
31	1.8476	0.5412	42.3794	0.0236	22.9377	0.0436	13.4257	307.9538	31
32	1.8845	0.5306	44.2270	0.0226	23.4683	0.0426	13.8230	324.4035	32
33	1.9222	0.5202	46.1116	0.0217	23.9886	0.0417	14.2172	341.0508	33
34	1.9607	0.5100	48.0338	0.0208	24.4986	0.0408	14.6083	357.8817	34
35	1.9999	0.5000	49.9945	0.0200	24.9986	0.0400	14.9961	374.8826	35
36	2.0399	0.4902	51.9944	0.0192	25.4888	0.0392	15.3809	392.0405	36
40	2.2080	0.4529	60.4020	0.0166	27.3555	0.0366	16.8885	461.9931	40
48	2.5871	0.3865	79.3535	0.0126	30.6731	0.0326	19.7556	605.9657	48
50	2.6916	0.3715	84.5794	0.0118	31.4236	0.0318	20.4420	642.3606	50
60	3.2810	0.3048	114.0515	0.0088	34.7609	0.0288	23.6961	823.6975	60
72	4.1611	0.2403	158.0570	0.0063	37.9841	0.0263	27.2234	1034.0557	72
80	4.8754	0.2051	193.7720	0.0052	39.7445	0.0252	29.3572	1166.7868	80
84	5.2773	0.1895	213.8666	0.0047	40.5255	0.0247	30.3616	1230.4191	84
90	5.9431	0.1683	247.1567	0.0040	41.5869	0.0240	31.7929	1322.1701	90
96	6.6929	0.1494	284.6467	0.0035	42.5294	0.0235	33.1370	1409.2973	96
100	7.2446	0.1380	312.2323	0.0032	43.0984	0.0232	33.9863	1464.7527	100
108	8.4883	0.1178	374.4129	0.0027	44.1095	0.0227	35.5774	1569.3025	108
120	10.7652	0.0929	488.2582	0.0020	45.3554	0.0220	37.7114	1710.4160	120
240	115.8887	0.0086	5744.4368	0.0002	49.5686	0.0202	47.9110	2374.8800	240
360	1247.5611	0.0008	62328.0564	0.0000	49.9599	0.0200	49.7112	2483.5679	360

Table C.9 Interest Rate Factors (3.0%)

	Single Payment		Equal Payment Series				Gradient Series		
N	Compound Amount Factor (F/P,i,N)	Present Worth Factor (P/F,i,N)	Compound Amount Factor (F/A,i,N)	Sinking Fund Factor (A/F,i,N)	Present Worth Factor (P/A,i,N)	Capital Recovery Factor (A/P,i,N)	Gradient Uniform Series (A/G,i,N)	Gradient Present Worth (P/G.i.N)	N
1	1.0300	0.9709	1.0000	1.0000	0.9709	1.0300	0.0000	0.0000	1
2	1.0609	0.9426	2.0300	0.4926	1.9135	0.5226	0.4926	0.9426	2
3	1.0927	0.9151	3.0909	0.3235	2.8286	0.3535	0.9803	2.7729	3
4	1.1255	0.8885	4.1836	0.2390	3.7171	0.2690	1.4631	5.4383	4
5	1.1593	0.8626	5.3091	0.1884	4.5797	0.2184	1.9409	8.8888	5
6	1.1941	0.8375	6.4684	0.1546	5.4172	0.1846	2.4138	13.0762	6
7	1.2299	0.8131	7.6625	0.1305	6.2303	0.1605	2.8819	17.9547	7
8	1.2668	0.7894	8.8923	0.1125	7.0197	0.1425	3.3450	23.4806	8
9	1.3048	0.7664	10.1591	0.0984	7.7861	0.1284	3.8032	29.6119	9
10	1.3439	0.7441	11.4639	0.0872	8.5302	0.1172	4.2565	36.3088	10
11	1.3842	0.7224	12.8078	0.0781	9.2526	0.1081	4.7049	43.5330	11
12	1.4258	0.7014	14.1920	0.0705	9.9540	0.1005	5.1485	51.2482	12
13	1.4685	0.6810	15.6178	0.0640	10.6350	0.0940	5.5872	59.4196	13
14	1.5126	0.6611	17.0863	0.0585	11.2961	0.0885	6.0210	68.0141	14
15	1.5580	0.6419	18.5989	0.0538	11.9379	0.0838	6.4500	77.0002	15
16	1.6047	0.6232	20.1569	0.0496	12.5611	0.0796	6.8742	86.3477	16
17	1.6528	0.6050	21.7616	0.0460	13.1661	0.0760	7.2936	96.0280	17
18	1.7024	0.5874	23.4144	0.0427	13.7535	0.0727	7.7081	106.0137	18
19	1.7535	0.5703	25.1169	0.0398	14.3238	0.0698	8.1179	116.2788	19
20	1.8061	0.5537	26.8704	0.0372	14.8775	0.0672	8.5229	126.7987	20
21	1.8603	0.5375	28.6765	0.0349	15.4150	0.0649	8.9231	137.5496	21
22	1.9161	0.5219	30.5368	0.0327	15.9369	0.0627	9.3186	148.5094	22
23	1.9736	0.5067	32.4529	0.0308	16.4436	0.0608	9.7093	159.6566	23
24	2.0328	0.4919	34.4265	0.0290	16.9355	0.0590	10.0954	170.9711	24
25	2.0938	0.4776	36.4593	0.0274	17.4131	0.0574	10.4768	182.4336	25
26	2.1566	0.4637	38.5530	0.0259	17.8768	0.0559	10.8535	194.0260	26
27	2.2213	0.4502	40.7096	0.0246	18.3270	0.0546	11.2255	205.7309	27
28	2.2879	0.4371	42.9309	0.0233	18.7641	0.0533	11.5930	217.5320	28
29	2.3566	0.4243	45.2189	0.0221	19.1885	0.0521	11.9558	229.4137	29
30	2.4273	0.4120	47.5754	0.0210	19.6004	0.0510	12.3141	241.3613	30
31	2.5001	0.4000	50.0027	0.0200	20.0004	0.0500	12.6678	253.3609	31
32	2.5751	0.3883	52.5028	0.0190	20.3888	0.0490	13.0169	265.3993	32
33	2.6523	0.3770	55.0778	0.0182	20.7658	0.0482	13.3616	277.4642	33
34	2.7319	0.3660	57.7302	0.0173	21.1318	0.0473	13.7018	289.5437	34
35	2.8139	0.3554	60.4621	0.0165	21.4872	0.0465	14.0375	301.6267	35
40	3.2620	0.3066	75.4013	0.0133	23.1148	0.0433	15.6502	361.7499	40
45	3.7816	0.2644	92.7199	0.0108	24.5187	0.0408	17.1556	420.6325	45
50	4.3839	0.2281	112.7969	0.0089	25.7298	0.0389	18.5575	477.4803	50
55	5.0821	0.1968	136.0716	0.0073	26.7744	0.0373	19.8600	531.7411	55
60	5.8916	0.1697	163.0534	0.0061	27.6756	0.0361	21.0674	583.0526	60
65	6.8300	0.1464	194.3328	0.0051	28.4529	0.0351	22.1841	631.2010	65
70	7.9178	0.1263	230.5941	0.0043	29.1234	0.0343	23.2145	676.0869	70
75	9.1789	0.1089	272.6309	0.0037	29.7018	0.0337	24.1634	717.6978	75
80	10.6409	0.0940	321.3630	0.0031	30.2008	0.0331	25.0353	756.0865	80
85	12.3357	0.0811	377.8570	0.0026	30.6312	0.0326	25.8349	791.3529	85
90	14.3005	0.0699	443.3489	0.0023	31.0024	0.0323	26.5667	823.6302	90
95	16.5782	0.0603	519.2720	0.0019	31.3227	0.0319	27.2351	853.0742	95
100	19.2186	0.0520	607.2877	0.0016	31.5989	0.0316	27.8444	879.8540	100

Table C.10 Interest Rate Factors (4.0%)

	Single Payment		Equal Payment Series				Gradient Series		
N	Compound Amount Factor (F/P,i,N)	Present Worth Factor (P/F,i,N)	Compound Amount Factor (F/A,i,N)	Sinking Fund Factor (A/F,i,N)	Present Worth Factor (P/A,i,N)	Capital Recovery Factor (A/P,i,N)	Gradient Uniform Series (A/G,i,N)	Gradient Present Worth (P/G.i.N)	N
1	1.0400	0.9615	1.0000	1.0000	0.9615	1.0400	0.0000	0.0000	1
2	1.0816	0.9246	2.0400	0.4902	1.8861	0.5302	0.4902	0.9246	2
3	1.1249	0.8890	3.1216	0.3203	2.7751	0.3603	0.9739	2.7025	3
4	1.1699	0.8548	4.2465	0.2355	3.6299	0.2755	1.4510	5.2670	4
5	1.2167	0.8219	5.4163	0.1846	4.4518	0.2246	1.9216	8.5547	5
6	1.2653	0.7903	6.6330	0.1508	5.2421	0.1908	2.3857	12.5062	6
7	1.3159	0.7599	7.8983	0.1266	6.0021	0.1666	2.8433	17.0657	7
8	1.3686	0.7307	9.2142	0.1085	6.7327	0.1485	3.2944	22.1806	8
9	1.4233	0.7026	10.5828	0.0945	7.4353	0.1345	3.7391	27.8013	9
10	1.4802	0.6756	12.0061	0.0833	8.1109	0.1233	4.1773	33.8814	10
11	1.5395	0.6496	13.4864	0.0741	8.7605	0.1141	4.6090	40.3772	11
12	1.6010	0.6246	15.0258	0.0666	9.3851	0.1066	5.0343	47.2477	12
13	1.6651	0.6006	16.6268	0.0601	9.9856	0.1001	5.4533	54.4546	13
14	1.7317	0.5775	18.2919	0.0547	10.5631	0.0947	5.8659	61.9618	14
15	1.8009	0.5553	20.0236	0.0499	11.1184	0.0899	6.2721	69.7355	15
16	1.8730	0.5339	21.8245	0.0458	11.6523	0.0858	6.6720	77.7441	16
17	1.9479	0.5134	23.6975	0.0422	12.1657	0.0822	7.0656	85.9581	17
18	2.0258	0.4936	25.6454	0.0390	12.6593	0.0790	7.4530	94.3498	18
19	2.1068	0.4746	27.6712	0.0361	13.1339	0.0761	7.8342	102.8933	19
20	2.1911	0.4564	29.7781	0.0336	13.5903	0.0736	8.2091	111.5647	20
21	2.2788	0.4388	31.9692	0.0313	14.0292	0.0713	8.5779	120.3414	21
22	2.3699	0.4220	34.2480	0.0292	14.4511	0.0692	8.9407	129.2024	22
23	2.4647	0.4057	36.6179	0.0273	14.8568	0.0673	9.2973	138.1284	23
24	2.5633	0.3901	39.0826	0.0256	15.2470	0.0656	9.6479	147.1012	24
25	2.6658	0.3751	41.6459	0.0240	15.6221	0.0640	9.9925	156.1040	25
26	2.7725	0.3607	44.3117	0.0226	15.9828	0.0626	10.3312	165.1212	26
27	2.8834	0.3468	47.0842	0.0212	16.3296	0.0612	10.6640	174.1385	27
28	2.9987	0.3335	49.9676	0.0200	16.6631	0.0600	10.9909	183.1424	28
29	3.1187	0.3207	52.9663	0.0189	16.9837	0.0589	11.3120	192.1206	29
30	3.2434	0.3083	56.0849	0.0178	17.2920	0.0578	11.6274	201.0618	30
31	3.3731	0.2965	59.3283	0.0169	17.5885	0.0569	11.9371	209.9556	31
32	3.5081	0.2851	62.7015	0.0159	17.8736	0.0559	12.2411	218.7924	32
33	3.6484	0.2741	66.2095	0.0151	18.1476	0.0551	12.5396	227.5634	33
34	3.7943	0.2636	69.8579	0.0143	18.4112	0.0543	12.8324	236.2607	34
35	3.9461	0.2534	73.6522	0.0136	18.6646	0.0536	13.1198	244.8768	35
40	4.8010	0.2083	95.0255	0.0105	19.7928	0.0505	14.4765	286.5303	40
45	5.8412	0.1712	121.0294	0.0083	20.7200	0.0483	15.7047	325.4028	45
50	7.1067	0.1407	152.6671	0.0066	21.4822	0.0466	16.8122	361.1638	50
55	8.6464	0.1157	191.1592	0.0052	22.1086	0.0452	17.8070	393.6890	55
60	10.5196	0.0951	237.9907	0.0042	22.6235	0.0442	18.6972	422.9966	60
65	12.7987	0.0781	294.9684	0.0034	23.0467	0.0434	19.4909	449.2014	65
70	15.5716	0.0642	364.2905	0.0027	23.3945	0.0427	20.1961	472.4789	70
75	18.9453	0.0528	448.6314	0.0022	23.6804	0.0422	20.8206	493.0408	75
80	23.0498	0.0434	551.2450	0.0018	23.9154	0.0418	21.3718	511.1161	80
85	28.0436	0.0357	676.0901	0.0015	24.1085	0.0415	21.8569	526.9384	85
90	34.1193	0.0293	827.9833	0.0012	24.2673	0.0412	22.2826	540.7369	90
95	41.5114	0.0241	1012.7846	0.0010	24.3978	0.0410	22.6550	552.7307	95
100	50.5049	0.0198	1237.6237	0.0008	24.5050	0.0408	22.9800	563.1249	100

Table C.11 Interest Rate Factors (5.0%)

	Single Payment		Equal Payment Series				Gradient Series		
N	Compound Amount Factor (F/P,i,N)	Present Worth Factor (P/F,i,N)	Compound Amount Factor (F/A,i,N)	Sinking Fund Factor (A/F,i,N)	Present Worth Factor (P/A,i,N)	Capital Recovery Factor (A/P,i,N)	Gradient Uniform Series (A/G,i,N)	Gradient Present Worth (P/G.i.N)	N
1	1.0500	0.9524	1.0000	1.0000	0.9524	1.0500	0.0000	0.0000	1
2	1.1025	0.9070	2.0500	0.4878	1.8594	0.5378	0.4878	0.9070	2
3	1.1576	0.8638	3.1525	0.3172	2.7232	0.3672	0.9675	2.6347	3
4	1.2155	0.8227	4.3101	0.2320	3.5460	0.2820	1.4391	5.1028	4
5	1.2763	0.7835	5.5256	0.1810	4.3295	0.2310	1.9025	8.2369	5
6	1.3401	0.7462	6.8019	0.1470	5.0757	0.1970	2.3579	11.9680	6
7	1.4071	0.7107	8.1420	0.1228	5.7864	0.1728	2.8052	16.2321	7
8	1.4775	0.6768	9.5491	0.1047	6.4632	0.1547	3.2445	20.9700	8
9	1.5513	0.6446	11.0266	0.0907	7.1078	0.1407	3.6758	26.1268	9
10	1.6289	0.6139	12.5779	0.0795	7.7217	0.1295	4.0991	31.6520	10
11	1.7103	0.5847	14.2068	0.0704	8.3064	0.1204	4.5144	37.4988	11
12	1.7959	0.5568	15.9171	0.0628	8.8633	0.1128	4.9219	43.6241	12
13	1.8856	0.5303	17.7130	0.0565	9.3936	0.1065	5.3215	49.9879	13
14	1.9799	0.5051	19.5986	0.0510	9.8986	0.1010	5.7133	56.5538	14
15	2.0789	0.4810	21.5786	0.0463	10.3797	0.0963	6.0973	63.2880	15
16	2.1829	0.4581	23.6575	0.0423	10.8378	0.0923	6.4736	70.1597	16
17	2.2920	0.4363	25.8404	0.0387	11.2741	0.0887	6.8423	77.1405	17
18	2.4066	0.4155	28.1324	0.0355	11.6896	0.0855	7.2034	84.2043	18
19	2.5270	0.3957	30.5390	0.0327	12.0853	0.0827	7.5569	91.3275	19
20	2.6533	0.3769	33.0660	0.0302	12.4622	0.0802	7.9030	98.4884	20
21	2.7860	0.3589	35.7193	0.0280	12.8212	0.0780	8.2416	105.6673	21
22	2.9253	0.3418	38.5052	0.0260	13.1630	0.0760	8.5730	112.8461	22
23	3.0715	0.3256	41.4305	0.0241	13.4886	0.0741	8.8971	120.0087	23
24	3.2251	0.3101	44.5020	0.0225	13.7986	0.0725	9.2140	127.1402	24
25	3.3864	0.2953	47.7271	0.0210	14.0939	0.0710	9.5238	134.2275	25
26	3.5557	0.2812	51.1135	0.0196	14.3752	0.0696	9.8266	141.2585	26
27	3.7335	0.2678	54.6691	0.0183	14.6430	0.0683	10.1224	148.2226	27
28	3.9201	0.2551	58.4026	0.0171	14.8981	0.0671	10.4114	155.1101	28
29	4.1161	0.2429	62.3227	0.0160	15.1411	0.0660	10.6936	161.9126	29
30	4.3219	0.2314	66.4388	0.0151	15.3725	0.0651	10.9691	168.6226	30
31	4.5380	0.2204	70.7608	0.0141	15.5928	0.0641	11.2381	175.2333	31
32	4.7649	0.2099	75.2988	0.0133	15.8027	0.0633	11.5005	181.7392	32
33	5.0032	0.1999	80.0638	0.0125	16.0025	0.0625	11.7566	188.1351	33
34	5.2533	0.1904	85.0670	0.0118	16.1929	0.0618	12.0063	194.4168	34
35	5.5160	0.1813	90.3203	0.0111	16.3742	0.0611	12.2498	200.5807	35
40	7.0400	0.1420	120.7998	0.0083	17.1591	0.0583	13.3775	229.5452	40
45	8.9850	0.1113	159.7002	0.0063	17.7741	0.0563	14.3644	255.3145	45
50	11.4674	0.0872	209.3480	0.0048	18.2559	0.0548	15.2233	277.9148	50
55	14.6356	0.0683	272.7126	0.0037	18.6335	0.0537	15.9664	297.5104	55
60	18.6792	0.0535	353.5837	0.0028	18.9293	0.0528	16.6062	314.3432	60
65	23.8399	0.0419	456.7980	0.0022	19.1611	0.0522	17.1541	328.6910	65
70	30.4264	0.0329	588.5285	0.0017	19.3427	0.0517	17.6212	340.8409	70
75	38.8327	0.0258	756.6537	0.0013	19.4850	0.0513	18.0176	351.0721	75
80	49.5614	0.0202	971.2288	0.0010	19.5965	0.0510	18.3526	359.6460	80
85	63.2544	0.0158	1245.0871	0.0008	19.6838	0.0508	18.6346	366.8007	85
90	80.7304	0.0124	1594.6073	0.0006	19.7523	0.0506	18.8712	372.7488	90
95	103.0347	0.0097	2040.6935	0.0005	19.8059	0.0505	19.0689	377.6774	95
100	131.5013	0.0076	2610.0252	0.0004	19.8479	0.0504	19.2337	381.7492	100

Table C.12 Interest Rate Factors (6.0%)

	Single Payment		Equal Payment Series				Gradient Series		
N	Compound Amount Factor (F/P,i,N)	Present Worth Factor (P/F,i,N)	Compound Amount Factor (F/A,i,N)	Sinking Fund Factor (A/F,i,N)	Present Worth Factor (P/A,i,N)	Capital Recovery Factor (A/P,i,N)	Gradient Uniform Series (A/G,i,N)	Gradient Present Worth (P/G.i.N)	N
1	1.0600	0.9434	1.0000	1.0000	0.9434	1.0600	0.0000	0.0000	1
2	1.1236	0.8900	2.0600	0.4854	1.8334	0.5454	0.4854	0.8900	2
3	1.1910	0.8396	3.1836	0.3141	2.6730	0.3741	0.9612	2.5692	3
4	1.2625	0.7921	4.3746	0.2286	3.4651	0.2886	1.4272	4.9455	4
5	1.3382	0.7473	5.6371	0.1774	4.2124	0.2374	1.8836	7.9345	5
6	1.4185	0.7050	6.9753	0.1434	4.9173	0.2034	2.3304	11.4594	6
7	1.5036	0.6651	8.3938	0.1191	5.5824	0.1791	2.7676	15.4497	7
8	1.5938	0.6274	9.8975	0.1010	6.2098	0.1610	3.1952	19.8416	8
9	1.6895	0.5919	11.4913	0.0870	6.8017	0.1470	3.6133	24.5768	9
10	1.7908	0.5584	13.1808	0.0759	7.3601	0.1359	4.0220	29.6023	10
11	1.8983	0.5268	14.9716	0.0668	7.8869	0.1268	4.4213	34.8702	11
12	2.0122	0.4970	16.8699	0.0593	8.3838	0.1193	4.8113	40.3369	12
13	2.1329	0.4688	18.8821	0.0530	8.8527	0.1130	5.1920	45.9629	13
14	2.2609	0.4423	21.0151	0.0476	9.2950	0.1076	5.5635	51.7128	14
15	2.3966	0.4173	23.2760	0.0430	9.7122	0.1030	5.9260	57.5546	15
16	2.5404	0.3936	25.6725	0.0390	10.1059	0.0990	6.2794	63.4592	16
17	2.6928	0.3714	28.2129	0.0354	10.4773	0.0954	6.6240	69.4011	17
18	2.8543	0.3503	30.9057	0.0324	10.8276	0.0924	6.9597	75.3569	18
19	3.0256	0.3305	33.7600	0.0296	11.1581	0.0896	7.2867	81.3062	19
20	3.2071	0.3118	36.7856	0.0272	11.4699	0.0872	7.6051	87.2304	20
21	3.3996	0.2942	39.9927	0.0250	11.7641	0.0850	7.9151	93.1136	21
22	3.6035	0.2775	43.3923	0.0230	12.0416	0.0830	8.2166	98.9412	22
23	3.8197	0.2618	46.9958	0.0213	12.3034	0.0813	8.5099	104.7007	23
24	4.0489	0.2470	50.8156	0.0197	12.5504	0.0797	8.7951	110.3812	24
25	4.2919	0.2330	54.8645	0.0182	12.7834	0.0782	9.0722	115.9732	25
26	4.5494	0.2198	59.1564	0.0169	13.0032	0.0769	9.3414	121.4684	26
27	4.8223	0.2074	63.7058	0.0157	13.2105	0.0757	9.6029	126.8600	27
28	5.1117	0.1956	68.5281	0.0146	13.4062	0.0746	9.8568	132.1420	28
29	5.4184	0.1846	73.6398	0.0136	13.5907	0.0736	10.1032	137.3096	29
30	5.7435	0.1741	79.0582	0.0126	13.7648	0.0726	10.3422	142.3588	30
31	6.0881	0.1643	84.8017	0.0118	13.9291	0.0718	10.5740	147.2864	31
32	6.4534	0.1550	90.8898	0.0110	14.0840	0.0710	10.7988	152.0901	32
33	6.8406	0.1462	97.3432	0.0103	14.2302	0.0703	11.0166	156.7681	33
34	7.2510	0.1379	104.1838	0.0096	14.3681	0.0696	11.2276	161.3192	34
35	7.6861	0.1301	111.4348	0.0090	14.4982	0.0690	11.4319	165.7427	35
40	10.2857	0.0972	154.7620	0.0065	15.0463	0.0665	12.3590	185.9568	40
45	13.7646	0.0727	212.7435	0.0047	15.4558	0.0647	13.1413	203.1096	45
50	18.4202	0.0543	290.3359	0.0034	15.7619	0.0634	13.7964	217.4574	50
55	24.6503	0.0406	394.1720	0.0025	15.9905	0.0625	14.3411	229.3222	55
60	32.9877	0.0303	533.1282	0.0019	16.1614	0.0619	14.7909	239.0428	60
65	44.1450	0.0227	719.0829	0.0014	16.2891	0.0614	15.1601	246.9450	65
70	59.0759	0.0169	967.9322	0.0010	16.3845	0.0610	15.4613	253.3271	70
75	79.0569	0.0126	1300.9487	0.0008	16.4558	0.0608	15.7058	258.4527	75
80	105.7960	0.0095	1746.5999	0.0006	16.5091	0.0606	15.9033	262.5493	80
85	141.5789	0.0071	2342.9817	0.0004	16.5489	0.0604	16.0620	265.8096	85
90	189.4645	0.0053	3141.0752	0.0003	16.5787	0.0603	16.1891	268.3946	90
95	253.5463	0.0039	4209.1042	0.0002	16.6009	0.0602	16.2905	270.4375	95
100	339.3021	0.0029	5638.3681	0.0002	16.6175	0.0602	16.3711	272.0471	100

Table C.13 Interest Rate Factors (7.0%)

	Single Payment		Equal Payment Series				Gradient Series		
N	Compound Amount Factor (F/P,i,N)	Present Worth Factor (P/F,i,N)	Compound Amount Factor (F/A,i,N)	Sinking Fund Factor (A/F,i,N)	Present Worth Factor (P/A,i,N)	Capital Recovery Factor (A/P,i,N)	Gradient Uniform Series (A/G,i,N)	Gradient Present Worth (P/G.i.N)	N
1	1.0700	0.9346	1.0000	1.0000	0.9346	1.0700	0.0000	0.0000	1
2	1.1449	0.8734	2.0700	0.4831	1.8080	0.5531	0.4831	0.8734	2
3	1.2250	0.8163	3.2149	0.3111	2.6243	0.3811	0.9549	2.5060	3
4	1.3108	0.7629	4.4399	0.2252	3.3872	0.2952	1.4155	4.7947	4
5	1.4026	0.7130	5.7507	0.1739	4.1002	0.2439	1.8650	7.6467	5
6	1.5007	0.6663	7.1533	0.1398	4.7665	0.2098	2.3032	10.9784	6
7	1.6058	0.6227	8.6540	0.1156	5.3893	0.1856	2.7304	14.7149	7
8	1.7182	0.5820	10.2598	0.0975	5.9713	0.1675	3.1465	18.7889	8
9	1.8385	0.5439	11.9780	0.0835	6.5152	0.1535	3.5517	23.1404	9
10	1.9672	0.5083	13.8164	0.0724	7.0236	0.1424	3.9461	27.7156	10
11	2.1049	0.4751	15.7836	0.0634	7.4987	0.1334	4.3296	32.4665	11
12	2.2522	0.4440	17.8885	0.0559	7.9427	0.1259	4.7025	37.3506	12
13	2.4098	0.4150	20.1406	0.0497	8.3577	0.1197	5.0648	42.3302	13
14	2.5785	0.3878	22.5505	0.0443	8.7455	0.1143	5.4167	47.3718	14
15	2.7590	0.3624	25.1290	0.0398	9.1079	0.1098	5.7583	52.4461	15
16	2.9522	0.3387	27.8881	0.0359	9.4466	0.1059	6.0897	57.5271	16
17	3.1588	0.3166	30.8402	0.0324	9.7632	0.1024	6.4110	62.5923	17
18	3.3799	0.2959	33.9990	0.0294	10.0591	0.0994	6.7225	67.6219	18
19	3.6165	0.2765	37.3790	0.0268	10.3356	0.0968	7.0242	72.5991	19
20	3.8697	0.2584	40.9955	0.0244	10.5940	0.0944	7.3163	77.5091	20
21	4.1406	0.2415	44.8652	0.0223	10.8355	0.0923	7.5990	82.3393	21
22	4.4304	0.2257	49.0057	0.0204	11.0612	0.0904	7.8725	87.0793	22
23	4.7405	0.2109	53.4361	0.0187	11.2722	0.0887	8.1369	91.7201	23
24	5.0724	0.1971	58.1767	0.0172	11.4693	0.0872	8.3923	96.2545	24
25	5.4274	0.1842	63.2490	0.0158	11.6536	0.0858	8.6391	100.6765	25
26	5.8074	0.1722	68.6765	0.0146	11.8258	0.0846	8.8773	104.9814	26
27	6.2139	0.1609	74.4838	0.0134	11.9867	0.0834	9.1072	109.1656	27
28	6.6488	0.1504	80.6977	0.0124	12.1371	0.0824	9.3289	113.2264	28
29	7.1143	0.1406	87.3465	0.0114	12.2777	0.0814	9.5427	117.1622	29
30	7.6123	0.1314	94.4608	0.0106	12.4090	0.0806	9.7487	120.9718	30
31	8.1451	0.1228	102.0730	0.0098	12.5318	0.0798	9.9471	124.6550	31
32	8.7153	0.1147	110.2182	0.0091	12.6466	0.0791	10.1381	128.2120	32
33	9.3253	0.1072	118.9334	0.0084	12.7538	0.0784	10.3219	131.6435	33
34	9.9781	0.1002	128.2588	0.0078	12.8540	0.0778	10.4987	134.9507	34
35	10.6766	0.0937	138.2369	0.0072	12.9477	0.0772	10.6687	138.1353	35
40	14.9745	0.0668	199.6351	0.0050	13.3317	0.0750	11.4233	152.2928	40
45	21.0025	0.0476	285.7493	0.0035	13.6055	0.0735	12.0360	163.7559	45
50	29.4570	0.0339	406.5289	0.0025	13.8007	0.0725	12.5287	172.9051	50
55	41.3150	0.0242	575.9286	0.0017	13.9399	0.0717	12.9215	180.1243	55
60	57.9464	0.0173	813.5204	0.0012	14.0392	0.0712	13.2321	185.7677	60
65	81.2729	0.0123	1146.7552	0.0009	14.1099	0.0709	13.4760	190.1452	65
70	113.9894	0.0088	1614.1342	0.0006	14.1604	0.0706	13.6662	193.5185	70
75	159.8760	0.0063	2269.6574	0.0004	14.1964	0.0704	13.8136	196.1035	75
80	224.2344	0.0045	3189.0627	0.0003	14.2220	0.0703	13.9273	198.0748	80
85	314.5003	0.0032	4478.5761	0.0002	14.2403	0.0702	14.0146	199.5717	85
90	441.1030	0.0023	6287.1854	0.0002	14.2533	0.0702	14.0812	200.7042	90
95	618.6697	0.0016	8823.8535	0.0001	14.2626	0.0701	14.1319	201.5581	95
100	867.7163	0.0012	12381.6618	0.0001	14.2693	0.0701	14.1703	202.2001	100

Table C.14 Interest Rate Factors (8.0%)

	Single Payment		Equal Payment Series				Gradient Series		
N	Compound Amount Factor (F/P,i,N)	Present Worth Factor (P/F,i,N)	Compound Amount Factor (F/A,i,N)	Sinking Fund Factor (A/F,i,N)	Present Worth Factor (P/A,i,N)	Capital Recovery Factor (A/P,i,N)	Gradient Uniform Series (A/G,i,N)	Gradient Present Worth (P/G.i.N)	N
1	1.0800	0.9259	1.0000	1.0000	0.9259	1.0800	0.0000	0.0000	1
2	1.1664	0.8573	2.0800	0.4808	1.7833	0.5608	0.4808	0.8573	2
3	1.2597	0.7938	3.2464	0.3080	2.5771	0.3880	0.9487	2.4450	3
4	1.3605	0.7350	4.5061	0.2219	3.3121	0.3019	1.4040	4.6501	4
5	1.4693	0.6806	5.8666	0.1705	3.9927	0.2505	1.8465	7.3724	5
6	1.5869	0.6302	7.3359	0.1363	4.6229	0.2163	2.2763	10.5233	6
7	1.7138	0.5835	8.9228	0.1121	5.2064	0.1921	2.6937	14.0242	7
8	1.8509	0.5403	10.6366	0.0940	5.7466	0.1740	3.0985	17.8061	8
9	1.9990	0.5002	12.4876	0.0801	6.2469	0.1601	3.4910	21.8081	9
10	2.1589	0.4632	14.4866	0.0690	6.7101	0.1490	3.8713	25.9768	10
11	2.3316	0.4289	16.6455	0.0601	7.1390	0.1401	4.2395	30.2657	11
12	2.5182	0.3971	18.9771	0.0527	7.5361	0.1327	4.5957	34.6339	12
13	2.7196	0.3677	21.4953	0.0465	7.9038	0.1265	4.9402	39.0463	13
14	2.9372	0.3405	24.2149	0.0413	8.2442	0.1213	5.2731	43.4723	14
15	3.1722	0.3152	27.1521	0.0368	8.5595	0.1168	5.5945	47.8857	15
16	3.4259	0.2919	30.3243	0.0330	8.8514	0.1130	5.9046	52.2640	16
17	3.7000	0.2703	33.7502	0.0296	9.1216	0.1096	6.2037	56.5883	17
18	3.9960	0.2502	37.4502	0.0267	9.3719	0.1067	6.4920	60.8426	18
19	4.3157	0.2317	41.4463	0.0241	9.6036	0.1041	6.7697	65.0134	19
20	4.6610	0.2145	45.7620	0.0219	9.8181	0.1019	7.0369	69.0898	20
21	5.0338	0.1987	50.4229	0.0198	10.0168	0.0998	7.2940	73.0629	21
22	5.4365	0.1839	55.4568	0.0180	10.2007	0.0980	7.5412	76.9257	22
23	5.8715	0.1703	60.8933	0.0164	10.3711	0.0964	7.7786	80.6726	23
24	6.3412	0.1577	66.7648	0.0150	10.5288	0.0950	8.0066	84.2997	24
25	6.8485	0.1460	73.1059	0.0137	10.6748	0.0937	8.2254	87.8041	25
26	7.3964	0.1352	79.9544	0.0125	10.8100	0.0925	8.4352	91.1842	26
27	7.9881	0.1252	87.3508	0.0114	10.9352	0.0914	8.6363	94.4390	27
28	8.6271	0.1159	95.3388	0.0105	11.0511	0.0905	8.8289	97.5687	28
29	9.3173	0.1073	103.9659	0.0096	11.1584	0.0896	9.0133	100.5738	29
30	10.0627	0.0994	113.2832	0.0088	11.2578	0.0888	9.1897	103.4558	30
31	10.8677	0.0920	123.3459	0.0081	11.3498	0.0881	9.3584	106.2163	31
32	11.7371	0.0852	134.2135	0.0075	11.4350	0.0875	9.5197	108.8575	32
33	12.6760	0.0789	145.9506	0.0069	11.5139	0.0869	9.6737	111.3819	33
34	13.6901	0.0730	158.6267	0.0063	11.5869	0.0863	9.8208	113.7924	34
35	14.7853	0.0676	172.3168	0.0058	11.6546	0.0858	9.9611	116.0920	35
40	21.7245	0.0460	259.0565	0.0039	11.9246	0.0839	10.5699	126.0422	40
45	31.9204	0.0313	386.5056	0.0026	12.1084	0.0826	11.0447	133.7331	45
50	46.9016	0.0213	573.7702	0.0017	12.2335	0.0817	11.4107	139.5928	50
55	68.9139	0.0145	848.9232	0.0012	12.3186	0.0812	11.6902	144.0065	55
60	101.2571	0.0099	1253.2133	0.0008	12.3766	0.0808	11.9015	147.3000	60
65	148.7798	0.0067	1847.2481	0.0005	12.4160	0.0805	12.0602	149.7387	65
70	218.6064	0.0046	2720.0801	0.0004	12.4428	0.0804	12.1783	151.5326	70
75	321.2045	0.0031	4002.5566	0.0002	12.4611	0.0802	12.2658	152.8448	75
80	471.9548	0.0021	5886.9354	0.0002	12.4735	0.0802	12.3301	153.8001	80
85	693.4565	0.0014	8655.7061	0.0001	12.4820	0.0801	12.3772	154.4925	85
90	1018.9151	0.0010	12723.9386	0.0001	12.4877	0.0801	12.4116	154.9925	90
95	1497.1205	0.0007	18701.5069	0.0001	12.4917	0.0801	12.4365	155.3524	95
100	2199.7613	0.0005	27484.5157	0.0000	12.4943	0.0800	12.4545	155.6107	100

Table C.15 Interest Rate Factors (9.0%)

	Single Payment		Equal Payment Series				Gradient Series		
N	Compound Amount Factor (F/P,i,N)	Present Worth Factor (P/F,i,N)	Compound Amount Factor (F/A,i,N)	Sinking Fund Factor (A/F,i,N)	Present Worth Factor (P/A,i,N)	Capital Recovery Factor (A/P,i,N)	Gradient Uniform Series (A/G,i,N)	Gradient Present Worth (P/G.i,N)	N
1	1.0900	0.9174	1.0000	1.0000	0.9174	1.0900	0.0000	0.0000	1
2	1.1881	0.8417	2.0900	0.4785	1.7591	0.5685	0.4785	0.8417	2
3	1.2950	0.7722	3.2781	0.3051	2.5313	0.3951	0.9426	2.3860	3
4	1.4116	0.7084	4.5731	0.2187	3.2397	0.3087	1.3925	4.5113	4
5	1.5386	0.6499	5.9847	0.1671	3.8897	0.2571	1.8282	7.1110	5
6	1.6771	0.5963	7.5233	0.1329	4.4859	0.2229	2.2498	10.0924	6
7	1.8280	0.5470	9.2004	0.1087	5.0330	0.1987	2.6574	13.3746	7
8	1.9926	0.5019	11.0285	0.0907	5.5348	0.1807	3.0512	16.8877	8
9	2.1719	0.4604	13.0210	0.0768	5.9952	0.1668	3.4312	20.5711	9
10	2.3674	0.4224	15.1929	0.0658	6.4177	0.1558	3.7978	24.3728	10
11	2.5804	0.3875	17.5603	0.0569	6.8052	0.1469	4.1510	28.2481	11
12	2.8127	0.3555	20.1407	0.0497	7.1607	0.1397	4.4910	32.1590	12
13	3.0658	0.3262	22.9534	0.0436	7.4869	0.1336	4.8182	36.0731	13
14	3.3417	0.2992	26.0192	0.0384	7.7862	0.1284	5.1326	39.9633	14
15	3.6425	0.2745	29.3609	0.0341	8.0607	0.1241	5.4346	43.8069	15
16	3.9703	0.2519	33.0034	0.0303	8.3126	0.1203	5.7245	47.5849	16
17	4.3276	0.2311	36.9737	0.0270	8.5436	0.1170	6.0024	51.2821	17
18	4.7171	0.2120	41.3013	0.0242	8.7556	0.1142	6.2687	54.8860	18
19	5.1417	0.1945	46.0185	0.0217	8.9501	0.1117	6.5236	58.3868	19
20	5.6044	0.1784	51.1601	0.0195	9.1285	0.1095	6.7674	61.7770	20
21	6.1088	0.1637	56.7645	0.0176	9.2922	0.1076	7.0006	65.0509	21
22	6.6586	0.1502	62.8733	0.0159	9.4424	0.1059	7.2232	68.2048	22
23	7.2579	0.1378	69.5319	0.0144	9.5802	0.1044	7.4357	71.2359	23
24	7.9111	0.1264	76.7898	0.0130	9.7066	0.1030	7.6384	74.1433	24
25	8.6231	0.1160	84.7009	0.0118	9.8226	0.1018	7.8316	76.9265	25
26	9.3992	0.1064	93.3240	0.0107	9.9290	0.1007	8.0156	79.5863	26
27	10.2451	0.0976	102.7231	0.0097	10.0266	0.0997	8.1906	82.1241	27
28	11.1671	0.0895	112.9682	0.0089	10.1161	0.0989	8.3571	84.5419	28
29	12.1722	0.0822	124.1354	0.0081	10.1983	0.0981	8.5154	86.8422	29
30	13.2677	0.0754	136.3075	0.0073	10.2737	0.0973	8.6657	89.0280	30
31	14.4618	0.0691	149.5752	0.0067	10.3428	0.0967	8.8083	91.1024	31
32	15.7633	0.0634	164.0370	0.0061	10.4062	0.0961	8.9436	93.0690	32
33	17.1820	0.0582	179.8003	0.0056	10.4644	0.0956	9.0718	94.9314	33
34	18.7284	0.0534	196.9823	0.0051	10.5178	0.0951	9.1933	96.6935	34
35	20.4140	0.0490	215.7108	0.0046	10.5668	0.0946	9.3083	98.3590	35
40	31.4094	0.0318	337.8824	0.0030	10.7574	0.0930	9.7957	105.3762	40
45	48.3273	0.0207	525.8587	0.0019	10.8812	0.0919	10.1603	110.5561	45
50	74.3575	0.0134	815.0836	0.0012	10.9617	0.0912	10.4295	114.3251	50
55	114.4083	0.0087	1260.0918	0.0008	11.0140	0.0908	10.6261	117.0362	55
60	176.0313	0.0057	1944.7921	0.0005	11.0480	0.0905	10.7683	118.9683	60
65	270.8460	0.0037	2998.2885	0.0003	11.0701	0.0903	10.8702	120.3344	65
70	416.7301	0.0024	4619.2232	0.0002	11.0844	0.0902	10.9427	121.2942	70
75	641.1909	0.0016	7113.2321	0.0001	11.0938	0.0901	10.9940	121.9646	75
80	986.5517	0.0010	10950.5741	0.0001	11.0998	0.0901	11.0299	122.4306	80
85	1517.9320	0.0007	16854.8003	0.0001	11.1038	0.0901	11.0551	122.7533	85
90	2335.5266	0.0004	25939.1842	0.0000	11.1064	0.0900	11.0726	122.9758	90
95	3593.4971	0.0003	39916.6350	0.0000	11.1080	0.0900	11.0847	123.1287	95
100	5529.0408	0.0002	61422.6755	0.0000	11.1091	0.0900	11.0930	123.2335	100

Table C.16 Interest Rate Factors (10.0%)

	Single Payment		Equal Payment Series				Gradient Series		
N	Compound Amount Factor (F/P,i,N)	Present Worth Factor (P/F,i,N)	Compound Amount Factor (F/A,i,N)	Sinking Fund Factor (A/F,i,N)	Present Worth Factor (P/A,i,N)	Capital Recovery Factor (A/P,i,N)	Gradient Uniform Series (A/G,i,N)	Gradient Present Worth (P/G.i.N)	N
1	1.1000	0.9091	1.0000	1.0000	0.9091	1.1000	0.0000	0.0000	1
2	1.2100	0.8264	2.1000	0.4762	1.7355	0.5762	0.4762	0.8264	2
3	1.3310	0.7513	3.3100	0.3021	2.4869	0.4021	0.9366	2.3291	3
4	1.4641	0.6830	4.6410	0.2155	3.1699	0.3155	1.3812	4.3781	4
5	1.6105	0.6209	6.1051	0.1638	3.7908	0.2638	1.8101	6.8618	5
6	1.7716	0.5645	7.7156	0.1296	4.3553	0.2296	2.2236	9.6842	6
7	1.9487	0.5132	9.4872	0.1054	4.8684	0.2054	2.6216	12.7631	7
8	2.1436	0.4665	11.4359	0.0874	5.3349	0.1874	3.0045	16.0287	8
9	2.3579	0.4241	13.5795	0.0736	5.7590	0.1736	3.3724	19.4215	9
10	2.5937	0.3855	15.9374	0.0627	6.1446	0.1627	3.7255	22.8913	10
11	2.8531	0.3505	18.5312	0.0540	6.4951	0.1540	4.0641	26.3963	11
12	3.1384	0.3186	21.3843	0.0468	6.8137	0.1468	4.3884	29.9012	12
13	3.4523	0.2897	24.5227	0.0408	7.1034	0.1408	4.6988	33.3772	13
14	3.7975	0.2633	27.9750	0.0357	7.3667	0.1357	4.9955	36.8005	14
15	4.1772	0.2394	31.7725	0.0315	7.6061	0.1315	5.2789	40.1520	15
16	4.5950	0.2176	35.9497	0.0278	7.8237	0.1278	5.5493	43.4164	16
17	5.0545	0.1978	40.5447	0.0247	8.0216	0.1247	5.8071	46.5819	17
18	5.5599	0.1799	45.5992	0.0219	8.2014	0.1219	6.0526	49.6395	18
19	6.1159	0.1635	51.1591	0.0195	8.3649	0.1195	6.2861	52.5827	19
20	6.7275	0.1486	57.2750	0.0175	8.5136	0.1175	6.5081	55.4069	20
21	7.4002	0.1351	64.0025	0.0156	8.6487	0.1156	6.7189	58.1095	21
22	8.1403	0.1228	71.4027	0.0140	8.7715	0.1140	6.9189	60.6893	22
23	8.9543	0.1117	79.5430	0.0126	8.8832	0.1126	7.1085	63.1462	23
24	9.8497	0.1015	88.4973	0.0113	8.9847	0.1113	7.2881	65.4813	24
25	10.8347	0.0923	98.3471	0.0102	9.0770	0.1102	7.4580	67.6964	25
26	11.9182	0.0839	109.1818	0.0092	9.1609	0.1092	7.6186	69.7940	26
27	13.1100	0.0763	121.0999	0.0083	9.2372	0.1083	7.7704	71.7773	27
28	14.4210	0.0693	134.2099	0.0075	9.3066	0.1075	7.9137	73.6495	28
29	15.8631	0.0630	148.6309	0.0067	9.3696	0.1067	8.0489	75.4146	29
30	17.4494	0.0573	164.4940	0.0061	9.4269	0.1061	8.1762	77.0766	30
31	19.1943	0.0521	181.9434	0.0055	9.4790	0.1055	8.2962	78.6395	31
32	21.1138	0.0474	201.1378	0.0050	9.5264	0.1050	8.4091	80.1078	32
33	23.2252	0.0431	222.2515	0.0045	9.5694	0.1045	8.5152	81.4856	33
34	25.5477	0.0391	245.4767	0.0041	9.6086	0.1041	8.6149	82.7773	34
35	28.1024	0.0356	271.0244	0.0037	9.6442	0.1037	8.7086	83.9872	35
40	45.2593	0.0221	442.5926	0.0023	9.7791	0.1023	9.0962	88.9525	40
45	72.8905	0.0137	718.9048	0.0014	9.8628	0.1014	9.3740	92.4544	45
50	117.3909	0.0085	1163.9085	0.0009	9.9148	0.1009	9.5704	94.8889	50
55	189.0591	0.0053	1880.5914	0.0005	9.9471	0.1005	9.7075	96.5619	55
60	304.4816	0.0033	3034.8164	0.0003	9.9672	0.1003	9.8023	97.7010	60
65	490.3707	0.0020	4893.7073	0.0002	9.9796	0.1002	9.8672	98.4705	65
70	789.7470	0.0013	7887.4696	0.0001	9.9873	0.1001	9.9113	98.9870	70
75	1271.8954	0.0008	12708.9537	0.0001	9.9921	0.1001	9.9410	99.3317	75
80	2048.4002	0.0005	20474.0021	0.0000	9.9951	0.1000	9.9609	99.5606	80
85	3298.9690	0.0003	32979.6903	0.0000	9.9970	0.1000	9.9742	99.7120	85
90	5313.0226	0.0002	53120.2261	0.0000	9.9981	0.1000	9.9831	99.8118	90
95	8556.6760	0.0001	85556.7605	0.0000	9.9988	0.1000	9.9889	99.8773	95
100	13780.6123	0.0001	137796.1234	0.0000	9.9993	0.1000	9.9927	99.9202	100

Table C.17 Interest Rate Factors (11.0%)

	Single Payment		Equal Payment Series				Gradient Series		
N	Compound Amount Factor (F/P,i,N)	Present Worth Factor (P/F,i,N)	Compound Amount Factor (F/A,i,N)	Sinking Fund Factor (A/F,i,N)	Present Worth Factor (P/A,i,N)	Capital Recovery Factor (A/P,i,N)	Gradient Uniform Series (A/G,i,N)	Gradient Present Worth (P/G.i.N)	N
1	1.1100	0.9009	1.0000	1.0000	0.9009	1.1100	0.0000	0.0000	1
2	1.2321	0.8116	2.1100	0.4739	1.7125	0.5839	0.4739	0.8116	2
3	1.3676	0.7312	3.3421	0.2992	2.4437	0.4092	0.9306	2.2740	3
4	1.5181	0.6587	4.7097	0.2123	3.1024	0.3223	1.3700	4.2502	4
5	1.6851	0.5935	6.2278	0.1606	3.6959	0.2706	1.7923	6.6240	5
6	1.8704	0.5346	7.9129	0.1264	4.2305	0.2364	2.1976	9.2972	6
7	2.0762	0.4817	9.7833	0.1022	4.7122	0.2122	2.5863	12.1872	7
8	2.3045	0.4339	11.8594	0.0843	5.1461	0.1943	2.9585	15.2246	8
9	2.5580	0.3909	14.1640	0.0706	5.5370	0.1806	3.3144	18.3520	9
10	2.8394	0.3522	16.7220	0.0598	5.8892	0.1698	3.6544	21.5217	10
11	3.1518	0.3173	19.5614	0.0511	6.2065	0.1611	3.9788	24.6945	11
12	3.4985	0.2858	22.7132	0.0440	6.4924	0.1540	4.2879	27.8388	12
13	3.8833	0.2575	26.2116	0.0382	6.7499	0.1482	4.5822	30.9290	13
14	4.3104	0.2320	30.0949	0.0332	6.9819	0.1432	4.8619	33.9449	14
15	4.7846	0.2090	34.4054	0.0291	7.1909	0.1391	5.1275	36.8709	15
16	5.3109	0.1883	39.1899	0.0255	7.3792	0.1355	5.3794	39.6953	16
17	5.8951	0.1696	44.5008	0.0225	7.5488	0.1325	5.6180	42.4095	17
18	6.5436	0.1528	50.3959	0.0198	7.7016	0.1298	5.8439	45.0074	18
19	7.2633	0.1377	56.9395	0.0176	7.8393	0.1276	6.0574	47.4856	19
20	8.0623	0.1240	64.2028	0.0156	7.9633	0.1256	6.2590	49.8423	20
21	8.9492	0.1117	72.2651	0.0138	8.0751	0.1238	6.4491	52.0771	21
22	9.9336	0.1007	81.2143	0.0123	8.1757	0.1223	6.6283	54.1912	22
23	11.0263	0.0907	91.1479	0.0110	8.2664	0.1210	6.7969	56.1864	23
24	12.2392	0.0817	102.1742	0.0098	8.3481	0.1198	6.9555	58.0656	24
25	13.5855	0.0736	114.4133	0.0087	8.4217	0.1187	7.1045	59.8322	25
26	15.0799	0.0663	127.9988	0.0078	8.4881	0.1178	7.2443	61.4900	26
27	16.7386	0.0597	143.0786	0.0070	8.5478	0.1170	7.3754	63.0433	27
28	18.5799	0.0538	159.8173	0.0063	8.6016	0.1163	7.4982	64.4965	28
29	20.6237	0.0485	178.3972	0.0056	8.6501	0.1156	7.6131	65.8542	29
30	22.8923	0.0437	199.0209	0.0050	8.6938	0.1150	7.7206	67.1210	30
31	25.4104	0.0394	221.9132	0.0045	8.7331	0.1145	7.8210	68.3016	31
32	28.2056	0.0355	247.3236	0.0040	8.7686	0.1140	7.9147	69.4007	32
33	31.3082	0.0319	275.5292	0.0036	8.8005	0.1136	8.0021	70.4228	33
34	34.7521	0.0288	306.8374	0.0033	8.8293	0.1133	8.0836	71.3724	34
35	38.5749	0.0259	341.5896	0.0029	8.8552	0.1129	8.1594	72.2538	35
40	65.0009	0.0154	581.8261	0.0017	8.9511	0.1117	8.4659	75.7789	40
45	109.5302	0.0091	986.6386	0.0010	9.0079	0.1110	8.6763	78.1551	45
50	184.5648	0.0054	1668.7712	0.0006	9.0417	0.1106	8.8185	79.7341	50
55	311.0025	0.0032	2818.2042	0.0004	9.0617	0.1104	8.9135	80.7712	55
60	524.0572	0.0019	4755.0658	0.0002	9.0736	0.1102	8.9762	81.4461	60

Table C.18 Interest Rate Factors (12.0%)

	Single Payment		Equal Payment Series				Gradient Series		
N	Compound Amount Factor (F/P,i,N)	Present Worth Factor (P/F,i,N)	Compound Amount Factor (F/A,i,N)	Sinking Fund Factor (A/F,i,N)	Present Worth Factor (P/A,i,N)	Capital Recovery Factor (A/P,i,N)	Gradient Uniform Series (A/G,i,N)	Gradient Present Worth (P/G.i.N)	N
1	1.1200	0.8929	1.0000	1.0000	0.8929	1.1200	0.0000	0.0000	1
2	1.2544	0.7972	2.1200	0.4717	1.6901	0.5917	0.4717	0.7972	2
3	1.4049	0.7118	3.3744	0.2963	2.4018	0.4163	0.9246	2.2208	3
4	1.5735	0.6355	4.7793	0.2092	3.0373	0.3292	1.3589	4.1273	4
5	1.7623	0.5674	6.3528	0.1574	3.6048	0.2774	1.7746	6.3970	5
6	1.9738	0.5066	8.1152	0.1232	4.1114	0.2432	2.1720	8.9302	6
7	2.2107	0.4523	10.0890	0.0991	4.5638	0.2191	2.5515	11.6443	7
8	2.4760	0.4039	12.2997	0.0813	4.9676	0.2013	2.9131	14.4714	8
9	2.7731	0.3606	14.7757	0.0677	5.3282	0.1877	3.2574	17.3563	9
10	3.1058	0.3220	17.5487	0.0570	5.6502	0.1770	3.5847	20.2541	10
11	3.4785	0.2875	20.6546	0.0484	5.9377	0.1684	3.8953	23.1288	11
12	3.8960	0.2567	24.1331	0.0414	6.1944	0.1614	4.1897	25.9523	12
13	4.3635	0.2292	28.0291	0.0357	6.4235	0.1557	4.4683	28.7024	13
14	4.8871	0.2046	32.3926	0.0309	6.6282	0.1509	4.7317	31.3624	14
15	5.4736	0.1827	37.2797	0.0268	6.8109	0.1468	4.9803	33.9202	15
16	6.1304	0.1631	42.7533	0.0234	6.9740	0.1434	5.2147	36.3670	16
17	6.8660	0.1456	48.8837	0.0205	7.1196	0.1405	5.4353	38.6973	17
18	7.6900	0.1300	55.7497	0.0179	7.2497	0.1379	5.6427	40.9080	18
19	8.6128	0.1161	63.4397	0.0158	7.3658	0.1358	5.8375	42.9979	19
20	9.6463	0.1037	72.0524	0.0139	7.4694	0.1339	6.0202	44.9676	20
21	10.8038	0.0926	81.6987	0.0122	7.5620	0.1322	6.1913	46.8188	21
22	12.1003	0.0826	92.5026	0.0108	7.6446	0.1308	6.3514	48.5543	22
23	13.5523	0.0738	104.6029	0.0096	7.7184	0.1296	6.5010	50.1776	23
24	15.1786	0.0659	118.1552	0.0085	7.7843	0.1285	6.6406	51.6929	24
25	17.0001	0.0588	133.3339	0.0075	7.8431	0.1275	6.7708	53.1046	25
26	19.0401	0.0525	150.3339	0.0067	7.8957	0.1267	6.8921	54.4177	26
27	21.3249	0.0469	169.3740	0.0059	7.9426	0.1259	7.0049	55.6369	27
28	23.8839	0.0419	190.6989	0.0052	7.9844	0.1252	7.1098	56.7674	28
29	26.7499	0.0374	214.5828	0.0047	8.0218	0.1247	7.2071	57.8141	29
30	29.9599	0.0334	241.3327	0.0041	8.0552	0.1241	7.2974	58.7821	30
31	33.5551	0.0298	271.2926	0.0037	8.0850	0.1237	7.3811	59.6761	31
32	37.5817	0.0266	304.8477	0.0033	8.1116	0.1233	7.4586	60.5010	32
33	42.0915	0.0238	342.4294	0.0029	8.1354	0.1229	7.5302	61.2612	33
34	47.1425	0.0212	384.5210	0.0026	8.1566	0.1226	7.5965	61.9612	34
35	52.7996	0.0189	431.6635	0.0023	8.1755	0.1223	7.6577	62.6052	35
40	93.0510	0.0107	767.0914	0.0013	8.2438	0.1213	7.8988	65.1159	40
45	163.9876	0.0061	1358.2300	0.0007	8.2825	0.1207	8.0572	66.7342	45
50	289.0022	0.0035	2400.0182	0.0004	8.3045	0.1204	8.1597	67.7624	50
55	509.3206	0.0020	4236.0050	0.0002	8.3170	0.1202	8.2251	68.4082	55
60	897.5969	0.0011	7471.6411	0.0001	8.3240	0.1201	8.2664	68.8100	60

Table C.19 Interest Rate Factors (13.0%)

	Single Payment		Equal Payment Series				Gradient Series		
N	Compound Amount Factor (F/P,i,N)	Present Worth Factor (P/F,i,N)	Compound Amount Factor (F/A,i,N)	Sinking Fund Factor (A/F,i,N)	Present Worth Factor (P/A,i,N)	Capital Recovery Factor (A/P,i,N)	Gradient Uniform Series (A/G,i,N)	Gradient Present Worth (P/G.i.N)	N
1	1.1300	0.8850	1.0000	1.0000	0.8850	1.1300	0.0000	0.0000	1
2	1.2769	0.7831	2.1300	0.4695	1.6681	0.5995	0.4695	0.7831	2
3	1.4429	0.6931	3.4069	0.2935	2.3612	0.4235	0.9187	2.1692	3
4	1.6305	0.6133	4.8498	0.2062	2.9745	0.3362	1.3479	4.0092	4
5	1.8424	0.5428	6.4803	0.1543	3.5172	0.2843	1.7571	6.1802	5
6	2.0820	0.4803	8.3227	0.1202	3.9975	0.2502	2.1468	8.5818	6
7	2.3526	0.4251	10.4047	0.0961	4.4226	0.2261	2.5171	11.1322	7
8	2.6584	0.3762	12.7573	0.0784	4.7988	0.2084	2.8685	13.7653	8
9	3.0040	0.3329	15.4157	0.0649	5.1317	0.1949	3.2014	16.4284	9
10	3.3946	0.2946	18.4197	0.0543	5.4262	0.1843	3.5162	19.0797	10
11	3.8359	0.2607	21.8143	0.0458	5.6869	0.1758	3.8134	21.6867	11
12	4.3345	0.2307	25.6502	0.0390	5.9176	0.1690	4.0936	24.2244	12
13	4.8980	0.2042	29.9847	0.0334	6.1218	0.1634	4.3573	26.6744	13
14	5.5348	0.1807	34.8827	0.0287	6.3025	0.1587	4.6050	29.0232	14
15	6.2543	0.1599	40.4175	0.0247	6.4624	0.1547	4.8375	31.2617	15
16	7.0673	0.1415	46.6717	0.0214	6.6039	0.1514	5.0552	33.3841	16
17	7.9861	0.1252	53.7391	0.0186	6.7291	0.1486	5.2589	35.3876	17
18	9.0243	0.1108	61.7251	0.0162	6.8399	0.1462	5.4491	37.2714	18
19	10.1974	0.0981	70.7494	0.0141	6.9380	0.1441	5.6265	39.0366	19
20	11.5231	0.0868	80.9468	0.0124	7.0248	0.1424	5.7917	40.6854	20
21	13.0211	0.0768	92.4699	0.0108	7.1016	0.1408	5.9454	42.2214	21
22	14.7138	0.0680	105.4910	0.0095	7.1695	0.1395	6.0881	43.6486	22
23	16.6266	0.0601	120.2048	0.0083	7.2297	0.1383	6.2205	44.9718	23
24	18.7881	0.0532	136.8315	0.0073	7.2829	0.1373	6.3431	46.1960	24
25	21.2305	0.0471	155.6196	0.0064	7.3300	0.1364	6.4566	47.3264	25
26	23.9905	0.0417	176.8501	0.0057	7.3717	0.1357	6.5614	48.3685	26
27	27.1093	0.0369	200.8406	0.0050	7.4086	0.1350	6.6582	49.3276	27
28	30.6335	0.0326	227.9499	0.0044	7.4412	0.1344	6.7474	50.2090	28
29	34.6158	0.0289	258.5834	0.0039	7.4701	0.1339	6.8296	51.0179	29
30	39.1159	0.0256	293.1992	0.0034	7.4957	0.1334	6.9052	51.7592	30
31	44.2010	0.0226	332.3151	0.0030	7.5183	0.1330	6.9747	52.4380	31
32	49.9471	0.0200	376.5161	0.0027	7.5383	0.1327	7.0385	53.0586	32
33	56.4402	0.0177	426.4632	0.0023	7.5560	0.1323	7.0971	53.6256	33
34	63.7774	0.0157	482.9034	0.0021	7.5717	0.1321	7.1507	54.1430	34
35	72.0685	0.0139	546.6808	0.0018	7.5856	0.1318	7.1998	54.6148	35
40	132.7816	0.0075	1013.7042	0.0010	7.6344	0.1310	7.3888	56.4087	40
45	244.6414	0.0041	1874.1646	0.0005	7.6609	0.1305	7.5076	57.5148	45
50	450.7359	0.0022	3459.5071	0.0003	7.6752	0.1303	7.5811	58.1870	50
55	830.4517	0.0012	6380.3979	0.0002	7.6830	0.1302	7.6260	58.5909	55
60	1530.0535	0.0007	11761.9498	0.0001	7.6873	0.1301	7.6531	58.8313	60

Table C.20 Interest Rate Factors (14.0%)

	Single Payment		Equal Payment Series				Gradient Series		
N	Compound Amount Factor (F/P,i,N)	Present Worth Factor (P/F,i,N)	Compound Amount Factor (F/A,i,N)	Sinking Fund Factor (A/F,i,N)	Present Worth Factor (P/A,i,N)	Capital Recovery Factor (A/P,i,N)	Gradient Uniform Series (A/G,i,N)	Gradient Present Worth (P/G.i.N)	N
1	1.1400	0.8772	1.0000	1.0000	0.8772	1.1400	0.0000	-0.0000	1
2	1.2996	0.7695	2.1400	0.4673	1.6467	0.6073	0.4673	0.7695	2
3	1.4815	0.6750	3.4396	0.2907	2.3216	0.4307	0.9129	2.1194	3
4	1.6890	0.5921	4.9211	0.2032	2.9137	0.3432	1.3370	3.8957	4
5	1.9254	0.5194	6.6101	0.1513	3.4331	0.2913	1.7399	5.9731	5
6	2.1950	0.4556	8.5355	0.1172	3.8887	0.2572	2.1218	8.2511	6
7	2.5023	0.3996	10.7305	0.0932	4.2883	0.2332	2.4832	10.6489	7
8	2.8526	0.3506	13.2328	0.0756	4.6389	0.2156	2.8246	13.1028	8
9	3.2519	0.3075	16.0853	0.0622	4.9464	0.2022	3.1463	15.5629	9
10	3.7072	0.2697	19.3373	0.0517	5.2161	0.1917	3.4490	17.9906	10
11	4.2262	0.2366	23.0445	0.0434	5.4527	0.1834	3.7333	20.3567	11
12	4.8179	0.2076	27.2707	0.0367	5.6603	0.1767	3.9998	22.6399	12
13	5.4924	0.1821	32.0887	0.0312	5.8424	0.1712	4.2491	24.8247	13
14	6.2613	0.1597	37.5811	0.0266	6.0021	0.1666	4.4819	26.9009	14
15	7.1379	0.1401	43.8424	0.0228	6.1422	0.1628	4.6990	28.8623	15
16	8.1372	0.1229	50.9804	0.0196	6.2651	0.1596	4.9011	30.7057	16
17	9.2765	0.1078	59.1176	0.0169	6.3729	0.1569	5.0888	32.4305	17
18	10.5752	0.0946	68.3941	0.0146	6.4674	0.1546	5.2630	34.0380	18
19	12.0557	0.0829	78.9692	0.0127	6.5504	0.1527	5.4243	35.5311	19
20	13.7435	0.0728	91.0249	0.0110	6.6231	0.1510	5.5734	36.9135	20
21	15.6676	0.0638	104.7684	0.0095	6.6870	0.1495	5.7111	38.1901	21
22	17.8610	0.0560	120.4360	0.0083	6.7429	0.1483	5.8381	39.3658	22
23	20.3616	0.0491	138.2970	0.0072	6.7921	0.1472	5.9549	40.4463	23
24	23.2122	0.0431	158.6586	0.0063	6.8351	0.1463	6.0624	41.4371	24
25	26.4619	0.0378	181.8708	0.0055	6.8729	0.1455	6.1610	42.3441	25
26	30.1666	0.0331	208.3327	0.0048	6.9061	0.1448	6.2514	43.1728	26
27	34.3899	0.0291	238.4993	0.0042	6.9352	0.1442	6.3342	43.9289	27
28	39.2045	0.0255	272.8892	0.0037	6.9607	0.1437	6.4100	44.6176	28
29	44.6931	0.0224	312.0937	0.0032	6.9830	0.1432	6.4791	45.2441	29
30	50.9502	0.0196	356.7868	0.0028	7.0027	0.1428	6.5423	45.8132	30
31	58.0832	0.0172	407.7370	0.0025	7.0199	0.1425	6.5998	46.3297	31
32	66.2148	0.0151	465.8202	0.0021	7.0350	0.1421	6.6522	46.7979	32
33	75.4849	0.0132	532.0350	0.0019	7.0482	0.1419	6.6998	47.2218	33
34	86.0528	0.0116	607.5199	0.0016	7.0599	0.1416	6.7431	47.6053	34
35	98.1002	0.0102	693.5727	0.0014	7.0700	0.1414	6.7824	47.9519	35
40	188.8835	0.0053	1342.0251	0.0007	7.1050	0.1407	6.9300	49.2376	40
45	363.6791	0.0027	2590.5648	0.0004	7.1232	0.1404	7.0188	49.9963	45
50	700.2330	0.0014	4994.5213	0.0002	7.1327	0.1402	7.0714	50.4375	50

Table C.21 Interest Rate Factors (15.0%)

	Single Payment		Equal Payment Series				Gradient Series		
N	Compound Amount Factor (F/P,i,N)	Present Worth Factor (P/F,i,N)	Compound Amount Factor (F/A,i,N)	Sinking Fund Factor (A/F,i,N)	Present Worth Factor (P/A,i,N)	Capital Recovery Factor (A/P,i,N)	Gradient Uniform Series (A/G,i,N)	Gradient Present Worth (P/G.i.N)	N
1	1.1500	0.8696	1.0000	1.0000	0.8696	1.1500	-0.0000	-0.0000	1
2	1.3225	0.7561	2.1500	0.4651	1.6257	0.6151	0.4651	0.7561	2
3	1.5209	0.6575	3.4725	0.2880	2.2832	0.4380	0.9071	2.0712	3
4	1.7490	0.5718	4.9934	0.2003	2.8550	0.3503	1.3263	3.7864	4
5	2.0114	0.4972	6.7424	0.1483	3.3522	0.2983	1.7228	5.7751	5
6	2.3131	0.4323	8.7537	0.1142	3.7845	0.2642	2.0972	7.9368	6
7	2.6600	0.3759	11.0668	0.0904	4.1604	0.2404	2.4498	10.1924	7
8	3.0590	0.3269	13.7268	0.0729	4.4873	0.2229	2.7813	12.4807	8
9	3.5179	0.2843	16.7858	0.0596	4.7716	0.2096	3.0922	14.7548	9
10	4.0456	0.2472	20.3037	0.0493	5.0188	0.1993	3.3832	16.9795	10
11	4.6524	0.2149	24.3493	0.0411	5.2337	0.1911	3.6549	19.1289	11
12	5.3503	0.1869	29.0017	0.0345	5.4206	0.1845	3.9082	21.1849	12
13	6.1528	0.1625	34.3519	0.0291	5.5831	0.1791	4.1438	23.1352	13
14	7.0757	0.1413	40.5047	0.0247	5.7245	0.1747	4.3624	24.9725	14
15	8.1371	0.1229	47.5804	0.0210	5.8474	0.1710	4.5650	26.6930	15
16	9.3576	0.1069	55.7175	0.0179	5.9542	0.1679	4.7522	28.2960	16
17	10.7613	0.0929	65.0751	0.0154	6.0472	0.1654	4.9251	29.7828	17
18	12.3755	0.0808	75.8364	0.0132	6.1280	0.1632	5.0843	31.1565	18
19	14.2318	0.0703	88.2118	0.0113	6.1982	0.1613	5.2307	32.4213	19
20	16.3665	0.0611	102.4436	0.0098	6.2593	0.1598	5.3651	33.5822	20
21	18.8215	0.0531	118.8101	0.0084	6.3125	0.1584	5.4883	34.6448	21
22	21.6447	0.0462	137.6316	0.0073	6.3587	0.1573	5.6010	35.6150	22
23	24.8915	0.0402	159.2764	0.0063	6.3988	0.1563	5.7040	36.4988	23
24	28.6252	0.0349	184.1678	0.0054	6.4338	0.1554	5.7979	37.3023	24
25	32.9190	0.0304	212.7930	0.0047	6.4641	0.1547	5.8834	38.0314	25
26	37.8568	0.0264	245.7120	0.0041	6.4906	0.1541	5.9612	38.6918	26
27	43.5353	0.0230	283.5688	0.0035	6.5135	0.1535	6.0319	39.2890	27
28	50.0656	0.0200	327.1041	0.0031	6.5335	0.1531	6.0960	39.8283	28
29	57.5755	0.0174	377.1697	0.0027	6.5509	0.1527	6.1541	40.3146	29
30	66.2118	0.0151	434.7451	0.0023	6.5660	0.1523	6.2066	40.7526	30
31	76.1435	0.0131	500.9569	0.0020	6.5791	0.1520	6.2541	41.1466	31
32	87.5651	0.0114	577.1005	0.0017	6.5905	0.1517	6.2970	41.5006	32
33	100.6998	0.0099	664.6655	0.0015	6.6005	0.1515	6.3357	41.8184	33
34	115.8048	0.0086	765.3654	0.0013	6.6091	0.1513	6.3705	42.1033	34
35	133.1755	0.0075	881.1702	0.0011	6.6166	0.1511	6.4019	42.3586	35
40	267.8635	0.0037	1779.0903	0.0006	6.6418	0.1506	6.5168	43.2830	40
45	538.7693	0.0019	3585.1285	0.0003	6.6543	0.1503	6.5830	43.8051	45
50	1083.6574	0.0009	7217.7163	0.0001	6.6605	0.1501	6.6205	44.0958	50

Table C.22 Interest Rate Factors (16.0%)

	Single Payment		Equal Payment Series				Gradient Series		
N	Compound Amount Factor (F/P,i,N)	Present Worth Factor (P/F,i,N)	Compound Amount Factor (F/A,i,N)	Sinking Fund Factor (A/F,i,N)	Present Worth Factor (P/A,i,N)	Capital Recovery Factor (A/P,i,N)	Gradient Uniform Series (A/G,i,N)	Gradient Present Worth (P/G.i,N)	N
1	1.1600	0.8621	1.0000	1.0000	0.8621	1.1600	0.0000	0.0000	1
2	1.3456	0.7432	2.1600	0.4630	1.6052	0.6230	0.4630	0.7432	2
3	1.5609	0.6407	3.5056	0.2853	2.2459	0.4453	0.9014	2.0245	3
4	1.8106	0.5523	5.0665	0.1974	2.7982	0.3574	1.3156	3.6814	4
5	2.1003	0.4761	6.8771	0.1454	3.2743	0.3054	1.7060	5.5858	5
6	2.4364	0.4104	8.9775	0.1114	3.6847	0.2714	2.0729	7.6380	6
7	2.8262	0.3538	11.4139	0.0876	4.0386	0.2476	2.4169	9.7610	7
8	3.2784	0.3050	14.2401	0.0702	4.3436	0.2302	2.7388	11.8962	8
9	3.8030	0.2630	17.5185	0.0571	4.6065	0.2171	3.0391	13.9998	9
10	4.4114	0.2267	21.3215	0.0469	4.8332	0.2069	3.3187	16.0399	10
11	5.1173	0.1954	25.7329	0.0389	5.0286	0.1989	3.5783	17.9941	11
12	5.9360	0.1685	30.8502	0.0324	5.1971	0.1924	3.8189	19.8472	12
13	6.8858	0.1452	36.7862	0.0272	5.3423	0.1872	4.0413	21.5899	13
14	7.9875	0.1252	43.6720	0.0229	5.4675	0.1829	4.2464	23.2175	14
15	9.2655	0.1079	51.6595	0.0194	5.5755	0.1794	4.4352	24.7284	15
16	10.7480	0.0930	60.9250	0.0164	5.6685	0.1764	4.6086	26.1241	16
17	12.4677	0.0802	71.6730	0.0140	5.7487	0.1740	4.7676	27.4074	17
18	14.4625	0.0691	84.1407	0.0119	5.8178	0.1719	4.9130	28.5828	18
19	16.7765	0.0596	98.6032	0.0101	5.8775	0.1701	5.0457	29.6557	19
20	19.4608	0.0514	115.3797	0.0087	5.9288	0.1687	5.1666	30.6321	20
21	22.5745	0.0443	134.8405	0.0074	5.9731	0.1674	5.2766	31.5180	21
22	26.1864	0.0382	157.4150	0.0064	6.0113	0.1664	5.3765	32.3200	22
23	30.3762	0.0329	183.6014	0.0054	6.0442	0.1654	5.4671	33.0442	23
24	35.2364	0.0284	213.9776	0.0047	6.0726	0.1647	5.5490	33.6970	24
25	40.8742	0.0245	249.2140	0.0040	6.0971	0.1640	5.6230	34.2841	25
26	47.4141	0.0211	290.0883	0.0034	6.1182	0.1634	5.6898	34.8114	26
27	55.0004	0.0182	337.5024	0.0030	6.1364	0.1630	5.7500	35.2841	27
28	63.8004	0.0157	392.5028	0.0025	6.1520	0.1625	5.8041	35.7073	28
29	74.0085	0.0135	456.3032	0.0022	6.1656	0.1622	5.8528	36.0856	29
30	85.8499	0.0116	530.3117	0.0019	6.1772	0.1619	5.8964	36.4234	30
31	99.5859	0.0100	616.1616	0.0016	6.1872	0.1616	5.9356	36.7247	31
32	115.5196	0.0087	715.7475	0.0014	6.1959	0.1614	5.9706	36.9930	32
33	134.0027	0.0075	831.2671	0.0012	6.2034	0.1612	6.0019	37.2318	33
34	155.4432	0.0064	965.2698	0.0010	6.2098	0.1610	6.0299	37.4441	34
35	180.3141	0.0055	1120.7130	0.0009	6.2153	0.1609	6.0548	37.6327	35
40	378.7212	0.0026	2360.7572	0.0004	6.2335	0.1604	6.1441	38.2992	40
45	795.4438	0.0013	4965.2739	0.0002	6.2421	0.1602	6.1934	38.6598	45
50	1670.7038	0.0006	10435.6488	0.0001	6.2463	0.1601	6.2201	38.8521	50

Table C.23 Interest Rate Factors (18.0%)

	Single Payment		Equal Payment Series				Gradient Series		
N	Compound Amount Factor (F/P,i,N)	Present Worth Factor (P/F,i,N)	Compound Amount Factor (F/A,i,N)	Sinking Fund Factor (A/F,i,N)	Present Worth Factor (P/A,i,N)	Capital Recovery Factor (A/P,i,N)	Gradient Uniform Series (A/G,i,N)	Gradient Present Worth (P/G.i.N)	N
1	1.1800	0.8475	1.0000	1.0000	0.8475	1.1800	0.0000	0.0000	1
2	1.3924	0.7182	2.1800	0.4587	1.5656	0.6387	0.4587	0.7182	2
3	1.6430	0.6086	3.5724	0.2799	2.1743	0.4599	0.8902	1.9354	3
4	1.9388	0.5158	5.2154	0.1917	2.6901	0.3717	1.2947	3.4828	4
5	2.2878	0.4371	7.1542	0.1398	3.1272	0.3198	1.6728	5.2312	5
6	2.6996	0.3704	9.4420	0.1059	3.4976	0.2859	2.0252	7.0834	6
7	3.1855	0.3139	12.1415	0.0824	3.8115	0.2624	2.3526	8.9670	7
8	3.7589	0.2660	15.3270	0.0652	4.0776	0.2452	2.6558	10.8292	8
9	4.4355	0.2255	19.0859	0.0524	4.3030	0.2324	2.9358	12.6329	9
10	5.2338	0.1911	23.5213	0.0425	4.4941	0.2225	3.1936	14.3525	10
11	6.1759	0.1619	28.7551	0.0348	4.6560	0.2148	3.4303	15.9716	11
12	7.2876	0.1372	34.9311	0.0286	4.7932	0.2086	3.6470	17.4811	12
13	8.5994	0.1163	42.2187	0.0237	4.9095	0.2037	3.8449	18.8765	13
14	10.1472	0.0985	50.8180	0.0197	5.0081	0.1997	4.0250	20.1576	14
15	11.9737	0.0835	60.9653	0.0164	5.0916	0.1964	4.1887	21.3269	15
16	14.1290	0.0708	72.9390	0.0137	5.1624	0.1937	4.3369	22.3885	16
17	16.6722	0.0600	87.0680	0.0115	5.2223	0.1915	4.4708	23.3482	17
18	19.6733	0.0508	103.7403	0.0096	5.2732	0.1896	4.5916	24.2123	18
19	23.2144	0.0431	123.4135	0.0081	5.3162	0.1881	4.7003	24.9877	19
20	27.3930	0.0365	146.6280	0.0068	5.3527	0.1868	4.7978	25.6813	20
21	32.3238	0.0309	174.0210	0.0057	5.3837	0.1857	4.8851	26.3000	21
22	38.1421	0.0262	206.3448	0.0048	5.4099	0.1848	4.9632	26.8506	22
23	45.0076	0.0222	244.4868	0.0041	5.4321	0.1841	5.0329	27.3394	23
24	53.1090	0.0188	289.4945	0.0035	5.4509	0.1835	5.0950	27.7725	24
25	62.6686	0.0160	342.6035	0.0029	5.4669	0.1829	5.1502	28.1555	25
26	73.9490	0.0135	405.2721	0.0025	5.4804	0.1825	5.1991	28.4935	26
27	87.2598	0.0115	479.2211	0.0021	5.4919	0.1821	5.2425	28.7915	27
28	102.9666	0.0097	566.4809	0.0018	5.5016	0.1818	5.2810	29.0537	28
29	121.5005	0.0082	669.4475	0.0015	5.5098	0.1815	5.3149	29.2842	29
30	143.3706	0.0070	790.9480	0.0013	5.5168	0.1813	5.3448	29.4864	30
31	169.1774	0.0059	934.3186	0.0011	5.5227	0.1811	5.3712	29.6638	31
32	199.6293	0.0050	1103.4960	0.0009`	5.5277	0.1809	5.3945	29.8191	32
33	235.5625	0.0042	1303.1253	0.0008	5.5320	0.1808	5.4149	29.9549	33
34	277.9638	0.0036	1538.6878	0.0006	5.5356	0.1806	5.4328	30.0736	34
35	327.9973	0.0030	1816.6516	0.0006	5.5386	0.1806	5.4485	30.1773	35
40	750.3783	0.0013	4163.2130	0.0002	5.5482	0.1802	5.5022	30.5269	40
45	1716.6839	0.0006	9531.5771	0.0001	5.5523	0.1801	5.5293	30.7006	45
50	3927.3569	0.0003	21813.0937	0.0000	5.5541	0.1800	5.5428	30.7856	50

Table C.24 Interest Rate Factors (20.0%)

	Single Payment		Equal Payment Series				Gradient Series		
N	Compound Amount Factor (F/P,i,N)	Present Worth Factor (P/F,i,N)	Compound Amount Factor (F/A,i,N)	Sinking Fund Factor (A/F,i,N)	Present Worth Factor (P/A,i,N)	Capital Recovery Factor (A/P,i,N)	Gradient Uniform Series (A/G,i,N)	Gradient Present Worth (P/G.i.N)	N
1	1.2000	0.8333	1.0000	1.0000	0.8333	1.2000	0.0000	0.0000	1
2	1.4400	0.6944	2.2000	0.4545	1.5278	0.6545	0.4545	0.6944	2
3	1.7280	0.5787	3.6400	0.2747	2.1065	0.4747	0.8791	1.8519	3
4	2.0736	0.4823	5.3680	0.1863	2.5887	0.3863	1.2742	3.2986	4
5	2.4883	0.4019	7.4416	0.1344	2.9906	0.3344	1.6405	4.9061	5
6	2.9860	0.3349	9.9299	0.1007	3.3255	0.3007	1.9788	6.5806	6
7	3.5832	0.2791	12.9159	0.0774	3.6046	0.2774	2.2902	8.2551	7
8	4.2998	0.2326	16.4991	0.0606	3.8372	0.2606	2.5756	9.8831	8
9	5.1598	0.1938	20.7989	0.0481	4.0310	0.2481	2.8364	11.4335	9
10	6.1917	0.1615	25.9587	0.0385	4.1925	0.2385	3.0739	12.8871	10
11	7.4301	0.1346	32.1504	0.0311	4.3271	0.2311	3.2893	14.2330	11
12	8.9161	0.1122	39.5805	0.0253	4.4392	0.2253	3.4841	15.4667	12
13	10.6993	0.0935	48.4966	0.0206	4.5327	0.2206	3.6597	16.5883	13
14	12.8392	0.0779	59.1959	0.0169	4.6106	0.2169	3.8175	17.6008	14
15	15.4070	0.0649	72.0351	0.0139	4.6755	0.2139	3.9588	18.5095	15
16	18.4884	0.0541	87.4421	0.0114	4.7296	0.2114	4.0851	19.3208	16
17	22.1861	0.0451	105.9306	0.0094	4.7746	0.2094	4.1976	20.0419	17
18	26.6233	0.0376	128.1167	0.0078	4.8122	0.2078	4.2975	20.6805	18
19	31.9480	0.0313	154.7400	0.0065	4.8435	0.2065	4.3861	21.2439	19
20	38.3376	0.0261	186.6880	0.0054	4.8696	0.2054	4.4643	21.7395	20
21	46.0051	0.0217	225.0256	0.0044	4.8913	0.2044	4.5334	22.1742	21
22	55.2061	0.0181	271.0307	0.0037	4.9094	0.2037	4.5941	22.5546	22
23	66.2474	0.0151	326.2369	0.0031	4.9245	0.2031	4.6475	22.8867	23
24	79.4968	0.0126	392.4842	0.0025	4.9371	0.2025	4.6943	23.1760	24
25	95.3962	0.0105	471.9811	0.0021	4.9476	0.2021	4.7352	23.4276	25
26	114.4755	0.0087	567.3773	0.0018	4.9563	0.2018	4.7709	23.6460	26
27	137.3706	0.0073	681.8528	0.0015	4.9636	0.2015	4.8020	23.8353	27
28	164.8447	0.0061	819.2233	0.0012	4.9697	0.2012	4.8291	23.9991	28
29	197.8136	0.0051	984.0680	0.0010	4.9747	0.2010	4.8527	24.1406	29
30	237.3763	0.0042	1181.8816	0.0008	4.9789	0.2008	4.8731	24.2628	30
31	284.8516	0.0035	1419.2579	0.0007	4.9824	0.2007	4.8908	24.3681	31
32	341.8219	0.0029	1704.1095	0.0006	4.9854	0.2006	4.9061	24.4588	32
33	410.1863	0.0024	2045.9314	0.0005	4.9878	0.2005	4.9194	24.5368	33
34	492.2235	0.0020	2456.1176	0.0004	4.9898	0.2004	4.9308	24.6038	34
35	590.6682	0.0017	2948.3411	0.0003	4.9915	0.2003	4.9406	24.6614	35
40	1469.7716	0.0007	7343.8578	0.0001	4.9966	0.2001	4.9728	24.8469	40
45	3657.2620	0.0003	18281.3099	0.0001	4.9986	0.2001	4.9877	24.9316	45

Table C.25 Interest Rate Factors (25.0%)

	Single Payment		Equal Payment Series				Gradient Series		
N	Compound Amount Factor (F/P,i,N)	Present Worth Factor (P/F,i,N)	Compound Amount Factor (F/A,i,N)	Sinking Fund Factor (A/F,i,N)	Present Worth Factor (P/A,i,N)	Capital Recovery Factor (A/P,i,N)	Gradient Uniform Series (A/G,i,N)	Gradient Present Worth (P/G.i.N)	N
1	1.2500	0.8000	1.0000	1.0000	0.8000	1.2500	0.0000	0.0000	1
2	1.5625	0.6400	2.2500	0.4444	1.4400	0.6944	0.4444	0.6400	2
3	1.9531	0.5120	3.8125	0.2623	1.9520	0.5123	0.8525	1.6640	3
4	2.4414	0.4096	5.7656	0.1734	2.3616	0.4234	1.2249	2.8928	4
5	3.0518	0.3277	8.2070	0.1218	2.6893	0.3718	1.5631	4.2035	5
6	3.8147	0.2621	11.2588	0.0888	2.9514	0.3388	1.8683	5.5142	6
7	4.7684	0.2097	15.0735	0.0663	3.1611	0.3163	2.1424	6.7725	7
8	5.9605	0.1678	19.8419	0.0504	3.3289	0.3004	2.3872	7.9469	8
9	7.4506	0.1342	25.8023	0.0388	3.4631	0.2888	2.6048	9.0207	9
10	9.3132	0.1074	33.2529	0.0301	3.5705	0.2801	2.7971	9.9870	10
11	11.6415	0.0859	42.5661	0.0235	3.6564	0.2735	2.9663	10.8460	11
12	14.5519	0.0687	54.2077	0.0184	3.7251	0.2684	3.1145	11.6020	12
13	18.1899	0.0550	68.7596	0.0145	3.7801	0.2645	3.2437	12.2617	13
14	22.7374	0.0440	86.9495	0.0115	3.8241	0.2615	3.3559	12.8334	14
15	28.4217	0.0352	109.6868	0.0091	3.8593	0.2591	3.4530	13.3260	15
16	35.5271	0.0281	138.1085	0.0072	3.8874	0.2572	3.5366	13.7482	16
17	44.4089	0.0225	173.6357	0.0058	3.9099	0.2558	3.6084	14.1085	17
18	55.5112	0.0180	218.0446	0.0046	3.9279	0.2546	3.6698	14.4147	18
19	69.3889	0.0144	273.5558	0.0037	3.9424	0.2537	3.7222	14.6741	19
20	86.7362	0.0115	342.9447	0.0029	3.9539	0.2529	3.7667	14.8932	20
21	108.4202	0.0092	429.6809	0.0023	3.9631	0.2523	3.8045	15.0777	21
22	135.5253	0.0074	538.1011	0.0019	3.9705	0.2519	3.8365	15.2326	22
23	169.4066	0.0059	673.6264	0.0015	3.9764	0.2515	3.8634	15.3625	23
24	211.7582	0.0047	843.0329	0.0012	3.9811	0.2512	3.8861	15.4711	24
25	264.6978	0.0038	1054.7912	0.0009	3.9849	0.2509	3.9052	15.5618	25
26	330.8722	0.0030	1319.4890	0.0008	3.9879	0.2508	3.9212	15.6373	26
27	413.5903	0.0024	1650.3612	0.0006	3.9903	0.2506	3.9346	15.7002	27
28	516.9879	0.0019	2063.9515	0.0005	3.9923	0.2505	3.9457	15.7524	28
29	646.2349	0.0015	2580.9394	0.0004	3.9938	0.2504	3.9551	15.7957	29
30	807.7936	0.0012	3227.1743	0.0003	3.9950	0.2503	3.9628	15.8316	30
31	1009.7420	0.0010	4034.9678	0.0002	3.9960	0.2502	3.9693	15.8614	31
32	1262.1774	0.0008	5044.7098	0.0002	3.9968	0.2502	3.9746	15.8859	32
33	1577.7218	0.0006	6306.8872	0.0002	3.9975	0.2502	3.9791	15.9062	33
34	1972.1523	0.0005	7884.6091	0.0001	3.9980	0.2501	3.9828	15.9229	34
35	2465.1903	0.0004	9856.7613	0.0001	3.9984	0.2501	3.9858	15.9367	35
40	7523.1638	0.0001	30088.6554	0.0000	3.9995	0.2500	3.9947	15.9766	40

Table C.26 Interest Rate Factors (30.0%)

	Single Payment		Equal Payment Series				Gradient Series		
N	Compound Amount Factor (F/P,i,N)	Present Worth Factor (P/F,i,N)	Compound Amount Factor (F/A,i,N)	Sinking Fund Factor (A/F,i,N)	Present Worth Factor (P/A,i,N)	Capital Recovery Factor (A/P,i,N)	Gradient Uniform Series (A/G,i,N)	Gradient Present Worth (P/G.i.N)	N
1	1.3000	0.7692	1.0000	1.0000	0.7692	1.3000	0.0000	0.0000	1
2	1.6900	0.5917	2.3000	0.4348	1.3609	0.7348	0.4348	0.5917	2
3	2.1970	0.4552	3.9900	0.2506	1.8161	0.5506	0.8271	1.5020	3
4	2.8561	0.3501	6.1870	0.1616	2.1662	0.4616	1.1783	2.5524	4
5	3.7129	0.2693	9.0431	0.1106	2.4356	0.4106	1.4903	3.6297	5
6	4.8268	0.2072	12.7560	0.0784	2.6427	0.3784	1.7654	4.6656	6
7	6.2749	0.1594	17.5828	0.0569	2.8021	0.3569	2.0063	5.6218	7
8	8.1573	0.1226	23.8577	0.0419	2.9247	0.3419	2.2156	6.4800	8
9	10.6045	0.0943	32.0150	0.0312	3.0190	0.3312	2.3963	7.2343	9
10	13.7858	0.0725	42.6195	0.0235	3.0915	0.3235	2.5512	7.8872	10
11	17.9216	0.0558	56.4053	0.0177	3.1473	0.3177	2.6833	8.4452	11
12	23.2981	0.0429	74.3270	0.0135	3.1903	0.3135	2.7952	8.9173	12
13	30.2875	0.0330	97.6250	0.0102	3.2233	0.3102	2.8895	9.3135	13
14	39.3738	0.0254	127.9125	0.0078	3.2487	0.3078	2.9685	9.6437	14
15	51.1859	0.0195	167.2863	0.0060	3.2682	0.3060	3.0344	9.9172	15
16	66.5417	0.0150	218.4722	0.0046	3.2832	0.3046	3.0892	10.1426	16
17	86.5042	0.0116	285.0139	0.0035	3.2948	0.3035	3.1345	10.3276	17
18	112.4554	0.0089	371.5180	0.0027	3.3037	0.3027	3.1718	10.4788	18
19	146.1920	0.0068	483.9734	0.0021	3.3105	0.3021	3.2025	10.6019	19
20	190.0496	0.0053	630.1655	0.0016	3.3158	0.3016	3.2275	10.7019	20
21	247.0645	0.0040	820.2151	0.0012	3.3198	0.3012	3.2480	10.7828	21
22	321.1839	0.0031	1067.2796	0.0009	3.3230	0.3009	3.2646	10.8482	22
23	417.5391	0.0024	1388.4635	0.0007	3.3254	0.3007	3.2781	10.9009	23
24	542.8008	0.0018	1806.0026	0.0006	3.3272	0.3006	3.2890	10.9433	24
25	705.6410	0.0014	2348.8033	0.0004	3.3286	0.3004	3.2979	10.9773	25
26	917.3333	0.0011	3054.4443	0.0003	3.3297	0.3003	3.3050	11.0045	26
27	1192.5333	0.0008	3971.7776	0.0003	3.3305	0.3003	3.3107	11.0263	27
28	1550.2933	0.0006	5164.3109	0.0002	3.3312	0.3002	3.3153	11.0437	28
29	2015.3813	0.0005	6714.6042	0.0001	3.3317	0.3001	3.3189	11.0576	29
30	2619.9956	0.0004	8729.9855	0.0001	3.3321	0.3001	3.3219	11.0687	30
31	3405.9943	0.0003	11349.9811	0.0001	3.3324	0.3001	3.3242	11.0775	31
32	4427.7926	0.0002	14755.9755	0.0001	3.3326	0.3001	3.3261	11.0845	32
33	5756.1304	0.0002	19183.7681	0.0001	3.3328	0.3001	3.3276	11.0901	33
34	7482.9696	0.0001	24939.8985	0.0000	3.3329	0.3000	3.3288	11.0945	34
35	9727.8604	0.0001	32422.8681	0.0000	3.3330	0.3000	3.3297	11.0980	35

Table C.27 Interest Rate Factors (35.0%)

	Single Payment		Equal Payment Series				Gradient Series		
N	Compound Amount Factor (F/P,i,N)	Present Worth Factor (P/F,i,N)	Compound Amount Factor (F/A,i,N)	Sinking Fund Factor (A/F,i,N)	Present Worth Factor (P/A,i,N)	Capital Recovery Factor (A/P,i,N)	Gradient Uniform Series (A/G,i,N)	Gradient Present Worth (P/G.i.N)	N
1	1.3500	0.7407	1.0000	1.0000	0.7407	1.3500	0.0000	0.0000	1
2	1.8225	0.5487	2.3500	0.4255	1.2894	0.7755	0.4255	0.5487	2
3	2.4604	0.4064	4.1725	0.2397	1.6959	0.5897	0.8029	1.3616	3
4	3.3215	0.3011	6.6329	0.1508	1.9969	0.5008	1.1341	2.2648	4
5	4.4840	0.2230	9.9544	0.1005	2.2200	0.4505	1.4220	3.1568	5
6	6.0534	0.1652	14.4384	0.0693	2.3852	0.4193	1.6698	3.9828	6
7	8.1722	0.1224	20.4919	0.0488	2.5075	0.3988	1.8811	4.7170	7
8	11.0324	0.0906	28.6640	0.0349	2.5982	0.3849	2.0597	5.3515	8
9	14.8937	0.0671	39.6964	0.0252	2.6653	0.3752	2.2094	5.8886	9
10	20.1066	0.0497	54.5902	0.0183	2.7150	0.3683	2.3338	6.3363	10
11	27.1439	0.0368	74.6967	0.0134	2.7519	0.3634	2.4364	6.7047	11
12	36.6442	0.0273	101.8406	0.0098	2.7792	0.3598	2.5205	7.0049	12
13	49.4697	0.0202	138.4848	0.0072	2.7994	0.3572	2.5889	7.2474	13
14	66.7841	0.0150	187.9544	0.0053	2.8144	0.3553	2.6443	7.4421	14
15	90.1585	0.0111	254.7385	0.0039	2.8255	0.3539	2.6889	7.5974	15
16	121.7139	0.0082	344.8970	0.0029	2.8337	0.3529	2.7246	7.7206	16
17	164.3138	0.0061	466.6109	0.0021	2.8398	0.3521	2.7530	7.8180	17
18	221.8236	0.0045	630.9247	0.0016	2.8443	0.3516	2.7756	7.8946	18
19	299.4619	0.0033	852.7483	0.0012	2.8476	0.3512	2.7935	7.9547	19
20	404.2736	0.0025	1152.2103	0.0009	2.8501	0.3509	2.8075	8.0017	20
21	545.7693	0.0018	1556.4838	0.0006	2.8519	0.3506	2.8186	8.0384	21
22	736.7886	0.0014	2102.2532	0.0005	2.8533	0.3505	2.8272	8.0669	22
23	994.6646	0.0010	2839.0418	0.0004	2.8543	0.3504	2.8340	8.0890	23
24	1342.7973	0.0007	3833.7064	0.0003	2.8550	0.3503	2.8393	8.1061	24
25	1812.7763	0.0006	5176.5037	0.0002	2.8556	0.3502	2.8433	8.1194	25
26	2447.2480	0.0004	6989.2800	0.0001	2.8560	0.3501	2.8465	8.1296	26
27	3303.7848	0.0003	9436.5280	0.0001	2.8563	0.3501	2.8490	8.1374	27
28	4460.1095	0.0002	12740.3128	0.0001	2.8565	0.3501	2.8509	8.1435	28
29	6021.1478	0.0002	17200.4222	0.0001	2.8567	0.3501	2.8523	8.1481	29
30	8128.5495	0.0001	23221.5700	0.0000	2.8568	0.3500	2.8535	8.1517	30

Table C.28 Interest Rate Factors (40.0%)

	Single Payment		Equal Payment Series				Gradient Series		
N	Compound Amount Factor (F/P,i,N)	Present Worth Factor (P/F,i,N)	Compound Amount Factor (F/A,i,N)	Sinking Fund Factor (A/F,i,N)	Present Worth Factor (P/A,i,N)	Capital Recovery Factor (A/P,i,N)	Gradient Uniform Series (A/G,i,N)	Gradient Present Worth (P/G.i.N)	N
1	1.4000	0.7143	1.0000	1.0000	0.7143	1.4000	0.0000	0.0000	1
2	1.9600	0.5102	2.4000	0.4167	1.2245	0.8167	0.4167	0.5102	2
3	2.7440	0.3644	4.3600	0.2294	1.5889	0.6294	0.7798	1.2391	3
4	3.8416	0.2603	7.1040	0.1408	1.8492	0.5408	1.0923	2.0200	4
5	5.3782	0.1859	10.9456	0.0914	2.0352	0.4914	1.3580	2.7637	5
6	7.5295	0.1328	16.3238	0.0613	2.1680	0.4613	1.5811	3.4278	6
7	10.5414	0.0949	23.8534	0.0419	2.2628	0.4419	1.7664	3.9970	7
8	14.7579	0.0678	34.3947	0.0291	2.3306	0.4291	1.9185	4.4713	8
9	20.6610	0.0484	49.1526	0.0203	2.3790	0.4203	2.0422	4.8585	9
10	28.9255	0.0346	69.8137	0.0143	2.4136	0.4143	2.1419	5.1696	10
11	40.4957	0.0247	98.7391	0.0101	2.4383	0.4101	2.2215	5.4166	11
12	56.6939	0.0176	139.2348	0.0072	2.4559	0.4072	2.2845	5.6106	12
13	79.3715	0.0126	195.9287	0.0051	2.4685	0.4051	2.3341	5.7618	13
14	111.1201	0.0090	275.3002	0.0036	2.4775	0.4036	2.3729	5.8788	14
15	155.5681	0.0064	386.4202	0.0026	2.4839	0.4026	2.4030	5.9688	15
16	217.7953	0.0046	541.9883	0.0018	2.4885	0.4018	2.4262	6.0376	16
17	304.9135	0.0033	759.7837	0.0013	2.4918	0.4013	2.4441	6.0901	17
18	426.8789	0.0023	1064.6971	0.0009	2.4941	0.4009	2.4577	6.1299	18
19	597.6304	0.0017	1491.5760	0.0007	2.4958	0.4007	2.4682	6.1601	19
20	836.6826	0.0012	2089.2064	0.0005	2.4970	0.4005	2.4761	6.1828	20
21	1171.3556	0.0009	2925.8889	0.0003	2.4979	0.4003	2.4821	6.1998	21
22	1639.8978	0.0006	4097.2445	0.0002	2.4985	0.4002	2.4866	6.2127	22
23	2295.8569	0.0004	5737.1423	0.0002	2.4989	0.4002	2.4900	6.2222	23
24	3214.1997	0.0003	8032.9993	0.0001	2.4992	0.4001	2.4925	6.2294	24
25	4499.8796	0.0002	11247.1990	0.0001	2.4994	0.4001	2.4944	6.2347	25
26	6299.8314	0.0002	15747.0785	0.0001	2.4996	0.4001	2.4959	6.2387	26
27	8819.7640	0.0001	22046.9099	0.0000	2.4997	0.4000	2.4969	6.2416	27
28	12347.6696	0.0001	30866.6739	0.0000	2.4998	0.4000	2.4977	6.2438	28
29	17286.7374	0.0001	43214.3435	0.0000	2.4999	0.4000	2.4983	6.2454	29
30	24201.4324	0.0000	60501.0809	0.0000	2.4999	0.4000	2.4988	6.2466	30

Table C.29 Interest Rate Factors (50.0%)

	Single Payment		Equal Payment Series				Gradient Series		
N	Compound Amount Factor (F/P,i,N)	Present Worth Factor (P/F,i,N)	Compound Amount Factor (F/A,i,N)	Sinking Fund Factor (A/F,i,N)	Present Worth Factor (P/A,i,N)	Capital Recovery Factor (A/P,i,N)	Gradient Uniform Series (A/G,i,N)	Gradient Present Worth (P/G.i,N)	N
1	1.5000	0.6667	1.0000	1.0000	0.6667	1.5000	0.0000	0.0000	1
2	2.2500	0.4444	2.5000	0.4000	1.1111	0.9000	0.4000	0.4444	2
3	3.3750	0.2963	4.7500	0.2105	1.4074	0.7105	0.7368	1.0370	3
4	5.0625	0.1975	8.1250	0.1231	1.6049	0.6231	1.0154	1.6296	4
5	7.5938	0.1317	13.1875	0.0758	1.7366	0.5758	1.2417	2.1564	5
6	11.3906	0.0878	20.7813	0.0481	1.8244	0.5481	1.4226	2.5953	6
7	17.0859	0.0585	32.1719	0.0311	1.8829	0.5311	1.5648	2.9465	7
8	25.6289	0.0390	49.2578	0.0203	1.9220	0.5203	1.6752	3.2196	8
9	38.4434	0.0260	74.8867	0.0134	1.9480	0.5134	1.7596	3.4277	9
10	57.6650	0.0173	113.3301	0.0088	1.9653	0.5088	1.8235	3.5838	10
11	86.4976	0.0116	170.9951	0.0058	1.9769	0.5058	1.8713	3.6994	11
12	129.7463	0.0077	257.4927	0.0039	1.9846	0.5039	1.9068	3.7842	12
13	194.6195	0.0051	387.2390	0.0026	1.9897	0.5026	1.9329	3.8459	13
14	291.9293	0.0034	581.8585	0.0017	1.9931	0.5017	1.9519	3.8904	14
15	437.8939	0.0023	873.7878	0.0011	1.9954	0.5011	1.9657	3.9224	15
16	656.8408	0.0015	1311.6817	0.0008	1.9970	0.5008	1.9756	3.9452	16
17	985.2613	0.0010	1968.5225	0.0005	1.9980	0.5005	1.9827	3.9614	17
18	1477.8919	0.0007	2953.7838	0.0003	1.9986	0.5003	1.9878	3.9729	18
19	2216.8378	0.0005	4431.6756	0.0002	1.9991	0.5002	1.9914	3.9811	19
20	3325.2567	0.0003	6648.5135	0.0002	1.9994	0.5002	1.9940	3.9868	20

Capital Tax Factors

Table D.1 Half Year Rule Capital Tax Factors

Tax Rate = 20%

Interest Rate (%)	Declining Balance CCA Rate (%)						
	4	8	10	20	25	30	40
1	0.8408	0.8231	0.8191	0.8105	0.8086	0.8074	0.8058
2	0.8680	0.8416	0.8350	0.8200	0.8166	0.8143	0.8114
3	0.8874	0.8567	0.8484	0.8286	0.8240	0.8208	0.8167
4	0.9019	0.8692	0.8599	0.8365	0.8309	0.8269	0.8217
5	0.9132	0.8799	0.8698	0.8438	0.8373	0.8327	0.8265
6	0.9223	0.8889	0.8785	0.8505	0.8433	0.8381	0.8310
7	0.9297	0.8968	0.8862	0.8567	0.8489	0.8431	0.8354
8	0.9358	0.9037	0.8930	0.8624	0.8541	0.8480	0.8395
9	0.9410	0.9098	0.8991	0.8678	0.8590	0.8525	0.8435
10	0.9455	0.9152	0.9045	0.8727	0.8636	0.8568	0.8473
11	0.9493	0.9200	0.9095	0.8774	0.8680	0.8609	0.8509
12	0.9527	0.9243	0.9140	0.8817	0.8721	0.8648	0.8544
13	0.9556	0.9282	0.9180	0.8858	0.8760	0.8685	0.8577
14	0.9583	0.9317	0.9218	0.8896	0.8797	0.8720	0.8609
15	0.9606	0.9350	0.9252	0.8932	0.8832	0.8754	0.8640
16	0.9628	0.9379	0.9284	0.8966	0.8865	0.8786	0.8670
17	0.9647	0.9406	0.9313	0.8997	0.8896	0.8816	0.8698
18	0.9664	0.9432	0.9340	0.9028	0.8926	0.8845	0.8726
19	0.9680	0.9455	0.9365	0.9056	0.8954	0.8873	0.8752
20	0.9694	0.9476	0.9389	0.9083	0.8981	0.8900	0.8778
21	0.9708	0.9496	0.9411	0.9109	0.9007	0.8926	0.8802
22	0.9720	0.9515	0.9431	0.9133	0.9032	0.8950	0.8826
23	0.9731	0.9532	0.9451	0.9157	0.9056	0.8974	0.8849
24	0.9742	0.9548	0.9469	0.9179	0.9078	0.8996	0.8871
25	0.9752	0.9564	0.9486	0.9200	0.9100	0.9018	0.8892
30	0.9792	0.9628	0.9558	0.9292	0.9196	0.9115	0.8989
35	0.9821	0.9676	0.9613	0.9367	0.9275	0.9197	0.9072
40	0.9844	0.9714	0.9657	0.9429	0.9341	0.9265	0.9143
45	0.9862	0.9745	0.9693	0.9480	0.9397	0.9324	0.9205
50	0.9877	0.9770	0.9722	0.9524	0.9444	0.9375	0.9259

Table D.2 Full Year Rule Capital Tax Factors

Tax Rate = 20%

Interest Rate (%)	Declining Balance CCA Rate (%)						
	4	8	10	20	25	30	40
1	0.8400	0.8222	0.8182	0.8095	0.8077	0.8065	0.8049
2	0.8667	0.8400	0.8333	0.8182	0.8148	0.8125	0.8095
3	0.8857	0.8545	0.8462	0.8261	0.8214	0.8182	0.8140
4	0.9000	0.8667	0.8571	0.8333	0.8276	0.8235	0.8182
5	0.9111	0.8769	0.8667	0.8400	0.8333	0.8286	0.8222
6	0.9200	0.8857	0.8750	0.8462	0.8387	0.8333	0.8261
7	0.9273	0.8933	0.8824	0.8519	0.8438	0.8378	0.8298
8	0.9333	0.9000	0.8889	0.8571	0.8485	0.8421	0.8333
9	0.9385	0.9059	0.8947	0.8621	0.8529	0.8462	0.8367
10	0.9429	0.9111	0.9000	0.8667	0.8571	0.8500	0.8400
11	0.9467	0.9158	0.9048	0.8710	0.8611	0.8537	0.8431
12	0.9500	0.9200	0.9091	0.8750	0.8649	0.8571	0.8462
13	0.9529	0.9238	0.9130	0.8788	0.8684	0.8605	0.8491
14	0.9556	0.9273	0.9167	0.8824	0.8718	0.8636	0.8519
15	0.9579	0.9304	0.9200	0.8857	0.8750	0.8667	0.8545
16	0.9600	0.9333	0.9231	0.8889	0.8780	0.8696	0.8571
17	0.9619	0.9360	0.9259	0.8919	0.8810	0.8723	0.8596
18	0.9636	0.9385	0.9286	0.8947	0.8837	0.8750	0.8621
19	0.9652	0.9407	0.9310	0.8974	0.8864	0.8776	0.8644
20	0.9667	0.9429	0.9333	0.9000	0.8889	0.8800	0.8667
21	0.9680	0.9448	0.9355	0.9024	0.8913	0.8824	0.8689
22	0.9692	0.9467	0.9375	0.9048	0.8936	0.8846	0.8710
23	0.9704	0.9484	0.9394	0.9070	0.8958	0.8868	0.8730
24	0.9714	0.9500	0.9412	0.9091	0.8980	0.8889	0.8750
25	0.9724	0.9515	0.9429	0.9111	0.9000	0.8909	0.8769
30	0.9765	0.9579	0.9500	0.9200	0.9091	0.9000	0.8857
35	0.9795	0.9628	0.9556	0.9273	0.9167	0.9077	0.8933
40	0.9818	0.9667	0.9600	0.9333	0.9231	0.9143	0.9000
45	0.9837	0.9698	0.9636	0.9385	0.9286	0.9200	0.9059
50	0.9852	0.9724	0.9667	0.9429	0.9333	0.9250	0.9111

Table D.3 Half Year Rule Capital Tax Factors

Tax Rate = 25%

Interest Rate (%)	Declining Balance CCA Rate (%)						
	4	8	10	20	25	30	40
1	0.8010	0.7789	0.7739	0.7631	0.7608	0.7593	0.7573
2	0.8350	0.8020	0.7937	0.7750	0.7708	0.7679	0.7642
3	0.8592	0.8208	0.8105	0.7858	0.7800	0.7760	0.7708
4	0.8774	0.8365	0.8249	0.7957	0.7886	0.7837	0.7771
5	0.8915	0.8498	0.8373	0.8048	0.7966	0.7908	0.7831
6	0.9028	0.8612	0.8482	0.8131	0.8041	0.7976	0.7888
7	0.9121	0.8710	0.8578	0.8209	0.8111	0.8039	0.7942
8	0.9198	0.8796	0.8663	0.8280	0.8176	0.8099	0.7994
9	0.9263	0.8872	0.8739	0.8347	0.8238	0.8156	0.8043
10	0.9318	0.8939	0.8807	0.8409	0.8295	0.8210	0.8091
11	0.9366	0.9000	0.8869	0.8467	0.8350	0.8261	0.8136
12	0.9408	0.9054	0.8925	0.8521	0.8401	0.8310	0.8180
13	0.9446	0.9102	0.8976	0.8572	0.8450	0.8356	0.8222
14	0.9479	0.9147	0.9022	0.8620	0.8496	0.8400	0.8262
15	0.9508	0.9187	0.9065	0.8665	0.8539	0.8442	0.8300
16	0.9534	0.9224	0.9105	0.8707	0.8581	0.8482	0.8337
17	0.9558	0.9258	0.9141	0.8747	0.8620	0.8520	0.8373
18	0.9580	0.9289	0.9175	0.8785	0.8657	0.8557	0.8407
19	0.9600	0.9318	0.9207	0.8820	0.8693	0.8592	0.8440
20	0.9618	0.9345	0.9236	0.8854	0.8727	0.8625	0.8472
21	0.9635	0.9370	0.9264	0.8886	0.8759	0.8657	0.8503
22	0.9650	0.9393	0.9289	0.8917	0.8790	0.8688	0.8533
23	0.9664	0.9415	0.9313	0.8946	0.8820	0.8717	0.8561
24	0.9677	0.9435	0.9336	0.8974	0.8848	0.8746	0.8589
25	0.9690	0.9455	0.9357	0.9000	0.8875	0.8773	0.8615
30	0.9740	0.9534	0.9447	0.9115	0.8995	0.8894	0.8736
35	0.9777	0.9595	0.9516	0.9209	0.9093	0.8996	0.8840
40	0.9805	0.9643	0.9571	0.9286	0.9176	0.9082	0.8929
45	0.9828	0.9681	0.9616	0.9350	0.9246	0.9155	0.9006
50	0.9846	0.9713	0.9653	0.9405	0.9306	0.9219	0.9074

Table D.4 Full Year Rule Capital Tax Factors

Tax Rate = 25%

Interest Rate (%)	Declining Balance CCA Rate (%)						
	4	8	10	20	25	30	40
1	0.8000	0.7778	0.7727	0.7619	0.7596	0.7581	0.7561
2	0.8333	0.8000	0.7917	0.7727	0.7685	0.7656	0.7619
3	0.8571	0.8182	0.8077	0.7826	0.7768	0.7727	0.7674
4	0.8750	0.8333	0.8214	0.7917	0.7845	0.7794	0.7727
5	0.8889	0.8462	0.8333	0.8000	0.7917	0.7857	0.7778
6	0.9000	0.8571	0.8438	0.8077	0.7984	0.7917	0.7826
7	0.9091	0.8667	0.8529	0.8148	0.8047	0.7973	0.7872
8	0.9167	0.8750	0.8611	0.8214	0.8106	0.8026	0.7917
9	0.9231	0.8824	0.8684	0.8276	0.8162	0.8077	0.7959
10	0.9286	0.8889	0.8750	0.8333	0.8214	0.8125	0.8000
11	0.9333	0.8947	0.8810	0.8387	0.8264	0.8171	0.8039
12	0.9375	0.9000	0.8864	0.8438	0.8311	0.8214	0.8077
13	0.9412	0.9048	0.8913	0.8485	0.8355	0.8256	0.8113
14	0.9444	0.9091	0.8958	0.8529	0.8397	0.8295	0.8148
15	0.9474	0.9130	0.9000	0.8571	0.8438	0.8333	0.8182
16	0.9500	0.9167	0.9038	0.8611	0.8476	0.8370	0.8214
17	0.9524	0.9200	0.9074	0.8649	0.8512	0.8404	0.8246
18	0.9545	0.9231	0.9107	0.8684	0.8547	0.8438	0.8276
19	0.9565	0.9259	0.9138	0.8718	0.8580	0.8469	0.8305
20	0.9583	0.9286	0.9167	0.8750	0.8611	0.8500	0.8333
21	0.9600	0.9310	0.9194	0.8780	0.8641	0.8529	0.8361
22	0.9615	0.9333	0.9219	0.8810	0.8670	0.8558	0.8387
23	0.9630	0.9355	0.9242	0.8837	0.8698	0.8585	0.8413
24	0.9643	0.9375	0.9265	0.8864	0.8724	0.8611	0.8438
25	0.9655	0.9394	0.9286	0.8889	0.8750	0.8636	0.8462
30	0.9706	0.9474	0.9375	0.9000	0.8864	0.8750	0.8571
35	0.9744	0.9535	0.9444	0.9091	0.8958	0.8846	0.8667
40	0.9773	0.9583	0.9500	0.9167	0.9038	0.8929	0.8750
45	0.9796	0.9623	0.9545	0.9231	0.9107	0.9000	0.8824
50	0.9815	0.9655	0.9583	0.9286	0.9167	0.9063	0.8889

Table D.5 Half Year Rule Capital Tax Factors

Tax Rate = 30%

Interest Rate (%)	Declining Balance CCA Rate (%)						
	4	8	10	20	25	30	40
1	0.7612	0.7347	0.7286	0.7157	0.7130	0.7111	0.7088
2	0.8020	0.7624	0.7525	0.7299	0.7249	0.7215	0.7171
3	0.8311	0.7850	0.7726	0.7429	0.7360	0.7312	0.7250
4	0.8529	0.8038	0.7898	0.7548	0.7464	0.7404	0.7325
5	0.8698	0.8198	0.8048	0.7657	0.7560	0.7490	0.7397
6	0.8834	0.8334	0.8178	0.7758	0.7649	0.7571	0.7465
7	0.8945	0.8452	0.8293	0.7850	0.7733	0.7647	0.7530
8	0.9037	0.8556	0.8395	0.7937	0.7811	0.7719	0.7593
9	0.9115	0.8647	0.8486	0.8016	0.7885	0.7788	0.7652
10	0.9182	0.8727	0.8568	0.8091	0.7955	0.7852	0.7709
11	0.9240	0.8799	0.8642	0.8160	0.8020	0.7914	0.7764
12	0.9290	0.8864	0.8709	0.8225	0.8082	0.7972	0.7816
13	0.9335	0.8923	0.8771	0.8286	0.8140	0.8027	0.7866
14	0.9374	0.8976	0.8827	0.8344	0.8195	0.8080	0.7914
15	0.9410	0.9025	0.8878	0.8398	0.8247	0.8130	0.7960
16	0.9441	0.9069	0.8926	0.8448	0.8297	0.8178	0.8005
17	0.9470	0.9110	0.8970	0.8496	0.8344	0.8224	0.8048
18	0.9496	0.9147	0.9010	0.8541	0.8389	0.8268	0.8089
19	0.9520	0.9182	0.9048	0.8584	0.8432	0.8310	0.8128
20	0.9542	0.9214	0.9083	0.8625	0.8472	0.8350	0.8167
21	0.9562	0.9244	0.9116	0.8664	0.8511	0.8388	0.8203
22	0.9580	0.9272	0.9147	0.8700	0.8548	0.8425	0.8239
23	0.9597	0.9298	0.9176	0.8735	0.8584	0.8461	0.8273
24	0.9613	0.9323	0.9203	0.8768	0.8618	0.8495	0.8306
25	0.9628	0.9345	0.9229	0.8800	0.8650	0.8527	0.8338
30	0.9688	0.9441	0.9337	0.8938	0.8794	0.8673	0.8484
35	0.9732	0.9514	0.9420	0.9051	0.8912	0.8795	0.8607
40	0.9766	0.9571	0.9486	0.9143	0.9011	0.8898	0.8714
45	0.9793	0.9617	0.9539	0.9220	0.9095	0.8986	0.8807
50	0.9815	0.9655	0.9583	0.9286	0.9167	0.9063	0.8889

Table D.6 Full Year Rule Capital Tax Factors

Tax Rate = 30%

Interest Rate (%)	Declining Balance CCA Rate (%)						
	4	8	10	20	25	30	40
1	0.7600	0.7333	0.7273	0.7143	0.7115	0.7097	0.7073
2	0.8000	0.7600	0.7500	0.7273	0.7222	0.7188	0.7143
3	0.8286	0.7818	0.7692	0.7391	0.7321	0.7273	0.7209
4	0.8500	0.8000	0.7857	0.7500	0.7414	0.7353	0.7273
5	0.8667	0.8154	0.8000	0.7600	0.7500	0.7429	0.7333
6	0.8800	0.8286	0.8125	0.7692	0.7581	0.7500	0.7391
7	0.8909	0.8400	0.8235	0.7778	0.7656	0.7568	0.7447
8	0.9000	0.8500	0.8333	0.7857	0.7727	0.7632	0.7500
9	0.9077	0.8588	0.8421	0.7931	0.7794	0.7692	0.7551
10	0.9143	0.8667	0.8500	0.8000	0.7857	0.7750	0.7600
11	0.9200	0.8737	0.8571	0.8065	0.7917	0.7805	0.7647
12	0.9250	0.8800	0.8636	0.8125	0.7973	0.7857	0.7692
13	0.9294	0.8857	0.8696	0.8182	0.8026	0.7907	0.7736
14	0.9333	0.8909	0.8750	0.8235	0.8077	0.7955	0.7778
15	0.9368	0.8957	0.8800	0.8286	0.8125	0.8000	0.7818
16	0.9400	0.9000	0.8846	0.8333	0.8171	0.8043	0.7857
17	0.9429	0.9040	0.8889	0.8378	0.8214	0.8085	0.7895
18	0.9455	0.9077	0.8929	0.8421	0.8256	0.8125	0.7931
19	0.9478	0.9111	0.8966	0.8462	0.8295	0.8163	0.7966
20	0.9500	0.9143	0.9000	0.8500	0.8333	0.8200	0.8000
21	0.9520	0.9172	0.9032	0.8537	0.8370	0.8235	0.8033
22	0.9538	0.9200	0.9063	0.8571	0.8404	0.8269	0.8065
23	0.9556	0.9226	0.9091	0.8605	0.8438	0.8302	0.8095
24	0.9571	0.9250	0.9118	0.8636	0.8469	0.8333	0.8125
25	0.9586	0.9273	0.9143	0.8667	0.8500	0.8364	0.8154
30	0.9647	0.9368	0.9250	0.8800	0.8636	0.8500	0.8286
35	0.9692	0.9442	0.9333	0.8909	0.8750	0.8615	0.8400
40	0.9727	0.9500	0.9400	0.9000	0.8846	0.8714	0.8500
45	0.9755	0.9547	0.9455	0.9077	0.8929	0.8800	0.8588
50	0.9778	0.9586	0.9500	0.9143	0.9000	0.8875	0.8667

Table D.7 Half Year Rule Capital Tax Factors

Tax Rate = 35%

Interest Rate (%)	Declining Balance CCA Rate (%)						
	4	8	10	20	25	30	40
1	0.7214	0.6904	0.6834	0.6683	0.6651	0.6630	0.6602
2	0.7690	0.7227	0.7112	0.6849	0.6791	0.6751	0.6699
3	0.8029	0.7492	0.7347	0.7001	0.6921	0.6865	0.6792
4	0.8284	0.7712	0.7548	0.7139	0.7041	0.6971	0.6879
5	0.8481	0.7897	0.7722	0.7267	0.7153	0.7071	0.6963
6	0.8640	0.8057	0.7874	0.7384	0.7257	0.7166	0.7043
7	0.8769	0.8194	0.8009	0.7492	0.7355	0.7255	0.7119
8	0.8877	0.8315	0.8128	0.7593	0.7447	0.7339	0.7191
9	0.8968	0.8421	0.8234	0.7686	0.7533	0.7419	0.7261
10	0.9045	0.8515	0.8330	0.7773	0.7614	0.7494	0.7327
11	0.9113	0.8599	0.8416	0.7854	0.7690	0.7566	0.7391
12	0.9172	0.8675	0.8494	0.7930	0.7762	0.7634	0.7452
13	0.9224	0.8743	0.8566	0.8001	0.7830	0.7699	0.7510
14	0.9270	0.8805	0.8631	0.8068	0.7894	0.7760	0.7567
15	0.9311	0.8862	0.8691	0.8130	0.7955	0.7819	0.7621
16	0.9348	0.8914	0.8747	0.8190	0.8013	0.7875	0.7672
17	0.9382	0.8961	0.8798	0.8246	0.8068	0.7928	0.7722
18	0.9412	0.9005	0.8845	0.8298	0.8120	0.7979	0.7770
19	0.9440	0.9046	0.8889	0.8348	0.8170	0.8028	0.7817
20	0.9465	0.9083	0.8931	0.8396	0.8218	0.8075	0.7861
21	0.9489	0.9118	0.8969	0.8441	0.8263	0.8120	0.7904
22	0.9510	0.9151	0.9005	0.8484	0.8306	0.8163	0.7946
23	0.9530	0.9181	0.9039	0.8524	0.8348	0.8204	0.7986
24	0.9548	0.9210	0.9070	0.8563	0.8387	0.8244	0.8024
25	0.9566	0.9236	0.9100	0.8600	0.8425	0.8282	0.8062
30	0.9636	0.9348	0.9226	0.8762	0.8593	0.8452	0.8231
35	0.9688	0.9433	0.9323	0.8892	0.8731	0.8594	0.8375
40	0.9727	0.9500	0.9400	0.9000	0.8846	0.8714	0.8500
45	0.9759	0.9554	0.9462	0.9090	0.8944	0.8817	0.8609
50	0.9784	0.9598	0.9514	0.9167	0.9028	0.8906	0.8704

Table D.8 Full Year Rule Capital Tax Factors

Tax Rate = 35%

Interest Rate (%)	Declining Balance CCA Rate (%)						
	4	8	10	20	25	30	40
1	0.7200	0.6889	0.6818	0.6667	0.6635	0.6613	0.6585
2	0.7667	0.7200	0.7083	0.6818	0.6759	0.6719	0.6667
3	0.8000	0.7455	0.7308	0.6957	0.6875	0.6818	0.6744
4	0.8250	0.7667	0.7500	0.7083	0.6983	0.6912	0.6818
5	0.8444	0.7846	0.7667	0.7200	0.7083	0.7000	0.6889
6	0.8600	0.8000	0.7813	0.7308	0.7177	0.7083	0.6957
7	0.8727	0.8133	0.7941	0.7407	0.7266	0.7162	0.7021
8	0.8833	0.8250	0.8056	0.7500	0.7348	0.7237	0.7083
9	0.8923	0.8353	0.8158	0.7586	0.7426	0.7308	0.7143
10	0.9000	0.8444	0.8250	0.7667	0.7500	0.7375	0.7200
11	0.9067	0.8526	0.8333	0.7742	0.7569	0.7439	0.7255
12	0.9125	0.8600	0.8409	0.7813	0.7635	0.7500	0.7308
13	0.9176	0.8667	0.8478	0.7879	0.7697	0.7558	0.7358
14	0.9222	0.8727	0.8542	0.7941	0.7756	0.7614	0.7407
15	0.9263	0.8783	0.8600	0.8000	0.7813	0.7667	0.7455
16	0.9300	0.8833	0.8654	0.8056	0.7866	0.7717	0.7500
17	0.9333	0.8880	0.8704	0.8108	0.7917	0.7766	0.7544
18	0.9364	0.8923	0.8750	0.8158	0.7965	0.7813	0.7586
19	0.9391	0.8963	0.8793	0.8205	0.8011	0.7857	0.7627
20	0.9417	0.9000	0.8833	0.8250	0.8056	0.7900	0.7667
21	0.9440	0.9034	0.8871	0.8293	0.8098	0.7941	0.7705
22	0.9462	0.9067	0.8906	0.8333	0.8138	0.7981	0.7742
23	0.9481	0.9097	0.8939	0.8372	0.8177	0.8019	0.7778
24	0.9500	0.9125	0.8971	0.8409	0.8214	0.8056	0.7813
25	0.9517	0.9152	0.9000	0.8444	0.8250	0.8091	0.7846
30	0.9588	0.9263	0.9125	0.8600	0.8409	0.8250	0.8000
35	0.9641	0.9349	0.9222	0.8727	0.8542	0.8385	0.8133
40	0.9682	0.9417	0.9300	0.8833	0.8654	0.8500	0.8250
45	0.9714	0.9472	0.9364	0.8923	0.8750	0.8600	0.8353
50	0.9741	0.9517	0.9417	0.9000	0.8833	0.8688	0.8444

Table D.9 Half Year Rule Capital Tax Factors

Tax Rate = 40%

Interest Rate (%)	Declining Balance CCA Rate (%)						
	4	8	10	20	25	30	40
1	0.6816	0.6462	0.6382	0.6209	0.6173	0.6148	0.6117
2	0.7359	0.6831	0.6699	0.6399	0.6333	0.6287	0.6228
3	0.7748	0.7133	0.6968	0.6572	0.6481	0.6417	0.6333
4	0.8038	0.7385	0.7198	0.6731	0.6618	0.6538	0.6434
5	0.8265	0.7597	0.7397	0.6876	0.6746	0.6653	0.6529
6	0.8445	0.7779	0.7571	0.7010	0.6865	0.6761	0.6620
7	0.8593	0.7936	0.7724	0.7134	0.6977	0.6863	0.6707
8	0.8716	0.8074	0.7860	0.7249	0.7082	0.6959	0.6790
9	0.8820	0.8195	0.7982	0.7355	0.7180	0.7050	0.6870
10	0.8909	0.8303	0.8091	0.7455	0.7273	0.7136	0.6945
11	0.8986	0.8399	0.8190	0.7547	0.7360	0.7218	0.7018
12	0.9054	0.8486	0.8279	0.7634	0.7442	0.7296	0.7088
13	0.9113	0.8564	0.8361	0.7715	0.7520	0.7370	0.7155
14	0.9166	0.8635	0.8436	0.7792	0.7593	0.7440	0.7219
15	0.9213	0.8699	0.8504	0.7863	0.7663	0.7507	0.7281
16	0.9255	0.8759	0.8568	0.7931	0.7729	0.7571	0.7340
17	0.9293	0.8813	0.8626	0.7995	0.7792	0.7632	0.7397
18	0.9328	0.8863	0.8680	0.8055	0.7852	0.7691	0.7452
19	0.9360	0.8909	0.8731	0.8112	0.7909	0.7747	0.7505
20	0.9389	0.8952	0.8778	0.8167	0.7963	0.7800	0.7556
21	0.9416	0.8992	0.8822	0.8218	0.8015	0.7851	0.7605
22	0.9440	0.9030	0.8863	0.8267	0.8064	0.7900	0.7652
23	0.9463	0.9064	0.8901	0.8313	0.8111	0.7948	0.7698
24	0.9484	0.9097	0.8937	0.8358	0.8157	0.7993	0.7742
25	0.9503	0.9127	0.8971	0.8400	0.8200	0.8036	0.7785
30	0.9584	0.9255	0.9115	0.8585	0.8392	0.8231	0.7978
35	0.9643	0.9352	0.9226	0.8734	0.8549	0.8393	0.8143
40	0.9688	0.9429	0.9314	0.8857	0.8681	0.8531	0.8286
45	0.9724	0.9490	0.9386	0.8960	0.8793	0.8648	0.8410
50	0.9753	0.9540	0.9444	0.9048	0.8889	0.8750	0.8519

Table D.10 Full Year Rule Capital Tax Factors

Tax Rate = 40%

Interest Rate (%)	Declining Balance CCA Rate (%)						
	4	8	10	20	25	30	40
1	0.6800	0.6444	0.6364	0.6190	0.6154	0.6129	0.6098
2	0.7333	0.6800	0.6667	0.6364	0.6296	0.6250	0.6190
3	0.7714	0.7091	0.6923	0.6522	0.6429	0.6364	0.6279
4	0.8000	0.7333	0.7143	0.6667	0.6552	0.6471	0.6364
5	0.8222	0.7538	0.7333	0.6800	0.6667	0.6571	0.6444
6	0.8400	0.7714	0.7500	0.6923	0.6774	0.6667	0.6522
7	0.8545	0.7867	0.7647	0.7037	0.6875	0.6757	0.6596
8	0.8667	0.8000	0.7778	0.7143	0.6970	0.6842	0.6667
9	0.8769	0.8118	0.7895	0.7241	0.7059	0.6923	0.6735
10	0.8857	0.8222	0.8000	0.7333	0.7143	0.7000	0.6800
11	0.8933	0.8316	0.8095	0.7419	0.7222	0.7073	0.6863
12	0.9000	0.8400	0.8182	0.7500	0.7297	0.7143	0.6923
13	0.9059	0.8476	0.8261	0.7576	0.7368	0.7209	0.6981
14	0.9111	0.8545	0.8333	0.7647	0.7436	0.7273	0.7037
15	0.9158	0.8609	0.8400	0.7714	0.7500	0.7333	0.7091
16	0.9200	0.8667	0.8462	0.7778	0.7561	0.7391	0.7143
17	0.9238	0.8720	0.8519	0.7838	0.7619	0.7447	0.7193
18	0.9273	0.8769	0.8571	0.7895	0.7674	0.7500	0.7241
19	0.9304	0.8815	0.8621	0.7949	0.7727	0.7551	0.7288
20	0.9333	0.8857	0.8667	0.8000	0.7778	0.7600	0.7333
21	0.9360	0.8897	0.8710	0.8049	0.7826	0.7647	0.7377
22	0.9385	0.8933	0.8750	0.8095	0.7872	0.7692	0.7419
23	0.9407	0.8968	0.8788	0.8140	0.7917	0.7736	0.7460
24	0.9429	0.9000	0.8824	0.8182	0.7959	0.7778	0.7500
25	0.9448	0.9030	0.8857	0.8222	0.8000	0.7818	0.7538
30	0.9529	0.9158	0.9000	0.8400	0.8182	0.8000	0.7714
35	0.9590	0.9256	0.9111	0.8545	0.8333	0.8154	0.7867
40	0.9636	0.9333	0.9200	0.8667	0.8462	0.8286	0.8000
45	0.9673	0.9396	0.9273	0.8769	0.8571	0.8400	0.8118
50	0.9704	0.9448	0.9333	0.8857	0.8667	0.8500	0.8222

Table D.11 Half Year Rule Capital Tax Factors

Tax Rate = 45%

Interest Rate (%)	Declining Balance CCA Rate (%)						
	4	8	10	20	25	30	40
1	0.6418	0.6020	0.5929	0.5736	0.5694	0.5667	0.5631
2	0.7029	0.6435	0.6287	0.5949	0.5874	0.5823	0.5756
3	0.7466	0.6775	0.6589	0.6144	0.6041	0.5969	0.5875
4	0.7793	0.7058	0.6848	0.6322	0.6195	0.6106	0.5988
5	0.8048	0.7297	0.7071	0.6486	0.6339	0.6235	0.6095
6	0.8251	0.7501	0.7267	0.6636	0.6474	0.6356	0.6198
7	0.8417	0.7679	0.7440	0.6776	0.6599	0.6471	0.6295
8	0.8556	0.7833	0.7593	0.6905	0.6717	0.6579	0.6389
9	0.8673	0.7970	0.7729	0.7025	0.6828	0.6681	0.6478
10	0.8773	0.8091	0.7852	0.7136	0.6932	0.6778	0.6564
11	0.8859	0.8199	0.7963	0.7241	0.7030	0.6870	0.6645
12	0.8935	0.8296	0.8064	0.7338	0.7122	0.6958	0.6724
13	0.9002	0.8384	0.8156	0.7430	0.7210	0.7041	0.6799
14	0.9061	0.8464	0.8240	0.7515	0.7293	0.7120	0.6871
15	0.9114	0.8537	0.8317	0.7596	0.7371	0.7196	0.6941
16	0.9162	0.8603	0.8389	0.7672	0.7445	0.7268	0.7007
17	0.9205	0.8665	0.8454	0.7744	0.7516	0.7336	0.7072
18	0.9244	0.8721	0.8515	0.7812	0.7583	0.7402	0.7133
19	0.9280	0.8773	0.8572	0.7877	0.7647	0.7465	0.7193
20	0.9313	0.8821	0.8625	0.7938	0.7708	0.7525	0.7250
21	0.9342	0.8866	0.8674	0.7995	0.7767	0.7583	0.7305
22	0.9370	0.8908	0.8721	0.8050	0.7822	0.7638	0.7359
23	0.9396	0.8947	0.8764	0.8103	0.7875	0.7691	0.7410
24	0.9419	0.8984	0.8805	0.8152	0.7926	0.7742	0.7460
25	0.9441	0.9018	0.8843	0.8200	0.7975	0.7791	0.7508
30	0.9532	0.9162	0.9005	0.8408	0.8191	0.8010	0.7725
35	0.9598	0.9271	0.9130	0.8576	0.8368	0.8192	0.7911
40	0.9649	0.9357	0.9229	0.8714	0.8516	0.8347	0.8071
45	0.9690	0.9426	0.9309	0.8830	0.8642	0.8479	0.8211
50	0.9722	0.9483	0.9375	0.8929	0.8750	0.8594	0.8333

Table D.12 Full Year Rule Capital Tax Factors

Tax Rate = 45%

Interest Rate (%)	Declining Balance CCA Rate (%)						
	4	8	10	20	25	30	40
1	0.6400	0.6000	0.5909	0.5714	0.5673	0.5645	0.5610
2	0.7000	0.6400	0.6250	0.5909	0.5833	0.5781	0.5714
3	0.7429	0.6727	0.6538	0.6087	0.5982	0.5909	0.5814
4	0.7750	0.7000	0.6786	0.6250	0.6121	0.6029	0.5909
5	0.8000	0.7231	0.7000	0.6400	0.6250	0.6143	0.6000
6	0.8200	0.7429	0.7188	0.6538	0.6371	0.6250	0.6087
7	0.8364	0.7600	0.7353	0.6667	0.6484	0.6351	0.6170
8	0.8500	0.7750	0.7500	0.6786	0.6591	0.6447	0.6250
9	0.8615	0.7882	0.7632	0.6897	0.6691	0.6538	0.6327
10	0.8714	0.8000	0.7750	0.7000	0.6786	0.6625	0.6400
11	0.8800	0.8105	0.7857	0.7097	0.6875	0.6707	0.6471
12	0.8875	0.8200	0.7955	0.7188	0.6959	0.6786	0.6538
13	0.8941	0.8286	0.8043	0.7273	0.7039	0.6860	0.6604
14	0.9000	0.8364	0.8125	0.7353	0.7115	0.6932	0.6667
15	0.9053	0.8435	0.8200	0.7429	0.7188	0.7000	0.6727
16	0.9100	0.8500	0.8269	0.7500	0.7256	0.7065	0.6786
17	0.9143	0.8560	0.8333	0.7568	0.7321	0.7128	0.6842
18	0.9182	0.8615	0.8393	0.7632	0.7384	0.7188	0.6897
19	0.9217	0.8667	0.8448	0.7692	0.7443	0.7245	0.6949
20	0.9250	0.8714	0.8500	0.7750	0.7500	0.7300	0.7000
21	0.9280	0.8759	0.8548	0.7805	0.7554	0.7353	0.7049
22	0.9308	0.8800	0.8594	0.7857	0.7606	0.7404	0.7097
23	0.9333	0.8839	0.8636	0.7907	0.7656	0.7453	0.7143
24	0.9357	0.8875	0.8676	0.7955	0.7704	0.7500	0.7188
25	0.9379	0.8909	0.8714	0.8000	0.7750	0.7545	0.7231
30	0.9471	0.9053	0.8875	0.8200	0.7955	0.7750	0.7429
35	0.9538	0.9163	0.9000	0.8364	0.8125	0.7923	0.7600
40	0.9591	0.9250	0.9100	0.8500	0.8269	0.8071	0.7750
45	0.9633	0.9321	0.9182	0.8615	0.8393	0.8200	0.7882
50	0.9667	0.9379	0.9250	0.8714	0.8500	0.8313	0.8000

Table D.13 Half Year Rule Capital Tax Factors

Tax Rate = 50%

Interest Rate (%)	Declining Balance CCA Rate (%)						
	4	8	10	20	25	30	40
1	0.6020	0.5578	0.5477	0.5262	0.5216	0.5185	0.5146
2	0.6699	0.6039	0.5874	0.5499	0.5416	0.5358	0.5285
3	0.7184	0.6417	0.6210	0.5715	0.5601	0.5521	0.5417
4	0.7548	0.6731	0.6497	0.5913	0.5773	0.5673	0.5542
5	0.7831	0.6996	0.6746	0.6095	0.5933	0.5816	0.5661
6	0.8057	0.7224	0.6963	0.6263	0.6082	0.5951	0.5775
7	0.8241	0.7421	0.7155	0.6417	0.6222	0.6079	0.5884
8	0.8395	0.7593	0.7325	0.6561	0.6352	0.6199	0.5988
9	0.8525	0.7744	0.7477	0.6694	0.6475	0.6313	0.6087
10	0.8636	0.7879	0.7614	0.6818	0.6591	0.6420	0.6182
11	0.8733	0.7999	0.7737	0.6934	0.6700	0.6523	0.6273
12	0.8817	0.8107	0.7849	0.7042	0.6803	0.6620	0.6360
13	0.8891	0.8205	0.7951	0.7144	0.6900	0.6712	0.6443
14	0.8957	0.8293	0.8045	0.7239	0.6992	0.6800	0.6524
15	0.9016	0.8374	0.8130	0.7329	0.7079	0.6884	0.6601
16	0.9069	0.8448	0.8210	0.7414	0.7161	0.6964	0.6675
17	0.9117	0.8516	0.8283	0.7494	0.7240	0.7040	0.6746
18	0.9160	0.8579	0.8350	0.7569	0.7315	0.7113	0.6815
19	0.9200	0.8637	0.8414	0.7641	0.7386	0.7183	0.6881
20	0.9236	0.8690	0.8472	0.7708	0.7454	0.7250	0.6944
21	0.9269	0.8740	0.8527	0.7773	0.7518	0.7314	0.7006
22	0.9300	0.8787	0.8578	0.7834	0.7580	0.7375	0.7065
23	0.9329	0.8830	0.8627	0.7892	0.7639	0.7434	0.7122
24	0.9355	0.8871	0.8672	0.7947	0.7696	0.7491	0.7177
25	0.9379	0.8909	0.8714	0.8000	0.7750	0.7545	0.7231
30	0.9480	0.9069	0.8894	0.8231	0.7990	0.7788	0.7473
35	0.9554	0.9190	0.9033	0.8418	0.8187	0.7991	0.7679
40	0.9610	0.9286	0.9143	0.8571	0.8352	0.8163	0.7857
45	0.9655	0.9362	0.9232	0.8700	0.8491	0.8310	0.8012
50	0.9691	0.9425	0.9306	0.8810	0.8611	0.8438	0.8148

Table D.14 Full Year Rule Capital Tax Factors

Tax Rate = 50%

Interest Rate (%)	Declining Balance CCA Rate (%)						
	4	8	10	20	25	30	40
1	0.6000	0.5556	0.5455	0.5238	0.5192	0.5161	0.5122
2	0.6667	0.6000	0.5833	0.5455	0.5370	0.5313	0.5238
3	0.7143	0.6364	0.6154	0.5652	0.5536	0.5455	0.5349
4	0.7500	0.6667	0.6429	0.5833	0.5690	0.5588	0.5455
5	0.7778	0.6923	0.6667	0.6000	0.5833	0.5714	0.5556
6	0.8000	0.7143	0.6875	0.6154	0.5968	0.5833	0.5652
7	0.8182	0.7333	0.7059	0.6296	0.6094	0.5946	0.5745
8	0.8333	0.7500	0.7222	0.6429	0.6212	0.6053	0.5833
9	0.8462	0.7647	0.7368	0.6552	0.6324	0.6154	0.5918
10	0.8571	0.7778	0.7500	0.6667	0.6429	0.6250	0.6000
11	0.8667	0.7895	0.7619	0.6774	0.6528	0.6341	0.6078
12	0.8750	0.8000	0.7727	0.6875	0.6622	0.6429	0.6154
13	0.8824	0.8095	0.7826	0.6970	0.6711	0.6512	0.6226
14	0.8889	0.8182	0.7917	0.7059	0.6795	0.6591	0.6296
15	0.8947	0.8261	0.8000	0.7143	0.6875	0.6667	0.6364
16	0.9000	0.8333	0.8077	0.7222	0.6951	0.6739	0.6429
17	0.9048	0.8400	0.8148	0.7297	0.7024	0.6809	0.6491
18	0.9091	0.8462	0.8214	0.7368	0.7093	0.6875	0.6552
19	0.9130	0.8519	0.8276	0.7436	0.7159	0.6939	0.6610
20	0.9167	0.8571	0.8333	0.7500	0.7222	0.7000	0.6667
21	0.9200	0.8621	0.8387	0.7561	0.7283	0.7059	0.6721
22	0.9231	0.8667	0.8438	0.7619	0.7340	0.7115	0.6774
23	0.9259	0.8710	0.8485	0.7674	0.7396	0.7170	0.6825
24	0.9286	0.8750	0.8529	0.7727	0.7449	0.7222	0.6875
25	0.9310	0.8788	0.8571	0.7778	0.7500	0.7273	0.6923
30	0.9412	0.8947	0.8750	0.8000	0.7727	0.7500	0.7143
35	0.9487	0.9070	0.8889	0.8182	0.7917	0.7692	0.7333
40	0.9545	0.9167	0.9000	0.8333	0.8077	0.7857	0.7500
45	0.9592	0.9245	0.9091	0.8462	0.8214	0.8000	0.7647
50	0.9630	0.9310	0.9167	0.8571	0.8333	0.8125	0.7778

French/English Concordance

acomptes provisionnel	tax paid by instalments
acquisition	acquisition
actif à court terme	current assets
actif	asset
actions	shares; stocks
actions ordinaires	common stock
actions privilégiées	preferred stock
actualisation	discounting
actualisation continue	continuous discounting
actualisation quotidienne	daily compounding
Agence de douanes et du revenu du Canada	Canada Customs and Revenue Agency
alternatives, options	alternatives
amélioration	betterment
améloriations locatives	leasehold improvements
amortissement linéaire; constant	straight line depreciation
amortissement avec solde dégressif	declining balance depreciation
amortissement	amortization
amortissement comptable	accounting depreciation
amortissement	depreciation
amortissement fiscal	tax depreciation, capital cost allowance
amortissement aux livres	book depreciation
analyse coûts-avantages	benefit-cost analysis
analyse de sensibilité	sensitivity analysis
analyse marginale, analyse différentielle	incremental analysis
analyse du seuil de rentabilité; du point mort	break-even analysis
analyse financière	financial analysis
année d'imposition	taxation year
annuité de début de période	annuity due
annuité	annuity
avoir minier	resource property
avoirs forestiers	timber resource property
bail financier	financial lease
bail d'exploitation	operating lease
bailleur	lessor
barème d'imposition	tax schedule
bénéfice par action	earnings per share

bénéfice durable	enduring benefit
bénéfices non répartis	retained earnings
biens d'équipement	capital goods
biens publics	public goods
biens prêts à être mis en service	available-for-use
biens amortissables	depreciable assets
bilan	balance sheet
bons du Trésor	Treasury bills
bourse des valeurs mobilières	stock market; stock exchange
budgét des investissements	capital budget
calendrier d'amortissement	depreciation schedule
capital fixe	fixed assets
capital amorti, capital recouvert	recovered capital
capital	capital
capitalisation continue	continuous compounding
capitalisé	capitalized
capitaux propres	net worth
capitaux propres	equity
catégorie de biens	asset class (pool)
certificat de placement garanti	guaranteed investment certificate (GIC)
charges fixes	fixed costs
choix, option	option
comptabilisation à la valeur d'acquisition	cost accounting
concessions forestières	timber limit
conditionnel; potentiel	contingent, conditional
contribution	contribution
courbe d'apprentissage	learning curve
courbe d'indifférence	indifference curve
courbe de valeur	present worth profile
coût annuel équivalent	equivalent annual cost
coût du revient de cycle de vie	life cycle costing
coût d'opportunité	opportunity cost
coût des produits vendus	cost of goods sold
coût capitalisé équivalent	capitalized equivalent
coût de recouvrement du capital	capital recover cost
coût du capital	cost of capital
coût variable	variable cost
coût en capital	cost basis
coût unitaire	unit cost
coût initial	first cost
coût irrécupérables, coûts perdus	sunk costs
coûts d'entretien	maintenance costs
coûts d'émission	floatation costs
coûts marginaux	marginal costs
coûts ventile; reparti	imputed cost
coûts d'exploitation	operating costs
crédit d'impôt à l'investissement	investment tax credit

crédit d'impôt remboursable	refundable tax credits
crédit d'impôt	tax credit
crédits d'impôt non remboursable	nonrefundable tax credit
date d'échéance	maturity date
date d'exigibilité	due date
débenture	debenture
débours	disbursements
décision à court terme	short-run decision
décision de fabriquer ou d'acheter	make-or-buy decision
décisions dans un contexte aléatoire	decisions under risk
décisions dans un contexte incertain	decisions under uncertainty
déduction accordée aux petites entreprises	small business deduction
déduction pour amortissement	capital cost allowance
défenseur	defender
déflation	deflation
délai de récupération	payback period
demande de dépense	request for expenditure
dépenses d'exploitation	operating expenses
dépôt à terme	term deposit (TD)
dépréciation physique	physical depreciation
dépréciation économique	economic depreciation
dépréciation fonctionnelle	functional depreciation
devis	estimate
diagramme du flux monétaire	cash flow diagram
différentiel	incremental
discret	discrete
disponibilité	availability
disposition	disposition
distribution cumulative	cumulative distribution
dividende	dividend
dollar constant	constant dollar
dollars constants	real dollars
dollars courants, dollars du moment	actual dollars
dotation aux amortissements	depreciation expense
DPA récupérée	recaptured CCA
DPA	CCA
durée amortissable	depreciable life
durée de vie utile	useful life
durée de vie économique	economic service life
durée du projet	project life
effet de levier	leverage
effet fiscal à la disposition	disposal tax effect
efficacité des coûts	cost effectiveness
efficacité	effectiveness
efficience	efficiency
emprunt hypothécaire non plafonné	open-end mortgage
entreprise	business

entretien	maintenance
épuisement	depletion
équivalence	equivalence
équivalence économique	economic equivalence
équivalent annuel	annual equivalent
état financier	financial statement
état des résultats	income statement
exercice	fiscal period
extrant	output
facteur de valeur accumulée	compound amount factor
facteur de recouvrement du capital	capital recovery factor
facteur d'annuité	annuity factor
facteur du coût en capital	capital cost tax factor
facteur d'impôt du coût en capital	capital cost tax factor
facteur d'intérêt	interest factor
financement par actions	equity financing
financement par emprunt	debt financing
flux monétaire irrégulier	irregular series
flux monétaires d'un gradient géométrique	geometric gradient series
flux monétaire d'un gradient arithmétique	linear gradient series
flux monétaire uniforme	uniform cash flow series
flux monetaire actualisé	discounted cash flow
flux monetaire net	net cash flow
fonds commun d'investissement	investment pool
fonds d'amortissement	sinking fund
fonds de roulement	working capital
fraction non amortie du coût en capital (FNACC)	undepreciated capital cost (UCC)
frais engagés	costs incurred
frais généraux	overhead costs
gains en capital	capital gains
gradient	gradient
graphique radar	radar chart
horizon de planification	planning horizon
hypothèque	mortgage
hypothèque mobilière	collateral mortgage
immobilisations	capital assets
impôt d'inflation	inflation tax
impôt sur le revenu	income tax
incertitude	uncertainty
incitations fiscales	tax incentives
incorporels	intangibles
indice des prix	price index
indice du volume	volume index
indice des prix industriels	industry price index
indice du coût de la vie	cost-of-living index
indice des prix à la consommation	consumer price index

inflation	inflation
intérêt simple	simple interest
intérêt	interest
intérêt composé	compound interest
interpolation linéaire	linear interpolation
intrant	input
investissement simple	simple investment
investissement	investment
investissement différé	deferred investment
l'économie de l'ingénieur	engineering economics
libération	working capital release
locataire	leasee
location	lease
loi des grands nombres	law of large numbers
marge bénéficiaire	profit margin
marginal	marginal
matrice des règlements	payoff matrix
mensuel	monthly
méthode de multiple commun	common-multiple method
méthode d'investissement total	total investment approach
méthode généralisée de flux monétaire	generalized cash flow approach
mines de minéraux industriel	industrial mineral mines
mise en service	commissioning
mode	mode
mouvement de trésorerie	cash flows
mutuellement exclusif	mutually exclusive
nombre aléatoire	random number
normes d'exploitation	standard operating procedures
obligation au porteur	bearer bond
obligation nominative	registered bond
obligation remboursable par anticipation	callable bond
obligation hypothécaire	mortgage bond
obligation avec prime d'émission	premium bond
obligation	bond
obligation avec escompte d'émission	discount bond
obligations négociables	marketable bonds
Obligations d'épargne du Canada	Canada savings bonds
opposant	challenger
option nulle	do-nothing alternative
paiement unique; somme forfaitaire	lump sum
passif	liabilities
passif à court terme	current liabilities
période d'analyse	analysis period
période d'actualisation	interest period
période d'étude	study period
période d'actualisation	compounding period
perte final	terminal loss

pertes en capital	capital losses
pertes de bénéfices	disbenefits
plan du occasions d'investissement	investment opportunity schedule
plus petit multiple commun	lowest common multiple
plus petit facteur commun	least common multiple
point de repère	benchmark
point mort	break-even point
politique fiscale	fiscal policy
politique monétaire	monetary policy
pouvoir d'achat	purchasing power
pouvoir de gain	earning power
prêt avec l'intérêt ajouté	add-on loan
prêt hypothécaire à rapport prêt-valeur élevé	high-ratio mortgage
prêt	loan
principal	principal
principe du rapprochement, d'appariement	matching principle
prix de base rajusté	adjusted cost basis
prix d'achat	purchase price
probabilité conditionnelle	conditional probability
probabilité marginale	marginal probability
probabilité objective	objective probability
probabilité conjointe	joint probability
probabilité subjective	subjective probability
produit de disposition	proceeds of disposition
profil de risque d'un investissement	investment risk profile
profit unitaire	unit profit
projet de service	service project
projet avec revenus	revenue project
propriétaire unique	sole proprietorship
protection fiscale	tax shield
quotidien	daily
ratio d'endettement	debt ratio
ratio d'exploitation	operating ratio
ratio cours-bénéfice	price-earnings ratio
ratio d'endettement	equity ratio
rationnement du capital	capital rationing
recettes	receipts
régime enregistré d'épargne-retraite (REER)	registered retirement savings plan (RRSP)
régime enregistré d'épargne-études	registered education savings plan
règle de la demi-année	half-year convention
règle du 50%	50% rule
reliquat	unrecovered capital
reliquat, solde	balance owing
reliquat	loan balance
remboursement	repayment

remplacement	replacement
rendement de l'investissement	return on investment
rendement	yield
rendement à l'échéance	yield to maturity
rentrée de fonds, encaissement	cash inflows
reports rétrospectifs	carrybacks
ressource minérale	mineral resource
revenu brut	gross income
revenu net	net income
revenus imposables	taxable income
risque	risk
risque financier	financial risk
rotation des stocks	inventory turnover
séries monétaires combinées	composite series
simulation	simulation
société	corporation
société de personnes	partnership
solde du projet	project balance
sortie de fonds, décaissement	cash outflows
structure du capital	capital structure
structure financière	financial structure
surtaxe	surtax
tableur	spreadsheet
taux de rendement acceptable minimum	minimum attractive rate of return
taux de rendement	rate of return
taux d'escompte social	social discount rate
taux de rendement interne	internal rate of return
taux d'intérêt	interest rate
taux d'intérêt réel	real interest rate
taux d'intérêt nominal	nominal interest rate
taux d'emprunt hypothécaire	mortgage rate
taux externe de rendement	external rate of return
taux d'actualisation	discount rate
taux composé annuel	annual percentage rate (APR)
taux d'intérêt externe	external interest rate
taux d'intérêt contractuel; nominal	coupon rate
taux du marché	market interest rate
taux de rendement multiples	multiple rates of return
taux de croissance	growth rate
taux de rendement courant	current yield
taux préférentiel	prime rate
taux d'intérêt effectif	effective interest rate
taux d'extinction	decay rate
test de l'investissement net	net investment test
TRAM	MARR
triage	screening
trimestrielle	quarterly

utilité	utility
valeur comptable	book value
valeur nominale	face value, par value
valeur de reprise	trade-in value
valeur espérée	expected value
valeur de l'argent dans le temps	time value of money
valeur actualisée, valeur présente	present equivalent, present worth, present value
valeur de récupération	salvage value
valeur capitalisée, valeur future	future value, future worth
valeur marchande	market value
valeur d'usage	going concern value
valeur amortissable	capital cost
valeur d'inventaire	inventories
valeurs disponibles	cash
valeurs réalisables à court terme	outstanding receipts
valorisation	valuation
variable aléatoire	random variable
variance	variance
venir à l'échéance	mature
versements	payments
vie de service	service life

Answers to Selected Problems

Chapter 2

$2.1 $24,481

2.2 (a)

$2.3 P = $5375

$2.4 C = $3741

$2.5 simple interest = $300, compound interest = $312.90

$2.6 F = $4422

$2.7 C = $8054

2.8 P = $30,000

2.9 (d)

$2.12 compound interest: $1191.02, simple interest: $1210

2.16 (a) P = $3532 (b) P = $2459 (c) P = $12,999 (d) P = $4627

$2.20 P = $7473.70

2.30 (a) P = $10,060.61 (b) $12,835.32 (c) $2036.45 (d) $22,346.49

$2.33 F = $4993.41

$2.38 i = 24.57%

2.40 C = $458.90

$2.43 (a) F = $18,231.52 (b) F = $19,872.35

2.57 A = $61.18

Chapter 3

$3.1 Bank A: 15.865%; Bank B: 15.948%; (c)

3.2 $F = \$2500(F/A, 2.2669\%, 40) = \$160,058$

$3.3 $A = \$10,000(A/P, 0.75\%, 60) = \$208, P = \$208(P/A, 0.75\%, 36) = \$6541$

$3.4 (b), Option 1: $F = \$1000(F/A, 1.5\%, 40) (F/P, 1.5\%, 60) = \$132,588$. Option 2: $F = \$6000(F/A, 6.136\%, 15) = \$141,111$

$3.5 (c)

3.6 $(1 + i)^2 = 1.12, i = 5.83\%$ semi-annually, $\$1000(P/A, 5.83\%, 6) = \$4944$

3.7 $2 = (1 + r/4)^{20}, r = 3.526\% \times 4 = 14.11\%$

3.8 $F = \$1000(F/A, 2.26692\%, 12) = \$13,615$

$3.9 $i, = (1 + 0.018)^{12} = 23.87\%$

3.10 $P = \$1000(P/A, 2.020\%, 20) = \$16,320$

3.11 $A = \$25,000(A/P, 0.75\%, 36) = \$795$

3.12 $\$70,000 = \$1000(P/A, 1\%, N), N = 121$ months

$3.13 $\$20,000 = \$922.90(P/A, i, 24), i = 0.83\%$ per month, $P = \$922.90(P/A, 0.83\%, 12) = \$10,500$

3.14 $\$20,000 = \$5548.19(P/A, i, 5), i = 12\%$

3.15 $2 = 1(F/P, 9.3083\%, N)$, $N = 8$ years

$3.16 $0.0887 = (1 + r/365)^{365} - 1$, $r = 8.5\%$

$3.17 $100(P/F, 8\%, 1) + \$300(P/F, 8\%, 2) + \$500(P/F, 8\%, 3) + X(P/F, 8\%, 4) =$ 1000; $X = \$345$

$3.24 24-month lease plan: PE = \$12,860.95; up-front lease plan: PE = \$12,026

$3.28 **(a)** \$116,678 **(b)** \$116,287 **(c)** \$116,089

3.38 **(a)** \$40,305.56 **(b)** \$76,054.66 **(c)** \$1,003,554.24

$3.45 A = \$471.03

$3.55 **(a)** \$127,646,246 **(b)** \$127, 655, 627

$3.58 **(a)** P = \$8875.42 **(b)** F = \$13,186 **(c)** A = \$2199.21

$3.67 A = \$1415.42

$3.75 **(a)** i_a = 24.6941% **(b)** \$1664.85

$3.77 **(a)** A = \$448.38 **(b)** A = \$438.88 **(c)** I_{Dealer} = \$727.23, I_{Credit} = \$3817.01

3.79 A = \$453.42

Chapter 4

4.1 n = 8 years

4.2 Payback periods: undiscounted = between 4 and 5 years; discounted = between 5 and 6 years

4.3 PE = \$866.51

4.4 PE = \$1386

4.10 FE's for projects A, B, C, D, E: \$3354.43, \$4741.10, \$23,006.14, \$7699.68, −\$831.87

4.13 \$1000

4.15 CR = \$3440

4.17 AE = \$4303.13

4.25 **(a)** X = \$2309.55 **(b)** AE = \$5265.46 >0

4.26 X = \$230

4.27 \$82,739

$4.29 F = \$904.67

4.35 i* = 10%

4.37 Payback method: undiscounted, (a); discounted, all statements are correct

$4.38 PE = \$218,420

4.39 PE = \$2,047,734

4.40 (d), project balances: n = 6, −\$2492; n = 7, \$2134

4.42 **(a)** PE = \$17,459.69 > 0 **(b)** PE = \$62,730 > 0 **(c)** i = 10.77%

4.45 **(a)** Cash flows n = 0 through n = 5: −\$10,000, − \$1000, −\$5000, \$8000, −\$6000, −\$3000

$4.59 (c)

Appendix 4A

4A.7 (a) A and B **(b)** C **(c)** project A IRR = 23.24%, project B IRR = 21.11%, project C IRR = 12.24% with MARR = 12% **(d)** All

4A.13 (c)

Chapter 5

5.1 PE_A (12%) = \$140.87; PE_B (12%) = \$197.68; choos

5.2 S_A = \$750 (after 2 years)

5.4 (a) PE_A (12%) = \$2103.23; PE_B(12%) = \$1019.20; choose A

   (b) FE_A(12%) = \$2954.48; PE_B(12%) = \$1431.90; choose A

5.5 PE_{BG} (10%) = \$13,582; PE_{EP} (10%) = −\$12,504; choose electric panels

5.7 (b)

5.8 AE_1(10%) = −\$9835; AE_2(10%) −\$9497; choose 2

5.9 (a)

5.11 Project C

5.14 Choose C

5.16 H ≥ 1018 hours

5.17 (a)

5.19 Annual energy cost: AE_A = (10/0.85)(0.7457)(1500)(0.07) = \$921.18;

   ∴ PE_A = −\$8669; AE_B = (10/0.90)(0.7457)(1500)(0.07) = \$869.97;

   ∴ PE_B = −\$8614; motor B is preferred

5.20 LCM = 6 years; PE_A = \$9,989; PE_B = \$15,056; choose B

5.22 Infinite analysis period: PE_A = −\$90,325; PE_B = −\$95,962; choose A

\$5.28 i_{s-a} = 3.04%; AE_{BOND} = \$79.59; AE_{STOCK} = \$107.17; AE_{LOAN} = \$126.60; choose loan

5.29 $AE(13\%)_A$ = \$260,083; $AE(13\%)_B$ = \$246,596; processing cost: C_A = \$35.63/ton; C_B = \$33.78; choose B

5.33 (a) IRR_A = 11.71% IRR_B = 19.15% (b) only B is acceptable (c) since A is not acceptable, don't use incremental analysis; accept B

5.36 IRR_A = 6.01%, IRR_B = 7.24%, IRR_{B-A} = 9.97%; Pick A

5.39 (a) IRR_B = 25.99% (b) $PE(15\%)_A$ = \$2,558 (c) IRR_{B-A} = 24.24% > 15%, select project B

Chapter 6

6.1 $CE(10\%)$ = \$400/0.1 + (\$100/0.1)(P/F, 10%, 10) = \$4386

6.2 $CE(10\%)$ = [\$100 + \$100(A/F, 10%,2)]/0.1 = \$1476

6.3 $AE(10\%)$ = \$1000 − (\$500/0.1)(P/F, 10%, 10) × 0.1 = \$807

6.4 $AE(10\%)$ = \$500 + (\$500/0.1)(P/F, 10%, 10) × 0.1 = \$693

6.5 (a) \$2500 (b) \$1500

6.6 AEC = − \$15,000(A/P, 15%, 2) − \$3000 + \$10,000(A/F, 15%,2) = \$7665.50

6.7 No. AEC^*_D < AEC^*_C. At least 2 more years

6.10 i_a = 6.136%, $CE(6.136\%)$ = \$195,567

6.14 $AEC(15\%)$ = \$5600 (A/P, 15%, 5) − \$6000 + \$5000 (A/F, 15%, 5) = \$21,964; machine cost per hour = \$21,964/2500 hrs = \$8.79/hour

6.15 $AEC(5\%)$ = \$800,000 × 0.05 + \$120,000 + \$13,000 − \$32,000 + \$50,000(A/F, 5%, 5) = \$150,049; Ticket cost = \$150,049/40,000 = \$3.75 per person

6.16 $AE(14\%)$ = [−\$100K + 30K(P/A_1, 3%, 14%, 5)](A/P, 14%, 5) = \$2482, savings per hour = \$2482/3000 hours = \$0.83/hour

6.17 (\$100,000 − \$12,000)(A/P, 12%, 8) + \$12,000(0.12) + \$10,000 = \$29,155, cost per book binding = \$29,155/1000 = \$29.16 per book; Annual volume of books needs = \$29,155/X = \$25, X = \$1166

6.18

	3200 kWh/month	6700 kWh/month
First 1500 kWh @ $0.025	$37.50	$37.50
next 1250 kWh @ $0.015	$18.75	$18.75
next 3000 kWh @ $0.009	$ 4.05	$27.00
All over 5750 kWh @ $0.008		$ 7.60
Total	$60.30	$90.85

Difference with an additional 3500 kWh = $90.85 − $60.30 = $30.55 per month

6.20 option 1: unit cost = $19.75; option 2: AEC = $496,776/yr, unit cost = $24.84; option 1 is better

6.22 option 1: unit cost = $0.25 per kilometre; option 2: AEC = $6221, unit cost = $0.52 per kilometre; option 1 is better

6.32 **(a)** sunk cost = $9000 **(b)** opportunity cost = $6000 **(c)** AEC = $6451 per year for two years **(d)** AEC = $6771 per year for five years

6.37 **(a)** remaining economic life = 2 years, AEC_D^* = $6435 **(b)** AEC_C^* = $6344 < AEC_D^*. Replace the defender now

6.42 **(a)** AEC_D = $30,000, AEC_C = $24,412 < AEC_D. Buy the proposed equipment now **(b)** AEC_D = $15,000, AEC_C = $22,014 > AEC_D. Don't buy the proposed equipment now

6.56 It appears that the $(j_0, 3)$, $(j, 3)$, $(j, 4)$ option is the least-cost replacement strategy with a PE value of −$17,221. In other words, we should keep the defender j_0 for 3 years, replace it with challenger j and keep it for 3 years, and replace it with another challenger j and keep it for 4 years.

6.59 convention method: unit cost = $23.49; laser blanking method: unit cost = $14.73. The company should go for the laser blanking method.

Chapter 7

7.1 B^2 = $25,000

7.2 d = $(1/4)* 2 = 0.5$; D_2 = $45,000(1 − 0.5)*0.5 = $11,250$

7.3 SOYD = $1 + 2 + 3 + 4 = 10$; D_1 = $(4/10)* (45,000 − 5000) = $16,000$; D_2 = $(3/10)*(40,000) = $12,000$

7.4 d = 0.2; U_n = $30,000*(1 − 0.2/2)*(1 − 0.2)^{n-1} = 11,059$; $(n−1)*1n(0.8) = 1n(0.4096)$; ∴ n = 5 years

7.7 CCA Class 1 (d = 4%); CCA_1 = $7500; CCA_2 = $14,700; CCA_3 = $14,112; CCA_4 = $13,548; CCA_5 = $13,005

7.8 Cost basis = $100,000

7.9 Cost basis = $31,000

7.10 d = $(1/8)*2.0 = 0.25$; D_3 = $0.25*150,000*(1 − 0.25)^2 = $21,094$

7.12 **(a)** 15 **(b)** $3333 **(c)** $2667

7.14 $D_{5000\ hours}$ = $[($60,000 − $8000/50,000)]*5000 = $5200

7.16 U_3 = $P(1 −d/2)(1 −d)^{3-1} = 100,000(1 −0.2/2)(1 −0.2)^2 = $57,600

7.17 CCA Class 43 (d = 30%); CCA_1 = $10,200; CCA_2 = $17,430; CCA_3 = $12,138; CCA_4 = $8497; CCA_5 = $5948

7.24 **(a)** D = $4167 **(b)** D_3 = $6713 **(c)** D_2 = $7051

7.27 **(a)** CCA_1 = $2400 (note: the land is not depreciable) **(b)** UCC_4 = $85,730

7.36 This limestone qualifies as an industrial mineral mine. Units-of-production depreciation rate: ($335,000 − 0)/475,000 = $0.7053/m³; CCA_1 = $31,737; CCA_2 = $33,324; CCA_3 = $34,990; CCA_4 = $36,739

7.38 **(a)** D = $32,000/year **(b)** B = $525,000 **(c)** Depreciation over the coming year: D = $23,333

Chapter 8

8.1 (b)

8.2 UCC = $10,204

8.3 taxable income = $200,000, net income = $78,400

8.4 net cash = $82,400

8.5 (a)

8.6 net income = $7,194,000

8.9 **(a)** loss = −$51,225, G = $17,146 **(b)** gain = $64,235, G = $21,839
(c) gain = $138,775, G = − $47,183

8.12 **(a)** marginal tax rate = 44.62% **(b)** average tax rates: without project 26.22%, with project 33.12%

$8.21 income taxes = $8,313, tax rates: average 21.8%, marginal 37.3%

$8.23 **(a)** income taxes = $21,146, tax rates: average 34.5%, marginal 50.8%
(b) income taxes = $24,791 (using 49.5% as the marginal tax rate above $65,000), tax rates: average 33.0%, marginal 49.5% **(c)** $78,8573

$8.25 income taxes with taxable income of $25,000, $45,000, $65,000 and $100,000: Ontario $3947, $9414, $17,518, $34,563; Newfoundland $4779, $12,639, $22,004 and $40,518

$8.27 capital gains tax = $12,607

Chapter 9

9.1 $75,040

9.2 $59,357

9.3 100% equity financing, PE(12%) = −$100,000 + $500,000(P/A, 12%, 5) + $600,000(P/F, 12%, 6) = $2,006,367; 100% debt financing, PE(12%) = 0 + ($500,000 − $7200)(P/A, 12%, 5) + ($600,000 − $7200 − $100,000) × (P/F, 12%, 6) = $2,026,102; difference = $19,735

9.4 $282,977

9.5 **(a)**, **(c)**, and **(d)**, For the incremental investment (Model B − Model A), we have PE = −$2121, AE = −$506, and IRR = 13%

9.13 PE(12%) = $295,929 >0. Yes, it is acceptable

9.19 with working capital requirement, PE(15%) = −$1,071,074; without working capital requirement, PE(15%) = −$958,152

9.27 net cash flows are −$60,000, $33,146, $52,224, $45,480, $52,495, $58,189; PE(18%) = $85,788; IRR = 66%

9.30 equity financing: PE(9%) = $182; debt financing: PE(9%) = $672; debt financing is the better option

9.35 **(a)** PE(15%) = $12,088,000 **(b)** PE(15%) = $12,601,000

Chapter 10

10.1 $920 = 100 (P/A, k_d, 10) + 100 (P/F, k_d, 10); k_d = 11.39\%$; after-tax k_d
$= (1 - 0.4)(11.39\%) = 6.83\%$

10.2 debt ratio = 30.09%; $800,000/0.3009 = $2,658,647

10.3 $k = (0.3)(0.075) + (0.2)(0.128) + (0.5)(0.20) = 14.81\%$ (Note: the debt interest is already given in after-tax basis.)

10.4 **(a)** 425,532 shares must be sold; flotation costs = $638,298 **(b)** 10,194 bonds must be sold; flotation costs = $193,680; annual interest = $1,223,280

10.6 **(a)** $(0.12)(1 - 0.25) = 0.09$; **(b)** $(0.14)(1 - 0.34) = 0.924$; **(c)** $(0.15)(1 - 0.40) = 0.09$

10.7 feasible combinations: (A1), (A2), (A2, B1), (A2, B1, C), (A2, B2) ∴ both (a) and (c) are correct

10.9 Amount of debt financing: $250,000 (0.40) = $100,000; amount of equity financing: $250,000 (0.60) = $150,000; $k = (0.40)(0.08) + (0.6)(0.20) = 15.20\%$ up to $250,000

10.10 $k = (0.40)(0.10) + (0.60)(0.23) = 17.8\%$ for any financing over $250,000

10.11

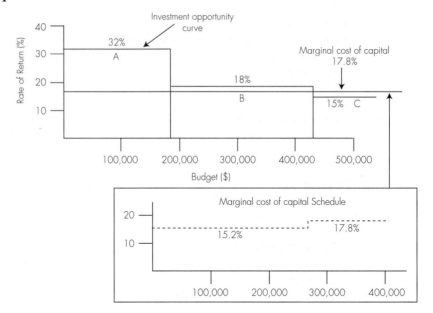

10.23

j	projects	Investment	PE(10%)
1	0	$ 0	$ 0
2	A	400	65
3	B	550	70
4	AC	1020	160
5	**ACD**	**1600**	**235**
6	BF	1150	1650
7	BEF	1530	210

with no capital limit, alternative 5 is the best

Chapter 11

$11.1 6.48% per year

$11.2 15.38% per year

11.3 $7200, $8640

11.4 $50,766

11.5 4 years, $18,473

11.6 3 years, $11,500

$11.7 **(a)** 6.677% per year **(b)** $1261.92 **(c)** 7.3586% per year

$11.16 12.96% per year

11.19 **(a)** Leasing: PEC = $23,554 **(b)** Buying: PEC = $21,521 **(c)** Buy option is better

11.27 **(a)** $247.50 **(b)** $6099 **(c)** AEC = $4193

11.36 **(a)** challenger: AEC = $17,389, defender: AEC = $18,193, replace the defender **(b)** challenger: AEC = $24,699, defender: AEC = $18,193, don't replace the defender

Chapter 12

12.1 P = $6000 $(P/A_i, 5\%, 6.73\%, 5)$ = $27,207, $i' = [(1 + i)/(1 + f)] - 1 = 6.73\%$

12.2 Constant series: P = $100 $(P/A, 1.9048\%, 3)$ = $288.92; Actual dollar series: P = $105 $(P/A, 7\%, 3)$ = $275.55

12.3 $108.6 = 102.0(F/P, f, 4)$, $f = 1.58\%$

12.4 Rule of 72: 72/9 = 8 years; Exact solution: $0.5 = 1 (P/F, 9\%, n)$ n = 8.03 years

12.5 b

12.6 Cash flow series: {– $1000, $95, $95 + $1080}; yield in actual dollars = 13.26%; yield in constant (real dollars) = $[(1 + 0.1326)/(1 + 0.04)] - 1 = 8.9\%$

12.7 c

12.12 $PE(12\%)$ = 4500(P/A, 12\%, 10)$ = $25,426

$12.14 actual dollars: A_{20} = $25,000$(A/P, 1\%, 48)$ = $658.35; real dollars: A'_{20} = 658.35(P/F, 0.5\%, 20)$ = $595.85

$12.16 $i' = (0.01 - 0.005)/(1 + 0.005) = 0.4975\%$; A' = $20,000$(A/P, 0.4975\%, 60)$ = $386.38

12.25 **(a)** A_0 = $210,000, A_1 = $61,600, A_2 = $72,456, A_3 = $162,830 **(b)** A'_0 = –$210,000, A'_1 = $58,113, A'_2 = $64,486, A'_3 = $136,715 **(c)** PE = –$14,120 ∴ reject

12.29 **(a)** A_0 = –$65,000, A_1 = $47,020, A_2 = $95,683 **(b)** ROR' = 50.72% **(c)** yes, it's profitable

Chapter 13

13.1 base: PE = $10,450(40 – 10) – 7890 = $305,610; 10% increase in X: PE = $10,450(44 – 10) – 7890 = $347,410; % change = $41,800/$305,610 = 13.68%

$13.2 Book value = $12,000 – $10,618 = $1382; gains tax = ($3500 – $1382) (?????? = $847; net proceeds from sale = $3500 – $847 = $2653; capital cost = ($12,000 – $2653) $(A/P, 15\%, 5)$ + $2653 (0.15) = $3186; equivalent annual depreciation tax credit = $912 (you obtain this figure by computing the depreciation tax credit each year, $0.4D_n$, finding the total Present Worth of those credits and annualizing the Present Worth amount over 5 years.); break-even equation: $3186 + 0.15 (1 − 0.40) X + 960 (1 − 0.4) − 912 = (1 − 0.4)0.25X$; solving for X yields $X = 47,506$

13.3 $E[FE] = 0.1 (−\$12,000) + 0.2 (\$4000) + (0.4) (\$12,000) + 0.2 (\$20,000) + 0.1 (\$30,000) = \$11,400$

13.4 $Var [FE] = (0.1)(−\$12,000 − \$11,400)^2 + \ldots + (0.1)(\$30,000 − \$111,400) = 115,240,000$; ??? $([FE]) = \$10,735$

$13.5 You lose 25 cents for each play, so the total expected loss will be $25.

13.8 useful life of the old bulb: $13,870/(19 \times 365) = 2$ years. Therefore, the new bulb would last for 4 years. Let X denote the price for the new light bulb. With an analysis period of 4 years: $PE(15\%)_{old} = (1 − 0.40)*\$61.90*[1 + (P/F, 15\%, 2)] = \$65.23$, $PE(15\%)_{new} = (1 − 0.40)(X + \$16)]$. The break-even price for the new bulb will be: $0.6X + 9.6 = \$63.23 \therefore X^* = \$92.72$. Since the new light bulb costs only $60, it is a good bargain

13.11 $E[PE]_1 = \$1900$; $E[PE]_2 = \$1850$; project 1 is preferred. $Var[PE]_1 = 1,240,000$; $Var[PE]_2 = 2,492,500$; since $Var_1 < Var_2$, project 1 is still preferred.

13.12 $U_S = \$4082$; $G = t(U − S) = 0.4(4082 − 5000) = −\$367$; net $SV = S + G = \$5000 − \$367 = \$4633$; $CR = (\$20,000 − 4633)(A/P, 15\%, 5) + 4633(0.15) = \$5279$; after-tax AE of CCA = $1326 (obtain this figure by computing the CCA tax credit each year, $0.4CCA_n$, find the PE of these credits, and annualize this amount over 5 years); break-even equation: $\$5279 + 0.14(1 − 0.40)X + 960(1 − 0.4) −1326 = (1 − 0.4)0.25X$; solving for X yields $X = 68,621$ km

13.13 **(a)** $PE(15\%) = \$3185 \therefore$ acceptable **(b)** $PE(15\%) = \$2238 \therefore$ acceptable **(c)** required annual savings (X): $\$85,000 = (0.6X)(P/A, 20\%, 4) + \$5100 (P/F, 20\%, 1) + \$8670(P/F, 20\%, 2) + \$6069(P/F, 20\%, 3) + \$32,161(P/F, 20\%, 4) \therefore X^* = \$35,865$

Chapter 14

14.1 (c)

14.2 $P = \$1,200,00 − \$50,000(P/F, 6\%, 25) = \$1,188,350$; $C' = \$100,000(P/A, 6\%, 25) = \$1,278,336$; $B = \$250,000(P/A, 6\%, 25) = \$3,195,834$; B/C ratio = $\$3,195,834/(\$1,188,350 + \$1,278,336) = 1.29$; net B/C ratio = $(\$3,195,834 − \$1,278,336)/\$1,188,350 = 1.61$

14.3 **(a)** $PE(8\%)_{A1} = \$398.954$, $PE(8\%)_{A2} = \$339,676$; B/C ratio: A1 = 1.22, A2 = 1.56

14.4 **(a)** some users' benefits: lower taxes for road maintenance, less rust on vehicles, less damage to utility lines, less damage to vegetation; some users' costs: higher taxes, possible environmental damage from CMA **(b)** experiment on sections of highway

14.9 $B_X = \$16,686,585$, $C_X = \$9,329,766$, $BC(10\%)_X = 1.79 > 1$, $B_Y = \$11,237,904$, $C_Y = \$12,462,207$, $BC(10\%)_Y = 0.90 < 1$. Choose X (incremental analysis not required

14.11 $BC_E = 1.24$, $BC_F = 1.65$, $BC_G = 1.25$, $BC_{G-F} = −5.7$, $BC_{E-F} = 0.37$, choose F

Index

Flow Type	Factor Notation	Formula	Cash Flow Diagram	Factor Relationship
SINGLE	Compound amount $(F/P, i, N)$	$F = P(1 + i)^N$		$(F/P,i,N) = i(F/A,i,N) + 1$
	Present worth $(P/F, i, N)$	$P = F(1 + i)^{-N}$		$(P/F,i,N) = 1 - (P/A,i,N)i$
EQUAL PAYMENT SERIES	Compound amount $(F/A,i,N)$	$F = A\left[\dfrac{(1+i)^N - 1}{i}\right]$		$(A/F,i,N) = (A/P,i,N) - i$
	Sinking fund $(A/F,i,N)$	$A = F\left[\dfrac{i}{(1+i)^N - 1}\right]$		
	Present worth $(P/A,i,N)$	$P = A\left[\dfrac{(1+i)^N - 1}{i(1+i)^N}\right]$		$(A/P, i, N) = \dfrac{i}{1 - (P/F,i,N)}$
	Capital recovery $(A/P,i,N)$	$A = P\left[\dfrac{i(1+i)^N}{(1+i)^N - 1}\right]$		
GRADIENT SERIES	Linear gradient Present worth $(P/G,i,N)$	$P = G\left[\dfrac{(1+i)^N - iN - 1}{i^2(1+i)^N}\right]$		$(F/G,i,N) = (P/G,i,N)\,(F/P,i,N)$ $(A/G,i,N) = (P/G,i,N)\,(A/P,i,N)$
	Geometric gradient Present worth $(P/A_1,g,i,N)$	$P = \begin{bmatrix} A_1\left[\dfrac{1 - (1+g)^N(1+i)^{-N}}{i - g}\right] \\ \dfrac{NA_1}{1+i} \quad (if\ i = g) \end{bmatrix}$		$(F/A_1,g,i,N) = (P/A_1,g,i,N)\,(F/P,i,N)$

Monster Meats

Comparing between 2 options:

options → find IRR for each

$$\frac{x^* - x_1}{x_2 - x_1} = \frac{y^* - y_1}{y_2 - y_1}$$

$x^* = IRR$ $y^* = (P/A, i^*, 8)$

$x_1 = $ lower ? $y_1 = $ at lower ? PE

$x_2 = $ higher? $y_2 = $ at higher?

find
then ~~IIR~~ IRR on incremental diff in PE. (challenger-defender)

if $i^*_{C-D} > MARR$ pick challenger as best as long as challenger is greater than MARR.

$i^*_{C-D} = MARR$

↳ pick either

$i^*_{C-D} < MARR$ pick defender as best as long as defender is greater than MARR

Declining Balance Depreciation (20%)

given

UCC = TAL − CCA

year	Asset Value	Total Asset value	CCA	UCC
03	10000	5000	$(10000\%)(0.2)$ = 1000	9000
04	20000	10000 + 9000	0.2(19000)	25200
05	—	25200	0.2(25200)	20160
06	—	20160	4032	16128
07	+3000	17628	3525.60	15602.4
08	−2000			

only depreciate $\frac{1}{2}$ 1st year!